BIOLOGY
living systems

AUTHOR
Raymond F. Oram
The Peddie School
Hightstown, NJ

CONSULTANTS
Paul J. Hummer, Jr.
Hood College
Frederick, MD

Robert C. Smoot
McDonogh School
McDonogh, MD

SIXTH EDITION

MERRILL
PUBLISHING COMPANY
Columbus, Ohio

A MERRILL SCIENCE PROGRAM

Biology: Living Systems, Student Edition
Biology: Living Systems, Teacher Annotated Edition
Biology: Living Systems, Teacher Resource Package
      Evaluation Program
      Reading and Study Guide, Teacher Edition
      Resource Master Book
Biology: Living Systems, Reading and Study Guide, Student Edition
Laboratory Biology: Investigating Living Systems, Student Edition and Teacher Annotated Edition
Probing Levels of Life: A Laboratory Manual, Student Edition and Teacher Annotated Edition
Biology: Living Systems, Transparency Package
Biology: Living Systems, Computer Test Bank

AUTHOR:

Raymond F. Oram is a teacher/administrator at The Peddie School, Hightstown, NJ, with over twenty-five years of teaching experience. He is the Science Department Head as well as a teacher at The Peddie School. He received his A.B. degree from Princeton University and his M.S.T. from Union College, Schenectady, NY. In 1976, Mr. Oram was recognized with the Princeton Prize for Distinguished Secondary School Teaching in the State of New Jersey.

CONSULTANTS:

Paul J. Hummer, Jr. taught science for twenty-eight years in the Frederick County, Maryland, schools. He is currently a biology teacher at Hood College, Frederick, Maryland. He received his B.S.Ed. from Lock Haven State University in Lock Haven, PA, and his M.A.S.T. from Union College in Schenectady, NY. He has taught various ability levels of biology as well as general science. He received the Presidential Award for Excellence in Science and Mathematics Teaching in 1984.

Robert C. Smoot is the Rollins Teaching Fellow in Science at the McDonogh School, McDonogh, MD. He earned his B.S. in Chemical Engineering from Pennsylvania State University and his M.A.T. from Johns Hopkins University. He is a fellow of the American Institute of Chemists. Mr. Smoot is author of Merrill's *Chemistry: A Modern Course* and a consultant on *Physics: Principles and Problems.*

Reviewers:
Cathie Banks, Science Division Chairperson, Wheeler High School, Marietta, GA
Carole Barsaloux, Science Department Chairperson, North Mid-High School, Edmond, OK
Catherine Bennett, Science Department Chairperson, Dunber High School, Dunber, WV
Dixie Duncan, Science teacher, Williams Township School, Whiteville, NC
Douglas Ramezane, Biology teacher, Gunderson High School, San Jose, CA
Louis Salvio, Biology teacher, Hall High School, West Hartford, CT
June Shealy, Science Department Chairperson, Mauldin High School, Mauldin, SC
George Tinker, Science Department Chairperson, School District 9, Coos Bay, OR

Project Editor: Carla J. Weiland; **Editors:** Linda Biggs Thornhill, Jennifer J. Whittingham; **Design and Project Coordinator:** Barton D. Hawkinberry; **Artists:** Jeffrey A. Clark, Terry D. Anderson; **Illustrators:** David M. Dennis, Peg Dougherty, Nancy A. Heim, Jean Helmer, Bill Robison, Don Robison, Jim Shough; **Photo Editor:** Barbara Buchholz; **Production Editors:** Kimberly Munsie, Janet B. Arledge; **Cover Photo:** Frank Cocco

ISBN 0-675-06484-8

Published by
Merrill Publishing Company

Columbus, Ohio 43216

Printed in the United States of America

7 8 9 10 11 12 13 14 15 VH/LP 00 99 98 97 96 95 94

To the Student

What common bond do you share with the more than two million other known kinds of organisms on Earth? You and the other organisms are living systems. What makes you and other life forms "alive"? What needs do you share? What activities do you all carry out? How are you all different? In what ways do you interact? How do you and other organisms affect the environment, and how does the environment affect you?

You are about to begin a course in biology that will enable you to answer some of these questions. You may or may not consider yourself a young "scientist." You may or may not ever take another course in biology. However, what you learn in this course will be valuable to you always. Knowledge of biology will be useful and important to you now as you learn about what makes you and other organisms "tick." It will also be important later as an aid to making responsible decisions as an adult.

BIOLOGY: Living Systems emphasizes the principles and features common to all life. It also provides information about the diversity of processes and structures. The content of the text is presented in a logical manner — from the simplest to the most complex levels of biological organization.

As a study of life, biology is a relevant and vital discipline. *BIOLOGY: Living Systems,* Sixth Edition, presents basic biological principles by using examples to illustrate them. Practical, applied, and familiar examples and illustrations help explain concepts and theories.

Understanding how to use your text will make your study of biology easier. Each unit begins with a photograph and a discussion that preview the material in the unit to follow. Each chapter begins with a photograph and introductory paragraph that convey the theme of the chapter. Major objectives are listed and will help you focus your study. Throughout the chapter, margin notes emphasize important ideas. Use these notes as a self-check to evaluate your understanding of facts and major ideas. Major terms are highlighted in boldface type. Each term is defined in the text material where boldface type appears.

Review questions and/or problems have been placed throughout the chapter. These, too, provide a means for you to check your understanding. At the end of each chapter, a *Summary* provides a list of the major points and ideas presented. Some of you will find reading this section useful both *before* and *after* studying the chapter. A word list, *Language of Biology,* reminds you of the most important terms so you may easily review each definition.

Also at the end of each chapter are sets of questions and problems. *Checking Your Ideas* is a set of questions stressing your understanding of major facts and terms. *Evaluating Your Ideas* contains questions and problems that are useful as a review of the chapter's concepts. *Applying Your Ideas* includes questions and problems that require you to apply what you have learned to new but related ideas. In addition, a set of project ideas is presented in *Extending Your Ideas*.

In each chapter an *Investigation* requires you to use the scientific method to solve a problem. Formulating hypotheses, using laboratory equipment, and gathering and analyzing data are skills stressed in the investigations. The appendices include *A Classification of Living Systems* and *Biology-Related Careers*. The Glossary contains over 1300 complete definitions that will aid your study of biology. The complete Index will be valuable to you in locating particular topics.

With these thoughts in mind, it is time to begin. Our hope is that *BIOLOGY: Living Systems* is your key to a successful year of study that will result in an increased awareness and understanding of yourself and your relationship with other life forms. Also, we hope that this understanding brings on a sense of excitement and wonderment that stimulates your continued questioning of and learning about the world around you.

TABLE OF CONTENTS

UNIT 1 CHARACTERISTICS OF LIFE 2

1 LIFE: COMMON CHARACTERISTICS 4
THE COMMUNITY 5
1:1   Organisms 6
1:2   Food................................ 7
1:3   Energy.............................. 8
1:4   Food Production and Energy Transfer 9
INVESTIGATION 1 12
THE USES OF ENERGY 14
1:5   Growth and Development.............. 14
1:6   Maintenance and Repair 15
1:7   Reproduction 16
1:8   Organization 16
ORGANISM AND ENVIRONMENT 18
1:9   Adaptation 18
1:10 Relationships Within the Environment 18

2   BIOLOGY AS A SCIENCE 22
METHODS OF SCIENCE.................... 24
2:1   What Is Science? 24
2:2   Observation 24
2:3   Interpretation 25
2:4   Formation of a Hypothesis 26
2:5   Experimentation..................... 26
PEOPLE IN BIOLOGY:
   DOROTHY CROWFOOT HODGKIN 29
2:6   Theory.............................. 29
2:7   Microscopes — Important Biological Tools 29
DIVERSITY OF ORGANISMS................. 31
2:8   Grouping Organisms 32
2:9   Kingdoms of Life 32
INVESTIGATION 2 34

3   MATERIALS OF LIFE 38
CLASSIFICATION OF MATTER 39
3:1   Atoms 39
3:2   Elements 41
3:3   Molecules 42
3:4   Compounds 42
3:5   Ions 44
3:6   Symbols and Formulas 45
3:7   Solutions 46
3:8   Acids and Bases..................... 48

CHANGE IN MATTER 50
3:9   Properties of Matter 50
3:10 Chemical Equations 51
BIOLOGICAL CHEMISTRY.................. 52
3:11 Carbon Compounds 52
3:12 Carbohydrates 54
3:13 Lipids.............................. 55
3:14 Proteins 55
3:15 Nucleic Acids 56
3:16 Reactions of Organic Compounds 56
INVESTIGATION 3 58

4   CELL STRUCTURE AND FUNCTION 62
DISCOVERING CELLS...................... 63
4:1   The Cell Theory...................... 63
4:2   Characteristics of Cells 64
THE CELL AND ITS ENVIRONMENT......... 65
4:3   Properties of the Cell Membrane 65
4:4   Structure of the Cell Membrane 66
4:5   Diffusion 68
4:6   Osmosis............................ 69
4:7   Passive Transport 70
4:8   Active Transport 71
4:9   Endocytosis and Exocytosis 72
INSIDE THE CELL 74
4:10 Protoplasm 74
4:11 Organization in Cells 75
PEOPLE IN BIOLOGY:
   EARNEST EVERETT JUST 75
4:12 Cell Parts of Eukaryotes 76
4:13 Other Eukaryotic Cell Parts........... 79
4:14 The Nucleus 81
ADVANCES IN BIOLOGY:
   Interaction of Cell Parts 82
INVESTIGATION 4 84

5   ENERGY FOR LIFE 88
ENERGY AND REACTIONS 89
5:1   Changes in Forms of Energy 89
5:2   Activation Energy 91
ENERGY FOR CELLULAR WORK 91
5:3   Enzymes 91
5:4   Properties of Enzymes 92
5:5   Lock and Key Hypothesis 92
5:6   Coenzymes 94
5:7   ATP 95
PRODUCTION OF ATP 96
5:8   Respiration With Oxygen 96
5:9   Respiration Without Oxygen 98
INVESTIGATION 5 100

UNIT 2 HEREDITY 104

6 THE CELLULAR BASIS OF HEREDITY 106
LIFE FROM LIFE 107
6:1 Spontaneous Generation: Prologue 107
6:2 Convincing Evidence 109
REPRODUCTION OF CELLS 111
6:3 The Cell Cycle 111
6:4 Mitosis 112
6:5 Analysis of Mitosis 115
6:6 Chromosome Number 115
6:7 Meiosis 117
6:8 Meiosis in Males 118
6:9 Meiosis in Females 119
6:10 Importance of Meiosis 121
INVESTIGATION 6 124

7 PRINCIPLES OF HEREDITY 128
ORIGIN OF GENETICS 129
7:1 Mendel's Experiments 129
7:2 Mendel's Results 130
7:3 Mendel's Hypothesis 131
7:4 The Test of Segregation 132
7:5 Terminology 133
SOLVING GENETICS PROBLEMS 134
7:6 Probability 134
7:7 Ratios 136
7:8 Incomplete Dominance 137
7:9 Two Traits 138
7:10 Multiple Alleles 141
INVESTIGATION 7 142

8 GENES AND CHROMOSOMES 146
THE CHROMOSOME THEORY OF HEREDITY 147
8:1 Sex Determination 148
8:2 Morgan's Discoveries 149
8:3 Solving Problems 150
8:4 Nondisjunction 151
PEOPLE IN BIOLOGY:
  HAR GOBIND KHORANA 152
OTHER GENETICS CONCEPTS 153
8:5 Gene Linkage 153
8:6 Crossing-Over 154
8:7 Many Genes — One Effect 156
8:8 Expression of Genes 157
HUMAN GENETIC DISEASE 159
8:9 Problem-Causing Genes 159
INVESTIGATION 8 162
8:10 Sex-linked Diseases 164
8:11 Chromosome Problems 165
ADVANCES IN BIOLOGY:
  Diagnosing Health Problems
  in the Fetus 166

9 THE GENETIC CODE 170
STRUCTURE OF DNA 171
9:1 Bacterial Transformation 171
9:2 A Model 173
9:3 Replication of DNA 175
THE ROLE OF DNA 176
9:4 Genes and Proteins 176
9:5 DNA Code 178
9:6 Protein Synthesis: Transcription 179
9:7 Protein Synthesis: Translation 181
INVESTIGATION 9 184
PEOPLE IN BIOLOGY:
  BARBARA McCLINTOCK 186
EXPRESSION OF GENES 186
9:8 Mutation 186
9:9 DNA Outside the Nucleus 188
9:10 DNA and Phenotypes 188
9:11 Genetic Control in Bacteria 189
9:12 Genetic Control in Eukaryotes 190

UNIT 3 CHANGES 194

10 CHANGE WITH TIME 196
EVOLUTION: EVIDENCE 197
10:1 Fossils: Formation and Dating 197
10:2 Fossil Record: Interpretation 199
10:3 Comparative Anatomy 201
10:4 Other Comparisons 202
10:5 Genetics 203
10:6 Evolution Observed 204
EVOLUTION: SOME EXPLANATIONS 205
10:7 Lamarck 206
10:8 Darwin: Gathering Evidence 207
10:9 Darwin's Explanation: Natural
  Selection 208
10:10 Darwin and Lamarck Compared 210
10:11 Population Genetics 211
ORIGIN OF LIFE 212
10:12 Formation of Organic Compounds 212
10:13 The First Organisms: Prokaryotes 213
ADVANCES IN BIOLOGY:
  Origin of Eukaryotic Cells 215
INVESTIGATION 10 216

11 ADAPTATION AND SPECIATION . 220
ADAPTATION 221
11:1 Origin of Adaptations 221
11:2 Evolution of Adaptations Observed 222
11:3 Types of Adaptations 224
11:4 Camouflage and Other "Tricks" 225
ORIGIN OF NEW GROUPS 228
11:5 Evolution of Species 228

11:6   The Tempo of Speciation 231
11:7   Adaptive Radiation 231
11:8   Convergence 232
HUMAN ORIGINS 233
11:9   Important Human Traits 234
11:10 African Origins 235
11:11 Earliest Humans 237
11:12 Modern Humans 238
INVESTIGATION 11 240

12   **CLASSIFICATION** 244
THEORY OF CLASSIFICATION 245
12:1   The Need for Classification 246
12:2   Binomial Nomenclature 247
12:3   Bases for Classification 248
SYSTEM OF CLASSIFICATION 250
12:4   Classification Groups 251
12:5   Some Examples 251
12:6   The Kingdom Problem 252
PEOPLE IN BIOLOGY:
    WILLIAM MONTAGUE COBB 253
INVESTIGATION 12 254

UNIT 4 DIVERSITY 258

13   MONERANS, PROTISTS,
       FUNGI, AND VIRUSES 260
KINGDOM MONERA 261
13:1   Characteristics of Prokaryotes 261
13:2   Structure of Bacteria 262
13:3   Heterotrophic Eubacteria 264
13:4   Autotrophic Eubacteria 264
13:5   Archaebacteria 265
13:6   Beneficial Bacteria 266
KINGDOM PROTISTA 267
13:7   Euglenoids 267
13:8   Golden Algae 268
13:9   Sarcodines 269
13:10 Ciliates 270
13:11 Flagellates 270
13:12 Slime Molds 271
KINGDOM FUNGI 273
13:13 Sporangium Fungi 273
13:14 Club Fungi 274
13:15 Sac Fungi 275
13:16 Lichens 276
VIRUSES 279
13:17 Characteristics 279
13:18 Some Questions 280
ADVANCES IN BIOLOGY:
    Recombinant DNA 280
INVESTIGATION 13 282

14   **PLANTS** 286
ALGAL PLANTS 287
14:1   Green Algae 287
14:2   Brown Algae 289
14:3   Red Algae 290
LAND PLANTS 290
14:4   From Water to Land 291
14:5   Bryophytes 291
14:6   Characteristics of Tracheophytes 293
14:7   Club Mosses and Horsetails 294
14:8   Ferns 294
14:9   Gymnosperms 296
14:10 Angiosperms 297
14:11 Monocots and Dicots 298
FEATURES OF ANGIOSPERMS 300
14:12 Root Types 300
14:13 The Stem 300
14:14 The Leaf 301
INVESTIGATION 14 304

15   ANIMALS: SPONGES
       THROUGH MOLLUSKS 308
THE SIMPLEST ANIMALS 309
15:1   Sponges 310
15:2   Coelenterates 312
15:3   Symmetry 314
WORMS AND MOLLUSKS 315
15:4   Flatworms 315
PEOPLE IN BIOLOGY:
    LIBBIE HENRIETTA HYMAN 316
15:5   The Tapeworm: A Parasitic Flatworm 317
15:6   Roundworms 318
15:7   Segmented Worms 319
15:8   Mollusks 321
INVESTIGATION 15 324

16   ANIMALS: ARTHROPODS
       THROUGH VERTEBRATES 328
ARTHROPODS AND ECHINODERMS 329
16:1   Arthropods 330
16:2   Centipedes and Millipedes 331
16:3   Crustaceans 332
16:4   Insects 334
16:5   Arachnids 335
16:6   Echinoderms 337
16:7   The Acorn Worms 337
CHORDATES 338
16:8   Characteristics of Chordates 338
16:9   Jawless Fish 340
16:10 Cartilage Fish 341
16:11 Bony Fish 342
16:12 Amphibians 343
16:13 Reptiles 344
16:14 Birds 346
16:15 Mammals 348
INVESTIGATION 16 352

UNIT 5 SIMPLE ORGANISMS ... 356

**17   SIMPLE ORGANISMS:
       REPRODUCTION** 358
ASEXUAL REPRODUCTION 359
17:1 Fission in Monerans 360
17:2 Fission in Eukaryotes 360
17:3 Budding and Fragmentation 361
17:4 Spore Formation 362
SEXUAL REPRODUCTION 363
17:5 Genetic Recombination
       in Bacteria......................... 364
17:6 Conjugation in Fungi 364
PEOPLE IN BIOLOGY:
    MAX DELBRÜCK...................... 366
REPRODUCTION IN VIRUSES 366
17:7 Bacteriophage Life Cycles 366
17:8 Transduction...................... 368
ADVANCES IN BIOLOGY:
    RNA Viruses 368
INVESTIGATION 17 370

**18   SIMPLE ORGANISMS: OTHER
       LIFE FUNCTIONS**.................. 374
NUTRITION 375
18:1 Nutrient Requirements 375
18:2 Digestion within Simple Autotrophs 376
18:3 Digestion within Simple Heterotrophs 377
18:4 Digestion by Fungi................... 378
OTHER LIFE PROCESSES 379
18:5 Transport 379
18:6 Gas Exchange...................... 379
18:7 Excretion......................... 380
18:8 Locomotion: Amoeboid Motion 381
18:9 Locomotion: Ciliary and Flagellar
       Motion 382
18:10 Response to Environment 383
INVESTIGATION 18 384

**19   SIMPLE ORGANISMS
       AND DISEASE** 388
BACTERIAL AND VIRAL DISEASES 389
19:1 Koch's Postulates 389
19:2 Bacterial Diseases and How
       They Are Spread 390
19:3 How Bacterial Diseases Occur 391
19:4 Viruses and Disease 392
OTHER AGENTS OF DISEASE 393
19:5 Fungal Diseases 393
19:6 Protozoan Diseases 395
DEFENDING AGAINST DISEASES 396
19:7 Barriers to Pathogens 396
19:8 The Immune System: Structures 397
19:9 The Immune System: Functions 397

19:10 Immunity and Treatment 399
19:11 AIDS 401
PEOPLE IN BIOLOGY:
    JANE C. WRIGHT 403
19:12 Cancer 403
ADVANCES IN BIOLOGY:
    New Techniques
    in Immunology...................... 404
INVESTIGATION 19 406

UNIT 6 PLANTS 410

**20   PLANT REPRODUCTION AND
       DEVELOPMENT** 412
LIFE CYCLES OF ALGAL PLANTS 413
20:1 *Spirogyra* and *Ulothrix* 413
20:2 Ulva 414
LIFE CYCLES OF HIGHER PLANTS 415
20:3 Mosses 416
20:4 Ferns 417
20:5 Conifers 418
20:6 Flowering Plants: Reproductive
       Structures 419
20:7 Flowering Plants: Sporophyte to
       Gametophyte 420
20:8 Flowering Plants: Fertilization
       and Seed Formation.................. 421
20:9 Flowering Plants: Fruit Formation
       and Seed Dispersal 422
20:10 Success Story...................... 423
20:11 Vegetative Reproduction 424
FLOWERING PLANTS: DEVELOPMENT 425
20:12 Germination 425
20:13 Further Development 426
INVESTIGATION 20 428

21   PLANT NUTRITION 432
TRAPPING ENERGY 433
21:1 The Leaf: Internal Structure 433
21:2 Light 435
21:3 Chlorophyll and Other Pigments 436
PHOTOSYNTHESIS 437
21:4 The Light Reactions 437
PEOPLE IN BIOLOGY:
    CHARLES STACY FRENCH 438
21:5 The Dark Reactions 439
21:6 Energy Relationships................. 439
DIGESTION IN PLANTS 441
21:7 Insectivorous Plants 441
21:8 Plant Nutrients 442
ADVANCES IN BIOLOGY:
    Fuel From Plants 442
INVESTIGATION 21 444

**22 PLANTS: OTHER LIFE
      FUNCTIONS** **448**
STRUCTURE OF ANGIOSPERMS 449
22:1   Roots: Internal Structure 449
22:2   Stems: Internal Structure 450
TRANSPORT AND GAS EXCHANGE 453
22:3   Simple Plants 453
22:4   Uptake of Water and Minerals in
       Vascular Plants 454
22:5   Transpiration-Cohesion Theory 454
22:6   Control of Stomata 456
22:7   Transport Through Phloem 456
22:8   Gas Exchange 457
RESPONSE TO ENVIRONMENT 459
22:9   Discovery of Auxins 459
PEOPLE IN BIOLOGY:
       HENRIK GUNNAR LUNDEGARDH 461
22:10 Other Effects of Auxins 462
22:11 Gibberellins 463
22:12 Control of Flowering 464
INVESTIGATION 22 466

UNIT 7 ANIMALS **470**

23   ANIMAL REPRODUCTION **472**
PATTERNS OF REPRODUCTION 473
23:1   Animal Cycles 473
23:2   Asexual Reproduction 474
FERTILIZATION 476
23:3   Conditions for Fertilization 476
23:4   External Fertilization 476
23:5   Internal Fertilization 478
HUMAN REPRODUCTION 479
23:6   Structures and Functions 479
23:7   Menstrual Cycle 481
INVESTIGATION 23 484

24   ANIMAL DEVELOPMENT **488**
PATTERNS OF DEVELOPMENT 489
24:1   External Development:
       Adaptations for Protection 489
24:2   External Development: Metamorphosis
       of a Frog 491
24:3   External Development: Metamorphosis
       in Insects 492
24:4   The Amniotic Egg 493
24:5   Partial Internal Development:
       The Marsupials 495
24:6   Complete Internal Development:
       The Placentals 496
24:7   Birth 499

MECHANISM OF DEVELOPMENT 501
24:8   Cleavage 501
24:9   Morphogenesis 505
24:10 Further Development 506
24:11 Control of Differentiation 507
24:12 Embryonic Induction 508
INVESTIGATION 24 510

25   FOOD GETTING AND DIGESTION 514
PATTERNS OF DIGESTION 515
25:1   Filter Feeding 515
25:2   Two-Way Traffic: Hydra 516
25:3   Two-Way Traffic: Planaria 518
25:4   One-Way Traffic: Earthworm 518
25:5   One-Way Traffic: Other Invertebrates 519
DIGESTION IN HUMANS 521
25:6   Nutrition 521
25:7   Diet and Health 522
PEOPLE IN BIOLOGY:
       LILLIE ROSA MINOKA-HILL 523
25:8   Mouth, Pharynx, and Esophagus 524
25:9   Stomach 525
25:10 Small Intestine 526
INVESTIGATION 25 528
25:11 Absorption of Food 530
25:12 Large Intestine 530

26   TRANSPORT **534**
CIRCULATORY SYSTEMS 535
26:1   Annelids 535
26:2   Arthropods 536
26:3   Fish 537
26:4   Amphibians and Reptiles 538
26:5   Birds and Mammals 539
PEOPLE IN BIOLOGY:
       FLORENCE RENA SABIN 541
26:6   The Heartbeat 542
26:7   Blood Pressure 543
26:8   Lymphatic System 544
INVESTIGATION 26 546
THE BLOOD 548
26:9   Gas Transport 548
26:10 Protection Against Injury and Infection . . . 549
26:11 Major Blood Types 551
26:12 Rh Factor 552
HEART DISEASE 554
26:13 Treatment and Prevention 554
26:14 Heart Transplants 556
ADVANCES IN BIOLOGY:
       Artificial Hearts 557

27   RESPIRATION AND EXCRETION . 562
RESPIRATION IN WATER 563
27:1   Annelids 564
27:2   Animals with Gills 564
RESPIRATION ON LAND 566
27:3   Tracheal System 566
27:4   Lung System 566

27:5   Control of Breathing 568
27:6   Respiratory Disease 569
27:7   Smoking and Its Effects 571
EXCRETION 573
27:8   Planaria 573
27:9   Earthworm and Grasshopper 574
27:10 Humans 575
27:11 Control of Osmotic Balance 577
27:12 Other Vertebrates 578
INVESTIGATION 27 580

28   CHEMICAL CONTROL 584
ROLES OF HUMAN HORMONES 585
28:1   Islets of Langerhans 586
28:2   Thyroid Gland 588
28:3   Adrenal Glands 589
28:4   Gonads 590
ENDOCRINE CONTROL 591
28:5   Pituitary Gland: Anterior Lobe 591
28:6   Feedback 592
PEOPLE IN BIOLOGY:
  CHOH HAO LI 593
28:7   Pituitary Gland: Posterior Lobe 594
28:8   Hormone Action 595
28:9   Hormones and Metamorphosis 597
INVESTIGATION 28 598

29   NERVOUS CONTROL 602
CONDUCTION 603
29:1   Neurons 604
PEOPLE IN BIOLOGY:
  HUMBERTO FERNANDEZ-MORAN 605
29:2   Membrane Potential 606
29:3   Action Potential 607
29:4   All-or-None Response 608
29:5   The Synapse 608
29:6   Effects of Drugs and Other Chemicals 609
NERVOUS SYSTEMS 611
29:7   Invertebrates 611
29:8   Vertebrates 612
29:9   Forebrain 613
29:10 Midbrain and Hindbrain 615
29:11 The Spinal Cord 615
29:12 Peripheral Nervous System 617
29:13 The Senses 618
ADVANCES IN BIOLOGY:
  Neuropeptides 620
INVESTIGATION 29 622

30   SUPPORT AND LOCOMOTION 626
SKELETAL SYSTEMS 627
30:1   The Exoskeleton 628
30:2   The Endoskeleton 629
30:3   Development of Bone 631
30:4   Structure of Bone 632
30:5   Joints 633
LOCOMOTION 634
30:6   Locomotion: Simple Animals 634

30:7   Locomotion: Mollusks and
      Echinoderms 635
30:8   Locomotion: Vertebrates and
      Arthropods 636
30:9   Types of Vertebrate Muscles 638
30:10 Contraction of Vertebrate Skeletal
      Muscle 639
30:11 Energy for Skeletal Muscle
      Contraction 641
INVESTIGATION 30 642

UNIT 8 ENVIRONMENT 646

31   BEHAVIOR 648
INNATE BEHAVIOR 649
31:1   Plant Behavior 650
31:2   Animal Behavior: Reflexes 650
31:3   Animal Behavior: Instincts 651
31:4   Bird Migration 652
31:5   Courtship Behavior 653
31:6   Biological Clocks 654
LEARNED BEHAVIOR 654
31:7   Trial and Error 655
INVESTIGATION 31 656
31:8   Conditioning 658
31:9   Imprinting 659
PEOPLE IN BIOLOGY:
  KONRAD ZACHARIAS LORENZ 660
31:10 Insight 661
COMMUNICATION 662
31:11 Honeybee "Talk" 662
31:12 Pheromones 664
31:13 Language 665
SOCIAL BEHAVIOR 666
31:14 Social Hierarchies 666
31:15 Territoriality 668

32   POPULATION BIOLOGY 672
GROWTH OF POPULATIONS 673
32:1   Biotic Potential 673
32:2   Population Growth Curves 674
LIMITING FACTORS 676
32:3   Predation and Food 677
32:4   Parasitism 678
32:5   Diseases and Populations 679
32:6   Interspecific Competition 680
32:7   Intraspecific Competition 681
32:8   Density-independent Limiting Factors 682
THE HUMAN POPULATION 683
32:9   Agriculture 684
32:10 Technology 684
32:11 The Population Explosion 685
INVESTIGATION 32 688

33 THE ECOSYSTEM 692
BIOTIC FACTORS OF ENVIRONMENT 693
33:1 Trophic Levels 694
33:2 Pyramid of Energy.................... 696
33:3 Pyramids of Numbers and Biomass 697
33:4 Recycling of Materials................. 698
33:5 Nitrogen Cycle...................... 699
33:6 Other Biotic Relationships 700
ABIOTIC FACTORS OF ENVIRONMENT 703
33:7 Water............................. 703
33:8 Soil............................... 704
33:9 Light 705
33:10 Temperature and Distribution of
        Organisms........................ 706
33:11 Temperature and Metabolism 707
INVESTIGATION 33 710

34 ORIGIN AND DISTRIBUTION
       OF COMMUNITIES 714
ECOLOGICAL SUCCESSION 715
34:1 Primary Succession 715
34:2 Climax Community.................... 717
34:3 Secondary Succession 717
34:4 From Lake to Forest................... 718
BIOMES.................................. 720
34:5 Tundra 720
34:6 Taiga 722
34:7 Temperate Forest 723
34:8 Tropical Rain Forest 724
34:9 Grassland 727
34:10 Desert 728
34:11 Ocean.............................. 730
34:12 Ocean: Littoral and Neritic Zones 731
34:13 Ocean: Abyssal Zone.................. 731
ADVANCES IN BIOLOGY:
  Unique Ocean Communities 732
INVESTIGATION 34 734

35 HUMANS AND THE
       ENVIRONMENT 738
PEST CONTROL AND POLLUTION 739
35:1 Pests and Pesticides 740
35:2 Pesticides: Some Problems 741
35:3 Pesticides: Effects on Other
        Organisms........................ 742
35:4 The Pesticide Dilemma................. 743
OTHER SOURCES OF POLLUTION.......... 744
35:5 Air Pollution 744
35:6 Acid Rain........................... 746
35:7 Pollution of Surface Waters 747

PEOPLE IN BIOLOGY:
  MICHIKO ISHIMURE..................... 749
INVESTIGATION 35 750
35:8 Pollution of Ground Water 752
CONSERVATION OF RESOURCES 753
35:9 Food............................... 753
35:10 Soil 755
35:11 Fuels 756
35:12 Wildlife and Forests................... 757
35:13 Cause for Optimism.................. 758
ADVANCES IN BIOLOGY:
  Alternative Solutions in Pest Control 759

APPENDICES 764

APPENDIX A: A Classification of
              Living Systems............... 764
APPENDIX B: SI Measurement............. 768
APPENDIX C: Respiration and
              Photosynthesis 769
APPENDIX D: Biology-Related
              Careers..................... 775

GLOSSARY 777

INDEX 806

PHOTO CREDITS 821

INVESTIGATIONS

1. Do living systems release carbon dioxide? **12**
2. How does one test a hypothesis? **34**
3. How does one determine if a solution is an acid or a base? **58**
4. How can cells be measured with the microscope? **84**
5. How does the amount of catalase enzyme compare in different tissues? **100**
6. How can you observe chromosomes in cells? **124**
7. How similar are traits of offspring to those of the parents? **142**
8. How does a genetic disease affect red blood cells? **162**
9. What is a method for extracting DNA from cells? **184**
10. How does comparative biochemistry support the theory of evolution? **216**
11. What evidence can be used to determine if an animal is or was a biped or a quadruped? **240**
12. What information can be gained from classifying living organisms? **254**
13. What are the traits of organisms in the Kingdoms Monera, Protista, and Fungi? **282**
14. What are the characteristics of gymnosperms? **304**
15. What does the pork worm *Trichinella spiralis* look like and what is its life cycle? **324**
16. How do poisonous and nonpoisonous snakes compare? **352**
17. How do different organisms reproduce asexually? **370**
18. How can you measure respiration rate in yeast? **384**
19. What are certain traits of bacteria? **406**
20. How can you tell if a seed is still alive? **428**
21. How can you estimate the number of stomata on a leaf? **444**
22. How do different environmental conditions alter transpiration rate? **466**
23. How does *Obelia* show sexual and asexual reproduction? **484**
24. How does sea urchin egg development compare to human egg development? **510**
25. How is the milk sugar lactose digested? **528**
26. How does one analyze pulse and heartbeat? **546**
27. How is urine used to help diagnose diseases? **580**
28. How does the body control calcium balance? **598**
29. What is the function of certain brain parts? **622**
30. How does muscle shorten when it contracts? **642**
31. How can you study the behavior of an animal? **656**
32. How do population changes alter population pyramids? **688**
33. How does one measure soil humus? **710**
34. How do biomes of North America differ? **734**
35. What evidence is there that the greenhouse effect is occurring? **750**

CHARACTERISTICS OF LIFE

iology, the study of life, is a vital and challenging field. Each year this field grows as old questions are answered and new questions are raised. What determines how an organism grows? What are the basic structures common to all living systems? Why do some life forms better resist disease than others? These and other questions and their answers affect you, for you are a part of the world of life.

Observe the characteristics of this great blue heron. What special features enabled it to capture the catfish? Notice that the structure of the heron's neck, beak, and legs distinguish it from other organisms. What features does it have in common with other organisms? Within the tremendous diversity of life, there is also unity. There are certain characteristics of life shared by all living systems.

Unit 1 focuses on the common characteristics of life. For example, all organisms require energy and interact with other life forms and the environment. You will learn about common features such as these, as well as some of the methods by which organisms are studied. You will also be introduced to some ideas of chemistry needed for today's study of biology.

Great Blue Heron with Catfish

LIFE: COMMON CHARACTERISTICS

I f you examine the world around you, you will find it full of living things. Most easy to observe may be plants and animals, but other types of life are also present. Although the life forms seem very different, they have some things in common. All living things need food. From the food they get energy. How does a living thing, such as this mouse, obtain and use energy? What is the major source of energy for living things on Earth?

chapter 1

You are about to begin a course in **biology,** the study of life. What will you be learning? One part of the course will explore the great diversity of living things. You will examine the major groups of living things that exist today. You will study the physical features that set one form of life apart from others.

There are more than two million known kinds of living things. Could you possibly learn about all or even most of their physical differences? Of course not. A modern course in biology does not stress differences among living things. Rather, it focuses on their similarities. Think about such different forms of life as an elephant, an earthworm, a pine tree, and a mushroom. It may be difficult to realize that they are similar in any way. But they are. How? They are very much alike in terms of their life processes — their activities and functions. As you study biology, you will come to see that these *life processes form the common bond shared by all organisms.* An **organism** is an individual capable of carrying on life processes. By the time you complete the course, you will have a better understanding of what is meant by the word "life."

Objectives:
You will
- describe the transfer of energy among organisms in a community.
- list the uses of energy common to all organisms.
- give examples of interactions among organisms and between organisms and their physical environment.

Living things are similar in their activities and functions.

THE COMMUNITY

What are the natural surroundings where you live? Is there a forest nearby? Are you close to a grassland or desert? Perhaps you are near a pond, lake, or stream. Have you ever taken the time to explore your natural surroundings? If so, doubtless you have discovered a great variety of life.

1:1 Organisms

Imagine that you are about to walk through a forest in early summer. As you near the edge of the forest, you see many organisms — plant life such as grasses, weeds, and small shrubs. Then, as you enter the forest, you are aware of the most striking forms of life, the trees.

As you move through the forest, you see much more. Ferns grow in the shade and mosses abound on the tree trunks. Leaves and fallen branches litter the forest floor. Mushrooms grow from some of the decaying wood. Under the leaves you discover insects and other small animals — ants, spiders, and millipedes. Other animal life is plentiful elsewhere. Birds sing their songs and a rabbit bounds quickly out of sight. In the distance, you see a deer nibbling leaves or a snake sliding off into the undergrowth. You look more closely at the trees and find many other animals. Bees swarm around their hives. A small butterfly, almost the same color as the bark, clings unmoving to the trunk of a tree. You see a caterpillar moving silently across a leaf, eating as it goes.

Your trip has been an eventful one. The many organisms you saw were part of the forest community. A **community** is a naturally occurring group of interacting organisms in a certain area. Yet, you saw but a fraction of the organisms in the community. Countless microscopic (mi kruh SKAHP ihk) organisms live in the soil, in puddles of water, and in or on other organisms. Soil is home for many tiny creatures such as mites, various insects, very small worms, and bac-

A community is a naturally occurring group of different kinds of interacting organisms.

FIGURE 1–1. A forest community may include organisms such as (a) red fox and (b) orange Indian pipe.

a

b

a

b

teria, among the smallest of organisms. Bacteria are plentiful in water, too, as are animallike forms such as *Amoeba* and *Paramecium* and tiny plantlike organisms called algae (AL jee).

Particular kinds of organisms vary from one community to another. Certainly the organisms in a desert or lake differ from those in a forest. However, study of any community reveals an important common feature of life. *Living things do not exist by themselves but exist in communities.* Can you suggest some reasons why this is so? As you consider possible reasons, other common features of organisms will become evident.

FIGURE 1–2. Organisms in different communities vary. (a) A unique Galapagos Island community may include penguins, iguanas, and sea lions. (b) A community in northern latitudes may include caribou and many plants.

bacteria (singular, bacterium)

1:2 Food

The study of different communities reveals many kinds of animal life. One striking similarity among all animals is their need for food. All animals must get their food from other living things. Some feed on other animals, some on plants, and some feed on both other animals and plants. However, in any community, there must be some organisms that can make their own food. These organisms are mainly the plants. Most other members of the community depend directly or indirectly on plants for food.

Plants are not the only living things that are able to make food. Several other forms of life, such as algae and some bacteria, can also make food. Biologists refer to all organisms that make their own food as **producers.** Most producers use sunlight to produce food by a process called **photosynthesis.** Photosynthesis means "put together with light."

Animals obtain their food from other organisms.

Every community includes some organisms, mainly plants, that can make food.

Producers manufacture food by the process of photosynthesis.

a

b

c

FIGURE 1–3. (a) A marmot is a consumer, eating grass, a producer. (b) A shelf fungus on a live tree is a parasite. (c) Coral fungus lives on dead matter and is a saprophyte.

Consumers depend directly or indirectly on producers for their food.

Parasites live in or on a host from which they obtain food.

Saprophytes (decomposers) feed on waste products of living things or upon dead organisms.

The food manufactured by producers is used by the producer itself and by all other members of the community. Organisms unable to produce their own food are called **consumers.**

Animals are consumers. Many are **predators** (PRED ut urz), animals that prey on other animals. Some animals feed upon plant life, grazing or browsing on a variety of producers. Most predators and plant-eating animals move around in search of food. But not all consumers are animals, and many of them obtain food by other means. Some are **parasites,** organisms that live in or on another living thing. Parasites obtain their food from a **host** and usually harm the host. A heartworm is an animal that is a parasite inside another animal—a dog. A rust is a fungus and is an example of a plantlike parasite that grows on wheat.

Some plant or plantlike consumers obtain their food from dead organisms or from the waste products of living organisms. These consumers are called **saprophytes** (SAP ruh fitez). A mushroom is a saprophyte that obtains food from soil materials. Saprophytes are also known as **decomposers,** organisms that cause decay.

The need for food is a basic feature of life. But there is another question to be answered. Why do all organisms need food?

1:3 Energy

In many ways a living organism is similar to an automobile engine. Both a living thing and an engine are complex structures. Each consists of many parts that work together so the whole may function smoothly.

In an automobile engine, all the parts have certain jobs or functions to perform. In order for them to perform, they must have fuel. Fuels are sources of energy. **Energy** is the ability to do work or cause motion. You cannot see energy, but you can see its effects. Just as gasoline is the source of energy in automobile engines, food is the source of energy in living things.

In an engine, the energy of burning gasoline sets in motion a chain of events. The result is the movement of the various parts. The energy is transferred to the wheels of the automobile. As long as fuel is available, the engine will run. As soon as the fuel is used up, the engine will stop. Similarly, each living thing must be provided with fuel if it is to function. When the fuel supply runs out, the organism will stop functioning.

Automobile engines can remain inactive for long periods of time. When provided with more fuel, an engine will run again. However, living "machines" need a continuous supply of fuel. If an organism is deprived of fuel for long periods of time, its parts will die. No amount of fuel supplied later will restore its ability to function.

Life, then, suggests a picture of constantly working machinery. There is a continuous need for fuel (in the form of food) to keep a living thing functioning. Every living thing uses the energy of food for its biological work. Obtaining and using energy is a feature of all living things.

A continuous supply of food provides energy for doing biological work.

1:4 Food Production and Energy Transfer

Sugar is the main food used by organisms.

The main type of food made by producers is sugar. It is not the same as the sugar you put on your cereal in the morning, but it is very similar. Sugar is the universal fuel for life.

Because the making of sugar involves work, energy is necessary. Most producers obtain energy for making food from sunlight. The energy of sunlight is used to combine water and a gas, carbon dioxide,

FIGURE 1–4. In photosynthesis, carbon dioxide and water are combined in plants to make sugar. Sometimes plants convert sugars to more complex substances that are stored in stems or roots.

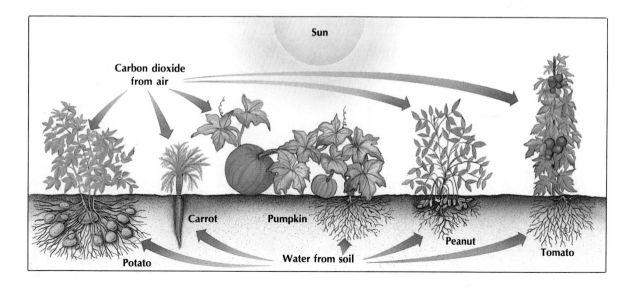

Sun

Carbon dioxide from air

Carrot

Pumpkin

Potato

Water from soil

Peanut

Tomato

Photosynthesis requires sunlight, carbon dioxide, water, and chlorophyll.

Cellular respiration produces carbon dioxide and water, raw materials for photosynthesis.

FIGURE 1–5. The processes of photosynthesis and cellular respiration result in recycling of certain materials in a community. Energy originally from the sun is used for biological work and cannot be recycled.

to make sugar. A special substance, chlorophyll, is also needed. **Chlorophyll** is the substance that *traps* the energy of sunlight for use in making food. Water and carbon dioxide are obtained from the environment, but chlorophyll is made by the producer. Chlorophyll is the substance that makes producers green.

Thus, part of the sun's energy is transferred to and stored in the sugar. The sugar is later "burned" by a living thing. "Burning" of sugar is called **cellular respiration.** Cellular respiration transfers energy first produced by the sun. This energy can be used for work by a living thing.

Carbon dioxide and water are not in endless supply. How can they continue to be available to producers? During cellular respiration, sugar is broken down into carbon dioxide and water. Carbon dioxide and water are the same substances needed for photosynthesis. These substances are constantly being returned to the environment. Thus, these substances are used over and over again. Your body is "burning" sugar right now. The carbon in that sugar might long ago have been part of a giant fern or a dinosaur.

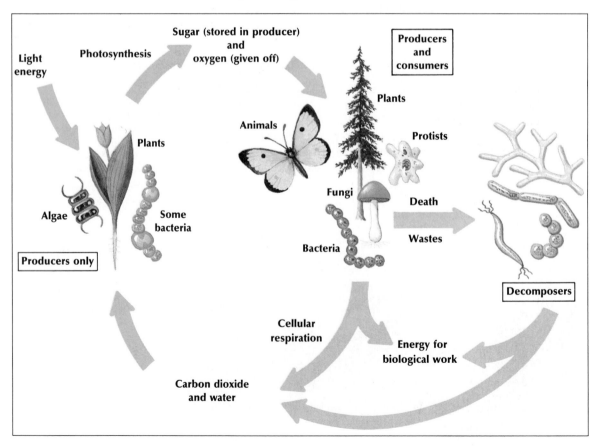

The process of cellular respiration releases materials needed for photosynthesis — water and carbon dioxide. Photosynthesis, in turn, produces sugar and releases an important substance, oxygen. Oxygen is needed by most organisms to "burn" the sugar. Can you see that the respiration process and photosynthesis are interdependent?

A producer, such as wheat, may "burn" sugar and use energy to do work. Or, a producer might be eaten by a consumer. Mice and deer, for example, are animals that feed on producers. Consumers "burn" the sugar in the producers they eat and use the energy for their own needs. A snake, in turn, may prey on the mouse, and a hawk may prey on the snake. The passage of energy from plants through a series of animals feeding on each other is a **food chain.** Wheat — mouse — snake — hawk is one food chain.

Living organisms constantly produce waste products. Eventually, the organisms die. Wastes and dead organisms are acted upon by other members in a food chain, the decomposers. Decomposers are organisms such as certain bacteria and fungi that cause decay. They are consumers. Decomposers obtain their needed energy from cellular respiration of the materials left in wastes or dead organisms. In so doing, they also release carbon dioxide and water that can be used by producers. Decomposers also return to the environment substances containing elements such as nitrogen, sulfur, and phosphorus. These elements will be used again by other organisms. *Thus, in any community, the basic materials of life are recycled.*

In a food chain, energy is passed from one organism to another. This energy is used for each organism's work. Once the energy is used for work, though, it cannot be used again. *Thus, energy cannot be recycled.* It must always be available in the form of food. There must be a constant source of energy for photosynthesis. That energy is sunlight. Without sunlight, life as we know it could not survive.

FIGURE 1–6. A food chain shows the direction energy moves from producer to consumers. Energy is not recycled in a food chain.

Energy is distributed to all organisms through food chains.

Materials are recycled through a community, but energy cannot be recycled.

REVIEWING YOUR IDEAS

1. Do all living things live in communities in nature? Why?
2. What is a producer? How are producers important to a community?
3. List several examples of consumers.
4. Why do living things need food?
5. What raw materials are necessary for photosynthesis? What does a producer need to combine these materials to form sugar?
6. What is a food chain? Why is it important? Give an example of a food chain involving yourself.

INVESTIGATION

Problem: Do living systems release carbon dioxide?

Materials:

apparatus shown in Figure 1–7   small flask
effervescent tablet   2 straws
4 test tubes   foil
bromothymol blue
test tube rack
dropper
glass marking pencil
2 stoppers to fit test tubes
Elodea
forceps
yeast

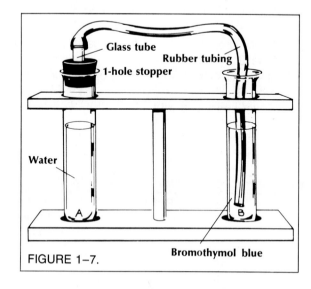

FIGURE 1–7.

Procedure:

Part A. Detecting Carbon Dioxide (CO$_2$) Gas

1. Make a table similar to the one shown in which to record your observations.
2. Obtain and set up the apparatus shown in Figure 1–7.
3. Observe the contents of each tube. Then, record the color of the liquid in tube B in Table 1–1.
4. Remove the rubber stopper from tube A. Drop ¼ of an effervescent tablet into the tube and *quickly* replace the stopper.
5. Note the production of gas (bubbles of carbon dioxide gas). When the bubbles stop, record the color of tube B in Table 1–1. NOTE: If bromothymol blue changes in color from blue to either blue-green, green, or yellow, carbon dioxide gas is present.

Part B. Do Producers Release Carbon Dioxide?

1. Fill two clean test tubes ⅔ full of bromothymol blue. Label one tube C and the other D.

2. Place a water plant, *Elodea,* into tube D. Stopper each tube. Label each tube with your name and with the contents of the tube.
3. Place both tubes in the dark for 24 hours.
4. After waiting the 24 hours, use forceps to remove the plant from tube D. Note the color of the liquid in each tube. Record the colors in Table 1–1.

Part C. Do Consumers Release Carbon Dioxide?

1. Fill a small flask ⅓ full with bromothymol blue. Label the flask E. Record the "before" color in the data table in the proper row.
2. Place two straws into the flask and exhale through the straws so that your breath bubbles through the liquid. Exhale through the straws for at least three minutes. **CAUTION:** *Do not drink the bromothymol blue.*
3. Record the "after" color of the liquid in the data table.

Part D. Do Decomposers Release Carbon Dioxide?

1. Fill two clean test tubes ⅓ full with bromothymol blue.
2. Label one test tube F and label the other tube G.
3. Using a dropper, add 5 drops of yeast to test tube G.
4. Cover each tube with foil.
5. Label each tube with your name and with the contents of the tube.
6. Place tube F and tube G in the dark for 24 hours.
7. After waiting the 24 hours, note the color of the liquid near the bottom of each of the tubes.
8. In the data table record the colors in each tube.

Data and Observations:

	TABLE 1–1. OBSERVATIONS OF BROMOTHYMOL BLUE			
	Part A Detecting CO_2	Part B Do Producers Release CO_2?	Part C Do Consumers Release CO_2?	Part D Do Decomposers Release CO_2?
Color of:	Tube B "before"	Tube C	Flask E "before"	Tube F
Color of:	Tube B "after"	Tube D	Flask E "after"	Tube G

Questions and Conclusion:

1. Define each of the following terms: (a) producer (b) consumer (c) decomposer.
2. Put each organism in this activity into one of the following categories: (a) producer (b) consumer (c) decomposer.
3. Explain how to experimentally detect the presence of carbon dioxide gas.
4. Do producers release carbon dioxide into their surroundings? What is your evidence?
5. Do consumers release carbon dioxide into their surroundings? What is your evidence?
6. Do decomposers release carbon dioxide into their surroundings? What is your evidence?
7. What life process is responsible for the release of carbon dioxide by organisms into their surroundings?
8. In Part B, why did you place a tube with no *Elodea* plant in the dark?
9. In Part D, why did you place a tube with no yeast in the dark?
10. If you repeated Part B, C, and D of the experiment, but this time tested to see if water was given off, what results might you expect? Explain.
11. If you repeated Part B, C, and D of the experiment but this time tested to see if oxygen was given off, what results might you expect? Explain.

Conclusion: Do living systems release carbon dioxide?

THE USES OF ENERGY

The ways in which energy is used are common to all organisms.

Sugar is made by producers and "burned" by all organisms. Thus, the materials and energy needed for proper functioning are made available to living things. What are these functions? They can be grouped into four broad areas: growth and development, maintenance and repair, reproduction, and organization. These functions are important because they are common features of all living things.

1:5 Growth and Development

Every organism has some period of growth during its lifetime. The overall growth may be slight, as of a bacterium; or it can be large, as of a whale or elephant. Whether great or small, growth is a feature of every living thing. But, what is growth, and how does it occur?

Suppose you place a wilted lettuce leaf in a bowl of water. After a period of time, the leaf will increase in mass and volume. Is this growth? No, **growth** is an increase in the amount of living material in an organism. The lettuce leaf does not add living material; it just absorbs water.

Growth is an increase in the amount of living material.

Consider another situation. Suppose you place a lettuce seed in some moist soil. In about a week, a lettuce plant will begin to appear. Water is taken in again, but this time, true growth occurs. The basic difference is that the seed has used the water (with carbon dioxide and minerals) to form living materials. Water was not just soaked up, or absorbed, in a physical manner. It was put together with the other

FIGURE 1–8. Growth, an increase in the amount of living material, of (a) lions, can be shown (b) in a graph of increase over time.

a

b

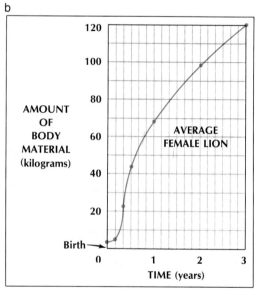

AMOUNT OF BODY MATERIAL (kilograms)

AVERAGE FEMALE LION

Birth→

TIME (years)

materials in a special way to become part of the living thing. Growth is not just an increase in amount. Growth is an increase in the amount of living material.

Not only does the lettuce seed grow, it becomes a certain kind of living thing. It becomes a lettuce plant with its specialized parts. How are materials put together in a lettuce seed to form a lettuce plant? Why does the seed not become a beet or carrot plant? The series of changes a living thing undergoes in reaching its mature form is called **development.** The lettuce seed develops into a lettuce plant. Perhaps you know that development involves heredity (huh RED ut ee). **Heredity** is the transmission of features from one generation to the next. Later you will study both heredity and development.

Development is a series of changes an organism undergoes in reaching its mature form.

1:6 Maintenance and Repair

Organisms are faced with many changes in the environment, including the activities of other organisms. Temperatures change during a day and from season to season, and the water and food supplies can also change. A resting animal suddenly may need to flee from a predator.

Organisms respond to changes in the environment.

If an organism is to survive such changes, somehow its internal operation must be adjusted. When you are outside in cold weather, you might shiver. A bat hibernates in the winter. A plant conserves water in dry spells. The heartbeat and breathing rates of a running animal increase. All features such as these help living things survive in their environments. How?

Homeostasis is the steady state of the internal operation of an organism.

These responses and activities use energy. They are controlled by the interaction of many parts and systems, such as chemicals, nerves, muscles, and the skeleton. Such responses maintain homeostasis (hoh mee oh STAY sus). **Homeostasis** is the steady state of the internal operation of an organism regardless of external changes. If homeostasis did not occur, organisms would not be able to live very long. Maintenance is a continual process.

FIGURE 1–9. Monitors on this woman detect her body's responses to her changed level of activity. Such responses maintain homeostasis.

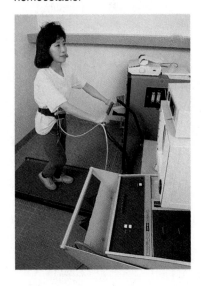

Other examples of homeostasis are repair of injuries and resistance to disease. Minor problems, such as cuts and scratches or infections, usually can be overcome by natural means. Scratches heal as a result of clotting blood that covers the wound and prevents infection while new skin is being formed. Special cells in an organism often attack and destroy bacteria in an infection. Sometimes repair of an injury can even involve replacement of body parts. A salamander can replace a lost limb by growing a new limb. It also can replace other body parts. A starfish can replace a lost "arm." Much energy goes into the repair and replacement of worn-out and damaged parts in living things.

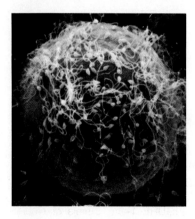

FIGURE 1–10. Coded instructions are passed in the sperm and egg from one generation to the next. A sea urchin egg covered with sperm is shown here magnified about 800 times.

During reproduction, a set of coded instructions is passed from one generation to the next.

Life is a continuous process.

Energy is needed for the parts of an organism to work together in an orderly way.

How is the degree of organization of your desk or locker related to the energy you spent organizing it?

Organisms have physical organization.

1:7 Reproduction

A feature that distinguishes organisms from nonliving things is **reproduction** (production of offspring). Only organisms can reproduce. Each new organism has certain features that set it apart from other kinds of organisms. Rabbits produce rabbits; oak trees produce oaks. "Like produces like" is a basic biological principle.

That all organisms can reproduce poses new questions. Many organisms begin life as a small, fertilized egg. How do you explain that from this small structure a fully grown, complex oak tree (or octopus, or orangutan) is produced?

A fertilized egg does not contain a small form of the living organism. What does it contain? Analysis shows that a fertilized egg is composed of many chemicals. One type of chemical directs the putting together of other materials so that a distinct kind of organism forms. In other words, the fertilized egg contains a "master code" for life. Reproduction is the transfer of the coded instructions that direct the formation of the new organism. Living organisms pass to each generation a "formula" for continuing life.

In many organisms, tiny sperm and eggs are living "messengers" that unite to form a fertilized egg, the beginning of a new life. As "messengers," sperm and eggs link one generation to the next so that the process of life continues. Individual organisms die and many groups of organisms have become extinct, but life continues without interruption.

1:8 Organization

You have now looked at some ways in which energy is used by organisms. Growth and development, maintenance and repair, and reproduction are all common functions of life that use energy. Although each has been discussed alone, all of them are ongoing processes in a living system. No one function of life is isolated from the others.

A common feature of organisms is their precise organization. **Organization** is the orderly functioning of a living system. It involves the coordination of all the complex processes and activities of an organism. Organization can only be maintained if energy is available. It is the main result of energy use. Without a steady supply of energy, organization soon disappears. What happens to an organism's organization once it dies? Why?

Organisms are not only organized in terms of complex processes and activities. They are organized physically as well. An organism is composed of interworking parts that differ in complexity. Thus, scientists have defined levels of organization.

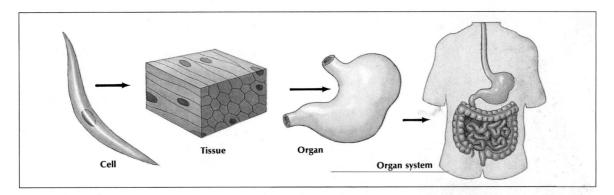

Cell   Tissue   Organ   Organ system

Most types of organisms are made of more than one kind of cell. You will study cells and their parts in Chapter 4. In organisms made of more than one kind of cell, similar cells are organized to do certain jobs. A group of cells having basically the same structure and function is called a **tissue.** Animal tissues include muscles, bones, skin, nerves, and blood. Plant tissues include protective areas such as bark and the outer surfaces of leaves, conductive tissue that transports water and food, and special regions of growth.

Groups of tissues are further organized into **organs.** The leaf of a plant, for example, is an organ of photosynthesis. It contains tissues for protection, water and mineral transport, food production, and transport of food. What are some other organs?

A group of organs working together is known as a **system.** The stomach is an organ that is only part of the entire digestive system. What other organs work with the stomach? What are some other animal systems and plant systems?

The total organism represents the next higher level of organization, but organism is not the most complex level. Life is more than individual organisms. Higher levels of organization, involving groups of organisms, also exist. For example, you have already learned about a community, which is composed of many kinds of organisms. Later you will learn about other ways in which organisms are grouped.

FIGURE 1–11. Levels of biological organization include cells, tissues, organs, and organ systems.

Biological levels of organization: cell, tissue, organ, system, organism.

REVIEWING YOUR IDEAS

7. What is development? How does it differ from growth?
8. Why are maintenance and repair important to living things?
9. What is reproduction? What is passed on to each generation?
10. List the levels of biological organization through organism.

ORGANISM AND ENVIRONMENT

When you think about a community, you think about the organisms living in it. A lake is inhabited by fish, snails, and algae. Cactus plants, scorpions, and lizards live in a desert. Crabs, jellyfish, and whales are found in the ocean. Deer, birds, and maple trees live in a forest.

1:9 Adaptation

Organisms are suited to particular environments.

Organisms are suited to particular environments and ways of life. A fish has gills that obtain oxygen from the water. A deer has teeth that grind plant material. Birds have wings. A cactus stores water. Maple trees have broad leaves that trap light during photosynthesis.

Adaptations promote survival.

These special parts and activities of organisms are examples of adaptations. An **adaptation** is a feature of a living thing that enables it to survive in its environment. Adaptations are another common feature of living things. You will learn more about adaptations in Chapter 11.

1:10 Relationships Within the Environment

Ecology is the study of the relationship between organisms and their environment.

Each living thing is a complex organization of interacting parts. Without the proper working of all of its parts, the organism could not function. So, too, the success of a community depends on the organisms in it. Living things do not exist alone; they live together and depend on each other. The study of the relationship between organisms and their environment is **ecology.** An ecologist studies how organisms interact with each other and with their physical surroundings.

FIGURE 1–12. (a) Water lilies and (b) a hermit crab have special adaptations that aid their survival. What are some of their adaptations?

a

b

A study of a community reveals many ways in which plants and animals interact. Many birds build nests in trees and other plants. Bees transfer pollen from flower to flower. Seeds of some plants are transported in the fur of mammals. Certain insects such as ants and termites live together in societies. They depend on one another for protection from enemies, information about the environment, or performance of special tasks. Many organisms require mates in order to reproduce, and many young organisms must be cared for by their parents. In Section 1:2, you learned how animals depend upon one another or other organisms for food.

Physical factors of the environment are also important to a community. Without water, carbon dioxide, and oxygen, most organisms could not live. The type of soil and amount of water determine the kinds of plants that can live in a certain place. The amount of shade the plants provide affects temperature, which in turn affects the kinds of plants and animals that can live there.

Can you see that a community, like an organism, is a well-organized system? Do you better understand that the community is a feature of life because no organism, regardless of its adaptations and complexity, can exist by itself? How do people depend on their environments? How do they affect their environments? What do you think is the responsibility of people concerning the environment?

FIGURE 1–13. Physical factors of the environment determine the kinds of organisms that can live in different places. (a) Trees cannot grow above a certain altitude due to colder temperatures and varying rainfall. (b) Because of special water-conserving adaptations, many plants can live in arid deserts.

Organisms depend on one another in many ways.

Physical factors of the environment determine the kinds of organisms living in a given community.

REVIEWING YOUR IDEAS

11. What is an adaptation? List three animal adaptations.
12. What would an ecologist likely want to find out about a community?
13. List several ways in which organisms depend on one another.

CHAPTER REVIEW

SUMMARY

1. Life processes are features common to all organisms. **p. 5**
2. Organisms live together in communities. **1:1**
3. All organisms need food. Producers make their own food; consumers depend on other organisms for food. **1:2**
4. Organisms depend upon the energy of food for their biological work. **1:3**
5. The energy in food comes from sunlight through photosynthesis. Cellular respiration or "burning" of food transfers energy. **1:4**
6. Materials in a community are recycled. Because energy cannot be recycled, a community must have a constant supply of energy. **1:4**
7. Energy from food is used in the growth and development of organisms. **1:5**
8. Homeostasis and repair enable organisms to cope with changes in their environment. **1:6**
9. Reproduction is a continuous process that links one generation to the next. **1:7**
10. The orderly functioning of a living system is maintained by organization at several different levels — cell, tissue, organ, system, organism, and community. **1:8**
11. Organisms are adapted for life in particular environments. **1:9**
12. Organisms interact with one another and with the physical aspects of their environment. **1:10**

LANGUAGE OF BIOLOGY

adaptation
biology
cellular respiration
chlorophyll
homeostasis
host
organ
organism
community
consumer
decomposer
development
ecology
energy
food chain
growth
heredity
organization
parasite
photosynthesis
predator
producer
reproduction
saprophyte
system
tissue

CHECKING YOUR IDEAS

On a separate paper, match each phrase from the left column with the proper term from the right column. Do not write in this book.

1. necessary for decay process
2. balance of internal processes
3. released during photosynthesis
4. characteristic promoting survival
5. interaction of organisms and environment
6. an increase in living material
7. obtains food from host
8. necessary for work
9. substance "burned" by all organisms
10. Like produces like.
11. tissues working together
12. makes own food
13. necessary for conversion of light to chemical energy
14. capable of carrying on life processes
15. group of organs working together

a. adaptation
b. decomposer
c. ecology
d. energy
e. growth
f. homeostasis
g. organ
h. oxygen
i. parasite
j. producer
k. reproduction
l. sugar
m. system
n. chlorophyll
o. organism

EVALUATING YOUR IDEAS

1. Distinguish between parasites and saprophytes. Are they producers or consumers?
2. What is energy? What is the source of energy in every community? How is it transferred to all members of a community?
3. How are decomposers important to a community? What would happen to a community in which there were no decomposers? Why?
4. How are photosynthesis and cellular respiration important to each other?
5. Compare what happens to the materials to what happens to the energy used by organisms.
6. What is homeostasis? Why is it important to living things?
7. What is meant by the phrase "like produces like"? Why is this true?
8. Why is it possible to say that the process of life is continuous?
9. In what ways are living things organized? Explain.
10. What is a tissue? What is an organ? What is a system? List some examples of each.
11. What are some physical factors of the environment?

APPLYING YOUR IDEAS

1. List several types of communities. What do they have in common? How do they differ?
2. Classify each of the following as a parasite or saprophyte.
   (a) tick on a dog
   (b) termites in dead wood
   (c) tapeworm in a cow
3. Give an example of a possible food chain in a lake and in a desert.

4. What would be the simplest possible food chain in a community? Explain.
5. A piece of potato immersed in water absorbs some of the water and swells. Has the potato grown? Explain.
6. How are availability of water, temperature, light, and type of soil important to a community?
7. List three adaptations of a tree. List three adaptations of a human. How do these adaptations help the organism survive in its environment?
8. Why are there few producers at the bottom of a deep lake or ocean?

EXTENDING YOUR IDEAS

1. Search the area near your home for different kinds of organisms. Classify each one as either producer or consumer. Identify any parasites or saprophytes you find. Determine how each organism fits into food chains of the community. List several adaptations of each organism.
2. Create your own miniature community by setting up a freshwater aquarium or by setting up a terrarium.
3. Make a list of five specific ways in which your body uses energy.

SUGGESTED READINGS

McCombs, Lawrence W., and Rosa, Nicholas, *What's Ecology?* Menlo Park, CA, Addison-Wesley Publishing Co., 1986.

Niegel, Joseph E., and Avise, John C., "On a Coral Reef, It's a Hard Knock Life." *Natural History,* December, 1984.

Nilson, Lisbet, "Focusing on Falling Leaves." *Science Digest,* November, 1984.

BIOLOGY AS A SCIENCE

Biology is the branch of science in which organisms are studied. To study something scientifically, a method must be followed and precise measurements must be made. Scientific research may be carried out either in the laboratory or outdoors. Here a marine biologist places a reference grid on a shallow reef so that she can study the grazing habits of fish. How is a scientific study done? What steps must be followed and what kinds of measurements are made?

umans have always wondered about themselves, other organisms, their world, and space. They have sought to explain nature and to make use of what they learned.

Early "biology" centered on matters vital to human life. Thousands of years ago humans learned about planting crops and taming animals. They were also concerned with the body and with health. Later, humans became curious about things other than their own needs.

Rather than question ideas themselves, most people relied on the beliefs of earlier scientists. Some of these beliefs were not correct. Thus, progress in biology (and other sciences) was slow. This pattern lasted until the 1500s. Then important findings based on new observations and study began to be made.

The late 1800s were an exciting time for biology. It was learned that all living things are made up of small units called cells, and that some cells called microbes can cause disease. It was shown that all life comes from life. The idea that organisms change with time was proposed and explained. The basic laws of heredity were discovered. These findings were important because they were general enough to apply to most organisms. Also, they provided a basis for later discoveries.

Today is an even more exciting time for biology. Diseases such as malaria and polio are now understood and can be prevented and treated. The structure of the "genetic message" has been worked out. Genes can be moved from one organism to another. Energy is being derived from many sources. Surgery now often includes the use of lasers. Crop yields are being improved. Computers are being used to gather, record, and analyze information.

Knowledge continues to grow at a rapid rate. Each year, new information is built on the foundation of prior knowledge. Much of this knowledge benefits humans and other organisms. How has science affected you personally?

Objectives:
You will
- describe how a scientist conducts an investigation.
- list the characteristics of the five kingdoms of life.

Many so-called "truths" remained untested until the sixteenth century when scientists began to experiment for themselves.

During the late 1800s some important general discoveries about biology were made.

METHODS OF SCIENCE

Early scientists did not know how to solve their problems. Often they did not study a problem carefully. Many times they did not know for what they were looking. One of the most important factors in good science is asking the right questions. Many discoveries made in the last few hundred years are valid because the scientists who made them were good detectives.

2:1 What Is Science?

The *methods* of science are unique. A scientist is a detective who must solve problems by asking questions and putting the answers together in a meaningful and conclusive way. Intelligent guessing is important to the scientist. But guessing alone is not enough. The guesses must be supported or rejected by evidence.

Science is a process that produces a body of knowledge about nature. Areas of study such as art, music, or history are no less scholarly than science because all of them involve creativity. But, the *manner* in which science studies nature makes it different from other subjects.

Science is carried out because people want to learn more about nature. Applying the knowledge to real problems is **technology** (or applied science). For example, the science of heredity can be used to solve real problems. Plants that make more food and bacteria that produce drugs have been produced. What practical problems have these technological advances helped to solve?

Scientists can solve problems in many ways. Each "case" is different. Usually certain parts such as observation, interpretation, hypothesis (hi PAHTH uh sus) formation, and experimentation are included. The relative place of each of these in scientific investigations may vary. However, they interact and are necessary in solving a problem.

2:2 Observation

Careful **observation** is very important to a scientist. No matter what the problem, scientists must observe carefully all they can about it. Often this process includes reading what is known already about the subject or related subjects. Careful and confirmed observations become facts. A **fact** is something about which there is no doubt. It is a fact that this paper is white.

In science, facts are often called **data.** Many data result from observations made during experiments. Some data are descriptive; for example, the color and shape of a tadpole. Other data involve numbers; for example, measurements of the growth of a tadpole.

A scientist must be a good detective.

Science differs from other areas of study in the methods used.

Technology is the application of scientific knowledge in a practical way.

FIGURE 2–1. Observations can be made with all of the senses. Here, a biologist identifies a tree by its smell.

How are numerical data recorded? Scientists worldwide use a universal "language" of measurements and their symbols, the **International System of Measurement (SI)**. SI is a modern form of the older metric system and is now used in most countries. One major advantage of SI is that it is a decimal system. It is based on tens, multiples of tens, and fractions of tens. Because it is a decimal system, it is much easier to use (and more logical) than the English system. See Appendix B for linear, mass, and volume measurements using SI.

Because science is a worldwide study, discoveries made in one place are important to scientists around the world. The transfer of information is hindered by the many languages in the world. Thus, reports must be translated in foreign journals to inform scientists worldwide of progress made.

Scientists must be careful not to overlook facts. They must know the importance of what they observe. Sometimes an observation may seem unimportant. However, it may end up being valuable.

In 1928, the British scientist Sir Alexander Fleming (1881–1955) made such an observation. He was studying a certain type of bacteria, *Staphylococcus* (staf loh KAHK us), in his laboratory. The bacteria were grown in culture dishes. Fleming noticed that a mold called *Penicillium* (pen uh SIHL ee um) grew in some of the culture dishes. He might have thrown out these dishes, but he observed something else. Around the *Penicillium* was a clear zone (Figure 2–3). This area was clear because the bacteria that had grown there had died. In dishes without the mold, no clear zone appeared because bacteria continued to grow and reproduce.

FIGURE 2–2. Observations of bird songs are made by hearing and are recorded for study.

Fleming made an unexpected observation that was the basis of an important discovery.

FIGURE 2–3. Penicillin from the soaked filter in the center of the dish produces a clear zone where no bacteria will grow. Notice the bacterial growth outside the clear zone.

2:3 Interpretation

This brief discussion of Fleming's discovery points out another important part of scientific procedures. Observing that bacteria were killed, Fleming *reasoned* that the mold must be producing a substance, a chemical, that killed the bacteria. He *assumed* that the chemical spread from the mold throughout the clear zone. He could not observe a substance either being produced by the mold or killing the bacteria. But this idea was the most *logical* explanation he could make of the observed facts.

Such reasoning is a key part of scientists' investigations. They must be able to *interpret* their observations of nature. These interpretations and explanations may not always be correct. What may seem to be a logical explanation might turn out completely or partly wrong. However, logical reasoning and interpretation are necessary to obtain a final answer to any scientific problem.

2:4 Formation of a Hypothesis

Facts have no meaning unless they can be tied together. A scientist must try to put together the pieces as if working a jigsaw puzzle without knowing what the total picture will look like.

If a scientist gains insight into the problem, an idea can be developed that may fit the pieces of the puzzle together. Such an idea or statement is called a **hypothesis.** The purpose of a hypothesis is to relate and explain observed facts. Fleming's reasoning about the action of *Penicillium* on bacteria could be stated as a hypothesis. Fleming's hypothesis was that some chemical produced by the *Penicillium* kills certain bacteria.

A good hypothesis not only *explains* the facts but also *predicts* new facts. Besides stating that a chemical killed the bacteria, Fleming also predicted that a certain chemical *alone* (not the entire mold) should be able to kill bacteria.

Forming a hypothesis is not the end of the problem. Many early scientists failed because they made a hypothesis their conclusion. A good hypothesis must be tested. Only if there is a test can a scientist be certain a hypothesis is correct. A hypothesis is a "tool" for further study of a problem.

A hypothesis relates and explains observations and predicts new facts.

2:5 Experimentation

Once a hypothesis has been presented, the next step is to test it. Such scientific testing is known as **experimentation.** You may have thought of experiments as involving smoking liquids in strange gadgets. Most experiments are fascinating but offer little of this type of excitement. Rather, they involve an orderly procedure and require careful work. The excitement comes from finding a way to check a hypothesis.

Experimentation is scientific testing.

A hypothesis must lend itself to testing.

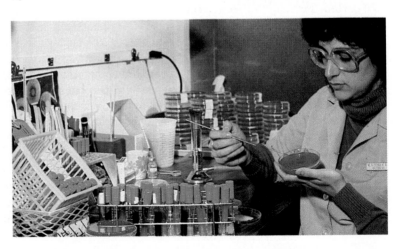

FIGURE 2–4. Experimentation is done to test hypotheses. This scientist is testing a hypothesis about the growth requirements of certain bacteria.

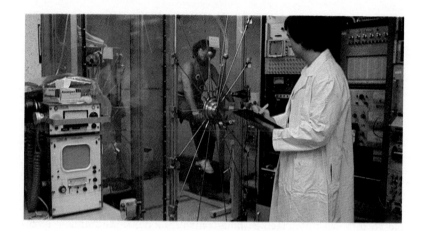

FIGURE 2–5. In this controlled experiment the man inside the chamber is undergoing a stress test in polluted air. The polluted air is the variable factor. All other factors are controlled.

Fleming had to isolate the substance he thought was killing the bacteria. Then he had to test the prediction that the substance would kill the bacteria. To do this test, he transferred some of the mold to nutrient broth solutions. Nutrient broth solutions contain the basic ingredients needed for the mold to grow and reproduce. Fleming assumed the mold would produce a chemical that would flow into the liquid broth. After he added the chemical and broth (minus the mold) to the bacteria, he observed that the bacteria were killed. Thus, he had isolated the chemical and also verified the prediction of his hypothesis. The chemical alone could kill the bacteria; the mold was not necessary. Has the hypothesis been verified beyond doubt?

A scientist must be sure of the cause of a certain effect. Fleming, for example, believed that the chemical in the broth killed the bacteria, but how could he be sure? Something in the broth itself might have killed the bacteria. When a test of the broth alone showed that it did not kill the bacteria, the doubt was removed. The hypothesis was supported. Fleming named the chemical that he discovered *penicillin*.

Scientists conduct their experiments in a carefully controlled manner. A **controlled experiment** is one in which all factors are the same except the one being tested. The factor being tested is called the **variable factor.** Suppose you were testing a substance suspected of killing bacteria. Like Fleming, you would have to run a controlled experiment to verify your idea. You would need two groups of bacteria that would be the same in all ways except that one would be exposed to the new substance and the other would not. Thus, the substance is the variable factor between the two groups. The group that received the chemical is the **experimental group;** the other group is the **control group.** Use of controls and using only one variable factor at a time are common features of scientific investigation because they reduce uncertainty.

Scientists carry out controlled experiments in which there is only one variable factor.

An experiment may indicate that a hypothesis is not correct. If so, a scientist, like a detective, must either revise or develop a new hypothesis to explain the facts.

An experiment may show that a hypothesis is partly or totally wrong. When this occurs, the facts must be looked at again and interpreted in some other way. As a result, the hypothesis may need to be changed or a completely new one formed. If a hypothesis is changed, new experiments must be carried out. Finally, to be acceptable to other scientists, the experiment that confirms a hypothesis must give the same results each time it is repeated.

Fleming's work with penicillin arose by chance during other research and led to a major *practical* breakthrough. Fleming realized the future use of penicillin as a drug and studied its effect on many common disease-causing bacteria. Penicillin's discovery has caused a search for similar drugs. Many drugs that kill bacteria continue to be discovered. These drugs, now called antibiotics, have been a great aid to modern medicine. Although many people take antibiotics for granted, it is not surprising that they were once called miracle drugs. Their use has saved millions of lives.

antibiotic (anti-, against; bio-, life)

This example illustrates two important points. First, much scientific research is carried on because of the natural curiosity of humans. They want to know the "hows" and "whys" of the world. As a result, there have been many benefits, from microwave ovens to antibiotic drugs. Fleming did not set out to discover a drug to cure bacterial infections, but when the unexpected happened, he did not ignore it. Use of the drug in medicine came about only after this "accidental" discovery.

Many practical benefits result from the natural curiosity of scientists.

Second, scientific knowledge builds upon the foundations of previous findings. Once penicillin was discovered and its value shown, widespread interest arose. A single, important finding led to many related discoveries. This pattern has been repeated often in the history of science — an important hypothesis provides a basis for further work.

Science builds upon previous knowledge.

FIGURE 2–6. While researching bacteria, Alexander Fleming discovered penicillin by chance.

PEOPLE IN BIOLOGY

About the time of Fleming's discovery of penicillin, Dorothy Crowfoot Hodgkin was becoming interested in a then new technique called X-ray crystallography (krihs tuh LAHG ruh fee). This technique involves the use of X rays to determine the structure of substances.

In 1937, Hodgkin received her doctorate from Oxford University in England where she continued X-ray research begun as a student. During World War II, the demand for penicillin became

Dorothy Crowfoot Hodgkin

(1910–   )

great. Penicillin could be made in large amounts if its structure were determined. Hodgkin set a goal to determine this structure. By 1948 she accomplished this goal. Through her work and that of her research team, the foundation was laid for mass production of antibiotics.

Among other accomplishments, Hodgkin has determined the structure of vitamin B_{12} and insulin. She has received many honors, including honorary degrees, the Royal Medal, and the Order of Merit. In 1964 she received the Nobel Prize in Chemistry for her contributions to biochemistry and medicine.

Hodgkin has three children and is now a grandmother, also. She lives in England with her husband and still works in her research lab in Oxford.

2:6 Theory

Scientists tend to shy away from the word "proof" in describing experimental results. It is possible that new data will cause revision of an accepted hypothesis.

In general, when a major hypothesis has survived testing, it is called a **theory.** Common usage of the word suggests an uncertainty, such as, "This is only a theory, but . . . " The opposite meaning is true in science. A theory is as close to a complete explanation as science has to offer and can be used to predict future outcomes.

All scientific theories are subject to revision and further analysis. New findings can alter an accepted explanation of observed facts.

A theory is a hypothesis supported by extensive experimentation.

2:7 Microscopes — Important Biological Tools

Observation, a key part of scientific methods, is especially important to a biologist. Because biology often calls for a study of extremely small objects — a tiny organism or the parts of a cell — observations often require use of microscopes.

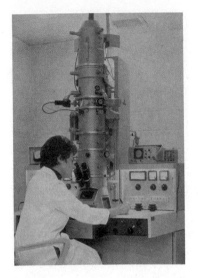

FIGURE 2–7. The electron microscope is a valuable tool in many fields of biology.

One nanometer is 0. 000 000 001 meter.

The TEM is useful in studying internal parts of cells.

The SEM produces a three-dimensional view of living specimens.

Magnification is an important property of microscopes. So, too, is resolving power. **Resolving power** is the ability to distinguish two objects as being separate. In your biology course you will be using a compound light microscope. Such a microscope makes use of glass lenses that focus light passing through an object. It can magnify, at the most, about 2000 times and has a resolving power of about 500 nanometers (nm). Two objects closer together than 500 nm will be seen as one object when viewed with a light microscope.

For greater magnification and resolving power, biologists now rely on electron microscopes. Such microscopes typically magnify hundreds of thousands of times and have a resolving power as great as 0.2 nm. Thus, they provide much greater detail and enable biologists to study inner and outer structures of even the smallest cells. As the name suggests, a beam of electrons, rather than light, strikes the object being viewed.

In order to view parts of very small structures, biologists use the **transmission electron microscope (TEM)**. The sample to be studied is frozen and then sliced into very thin sections. Thus, living specimens cannot be viewed. The electrons are focused by magnets instead of glass lenses. The electrons form an image that can be photographed or seen on a screen. The TEM has played an important role in the study of inner cell parts. It reveals details that aid in understanding how such parts function.

A new device, the **scanning electron microscope (SEM)**, permits a view of surfaces not possible with the TEM. Although its magnification is lower, it produces a three-dimensional image. This is possible because the beam of electrons does not pass through the object. Rather, it sweeps across it and bounces off. As it does so, it causes electrons to be knocked loose from the object. These electrons produce patterns seen as an image on a television-like screen.

FIGURE 2–8. Both the light microscope and the electron microscope direct a beam through a specimen, focus it with lenses, and enlarge the final image. They differ in that the light microscope focuses a beam of light with glass lenses whereas the electron microscope focuses a beam of electrons with magnets.

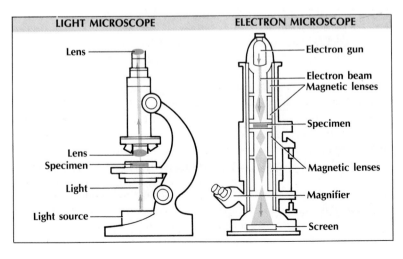

a

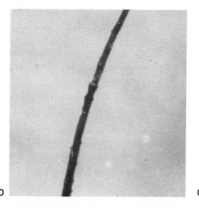
b

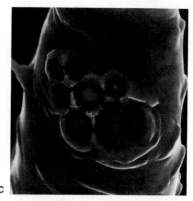
c

Because thin sections are not used, living organisms can be viewed and details about their surfaces can be learned. Knowledge of the surfaces of cells, for example, is important in studying cancer. Interactions of cells with each other and with chemicals can also be studied using the SEM.

FIGURE 2–9. (a) Aphids are small insects that can be seen with the unaided eye. Details of an aphid can be seen by use of microscopes. (b) The antenna can be seen enlarged with the light microscope and (c) sensors on the antenna can be seen with the electron microscope.

REVIEWING YOUR IDEAS

1. Compare science to other areas of study.
2. What is a fact? What are data?
3. What elements are necessary for interpretation?
4. What are the functions of a hypothesis?
5. What is a controlled experiment? Why is it necessary?
6. Distinguish between magnification and resolving power.

DIVERSITY OF ORGANISMS

You learned in Chapter 1 that there are more than two million different forms of life. No biologist is familiar with all, or even most, of these different kinds of organisms. Fortunately, though, these many kinds of organisms can be placed together in major groups based on certain common characteristics. Every biologist, regardless of his or her special field of interest or current research, does know about these major groups of organisms and their features. As you proceed with your biology course, it will be useful for you to have a broad knowledge of the major groups of organisms alive today. In Chapter 12 you will study in greater detail the basis for grouping and classifying living things.

Biologists group organisms together on the basis of common characteristics.

2:8 Grouping Organisms

Think back to your imaginary trip through a forest community (Section 1:1). Recall some of the organisms in such a community. There are mosses, ferns, shrubs, and trees. There are snakes, insects, deer, and birds. Other forms include bacteria and *Amoeba;* algae and mushrooms. How do biologists make sense out of the great diversity of life? Is there a way to group organisms that are so different?

For now, you will learn about the most general way in which organisms are grouped. All the various forms of life can be placed in five different categories, called kingdoms. A **kingdom** is the broadest division in the classification of organisms.

If you have a coin or stamp collection, you know that there must be a basis for organizing that collection. You arrange your coins or stamps by groups according to value or country of origin, for example. One basis upon which biologists group organisms is by similarity in structure. In Chapter 12 you will learn more about this and other bases for classification.

Kingdoms are the broadest groups into which organisms are classified.

Similarity of structure is one important basis for grouping organisms.

2:9 Kingdoms of Life

Below is a list of the five kingdoms. The major characteristics of organisms in each kingdom are given as well as examples of organisms.

• Kingdom **Monera** (muh NIHR uh). This is the smallest kingdom. It consists of two major groups of bacteria. Organisms are grouped in this kingdom based upon their primitive cell structure. Monerans are usually **unicellular** (composed of just one cell).

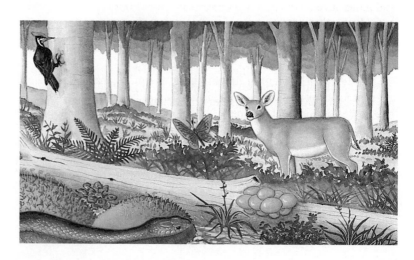

FIGURE 2–10. There is great diversity in the organisms in a community. What common characteristics would you use to group them?

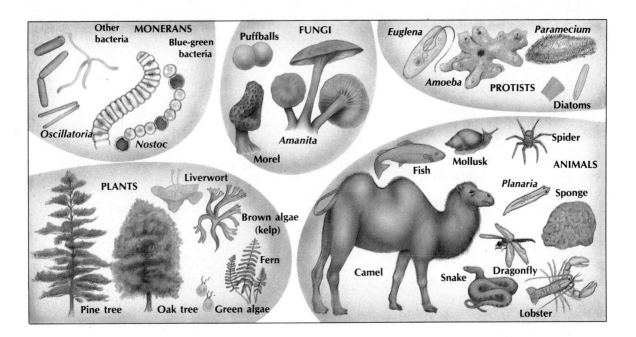

Kingdom **Protista** (pruh TIHS tuh). Most protists are unicellular. The kingdom includes both producers and consumers. In this group are animallike forms such as *Amoeba* and *Paramecium* and certain simple algae.

Kingdom **Fungi.** Most organisms in this kingdom are **multicellular** (composed of more than one cell). All fungi are consumers that absorb small food molecules. Included in this kingdom are mushrooms, molds, mildews, and yeast.

Kingdom **Plantae.** Most plants are multicellular, but some unicellular algae are included in this kingdom. All plants are producers and most have a complex structure. In addition to green, brown, and red algae, this group includes more familiar land forms such as mosses, ferns, conifers, and flowering plants.

Kingdom **Animalia.** Animals are multicellular organisms that are usually mobile at some stage of their life. They are consumers and ingest their food. **Ingestion** is the taking in of solid particles. Except for sponges, all animals have nerves. Animals include such diverse forms as jellyfish, worms, snails, starfish, crabs, centipedes, sharks, frogs, eagles, and humans.

FIGURE 2–11. The classification system used in this text has five kingdoms of organisms.

Notice that algae are found in three of the five kingdoms. The word algae is not a technical term. It refers to a variety of simple organisms possessing chlorophyll.

REVIEWING YOUR IDEAS

7. Identify one basis for grouping organisms.
8. What is the broadest division for grouping organisms?
9. Identify the five kingdoms of life.

INVESTIGATION

Problem: How does one test a hypothesis?

Materials:

2 beakers
paper toweling
40 radish seeds
2 plastic bags

4 labels
aspirin tablet
water
graduated cylinder

Procedure:

Part A. Forming a Hypothesis

1. It has been said that adding aspirin to plant seeds will help them grow faster than seeds not treated with aspirin. You will test this hypothesis.
2. Prepare a hypothesis for the problem stated above. Your hypothesis should be stated as an "if" and "then" statement. For example, *if* animals receive vitamin D in their diet, *then* they will not develop rickets.
3. Record your hypothesis in the space at the top of a copy of Table 2–1.

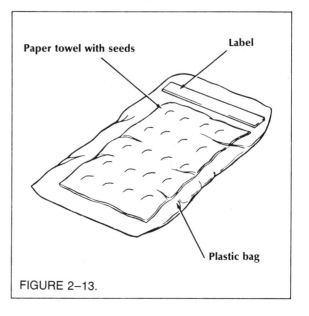

FIGURE 2–13.

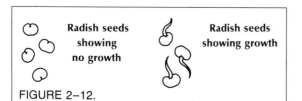

Radish seeds showing no growth

Radish seeds showing growth

FIGURE 2–12.

Part B. Testing Your Hypothesis

1. Prepare your seeds for testing by following the steps provided.
2. Count out 20 radish seeds.
3. Place these seeds into a beaker. Use a graduated cylinder to measure 20 mL of water. Add this water to the beaker.
4. Label the beaker with your name and the word "water."
5. Count out 20 radish seeds.

6. Place these seeds into a beaker. Use a graduated cylinder to measure 20 mL of water. Add this to the beaker along with one aspirin tablet.
7. Label the beaker with your name and the word "aspirin."
8. Allow the seeds to soak overnight.
9. The following day remove the seeds from the beaker marked "water" and place them between two wet paper towels. Fold the edges of the paper towel and slide the towel into a plastic lunch bag. Use Figure 2–13 as a guide.
10. Label the bag with your name and the word "water."
11. Also remove the seeds from the beaker marked "aspirin" and place them between two wet paper towels. Fold the edges of the paper towel and slide the towel into a plastic lunch bag.

34

12. Label the bag with your name and with the word "aspirin."
13. Allow both bags to remain undisturbed for two days.
14. After two days, open both bags and examine the seeds for evidence of growth. Those seeds that are growing will show signs of a small root extending from the seed. Those not growing will show no root. Use Figure 2–12 as a guide.
15. Count the number of seeds from the bag marked "water" that are growing, and those that are not. Record these numbers in Table 2–1.
16. Count the number of seeds from the bag marked "aspirin" that are growing, and those that are not. Record these numbers in Table 2–1.
17. Use class totals of all seeds used. Record the numbers growing and not growing.

Part C. Examining Your Hypothesis

1. Read your hypothesis written at the top of Table 2–1.
2. Considering the results of the experiment, does your hypothesis seem valid or must it be revised or modified?

Data and Observations:

Treatment	Individual Results				Class Results		
	Number of Seeds Used	Number of Seeds Growing	Number of Seeds Not Growing		Number of Seeds Used	Number of Seeds Growing	Number of Seeds Not Growing
Water	20						
Water and aspirin	20						

TABLE 2-1. RADISH SEED GROWTH

Questions and Conclusion:

1. Define hypothesis, control, and variable.
2. Which part of this experiment represents the: (a) data (b) control (c) variable?
3. Based on your original hypothesis, does class data appear to support it? Explain.
4. Restate a new hypothesis based on experimental findings for the class. Use the "if-then" format.
5. Explain why testing or experimentation is needed before accepting a hypothesis.
6. Suggest a reason why it was best to use class totals rather than only your data when drawing conclusions about your hypothesis.
7. A hypothesis helps to predict new facts. Determine if the following could possibly be predicted based on your corrected hypothesis regarding aspirin and seed growth. Explain.
   (a) Aspirin added to pea seeds will reduce the amount of growth.
   (b) Aspirin added to soil will help houseplants to form greener leaves.
   (c) Aspirin added to corn seeds will increase the amount of growth.

Conclusion: How does one test a hypothesis?

CHAPTER REVIEW

SUMMARY

1. An interest in themselves and their world led humans to study nature. **p. 23**
2. Science involves methods that produce a body of knowledge about nature. **2:1**
3. Observation leads to discovery of facts or data. **2:2**
4. Scientists logically interpret facts in order to explain observations of nature. **2:3**
5. A hypothesis is a statement that explains and relates facts. **2:4**
6. A controlled experiment is used to determine whether or not a hypothesis is correct. The experimental group is compared to the control group. **2:5**
7. Theories are major hypotheses that have survived repeated testing. **2:6**
8. Use of electron microscopes has yielded much information about the structures and functions of organisms and their parts. **2:7**
9. Biologists group all organisms into five broad groups called kingdoms. **2:8**
10. Monera, Protista, Fungi, Plantae, and Animalia are the five kingdoms of life. **2:9**

CHECKING YOUR IDEAS

On a separate paper, complete each of the following statements with the missing terms. Do not write in this book.

1. Applied science is also called _____.
2. A statement that explains and relates observations and also predicts new facts is a(n) _____.
3. Biologists group organisms into five very broad categories called _____.
4. Scientists use a universal set of measurements known as _____.
5. A controlled experiment has only one _____ factor.
6. The _____ electron microscope is used to view the surfaces of cells.
7. A(n) _____ is made by using one of your senses.
8. _____ are organisms having the simplest cell structure.
9. Facts are also known as _____.
10. A major hypothesis that has survived extensive testing over a long period of time is called a(n) _____.

LANGUAGE OF BIOLOGY

control group
controlled experiment
data
experimental group
experimentation
fact
hypothesis
ingestion
International System
  of Measurement
kingdom
multicellular

observation
resolving power
scanning electron
  microscope (SEM)
science
technology
theory
transmission electron
  microscope (TEM)
unicellular
variable factor

EVALUATING YOUR IDEAS

1. Why did some of the unconfirmed conclusions of early scientists prevail for more than a thousand years?
2. List some of the major biological discoveries made since the late 1800s.
3. Is a "good" hypothesis necessarily correct? Explain.
4. What hypothesis did Fleming attempt to test? Describe his test of that hypothesis, identifying the control group, experimental group, and variable factor.

5. What benefits were derived from Fleming's work? What point does this illustrate about science in general?

6. Why do scientists prefer not to use the word "proof"?

7. List some differences between light microscopes and electron microscopes.

8. Describe the purposes for which the TEM and SEM are used.

9. Fungi and animals are both consumers. How do they differ in terms of obtaining their food?

10. To what kingdom of organisms does each of the following descriptions apply?
    (a) mostly unicellular; advanced cell structure; both producers and consumers
    (b) generally multicellular, but some unicellular forms; all are producers
    (c) usually unicellular; both producers and consumers; primitive cell structure

11. To what kingdom does each of the following organisms belong?
    (a) bread mold
    (b) cat
    (c) bacterium
    (d) apple tree
    (e) moss
    (f) earthworm
    (g) *Paramecium*

3. A student wishes to test the hypothesis that plants grow taller if a certain mineral is added to their soil. The student uses two groups of plants, cacti and geraniums. The cacti, since they are desert plants, are given little water and are grown in a hot environment. The geraniums are given more water and grown in a cooler environment. The mineral is added only to the soil in which the geraniums are growing. Is the student conducting this experiment properly? Explain.

EXTENDING YOUR IDEAS

1. Prepare a report on several kinds of antibiotics. Include in your report the names of the organisms from which the antibiotics are prepared and the diseases against which they are useful.

2. Many molds growing in soil have antibiotic properties. Collect soil samples from different areas and place them on agar in different culture dishes. Transfer the different molds that appear to agar plates on which bacteria are growing. Determine whether any of these molds produce antibiotics.

SUGGESTED READINGS

Binnig, Gerd, and Rohrer, Heinrich, "The Scanning Tunneling Microscope." *Scientific American*, August, 1985.

Hawking, Stephen, "Handicapped People and Science." *Science Digest,* September, 1984.

Rensberger, Boyce, *How the World Works: a Guide to Science's Greatest Discoveries.* New York, New York, Morrow, 1986.

Winerip, Michael, "A Serendipitous Sort of Scientist." *Discover,* July, 1986.

APPLYING YOUR IDEAS

1. How would you conduct an experiment to test the hypothesis that the rate of photosynthesis increases as the amount of available light increases?

2. The micrometer (0. 000 001 m) and nanometer (0. 000 000 001 m) are units of linear measurement. Why are these units important to some biologists?

MATERIALS OF LIFE

Water, one of the most common and most vital substances on Earth, is the medium in which many chemical reactions take place. It is also involved in dissolving and transporting substances for use by the body. Water is a raw material needed by plants during the process of making their food. Of what is water composed? What are its properties? What other substances are considered to be basic materials of life?

chapter 3

For a long time biology was mainly a study of different organisms. Organisms were described, grouped, and named by their features. Early biologists were curious about the **anatomy** (study of internal and external structure) of organisms. They then became curious about *how* the parts worked. Study of functions of organisms is known as **physiology** (fihz ee AHL uh jee). The beating of the heart and the flowering of a plant are examples of physiological processes. How does a nerve transmit an impulse? What causes a muscle to contract? How are minerals moved from the roots of trees to the leaves?

In order to understand the life processes of organisms, biologists must have a knowledge of chemistry and physics. They must understand the properties of matter and be able to relate those properties to living systems. Such knowledge is needed for the understanding of many biological functions.

Objectives:
You will
- define matter and its properties.
- describe changes in matter.
- name important biological compounds and their roles.

Physiology is the study of the functions of organisms.

A knowledge of matter and its properties is needed to understand physiological processes.

CLASSIFICATION OF MATTER

You live in a world of both living and nonliving things. In Chapter 1 you learned about several features that distinguish them. It may surprise you to realize that even though living and nonliving things are quite different, they are composed of the same basic building blocks.

3:1 Atoms

Matter is anything that has mass and takes up space. All matter is composed of building blocks called **atoms.** Atoms are very small, most of them being from 0.1 to 0.5 nm in diameter. Today only the most powerful microscopes can photograph individual atoms. Much detail still cannot be seen.

Matter has mass and takes up space.

39

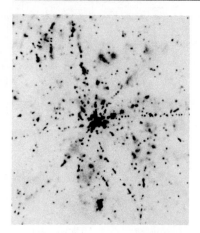

FIGURE 3–1. Because atoms are so small, not much detail can be seen in photographs of them. The phosphorus atoms found in the DNA of a bacteriophage are shown here.

Electrons form electron clouds around the nucleus of an atom. Some electrons are farther from the nucleus than others.

Different energy levels, when filled, have different numbers of electrons.

Atoms are composed of three major particles—protons, neutrons, and electrons. **Protons** and **neutrons** are bunched tightly together in the center to form the **nucleus** of an atom. Moving rapidly outside this nucleus are **electrons.** Protons are positively charged, neutrons have no charge, and electrons are negatively charged. Electrons are sometimes indicated as e^-. In an atom, the number of protons equals the number of electrons. Thus, the atom as a whole is neutral; it has no electric charge.

Diagrams have been drawn based on information gathered about atoms. A diagram of an atom is just another way of showing its parts. Such diagrams show protons and neutrons as tightly packed particles at the center of the atom. Electrons move about the nucleus at certain *probable* distances from the center. The space that electrons fill forms an electron cloud around the nucleus. The **electron cloud** represents the space around the nucleus that the electrons *can occupy.*

Hydrogen, with one proton and one electron, is the simplest atom. A helium atom has two protons, two electrons, and two neutrons. In a helium atom, the two electrons are about the same distance from the nucleus. Lithium (Figure 3–2) has three electrons per atom. One of the electrons is usually farther from the nucleus than the other two.

The different regions in which electrons travel about the nucleus are called **energy levels.** The first energy level holds a maximum of two electrons. The second energy level has a maximum of eight electrons, and the third energy level has eighteen electrons. Aluminum (Figure 3–2) has thirteen electrons per atom. Electrons of each aluminum atom would usually be arranged as follows: two in the first level, eight in the second level, and three in the third level. How would electrons be arranged in an atom with seven electrons? Eighteen electrons?

FIGURE 3–2. Diagrams of a hydrogen, lithium, and an aluminum atom show the energy levels at which electrons move around the nuclei of the atoms. The best description of the probable location of electrons in an atom is an electron cloud.

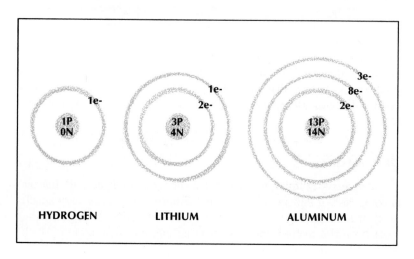

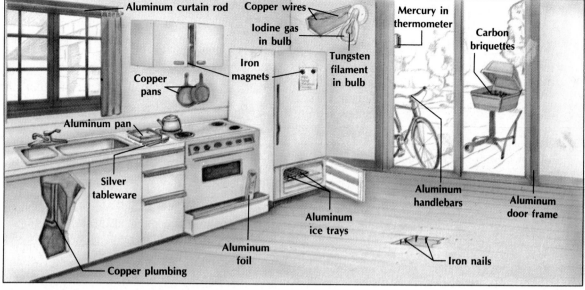

FIGURE 3–3. Many common objects are composed of one element each.

3:2 Elements

Hydrogen, helium, lithium, and aluminum are all **elements,** substances composed of only one type of atom. The number of protons in the nucleus determines the type of atom. For example, a piece of copper has only copper atoms. Each copper atom has the same number of protons, 29. Each atom in an iron bar has 26 protons.

Any given proton, neutron, or electron in a substance is the same as that in any other substance. The *combination* of these particles makes one kind of atom different from another. All copper atoms are identical to one another in the number of protons and electrons. But iron atoms differ from copper atoms in the number of protons and electrons each type has.

Atoms of the same element can differ in their number of neutrons. Such atoms are called **isotopes** (I suh tohps) of each other. Atoms of hydrogen usually contain no neutrons, but one isotope (heavy hydrogen) has one neutron, and another isotope (heavy, heavy hydrogen) has two neutrons. All three isotopes of hydrogen have one proton and one electron.

All atoms of a given element have the same number of protons in their nuclei.

Isotopes are atoms that differ from other atoms of the same element only in the number of neutrons in their nuclei.

REVIEWING YOUR IDEAS

1. What is matter? Of what is matter composed?
2. List the three major kinds of particles of an atom. Where is each located and what is its charge?
3. Why does a free atom have no electric charge?
4. What is an element? What is an isotope?

3:3 Molecules

Although only ninety elements exist naturally on Earth, there are thousands of different substances. Atoms combine with one another to form different substances. Often two or more atoms of the same element are combined. For example, the hydrogen atom never occurs alone in nature. It always combines with other atoms, including other hydrogen atoms.

Most atoms combine in such a way that their outer energy levels acquire a total of eight electrons. This principle is called the **octet rule.** (In the case of atoms with only one energy level, that energy level acquires a total of two electrons — the **duet rule.** For example, a hydrogen atom has only one electron in its outer level. In combining with other atoms, it must acquire one more electron.) An atom of nitrogen has five electrons in its outer level. When it combines with other atoms, it must acquire three more electrons.

Two hydrogen atoms can join by *sharing* their electrons (Figure 3–4) so that each hydrogen atom has two electrons in its outer energy level. When this sharing occurs, the electrons occupy the space between and around the two nuclei.

Atoms that have joined are held together by forces called *chemical bonds.* The forces involved in the sharing of electrons, such as between hydrogen atoms, are known as **covalent** (koh VAY lunt) **bonds.** Such a combination of two or more atoms joined by a covalent bond is called a **molecule.**

When simple atoms combine, their outer energy levels acquire an outer octet of electrons.

FIGURE 3–4. Two hydrogen atoms join to form a hydrogen molecule by sharing electrons.

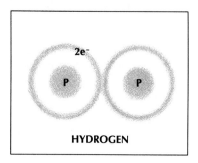

HYDROGEN

3:4 Compounds

Suppose you had a small, covered bottle of hydrogen gas (hydrogen molecules). The air in the room would contain oxygen molecules. Because the bottle is covered, the hydrogen and oxygen molecules cannot meet. If you removed the cover, oxygen would enter the bottle and collide with the hydrogen, but the two types of molecules would not combine. However, if an outside source of energy is applied, drops of water would form on the inside of the bottle (Figure 3–5). What has happened?

The energy source weakened the force of attraction within the two kinds of molecules. Bonds between the atoms in hydrogen molecules and between the atoms in oxygen molecules broke. The atoms then recombined to form water.

In order to form water, two hydrogen atoms must combine with one oxygen atom. Each hydrogen atom *shares* its electron with the oxygen atom. Thus, the oxygen atom gains the two electrons it needs — one from each hydrogen atom. Each hydrogen atom obtains its needed electron by sharing electrons with the oxygen atom. Thus, the duet and octet rules are obeyed because each hydrogen atom

Energy increases the force of collisions between particles.

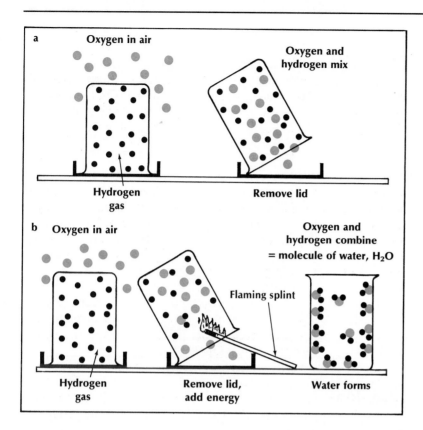

FIGURE 3–5. (a) Hydrogen and oxygen mix when no energy source is added. (b) With an added energy source, hydrogen and oxygen combine to form water.

requires one electron and each oxygen atom requires two electrons. Each atom obtains its required number of electrons by sharing. As a result, the atoms are bonded together forming a new substance. The bond is a covalent bond because it is formed by the sharing of electrons.

This combination of two hydrogen atoms and one oxygen atom is also a molecule, a molecule of water. Water, carbon dioxide, alcohol, and sugar are substances made of two or more kinds of atoms. Substances composed of different kinds of atoms chemically bonded together are known as **compounds.**

When atoms combine with one another, the new substances formed have a definite composition. A molecule of hydrogen always contains two hydrogen atoms. Water molecules always consist of two hydrogen atoms and one oxygen atom. Definite composition is explained by the octet or duet rule.

Biologists and biochemists are interested in the chemical reactions that occur in organisms. Many of these reactions involve the synthesis (making) of compounds from atoms of other compounds or elements. The properties and processes of living things depend on the structures and reactions of compounds.

In water, covalent bonds exist between the hydrogen and oxygen atoms.

FIGURE 3–6. A water molecule is formed and outer energy levels are filled when two hydrogen atoms and one oxygen atom share electrons.

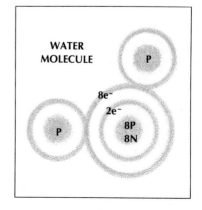

WATER MOLECULE

FIGURE 3–7. Sodium chloride forms when one electron is transferred from a sodium atom to a chlorine atom. The outer energy levels of both atoms are filled in this way.

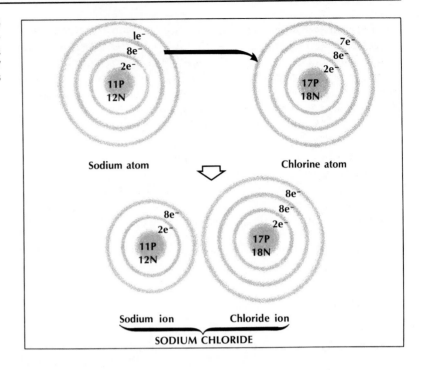

Sodium atom   Chlorine atom

Sodium ion   Chloride ion

SODIUM CHLORIDE

3:5 Ions

Some chemical bonds form by the transfer of electrons.

Sometimes electrons are *transferred* to form an outer octet or duet of electrons. Compounds can be formed as a result of electrons being added or taken away from atoms. Two simple atoms, sodium and chlorine, can combine in this way (Figure 3–7). They combine to form the compound sodium chloride, which is better known as table salt. A sodium atom has eleven electrons — two in the first energy level, eight in the second, and only one in its outer energy level. The outer level could be filled by "picking up" seven electrons or by "giving up" its one outer electron. Either way would give it an outer octet. A chlorine atom has seventeen electrons — two in the first energy level, eight in the second, and seven in the third. Because the atom has seven electrons in its outer energy level, it needs only one more. As sodium and chlorine atoms combine, each sodium atom "gives up" one electron, which is "picked up" by a chlorine atom. In this way, each atom obtains an outer octet of electrons.

Ions are charged atoms or groups of atoms.

A loss or gain of electrons results in an unbalanced electric charge on an atom. Because a sodium atom has eleven positively charged protons and ten negatively charged electrons after losing an electron, it is positive by one charge, 1+. After gaining an electron, a charged chlorine atom has an excess of one negative charge, 1−. Such charged atoms are known as **ions** (I ahnz). The electrons of each ion move about the nucleus of that ion.

Because sodium and chloride ions are oppositely charged, they attract each other. The attraction between ions is called an **ionic bond.** The number and kind of charge on each ion determine the composition of a compound. In order for a compound to form ionic bonds, the positive and negative charges must balance each other so the net charge is zero. Because sodium ions have a 1+ charge and chloride ions have a 1− charge, one of each ion gives a neutral compound. How would ions with a 2+ charge combine with ions with a 1− charge? Could two positive ions combine?

Instead of being called molecules, compounds of ions are called **ionic compounds.** Water is composed of water molecules in which each molecule contains two hydrogen atoms and one oxygen atom. However, a crystal of salt is made of a continuous network of alternating sodium and chloride ions. These ions are not paired with each other (Figure 3–8) to form a molecule. Salt is an ionic compound.

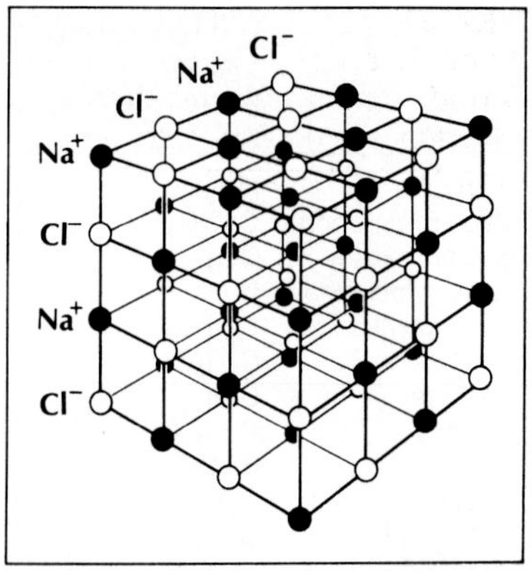

FIGURE 3–8. Ionic compounds do not exist as individual molecules. Salt, for example, is a crystal of alternating sodium and chloride ions.

3:6 Symbols and Formulas

Chemists have developed a shorthand system for representing elements. Each element is given a **symbol** agreed on by all scientists. This symbol consists of either one or two letters. Hydrogen has the symbol H, iron is Fe, and copper is Cu. Notice that when a symbol contains two letters, the second is not capitalized. *In addition to hydrogen, elements most commonly found in living systems are carbon (C), oxygen (O), and nitrogen (N).*

Each element can be represented by a symbol of one or two letters.

Carbon, hydrogen, oxygen, and nitrogen are the elements most abundant in organisms.

Chlorophyll
$C_{55}H_{72}MgN_4O_5$

Carbon dioxide
CO_2

Salt
NaCl

Baking soda
$NaHCO_3$

Hemoglobin
$(C_{738}H_{1166}FeN_{208}S_2)_4$

FIGURE 3–9. Chemical formulas can be written using chemical symbols for the atoms in the compounds and mixtures. Formulas of the items in this photo are shown.

TABLE 3-1. SOME COMMON COMPOUNDS

Name	Formula
ammonia	NH_3
calcium carbonate	$CaCO_3$
carbon dioxide	CO_2
carbon monoxide	CO
chlorophyll-a	$C_{55}H_{72}O_5N_4Mg$
ethyl alcohol	C_2H_5OH
fructose	$C_6H_{12}O_6$
glucose	$C_6H_{12}O_6$
glycerol	$C_3H_8O_3$
glycine	$C_2H_5O_2N$
lactic acid	$CH_3CHOHCOOH$
iron oxide (rust)	Fe_2O_3
sodium hydroxide	$NaOH$
sucrose	$C_{12}H_{22}O_{11}$
table salt	$NaCl$
water	H_2O

Chemical formulas are determined by the bonding requirements of the atoms of which the compound is composed.

A group of symbols, called a **chemical formula,** is used to show the number and kind of each atom in a compound. A molecule of water is composed of two hydrogen atoms and one oxygen atom. The chemical formula of water is H_2O. The subscript 2 refers to the element it follows. Subscripts in a formula represent the number of atoms of the element. When a symbol is not followed by a number, it is understood to represent one atom or ion.

In sodium chloride, the simplest ratio is one sodium ion to one chloride ion. Its formula is written $NaCl$. Notice that no subscripts are written because only one of each ion is involved.

The formulas H_2O and $NaCl$ represent one molecule of water and one unit of salt. Five water molecules would be indicated by a number in front of the formula. It would be written $5H_2O$.

How many atoms are there in $5H_2O$? The 5 refers to the entire molecule, H_2O. Thus, there are 5 times 2 atoms of hydrogen and 5 times 1 atom of oxygen. The total atoms of $5H_2O$ is $10 + 5 = 15$. How many atoms are there in $3NaCl$?

Chemists determine formulas by analyzing compounds in the laboratory. Some formulas are listed in Table 3–1. You do not need to learn many formulas in this course, but you should remember that the formula of a compound is determined by the bonding requirements of its atoms. Each compound has a definite composition that does not vary. You will be seeing formulas of biological substances throughout your study of biology.

A homogeneous substance is the same throughout.

FIGURE 3–10. Potassium permanganate, $KMnO_4$, is a solute that can be dissolved in water, a solvent. An aqueous (water) solution of potassium permanganate results.

3:7 Solutions

Hydrogen molecules and compounds, such as water and salt, are examples of homogeneous (hoh muh JEE nee us) substances. **Homogeneous** means the same throughout. Pure water in a beaker has definite composition and is homogeneous. Any water molecule is the same as any other (H_2O) and the molecules are evenly distributed as they move about.

Some homogeneous materials do not have a definite composition. Such materials are formed as a result of a physical combination of different substances. Suppose you dissolve one gram of sugar (a compound) in 100 mL of water (a compound). You would obtain a material that would be the same throughout. It would be homogeneous. If you added two grams of sugar to another 100 mL of water, you would again get a homogeneous material. The two materials would be the same, sugar water, but they would differ in composition. The second sample would contain a greater amount of sugar in the same amount of water.

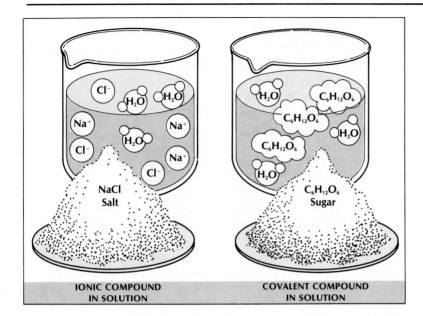

FIGURE 3–11. In solution, ionic compounds, such as salt, dissociate into ions. Covalent compounds, such as sugar, dissolve as separate molecules.

This type of homogeneous material is called a **solution.** True solutions are homogeneous, but they have variable compositions depending on how much of one substance is dissolved in the other. The portion of a solution in greater quantity is called the **solvent** (SAHL vunt); the part in lesser quantity is the **solute** (SAHL yewt). In the example, the solute, sugar, was dissolved in the solvent, water. In a water-alcohol solution in which there is more alcohol present than water, the alcohol is the solvent and the water is the solute.

Sugar is a covalent compound. As sugar crystals dissolve in water, the sugar molecules separate. The sugar molecules become dispersed (scattered) equally throughout the solvent. Ionic compounds do not behave in this way. When placed in a solvent, positive and negative ions of the compound undergo **dissociation,** a process whereby the ions separate. When salt dissolves in water, the sodium and chloride ions move independently through the water. Thus, a solution of sodium chloride in water consists of water molecules, sodium ions, and chloride ions (Figure 3–11). The units of which ionic compounds are composed behave quite differently in solution than do those of covalent compounds.

Water is an excellent solvent. Dissociation of ionic compounds within living systems is common because organisms are composed of about 70 percent water. Also, many ions that enter an organism are dissolved in water. Ions are important in many biological processes. For example, sodium (Na^+) and potassium (K^+) are essential to conduction of nerve impulses. Other ions are involved in processes such as muscle contraction. Ions are also parts of bone, blood, and other tissues.

A solution is a homogeneous substance that can have different compositions.

A solute dissolves in a solvent.

Covalent compounds dissolve as separate molecules.

When ionic compounds dissolve, their ions separate and move independently in the solution.

Water is an important solvent in living things.

3:8 Acids and Bases

Each water molecule contains two hydrogen atoms and one oxygen atom. Very rarely (about two molecules per billion) one hydrogen atom breaks away from the rest of a molecule. The formation of a positively charged hydrogen ion (H^+) and a negatively charged hydroxide ion (OH^-) results.

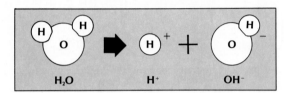

Therefore, in a given volume of pure water, most of the water exists as water molecules, but there are some hydrogen ions and some hydroxide ions. The number of hydrogen and hydroxide ions is equal because one of each kind of ion is produced from each water molecule that dissociates. A solution (or pure water) in which the concentration of hydrogen ions equals the concentration of hydroxide ions is said to be **neutral.**

In a neutral solution, the hydrogen ions and hydroxide ions are equal in concentration.

Hydrogen chloride is a gaseous compound that dissolves in water to form hydrogen ions and chloride ions. Addition of hydrogen chloride increases the concentration of hydrogen ions in the water (Figure 3–12). If the concentration of hydrogen ions is greater than the concentration of hydroxide ions, the solution is called an **acid.** A solution of hydrogen chloride in water is called hydrochloric acid. Acid solutions have a sour taste. Lemons and grapefruit are sour because they contain citric acid.

In an acid solution, the concentration of hydrogen ions is greater than the concentration of hydroxide ions.

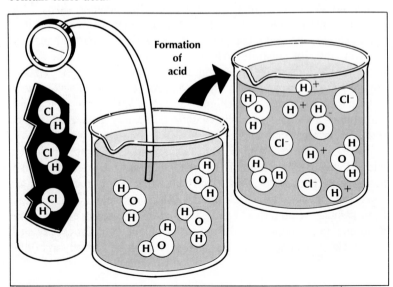

FIGURE 3–12. When an acid is formed, the concentration of hydrogen ions is greater than the concentration of hydroxide ions.

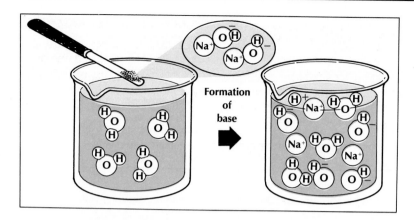

FIGURE 3–13. When a base is formed, the concentration of hydroxide ions is greater than the concentration of hydrogen ions.

If the concentration of hydroxide ions in a solution exceeds the concentration of hydrogen ions, the solution is a **base.** Basic solutions have a sharp, bitter taste, and they feel slippery. In water, sodium hydroxide dissociates into a solution of sodium ions and hydroxide ions. The solution is basic because of an excess of hydroxide ions (Figure 3–13).

In a basic solution, the concentration of hydroxide ions is greater than the concentration of hydrogen ions.

Chemists have devised a system for indicating the relative concentration of ions in acids and bases. The system is a scale of numbers, called the **pH scale.** The scale runs from 1 to 14 with 7 representing a neutral solution. Numbers below 7 represent acids. The lower the number, the greater the concentration of hydrogen ions and the more acidic the solution. A solution with a pH of 2 has a greater concentration of hydrogen ions and is more acidic than a solution with a pH of 6. Numbers above 7 represent bases. The higher the number, the greater the concentration of hydroxide ions and the more basic the solution. Therefore, a solution with a pH of 12 has a greater concentration of hydroxide ions and is more basic than a solution with a pH of 9.

Concentrations of ions in acids and bases are represented by a set of numbers called the pH scale.

FIGURE 3–14. (a) When the pH of substances is determined, it is a number between 1 and 14. (b) The pH can be determined with pH paper or with a pH meter. The pH paper changes color indicating pH. The pH meter indicates pH directly on a scale. Is the solution tested here acidic or basic?

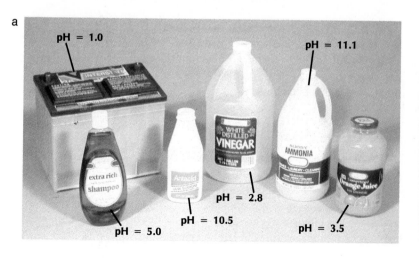

TABLE 3-2. RELATIVE CONCENTRATIONS OF HYDROGEN AND HYDROXIDE IONS IN SOLUTIONS			
pH	Concentration of Hydrogen Ions	Concentration of Hydroxide Ions	Kind of Solution
<7	Greater than concentration of hydroxide ions	Less than concentration of hydrogen ions	Acid
7	Equal to concentration of hydroxide ions	Equal to concentration of hydrogen ions	Neutral
>7	Less than concentration of hydroxide ions	Greater than concentration of hydrogen ions	Base

The table above summarizes the meaning of pH scale measurements.

pH influences the habitats and chemical reactions of organisms.

The pH of solutions is important to living systems. The pH of water determines which organisms will live in the water, and the pH of soil determines which organisms will grow in the soil. Also, the pH of solutions influences many of the chemical reactions that take place in organisms. For example, in order for digestion to occur properly, the pH of the stomach must be about 2.

REVIEWING YOUR IDEAS

5. What is a molecule? A compound?
6. What are ions? How are they formed?
7. What information does a chemical formula provide?
8. What is a solution? What are the two parts of a solution?
9. Compare neutral, acidic, and basic solutions. What is pH?

CHANGE IN MATTER

Each of the various forms of matter—element, molecule, compound, solution—has certain properties. These properties describe the characteristics of the substance itself or how it reacts in the presence of other substances.

3:9 Properties of Matter

A piece of wood has a definite mass, volume, shape, color, hardness, and texture. Such features that describe a piece of matter are called **physical properties.**

Physical properties and appearance may change. The size, shape, and mass of a piece of wood can be changed by carving it, but after the piece is carved, it is still wood. Any change in the physical properties

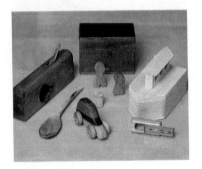

FIGURE 3–15. The physical properties of a wooden log are changed when the wood is carved (a physical change).

of a substance that does not change the arrangement of atoms is a **physical change.** Other examples of physical changes include stretching a rubber band and molding clay.

What happens when you burn a piece of wood? Many of its properties are changed. When wood is burned completely, the only visible remains are ashes that do not resemble the wood.

Burning is a **chemical change,** a change that results from rearranging the atoms or molecules of the materials involved. A chemical change may involve the formation of new bonds between two or more substances or the breaking of bonds between atoms. When wood burns, it combines with oxygen in the air. The carbon, hydrogen, and oxygen atoms of the molecules in wood are rearranged to form new substances with new properties. Among the new substances formed are the invisible gases carbon dioxide and water vapor. Such a change is called a **chemical reaction.**

A property of the wood is that it can be burned—wood can combine with oxygen to form new substances. A **chemical property** is a property that describes how a substance changes into a new substance or substances. For instance, it is a chemical property of sodium atoms to combine with chlorine atoms to form a new substance, sodium chloride (table salt). The properties of salt are entirely different from the properties of either sodium or chlorine.

FIGURE 3–16. When wood is burned, the chemical properties are changed (a chemical change).

A chemical change involves the rearranging of atoms or molecules as new substances are formed.

Chemical properties describe how a substance reacts with other substances.

3:10 Chemical Equations

Scientists use chemical shorthand to describe chemical reactions. When iron rusts, it reacts chemically with oxygen. The chemical reaction of iron and oxygen to form iron oxide can be expressed in words:

Write a word equation for the formation of table salt.

Iron plus oxygen yields iron oxide.

This statement can be shortened. A + can be used for "plus" and an → can be used for "yields":

iron + oxygen ⟶ iron oxide

This reaction can also be shown using symbols and formulas rather than words to represent the elements and compounds involved. A statement describing a chemical reaction with symbols and formulas is a **chemical equation.** All chemical reactions can be written as chemical equations. In the equation for the formation of iron oxide, iron and oxygen are called **reactants,** and iron oxide is the **product** of the reaction.

A chemical equation describes a chemical reaction.

A chemical equation is like an algebraic equation in that both sides must represent the same quantity. *The number of atoms of each element on each side of the arrow must be equal.* Why is this so? It is because of the **law of conservation of mass:** *during a chemical reaction, mass is neither created nor destroyed.* Therefore, a chemist would write the equation for the formation of iron oxide as follows:

$$4Fe + 3O_2 \longrightarrow 2Fe_2O_3$$

The sum of each kind of atom is the same on both sides of a balanced equation.

This statement is a **balanced equation** because the number of atoms of each element is the same on both sides. Four iron atoms must combine with three molecules of oxygen to produce two units of iron oxide.

REVIEWING YOUR IDEAS

10. What is a physical change?
11. What occurs during a chemical change? How do physical changes and chemical changes differ?
12. What is a chemical equation? What is a balanced equation? Why must an equation be balanced?
13. How are chemical equations and reactions related?

BIOLOGICAL CHEMISTRY

Scientists have been aware for many years of complex compounds being produced by living things. Because they were found in organisms, these substances were named organic compounds. At first scientists thought organic compounds could be produced only by living systems and not in the laboratory. Chemists have since learned to make them, but the name organic compound has remained.

FIGURE 3–17. An atom of carbon has four electrons in its outer energy level. It forms bonds by sharing electrons with other atoms.

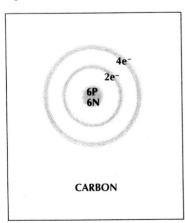

CARBON

3:11 Carbon Compounds

Analysis of **organic compounds** shows that they all contain the element carbon. They now are also called carbon compounds. The study of these compounds is called **organic chemistry.**

Carbon atoms are relatively simple in structure with six electrons. Because carbon has four electrons in its outer level (Figure 3–17), a carbon atom "needs" four more electrons in forming bonds with other atoms (Section 3:3). Carbon atoms obtain these electrons by forming covalent bonds.

A carbon atom can join with other kinds of atoms and with other carbon atoms. Many of the complex molecules found in living systems are composed of long chains or rings of carbon atoms. Attached to the carbon atoms are atoms such as hydrogen, oxygen, and nitrogen. Inorganic compounds (those not containing carbon) consist of small numbers of atoms and are not nearly so complex.

Chemical formulas such as H_2O and NaOH are simple formulas. A **simple formula** shows the kinds and numbers of atoms in a molecule or unit of an ionic compound. Simple formulas can also be written for organic compounds. Usually, though, chemists use structural formulas to show organic compounds. A **structural formula** gives the same information as a simple formula. In addition, it shows how the atoms are arranged in space.

Methane is a compound composed of one carbon atom and four hydrogen atoms per molecule. Its simple formula is CH_4, but its structural formula shows more about the compound. The formula shown in Figure 3–18 is a two-dimensional drawing. It shows that each of the four hydrogen atoms is bonded to the central carbon atom. (If the formula were drawn in three dimensions, the true shape of the molecule, it would resemble a pyramid.) Each line drawn between atoms represents a covalent bond, the sharing of one pair of electrons.

In some structural formulas you will see two or even three lines between atoms. One line between two atoms is a single covalent bond (C—C). Two lines between two atoms is a double covalent bond (C=C), and represents the sharing of two pairs of electrons. Three lines would be a triple bond (C≡C) in which three pairs of electrons are shared. Because a hydrogen atom shares only one electron, each hydrogen atom has only one bond.

Figure 3–19 shows two possible structural formulas for C_4H_{10}, butane. In each formula, each carbon atom has four bonds and each hydrogen atom has one bond as required by electron numbers. Compounds that have the same simple formula but different structures are called **isomers.** Isomers have different physical and chemical properties. The existence of isomers is another reason chemists use structural formulas to represent organic chemicals. With structural formulas, there is usually no doubt about which isomer is meant.

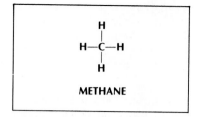

FIGURE 3–18. A structural formula of methane shows the arrangement of the atoms.

Structural formulas show the arrangement of atoms in space.

FIGURE 3–19. Each of these structures has the formula C_4H_{10}. Because the arrangement of the atoms differs, the structures are isomers.

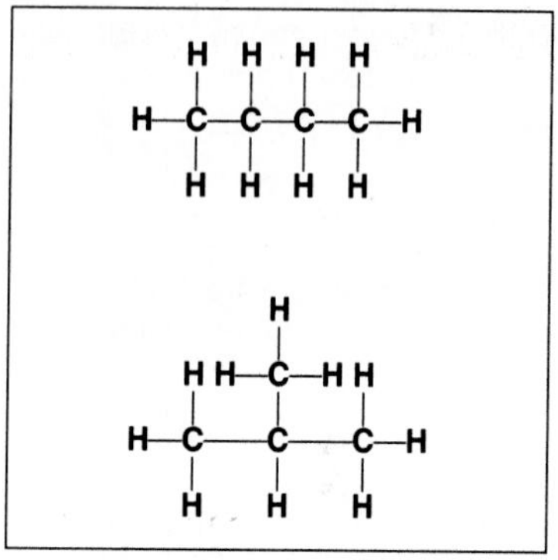

REVIEWING YOUR IDEAS

14. How is the term "organic compound" defined today? How is this definition better than the original meaning?
15. What information is given by a structural formula that is not given by a simple formula?
16. Which type of formula, structural or simple, gives more information about a compound?
17. What is an isomer? Which of the following are isomers?

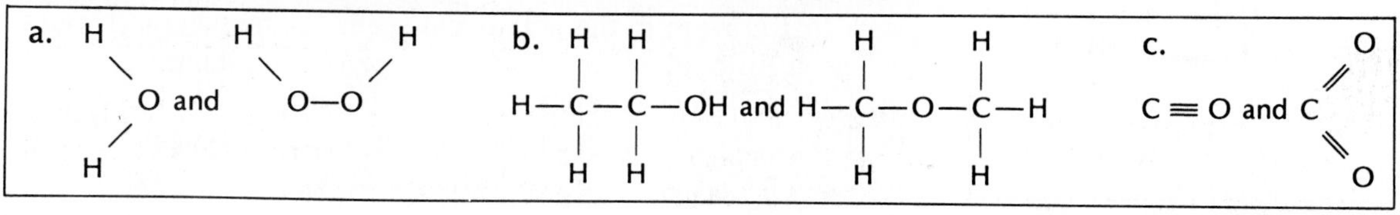

a   b   c

FIGURE 3–20. There are various sources of (a) carbohydrates, (b) fats, and (c) proteins.

Carbohydrates, composed of carbon, hydrogen, and oxygen, are used mainly as energy sources.

Monosaccharides are the basic building blocks of carbohydrates.

FIGURE 3–21. The structural formula of glucose, a monosaccharide, shows a ring made of one oxygen and five carbon atoms.

Glucose ($C_6H_{12}O_6$)

3:12 Carbohydrates

One of the most important groups of organic compounds of living systems is the carbohydrates. **Carbohydrates** are composed of carbon, hydrogen, and oxygen with a ratio of two hydrogen atoms to one oxygen atom. Glucose, the sugar produced by plants, is an important carbohydrate (Figure 3–21). Glucose and most other carbohydrates are energy sources for organisms. What would happen if all plants suddenly stopped producing glucose?

Glucose, $C_6H_{12}O_6$, is an example of the simplest type of carbohydrate, a **monosaccharide** (mahn uh SAK uh ride), or single sugar. Monosaccharides are the building blocks of complex carbohydrates.

Two monosaccharides can join to form a **disaccharide,** or double sugar. Table sugar (sucrose) is a disaccharide formed from joining one molecule each of glucose and fructose (another monosaccharide). Maltose is a disaccharide formed from two molecules of glucose. Both of these disaccharides have the simple formula $C_{12}H_{22}O_{11}$. Lactose, another isomer, is formed from one molecule each of glucose and galactose.

Larger carbohydrate molecules, **polysaccharides,** are also found in living systems. The most complex of all carbohydrates, the **starches,** are composed of hundreds of monosaccharides linked together. Plants produce more sugar than they use. They "store" excess sugar in the form of starch. A potato, kernels of corn, and beans are rich in starch. This starch will be used as food for the new plants that grow from the potato and seeds. Some polysaccharides, such as **cellulose,** are used by plants for structural purposes (Section 4:13). Cellulose is similar to starch. **Glycogen** (GLI kuh jun) is a starchlike carbohydrate in animals. When needed, glycogen is broken down into monosaccharides, which are "burned" for energy.

3:13 Lipids

Lipids are a class of organic compounds that includes fats, waxes, and oils. They are often used for energy and stored as energy reserves. Usually, solid fats are produced by animals, and liquid oils are produced by plants. Waxes are produced by both plants and animals. Like carbohydrates, lipids are composed of carbon, hydrogen, and oxygen, but the number of carbon and hydrogen atoms per molecule is much greater than the number of oxygen atoms. Thus, most lipid molecules are more complex than those of carbohydrates.

A fat molecule is formed from two less complex molecules: **fatty acids** and **glycerol** (GLIHS uh rawl). There are many different fatty acid molecules, but all of them have a certain part in common, a **carboxyl** (kar BAHK sul) **group** (—COOH). In the structural formula of a fatty acid (Figure 3–22), R represents a long chain of carbon and hydrogen atoms. The exact number, arrangement, and bonding of these atoms accounts for the differences among fatty acids.

Notice that there is a double bond (Section 3:11) between the carbon atom and one oxygen atom in the carboxyl group of each fatty acid. The other oxygen atom also shares two electrons, but with two different atoms. Note also that each carbon atom shares four electrons and each hydrogen atom shares one.

Not all fats are used for energy. Some have structural uses. For example, both cells and some of their inner parts have membranes composed of fats (Section 4:4). Cholesterol, a different type of lipid, is also part of some cell membranes.

3:14 Proteins

Proteins are the building blocks of living material. They make up much of the structure of living things. Proteins are important in the growth, maintenance, and repair of living material. Certain proteins are also essential in running the chemical reactions in living organisms (Section 4:3).

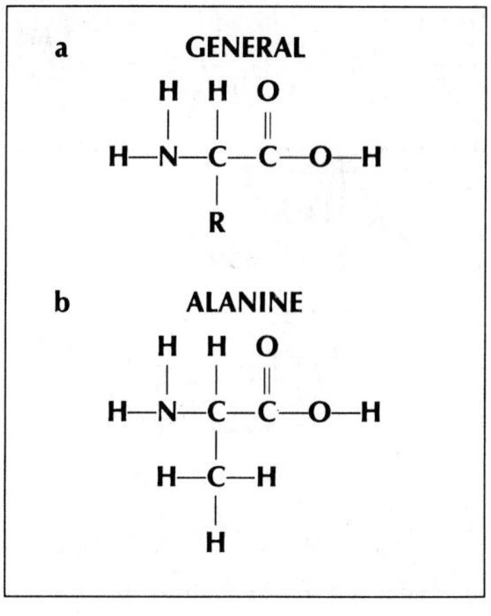

Like lipids and carbohydrates, proteins contain carbon, hydrogen, and oxygen; but, they also contain nitrogen. Proteins also can contain other elements such as sulfur. Proteins are much larger and more complex than either lipids or carbohydrates.

Proteins are made of hundreds or thousands of compounds called **amino** (uh MEE noh) **acids.** Most of the twenty amino acids important to living things contain a central carbon atom to which is attached a carboxyl group (—COOH), a hydrogen atom, and an **amino group** (—NH$_2$). Also attached to the carbon atom is the rest of the molecule, the part that makes each amino acid different. It is represented by the letter R.

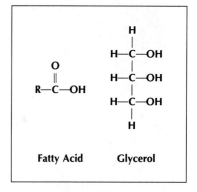

FIGURE 3–22. A fat molecule is formed from fatty acid and glycerol molecules.

Fats, energy reserves, are composed of fatty acids and glycerol.

Some fats play a structural role as parts of membranes.

FIGURE 3–23. (a) The general formula for an amino acid shows a central carbon atom to which are attached the amino group, the carboxyl group, and the R group. (b) Find each of these groups in alanine.

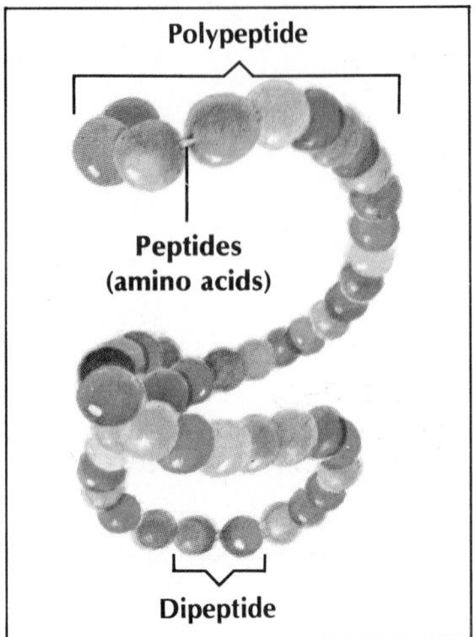

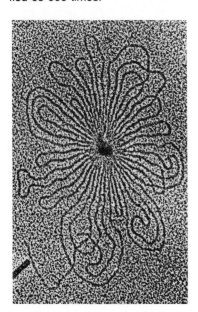

Polypeptide

Peptides
(amino acids)

Dipeptide

FIGURE 3–24. A protein is made of amino acids held together by peptide bonds. Proteins differ in the kind, number, sequence, and arrangement of amino acids.

Polypeptides vary in terms of the number, sequence, and kind of amino acids they contain.

FIGURE 3–25. A nucleic acid, DNA, from a virus is shown magnified 65 000 times.

As amino acids join together, they become known as **peptides.** Two amino acids joined together form a **dipeptide.** Many amino acids joined together form a **polypeptide.** One polypeptide differs from another in its kind, number, and sequence of amino acids. In the same way that the twenty-six letters of the alphabet form many words, the twenty amino acids can form an almost endless variety of polypeptides.

A protein molecule consists of one polypeptide or two or more polypeptides bonded together. Each protein has a distinct three-dimensional shape. That shape is crucial to the function of the protein. Some proteins, such as those in hair and nails, have a fibrous shape. Other proteins, such as hemoglobin in the blood, have a globular shape.

TABLE 3-3. IMPORTANT BIOLOGICAL MOLECULES			
Compound	Elements Present	Building Blocks	Use
Carbohydrate	Carbon Hydrogen Oxygen	Monosaccharides	Energy
Fat	Carbon Hydrogen Oxygen	Fatty Acids Glycerol	Energy reserve
Protein	Carbon Hydrogen Oxygen Nitrogen Sulfur	Amino Acids	Building living material Regulating chemical reactions

3:15 Nucleic Acids

Nucleic (noo KLAY ihk) **acids** are the most complex of all biological compounds. They are extremely large molecules that may consist of hundreds of thousands of atoms. Two very important nucleic acids are **deoxyribonucleic** (dee AHK sih ri boh noo KLAY ihk) **acid (DNA)** and **ribonucleic** (ri boh noo KLAY ihk) **acid (RNA)** that control heredity. DNA is the material that usually contains the "genetic message." RNA works with DNA in carrying out the instructions of the DNA code. You will study this process in Chapter 9.

3:16 Reactions of Organic Compounds

You have seen that organic compounds may combine to form larger molecules. Also, large compounds can be changed to smaller ones. As in any chemical reaction, old bonds must be broken and new bonds formed.

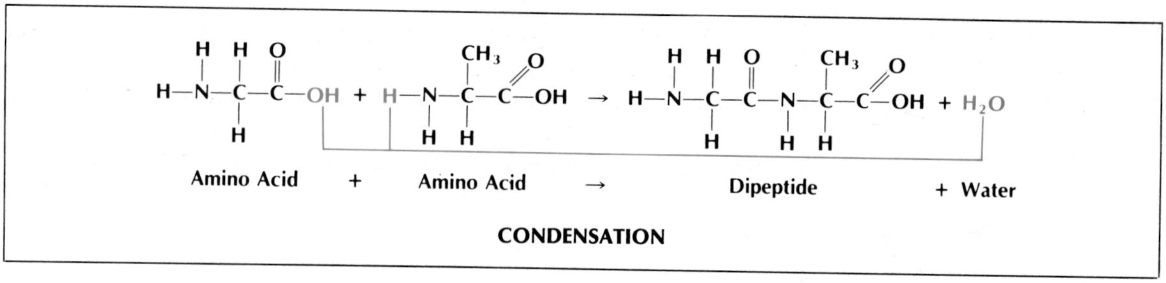

CONDENSATION

Consider the linking of two amino acids to form a dipeptide. The bonds that are broken involve oxygen and hydrogen atoms. A hydrogen atom is removed from the amino group of one amino acid. The other amino acid loses an oxygen atom and a hydrogen atom from its carboxyl group (Figure 3–26). When these atoms are removed, electrons become available for sharing and the two amino acids join together. The two hydrogen atoms and one oxygen atom removed during the reaction join to form a molecule of water. Because the new molecule is made by removing a molecule of water, the process is called **condensation.**

Condensation is common in all cells. It occurs in synthesizing fats from fatty acids and glycerol. It also occurs in constructing larger carbohydrate molecules from monosaccharides.

Condensation is reversed when large molecules are split into smaller molecules. One or more bonds holding the larger molecule together are broken. A molecule of water is added where each bond is broken to form two or more smaller molecules. This process is called **hydrolysis** (hi DRAHL uh sus) because water is added (Figure 3–27). Like condensation, hydrolysis is a common type of cell reaction.

FIGURE 3–26. When water is removed (condensation), a dipeptide is formed from amino acids.

Large molecules commonly are made in a cell by condensation.

Large molecules are converted to smaller ones in cells by hydrolysis.

FIGURE 3–27. Water is added in the breaking (hydrolysis) of a disaccharide into monosaccharides.

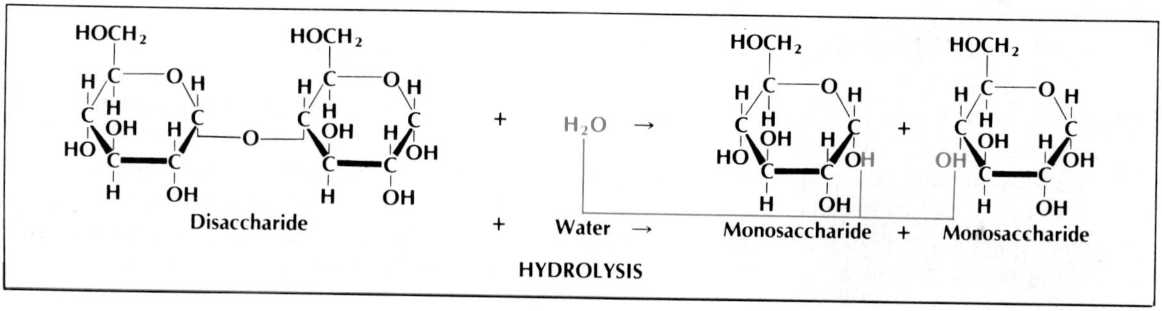

HYDROLYSIS

REVIEWING YOUR IDEAS

18. How are carbohydrates important to organisms?
19. How are proteins used by organisms?
20. Name two nucleic acids. What are their functions?

INVESTIGATION

Problem: How does one determine if a solution is an acid or a base?

Materials:

known acid solution with dropper
known base solution with dropper
known neutral solution with dropper
12 sample chemical solutions
pH paper strips
12 droppers

Procedure:

Part A. Understanding the pH Scale

1. Examine the pH scale shown in Figure 3–28. This scale allows you to measure the pH of chemical solutions.
2. Note that the various colors and numbers on the scale correspond to certain pH values.
3. Answer the following questions before going on to Part B.
   (a) Which numbers on the pH scale indicate an acid solution?
   (b) Which numbers on the pH scale indicate a basic solution?
   (c) Which number on the pH scale indicates a neutral solution?
   (d) Which number on the pH scale indicates the strongest acid solution?
   (e) Which number on the pH scale indicates the strongest basic solution?
   (f) Which number on the pH scale indicates the weakest acid solution?
   (g) Which number on the pH scale indicates the weakest basic solution?

Part B. Testing the pH of a Known Acid, Base, and Neutral Solution

1. Using a dropper, place one drop of known acid on a piece of pH paper. Wait 5 seconds.

   CAUTION: *Acids and bases are harmful to skin and clothing. Rinse with water if spillage occurs. Call your teacher.*

2. Observe the color change on the pH paper. Match the color on the pH paper with the pH color scale.
3. In a copy of Table 3–4, record the pH indicated by the color.
4. Repeat steps 1 and 2 using one drop of the known base and a new piece of pH paper. Record the pH in Table 3–4.
5. Repeat steps 1 and 2 using one drop of the neutral solution and a new piece of pH paper. Record the pH in Table 3–4.

Part C. Testing the pH of Unknown Solutions

1. Complete column A in a copy of Table 3–5. Your teacher will supply the list of items to be tested.
2. Make a guess, based on previous knowledge, of whether each substance is an acid, a base, or a neutral solution. Complete column B of Table 3–5.

FIGURE 3–28.

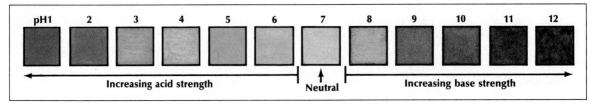

3. Place 1 drop of sample 1 onto a piece of pH paper. Wait 5 seconds. Observe the color and record the pH in column C of Table 3–5. NOTE: If the color does not match exactly, make an estimate of the pH.
4. Repeat step 3 for each of the remaining samples. Use a new piece of pH paper each time.
5. Using the results in column C, determine whether each solution is an acid, a base, or a neutral solution.

| TABLE 3-4. pH OF KNOWN SUBSTANCES ||
Solution	pH
Known Acid	
Known Base	
Known Neutral	

6. Compare your guess (column B), and the actual results. Use yes or no to complete column E of Table 3–5.

Data and Observations:

| TABLE 3-5. pH OF UNKNOWN SUBSTANCES ||||||
| | A | B | C | D | E |
Sample Number	Sample Name	Your Guess Acid, Base, Neutral	pH	Actual Results Acid, Base, Neutral	Did your guess match results?
1					
2					
3					
4					
5					
6					
7					
8					
9					
10					
11					
12					

Questions and Conclusion:

1. Define acid, base, neutral solution, pH scale, and ion.
2. Using numbers of the pH scale, describe how you can indicate if a solution is an acid, a base, or a neutral solution.
3. Identify the symbols OH^- and H^+.
4. Of the solutions you tested, name
   (a) the weakest acid
   (b) the strongest acid
   (c) the weakest base
   (d) the strongest base
   (e) any that were neutral
5. (a) What was the expected pH of drinking water? What was the actual pH?
   (b) List several factors that may affect the pH of drinking water.

Conclusion: How does one determine if a solution is an acid or a base?

SUMMARY

1. Matter is composed of atoms that are made of protons, neutrons, and electrons. **3:1**
2. Elements are substances composed of one type of atom. **3:2**
3. Molecules are composed of two or more atoms joined by a covalent bond. **3:3**
4. Compounds are substances composed of two or more different atoms joined by a chemical bond. **3:4**
5. The attraction between ions, charged atoms, is an ionic bond. **3:5**
6. Elements are represented by symbols. A chemical formula represents the number and kind of each atom in a compound. **3:6**
7. Solutions form when molecules dissolve and ionic compounds dissociate in a solvent. **3:7**
8. The concentration of hydrogen ions in acids and bases is measured by the pH scale. **3:8**
9. Matter has both physical and chemical properties and can undergo both physical and chemical changes. **3:9**
10. In a chemical equation, symbols and formulas describe a chemical reaction. **3:10**
11. Living systems contain complex organic compounds usually represented by structural formulas. **3:11**
12. Carbohydrates are organic compounds used primarily as energy sources. **3:12**
13. Lipids include fats, waxes, and oils. Some fats are used as energy reserves; others are part of membranes of cells. **3:13**
14. Proteins are complex organic compounds that play a variety of structural and functional roles in organisms. **3:14**
15. Nucleic acids, DNA and RNA, are chemicals controlling heredity. **3:15**
16. Condensation and hydrolysis are important reactions of organic compounds and are common in cells. **3:16**

LANGUAGE OF BIOLOGY

acid	element	monosaccharide
amino acid	energy level	neutron
atom	fatty acid	nucleic acid
base	formula	nucleus
carbohydrate	glycerol	peptide
cellulose	hydrolysis	pH scale
compound	ion	polypeptide
condensation	ionic bond	protein
covalent bond	isomer	proton
dissacharide	isotope	reactant
duet rule	lipid	solution
electron	matter	solvent
electron cloud	molecule	starch

CHECKING YOUR IDEAS

On a separate paper, complete each of the following statements with the missing term(s). Do not write in this book.

1. In an atom with no overall charge, the number of protons equals the _____.
2. Ions are formed from the _____ of electrons.
3. To show arrangement of atoms, organic compounds are shown by _____ formulas.
4. A(n) _____ is a homogeneous material of a solute in a solvent.
5. The stretching of a spring is an example of a(n) _____ change.
6. A(n) _____ is composed of atoms joined by covalent bonds.
7. _____ and _____ are the main energy sources of living systems.
8. A molecule of fat is changed to fatty acids and glycerol by the process of _____.
9. During a(n) _____ change, new substances with new properties are formed.
10. The part of an atom involved in bond formation is the _____.

EVALUATING YOUR IDEAS

1. Why is knowledge of chemistry and physics important to a biologist?
2. List the four elements most common in living things. What are their symbols?
3. What happens to the outer energy level when an atom joins with another atom?
4. Why must a molecule of water be composed of two hydrogen atoms and one oxygen atom?
5. What is the difference between a chlorine atom and a chloride ion?
6. How are ionic compounds different from molecular compounds?
7. What is meant by the expression $3H_2O$? How many atoms are there in $3H_2O$?
8. Describe the difference between a solution of sugar in water and a solution of salt in water.
9. Why is a solution of hydrogen chloride in water an acid? Why is a solution of sodium hydroxide in water a base?
10. A chemist tests two solutions. One has a pH of 6. The other has a pH of 2. Which is more acidic? Explain.
11. What properties of carbon make it suitable as a part of complex molecules?
12. What is a double bond? What is a triple bond? Why does hydrogen not form double or triple bonds?
13. What group of atoms is common to both fatty acids and amino acids?
14. What element found in proteins is not found in either lipids or carbohydrates?
15. How does one protein differ from another?
16. With what important biological function are the nucleic acids associated?
17. Classify each reaction below as either hydrolysis or condensation.
    (a) conversion of starch to glucose
    (b) synthesis of protein from amino acids
    (c) production of a disaccharide from monosaccharides

APPLYING YOUR IDEAS

1. A certain atom has 8 protons and 10 neutrons. How many electrons does it have? Draw a diagram of this atom.
2. One ion has a charge of 3+. Another ion has a charge of 1−. How would the two ions combine to form a neutral compound?
3. Helium, argon, and neon are atoms whose outer energy levels are filled. They are known as noble (inactive) gases. Why are they inactive?
4. Which of the following changes are physical and which are chemical?
    (a) tearing a sheet of paper
    (b) melting of ice
    (c) joining of two monosaccharides
5. Pentane, C_5H_{12}, is an organic compound containing single covalent bonds. Draw structural formulas of 2 isomers of C_5H_{12}.
6. Ecologists are concerned about the effects of acid rain. What pH range could acid rain have? How might acid rain affect the environment?

EXTENDING YOUR IDEAS

1. List ten common chemical changes.
2. How would you separate a mixture of sand and iron filings? Of sand and salt?
3. Using reference materials, determine how the pH of the blood is kept constant. What happens if the blood pH varies too much?
4. Prepare a report on cholesterol and its possible effects on humans.

SUGGESTED READINGS

Hopson, Janet, "Carbohydrates: A Key to Health and Performance." *Science Digest*, Jan., 1984.

Preuss, Paul, "The New Proteins." *Science 83*, Dec., 1983.

CELL STRUCTURE AND FUNCTION

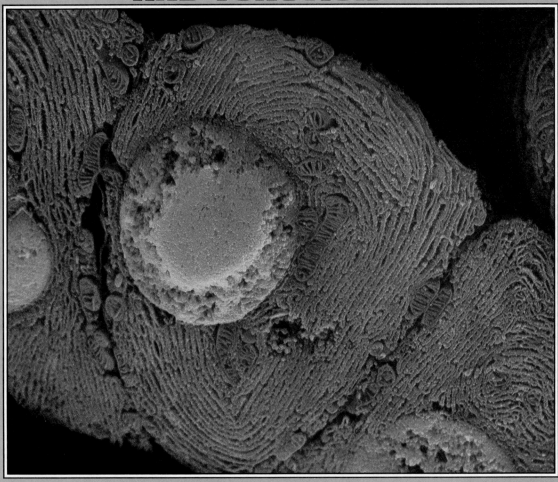

The cell is the basic unit of life. The various cell parts function in different ways resulting in life processes such as fluid balance and protein production. This human cell is shown magnified about 10 000 times. What cell structures can you identify in this cell? How do the cell parts function? How can cells and cell parts be studied?

chapter 4

In Chapter 1 you learned that organisms are organized in two ways. They are organized so that all their various functions are carried out smoothly in a coordinated fashion. Living things are also organized with respect to structure. From a single bacterium to a community, life is characterized by more and more complex levels of organization. In this chapter you will study cells, the simplest units of structure and function. The cell is the most basic level of biological organization.

Objectives:
You will
- describe the discovery of cells and their basic characteristics.
- explain the structure of the cell membrane and various ways in which substances pass across it.
- list the specialized parts of cells and their functions.

DISCOVERING CELLS

In 1665, Robert Hooke, an English scientist, made important observations. Using an early microscope, he examined razor-thin slices of cork, a plant material. He saw that the cork is composed of many tiny units. Because they reminded him of a honeycomb, he named these units **cells.** Although Hooke did not realize it, he was the first to observe this feature of life.

Robert Hooke, who first observed and named cells, saw only the walls of dead cork cells.

4:1 The Cell Theory

Hooke's discovery of cork cells was the foundation of a major biological theory. The development of that theory, however, took almost two hundred years. Study was limited by poor microscopes. Although later scientists reported finding cells in many plants and animals, they did not realize the importance of the observations. They made no general statements about cells.

FIGURE 4–1. Schleiden determined that all plants are composed of cells. Schwann determined that animals are composed of cells. These plant cells (a) and animal cells (b) are units of structure and function.

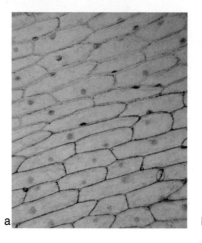

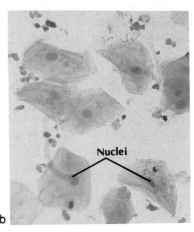

When Hooke looked at cork, he saw only dead cells. Only the outer parts, or walls, of the cells were visible. He observed, then, only the borders of hollow compartments. There was certainly no hint that living cells are active and are composed of still smaller parts.

Later discoveries by a number of scientists led to a more accurate picture of a living cell. In the 1830s Robert Brown (1773–1858) determined that a cell contains a central part, or nucleus (NEW klee us). It was later reported that cells are not hollow structures. Cells are filled with a thick, jellylike fluid. In 1838, Matthias Schleiden (SHLI dun) determined that all plants are composed of cells and that cells are the basis of a plant's function. A year later, Theodor Schwann (SHVAHN) determined the same things about animals. In 1858, Rudolf Virchow (FIHR koh) concluded that all cells come from other living cells.

Most cells have one nucleus and are filled with a jellylike fluid.

The conclusions made by Schleiden, Schwann, and Virchow are extremely important because they apply to all organisms. They have been supported by later findings. These statements form the basis of one of biology's most important concepts, the **cell theory.**

All organisms are composed of cells.

1. All organisms are composed of cells or cell fragments that are the basic units of structure and function.
2. All cells are produced from other cells.

The development of the cell theory is a good example of the interaction of careful observation, interpretation, and curiosity. Why is it called a theory rather than a hypothesis?

4:2 Characteristics of Cells

Cells exist in many shapes, such as spherical and cubical. Some cells, such as the amoeba, have shapes that can change. The surfaces of the cells vary too. Some have rough surfaces; some have extensions and other cell parts coming from their surfaces. A sperm cell has a "tail." Cells also vary in size. Ostrich eggs are as wide as 10 cm; some bacterial cells are smaller than half a micrometer. Regardless of shape, surface traits, and size, all cells are alive and are the basic units of life.

Many organisms such as bacteria, amoebae, certain algae, and yeasts are unicellular (Figure 4–2). More often, however, an organism is multicellular. A full grown human contains trillions of these units of life. Both unicellular organisms and cells of multicellular organisms perform the following functions. Most cells "burn" food for energy, rid themselves of wastes (dangerous by-products of chemical reactions), exchange gases, and synthesize (SIHN thuh size) (build up) new living material. They also take in necessary ions, regulate water balance, react to changes in the environment, and produce other cells.

How are all these life activities and processes carried out? The answer can be found by examining the composition of cells and learning about those parts common to most cells.

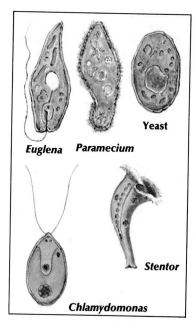

FIGURE 4–2. A unicellular organism is one cell. It performs all life functions. Many organisms are unicellular.

REVIEWING YOUR IDEAS

1. Why did Hooke see only the outer walls of cells?
2. Describe the findings of Schleiden, Schwann, and Virchow.
3. State the cell theory.
4. List some functions of cells.
5. Describe the various shapes and sizes of cells.

THE CELL AND ITS ENVIRONMENT

Each cell, whether an entire organism or a small part of a larger living system, lives in a fluid environment. An amoeba may be in a pond. A kidney cell is surrounded by blood. Wherever the cell is, it must obtain materials from and release substances into its moist environment.

4:3 Properties of the Cell Membrane

Each cell has a boundary that separates it from neighboring cells or its liquid environment. This boundary is called the **cell membrane.** The cell membrane holds the cellular contents together and determines the shape of the cell. Along with being a structural boundary, the cell membrane is also the "gatekeeper" of a cell. The membrane is the structure that particles must cross as they move into or out of the cell. Many cell functions (Section 4:2) depend upon passage

The cell membrane acts as a boundary and is the "gatekeeper" of the cell.

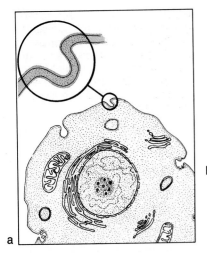

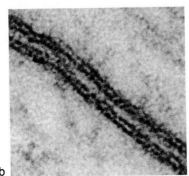

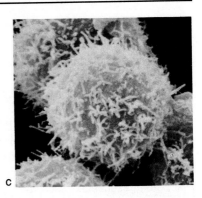

a

b

c

FIGURE 4–3. (a) The two layers of the cell membrane are shown in relation to the cell. (b) The cell membranes of two adjoining cells have been magnified about 134 000 times. Each cell membrane can be seen as having two layers. (c) A scanning electron micrograph of a hamster cell magnified 3500 times shows surface properties of the cell membrane.

Different cells are permeable to different substances.

Small particles, lipids, lipid-soluble substances, and neutral particles are those that cross membranes most easily.

of materials across the membrane. For example, amino acids and sugar molecules move from the environment, across the membrane, and into the cell. Once inside, the amino acids can be combined to form protein and the sugar can be "burned" for energy. Carbon dioxide produced during the respiration process leaves by passing across the cell membrane into the environment.

Long before much was known about the structure of the cell membrane, biologists had learned a great deal about its properties. One important property of the cell membrane is that it is semipermeable. A **semipermeable** (sem ih PUR mee uh bul) membrane is one that allows only *certain* particles to pass across (permeate) it. The fact that cell membranes are semipermeable is important. Only those materials that are needed or must leave a certain cell should cross the membrane. Cells differ in terms of particles to which they are permeable. Also, a given cell may be more or less permeable to a certain particle at different times.

What kinds of particles most freely cross cell membranes? Smaller particles pass through more easily than larger ones. Lipids and particles soluble in lipids cross the membrane more easily, too. In general, small ions dissolved in water do not cross the membrane as easily as uncharged particles of about the same size. Also, some ions pass through more readily than others.

4:4 Structure of the Cell Membrane

Scientists often propose models to explain known properties. Biologists have come up with several models of cell membrane structure over the last fifty years. Since the 1950s, electron microscopy and improved biochemical techniques have been especially important. The models have been revised over the years as more has been learned.

When viewed through a light microscope, a cell membrane appears as a thin, single line. But the electron microscope shows the

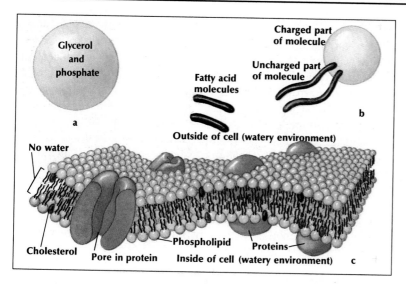

FIGURE 4–4. (a) Phosphate, glycerol, and fatty acid molecules combine to form phospholipids which have charged and uncharged parts (b). In addition to phospholipids, cell membranes contain proteins that can move in the bilayer (c).

Scientists use models to help explain known properties.

membrane to consist of two lines (Figure 4–3). Chemical analysis shows that the membrane is composed mainly of a bilayer (two layers) of molecules called phospholipids. A phospholipid is like a fat molecule, but one of the fatty acids is replaced by another part containing a combination of atoms called a phosphate group. This other part has an electric charge (Figure 4–4 a and b). The charged part of a phospholipid is soluble in water. The two fatty acids are uncharged and not soluble in water. These properties determine the way the two layers of phospholipids are arranged. The charged parts of each molecule, because they are soluble in water, are located toward the outside and inside of the cell (where there is water). The uncharged fatty acids are located away from water. Thus, the two layers of phospholipids face one another (Figure 4–4 c).

Different charge characteristics determine the arrangement of the phospholipids in the bilayer.

Phospholipids in one layer are different from those in the other. Within each layer the fatty acid "tails" can move back and forth. Also, an entire phospholipid can move sideways through its layer. These properties make the bilayer like a very light oil. Another type of lipid, cholesterol (Section 3:13), is embedded in the bilayer of animal cells. Cholesterol has a rigid structure and reduces movement of the lipids. Thus, the membrane is both flexible and strong.

The cell membrane is like a fine oil.

It had been known for many years that the cell membrane also contains proteins. Until the 1970s it was thought that the lipid bilayer was coated on either side by proteins. That is now known to be untrue. Some protein molecules are partly or totally embedded in the bilayer. Others lie on the surface of each lipid layer. The proteins, too, can move sideways within the bilayer. Because the lipid bilayer is a liquid and has proteins scattered through it, this current model of membrane structure is the **fluid mosaic model** (Figure 4–4).

Proteins lie on the surface of the lipid layers or are partly or totally embedded in the bilayer.

Proteins in the cell membrane play a variety of functional roles.

The lipid portion of the cell membrane forms the basic boundary of the cell. Its chemical characteristics allow it to act as a barrier across which most particles cannot pass. It is the proteins of the membrane that play a functional role. Some proteins are bonded together to form pores through which certain particles may pass. Many proteins (Section 4:7) help in passage of different particles across the membrane. Others act as "markers" that are "recognized" by chemicals from inside and outside the organism. Many of the proteins regulate chemical reactions that take place on the membrane itself.

Many parts of cells are also enclosed by membranes. These membranes also have fluid mosaic structure. Because the kinds and arrangements of the lipids and proteins vary from one membrane to another, each membrane has its own permeability characteristics.

4:5 Diffusion

If you place some crystals of blue copper sulfate in a beaker of water, you will see evidence of an important physical process. At first, the water will be colorless and the crystals will settle to the bottom of the beaker. Several hours later you will see that the water has turned blue and the crystals have disappeared.

When copper sulfate is placed in water, its ions begin to separate from each other. They move about randomly. As a result, they collide with each other, the molecules of water, and the sides and bottom of the beaker. Each time a collision occurs, ions bounce off and move in new directions. Random movement results in a uniform distribution of ions throughout the water.

Diffusion, the random motion of particles, results in the net movement of particles from a region of greater to lesser concentration.

The water appears blue because there are equal numbers of copper ions in any given region at any given time. This random movement of ions (and other particles) is called **diffusion** (dihf YEW zhun). Diffusion results in the spreading out of particles. The particles become distributed evenly in all the available space. After the ions become distributed evenly, they continue to move, but the same number of ions enters a certain area as leaves the area.

FIGURE 4–5. Copper sulfate, a blue crystal, diffuses through water distributing copper and sulfate ions evenly throughout. The solution turns bluer as diffusion proceeds.

Diffusion is a process that explains how *some* (but not most) particles pass across a membrane. Such particles include lipids and lipid-soluble molecules as well as gases such as oxygen and carbon dioxide. Suppose a cell is surrounded by oxygen molecules and the membrane is permeable to oxygen. Because the oxygen molecules are in random motion, they bump against each other and the membrane of the cell. The membrane is permeable to oxygen. Thus, the molecules can pass through it into the cell. As oxygen molecules enter the cell, their movement results in more collisions. Random movement across the membrane occurs in both directions. At first more oxygen molecules enter the cell than leave the cell. Later the oxygen molecules become distributed evenly inside and outside the cell. This balance is called a **dynamic equilibrium** (di NAM ihk • ee kwuh LIHB ree um). After dynamic equilibrium is reached, molecular motion continues, but the number of molecules entering and leaving the cell is equal.

This random movement of molecules shows the importance of diffusion in living systems. Some needed materials may enter a cell by diffusion across the cell membrane. Also, as some by-products of cellular activities build up in a cell, their random motion may move them out. *In either case, a substance moves from a region of greater concentration to a region of lesser concentration until equilibrium is reached.* Once equilibrium occurs, the substance remains in equal concentration on both sides of the membrane although movement across the membrane still occurs.

4:6 Osmosis

Movement of water into and out of cells is so common that it is given a special name — osmosis (ahs MOH sus). **Osmosis** is the diffusion of water through a semipermeable membrane. Water can pass in either direction across the membrane. In order for any cell to survive, it must be in osmotic (ahs MAHT ihk) balance with its watery environment. **Osmotic balance** is a kind of dynamic equilibrium that occurs when the amount of water entering and leaving a cell is equal.

Consider a red blood cell. The cell is normally in osmotic balance with its surrounding blood. The amount of water entering and leaving the cell is equal. However, if the cell is removed from the blood and placed in pure water, equilibrium is destroyed. The blood cell contains water and other substances while the surrounding water has no other substances in it. Thus, the concentration of water inside the cell is less than the concentration of water outside the cell. As a result, more water molecules enter the cell than leave the cell. So much water enters the red blood cell that it swells and finally bursts. The cell cannot survive because it cannot maintain an osmotic balance with its pure water environment.

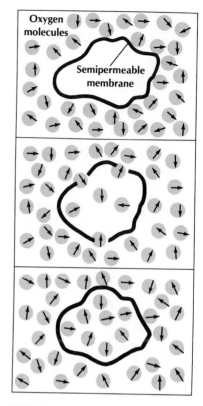

Oxygen molecules

Semipermeable membrane

FIGURE 4–6. Diffusion of oxygen molecules into a cell occurs as the oxygen molecules move from a region of greater concentration to a region of lesser concentration. Dynamic equilibrium results.

Diffusion of water through a semipermeable membrane is called osmosis.

a

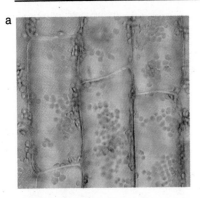

b

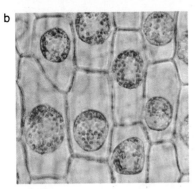

FIGURE 4–7. (a) In fresh water, the *Elodea* cells are in osmotic balance with the environment. (b) In salt water, water moves out of the cells. As a result, the contents shrink to the center of the cell.

In passive transport the cell expends no energy in moving particles across its membrane.

In facilitated diffusion, permease molecules help move particles from a region of greater to lesser concentration.

Pores are simple carriers that allow certain ions to cross a membrane.

Elodea (ih LOHD ee uh) is a freshwater plant that is often seen in home aquaria. The cells of *Elodea* contain water and dissolved materials while the freshwater environment consists mainly of water molecules. Therefore, in their environment, the cells of this plant tend to take in water. These cells do not burst, though. They have a tough cell wall (Section 4:13) that balances the pressure of the water that enters the cell. The cell wall pressure stops more water from entering. Thus, osmotic balance between the cell and the environment is reached and the plant can survive.

What would happen if an *Elodea* leaf were placed in a saltwater solution? The concentration of water molecules inside the cells would then be greater than outside the cells. Water molecules would leave the cells faster than they would enter. Such loss of water in a cell causes the cell contents to shrink (Figure 4–7) and is called **plasmolysis** (plaz MAHL uh sus). Prolonged plasmolysis or a large water loss results in death to cells and can result in death to an entire plant.

4:7 Passive Transport

Osmosis and diffusion are processes that explain how water, lipids, lipid-soluble molecules, and gases permeate membranes. These particles either dissolve in the lipid bilayer or are small enough to pass between the lipid molecules. Passage across the membrane is due to their random motion from a region of greater to lesser concentration. The cell plays no active role (does no work) in moving the particles. This type of movement is called **passive transport** because the cell uses no energy for osmosis or diffusion to occur.

Many particles, such as certain ions, sugars, and amino acids, are in greater concentration outside the membrane than inside. The cell needs such particles but they cannot dissolve in or pass between the lipids in the bilayer. How do such important particles pass across the membrane? Recall that many of the proteins of the membrane play a role in passage of particles. Use of proteins to aid (facilitate) the passage of particles from a region of greater to lesser concentration is called **facilitated diffusion.** Facilitated diffusion is a special form of passive transport.

In general, proteins involved in facilitated diffusion are known as **permeases.** A single permease is composed of two or more proteins. One of the simplest types of permeases is the protein pore. These pores extend through the bilayer and form a channel that allows certain ions to pass through. Different ions pass through different pores. Which ion passes through a given pore depends upon the size and charge of the ion and the inside diameter of the pore.

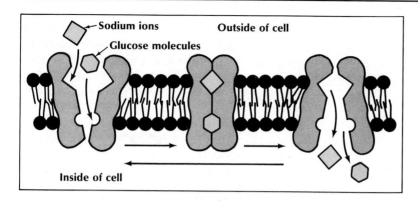

FIGURE 4–8. When sodium binds with the permease, it causes a shape change that allows glucose to also bind. Once both are picked up, the shape changes again letting the glucose and sodium enter the cell.

Some permeases are more complex. They aid in the passage of two particles at once. Glucose molecules and sodium ions are passed across a membrane in this way. A sodium ion outside the cell combines with a protein in the permease and is thought to change the permease's shape. This change in shape allows the permease to pick up a glucose molecule. Once the glucose is picked up, the shape of the permease changes again. It then becomes closed to the outside and open to the inside. The permease, in this shape, "loses its hold" on glucose and sodium, and the two particles then enter the cell. Once the particles have entered the cell, the permease resumes its original shape. It is once again open to the outside (Figure 4–8).

Some permeases are constructed in the forms of "gates." When a particle passes by, it combines with the gate. The shape of the permease changes such that the gate opens, allowing another particle to pass through.

Whether passive transport occurs by osmosis, simple diffusion, or facilitated diffusion, it always involves a net movement of particles from a region of greater to lesser concentration. The particles pass across the membrane as a result of their own energy of motion.

Passage of some particles across a membrane involves a change in the shape of permease.

4:8 Active Transport

Sometimes a cell must move particles from a region of lesser to a region of greater concentration. To do so the cell must use energy to counteract the random motion of the particles. (The cell's energy causes the movement of the particles to be less random.) The cell plays an active role and does work. Because the cell's energy is required to move the particles, this type of movement is called **active transport.** Active transport also involves permease molecules. Permeases involved in active transport are known as pumps. Energy from the cell is used to change the shape of the pump. The change in shape results in pulling a particle into (or pushing it out of) a cell.

In active transport, cells expend energy and use permeases called pumps to move particles from a region of lesser concentration to a region of greater concentration.

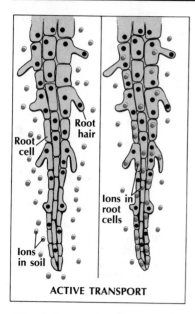

ACTIVE TRANSPORT

FIGURE 4–9. By active transport, root cells accumulate ions in concentrations greater than in the soil. Active transport requires energy.

The fluid mosaic model explains the known properties of cell membranes.

A cell can rid itself of harmful particles by active transport. For example, some cells of the kidney actively transport wastes out. Waste materials must be removed. Some wastes (even in small amounts) would poison an organism. Active transport maintains a greater concentration of the waste materials outside the kidney cells than inside the cells.

Active transport also may work to bring extra particles into the cell. For example, root cells of plants bring in and retain needed mineral ions from the soil. Even though there may be more ions in the root cells than in the soil, ions continue to enter the cells by active transport. Use of energy by these cells prevents the natural diffusion of these ions out of the cells (Figure 4–9).

Now that you have studied the structure of the cell membrane and the passage of particles across it, think again about the fluid mosaic model. The model most certainly explains the properties of the membrane. Lipids and lipid-soluble particles permeate more easily because they dissolve in the lipid bilayer. Smaller particles pass across more readily because they can fit between the lipids or through various pores. Larger particles, or those not soluble in lipids, can pass across, however, with the aid of permeases. Small neutral particles cross more freely than ions because they are neither attracted to nor repelled by the charged parts of the membrane. Membranes of different cells are permeable to different particles because they have different permeases. Also, a given cell may be more or less permeable to a certain particle depending on the position of its lipids and proteins.

4:9 Endocytosis and Exocytosis

Some materials that are too large to pass into a cell by passive or active transport can enter a cell in another way. **Endocytosis** (en duh si TOH sus) is a process in which a cell uses energy to surround and take in large particles. In endocytosis a substance does not pass *through* the bilayer as happens in passive or active transport.

Many unicellular organisms, such as an amoeba, depend on smaller organisms for food (Figure 4–10). After detecting food, an amoeba's cell membrane engulfs (flows over and encloses) the food organism. That portion of the membrane then breaks away and moves to the cell's interior. The membrane and its contents form a small sac or vesicle. The food contained within it will later be digested (Section 4:12). The trapping of a solid in this way is called **phagocytosis** (fag uh si TOH sus). Phagocytosis requires energy because work is done in capturing the food.

Some cells of multicellular organisms also use phagocytosis. For example, certain cells in your blood (white cells) are very much like

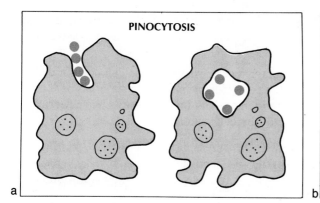

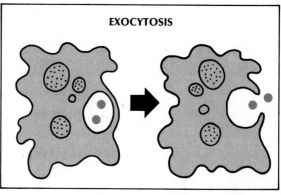

amoebae. These cells engulf and digest bacteria and other foreign particles that invade the body. Thus, the white cells are a line of defense against disease.

Another form of endocytosis is pinocytosis. **Pinocytosis** (pihn uh si TOH sus) is the taking in of liquid droplets or very small particles. In one form of pinocytosis, the substance to be taken in first collects on the outer surface of the membrane. The substance seems to collect where certain membrane proteins are located. A vesicle then forms, bringing the substance into the cell. In another form of pinocytosis, specific particles combine with membrane proteins called receptors. A vesicle forms around both the particles and the receptors and moves into the cell. The particles are released from the vesicle and used by the cell. The vesicle, still containing the receptors, returns to the outer part of the cell where it and the receptors become part of the cell membrane once again.

A reverse process is used to rid some cells of wastes or to expel useful substances needed elsewhere. The substances, enclosed in a vesicle, move toward the cell membrane. The two membranes fuse and the vesicle bursts. Thus, materials are expelled from the cell. This process is called **exocytosis** (ek soh si TOH sus).

FIGURE 4–10. (a) In pinocytosis, the cell forms channels through which liquids and small particles are taken in. (b) Exocytosis is a process in which substances are expelled from a cell.

Phagocytosis and pinocytosis are both forms of endocytosis.

REVIEWING YOUR IDEAS

6. What is the cell membrane? Of what is it composed?
7. What is diffusion? What is the name for the balance that results from diffusion?
8. What is osmosis?
9. Compare diffusion and facilitated diffusion.
10. Distinguish between passive and active transport.
11. Distinguish between endocytosis and exocytosis.

INSIDE THE CELL

Once substances enter a cell, what does the cell do with them? How and where does the cell assemble new living materials or "burn" food for energy? Where in the cell do wastes arise? The answers are to be found in a study of the cell's interior.

4:10 Protoplasm

The "living stuff" of cells was first described in the 1830s (Section 4:1). Hooke did not see it because the cork cells he saw were not alive. This "living stuff" of cells, the thick, jellylike substance that makes up most of each cell, is **protoplasm** (PROHT uh plaz um).

Protoplasm consists of two regions. Inside the nucleus it is called **nucleoplasm** (NEW klee uh plaz um) and outside the nucleus it is called **cytoplasm** (SITE uh plaz um). These terms are based only on location in the cell.

Protoplasm is about 70 percent water and about 30 percent proteins, fats, carbohydrates, nucleic acids, and small amounts of mineral ions. The composition of protoplasm varies. Thousands of different substances are brought into or made by a cell and are changed from second to second. A molecule of sugar now existing in one of your cells easily could be converted to carbon dioxide by the time you finish reading this sentence. Protoplasm changes constantly. Chemical reactions within a cell both build up and tear down many complex molecules. The total of the reactions is called **metabolism** (muh TAB uh lihz um).

Metabolism involves a variety of reactions. Some of them, such as condensation or hydrolysis (Section 3:16), occur in one or a few steps. Often a complex set of reactions is involved in a metabolic change. Cellular respiration and photosynthesis are complex reactions involving many steps.

Not all the reactions of metabolism occur in the cytoplasm. Many take place in specialized structures within the cytoplasm.

Protoplasm is the living, jellylike portion of cells.

Two regions of protoplasm exist in most cells — cytoplasm and nucleoplasm.

The composition of protoplasm is constantly changing.

Metabolism is the total of a cell's chemical reactions.

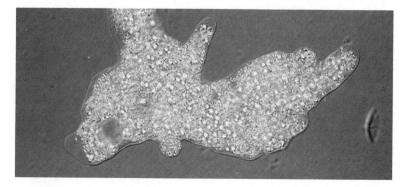

FIGURE 4–11. Protoplasm, the "living stuff" of cells — the thick jellylike substance that makes up most of each cell — is clearly visible in this amoeba.

4:11 Organization in Cells

In Chapter 1, you learned how cells are organized into tissues and organs. But, for cells to perform their many functions, they too must be organized. Certain jobs are carried out by specific cell parts called **organelles** (or guh NELZ), located in the cytoplasm. An organization of parts, each doing specific jobs, is known as **division of labor.** Division of labor occurs among the cell's organelles. However, a cell survives only if all its parts work together.

As you know from studying membrane structure, the electron microscope has proved very important in learning about cell structures. Chemical analysis has played a key role in study of the functions of those structures. The descriptions of cell parts given here are based on these methods. Keep in mind that many structures are either not clear or not visible with the light microscope. So do not expect to see all the parts described in this chapter when you view cells in the laboratory.

Organelles are specialized for cellular functions. Division of labor occurs among the organelles.

PEOPLE IN BIOLOGY

As a teenager, Earnest Just left the South with few academic skills other than reading. He completed high school and college in New Hampshire, graduating with honors, and soon began teaching in Washington, D.C. His interest in cell biology led him to the University of Chicago, where he completed his doctorate in 1916. The rest of his life was devoted to both research in cell biology and embryology, and to teaching.

Earnest Everett Just

(1883–1941)

In the 1920s and 1930s, Just took some unpopular stands concerning the roles of various cell parts. He thought that cell membranes were active parts of cells, not just coverings. He also felt that the cytoplasm was as important as the nucleus in determining the cell's activities. (Remember that much of the information we have today about cells was not known at that time.) Because of his unpopular views, even though later shown to be correct, Just and his work were not fully appreciated in his lifetime. He did, however, publish numerous articles and authored two books. In 1915, he was the first person to receive the NAACP Spingarn Medal for "highest achievement . . . in an honorable field of human endeavor."

FIGURE 4–12. The biggest division among cells is between (a) eukaryotic cells and (b) prokaryotic cells. A eukaryotic cell has a nucleus and membrane-bound organelles. A prokaryotic cell has nuclear material spread throughout the cytoplasm and lacks membrane-bound organelles. The eukaryote shown here is magnified about 4700 times. The prokaryotic cell shown here is magnified about 40 000 times.

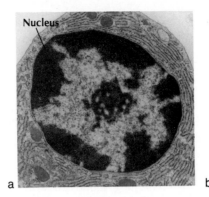
Nucleus

a

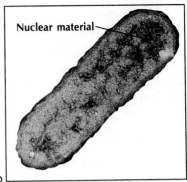
Nuclear material

b

4:12 Cell Parts of Eukaryotes

Eukaryotic cells have nuclei and membrane-bound organelles. Prokaryotic cells lack nuclei and membrane-bound organelles.

Bacteria are prokaryotes.

Cells of all organisms fall into one of two broad groups. They are either eukaryotes (yew KER ee ohtz) or prokaryotes (proh KER ee ohtz). **Eukaryotes** are cells that are more complex and contain a distinct nucleus surrounded by a membrane. Eukaryotes also contain organelles enclosed by membranes. **Prokaryotes** lack nuclei and other membrane-bound organelles. Bacteria are prokaryotes. All other cells are eukaryotic. Specialized cell parts presented in this section are parts of eukaryotic cells. Structure of prokaryotes will be studied in Chapter 13.

Mitochondria (singular, mitochondrion) are the "powerhouses" of the cell.

Mitochondria (mite uh KAHN dree uh) are organelles scattered through the cytoplasm. They vary in size, often are cigar-shaped, and have a double-layered membrane with the inner layer folded inward. The folds are called **cristae** (KRIS tee). The central cavity is filled with a fluid (Figure 4–13).

More mitochondria are found in cells requiring great amounts of energy than are found in other cells. Mitochondria are involved with energy. Because most of the many steps of aerobic respiration occur in the mitochondria, they often are called the "powerhouses" of the cell.

FIGURE 4–13. A mitochondrion is a double-membraned organelle that releases energy in a usable form (ATP) in a cell. This mitochondrion is shown magnified about 40 000 times.

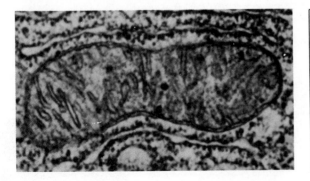

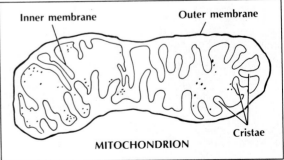
Inner membrane   Outer membrane

Cristae

MITOCHONDRION

Endoplasmic reticulum (en duh PLAZ mihk • rih TIHK yuh lum) is a network of fluid-filled, tubelike structures found in most eukaryotic cells. It is often called ER. These tubes connect with one another, often ending in "blind alleys." In some cells, portions of the network are connected with the nuclear membrane. The endoplasmic reticulum acts as a cellular "subway" or "canal system" that transports certain proteins within the cell.

Some of the cell's ER has a smooth surface and is called *smooth ER*. The structure of the ER's membrane is similar to that of the cell membrane. Some of the proteins in smooth ER membrane function in making phospholipids that make up the cell membrane.

Ribosomes (RI buh sohmz) are tiny spherical organelles most often located on the surface of ER. (Such ER, because of its "bumpy" surface, is called *rough ER*.) Some ribosomes are scattered through the cytoplasm. Ribosomes are not enclosed by a membrane. These bodies, composed mainly of RNA, are the sites of protein synthesis. Proteins synthesized on the ribosomes of rough ER enter channels in the ER membrane. They then travel through the ER and are later sent to the part of the cell in which they will be used. Proteins made on free ribosomes probably do not enter the ER. Rather, they pass directly to the cytoplasm. Some proteins in the membrane of rough ER probably function to direct a variety of chemical reactions in the cell.

Golgi (GAWL jee) bodies resemble ER. They are sets of flattened, slightly-curved sacs. Small vesicles are often seen near the ends of the sacs. Unlike ER, the sacs of the Golgi body are not connected to each other. **Golgi bodies** are involved in the storage and **secretion** (sih KREE shun) (pouring out) of chemicals from the cell. These chemicals may affect other parts of the body or other organisms. Digestive juices are cell secretions that affect other parts of the organism. Snake venom is a secretion that affects another organism. Many such chemicals are proteins that have been produced on ribosomes and channelled through the ER. Thus, the Golgi body and ER work together in directing transport of various chemicals around the cell.

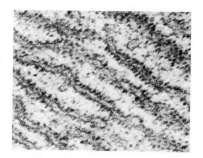

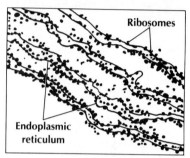

FIGURE 4–14. Cellular materials are thought to be transported in the tubes of endoplasmic reticulum. Proteins are synthesized on ribosomes. This ER is shown magnified about 40 000 times.

Some proteins synthesized on the ribosomes pass into the rough ER for transport. Others enter the cytoplasm directly.

Golgi bodies function in the storage and secretion of cellular chemicals.

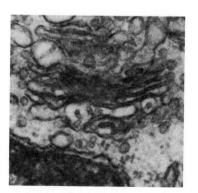

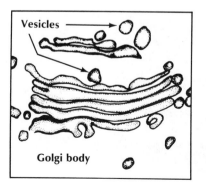

Vesicles

Golgi body

FIGURE 4–15. Golgi bodies aid in manufacturing and storing molecules that later are secreted from cells. Golgi bodies are shaped like stacked plates. This Golgi body is shown magnified about 54 000 times.

FIGURE 4–16. Microtubules are long, thin structures composed of protein. They are hollow rods that help maintain the cell's structure.

Microfilaments are made of protein and are involved in movement.

Microtubules are hollow rods of protein that provide support and shape to the cell.

Vacuoles (VAK yuh wohlz) are fluid-filled areas having a membrane. Many vacuoles are used for the storage of food, water, and minerals. In some plant cells small vacuoles join to form a large one that may be the prominent structure. Certain unicellular freshwater organisms have vacuoles that serve as pumps for removing excess water. Such water-removing structures are called **contractile** (kun TRAK tul) **vacuoles** (Section 18:7).

Lysosomes (LI suh sohmz) are small, membrane-bound spherical organelles that contain digestive chemicals. They are found mostly in animal cells. The chemicals break down molecules brought into the cell by endocytosis.

Lysosomes sometimes digest other cell parts. "Worn out" cell parts are broken down and removed from cells. Also, some parts may be digested to provide energy during starvation. If lysosomes break, the chemicals may even destroy the cell itself. "Worn out" cells may be removed in this way. Normally, the cell is protected from the chemicals by the lysosome's membrane.

Microfilaments are long, thin threadlike structures that have been found in a variety of cells. They are composed of protein and are involved with movement. Contraction of skeletal muscle (Chapter 30) depends upon microfilament movement. They also function in other ways such as locomotion of an amoeba or in causing changes in a cell's shape. Microfilaments also give strength to certain cells or parts of cells.

Microtubules are long, thin structures that provide support and shape to the cell. They are hollow rods composed of protein. Microtubules extend out from a region near the cell's nucleus to just beneath the cell membrane. There they help maintain the cell's structure. Microtubules also help with the movement of cell parts. For example, vesicles and mitochondria move along microtubules as they travel from one part of a cell to another. Microtubules are important structures in moving genetic material in cell division (Section 6:4) and are also involved in locomotion of some cells (Section 4:13).

Some microtubules are permanent parts of cells. Most are built up from protein within the cell when needed. They are then torn down

FIGURE 4–17. (a) Most vacuoles function in storage. In this cell, one vacuole takes up most of the space inside. (b) Lysosomes contain chemicals that hydrolyze food molecules. These lysosomes are shown magnified 25 000 times.

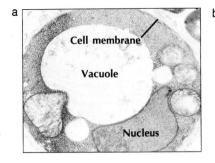

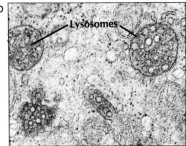

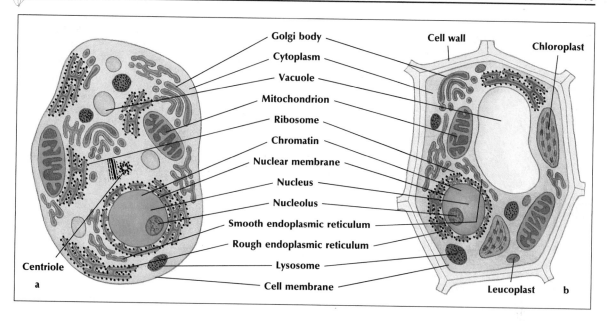

Golgi body

Cytoplasm

Vacuole

Mitochondrion

Ribosome

Chromatin

Nuclear membrane

Nucleus

Nucleolus

Smooth endoplasmic reticulum

Rough endoplasmic reticulum

Lysosome

Cell membrane

Centriole

a

Cell wall

Chloroplast

Leucoplast b

FIGURE 4–18. A generalized (a) animal cell and (b) plant cell are shown. There are basic differences between these types. However, in each type of cell, there is interaction of cell parts for proper metabolism of the cell.

and their protein can be reused later in the formation of other microtubules.

Microfilaments, microtubules, and certain other kinds of filaments provide shape and support for the cell. For this reason, they are referred to together as the cytoskeleton (cyto = cell).

4:13 Other Eukaryotic Cell Parts

The cell parts studied so far are found in most eukaryotic cells. Some other parts, though, are found in only some eukaryotes.

Plant cells and algae contain structures called **plastids** (PLAS tudz). The most common plastids are **chloroplasts** (KLOR uh plastz) that contain the chlorophyll needed for photosynthesis. A chloroplast has many stacks of membranes called **grana** (GRAY nuh) (Figure 4–19) located in a liquid. The liquid is called **stroma.** Light energy for photosynthesis is trapped by the grana. Glucose is made in the stroma (Chapter 21).

Other types of plastids contain pigments that aid chlorophyll in trapping solar energy. Some plastids produce and store starch and fats.

Cells of plants, fungi, and algae have a boundary called the **cell wall** outside the cell membrane. The cell wall of plants and certain algae is composed of cellulose (Section 3:12). A cell wall gives structure and shape to individual cells. In land plants the cell wall is important in support. It allows the plant to grow upright.

FIGURE 4–19. Chloroplasts are organelles that function in photosynthesis. The stacks of membranes in chloroplasts are grana; the liquid is stroma.

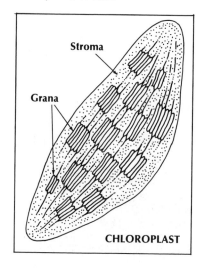

Stroma

Grana

CHLOROPLAST

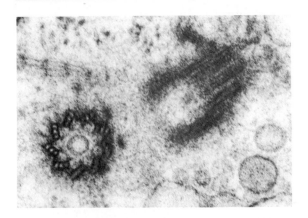

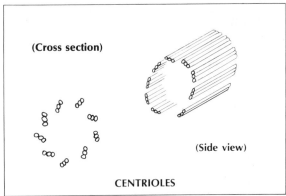

FIGURE 4–20. Centrioles, mainly important for cell division in animal cells, lie at right angles to each other. In the photograph the centriole at the left was cut across as the cell was sliced thinly for observation.

Centrioles function in the division of animal cells.

Cilia and flagella aid locomotion in unicellular organisms and some specialized cells of multicellular organisms.

Cells of animals and of some algae and fungi contain a pair of organelles called **centrioles** (SEN tree olz) that play an important role in cell reproduction. Centrioles exist in pairs and are found near the nucleus. Each centriole is composed of nine sets of microtubules arranged to form a circle. Each of the nine sets is composed of three tubules and the sets are connected (Figure 4–20).

Many cells have one or more hairlike structures. There are two kinds of these structures, flagella (fluh JEL uh) and cilia (SIHL ee uh). **Flagella** are long structures that usually appear singly or in pairs. **Cilia** are much shorter than flagella and appear in greater numbers. Both flagella and cilia are similar in structure to a centriole except that they have two microtubules in the center of the circle. Also, each outer set contains only two tubules. An extension of the cell membrane forms the borders of flagella and cilia. Flagella and cilia are attached to the interior of a cell by a **basal body** that has the same structure as a centriole. In some cells centrioles may become basal bodies and sometimes basal bodies become centrioles. They seem to be one and the same.

Movement of flagella and cilia by unicellular organisms results in locomotion of the organism. These structures also are present on some cells of multicellular organisms. A sperm cell, for example, moves by means of a flagellum. Also, cells lining your windpipe have cilia. The cells do not move but the movement of the cilia aids in moving mucus up from your lungs (Section 19:9).

FIGURE 4–21. Cilia are movable hairlike structures that usually are smaller and exist in greater numbers than flagella. Note the two microtubules in the centers of the cross-sections of cilia.

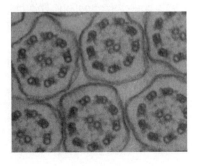

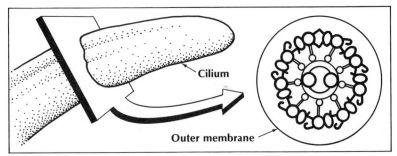

TABLE 4-1. EUKARYOTIC CELL STRUCTURES

Structure	Found in Cells of	Main Functions
CELL WALL	Plants; algae; fungi	Support; protection
CELL MEMBRANE	All eukaryotes	Boundary; "gatekeeper"
CYTOPLASM	All eukaryotes	Site of most metabolism
CENTRIOLES	Animals; some algae and fungi	Cell reproduction
CHLOROPLASTS	Plants; algae	Photosynthesis
ENDOPLASMIC RETICULUM	All eukaryotes	Intracellular transport of proteins
GOLGI BODIES	All eukaryotes	Storage; secretion
LYSOSOMES	Mainly animals	Intracellular digestion
MICROFILAMENTS	All eukaryotes	Movement
MICROTUBULES	All eukaryotes	Support; movement
MITOCHONDRIA	All eukaryotes	Cellular respiration
NUCLEOLUS	All eukaryotes	Production of ribosomes
NUCLEUS	All eukaryotes	Control; heredity
RIBOSOMES	All eukaryotes	Protein synthesis
VACUOLES	All eukaryotes	Storage

4:14 The Nucleus

Almost every kind of eukaryotic cell contains a central, spherical body called the **nucleus.** The nucleus is the control center of the cell and is important in its reproduction. It is separated from the rest of the cell by a double membrane. Definite pores or openings exist in this double membrane. Certain materials may pass through these pores when they travel between the nucleus and cytoplasm. Other materials move from the nucleus to the cytoplasm by way of the endoplasmic reticulum. Inside the nuclear membrane is nucleoplasm (Section 4:10). Also within the nucleus is a substance called **chromatin** (KROH mut un). In a nondividing cell, chromatin appears as a dense mass of material. However, as cell reproduction occurs, chromatin can be seen to consist of distinct threadlike bodies called **chromosomes** (KROH muh sohmz). Chromosomes are composed of protein and DNA. It is the DNA that carries the "genetic code." DNA thus controls the reproduction and activities of the cell.

One or more smaller bodies called **nucleoli** (new KLEE uh li) are also in the nucleus. Nucleoli are special parts of certain chromosomes. The DNA of nucleoli makes the RNA of ribosomes. The role of the nucleus in reproduction will be studied in Chapter 6 and again in Chapter 9.

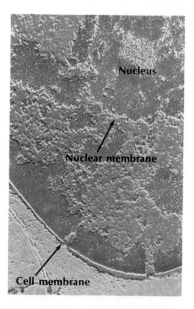

Nucleus

Nuclear membrane

Cell membrane

FIGURE 4–22. The nucleus is the cell's control center. It contains the cell's coded information. This cell has a large central nucleus.

REVIEWING YOUR IDEAS

12. Of what is protoplasm composed? What is its function?
13. What is division of labor? How is division of labor accomplished in cells?
14. How are prokaryotes and eukaryotes different?
15. What are the general functions of a nucleus?

Interaction of Cell Parts

Organelles serve as enclosed areas where specialized chemical reactions can occur.

Organelles are specialized regions where specific functions are carried out in cells. Because most organelles are separated from other parts of the cell by membranes, they can carry out their chemical activities independently and efficiently. Each organelle is a compartment where only certain chemicals react. While isolation of functions in distinct organelles is efficient, it is the entire cell that is the unit of life. The organelles must be specialized areas, but they must interact with one another for the cell to function as a whole.

Cell parts must interact for the entire cell to function properly.

Study of production of lysosomes and their functions shows the interdependence of cell parts. The digestive chemicals in lysosomes are proteins. These proteins are manufactured on the ribosomes of rough ER. The proteins pass from the ribosomes and into the ER.

How do these proteins become part of lysosomes? The proteins travel through the ER. A part of the membrane of the ER surrounds the proteins and pinches off, forming a vesicle. The vesicle then moves toward the membrane of a sac of a nearby Golgi body and fuses with it. The proteins of the transport vesicle are then inside the Golgi body sac. Once in the Golgi body, the proteins are transported from one sac to another by means of vesicles formed from the membrane of the Golgi body. (This description explains why vesicles are seen when Golgi bodies are viewed.) Finally a vesicle buds from the Golgi body sac farthest from the ER. That vesicle, containing the digestive proteins, becomes a lysosome.

Proteins travel from ER to Golgi bodies in vesicles formed from the ER membrane. They then travel from one Golgi sac to the next via vesicles formed from the membranes of the Golgi. The last vesicle to bud from the Golgi is a lysosome.

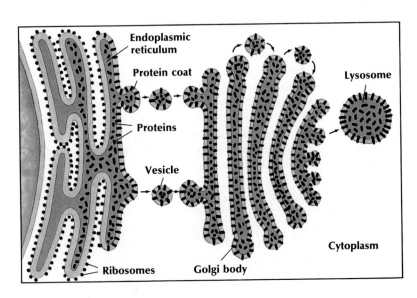

FIGURE 4–23. In the ER a membrane surrounds proteins and forms a vesicle. The vesicle fuses with a sac of a Golgi body. The proteins are passed along the Golgi body. Finally, a vesicle containing the proteins (enzymes) buds from the Golgi sac farthest from the ER and becomes a lysosome.

When a cell traps food by phagocytosis (Section 4:9), the food particles are also inside a vesicle. Often that vesicle is called a food vacuole. The food must be digested into small molecules the cell can use. The lysosome contains the chemicals needed for this digestion, but it cannot occur in the cytoplasm. The lysosome and the food vacuole fuse. Then the food can be safely digested within a membrane that protects the rest of the cell. The large molecules of food are digested into molecules small enough to diffuse across the membrane and into the cytoplasm. Some might be amino acids. They would be synthesized into new proteins on the ribosomes. Fuel molecules such as sugars would enter the mitochondria.

Undigested food particles must be removed from the cell. Such particles remain in the membrane after digestion occurs. The vacuole then moves to the cell membrane, and its contents are removed by exocytosis (Section 4:9).

Formation of vesicles by endocytosis and from membranes of ER and Golgi bodies as well as the process of exocytosis reveals another important interaction of cells. These processes show that the many membranes of cells are "interchangeable." Membranes of lysosomes and vesicles can fuse and may become parts of the cell membrane. A part of the membrane of the ER might become part of the Golgi membrane and then the cell membrane! Membranes seem to get recycled from one cell part to another. This shuttling of membranes is possible because they all have the same basic fluid mosaic structure.

Secretion of chemicals from the cell is similar to production of lysosomes. Many secretions contain proteins made on ribosomes. They enter the ER and move to the Golgi body from which they are pinched off in a vesicle. This kind of vesicle moves to the cell membrane and empties its secretion by means of exocytosis.

Many different proteins enter the ER and pass into the Golgi bodies. Each protein must make its way to the proper part of the cell. How does the cell package certain proteins and then ensure that each package reaches its proper destination?

The ER and Golgi bodies work together to sort out the various proteins. In the ER, an identical chain of sugar molecules is added to every protein. The proteins, all having the same sugar chain, then pass into the Golgi body. (The sugars act as markers to identify proteins.) What happens to the markers inside the Golgi body varies with different proteins. The markers become increasingly different depending on what sugars and other substances are added or removed. It is now known that each sac modifies a protein in a certain way. Thus, the passage of proteins through the sacs must occur in a definite order. When the proteins, with their specific markers, reach the Golgi sac farthest from the ER, they are packaged into vesicles that break from the Golgi body.

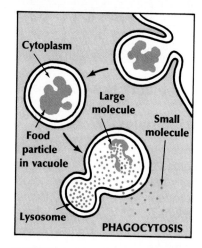

FIGURE 4–24. When a cell traps food by phagocytosis, the food is in a food vacuole. The vacuole and a lysosome fuse. Digestion occurs. Small molecules diffuse across the membrane.

Because they all have fluid mosaic structure, parts of different membranes are interchangeable.

Based on their sugar markers, different proteins are sorted and packaged for transport by Golgi bodies.

INVESTIGATION

Problem: How can cells be measured with the microscope?

Materials:

microscope
glass slide
coverslip
dropper
Elodea

water
prepared slide of frog
    blood
prepared slide of *Para-*
    mecium

Procedure:

1. Look through the eyepiece of your microscope using low power magnification. The circle of light you see is called the field of view. It has a diameter of about 1.5 mm. If you convert millimeters (mm) to micrometers (μm), the diameter of the field of view equals 1500 μm (Figure 4–25).

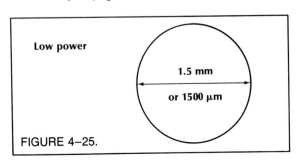

Low power

1.5 mm

or 1500 μm

FIGURE 4–25.

2. Look through the microscope using high power magnification. The field of view is now about 0.375 mm or 375 μm in diameter (Figure 4–26).

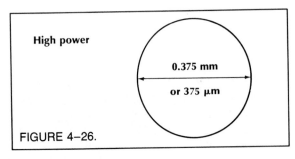

High power

0.375 mm

or 375 μm

FIGURE 4–26.

3. Convert the following measurements. NOTE: 1000 μm = 1 mm
   (a) 1.35 mm = _____ μm
   (b) 14.2 mm = _____ μm
   (c) 400 μm = _____ mm
   (d) 12 μm = _____ mm

4. To estimate the size of a cell while looking through the microscope under low power, follow these steps:
   (a) Locate the cell under low power.
   (b) Draw a circle that represents your field of view.
   (c) Within the circle, draw the cell the size it appears to be in relation to your field of view. This is called drawing to scale.
   (d) Estimate the number of same-sized cells that could fit side-by-side across the diameter of the field of view (shown as dashed outlines in Figure 4–27).

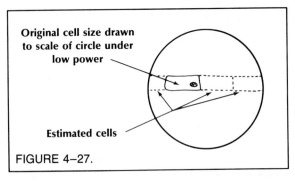

Original cell size drawn to scale of circle under low power

Estimated cells

FIGURE 4–27.

   (e) To determine cell size on low power, divide 1500 μm by the total number of cells that would fit across the circle's diameter. For example: the size of the cell in Figure 4–27 is 428 μm (1500 μm ÷ 3.5 = 428).

5. To estimate the size of a cell under high power, follow steps 4a–d. Divide 375 μm by the total number of cells that would fit across the field of view's diameter.

6. Examine the prepared slide of frog red blood cells. Determine if this slide is best viewed under low or high power. Complete the first column in a copy of Table 4–2.
7. Examine the prepared slide of *Paramecium*. Determine if this slide is best viewed under low or high power. Complete the second column of Table 4–2.
8. Prepare a wet mount of an *Elodea* leaf and observe it through the microscope. Determine if this slide is best viewed under low or high power. Complete the third column of Table 4–2.
9. Examine a prepared slide of frog blood cells under low power of your microscope. Locate the red blood cells that will appear oval in shape with a purple oval nucleus.
10. Trace Figure 4–28 and then make a drawing to scale in it. Diagram only one cell.

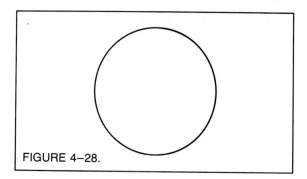

FIGURE 4–28.

11. Repeat steps 9 and 10 using high power.
12. Use your diagrams to make and record your calculations of cell size.
13. Repeat steps 9–12 using a prepared slide of *Paramecium*. Diagram only one cell.
14. Prepare a wet mount of an *Elodea* leaf.
15. Repeat steps 9–12 for *Elodea*. Diagram only one cell.

Data and Observations:

TABLE 4-2. ESTIMATING CELL SIZE						
	Frog red blood cells		Paramecium		Elodea	
Viewing under low or high power	Low power	High power	Low power	High power	Low power	High power
Number of cells that fit side-by-side						
Diameter of field of view						
Cell size in micrometers						

Questions and Conclusion:

1. Define millimeter, micrometer, and field of view.
2. Compare the cell size of your frog blood cells under low and high power. Are they very close or far apart? Explain.
3. Repeat question 2, comparing *Paramecium* measurements. Then compare *Elodea* measurements.

Conclusion: How can cells be measured with the microscope?

CHAPTER REVIEW

SUMMARY

1. After Robert Hooke's discovery of the cell, later discoveries led to the development of the cell theory. **4:1**
2. Cells can exist in many sizes and shapes and can exist independently as organisms or as parts of multicellular organisms. **4:2**
3. Cells have a semipermeable membrane that provides structure and regulates the passage of materials between the cell and its environment. **4:3**
4. The fluid mosaic model describes the cell membrane as a lipid bilayer in which proteins are embedded. **4:4**
5. Some small particles, lipids, and lipid-soluble particles may cross a membrane by means of diffusion. **4:5**
6. Water enters and leaves a cell by osmosis, a special type of diffusion. Cells must maintain an osmotic balance with their liquid environment in order to survive. **4:6**
7. Many kinds of particles pass across a membrane by means of a special form of passive transport called facilitated diffusion. **4:7**
8. In active transport cells expend energy to move particles from a region of lesser to greater concentration. **4:8**
9. Some cells obtain very large particles from their environment by endocytosis. Certain particles leave a cell by exocytosis. **4:9**
10. Protoplasm is the site of much of a cell's metabolism. **4:10**
11. Organelles contribute to division of labor in cells. **4:11**
12. Eukaryotes, unlike prokaryotes, have a true nucleus and membrane-bound organelles. Organelles common to eukaryotes include mitochondria, ER, ribosomes, Golgi bodies, vacuoles, and lysosomes. **4:12**
13. Plastids, centrioles, cilia, and flagella are cell parts not found in all eukaryotes. **4:13**
14. The nucleus controls a cell. **4:14**
15. Specialized organelles must function with one another for the cell to function as a whole. **Advances in Biology**

LANGUAGE OF BIOLOGY

active transport
cell
cell membrane
cell theory
cell wall
chromatin
centriole
cilium
diffusion
division of labor
dynamic equilibrium
endocytosis
endoplasmic reticulum
eukaryote
exocytosis
facilitated diffusion
flagellum
fluid mosaic model
Golgi body
lysosome
metabolism
microfilament
microtubule
mitochondrion
nucleolus
nucleus
organelle
osmosis
passive transport
permease
phagocytosis
pinocytosis
prokaryote
protoplasm
ribosome
vacuole

CHECKING YOUR IDEAS

On a separate paper, match the phrase from the left column with the proper term from the right column. Do not write in this book.

1. random movement of particles
2. controls reproduction
3. requires energy
4. semipermeable
5. cell containing definite nucleus
6. total of cell's reactions
7. all organisms are composed of cells

a. active transport
b. cell membrane
c. cell theory
d. diffusion
e. division of labor
f. eukaryote
g. metabolism
h. nucleus
i. osmosis
j. phagocytosis

EVALUATING YOUR IDEAS

1. Why is the cell theory important?
2. Identify the particles that most readily cross a cell membrane.
3. Describe the structure and properties of the phospholipid bilayer in the cell membrane. What is the role of cholesterol?
4. How are the proteins of the cell membrane arranged? What are their functions?
5. What kinds of particles can cross a cell membrane by diffusion?
6. Explain how carbon dioxide leaves your cells and enters the blood.
7. Why is osmotic balance important to cells?
8. Describe the pores of the cell membrane. What passes through them?
9. Describe how glucose is thought to enter cells.
10. Explain how a cell uses active transport to move particles.
11. Explain how each of the following properties of cell membranes is explained by the fluid mosaic model:
    (a) Lipids and lipid-soluble particles pass through freely.
    (b) Smaller particles pass through more readily than larger ones.
    (c) Membranes of different cells are permeable to different particles.
12. Contrast phagocytosis and pinocytosis.
13. State the main functions of each of the following. In what kind(s) of cells is each usually found: (a) cell wall (b) centriole (c) chloroplast (d) cilium, flagellum (e) ER (f) Golgi body (g) lysosome (h) microfilament (i) microtubule (j) mitochondrion (k) nucleolus (l) nucleus (m) ribosome (n) vacuole
14. Distinguish between chromatin and chromosomes. Of what are chromosomes composed? What part of a chromosome carries the "genetic code"?
15. Why is it important that different cell functions be carried out in distinct organelles?
16. Explain how a lysosome and its contents are produced and how a lysosome functions in digestion.
17. How does the cell sort out and package the many proteins it produces?

APPLYING YOUR IDEAS

1. *Paramecium* is a single-celled, freshwater organism. Its cytoplasm contains water and many other substances. What "problem" does a paramecium have? What structure might a paramecium have to deal with this problem? What might happen to a paramecium in salt water? Why?
2. Oxygen from blood continually moves into cells by diffusion. What explains the fact that the net movement of oxygen is always into the cell? (Hint: What happens to the oxygen inside the cell?)
3. Why are there many mitochondria in a muscle cell?
4. Hormones are chemicals secreted by specialized gland cells. Name a structure that would be abundant in these cells. Explain.

EXTENDING YOUR IDEAS

1. Prepare a report, with drawings, on five different specialized animal or plant cells.
2. Build a model of a "typical" cell. Is there really such a cell?

SUGGESTED READINGS

DeDuve, Christian, *A Guided Tour of the Living Cell.* NY, W. H. Freeman and Co., 1984.
Hellerstein, David, "The Promise of Artificial Skin." *Science Digest,* Sept., 1985.

ENERGY FOR LIFE

Organisms need energy to maintain life processes. Energy must not only be present, it must be in a form usable within the cells of an organism. Here, chemicals react within the tissues of the firefly to release energy causing the firefly to glow. What causes the chemicals to react? What form of energy do they release when they react? What is the source of that energy?

iving systems require a supply of energy for growth and development, maintenance and repair, reproduction, and overall organization. Without a constant input of energy, no single organism or community could survive for very long. Energy is necessary to maintain life.

What is energy? The best way to define energy is to describe what it does. The movement of a car, the running of an air conditioner, and the changing of ice to water all require energy. Each of these examples involves motion and can be considered work. **Energy** is the ability to do work or cause motion.

Objectives:
You will
- explain the importance of energy transformation in living systems.
- describe how specific cellular reactions occur.
- explain how the process of cellular respiration releases energy needed for certain cellular reactions.

ENERGY AND REACTIONS

Think about examples of energy familiar to you. Light shines from a bulb, enabling you to read. The blades of a fan turn rapidly to cool you. Heat in an oven cooks your food. Light, movement, and heat are different forms that energy may take. Not only does energy exist in different forms, one form of energy may change to another. Changes in forms of energy are important to organisms.

Energy is the ability to do work or cause motion.

5:1 Changes in Forms of Energy

Consider a ball at the top of a flight of stairs. The ball could be pushed to bounce slowly down the stairs. It could also be dropped directly over the railing to fall quickly to the floor. In either case, as the ball moves downward, it could do work or cause motion such as breaking a nearby vase or lamp.

Now think about burning wood. As the wood burns, heat and light (radiant energy) are released. The energy could be used for work. For example, the heat could be used to warm a room. When the wood is all burned, no more work can be done.

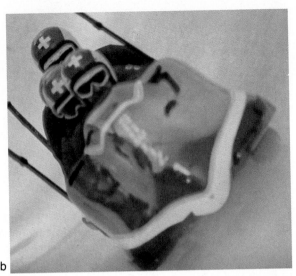

a

b

FIGURE 5-1. (a) A bobsled team at the top of the run has high potential energy and low kinetic energy. (b) While on the run, the potential energy is converted to kinetic energy.

Potential energy can be converted to kinetic energy.

Both the ball at the top of the stairs and the wood have high potential energy. **Potential energy** is energy of position or stored energy. As the ball bounces or falls down the stairs or as the wood burns, the potential energy is converted to kinetic energy. **Kinetic energy** is energy of motion. The ball at the bottom of the stairs and the products formed from burning the wood have low potential energy. The difference between the high and low potential energies equals the energy released.

Energy, either potential or kinetic, can be in many forms and can change from one form to another. The potential energy in the wood molecules is called **chemical energy.** That energy is "stored" in the bonds of the molecules in wood cells. (Where did that energy come from originally?) When wood is burned, old bonds are broken and new bonds are formed. The potential chemical energy is changed to heat and light (radiant energy).

Energy is conserved as it changes form.

Anytime energy is transformed, it is conserved so that the total amount of energy in a system is constant. For example, in the burning of wood, some energy is released and can be used for work. Some of the original energy exists in the bonds of the new compounds formed, carbon dioxide and water. The total amount of the energy released plus the energy in the bonds is equal to the amount of energy in the wood and the oxygen with which it reacted. This idea can be stated as the **law of conservation of energy:** *energy is neither created nor destroyed, but it can be transformed.*

In cells, potential chemical energy is converted to kinetic energy used for work.

Transformation of energy is important to living systems. *Cells depend upon potential chemical energy in the bonds of energy-rich molecules. When the bonds are broken, the potential energy is converted to kinetic energy that can be used by cells for necessary biological work.*

5:2 Activation Energy

In order for any chemical reaction to occur, collisions between reacting molecules must be forceful enough for old bonds to break. The force of collisions depends upon the kinetic energy of the molecules. The kinetic energy, in turn, increases as temperature increases. The amount of energy required to start any chemical reaction is called **activation energy.** (Activation energy is like the push needed to get a ball rolling down the stairs.) The amount of energy needed to start a reaction varies with the reaction.

Consider sugar molecules in a test tube in a laboratory. Even though oxygen molecules in the air collide with the sugar molecules, the sugar does not burn. A reaction does not occur. If the test tube is heated though, the sugar will burn. The increase in temperature raises the average kinetic energy of sugar and oxygen molecules so that a reaction can occur. In this reaction, additional heat is the source of activation energy.

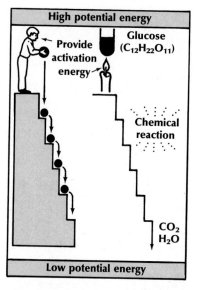

High potential energy

Glucose ($C_{12}H_{22}O_{11}$)

Provide activation energy

Chemical reaction

CO_2
H_2O

Low potential energy

FIGURE 5–2. Activation energy is required to start a ball bouncing down stairs and to start chemical reactions. In both cases, potential energy is converted to kinetic energy as the processes occur.

REVIEWING YOUR IDEAS

1. Distinguish between potential and kinetic energy.
2. How do cells obtain their usable energy?

ENERGY FOR CELLULAR WORK

In organisms, energy is used for biological work or motion. Without energy, muscles would not contract, active transport would not occur, and new living tissue would not be built.

5:3 Enzymes

In some ways, the chemical reactions of organisms and nonliving things are alike. For example, sugar in a test tube burns by combining with oxygen, and a mouse needs oxygen to "burn" its food in order to stay alive. But there are differences. The test-tube reaction gives off heat and light while the mouse gives off a little heat but no light. Burning in the test tube is "explosive"; the energy is released quickly. In the mouse, the energy is released more slowly and the "burning" of food is controlled.

Heat energy is the activation energy used to burn the sugar in the test tube. But heat cannot be the activation energy for chemical reactions in organisms. The amount of heat needed to start most cellular chemical reactions would be so great it would kill the cell. Even if heat

In a cell, sugar is "burned" so that energy is released slowly.

Heat is not a suitable source of activation energy for organisms.

could be used, it would cause most reactions to occur at the same time. For a cell to function smoothly, *specific* reactions must occur at definite times.

If heat is not the source of activation energy in cells, how do cellular reactions get started? Cells contain special chemicals called enzymes (EN zimes). **Enzymes** are chemicals that *lower* the amount of activation energy required and allow reactions to occur at cellular temperatures. Thus, no additional heat is required.

Enzymes lower the activation energy necessary to start biological reactions.

5:4 Properties of Enzymes

All enzymes have certain properties. Unlike heat, each enzyme is specific in that it boosts only one or one type of a cell's many reactions. Also, enzymes themselves are not changed or used up during a reaction.

Enzymes are globular proteins. Recall that each protein has a particular three-dimensional shape (Section 3:14). The fact that each enzyme has a unique 3-D shape is related to the fact that each enzyme governs a specific reaction.

Enzymes are proteins. They are specific and reusable.

The substance upon which a certain enzyme acts is called a **substrate.** An enzyme often is named by adding the suffix *-ase* to part of the name of the substrate. Thus, the enzyme that converts maltose to simpler sugars is called maltase. The enzymes that break down proteins are called proteases (PROHT ee ays uz). Some enzymes are named on the basis of their function. For example, an enzyme that removes hydrogen is called a dehydrogenase (dee hi DRAHJ uh nays). An enzyme that transfers an amino group is called a transaminase (tranz AM uh nays). Some names of enzymes, such as pepsin, an enzyme in your stomach, do not end in *-ase*. These enzymes were named before a system of naming was used, although some have since been renamed according to the system.

What is the name of the enzyme that breaks down sucrose?

FIGURE 5-3. According to the lock and key hypothesis, an enzyme fits a specific substrate just as a key fits a specific lock.

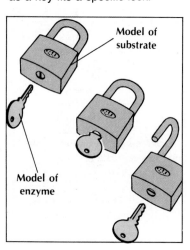

Model of substrate

Model of enzyme

5:5 Lock and Key Hypothesis

How does an enzyme lower the activation energy needed to start a reaction? The properties of enzymes have been studied in detail. The **lock and key hypothesis** is an accepted explanation of enzyme function. It explains many properties of enzymes.

Every lock has a specific key that will open it. The key is cut so that it matches the tumblers of the lock. Only if the two surfaces fit together will the lock open.

Each substrate (lock) in a given reaction can be acted upon only by a specific enzyme (key). The surface shapes of the enzyme and substrate must match each other and fit together. Part of the enzyme has a special shape that matches (fits with) a part of the corresponding substrate.

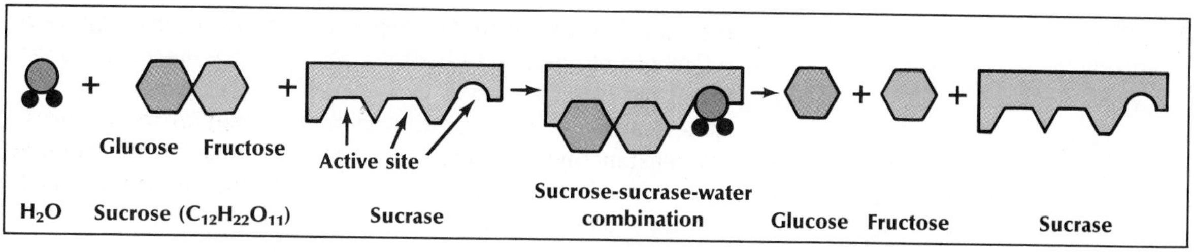

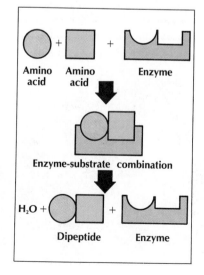

Glucose   Fructose   Active site

H_2O Sucrose $(C_{12}H_{22}O_{11})$ Sucrase   Sucrose-sucrase-water combination   Glucose   Fructose   Sucrase

FIGURE 5–4. The enzyme sucrase is involved in the breakdown of sucrose. The active site of the enzyme fits the shape of the substrates (sucrose and water).

The active site is the "key" that fits the substrate "lock."

The enzyme and substrate fit just as a key fits a lock or as two pieces of a jigsaw puzzle fit together. The part of the enzyme molecule that fits the substrate is called the **active site.** This explains the fact that each enzyme is specific for its substrate.

Sometimes other substances alter the shapes of active sites (Section 5:6). In some cases the fit between enzyme and substrate is not exact, but when they join, the shape of the enzyme changes to make a better fit.

Now consider a common biological reaction. Sucrose (table sugar), a disaccharide, breaks down into two monosaccharides, glucose and fructose (Section 3:12). This hydrolysis reaction (Section 3:16) occurs during digestion. Although sucrose and water molecules collide, the force of such collisions is not great enough to cause the hydrolysis reaction to occur. The bond holding the two monosaccharides together is too strong.

Now suppose a molecule of sucrose and a molecule of water collide with a molecule of the enzyme sucrase (Figure 5–4). According to the lock and key hypothesis, the sucrase molecule has an active site that fits a certain part of both the sucrose molecule and the water molecule. When the water and sucrose molecules collide with a sucrase molecule, the three molecules may join for a short time. *When an enzyme and substrates are joined, the amount of activation energy needed for a reaction to occur is lowered. The reaction can then occur at cellular temperatures.* While sucrose, water, and sucrase are joined, bonds holding the two monosaccharides of sucrose together and bonds in the water molecule are somehow weakened and then broken. As this occurs, new bonds are formed, and the reaction is completed. The enzyme separates as the products, glucose and fructose, are formed.

Enzymes also work in condensation reactions (Section 3:16). For example, dipeptides are formed from two amino acids. An enzyme that aids the formation of a dipeptide has an active site into which both of the amino acids can fit. Again, while the enzyme and substrates are joined, the activation energy required is lowered. As the two amino acids lie side by side on the surface of the enzyme, they are in a position that increases the chances for bonds to form between them. A molecule of water is removed and a dipeptide is formed as the two amino acids join (Figure 5–5).

FIGURE 5–5. In the synthesis of a dipeptide, two amino acids must fit into the active site of the enzyme.

Amino acid   Amino acid   Enzyme

Enzyme-substrate combination

H_2O +   Dipeptide   Enzyme

The lock and key hypothesis is supported by experimentation.

Experimental evidence supports the lock and key hypothesis. If the active site of an enzyme is destroyed, the enzyme cannot direct its normal activities. However, if the active site remains and another region is destroyed, the enzyme can sometimes perform properly. Also, a substance with a shape similar to the normal substrate can compete with the substrate. It joins with some of the available enzyme molecules so that the enzyme cannot do its job.

Other molecules may sometimes combine with an enzyme, altering its shape and thus controlling when a reaction occurs.

Using computers, scientists are identifying the active sites of some enzymes. They have learned which part of the substrate the enzyme joins. The shape of the active sites of some enzymes may be altered by molecules that either produce the proper active site or distort it. These molecules can control when a particular reaction occurs. Such control is important because a cell must carry out only certain reactions at certain times.

5:6 Coenzymes

Some enzymes join temporarily with other molecules, **coenzymes** (koh EN zimes), during a reaction. In many cases, a coenzyme alters the shape of the enzyme's active site. The enzyme then fits its substrate better and the reaction occurs.

Some coenzymes change the shape of an enzyme's active site.

A well-balanced diet provides all the vitamins required as coenzymes or for the synthesis of necessary coenzymes.

Coenzymes are organic molecules, smaller than enzymes. Like enzymes, they are not consumed during a reaction and can be reused. Coenzymes often serve as transfer agents of atoms, parts of atoms, or groups of atoms during a reaction. For example, an enzyme might remove a hydrogen atom from a molecule. The coenzyme would pick up the atom and donate it to some other substance. The reaction could not occur without the coenzyme to accept the hydrogen atom.

FIGURE 5–6. These foods are high in vitamin B_1. A disease called beriberi can result if the diet does not include enough vitamin B_1.

Some essential coenzymes are made from vitamins or vitamin fragments. If an organism's diet does not include enough of a needed vitamin, the coenzyme cannot be made. Certain cellular reactions cannot occur and symptoms of a disease result. For example, lack of Vitamin B_1 causes beriberi. Beriberi involves weakening of the muscles and paralysis. Vitamin B_1 functions as a coenzyme. It activates an enzyme. The enzyme is involved in the process of cellular respiration.

a   b

5:7 ATP

Reactions require activation energy to get started. Once started, some reactions release energy. A reaction that gives off energy is called **exergonic** (ek sur GAHN ihk). Burning wood (wood combining with oxygen) and a bouncing ball are exergonic reactions. Activation energy is required to start the wood burning and the ball bouncing.

Some reactions use energy once started. Reactions that use energy are called **endergonic** (en dur GAHN ihk). Compresses sometimes used on sprained ankles contain chemicals that react with each other when mixed. The reaction uses up energy. As the chemicals react and the energy is used, the compress gets colder. Heat energy is absorbed. Many cellular reactions are endergonic because energy is used up as each reaction continues.

All cellular reactions require an enzyme to lower activation energy. But, endergonic reactions need more than this enzyme "spark" to occur. (Consider the ball and stairs. The ball cannot return to the top of the stairs with simply a push. Moving the ball back to the top is an endergonic reaction.) Recall that cells rely on chemical energy for biological work (Section 5:1). The source of chemical energy for most endergonic reactions in a cell is a special kind of molecule. The molecule is **adenosine triphosphate** (uh DEN uh seen • tri FAHS fayt). This source of chemical energy is commonly called **ATP.**

ATP is a complex molecule that often is represented as A—Ⓟ ⟿Ⓟ⟿Ⓟ. The A stands for the adenosine part of the molecule, and the Ⓟs represent groups of atoms called phosphate groups. The wavy lines between phosphate groups represent special bonds called **high-energy bonds.** When these bonds are broken (and new bonds are formed), a large amount of chemical energy is released. The high-energy phosphate bonds liberate about twice as much energy as normal phosphate bonds. *The chemical energy released from ATP is used in endergonic reactions.*

In an endergonic reaction in a cell, an enzyme is used to lower the amount of activation needed to get the reaction started. In addition,

FIGURE 5–7. (a) The explosion of fireworks is an exergonic reaction. Heat and light are given off. (b) A cold compress has chemicals undergoing an endergonic reaction. Heat is being absorbed.

An exergonic reaction liberates energy.

An endergonic reaction requires energy.

ATP provides energy for most endergonic reactions in cells.

ATP contains high-energy bonds that, when broken, release chemical energy for endergonic reactions.

FIGURE 5–8. Energy is released and work is done when ATP is converted to ADP and a low-energy phosphate group. Energy is required to produce ATP.

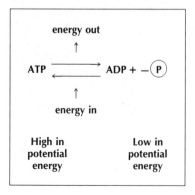

energy out

↑

ATP ⇄ ADP + —Ⓟ

↑

energy in

High in
potential
energy

Low in
potential
energy

Energy is needed to change ADP and low-energy phosphate to ATP.

cells must use ATP because there is more energy in the bonds of the products than in the bonds of the reactants. Another enzyme acts upon ATP, breaking the bond that holds the third phosphate group. Then the molecule contains only two phosphates and is called **adenosine diphosphate** (di FAHS fayt) or **ADP.** A high-energy phosphate group, represented as ∿Ⓟ, is also a product. When the high-energy phosphate group forms a bond with another substance, chemical energy is released and work is done. As the work is done, the high-energy phosphate group is changed to a low-energy phosphate group, —Ⓟ (Figure 5–10).

ATP is used in many cellular jobs. For example, energy released from a high-energy phosphate group is necessary for work such as contracting a muscle, or lighting a firefly. It is also used to build proteins, starches, and fats.

Recall that for each molecule of ATP used, the molecule of ADP and a low-energy phosphate group are produced. Cells can maintain their supply of ATP by rejoining ADP and a low-energy phosphate group. But, the total energy potential of ATP is greater than the energy of an individual ADP plus a low-energy phosphate group. Therefore, energy is required to produce more ATP.

REVIEWING YOUR IDEAS

3. List three characteristics of enzymes.
4. How are enzymes named?
5. Of what are coenzymes made? How do they function?
6. Why is ATP necessary for many cellular reactions? What property of ATP makes it suitable for its function?

PRODUCTION OF ATP

Sugar can be "burned" to supply energy for living systems (Chapter 1). The sugar that organisms most commonly use as an energy source is glucose, $C_6H_{12}O_6$. It is the source of energy for making ATP. Why is glucose such a good energy source?

5:8 Respiration With Oxygen

The process in which a cell releases the energy of glucose is called **cellular respiration.** During this process, the energy in the bonds of glucose is transferred to the bonds of ATP. Respiration in most cells makes use of oxygen and is called **aerobic** (er ROH bihk) **respiration.** ("Aerobic" means requiring oxygen.)

Aerobic respiration requires oxygen.

FIGURE 5–9. Oxygen taken in by the body is used in cellular respiration. During exercise, more oxygen is taken in.

The process of aerobic respiration involves many separate chemical reactions that occur in a definite sequence. Most of the reactions occur in the mitochondria (Section 4:12). Each reaction requires an enzyme. Energy is used in some of these steps and released during others. Because more energy is released than used, the overall reaction is exergonic. The following equation summarizes the many reactions of aerobic respiration of glucose:

$$C_6H_{12}O_6 + 6O_2 \xrightarrow[\text{2 ATP}]{\text{enzymes}} 6CO_2 + 6H_2O + \begin{array}{l}\text{energy for} \\ \text{many ATP}\end{array}$$

In the series of steps glucose is converted to carbon dioxide and water. *Energy of glucose is released and is used in the formation of ATP from ADP and a low-energy phosphate group.* Thus, ATP contains energy originally in glucose.

Note that two ATP molecules are "invested" in the process. This "investment" is needed because certain early steps require energy. However, later steps release energy. For many years it was thought that each molecule of glucose releases enough energy to produce a net yield of 36 ATP molecules. However, recent studies show that the actual number may be fewer than 36.

ATP molecules are the major *usable* energy source in a cell. The step-by-step process of cellular respiration slowly releases the potential energy of glucose and transfers it to ATP. Thus, the process is like the ball bouncing down the stairs. If the energy were released more rapidly (as when glucose is burned in a test tube), it could not be trapped in chemical form. Rapid burning of glucose is similar to the ball falling quickly over the railing. The amount of energy released is the same whether the ball is quickly falling or slowly bouncing. Similarly, the amount of energy released from glucose is the same whether it is burned in a test tube or "burned" in a cell. However, cells can only trap and use energy released slowly from glucose. The energy is transferred to the bonds of ATP. The ATP can then be used for biological work.

During aerobic respiration, some of the energy in the bonds of glucose is transferred to the bonds of ATP.

FIGURE 5–10. When ATP is converted to ADP and (P), energy is released that is used in cells for different purposes.

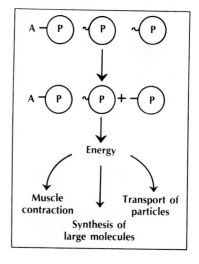

FIGURE 5–11. In cellular respiration, the energy of glucose converts ADP and (P) to ATP. The ATP is broken down later, releasing energy that is used in cells.

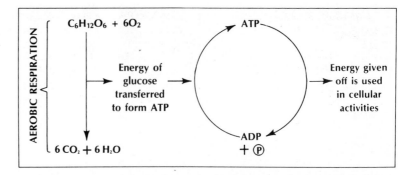

Figure 5–11 summarizes the interrelationships among aerobic cellular respiration, ATP, ADP, and the work of the cell. This series of chemical events is called the ATP-ADP cycle. The cycle shows that during aerobic respiration, the energy of glucose is released. This energy is used to join ADP and a low-energy phosphate group to form ATP that can then be used as an energy source for cellular work.

Anaerobic respiration occurs in the absence of oxygen.

FIGURE 5–12. (a) Anaerobic respiration of yeast produces carbon dioxide that makes bread rise. (b) Anaerobic respiration also occurs in muscle cells during times of heavy exercise when oxygen is lacking.

5:9 Respiration Without Oxygen

Most cells undergo aerobic respiration. However, some of these cells may be able to change to **anaerobic respiration** if deprived of oxygen. ("Anaerobic" means not requiring oxygen.) In some cells, anaerobic respiration may occur even if oxygen is present. Some microbes rely only on anaerobic respiration for energy because they cannot live in the presence of oxygen.

Most plant cells and microbes commonly undergo a type of anaerobic respiration called alcoholic **fermentation.** Glucose is broken down as shown in the following equation:

$$C_6H_{12}O_6 \xrightarrow[\text{2 ATP}]{\text{enzymes}} \underset{\text{(ethyl alcohol)}}{2C_2H_5OH + 2CO_2} + \underset{\text{4 ATP}}{\text{energy for}}$$

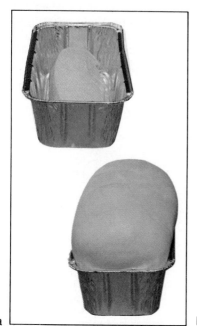

a

b

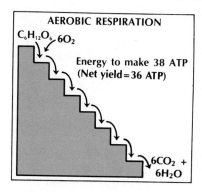

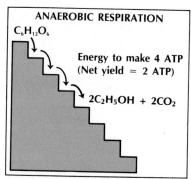

FIGURE 5–13. Aerobic respiration and anaerobic respiration both produce ATP. What are the main differences in the two processes?

As in aerobic respiration, enzymes and a small amount of ATP are used. Also, the energy of glucose is released for making more ATP. Notice, though, that oxygen is not a reactant. Without oxygen, glucose cannot be broken down to CO_2 and water. Glucose in anaerobic respiration is like a ball that bounces down just a few steps and stops. Only a small amount of its potential energy is released. Much of the energy remains "locked" in the bonds of ethyl alcohol. Thus, far fewer ATP molecules are produced.

Anaerobic respiration releases much less energy than aerobic respiration.

Yeasts are useful in industry because of their fermentation products—ethyl alcohol and carbon dioxide. Ethyl alcohol is used in some beverages. The carbon dioxide formed when yeast cells break down the carbohydrates in dough causes bread and cakes to rise.

Fermentation is useful in the baking and brewing industries.

Many animals and some microbes carry on anaerobic respiration at certain times. For example, during times of heavy physical activity, oxygen may not be supplied fast enough to muscle cells for aerobic respiration to take place. When oxygen is low, the muscle cells "switch" to anaerobic respiration. At these times, glycogen (Section 3:12), which is stored in muscles, is converted to glucose. Glucose is then broken down without oxygen to produce lactic acid. Energy is released and used to form ATP molecules. Thus, energy is supplied for muscle contraction even in the absence of oxygen.

Muscle cells undergo anaerobic respiration when oxygen is unavailable.

Why would oxygen not be supplied quickly enough during heavy exercise?

$$C_6H_{12}O_6 \xrightarrow[\text{2 ATP}]{\text{enzymes}} \underset{\text{(lactic acid)}}{2CH_3CHOHCOOH} + \underset{\text{4 ATP}}{\text{energy for}}$$

Figure 5–13 compares the events of aerobic and anaerobic respiration. Examine the figure closely.

REVIEWING YOUR IDEAS

7. What is cellular respiration? Why is it necessary? How is it related to breathing?
8. Contrast aerobic and anaerobic respiration.

INVESTIGATION

Problem: How does the amount of catalase enzyme compare in different tissues?

Materials:

8 test tubes
hydrogen peroxide
8 droppers
tweezers
chunks of:
  unboiled carrot, potato, liver, muscle
  boiled carrot, potato, liver, muscle
chopped:
  unboiled carrot, potato, liver, muscle
  boiled carrot, potato, liver, muscle

test tube rack
graduated cylinder
glass-marking pencil

Procedure:

Part A. What Is Catalase?

1. Most living cells contain an enzyme called catalase. This enzyme changes a chemical called hydrogen peroxide (our substrate) into oxygen gas and water.
2. A shorthand way of expressing this idea is as follows:

$$\text{hydrogen peroxide} \xrightarrow[\text{(enzyme)}]{\text{catalase}} \text{oxygen} + \text{water}$$
$$\text{(substrate)}$$

3. Fill a test tube half way with hydrogen peroxide. Add a small chunk of unboiled liver to the tube. Note the change that takes place.
4. The bubbling or foam that occurs in the test tube is caused by the oxygen gas given off when the catalase in the liver changed the hydrogen peroxide into water and oxygen.
5. Repeat step 3, this time adding no liver to the test tube. Observe the tube for at least one minute.

6. Answer these questions before going on:
  (a) Name the enzyme used in this activity.
  (b) What is the source of the enzyme?
  (c) Name the substrate used.
  (d) What chemicals are formed when catalase reacts with its substrate?
  (e) What evidence do you have that the enzyme is responsible for changes in the substrate?

Part B. Testing Unboiled Cells for Catalase

1. Number the test tubes 1–8. Place them in the test tube rack.
2. Using tweezers, add the following:
  Tube 1—small chunk of unboiled carrot
  Tube 2—small chunk of unboiled potato
  Tube 3—small chunk of unboiled liver
  Tube 4—small chunk of unboiled muscle
3. Using a different dropper for each tube, add the following:
  Tube 5—add 3 drops of chopped unboiled carrot
  Tube 6—add 3 drops of chopped unboiled potato
  Tube 7—add 3 drops of chopped unboiled liver
  Tube 8—add 3 drops of chopped unboiled muscle
4. Using a graduated cylinder, add 5 mL of hydrogen peroxide to each tube.
5. Observe the amount of bubbling or foam (oxygen gas being given off) in each tube. Using a scale of 0–10, record the amount in a copy of Table 5–1. NOTE: 0 = no bubbles or no enzyme present in cells being used; 10 = most bubbles or most enzyme present in cells being used.

Part C. Testing Boiled Cells for Catalase

1. Number the test tubes 1–8. Place them in the test tube rack.
2. Using tweezers, add the following:
   Tube 1—small chunk of boiled carrot
   Tube 2—small chunk of boiled potato
   Tube 3—small chunk of boiled liver
   Tube 4—small chunk of boiled muscle
3. Using a different dropper for each tube, add the following:
   Tube 5—3 drops of chopped boiled carrot

Tube 6—3 drops of chopped boiled potato
Tube 7—3 drops of chopped boiled liver
Tube 8—3 drops of chopped boiled muscle
4. Using a graduated cylinder, add 5 mL of hydrogen peroxide to each tube.
5. Observe the amount of bubbling or foam (oxygen gas) in each tube. Using a scale of 0–10, record the amount in your copy of Table 5–1. NOTE: 0 = no bubbles or no enzyme; 10 = most bubbles or most enzyme.

Data and Observations:

TABLE 5-1. COMPARISON OF THE AMOUNTS OF CATALASE IN DIFFERENT TISSUES				
Cell Type	Unboiled		Boiled	
	Chunk	Chopped	Chunk	Chopped
Carrot				
Potato				
Liver				
Muscle				

Questions and Conclusion:

1. List four properties of all enzymes.
2. How were you able to judge if the enzyme was or was not working in this activity?
3. In general, do plant or animal cells contain more catalase? Give evidence from your data.
4. (a) In general, do chunk unboiled cells release more, less, or about the same amount of catalase as chopped unboiled cells? Give evidence from your data.
   (b) Offer a possible hypothesis to explain your data.
5. In general, do chunk unboiled cells release more, less, or about the same amount of catalase as chunk boiled cells? Give evidence from your data.
6. In general, do chunk boiled cells release more, less, or about the same amount of catalase as chopped boiled cells? Give evidence from your data.
7. Offer possible hypotheses to explain the data you obtained in questions 5 and 6.
Conclusion: How does the amount of catalase enzyme compare in different tissues?

CHAPTER REVIEW

SUMMARY

1. Living systems require a supply of energy for biological work. **p. 89**
2. Organisms obtain usable energy by breaking down high-energy molecules. Chemical potential energy is converted to kinetic energy. **5:1**
3. All chemical reactions require activation energy to get started. **5:2**
4. Enzymes lower the required activation energy in cells. **5:3**
5. Enzymes are reusable proteins. Each enzyme guides a specific cellular reaction. **5:4**
6. The lock and key hypothesis explains how a given enzyme guides a specific cellular reaction. **5:5**
7. Coenzymes work along with many enzymes. A coenzyme may combine with an enzyme to alter the enzyme's active site. Coenzymes also act as transfer agents. **5:6**
8. When the high-energy bonds of ATP are broken, energy is released for biological work. ATP is used in endergonic reactions. **5:7**
9. Each time a molecule of ATP is used, the cell is left with a molecule of ADP and a low-energy phosphate group. ADP and a low-energy phosphate group can rejoin in a cell when energy is added. **5:7**
10. Most cells break the bonds of glucose by the process of aerobic respiration. Much energy is released from the bonds of glucose during aerobic respiration and transferred to the bonds of ATP. The process of aerobic respiration involves many separate chemical reactions that occur in a definite sequence. **5:8**
11. Transfer of energy from glucose to ATP is necessary because ATP is the major usable energy source in cells. **5:8**
12. Fewer ATP molecules are formed from the anaerobic breakdown of one molecule of glucose than from the aerobic breakdown. **5:9**

LANGUAGE OF BIOLOGY

activation energy
active site
ADP
aerobic respiration
anaerobic respiration
ATP
cellular respiration
chemical energy
coenzyme
endergonic
energy
enzyme
exergonic
fermentation
high-energy bond
kinetic energy
law of conservation of energy
lock and key hypothesis
potential energy
substrate

CHECKING YOUR IDEAS

On a separate paper, indicate whether each of the following statements is true or false. Do not write in this book.

1. Cellular respiration involves an endergonic set of reactions.
2. Enzymes lower the amount of activation energy required in cells.
3. All cellular reactions require the addition of energy.
4. Energy for making ATP can come from glucose.
5. The part of the enzyme that combines with the substrate is the active site.
6. More energy is released during anaerobic respiration than during aerobic respiration.
7. The total amount of energy in a system before a chemical change is always greater than the total amount of energy after the change.
8. Enzymes are reusable.
9. ATP is a usable source of chemical energy in living cells.
10. A battery in a flashlight that is not turned on has potential energy.
11. Heat is a good source of activation energy for cellular reactions.

EVALUATING YOUR IDEAS

1. Identify several functions for which organisms must have a supply of energy.
2. What are some forms that energy may take? Give an example of transformation of energy. Which form of energy is most important to organisms?
3. What is meant by "conservation of energy"? Give an example.
4. Why do chemical reactions require activation energy? Explain why wood in a fireplace does not just burst into flame. Some reactions occur at room temperature. Explain how such reactions occur.
5. Why is heat not a suitable source of activation energy for cellular reactions?
6. Describe the lock and key hypothesis. What properties of enzymes does it explain?
7. Using diagrams, show how an enzyme would operate in (a) the synthesis of a disaccharide and (b) the conversion of a dipeptide to two amino acids.
8. Distinguish between an endergonic reaction and an exergonic reaction. Which of these requires an enzyme in cells?
9. Explain why ATP is needed and how it is used by cells.
10. What remains after a molecule of ATP has been used? What is needed to produce ATP?
11. What are the reactants and products of aerobic cellular respiration?
12. How efficient would cells be if the energy from glucose were not transferred to ATP?
13. Describe the ATP-ADP cycle.
14. How is cellular respiration similar to a ball bouncing down the stairs? How is it different?
15. Is anaerobic respiration less "profitable" than aerobic respiration? Why?
16. How is fermentation in yeasts useful?

APPLYING YOUR IDEAS

1. Assume that the amount of enzyme in a reaction is kept constant. How would you expect the rate of a reaction to be influenced by adding more substrate? Why?
2. ATP may be considered as the "coins" that the cell uses to pay its debts. Using this analogy, how would you describe glucose? Why?
3. Yeasts can respire both with and without oxygen. Explain why yeasts respiring with oxygen grow and divide more rapidly than yeasts respiring without oxygen.
4. Which has the most potential energy — a diver on a high board, a diver on a low board, or a diver entering the water? Explain your answer in terms of kinetic energy.
5. What would be the probable effect of very cold temperatures on most chemical reactions? Explain.
6. Glucose is made by producers. How does the producer use that glucose?
7. How is photosynthesis an example of energy transformation?

EXTENDING YOUR IDEAS

1. Conduct an experiment to determine the effect of temperature or pH on the rate at which an enzyme operates.
2. Louis Pasteur was a versatile scientist. What role did he play in solving the problems of the wine industry?

SUGGESTED READINGS

Baker, Jeffrey W. and Allen, Garland E., *Matter, Energy and Life: An Introduction for Biology Students,* 4th ed. Reading, MA, Addison-Wesley Publishing Co., Inc., 1981.

HEREDITY

When organisms reproduce, what ensures that their offspring will resemble them? What determines that alligators will produce only alligators? How are the features of these and other living systems determined? The results of reproduction are determined by genetic messages that are copied and passed on to the offspring. The study of genetic messages and the traits that result from them is called genetics or heredity.

In this unit, you will learn about cell reproduction and the "vehicles" that carry genetic messages to new cells. Knowledge of how genetic information is passed on will enable you to predict some of the traits new organisms will have. You also will investigate the chemical basis of heredity. You will study the materials of which genetic messages are composed and learn how those messages are decoded to produce an organism's feature.

American Alligator with Young

THE CELLULAR BASIS OF HEREDITY

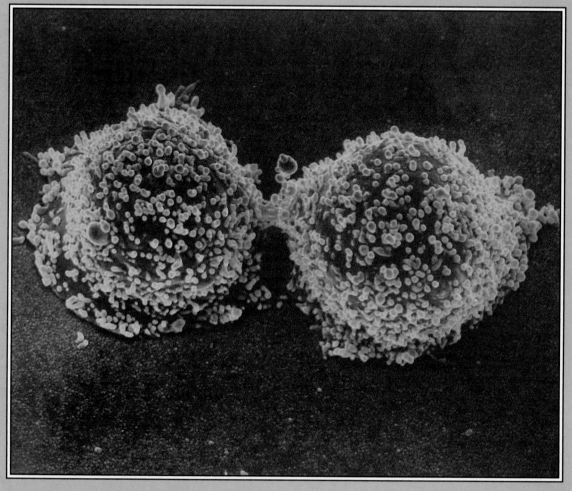

New cells are produced by other living cells. Dividing human body cells are shown here magnified about 8000 times. Notice that the two daughter cells are still connected by spindle fibers. How do cells divide? Are all cells produced in the same way? How do body cells differ from sperm and eggs? Are all cells produced by cell division identical to the parent cell? What are some of the ways cell division is important in your body?

ristotle (AR uhs taht uhl), 384–322 B.C., a Greek philosopher and scientist, often is called the "Father of Biology." He made important studies in anatomy and classification. Other scientists thought his reports true without question, but some of his findings and conclusions were not correct.

One of Aristotle's mistaken beliefs concerned the origin of organisms. He thought that some forms of life could be produced from nonliving sources. For example, Aristotle believed that eels came from the slime at the bottoms of rivers or oceans. Other popular notions included the ideas that maggots were produced from decaying meats and that a horse hair in water became a worm.

Objectives:
You will
- describe spontaneous generation and the debate surrounding it.
- explain the cellular events by which body cells and sex cells are produced.
- analyze the significance of mitosis and meiosis with respect to genetic continuity and adaptation.

LIFE FROM LIFE

Today most people know that the beliefs described above are false. Why were they accepted as facts for so long? Careful observation and experimentation were not used to test these ideas. Eels were thought to have arisen from mud because they were found in bodies of water and were slimy like the bottom of an ocean or river. But no one had ever studied their origin carefully. People just concluded that eels came from the slime.

Lack of experimentation led to false beliefs about reproduction.

6:1 Spontaneous Generation: Prologue

The concept that organisms come from nonliving things is called **spontaneous** (spahn TAY nee us) **generation.** For hundreds of years, people believed that maggots were spontaneously generated from decaying meat.

Spontaneous generation is the idea that organisms can come from nonliving things.

FIGURE 6–1. Redi showed in a controlled experiment that maggots did not spontaneously generate from decaying meat.

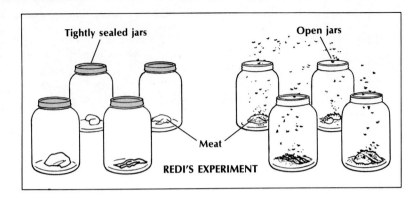

REDI'S EXPERIMENT

Redi assumed that maggots are produced by flies.

Francesco Redi (RAY dee), an Italian physician and biologist, was not convinced that spontaneous generation occurred. Thus, he tried to show that maggots come from other organisms. In 1668, he performed a series of important experiments. Redi noticed that adult flies hovered over decaying meat. He assumed that the flies produced the maggots and that if the flies could be kept away from the meat, no maggots would appear.

A controlled experiment was done to learn about the origin of maggots.

To test his hypothesis, Redi prepared several jars with various types of meat. He left some of the jars open and sealed the other jars. After a few days, the open jars were covered with maggots and swarming with adult flies, but the sealed jars revealed no life at all! Redi then put some maggots in a container and observed them. He noted that each maggot went through a series of changes and developed into an adult fly.

Redi's experiment did not remove all doubt about the validity of spontaneous generation.

What was the importance of Redi's investigations? By carrying out *controlled* experiments, Redi showed that maggots do not come from decaying meat. He did not, however, completely disprove the concept of spontaneous generation. His work was not enough to fully convince those who accepted the concept; nevertheless it did cast

FIGURE 6–2. Maggots develop from fly eggs. The maggots go through a pupal stage during which they change. They emerge from the pupal stage as flies.

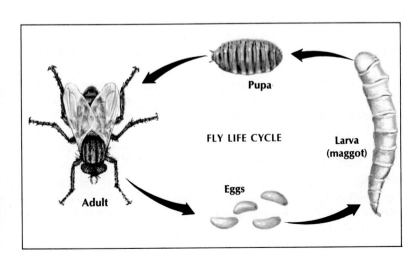

FLY LIFE CYCLE

Pupa

Larva (maggot)

Eggs

Adult

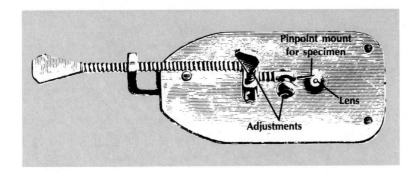

FIGURE 6–3. This paddlelike instrument is one of the many microscopes made by Anton van Leeuwenhoek. Specimens were mounted on the pinpoint and enlarged with the lens. This drawing is one-third larger than the actual microscope.

doubt on the idea. Redi was the first to use experimental procedures to test the hypothesis. Thus, he opened the door for further investigation of the problem.

6:2 Convincing Evidence

The conflict regarding spontaneous generation of maggots and other familiar organisms gradually subsided. Then, in 1675, Anton van Leeuwenhoek (LAY vun hook) discovered the world of microorganisms. A major event in itself, this discovery renewed the battle between those who believed in spontaneous generation and those who did not believe in it. These small organisms had never been seen before, so many people were quick to conclude that they were spontaneously generated.

Leeuwenhoek's discovery of microorganisms revived the debate about spontaneous generation.

The debate continued for almost two hundred years. Some experiments seemed to favor spontaneous generation. Others did not. Many experiments were done to test whether microorganisms would appear in nutrient solutions called **infusions** (ihn FYEW zhunz). Slight differences among the experiments affected the results. If air were present in the flasks containing the infusions, microorganisms usually appeared after several days. If no air were present, no life appeared. Those who believed in spontaneous generation concluded that air was an essential ingredient in spontaneous generation. They would not accept the conclusion that spontaneous generation does not exist.

Louis Pasteur (1822–1895), a great French scientist, opposed spontaneous generation. From his early experiments, he believed that microorganisms came from cells called **spores.** Spores are reproductive cells of certain plants and fungi. Pasteur thought that spores were carried on dust particles in the air. He believed that spores were inactive in air, but developed into new organisms when nutrients were available (as in an infusion).

FIGURE 6–4. These spores from a fungus are cells carried in the air. The spores are shown here magnified about 40 500 times.

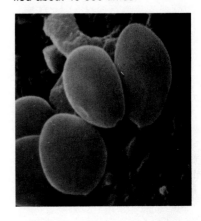

To disprove spontaneous generation, Pasteur had to keep spores from entering his infusions. Pasteur's opponents insisted that air must be present. Thus, he was faced with the problem of letting air into an infusion while keeping spores out.

FIGURE 6–5. Pasteur used S-shaped flasks in his experiments to disprove spontaneous generation. Although air entered the flasks, spores were trapped in the curved neck. No growth occurred in the broth.

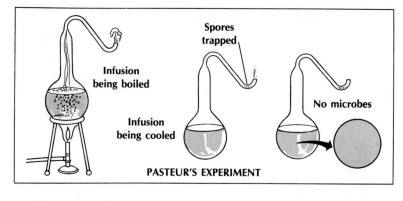

Pasteur devised an experiment that allowed air to enter an infusion but kept spores out.

Pasteur's experiment showed that microorganisms, like other life, are produced from other organisms.

The theory of biogenesis states that organisms are produced by other organisms.

The cell theory and the theory of biogenesis fit together.

Pasteur devised an experiment that allowed infusions to be in contact with air. He prepared infusions and boiled them thoroughly in flasks. The flasks were specially made and had long, S-shaped necks (Figure 6–5). Pasteur thought the boiling would kill any organisms present in the infusion. The steam produced would destroy any spores clinging to the glass. Air entering the flask would contain dust and spores, but the spores would be trapped in the curved portion of the neck. Thus, the infusion would remain **sterile,** free of life, for a long time.

Pasteur's prediction that the infusions would remain clear proved to be correct. As a result, the major objection of his critics was overcome. He showed that microorganisms, like other organisms, do not come from nonliving things. A belief held for over two thousand years was finally disproved.

Pasteur's work led to a principle that applies to all life. Biologists refer to this idea as the **theory of biogenesis** (bi oh JEN uh sus): *At the present time and under present conditions on Earth, all organisms are produced from other organisms.*

Pasteur's convincing evidence about the reproduction of organisms appeared in 1864. Recall that the announcement came in 1858 that all cells come from other cells (Section 4:1). These two discoveries were significant as biological principles, but more importantly, they fit together. Cells are the basic units of structure and function, and all cells are reproduced from other cells. Thus, all life must come from life.

REVIEWING YOUR IDEAS

1. Why was the idea of spontaneous generation believed for a long time?
2. Why did Redi seal only some of the jars in his test? Why did he observe fly development?
3. State the theory of biogenesis.

REPRODUCTION OF CELLS

Once it had been learned that all cells arise from previously existing cells, biologists began to study how cells reproduce. Improvement in lab techniques, such as the discovery of special dyes and stains and the development of better microscopes helped biologists learn about the process of cell reproduction.

6:3 The Cell Cycle

Recall that the nucleus (Section 4:14) controls cell reproduction. The chromosomes in the nucleus contain DNA, which carries the "genetic code." The genetic code instructs the cell in carrying out its chemical reactions. The genetic code also carries information about the building of new cell parts. To survive, each new cell must have the same genetic code as its parent cell. Cells must reproduce in such a way that each new cell ends up with the same number of chromosomes as the parent cell.

Each new cell has the same genetic information as its parent cell.

For each new cell to have the same genetic information as its parent, chromosomes must be replicated and the copies must be distributed equally. In eukaryotes, replicated chromosomes are distributed equally to new nuclei by a process called **mitosis** (mi TOH sus). Mitosis usually, but not always, occurs along with division of the cytoplasm (cell division) so that two cells are produced from one. However, the term mitosis refers only to distribution of the chromosomes to new nuclei.

Mitosis involves the equal distribution of replicated chromosomes to new nuclei.

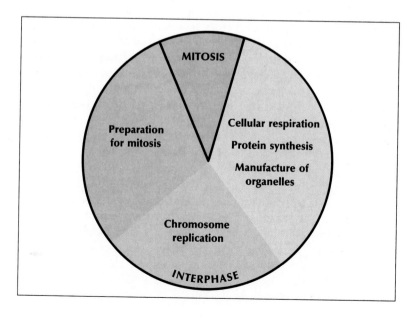

FIGURE 6–6. The cell cycle is made up of interphase and the phases of mitosis. The duration of interphase is longer than all of the phases of mitosis together.

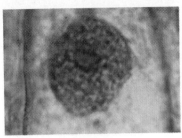

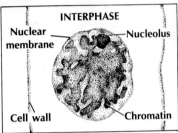

FIGURE 6–7. Although they appear inactive, interphase cells are actively taking in and making materials. Chromosomes are not distinct in interphase but are undergoing duplication.

FIGURE 6–8. In prophase, the nuclear membrane disappears and chromosomes become visible.

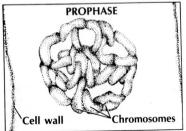

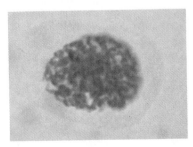

Between cell divisions, a cell is in a nonreproductive phase. This part of a cell's life is known as **interphase** (IHNT ur fayz). Although cells in interphase are not active in terms of mitosis, they are active in other ways. In the early stages of interphase, the cell is carrying out a variety of functions. It undergoes cellular respiration and builds new proteins and whole new organelles as it exchanges materials with its environment.

In the late stages of interphase, each chromosome replicates. Replication cannot be seen directly, but chemical tests show that it occurs. The replicated chromosomes become visually apparent when the cell is undergoing mitosis. Replication is followed by a brief period of time and then mitosis begins. The stages of interphase, together with mitosis, make up the **cell cycle** (Figure 6–6).

The duration of the cell cycle varies from one type of cell to another. In some cell types it lasts less than an hour, while in others it lasts for several weeks. Some cells, once produced, remain in interphase. Examples of such cells are nerve cells and certain muscle cells. In general, most of the cell cycle is spent in interphase. Thus, when you observe living cells through a microscope, you most often see them in this stage.

During interphase, the nuclear membrane is clearly visible. Inside the membrane, chromatin and nucleoli (Section 4:14) can be readily seen. Chromatin occupies most of the volume of the nucleus. The chromatin appears as a mass of long, thin, intertwined chromosomes (Figure 6–7).

6:4 Mitosis

Although nuclear division is a continuous series of events, there are four main phases of mitosis. The four phases are **prophase** (PROH fayz), **metaphase** (MET uh fayz), **anaphase** (AN uh fayz), and **telophase** (TEL uh fayz). Each of the four phases can be thought of as frames of a movie that continues to run without stopping at each frame. The description that follows applies to *most* of the cells of complex eukaryotes.

● **Prophase.** Prophase is a series of events that set the stage for later events. Those events lead to the equal distribution of chromosomes to new nuclei. Prophase follows interphase. In early prophase, the nucleolus (or nucleoli) begins to disintegrate. Chromatin becomes thicker and shorter, but separate chromosomes cannot be seen clearly. By middle prophase, the nucleolus has disappeared completely. Chromosomes are clearly visible, and the nuclear membrane begins to break down. In animal cells, centrioles (Section 4:13) begin to separate and migrate toward opposite poles of the cell. Microtubules (Section 4:12) then form and become assembled into tiny fibers between the poles.

By late prophase, the nuclear membrane is completely absent. Chromosomes appear even more distinct. The fact that replication has occurred is noticeable because the chromosomes appear as double-stranded structures. Each strand is a replica of the other and is called a **chromatid** (KROH muh tud). The two chromatids of a chromosome are joined at a special region, the **centromere** (SEN truh mihr). Short microtubules extend from each centromere. Fibers between the poles form an oval-shaped structure called the **spindle.** Still other fibers form the **aster,** a structure that radiates outward from each centriole within most animal cells. Most other eukaryotic cells lack centrioles and asters, but they have spindles.

● **Metaphase.** During metaphase, a short stage, chromosomes move toward the center of the spindle. The short microtubules extending from each centromere are joined to the microtubules that make up the spindle fibers. Each centromere attaches to a separate spindle fiber near the center, or equator, of the spindle apparatus. The two chromatids may face in any direction.

● **Anaphase.** At the beginning of anaphase, the centromeres split. The two chromatids begin to separate from one another and move apart, each one moving toward a pole of the spindle. Movement of chromatids is related to microtubule activity. The microtubules of the centromeres shorten, thus pulling the chromatids toward opposite poles. At the same time, the microtubules of the spindle fibers push the poles of the cell farther apart.

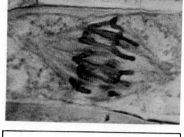

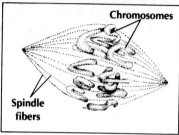

FIGURE 6-9. A spindle is an oval-shaped structure made of microtubules. Chromosomes attach to the spindle fibers.

By the end of anaphase, one set of single-stranded chromosomes has been pulled to each end of the spindle.

a
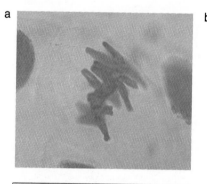

b
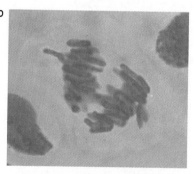

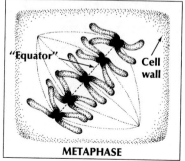

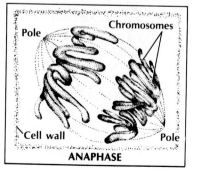

FIGURE 6-10. (a) During metaphase, chromosomes line up at the center ("equator") of the cell. (b) During anaphase, the chromatids of each chromosome separate and begin moving to opposite ends (poles) of the cell.

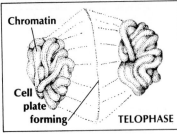

FIGURE 6–11. During telophase, chromosomes finish moving toward opposite poles of the cell and the formation of two daughter cells takes place.

At the end of anaphase, there is one set of single-stranded chromosomes at each end of the cell. The cell membrane of animal cells begins to pinch together at the cell's center. Microfilaments (Section 4:12) seem to be involved. In most plants, the **cell plate** begins to appear midway across the cell. It grows from the middle of the cell outwards. The cell plate may form from membranes of Golgi bodies or ER. The cell plate will become the membranes of the two new cells. Later, new cell walls form between the membranes.

• **Telophase.** Telophase ends mitosis with events opposite to those of prophase. The nucleolus reappears, and a nuclear membrane forms around each set of chromosomes. The nuclear membrane forms from ER. Meanwhile, chromosomes lose their distinct form so they once again appear as a mass of chromatin. In animal cells, when cell division is completed the cell membrane pinches completely together so that the single cell is separated into two **daughter cells.** In plants, the cell plate is completed to form the daughter cells. In late telophase of animal cells, the centrioles also replicate.

FIGURE 6–12. Mitosis ensures that each daughter cell has the same number and kinds of chromosomes as the parent cell. Follow the movement of the chromosomes through the stages of mitosis in an animal cell.

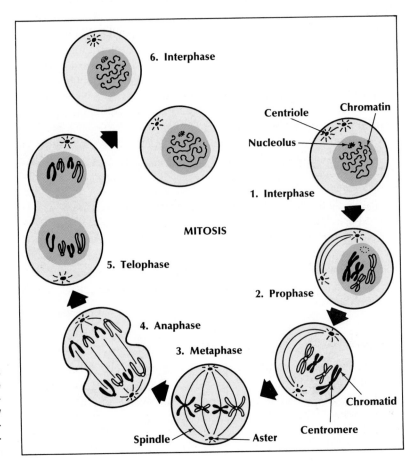

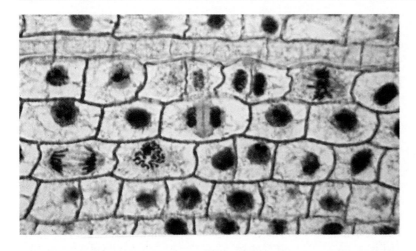

FIGURE 6–13. Stained cells from an onion root tip show different stages of mitosis. Which stages can you identify?

Each daughter cell has the same number and kinds of chromosomes as the parent cell that produced it. After mitosis and cell division, the daughter cells enter interphase and a new cell cycle begins. As the new cells mature and grow, their single-stranded chromosomes eventually will replicate. Later, mitosis will begin again.

6:5 Analysis of Mitosis

Mitosis guarantees genetic continuity (kahnt un EW ut ee) when it occurs properly. **Genetic continuity** results in the reproduction of offspring with the same features as the parent.

Replication of chromosomes contributes to genetic continuity. If exact replication does not occur, each daughter cell does not receive the proper genetic instructions. Thus, it will not function correctly and probably will die.

Equal distribution of replicated chromosomes is another important factor. During metaphase each double-stranded chromosome must line up on a *separate* spindle fiber. Proper alignment of chromosomes is crucial if each new nucleus is to receive an identical set of chromosomes. What would happen if the chromosomes were to line up randomly in metaphase? Although errors do occur, mitosis almost always results in genetic continuity.

The process of mitosis guarantees genetic continuity.

Replication of chromosomes during interphase is essential.

Because each chromosome attaches to a separate spindle fiber during metaphase of mitosis, each daughter cell receives one of each kind of chromosome.

6:6 Chromosome Number

The number of chromosomes per cell varies from one kind of organism to another. Humans have 46 chromosomes per cell and fruit flies have eight. The cell in Figure 6–12 has a total of four chromosomes. Some types of organisms have several hundred chromosomes per cell. A certain chromosome from one cell of an organism carries the same kinds of information as its counterpart in another cell.

Each kind of organism has a specific number of chromosomes in its cells.

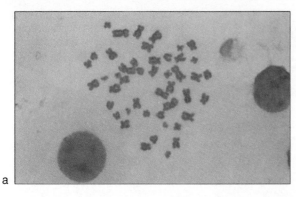

a

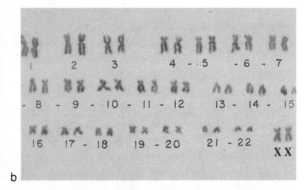

b

FIGURE 6–14. (a) Chromosomes can be photographed and then (b) paired in a karyotype. This karyotype shows the chromosomes of a normal human female. Humans have 23 pairs of chromosomes.

Diploid cells contain two of each kind of chromosome.

The two chromosomes of each pair are called homologs.

Chromosomes can be isolated for detailed study. Features such as the shape and number of chromosomes make up a cell's **karyotype** (KER ee uh tipe). Karyotypes show that all body cells of a given kind of organism have the same numbers and kinds of chromosomes. Cells of animals usually have *paired* chromosomes while cells of many plants and simple organisms do not. Human cells have 23 pairs; fruit fly cells have four pairs. Cells having two of each chromosome are said to be **diploid** (DIHP loyd). The diploid number of chromosomes is represented by *2n,* where *n* = the number of different pairs. Humans have 23 different pairs, so *2n* = 46. How many chromosomes does a fruit fly have?

In diploid cells, each chromosome of a pair has the same basic structure. The two members of each pair are said to be **homologous** (huh MAHL uh gus) and are called **homologs** (HOH muh lawgz). Each homolog carries messages for the same traits as its partner although the specific codes may differ. For example, each homolog may carry the message for eye color. The code on one homolog may be for brown eyes; the code on the other homolog may be for blue eyes. The combination of the two codes determines the eye color of the offspring. One homologous pair of chromosomes may carry codes for thousands of inherited traits.

REVIEWING YOUR IDEAS

4. What is mitosis?
5. What is the cell cycle? During what part of the cell cycle does replication of chromosomes occur?
6. What are the phases of mitosis?
7. What are chromatids? What is a centromere?
8. Why is genetic continuity important? How does mitosis contribute to genetic continuity?
9. What does a karyotype show?
10. What is meant by diploid number? What is a human's diploid number?

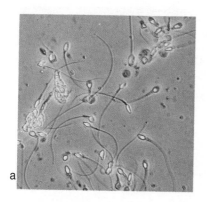
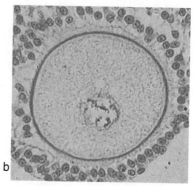

a

b

FIGURE 6–15. The products of meiosis are (a) sperm cells and (b) egg cells. (Only one egg cell is shown here.) A sperm and an egg each carry half the genetic message.

6:7 Meiosis

What is the source of the two homologs of each pair of chromosomes in a diploid animal cell? Development in animals usually starts with a fertilized egg or **zygote** (ZI gote). Genetic information in the zygote guides development as mitosis and cell division produce a multicellular animal. The zygote results from the union of two different kinds of cells, the sex cells. The sex cells, also called **gametes** (GAM eets), are a **sperm** from the male parent and an **egg,** or **ovum** (OH vum), from the female parent. All your body cells were produced as a result of mitosis and cell division in a zygote. Your body cells each contain 46 chromosomes. Cells of your parents also contain 46 chromosomes, and of their parents, too. Sex cells of animals, unlike their body cells, do not have the diploid number of chromosomes. They contain only *one* chromosome from each homologous pair. Cells, such as animal gametes, that contain one chromosome of each homologous pair are said to be **haploid** (HAP loyd). The haploid number is represented by *n*. In human gametes, *n* = 23.

Haploid cells are not produced by mitosis. Instead, they are produced by a process called **meiosis** (mi OH sus). Meiosis is the process by which haploid nuclei are produced from diploid nuclei. Meiosis usually is accompanied by cell division. In animals, meiosis results in production of gametes.

Unlike mitosis, which occurs in most kinds of tissues, meiosis occurs only in certain reproductive tissues. Correct meiosis ensures that animal gametes will have the right *number* and *kinds* of chromosomes. As a result, after fertilization occurs, the zygote will also contain the proper *number* and *kinds* of chromosomes.

Meiosis also occurs in the life cycles of plants, algae, fungi, and some unicellular organisms. In plants and some algae and fungi, meiosis results in the production of spores, not gametes. Meiosis in unicellular organisms often produces haploid nuclei rather than spores or gametes. In any case, meiosis occurs in two major parts — **meiosis I** and **meiosis II.**

FIGURE 6–16. (a) In meiosis I, the members of each pair of chromosomes separate and move apart forming two cells. (b) In meiosis II, chromosomes line up at the center of each cell and the chromatids separate. Four cells result, each with half the chromosome number of the original cell.

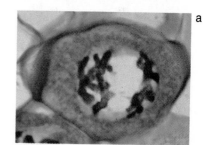

a

b

FIGURE 6–17. Meiosis in males occurs in tubules in the testes. Part of one of the tubules is shown magnified about 1800 times. Cells near the outside of the tubule (left) move toward the tubule's center (right) as they undergo meiosis. The dark lines at the right are mature sperm.

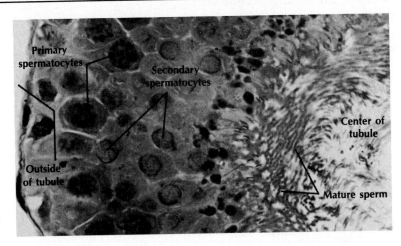

6:8 Meiosis in Males

The main events of meiosis as it occurs in a male animal are diagrammed in Figure 6–18. In early prophase of meiosis I, chromosomes do not appear to be double-stranded. However, studies show that the chromosomes replicate during interphase (as in mitosis). In middle prophase I of meiosis, homologous chromosomes move close together by a process called **synapsis** (suh NAP sus). Not only do the chromosomes come together, but similar parts lie side by side. During this time the chromosomes may be wound around one another. Finally, in late prophase I, the replicated chromosomes are seen clearly. The cell now is called a **primary spermatocyte** (spur MAT uh site). Each pair of chromosomes is called a **tetrad** (TEH trad) because four strands (chromatids) lie close together.

While these chromosome activities are occurring, the cell undergoes some events similar to mitosis. The nucleolus (if present) and nuclear membrane disappear, the centrioles move apart, and the spindle and aster form. The major difference between this prophase and prophase of mitosis is that the homologous chromosomes are paired in meiosis I.

There is another main difference between mitosis and meiosis that occurs in metaphase I. Each chromosome does not attach to a separate spindle fiber in meiosis. Instead, each *tetrad*, or pair of chromosomes, attaches to a separate fiber. During anaphase I, which follows, the centromeres do not split, and *one pair of chromatids of each tetrad is pulled to each pole.*

Then, the cell membrane begins to pinch in and telophase I begins. The original cell divides to form two **secondary spermatocytes** in which each cell has one member of each homologous pair *(n)*. However, each chromosome still consists of two chromatids that are joined together.

Chromosomes replicate prior to meiosis I.

During prophase I, homologous chromosomes undergo synapsis.

In metaphase I, tetrads attach to spindle fibers.

During anaphase I, one double-stranded chromosome of each tetrad is pulled to each pole.

In males, meiosis I produces two secondary spermatocytes, each of which is haploid.

After a short interphase, meiosis II begins. Meiosis II is much like mitosis. The secondary spermatocytes may skip prophase and go into metaphase directly. During anaphase, the centromere of each chromosome splits, so each new nucleus contains single-stranded chromosomes (Figure 6–18).

Cellular events of meiosis II are like those of mitosis.

TABLE 6-1. COMPARISON OF MITOSIS AND MEIOSIS		
	Mitosis	Meiosis
Occurs in	all body cells	certain cells of reproductive organs
Number of cells produced per parent cell	two	four (three may die)
Chromosome number of parent cell	diploid (*2n*) or haploid (*n*)	diploid (*2n*)
Chromosome number of daughter cells	same as parent cell	haploid (*n*)
Kind of cell produced	various cells	animal gametes or certain spores
Function	genetic continuity from cell to cell	genetic continuity between generations; promotes variation

A total of four haploid cells, called **spermatids,** are produced by meiosis II. A flagellum (Section 4:13) then forms from one of the centrioles of each spermatid. The spermatid with its tail becomes a sperm. The flagellum helps the sperm swim toward an egg, thus aiding fertilization.

Four spermatids are produced from one primary spermatocyte.

6:9 Meiosis in Females

Production of eggs by meiosis in a female animal follows the same general pattern as sperm production, but there are interesting differences. Meiosis I in human females begins before birth! As early as the third month of development, **primary oocytes** (OH uh sites) begin to form in a human female fetus. (Oocytes correspond to the spermatocytes of males.) Meiosis then stops until sexual maturity is reached many years later. At that point, several primary oocytes continue meiosis I on a cyclic basis. Usually, however, only one oocyte survives per cycle. A primary oocyte divides to produce two cells of unequal size. The larger cell is the **secondary oocyte,** and the smaller cell is the **first polar body.** The first polar body may divide again, but cells produced from the first polar body do not survive.

The secondary oocyte enters meiosis II and divides unequally to form a small **second polar body** and a large **ootid** (OH uh tihd). The second polar body dies, but the ootid develops into a mature ovum as shown by Figure 6–18.

In females, meiosis I begins before birth occurs.

A fetus is a developing organism.

Unequal cell division of a primary oocyte produces a secondary oocyte and the first polar body.

An ootid develops into a mature egg, or ovum.

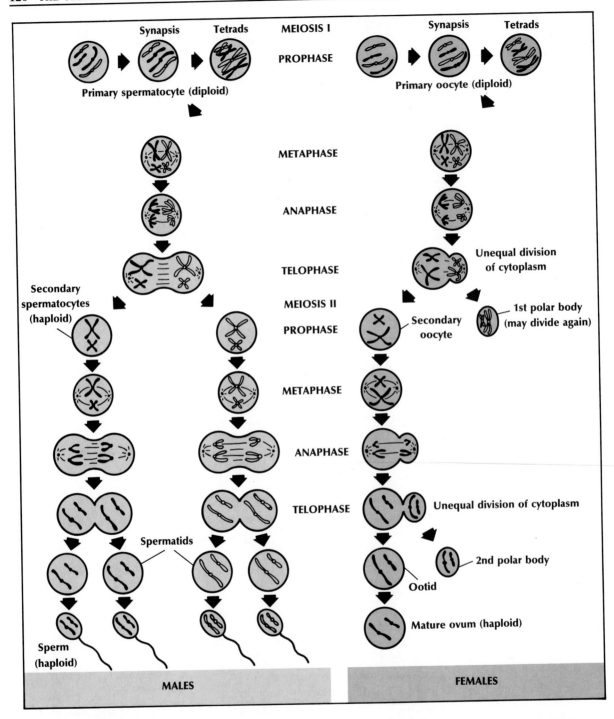

FIGURE 6–18. Meiosis yields cells with half the number of chromosomes as the parent cell. Compare and contrast meiosis in males with meiosis in females.

A sperm must be in final form before fertilization can occur. The egg may be fertilized before it is fully mature. Sometimes fertilization triggers the final events of meiosis in a female. In a male, four sperm develop from each cell (primary spermatocyte) that undergoes meiosis. Only one egg develops from each cell (primary oocyte) that undergoes meiosis. In what way would unequal cell division in a female be adaptive?

6:10 Importance of Meiosis

Meiosis may produce sperm or eggs in animals or spores in plants, algae, and fungi. Each of these types of reproductive cells contains the haploid number of chromosomes. Thus, each contains genetic information in a form that can be passed to the next generation. If meiosis did not occur, the life cycle could not continue.

Meiosis provides for more than the transfer of the genetic code. Meiosis and fertilization, needed for sexual reproduction, also provide the means for variation among offspring. Meiosis and fertilization have a special advantage for organisms. Reproduction by mitosis alone produces little variety in offspring.

In playing a game of cards, the deck is shuffled before each new hand is played. Shuffling rearranges the cards so you are dealt a new set of cards with which to play. Some hands are better than others; they are the winning ones. Other combinations are not so good, so sometimes you lose.

Meiosis is a process that reshuffles genetic information (chromosomes). In this case, the hands being played are the offspring produced as a result of the information they receive. In any population there are variations among individual organisms. Such differences are due to various combinations of chromosomes.

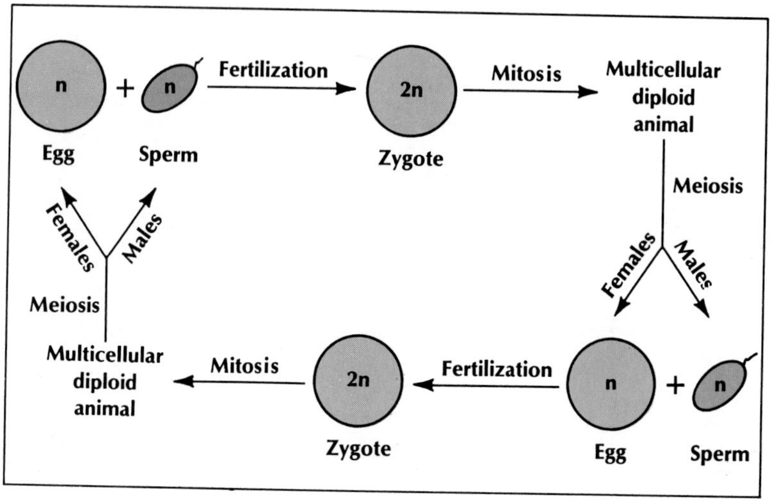

Meiosis assures that the genetic code is passed from one generation to the next.

Meiosis and fertilization contribute to variety among offspring.

FIGURE 6–19. Animals develop by mitosis and produce offspring by the union of eggs and sperm. Eggs and sperm are produced by meiosis.

How is this variation important? *Variation promotes a better chance for survival of some organisms should changes in the environment occur over time.* By keeping many different "hands" in the game, nature decreases the chances of too many losers. Adaptations of organisms result from genetically controlled variations that enable them to survive in a particular environment.

How does meiosis increase variety? Consider an animal whose diploid number is four. Such an animal will produce gametes having two chromosomes, one of each pair. How many possible combinations of chromosomes may appear among the gametes? Look at Figure 6–20. The parent cell, shown in metaphase I, has two pairs of homologous chromosomes—*A* and *a; B* and *b*.

When the cell at the left completes meiosis, the gametes produced from it contain two possible chromosome combinations—*A* with *B* or *a* with *b*. The cell at the right produces gametes with different combinations—*a* with *B* or *A* with *b*. (Notice that there is always one member of each homologous pair in each gamete.) The particular combination of chromosomes in a gamete *depends on how they line up at metaphase I.* In this example, four combinations of chromosomes are possible among the gametes.

Now consider the possible combinations of chromosomes in zygotes produced by the union of these gametes. Figure 6–21 shows that 16 different combinations of chromosomes may be produced! Some zygotes contain the same chromosome combinations as the parent cells, but many are different. Recall that homologous chromo-

Different combinations of chromosomes are possible in gametes, depending upon how homologs line up at metaphase I.

FIGURE 6–20. How chromosomes are combined in gametes depends on how homologs line up during metaphase I of meiosis.

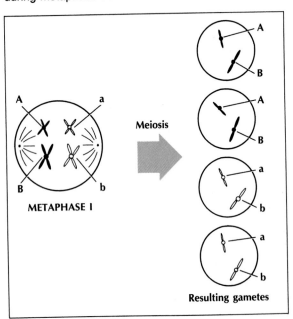
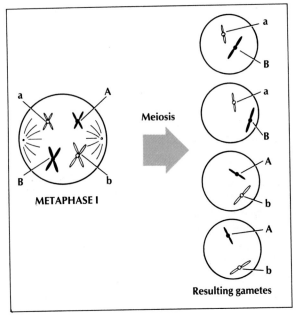

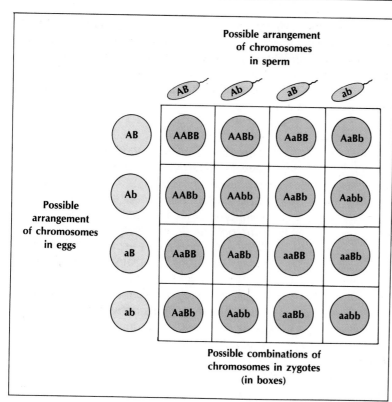

Possible arrangement
of chromosomes
in sperm

Possible
arrangement
of chromosomes
in eggs

Possible combinations of
chromosomes in zygotes
(in boxes)

FIGURE 6–21. When gametes contain different combinations of chromosomes, fertilization leads to genetic recombination and increased variety among offspring.

somes carry information for the same traits, but that the information may be in a different form. For example, one homolog may carry information for brown eyes; the other may carry information for blue eyes. Also, a single chromosome may carry information for thousands of traits. Thus, many different features are possible among offspring depending on the combination of chromosomes they have. Some combinations may be more useful to an organism's survival than others. This reshuffling of chromosomes and the genetic information they carry is called **genetic recombination** (ree kahm buh NAY shun). Another form of recombination occurs between parts of chromosomes during prophase I (Section 8:6).

REVIEWING YOUR IDEAS

11. What is a haploid cell? How do gametes differ from other body cells in animals?
12. What is meiosis? Where does it occur?
13. What is synapsis? What is a tetrad?
14. How does the number of gametes produced from one cell differ between male and female animals?
15. What is genetic recombination? Why is it important?

INVESTIGATION

Problem: How can you observe chromosomes in cells?

Materials:

horse bean roots
metric ruler
microscope
Bunsen burner
2 glass slides
2 coverslips
aceto-orcein stain
single-edged razor blade
prepared slide of *Drosophila* chromosome

3 droppers
hydrochloric acid
test tube
test tube holder
test tube rack
paper towel
water

Procedure:

Part A. Observing Chromosomes from a Root Tip

1. Use a single-edged razor to cut off two root tips from a bean plant. The tips should be no more than 3 mm. **CAUTION:** *Use extreme care with the razor blade.*
2. Place the tips into a test tube. Add 10 drops of aceto-orcein stain and 1 drop of hydrochloric acid to the tube. **CAUTION:** *If spillage of the stain or acid occurs, rinse with water and call your teacher.*
3. Place the tube in a rack for 10 minutes.
4. Holding the tube with a test tube holder, heat the stain by *rapidly* passing the tube through a Bunsen burner flame several times. Do not

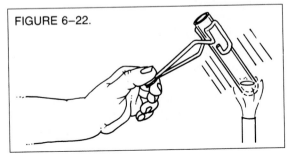

FIGURE 6-22.

boil the liquid. See Figure 6-22. **CAUTION:** *Secure all loose hair and clothing from the open flame. Do not direct the open end of the tube toward anyone.*

5. Allow the root tips to cool and remain in the stain for another 10 minutes.
6. Pour out the stain without losing the tips. Place each root tip on a separate, clean glass slide. Add 1 to 2 drops of water to each tip.
7. Add a coverslip to each tip. Place a piece of paper towel over each coverslip. Squash each root tip by firmly pressing down on the paper towel with your thumb. *Do not rotate your thumb.* Use Figure 6-23 as a guide. Remove the paper towel.

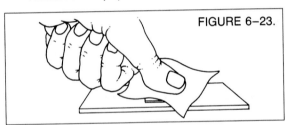

FIGURE 6-23.

8. Observe your slide under low power and then high power. Look for cells that show well-stained chromosomes.
9. Copy Table 6-2. Complete the top two rows of the table by making diagrams of those cells that are undergoing mitosis. Name the phases of mitosis that you see.

Part B. Observing *Drosophila* Chromosomes

1. Obtain a prepared slide of tissue from the salivary glands of *Drosophila*. Observe the slide under low and then high power.
2. Locate the chromosomes. They will look like long, red, twisted threads with bands.
3. Complete the bottom row of Table 6-2 by drawing several fruit fly chromosomes.

Data and Observations:

TABLE 6-2. CHROMOSOME APPEARANCE		
Organism	Diagram of Chromosomes	Phases of Mitosis Seen
Bean root tip		
Bean root tip		
Drosophila salivary gland		

Questions and Conclusion:

1. Define the following terms:
   (a) chromosome
   (b) mitosis
   (c) interphase
   (d) cell cycle
2. (a) Using your data from Table 6–2, list the phases of mitosis you observed in the bean root tip.
   (b) Were there more cells undergoing a particular phase of the *cell cycle* than any other phase? If so, which phase did you observe most often?
3. How did the following help in the observations of the chromosomes:
   (a) staining
   (b) squashing?
4. Why did you use the tip of the root and not the entire root when preparing your slide?

5. The term chromosome is derived from *chromo-* (meaning color) and *-some* (meaning body). How is the term chromosome an appropriate one?
6. If human skin tissue instead of root tips had been used in Part A:
   (a) how many chromosomes would you expect to find in each cell?
   (b) would the skin cells have been undergoing mitosis?
   (c) explain your answer to question b.
7. (a) Consider the tissue from the salivary glands of *Drosophila*. How do the chromosomes from one cell compare to those from any other cell?
   (b) How does the situation described in 7(a) illustrate genetic continuity?

Conclusion: How can you observe chromosomes in cells?

CHAPTER REVIEW

SUMMARY

1. For several centuries, some biologists believed that certain organisms were produced by spontaneous generation. **6:1**
2. Pasteur conducted experiments in 1864 that showed that microorganisms are not spontaneously generated. His conclusions led to the theory of biogenesis. **6:2**
3. The cell cycle consists of the various stages of interphase and mitosis. Chromosomes replicate during interphase and are equally distributed during mitosis. **6:3**
4. The four phases of mitosis are prophase, metaphase, anaphase, and telophase. **6:4**
5. Mitosis results in genetic continuity. **6:5**
6. Cells having two of each chromosome are said to be diploid. The two members of each chromosome pair are homologous. **6:6**
7. Fertilization of an egg by a sperm results in the formation of a diploid zygote. **6:7**
8. Meiosis and cell division result in haploid reproductive cells. **6:7**
9. In male animals each cell undergoing meiosis produces four haploid sperm. **6:8**
10. In female animals only one cell, the haploid egg, survives meiosis. **6:9**
11. Meiosis and fertilization contribute to variation among offspring. **6:10**

LANGUAGE OF BIOLOGY

anaphase	genetic	prophase
biogenesis	recombination	sperm
cell cycle	haploid	spindle
cell plate	homologs	spontaneous
centromere	interphase	generation
chromatid	meiosis	spore
daughter cell	metaphase	telophase
diploid	mitosis	tetrad
gamete	ovum	zygote

CHECKING YOUR IDEAS

On a separate paper, complete each of the following statements with the missing term(s). Do not write in this book.

1. Mitosis of a *2n* cell results in the production of cells with a(n) _____ number of chromosomes.
2. Pasteur's experiments led to the theory of _____.
3. Meiosis in animals produces cells called _____.
4. Replication of chromosomes occurs during _____.
5. Cells that do not survive meiosis in female animals are _____.
6. During telophase a(n) _____ forms between the two poles of plant cells.
7. Meiosis and fertilization promote _____ among organisms.
8. Cells produced from a *2n* cell by meiosis have a(n) _____ number of chromosomes.
9. _____ is the idea that organisms arise from nonliving things.
10. The chromosomes of a matching pair are called _____.

EVALUATING YOUR IDEAS

1. What was Aristotle's mistaken belief about the reproduction of organisms?
2. Did Redi disprove the theory of spontaneous generation? Explain. What important role did Redi play in the debate over this theory?
3. Write an essay about Pasteur's experiment that disproved spontaneous generation. Discuss his equipment, the reasons for his procedure, and his results and their significance.
4. Why is the theory of biogenesis important?
5. How are the cell theory and the theory of biogenesis related?

6. Describe the events of interphase and mitosis: (a) prophase, (b) metaphase, (c) anaphase, and (d) telophase.
7. How does mitosis in plant cells differ from mitosis in animal cells?
8. What is genetic continuity? How does mitosis contribute to genetic continuity?
9. What are homologous chromosomes? How are homologs the same? How are homologous chromosomes different?
10. In a given animal, the diploid number of chromosomes is 12. How many chromosomes will a gamete produced by this animal contain? Explain your answer.
11. What is the origin of each member of a homologous pair of chromosomes?
12. Why is it important that a human egg or sperm contain one of each pair of chromosomes rather than just any 23?
13. Describe the events of meiosis: (a) prophase I, (b) metaphase I, (c) anaphase I, and (d) telophase I as they occur in a male animal.
14. Compare meiosis II with mitosis.
15. How does meiosis in a female animal differ from meiosis in a male animal?
16. In terms of genetic continuity, summarize the relationship among meiosis, fertilization, and mitosis in animals.
17. How does meiosis and fertilization contribute to genetic recombination?
18. Why does asexual reproduction (mitosis) offer little opportunity for variety among offspring? How may mitosis lead to variation?

APPLYING YOUR IDEAS

1. Supporters of spontaneous generation claimed that air must be present for the process to occur. Do you think this idea was good science? Explain.
2. Explain how Redi's experiments made use of the elements of the scientific method.

3. Damage to nerves is often permanent. Why is it sometimes not possible to recover from such damage?
4. Make a series of labeled drawings illustrating mitosis in an animal cell with a *2n* of six.
5. Make a series of labeled drawings illustrating meiosis in a male animal cell with a *2n* of six. Compare your drawings with the drawings in question 4.
6. How is the structure of a sperm cell adapted to its function? An egg?
7. List several examples of variations among organisms. Choose one feature and explain how a variation in this trait might be important if the environment were to change.
8. Suppose a cell had four chromosomes of each type instead of two. How many chromosomes of each type would be in the daughter cells produced by mitosis?

EXTENDING YOUR IDEAS

1. Prepare your own mitosis slides from onion root tips that you have grown.
2. Salivary glands of fruit flies have giant chromosomes. Describe the structure of these chromosomes by using library sources or by making and examining slides of them.
3. Design and conduct an experiment about spontaneous generation of microorganisms. Consider variable factors such as presence or absence of air, thorough boiling of infusions, and types of nutrients.

SUGGESTED READINGS

Angyal, Jennifer, *Mitosis and Meiosis Illustrated,* Burlington, NC, Carolina Biological Supply Co., 1980.

John, B., *The Meiotic Mechanism,* 2nd ed. Carolina Biology Reader. Burlington, NC, Carolina Biological Supply Co., 1984.

PRINCIPLES OF HEREDITY

Offspring inherit traits from their parents. This inheritance of traits gives parents and offspring many features in common. Some offspring are exactly like their parents while others are not. Differences often exist among parents and offspring, among offspring, and among different types of organisms. How do organisms of the same type differ? How do you differ from your brothers and sisters, your cousins, or your friends?

chapter 7

Genetic information is passed from cell to cell and from parents to offspring. That information is encoded within molecules of DNA. DNA is a part of chromosomes, and, in most organisms, replicated chromosomes are distributed during mitosis and meiosis. Such knowledge has developed mainly in this century. It is the basis of the current study of **genetics** (juh NET ihks), the science of heredity.

Objectives:
You will
- discuss the principles of genetics set forth by Mendel.
- explain how genotype and phenotype are related.
- solve genetics problems.

ORIGIN OF GENETICS

The discipline of genetics developed as a result of the work of Gregor Mendel (1822–1884). In 1866, Mendel published the results of eight years of experiments and analysis. His work was ignored until 1900, when it was finally recognized as a major new development in biology. As you read about Mendel, think about his experiments and conclusions. Keep in mind the methods of good science. All of them are shown in Mendel's creative work.

7:1 Mendel's Experiments

Mendel worked with the common pea plant. Pea plants are grown easily and produce large numbers of offspring in a short time. Mendel found that traits (characteristics) are hereditary and that they are transmitted (passed) from generation to generation. He also found that many traits exist in one of two forms. For example, pea seeds are either round or wrinkled. The stems of pea plants are either tall or short. Mendel set out to determine how traits are transmitted from parents to offspring.

In pea plants, several traits exist in one of two forms.

DOMINANT	RECESSIVE
SEEDS	
Round	Wrinkled
Yellow	Green
COATS	
Colored	White
PODS	
Smooth	Wrinkled
Green	Yellow
STEMS	
Long	Short
FLOWERS	
Axial	Terminal

FIGURE 7–1. Mendel worked with seven pea plant traits in his first experiments. Each trait could appear in one of two forms.

In the F$_2$ generation, a ratio of three round-seeded plants to one wrinkled-seeded plant appeared.

It is a good thing that Mendel chose to work with a large number of plants. Perhaps he realized that the greater the number of plants, the greater the chances that his results would be more meaningful. He was looking for general trends because he wanted to form a basic set of rules about the transmission of traits. Mendel worked with a total of seven traits in his first experiments on pea plants. Those traits are shown in Figure 7–1.

7:2 Mendel's Results

Pea plants reproduce sexually. Both male and female sex organs are in the same flower. Normally, male gametes fertilize eggs of the same flower. After many generations, offspring still have the same features as the parents. Such plants are said to be pure.

Mendel wished to cross a plant having a certain trait with a plant having the opposite trait. For example, he wanted to cross plants that produced round seeds with plants that produced wrinkled seeds. Therefore, he transferred the male gametes (pollen) of one plant to another plant (Figure 7–2).

In some cases, the pollen was obtained from round-seeded plants and transferred to wrinkled-seeded plants. The opposite combination also was done. Mendel found that in every case, these **parental crosses (P)** yielded offspring that all had round seeds. The offspring of a parental cross are called the **first filial** (FIHL ee ul), or **F$_1$, generation.**

Mendel was impressed that there were no wrinkled-seeded plants in the F$_1$ generation. He then allowed members of the F$_1$ generation to reproduce in the usual fashion.

The offspring of this second cross are called the **second filial,** or **F$_2$, generation.** Mendel's results for this cross were very revealing. Of the 7324 offspring produced, 5474 plants had round seeds and 1850

FIGURE 7–2. Mendel cross-pollinated pea plants by transferring pollen from a plant that showed one form of a trait to a plant that showed a different form of the trait. In this way, he could determine what types of offspring are produced from these parents.

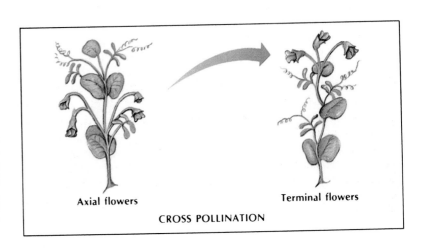

Axial flowers   Terminal flowers

CROSS POLLINATION

had wrinkled seeds. These numbers give a ratio very close to three round-seeded plants to one wrinkled-seeded plant. Mendel repeated both these experiments with the other pairs of traits and obtained a 3:1 F_2 ratio in each case.

Mendel noted that for each trait there is one form that "dominates" the other. Based on the results for the F_1 generation, the trait for round seeds is the **dominant** (DAHM uh nunt) **trait.** The other trait, in this case wrinkled seeds, disappears in the F_1 generation and is called a **recessive** (rih SES ihv) **trait.** Mendel generalized this idea in his **law of dominance** (DAHM uh nunts): *one form of a hereditary trait, the dominant trait, dominates or prevents the expression of the recessive trait.*

7:3 Mendel's Hypothesis

In Mendel's crosses the recessive trait disappeared in the F_1 generation but reappeared in the F_2 generation. To explain this result, Mendel made certain assumptions. Each pea plant is produced as a result of the union of a sperm and an egg. Thus, he reasoned that for every trait there must be two governing factors. One must come from the sperm and one must come from the egg. He called the factors characters, but we now call them **genes.** Mendel represented the genes by symbols. A dominant gene is represented by a capital letter. A recessive gene is represented by the same letter as the dominant gene but in lowercase. For example, if the gene for round seed shape (dominant) were R, then the gene for wrinkled seeds (recessive) would be r. The two genes combine in the fertilized egg so that each new pea plant contains two genes for each trait as shown in Figure 7–3.

In Mendel's experiments, the pure parental plants each had duplicate genes. Round-seeded plants (dominant) were RR. Wrinkled-seeded plants (recessive) were rr. Mendel reasoned that the two genes segregate (separate) during gamete formation. Sperm and eggs would have just one gene for each trait. Thus, the gametes of round-seeded parents should have the R gene, and gametes of wrinkled-seeded parents should have the r gene. As a result, all F_1 plants should be Rr, the only possible combination. Because the R gene is dominant to the r gene, all F_1 plants would have round seeds.

If Mendel's assumptions were correct, all F_1 plants *(Rr)* should produce half their gametes with the R gene and half with the r gene. There would be three possible combinations in the F_2 generation — RR, Rr, and rr. The chances of R and r gametes combining would be twice as great as other possible combinations. These possibilities can be expressed as a ratio — $1\ RR:2\ Rr:1\ rr$. Both RR and Rr plants will produce round seeds, while rr plants will produce wrinkled seeds. This result represents a ratio of 3 round-seeded plants:1 wrinkled-seeded plant (Figure 7–3).

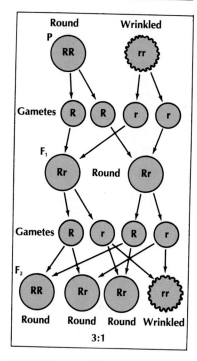

FIGURE 7–3. Mendel's analysis of his results can be set up in a flow chart that shows parents, gametes, and possible offspring.

Mendel reasoned that the two genes governing each trait segregate from each other during gamete formation.

FIGURE 7–4. Combinations of characters among offspring can be determined by using Punnett squares. Sperm and egg possibilities are listed along the top and left side of the square. Possible offspring are shown inside the Punnett square.

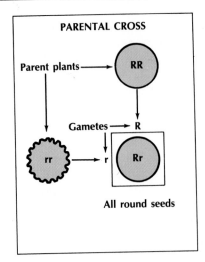
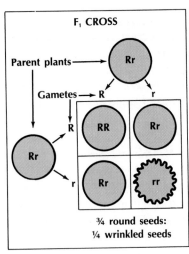

Punnett squares can be used to predict results of crosses.

Results of crosses can be predicted with Punnett (PUN ut) squares (Figure 7–4). A **Punnett square** is a chart used to determine the possible combinations of genes among offspring. Across the top of the square are written the gametes produced by one parent. The gametes produced by the other parent are listed along the left side of the square. The square is then filled in by listing all the possible combinations of gametes. Each square in the chart represents a *possible* zygote.

Law of segregation: During gamete formation, the pair of genes responsible for each trait segregates so that each gamete contains only one gene for each trait.

7:4 The Test of Segregation

Mendel was able to explain what he observed in his experiments. His laws of dominance and his hypothesis that genes segregate when gametes are formed grew from his experiments. His first experiments alone did not mean that his explanation was correct; another test was needed to confirm his hypothesis. Mendel did this test after predicting the outcome of yet another cross.

Mendel predicted that if F_1 generation round-seeded plants were crossed with wrinkled-seeded plants, he would get ratios different from any of his previously obtained ratios. According to his hypothesis, F_1 round-seeded plants are *Rr* and wrinkled-seeded plants are *rr*. Round-seeded plants produce both *R* and *r* gametes while wrinkled-seeded plants produce only *r* gametes. Mendel predicted that the possible combinations in offspring resulting from a cross between *Rr* and *rr* parents would be *Rr* (round) and *rr* (wrinkled). He predicted also that the numbers of each would be about the same (Figure 7–5). When Mendel performed the experiment, he obtained about half (50%) round-seeded and half (50%) wrinkled-seeded plants. The same ratio was observed when he did the same experiment using plants having other traits.

FIGURE 7–5. Mendel predicted the results of this cross *(Rr × rr)*. His results matched his predictions and verified his assumptions.

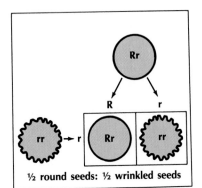

As a result of his experiments, Mendel assumed his hypothesis was correct. His hypothesis and data led him to form the **law of segregation** (seg rih GAY shun): *during gamete formation the pair of genes responsible for each trait separates so that each gamete receives only one gene for each trait.* The gametes unite to produce predictable ratios of traits among the offspring.

7:5 Terminology

Recall that Mendel's "characters" (genes) took different forms, such as dominant or recessive. Each form of a gene is an **allele** (uh LEEL). Thus, the trait expressed as seed shape is determined by two alleles, R and r. The combination of alleles that an organism has is called its **genotype** (JEE nuh tipe). The genotype for any trait includes both alleles. For seed shape in pea plants, the possible genotypes are RR, Rr, and rr. The physical or visible feature that each genotype determines is called the **phenotype** (FEE nuh tipe). Round seeds is the phenotype determined by the genotypes RR and Rr. Wrinkled seeds is the phenotype determined by the genotype rr. Notice that different genotypes in the above example may result in the same phenotype.

When each cell of an organism contains two alleles that are the same, the organism is **homozygous** (hoh muh ZI gus) for that trait. RR and rr are both homozygous genotypes. To avoid confusion, the terms homozygous dominant and homozygous recessive are used. When each cell of an organism contains one of each kind of allele, it is **heterozygous** (het uh roh ZI gus). A plant with the genotype Rr is heterozygous. You should learn these terms in order to do your work in genetics.

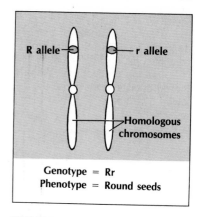

FIGURE 7–6. Terminology: A genotype represents the alleles on homologous chromosomes that result in a certain phenotype.

A genotype indicates the combination of alleles an organism has for a given trait.

A phenotype is the trait determined by a particular genotype.

REVIEWING YOUR IDEAS

1. What did Mendel notice about the appearance of hereditary traits? What did Mendel try to find? Did he succeed?
2. Why did Mendel transfer the male gametes of one plant to another plant?
3. State the law of dominance.
4. How are dominant and recessive traits represented?
5. What led Mendel to form his law of segregation? How did Mendel test the law of segregation?
6. Define the following terms: allele, genotype, phenotype, homozygous, and heterozygous.
7. In a certain organism, the gene for tall, *T,* is dominant to the gene for short, *t.* Write the genotype and phenotype for an organism that is (a) heterozygous, (b) homozygous dominant, and (c) homozygous recessive.

FIGURE 7–7. Each toss of a coin has a 50% probability of being a head (or a tail). To find the probability of combinations of tosses in a row, use the product rule. What is the probability of tossing a tail followed by a head?

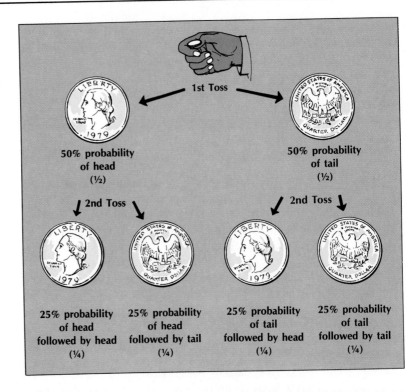

1st Toss

50% probability of head (½)

50% probability of tail (½)

2nd Toss   2nd Toss

25% probability of head followed by head (¼)

25% probability of head followed by tail (¼)

25% probability of tail followed by head (¼)

25% probability of tail followed by tail (¼)

SOLVING GENETICS PROBLEMS

Biologists use Mendel's laws to predict the outcomes of certain crosses between organisms. If the genotype of each parent is known, the different genotypes and phenotypes of the offspring can be determined. Such predictions, though, are based on probability. Predictions based on probability are *expected* results. Expected results are not necessarily the same as actual results.

Expected results and actual results may differ.

7:6 Probability

When you toss a coin, the chances are equal, or about fifty-fifty, that you will get a head or a tail. Stated another way, there is an equal probability of tossing either a head or a tail. This probability also can be written 1 head: 1 tail, or ½ heads: ½ tails. These expressions say the same thing. It is simply a matter of chance whether a head or a tail appears.

What are the chances of tossing two heads in a row? The answer is the product of the individual probabilities. This rule is called the **product rule.** Each time the coin is tossed, the chance of getting a head is ½. To find the probability of tossing two heads in a row, multiply ½ times ½. The answer is ¼. *Previous tosses of the coin do not influence any other tosses.* Even if you tossed heads five times in a row, the chance that the next toss would be a head is still ½. But, what are the chances of tossing five heads in a row? Using the product rule, the answer is ½ × ½ × ½ × ½ × ½, or ⅟₃₂.

Probability is important in genetics and in solving genetics problems (making predictions about the outcomes of certain crosses). The law of segregation states that alleles separate during gamete formation. Each gamete receives its allele for a given trait purely by chance. Consider the F_1 generation in Mendel's experiments in which each plant has the genotype *Rr*. Gametes produced by these plants must have had either the *R* or *r* allele. Like a two-sided coin, the chances are equal for *R* or *r* to appear in any one gamete, so the ratio of alleles produced by all the F_1 plants taken together should be ½ *R*: ½ *r*. One half the gametes, by chance, will contain the *R* allele and one half will contain the *r* allele.

How can this idea be used to predict the results of breeding plants or animals? Consider ½ *R* + ½ *r* as the chance that half the gametes produced by a parent with the *Rr* genotype will have the *R* allele and the other half will have the *r* allele. The probable ratio of genotypes and the probable ratio of phenotypes for the offspring of two *Rr* parents can be determined by using the product rule in the following way:

(1) Probable distribution of alleles:
  ½ *R* + ½ *r* (for both parents)
(2) Probable combinations of alleles:

$$½\ R + ½\ r$$
$$× ½\ R + ½\ r$$
$$\overline{\qquad\qquad\qquad}$$
$$¼\ Rr + ¼\ rr$$
$$¼\ RR + ¼\ Rr$$
$$\overline{\qquad\qquad\qquad}$$
$$¼\ RR + ½\ Rr + ¼\ rr$$

(3) Genotypes:
  ¼ *RR* + ½ *Rr* + ¼ *rr*
(4) Phenotypes:
  ¾ round: ¼ wrinkled

This answer is the same ratio that Mendel obtained. Do you see why it was good that he worked with many plants? For example, suppose Mendel had examined only four offspring of the *Rr* × *Rr* cross. It is possible that the four offspring would have all been *rr,* all wrinkled. This answer would have been as far as possible from the expected 3:1 ratio. As the number of offspring examined increases, the chance that the actual results will be close to the expected results

The product rule: The probability of two or more independent events occurring together is the product of the individual probabilities of each event occurring alone.

The laws of probability are used widely in the study of genetics.

Note that the genotypes *RR* and *Rr* produce the same phenotype, round seeds.

Both the genotypic ratio and the phenotypic ratio should add up to a total of one. Check your work.

Expected results are more likely to be obtained when large numbers of individuals are considered.

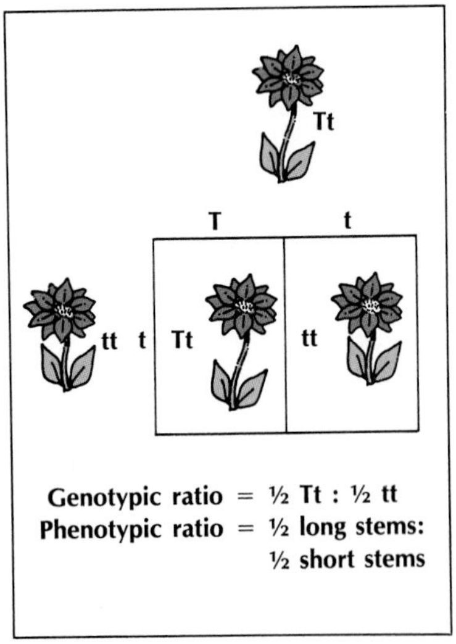

Genotypic ratio = ½ Tt : ½ tt
Phenotypic ratio = ½ long stems:
½ short stems

FIGURE 7–8. In this Punnett square solution to Sample Problem 1, notice that half the expected offspring are *Tt* and half are *tt*.

also increases. Mendel examined hundreds of offspring to get his results. Segregation and combination occur by chance, so the larger the number of offspring observed, the closer the result will be to the expected 3:1 ratio.

Compare this method with the Punnett square method. Listing the possible gametes across the top and side of the square is like following the rules of probability. By using the Punnett square method, you actually are multiplying probabilities (the product rule).

7:7 Ratios

Often a geneticist (juh NET uh sust) is interested in finding probable ratios of genotypes and phenotypes among the offspring of a particular cross. The genotypic (jee nuh TIHP ihk) ratio is the result obtained by the product rule. The genotypic ratio tells how many offspring are expected to have a particular gene combination. The phenotypic (fee nuh TIHP ihk) ratio is found by interpreting the genotypes. Study the following sample problem.

Sample Problem 1: In a certain plant, long stems *(T)* are dominant to short stems *(t)*. A farmer crosses a short-stemmed plant with a heterozygous long-stemmed plant. What are the expected genotypic and phenotypic ratios?

Solution:
(1) Write the genotype of each parent.
$$\text{short stem: } tt$$
$$\text{long stem: } Tt$$
(2) Determine the probable distribution of alleles in the gametes.
$$\text{short} = tt \rightarrow \tfrac{1}{1}\,t$$
$$\text{long } = Tt \rightarrow \tfrac{1}{2}\,T + \tfrac{1}{2}\,t$$
(3) Using the product rule, multiply to find the probable combination of alleles (genotypic ratio).
$$\tfrac{1}{2}\,T + \tfrac{1}{2}\,t$$
$$\underline{\times \tfrac{1}{1}\,t}$$
$$\tfrac{1}{2}\,Tt + \tfrac{1}{2}\,tt$$
(4) Interpret the genotypes to find the phenotypic ratio.
$$\tfrac{1}{2}\,Tt = \text{long stems}$$
$$\tfrac{1}{2}\,tt = \text{short stems}$$

Therefore, the expected phenotypic ratio is ½ long stems: ½ short stems. Figure 7–8 shows the solution to this problem by the Punnett square method.

REVIEWING YOUR IDEAS

8. What is the probability of tossing a coin and getting heads three times in a row?
9. What is the chance that the next toss will be heads? What is the chance that the next toss will be tails?
10. What is the probable ratio of alleles in the gametes of an organism that is (a) *MM,* (b) *Mm,* or (c) *mm?*
11. Why was it good that Mendel worked with very large numbers of plants?
12. In a certain plant, yellow fruit, *Y,* is dominant to white fruit, *y.* A heterozygous plant with yellow fruit is crossed with a plant with white fruit. Find the probable genotypic and phenotypic ratios resulting from this cross.
13. Find the probable genotypic and phenotypic ratios expected from crossing two heterozygous plants of Problem 12.

7:8 Incomplete Dominance

For some traits, one allele of a pair is not dominant to the other. For example, certain plants have alleles for red flowers and white flowers, but neither of these colors is dominant. In these flowers, heterozygous genotypes produce pink flowers. This phenomenon is called incomplete dominance. **Incomplete dominance** of two alleles results in the possibility of three different phenotypes because neither allele is dominant to the other. The third phenotype, shown when the two alleles are together, is an intermediate form of the other two

In incomplete dominance, neither allele is dominant to the other and three completely different phenotypes are possible.

a

b

FIGURE 7–9. (a) In four o'clocks, the alleles for red flowers and white flowers show incomplete dominance. (b) The heterozygote has pink flowers.

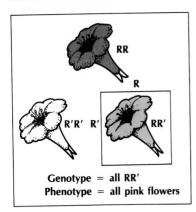

Genotype = all RR′
Phenotype = all pink flowers

FIGURE 7–10. The Punnett square solution to Sample Problem 2 illustrates how incomplete dominance works.

Any combination of alleles may appear in a gamete. There will be, however, only one allele for each trait.

FIGURE 7–11. The F_1 generation of a parental cross involving two traits is shown in this Punnett square.

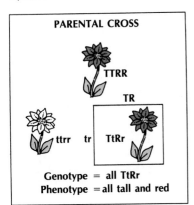

PARENTAL CROSS

Genotype = all TtRr
Phenotype = all tall and red

phenotypes. In this example, the two alleles for flower color are represented as R (red) and R' (white). To solve a genetics problem involving incomplete dominance, proceed in the same manner as for Sample Problem 1.

Sample Problem 2: In four-o'clocks the alleles for red flowers, R, and white flowers, R', show incomplete dominance. The heterozygous condition results in pink flowers. A gardener crosses a red four-o'clock with a white one. What are the expected genotypes and phenotypes of the offspring?

Solution: (1) The parental genotypes must be RR for the red flower and $R'R'$ for the white. Why? (2) The RR flowers produce all R gametes and the $R'R'$ flowers produce all R' gametes. (3) Therefore, all offspring are expected to have the genotype RR' and the phenotype pink flower.

7:9 Two Traits

After Mendel's first set of experiments, he studied the inheritance of two traits at once. For example, what types of plants would develop from a cross between a tall, red-flowered plant *(TTRR)* and a short, white-flowered plant *(ttrr)*? Remember, *each gamete must contain one allele for each trait* (law of segregation). Therefore, the gametes of the tall, red parent should have the alleles T and R and the short, white plant should produce gametes with t and r alleles. If they do, F_1 plants would all have the genotype *TtRr* and the phenotype tall and red (Figure 7–11). Mendel found these results.

The F_1 cross presented another problem to Mendel. The F_1 plants had the genotype *TtRr*. He knew that gametes produced by these plants must contain one of each kind of allele. But, do the alleles segregate independently of one another during gamete formation? Or do they stay in the same combination *(TR or tr)*? Mendel hypothesized that they segregate independently and that the possible gametes would be *TR, Tr, tR,* and *tr*. Given these possible gametes, the F_2 generation should show a phenotypic ratio of 9:3:3:1, or 9 tall and red: 3 tall and white: 3 short and red: 1 short and white (Figure 7–12). Again, experiments verified this prediction. The **law of independent assortment** states that *genes for different traits segregate independently during gamete formation.*

Genetics problems involving two traits are solved easily. The problem is a little longer, but no more difficult. Each trait is treated separately. Then, using the product rule again, the probabilities that the four alleles will combine in a given way can be determined.

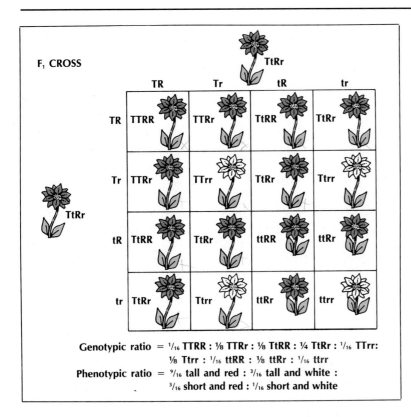

F₁ CROSS

Genotypic ratio = ¹/₁₆ **TTRR** : ⅛ **TTRr** : ⅛ **TtRR** : ¼ **TtRr** : ¹/₁₆ **TTrr** :
⅛ **Ttrr** : ¹/₁₆ **ttRR** : ⅛ **ttRr** : ¹/₁₆ **ttrr**

Phenotypic ratio = ⁹/₁₆ **tall and red** : ³/₁₆ **tall and white** :
³/₁₆ **short and red** : ¹/₁₆ **short and white**

FIGURE 7–12. The result of an F₁ cross involving two traits is shown in this Punnett square. You can see how Punnett square solutions become more involved as more traits are considered in a cross.

Sample Problem 3: In guinea pigs, rough coat *(R)* is dominant to smooth coat *(r)*. Black color *(B)* is dominant to albino *(b)*. A heterozygous black, smooth male is bred to a heterozygous black, heterozygous rough female. What are the probable genotypic and phenotypic ratios among their offspring?

Solution:

(1) Genotypes:

male: *Bbrr*

female: *BbRr*

(2) Gametes produced for color (*B* and *b* alleles):

male = *Bb* → ½ *B* + ½ *b*

female = *Bb* → ½ *B* + ½ *b*

(3) Multiply to find the probable combinations of *B* and *b* alleles among offspring.

$$
\begin{array}{r}
\tfrac{1}{2}\,B + \tfrac{1}{2}\,b \\
\times\ \tfrac{1}{2}\,B + \tfrac{1}{2}\,b \\
\hline
\tfrac{1}{4}\,Bb + \tfrac{1}{4}\,bb \\
\tfrac{1}{4}\,BB + \tfrac{1}{4}\,Bb \\
\hline
\tfrac{1}{4}\,BB + \tfrac{1}{2}\,Bb + \tfrac{1}{4}\,bb
\end{array}
$$

FIGURE 7–13. A cell that is heterozygous for two traits could produce four different combinations of alleles depending on how the chromosomes assort.

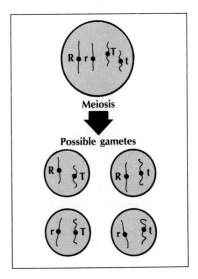

Meiosis

Possible gametes

Use the product rule to determine the second trait.

(4) Repeat steps 2 and 3 for the other trait (R and r alleles).

$$\text{male} = rr \rightarrow \frac{1}{1}\,r$$
$$\text{female} = Rr \rightarrow \frac{1}{2}\,R + \frac{1}{2}\,r$$
$$\frac{1}{2}\,R + \frac{1}{2}\,r$$
$$\times \frac{1}{1}\,r$$
$$\overline{\frac{1}{2}\,Rr + \frac{1}{2}\,rr}$$

Use the product rule again to determine the probable combinations of all four alleles.

(5) Multiply the two products to find the probable combinations of all four alleles (genotypic ratio).

$$\frac{1}{4}\,BB + \frac{1}{2}\,Bb + \frac{1}{4}\,bb$$
$$\times \frac{1}{2}\,Rr + \frac{1}{2}\,rr$$

$$\overline{\frac{1}{8}\,BBRr + \frac{1}{4}\,BbRr + \frac{1}{8}\,bbRr + \frac{1}{8}\,BBrr + \frac{1}{4}\,Bbrr + \frac{1}{8}\,bbrr}$$

(6) Interpret genotypes to find the phenotypic ratio.

$\frac{1}{8}\,BBRr$ = black and rough
$\frac{1}{4}\,BbRr$ = black and rough
$\frac{1}{8}\,bbRr$ = albino and rough
$\frac{1}{8}\,BBrr$ = black and smooth
$\frac{1}{4}\,Bbrr$ = black and smooth
$\frac{1}{8}\,bbrr$ = albino and smooth

The phenotypic ratio is $\frac{3}{8}$ black and rough: $\frac{1}{8}$ albino and rough: $\frac{3}{8}$ black and smooth: $\frac{1}{8}$ albino and smooth. Check the accuracy of the calculations used to find the phenotypic ratio by adding the fractions. The sum should equal one: $\frac{3}{8} + \frac{1}{8} + \frac{3}{8} + \frac{1}{8} = \frac{8}{8} = 1$.

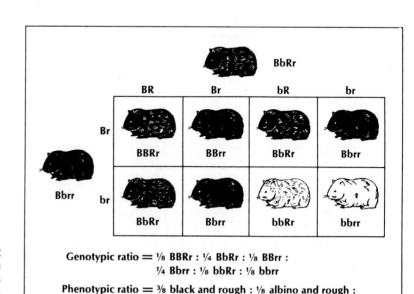

FIGURE 7–14. The Punnett square solution to Sample Problem 3 shows the genotypes of a cross involving two traits. The phenotypes can be determined from the genotypes.

Genotypic ratio = $\frac{1}{8}$ BBRr : $\frac{1}{4}$ BbRr : $\frac{1}{8}$ BBrr : $\frac{1}{4}$ Bbrr : $\frac{1}{8}$ bbRr : $\frac{1}{8}$ bbrr

Phenotypic ratio = $\frac{3}{8}$ black and rough : $\frac{1}{8}$ albino and rough : $\frac{3}{8}$ black and smooth : $\frac{1}{8}$ albino and smooth

7:10 Multiple Alleles

You have learned that two alleles act together to produce a phenotype. Although each *organism* normally has just two alleles for a trait, there may be more than two alleles possible for that trait in the *population*. In the human population, for example, there are three alleles that govern blood type: I^A, I^B, and I^O. Human blood types are said to be controlled by **multiple alleles,** a set of three or more different alleles controlling a trait. In humans, both I^A and I^B are dominant to I^O. But, I^A and I^B are codominant. Like incomplete dominance, **codominance** results when one allele is not dominant to the other. Unlike incomplete dominance, in which presence of both alleles results in an intermediate third phenotype, codominance results in both alleles being expressed equally.

From Table 7–1 you can see that for human blood types six genotypes and four phenotypes (blood types) are possible. The blood types are A, B, AB, and O.

Sample Problem 4: A woman who is known to have type B blood marries a man who has type A blood. They have five children, all with type AB blood. What are the most *probable* parental genotypes?

Solution: Consider the genotype of the children. Each of them has type AB blood. Therefore, their genotype must be $I^A I^B$. Now consider the parental possibilities. The mother could be either $I^B I^B$ or $I^B I^O$. The father could have the genotype $I^A I^A$ or $I^A I^O$. That all their children have $I^A I^B$ indicates that both parents are *probably* homozygous. If either of the parents were heterozygous, then some other blood types would probably occur among the five children. The mother's genotype is probably $I^B I^B$ and the father's is probably $I^A I^A$.

If these parents have a sixth child whose blood type is O $(I^O I^O)$, would you change your answer? Why or why not?

TABLE 7-1. BLOOD TYPES

Phenotype	Genotype(s)
A	$I^A I^A$, $I^A I^O$
B	$I^B I^B$, $I^B I^O$
AB	$I^A I^B$
O	$I^O I^O$

More than two alleles for a trait may be present in a population, but an individual diploid organism has only two alleles for the trait.

REVIEWING YOUR IDEAS

14. In four-o'clocks the genes for red flowers, *R*, and white flowers, *R'*, show incomplete dominance. The heterozygous condition results in pink flowers. A gardener crosses two red four-o'clocks. What are the expected genotypes and phenotypes of the offspring? What would be the expected genotypes and phenotypes if a second cross were done using the offspring?
15. State the law of independent assortment.
16. What are multiple alleles?

INVESTIGATION

Problem: How similar are traits of offspring to those of the parents?

Materials:

scissors
3 × 5 file cards

Part B. Determining Children's Phenotypes

1. Take the four chromosomes marked "father" and turn them over. Move them to the left side of your desk.

Procedure:

Part A. Determining Parent Phenotypes

1. Examine the models of eight chromosomes shown in Figure 7–15. Four of the chromosomes are marked "mother" and four are marked "father." Chromosomes of equal length are to be considered homologs.
2. Trace the chromosome outlines and copy the words "mother" or "father" and the letters onto a 3 × 5 file card.
3. Use a pair of scissors to cut out the eight chromosome outlines from the file card.
4. The letters written on each representative chromosome stand for alleles. Examine Table 7–2. It shows what traits are represented by the letters.
5. Make a copy of Table 7–3.
6. Complete Table 7–3 for both the mother and father. Describe the nine different phenotypes for each parent based on the alleles present on their chromosomes.

TABLE 7-2. REPRESENTED TRAITS

Allele	Phenotype
F f	six fingers five fingers
B b	black hair blond hair
S s	space between top front teeth no space between top front teeth
E e	free earlobes attached earlobes
R r	Rh positive blood Rh negative blood
T t	can roll tongue cannot roll tongue
H h	hair present on middle digit of hand hair absent on middle digit of hand
D d	skeleton development resulting in average height skeleton development resulting in dwarfism
C c	presence of Huntington's disease absence of Huntington's disease

FIGURE 7–15.

Mother	Mother	Mother	Mother	Father	Father	Father	Father
F	f	t	t	f	f	T	t
b	b	h	H	b	B	h	H
S	s	D	d	s	s	D	d
e	E	c	c	e	e	C	c
R	r			R	r		

142

2. Take the four chromosomes marked "mother" and turn them over. Move them to the right side of your desk.

3. Select only one long and one short chromosome from the father's pile and one long and one short chromosome from the mother's pile. This selecting simulates chromosomes that are passed to the first child during fertilization. Turn the chromosomes over. Read the allele combinations and decide to which phenotypes they correspond. Record the phenotypes in Table 7–3 under "1st child."

4. Return the four chromosomes face down to the correct piles.

5. Mix up the chromosomes in each pile.

6. Repeat steps 3–5 twice to simulate chromosomes passed to a second child and then to a third child.

Data and Observations:

TABLE 7-3. PHENOTYPES OF PARENTS AND CHILDREN					
Trait	Father	Mother	1st Child	2nd Child	3rd Child
Fingers					
Hair color					
Teeth					
Earlobes					
Rh blood type					
Tongue rolling					
Mid-digit hair					
Height					
Huntington's disease					

Questions and Conclusion:

1. (a) In this simulation, how many chromosomes did each child receive from each of the parents?
   (b) If this were an actual situation, how many chromosomes would each child receive from each parent?

2. Explain how the element of chance enters into inheritance of traits.

3. Using the data you generated for the first child, determine:
   (a) the number of traits it inherited that were exactly like those of the father

   (b) the number of traits it inherited that were exactly like those of the mother.

4. (a) Using your data from the second child, answer questions 3(a) and 3(b).
   (b) How many traits does the second child have in common with the first child?

5. (a) Using your data from the third child, answer questions 3(a) and 3(b).
   (b) How many traits does the third child have in common with the second child?

Conclusion: How similar are traits of offspring to those of the parents?

CHAPTER REVIEW

SUMMARY

1. Mendel studied the transmission of traits in pea plants and noted that traits can exist in two possible forms. **7:1**
2. The law of dominance states that one form of a hereditary trait, dominant, prevents the expression of another form, recessive. **7:2**
3. Mendel proposed that each pea plant has two genes for each trait and that each sex cell has just one of the two genes. **7:3**
4. The law of segregation states that the two alleles for each trait separate during gamete formation and are distributed to different gametes. **7:4**
5. Alleles determine genotypes, which determine phenotypes. **7:5**
6. Laws of probability can be used to solve genetics problems. **7:6**
7. Expected genotypic and phenotypic ratios can be predicted using laws of probability. **7:7**
8. Incomplete dominance occurs when one allele for a trait is not dominant to another. **7:8**
9. The law of independent assortment states that segregation of one allele occurs independently of segregation of other alleles. **7:9**
10. Some traits are controlled by multiple alleles, sets of three or more genes in a population. **7:10**

LANGUAGE OF BIOLOGY

allele	law of dominance
codominance	law of
dominant	independent assortment
gene	law of segregation
genetics	multiple alleles
genotype	phenotype
heterozygous	product rule
homozygous	Punnett square
incomplete dominance	recessive

CHECKING YOUR IDEAS

On a separate paper, indicate whether each of the following statements is true or false. Do not write in this book.

1. If the gene for yellow seeds, Y, is dominant to the gene for green seeds, y, then an organism with the Yy genotype will have the phenotype green seeds.
2. The organism described in question 1 is considered heterozygous.
3. Alleles segregate independently during gamete formation.
4. Actual results obtained from a cross are always the same as expected results.
5. By chance, one fourth of the gametes of an Aa organism will contain an A allele.
6. If two heterozygotes are crossed, half their offspring are expected to be heterozygous.
7. If two heterozygotes are crossed and one allele is dominant to the other, then one-half the offspring are expected to have the dominant trait.
8. The chances of tossing three coins and getting all heads is $\frac{1}{6}$.
9. A gamete produced from an organism $AABB$ contains two A genes and two B genes.
10. If two organisms with the genotype $AaBb$ are crossed, a $9:3:3:1$ phenotypic ratio will be expected.
11. Parents with the genotype $I^A I^O$ and $I^B I^B$ can produce a child with type O blood.
12. Type AB blood results because the genes I^A and I^B show codominance.

EVALUATING YOUR IDEAS

1. How were the results of Mendel's parental crosses unusual? How did Mendel interpret these results?
2. What happened when Mendel crossed members of the F_1 generation?

3. Why did Mendel assume that each organism has two characters for each trait?
4. How did Mendel test segregation?
5. Distinguish among the laws of dominance, segregation, and independent assortment.

APPLYING YOUR IDEAS

1. In a certain animal, black fur, *B,* is dominant to white fur, *b.* Determine the expected genotypic ratios and phenotypic ratios resulting from crosses between (a) homozygous black × white, (b) two heterozygous blacks, and (c) heterozygous black × white.
2. Suppose that in outer space there exist creatures whose traits are inherited by Mendel's laws. You find that purple eyes, *P,* are dominant to yellow eyes, *p.* Two purple-eyed creatures mate and produce six offspring. Four of them have purple eyes and two have yellow eyes. What are the genotypes of the parents? The phenotypes? What are the genotypes of the offspring?
3. In fruit flies, long wing, *L,* is dominant to short wing, *l.* Two long-wing flies produced 49 short-wing and 148 long-wing offspring. What were the probable genotypes of the parents? About how many of the long-wing offspring should be heterozygous?
4. In humans, brown eyes, *B,* are dominant to blue eyes, *b.* A brown-eyed man marries a blue-eyed woman. They have eight children; all are brown-eyed. What are the possible genotypes of each person in the family?
5. In Andalusian fowl, *B* is the gene for black plumage. *B'* is the gene for white plumage. The genes show incomplete dominance. The heterozygous condition results in blue plumage. List the genotypic and phenotypic ratios expected from the crosses (a) black × blue, (b) blue × blue, (c) blue × white.

6. A black, smooth guinea pig was mated with an albino, rough guinea pig. Their offspring were black rough and black smooth. These were the only types produced over a period of years in a number of matings. Black and rough are dominant traits. What was the probable genotype of each parent?
7. What is the probable genotypic ratio among children born to a mother having the genotype $I^A I^O$ and a father with blood type AB?
8. One parent has type A blood and the other parent has type B blood. What are their genotypes if they produce a large number of children whose blood types are (a) all AB, (b) ½ AB and ½ B, (c) ½ AB and ½ A, and (d) ¼ AB, ¼ A, ¼ B, ¼ O?

EXTENDING YOUR IDEAS

1. An organism *AaBBCcDd* is crossed with an organism *AaBbCcDD.* Without working out the entire cross, find the expected frequency of an offspring with the genotype *aaBbCcDd.*
2. A plant is wanted with the genotype *AAbb.* Given parental strains *AABB* and *aabb,* outline the methods you would use to obtain the plant. How would you test the genotype of the plant once you thought you had obtained it?
3. Radishes may be long, round, or oval. Crosses of long and oval gave 159 long and 156 oval. Crosses of oval and round produced 203 oval and 199 round. Crosses of long and round gave 576 oval. Explain.

SUGGESTED READINGS

Bornstein, Jerry and Bornstein, Sandy, *What Is Genetics?* New York, Messner Pub., 1979.
Raab, Carl and Raab, Joan, *The Student Biologist Explores Genetics.* New York, Rosen Press, 1979.

GENES AND CHROMOSOMES

Within each cell of your body are structures called chromosomes that contain thousands of genes. The genes of the tuber begonia are responsible for the various colors of the flowers. How do genes control development of traits? What can happen if genes are not replicated normally during cell division? How can abnormal genes and chromosomes be detected in humans?

chapter 8

I mport observations of chromosomes in the early 1900s linked the events of meiosis (Section 6:7) with some of Mendel's findings. Mendel had concluded that there are two characters (genes) for each trait, both of which were transmitted to the offspring. He had concluded also that genes segregate during gamete formation — each gamete has just one gene per trait. Observation of chromosomes revealed that each diploid cell has one homologous pair of chromosomes of each type and that a zygote inherits both homologs. Also, it was noted that homologs segregate during meiosis — each gamete has only one homolog of a pair.

These observations about chromosomes and meiosis led to a hypothesis: genes are located on chromosomes. It was reasoned that one gene for a trait is carried on one homolog and the corresponding gene is on the other homolog. Later experiments verified this hypothesis.

Objectives:
You will
- trace the development of the chromosome theory of heredity.
- discuss modern concepts of genetics.
- give examples of human genetic diseases and their causes.

In the early 1900s, it was hypothesized that genes are located on chromosomes.

THE CHROMOSOME THEORY OF HEREDITY

Drosophila melanogaster (droh SAHF uh luh • mel uh NOH gas tur), fruit flies, are often used in genetic studies. They are small (about 2 mm in length), easily handled, and they produce many offspring in about two weeks. *Drosophila* played an important role in genetics during the first half of this century.

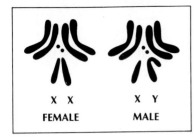

FIGURE 8–1. Male and female *Drosophila* have eight chromosomes. Three pairs are autosomes. The other two chromosomes are sex chromosomes. Females have two X chromosomes. Males have one X chromosome and one Y chromosome.

Cells of *Drosophila* have three pairs of autosomes and one pair of sex chromosomes. X and Y are the sex chromosomes.

Using probability shows that half of the *Drosophila* offspring would be male and half would be female.

8:1 Sex Determination

Thomas Hunt Morgan (1866–1945) was a pioneer in the use of fruit flies to study genetics. Early in his research he learned that chromosomes of male and female *Drosophila* cells are slightly different. Female flies have four pairs of homologous chromosomes, but males have three homologous pairs plus one pair of two different chromosomes (Figure 8–1). One of the two chromosomes looks like those of the fourth pair of female chromosomes. It is called the **X chromosome.** The other chromosome, which has a different shape, is called the **Y chromosome.** The female fruit fly has two X chromosomes, while a male has one X and one Y chromosome.

Because the X and Y chromosomes differ between the sexes, they are called **sex chromosomes.** The other chromosomes are called **autosomes** (AWT uh sohmz). Thus, both female and male *Drosophila* have three pairs of autosomes and one pair of sex chromosomes. A female normally has three pairs of autosomes and two X chromosomes. A male normally has three pairs of autosomes, an X chromosome, and a Y chromosome.

All the gametes produced by the female fruit fly will contain an X chromosome. The male fruit fly's X and Y chromosomes are not homologous, but they behave as a homologous pair during meiosis. They are separated from each other during gamete formation. Thus, half the male's gametes will contain the X chromosome, and half will contain the Y chromosome. The possible combinations of X and Y chromosomes among offspring can be determined like other traits (Figure 8–2). Using probability methods, it can be calculated that half the offspring would have the genotype XX (female). The other half would have XY (male).

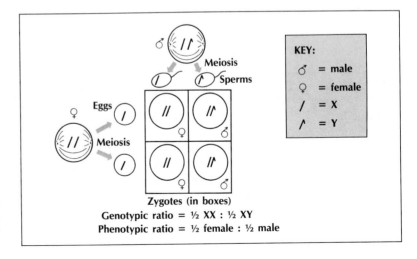

FIGURE 8–2. This Punnett square shows that half the possible offspring are female and half are male.

Zygotes (in boxes)
Genotypic ratio = ½ XX : ½ XY
Phenotypic ratio = ½ female : ½ male

KEY:
♂ = male
♀ = female
/ = X
∧ = Y

Such a pattern of sex determination is evident in many organisms, including humans. However, the pattern is different in some organisms. In grasshoppers, for example, males have only one X chromosome, while females have two.

8:2 Morgan's Discoveries

In 1910, Morgan made an unexpected discovery about eye color in fruit flies. He found a male with white eyes. This fly was produced from a pure line having only red eyes.

He decided to breed this white-eyed male to red-eyed females. Assuming that red eye color was dominant to white eye color, Morgan expected to find all red-eyed offspring in the F_1 generation. He also expected the F_1 flies, when interbred, to produce a ratio of three red-eyed flies to one white-eyed fly. He did, in fact, get these results.

Morgan also noted that the only white-eyed flies in the F_2 generation were males. There were no white-eyed females. Because white eye color seemed to be linked with sex, it was called a **sex-linked characteristic.**

Morgan's hypothesis was that the alleles for eye color are carried only on the X chromosomes, and that there are no alleles for eye color on the Y chromosome. Thus, the parental cross had been between red-eyed females ($X^R X^R$) and white-eyed males ($X^r Y$) (Figure 8–3). R represents the red eye color allele, and r stands for the white eye

Traits associated with sex are called sex-linked characteristics.

FIGURE 8–3. Assuming eye color is sex-linked, results predicted with Punnett squares agree with Morgan's experimental results. The parental and F_1 crosses he did are shown.

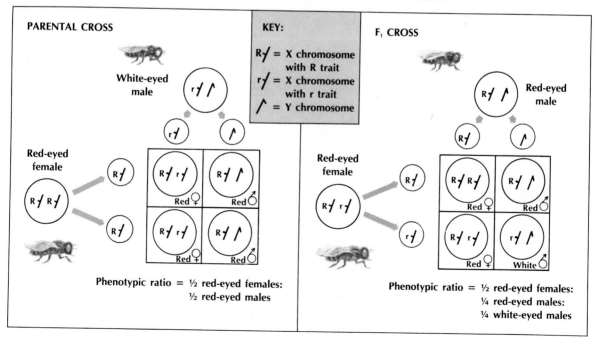

color allele. The results of the cross (F$_1$ generation) had been red-eyed males (X^{R}Y) and red-eyed females (X^{R}X^r). The F$_2$ generation would have been red-eyed females (X^{R}X^R) or (X^{R}X^r), red-eyed males (X^{R}Y), and white-eyed males (X^{r}Y). Morgan's hypothesis explained his previous data by showing how all white-eyed flies in the F$_2$ generation turned out to be males.

But, a hypothesis should predict new facts. Morgan predicted that he could produce white-eyed females by breeding an F$_1$ red-eyed female (X^{R}X^r) with a white-eyed male (X^{r}Y). When such a cross was done, Morgan's prediction was confirmed. There were white-eyed females among the offspring. This evidence was also important in supporting the hypothesis that genes are carried on chromosomes.

8:3 Solving Problems

Problems with sex-linked traits can be solved in the same way you solve other genetics problems.

Represent the alleles and the chromosomes in problems involving sex-linked traits.

Sample Problem 1: In a certain animal, the gene for black coat color (*B*) is dominant to the gene for orange coat color (*b*). The characteristic is sex-linked, the gene being carried on the X chromosome. Determine the probable genotypic and phenotypic ratios among the offspring produced by a heterozygous female and a black male.

Solution:

(1) Write the genotype of each parent.
$$\text{female: } X^B X^b$$
$$\text{male: } X^B Y$$

(2) Determine the probable distribution of chromosomes and alleles in the gametes.
$$\text{female} = X^B X^b \rightarrow \tfrac{1}{2} X^B + \tfrac{1}{2} X^b$$
$$\text{male} = X^B Y \rightarrow \tfrac{1}{2} X^B + \tfrac{1}{2} Y$$

(3) Using the product rule, multiply to find the possible combinations of chromosomes and alleles (genotypic ratio).
$$\begin{array}{r} \tfrac{1}{2} X^B + \tfrac{1}{2} X^b \\ \times\ \tfrac{1}{2} X^B + \tfrac{1}{2} Y \\ \hline \tfrac{1}{4} X^B X^B + \tfrac{1}{4} X^B X^b + \tfrac{1}{4} X^B Y + \tfrac{1}{4} X^b Y \end{array}$$

(4) Interpret the genotypes to find the phenotypic ratio.
$\tfrac{1}{4} X^B X^B$ = black female $\quad \tfrac{1}{4} X^B Y$ = black male
$\tfrac{1}{4} X^B X^b$ = black female $\quad \tfrac{1}{4} X^b Y$ = orange male
Therefore, the expected phenotypic ratio is $\tfrac{1}{2}$ black females: $\tfrac{1}{4}$ black males: $\tfrac{1}{4}$ orange males.

The Punnett square answer to Sample Problem 1 is shown in Figure 8–4.

FIGURE 8–4. The Punnett square solution to Sample Problem 1 shows the inheritance of coat color which, in this example, is sex-linked.

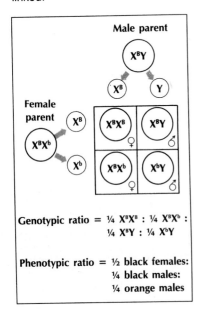

Genotypic ratio = $\tfrac{1}{4}$ X^{B}X^B : $\tfrac{1}{4}$ X^{B}X^b : $\tfrac{1}{4}$ X^{B}Y : $\tfrac{1}{4}$ X^{b}Y

Phenotypic ratio = $\tfrac{1}{2}$ black females: $\tfrac{1}{4}$ black males: $\tfrac{1}{4}$ orange males

Sample Problem 2: What genotypes and phenotypes must each parent have to produce an orange female?

Solution: The genotype of the orange female must be $X^b X^b$ because the gene for orange is recessive. One of the X chromosomes is received from the mother. The other is received from the father. Therefore, the father's genotype must be $X^b Y$. He is orange. The mother must have the genotype $X^B X^b$ (black) or $X^b X^b$ (orange).

8:4 Nondisjunction

Calvin Bridges (1889–1938) made another unexpected discovery working with fruit flies. He found female flies that had two X chromosomes and a Y chromosome. Also, some male flies had just an X chromosome and no Y chromosome at all. These unusual patterns occurred because the homologous sex chromosomes of the parents failed to segregate properly during meiosis (Figure 8–5). He called this failure of chromosomes to segregate **nondisjunction** (nahn dihs JUNK shun). Nondisjunction of autosomes can also occur. Bridges' discovery of nondisjunction was associated with abnormal genetic results and was important evidence of the **chromosome theory of heredity:** *Genes are located on chromosomes.*

Bridges' discovery of nondisjunction changed the concept of sex determination. It had been thought that the presence of the Y chromosome in fruit flies determines sex. Bridges found that in *Drosophila* the chromosome arrangement XXY will produce a female and a single X will produce a male. Thus, it is now known that the presence of *two or more* X chromosomes produces females and the absence of the second X chromosome produces males. Sex in fruit flies is determined by the number of X's.

Chromosome theory of heredity: Genes are located on chromosomes.

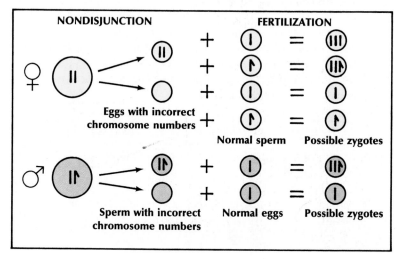

FIGURE 8–5. Nondisjunction is one explanation for the appearance of extra chromosomes that may result in unique phenotypes. In this example, the sex cell formed through nondisjunction joins with a sex cell having the usual number of sex chromosomes.

PEOPLE IN BIOLOGY

In 1970, in front of a small group of scientists gathered at the University of Wisconsin, Har Gobind Khorana announced that he and his research team had made the first synthetic gene. This announcement was the result of many years of research in chemical genetics.

Har Gobind Khorana

(1922–)

Khorana was born and reared in India. After he earned his degrees at Punjab University, he moved to England where he began his lifelong study of nucleic acids and proteins. From there his research led him to a post in Canada. Later he became Co-director of the Institute for Enzyme Research at the University of Wisconsin. It was here that Khorana began to decipher the genetic code of yeast. In 1968 he received the Nobel Prize for Medicine for interpreting the code and explaining how genes determine the function of cells. Only two years later, Khorana reported his synthesis of a yeast gene.

Dr. Khorana, who has been a United States citizen since 1966, now lives in Massachusetts with his wife. He has three children, and in his spare time he enjoys hiking, swimming, and listening to music.

Determination of sex involves genes on both sex chromosomes and autosomes.

In humans, unlike fruit flies, the Y chromosome determines sex. If a Y chromosome is present, a male will be produced. Cells with X, XX, and XXX chromosomes produce females, while cells with XY, XXY, and XXXY chromosomes produce males.

How combinations of X and Y chromosomes result in a particular sex is not known. Sex determination seems to result from the actions of many genes, some of which are on autosomes.

The X chromosome carries many more genes than the Y chromosome. Some Y genes have no corresponding genes on the X chromosome. For example, the gene that causes the rims of the ears to be hairy is on the Y chromosome only.

REVIEWING YOUR IDEAS

1. Why are *Drosophila* useful for genetic studies?
2. Compare the chromosomes of male *Drosophila* to those of female *Drosophila*.
3. Why are the chances of a couple having a girl 50:50?
4. What was unusual about Morgan's F_2 3:1 ratio?
5. What is nondisjunction?

OTHER GENETICS CONCEPTS

Discoveries made by Mendel, Morgan, Bridges, and others provided an understanding of basic patterns of inheritance. Work of later geneticists has, in turn, led to knowledge of more complex ways by which hereditary information is transferred.

8:5 Gene Linkage

Today it is known that the total number of genes in a cell is much greater than the total number of chromosomes. Thus, each chromosome must contain many different genes. Each gene influences certain traits. Genes that occur on the same chromosome are said to be linked. This phenomenon is called **gene linkage.**

Think of the cross between tall, red-flowered plants with a genotype of *TTRR* and short, white-flowered plants with a genotype of *ttrr.* A 9:3:3:1 ratio would be expected in the F$_2$ generation (see Figure 7–12). The expected results would be different, though, if the genes were linked (Figure 8–6).

Assume that in a certain plant green seeds, *G,* are dominant to yellow seeds, *g,* and that round seeds, *R,* are dominant to wrinkled seeds, *r.* Also, assume that the alleles for seed color and seed shape are on the same pair of homologous chromosomes. As a result, they do not assort independently (Section 7:9) when meiosis takes place. The alleles stay together during meiosis and, consequently, enter the gametes in the same combinations each generation.

A single chromosome has many genes.

If genes are linked, they do not assort independently during gamete formation.

FIGURE 8–6. A Punnett square shows that linked genes do not assort independently.

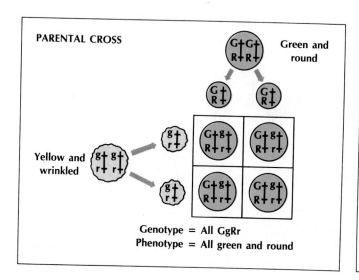

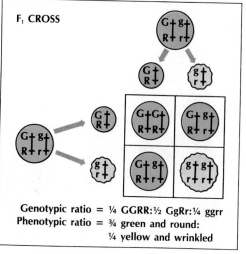

When *GGRR* and *ggrr* are crossed, the F₁ offspring have green, round seeds *(GgRr)*. These results are the same as those produced when genes are not linked. But, there are only three possible genotypes among the F₂ offspring—*GGRR, GgRr,* and *ggrr*. The phenotypic ratio is three green, round to one yellow, wrinkled. Without gene linkage, the F₂ phenotypic ratio is 9:3:3:1 due to independent assortment. *Thus, gene linkage has the potential for reducing the chances for genetic recombination and variety among offspring.*

Gene linkage reduces variety among offspring.

8:6 Crossing-Over

Geneticists have learned that linked genes can sometimes be separated. Recall that pairs of double-stranded chromosomes (tetrads) exist in prophase I of meiosis. During this time, part of a chromatid on one homolog may break and rejoin with one of the chromatids belonging to the second homolog in the same tetrad. This process is called **crossing-over,** an exchange of segments of chromosomes between two homologous chromatids (Figure 8–7). Crossing-over may occur at more than one place on chromosomes.

Crossing-over, exchange of alleles between homologous chromosomes, separates linked genes.

When homologous chromosomes pair during prophase I, the four chromatids lie side by side. They are positioned so that genes on one homolog line up with corresponding genes on the second homolog. When the homologous pairs are pulled apart later in meiosis I, the chromatids can have different combinations of genes as a result of exchange.

FIGURE 8–7. The differences in the genetic makeup of gametes produced with no cross-over, a single cross-over, and a double crossover can be easily compared.

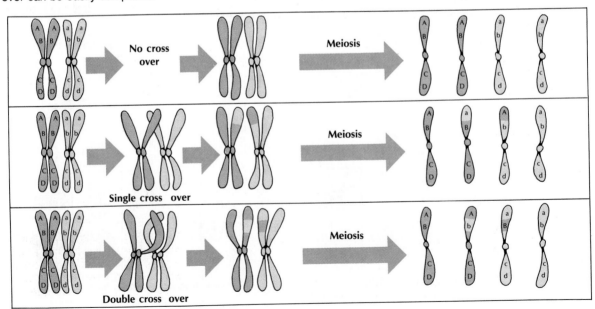

Although many gametes have the same sets of linked genes as the parent, some do not because of crossing-over. Gametes with a set of linked genes different than that found in the parent are called **recombination gametes.** The union of recombination gametes with other gametes results in offspring with a wider variety of traits than would be possible if crossing-over were not to take place. Thus, in a population, crossing-over aids in restoring some of the variation that would otherwise be prevented by gene linkage. Recall that variety in a population may be important if the environment changes (Section 6:10).

Crossing-over can serve as a tool for determining the locations of genes on chromosomes. First, the expected phenotypic ratio is determined for the set of linked genes being studied. Then, the expected ratio is compared to the observed phenotypic ratio. The degree of variation from the expected ratio is a measure of how often the two genes cross-over. The frequency or amount of cross-over can then be used to determine the locations of the genes on the chromosomes.

The linear arrangement of genes on chromosomes has been worked out for many organisms. The location of genes on a chromosome is called a **genetic map.** The genetic map is determined by the relative amounts of crossing-over between different groups of linked genes. The farther apart two genes are, the greater the chance of their crossing-over.

As an example, it was determined that in *Drosophila* the ruby-eye gene was farther from the bar-eye gene than it was from the cut-wing gene (Figure 8–8). More crossing-over occurred between the ruby-eye and bar-eye than between the ruby-eye and cut-wing genes. The distances between the genes are given in genetic map units, which are what the numbers on Figure 8–8 show. Genetic map units are an indication of the amount of crossing-over between genes. One unit equals 1% crossing-over. Finding locations of human genes is an important step in understanding, treating, or preventing certain genetic diseases.

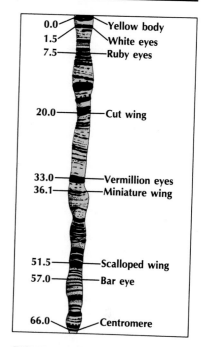

FIGURE 8–8. A genetic map of the *Drosophila* X chromosome shows the linear arrangement of genes. Many of the genes that have been mapped were omitted from this drawing for simplification. The numbers indicate relative distances between genes in genetic map units.

Frequency of crossing-over is an indication of the distance between genes on a given chromosome.

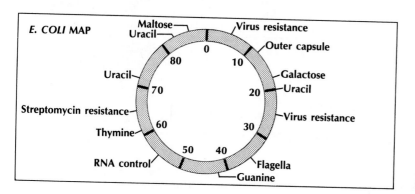

FIGURE 8–9. Chromosomes in bacteria are circular. This genetic map of the *E. coli* chromosome, a bacterial chromosome, shows some of the genes whose locations are known.

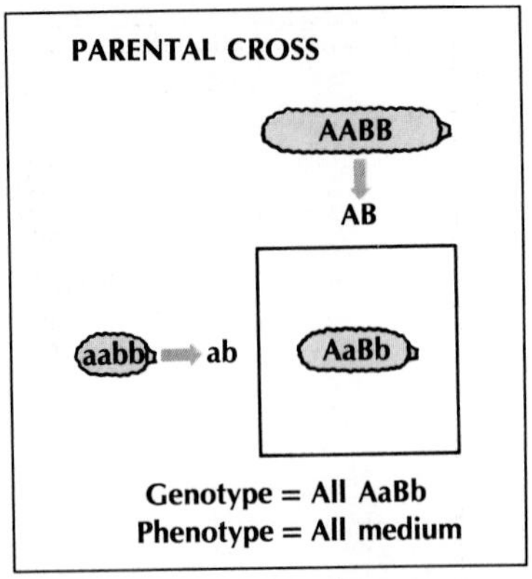

PARENTAL CROSS

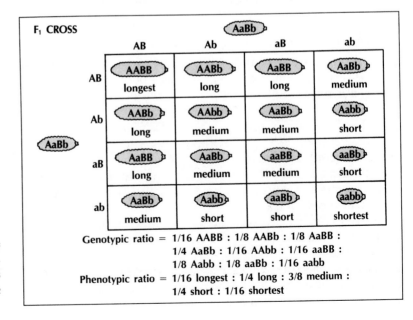

AABB

AB

aabb → ab   AaBb

Genotype = All AaBb
Phenotype = All medium

FIGURE 8–10. This Punnett square shows a parental cross of corn in which two pairs of genes determine one trait, ear length.

The range of phenotypes seen in continuous variation is controlled by several pairs of genes.

8:7 Many Genes — One Effect

The hereditary traits studied so far are explained by Mendel's laws, which state that there are single pairs of genes, the expressions of which are governed by dominance or incomplete dominance (Section 7:8). Examples are red flowers versus pink or white flowers, normal color vision versus color blindness, and round seeds versus wrinkled seeds. All of these traits are determined by the manner in which two alleles on a single pair of chromosomes are combined.

Mendel's laws, however, do not explain all genetic traits. For example, there is a wide range of ear lengths in corn. If the expression of ear length in corn were governed by simple dominance or recessiveness, ears of corn would be one of two lengths, long or short. Ears of corn are not just long or short. Long and short are limits between which many lengths exist. If length were controlled by a single pair of genes showing incomplete dominance, there would be only three lengths — long, medium, and short.

The presence of many ear lengths in a population of corn plants is an example of **continuous variation.** Continuous variation can be explained by assuming that more than one pair of genes is involved. Suppose that two pairs of genes (*A* and *a, B* and *b*) are responsible, and they are on different pairs of homologs. Suppose also that *AABB* represents the genotype of the longest ears of corn, and *aabb* represents the genotype of the shortest ears of corn. A cross between *AABB* and *aabb* parents results in F₁ offspring that are *AaBb* (intermediate length) (Figure 8–10).

FIGURE 8–11. When two pairs of genes determine a trait, many different phenotypes (continuous variation) are possible in the F₂ generation.

F₁ CROSS   AaBb

		AB	Ab	aB	ab
	AB	AABB longest	AABb long	AaBB long	AaBb medium
AaBb	Ab	AABb long	AAbb medium	AaBb medium	Aabb short
	aB	AaBB long	AaBb medium	aaBB medium	aaBb short
	ab	AaBb medium	Aabb short	aaBb short	aabb shortest

Genotypic ratio = 1/16 AABB : 1/8 AABb : 1/8 AaBB :
1/4 AaBb : 1/16 AAbb : 1/16 aaBB :
1/8 Aabb : 1/8 aaBb : 1/16 aabb

Phenotypic ratio = 1/16 longest : 1/4 long : 3/8 medium :
1/4 short : 1/16 shortest

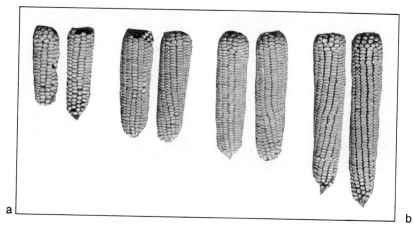

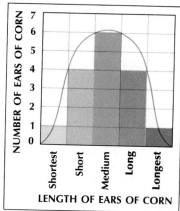

FIGURE 8–12. (a) Continuous variation in length can be seen in corn ears. (b) A graph of the distribution of ear lengths in corn is bell shaped. Medium length ears occur most often.

If these F_1 organisms were interbred, a broad range of genotypes and phenotypes would be produced in the F_2 offspring (Figure 8–11). The genotype *AaBb,* medium length ears of corn, is the one most often expected. *AABB* (longest) and *aabb* (shortest) are very seldom expected. The rest of the expected genotypes correspond to various intermediate lengths. Some ears would be longer than medium; some ears would be shorter than medium. When graphed, these expected results show a distribution such as that in Figure 8–12. When the actual crosses are made, the results agree with those predicted by the Punnett square. Geneticists explain many such traits in terms of **multiple genes** in which many genes may affect a single trait. Human skin color shows continuous variation and is thought to be determined by multiple genes.

8:8 Expression of Genes

Genes interact to control various patterns of inheritance. For example, consider human eye color. Brown eye color is due to presence of a pigment called melanin. People with blue eyes have no melanin in the eyes. In the last chapter, you worked genetics problems dealing with the way in which the phenotypes brown eyes and blue eyes are inherited. You know, however, that many other eye colors exist. Those other colors are due to the presence of other genes inherited along with the *B* and *b* alleles. These other **modifier genes** affect eye color by influencing the amount, intensity, and distribution of melanin in cells of the eye.

It is the rule, not the exception, that *many genes interact to control a phenotype.* Because most traits are controlled by several genes working together, an individual may have alleles for a particular trait, but not show that trait. For example, a person may have the genotype *Bb,* but have blue eyes. This may happen when a modifier gene prevents melanin from being produced.

Modifier genes may affect the expression of a genotype for a particular trait.

A particular phenotype most often depends upon many interacting genes.

FIGURE 8–13. (a) Curly-winged *Drosophila* bred at 16°C will produce (b) straight-winged offspring. Temperature is an environmental factor that influences wing shape in this organism.

Phenotypes are influenced by genes and environment.

Another factor that determines the expression of genes is the environment in which an organism develops. In fruit flies, for example, the gene for curly wings is expressed differently at different temperatures. Curly-winged flies bred at 25°C will have offspring with curly wings. But curly-winged flies bred at 16°C will have offspring with straight wings. Both sets of offspring, if interbred at 25°C, will produce curly-winged flies.

In general, the development of any organism is affected by the environment. A given genotype may be for long ears of corn. But, the soil in which the corn is growing may be poor, or there may be a drought during the growing season. Each of these may affect the phenotype so that there may be short ears of corn or even no ears.

In summary, expression of traits (phenotypes) is certainly more complicated than early biologists thought. Whether or not a certain trait is expressed depends upon the particular alleles for that trait, all other alleles carried by the organism, and the environment in which the organism develops and lives.

FIGURE 8–14. Environmental factors influence traits in many ways. (a) A soybean plant with ample iron (above) grows much better than an iron-deficient soybean plant (below). (b) Sun exposure can darken skin color and lighten hair color.

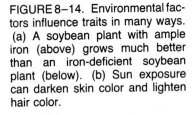

REVIEWING YOUR IDEAS

6. What is gene linkage? How does it affect variety among offspring?
7. What is crossing-over? How does it affect variety among offspring?
8. What is continuous variation? How is it explained?

HUMAN GENETIC DISEASE

Humans inherit traits the same way other organisms do. Sometimes humans develop diseases that are genetic in origin. Genetic diseases are not really rare. It is estimated that in the United States, genetic diseases cause about 40% of all miscarriages and 40% of all infant deaths. It is also thought that about 80% of all mentally retarded people and about ⅓ of all children in hospitals have problems of genetic origin.

Genetic disease affects many humans in a variety of ways.

Genetic counseling is now available to prospective parents such as those from families with a history of hereditary disease. In counseling sessions, genetic counselors discuss with the couples their chances of having children with certain genetic diseases. The counselors may discuss results of tests with the prospective parents. They also help the couples make decisions about having children when children with serious genetic defects are likely.

Genetic counseling helps couples weigh the risks of having children.

Some genetic diseases are caused by the presence of certain alleles. Others may result from abnormal inheritance of chromosomes. Examples of both types are given here.

8:9 Problem-Causing Genes

Genes may exist in two or more forms, or alleles. It is possible that one allele for a particular trait may be harmful or even lethal. A **lethal gene** causes death. It is estimated that every person has five to eight harmful or lethal genes. Yet, most people show no signs of disease. Usually the harmful alleles are recessive, and their effects are masked by dominant alleles. When a person is homozygous for a disease-causing recessive allele, the symptoms of the disease appear. Some examples of human diseases caused by harmful genes are discussed below.

● **Sickle-Cell Anemia.** This disease affects mostly black people. A person with the disease usually is homozygous recessive for a gene that codes for hemoglobin. Hemoglobin is a protein in red blood cells that combines with oxygen and transports it to the cells. In persons

FIGURE 8–15. (a) Normal red blood cells are disc-shaped. (b) Red blood cells of a person with sickle-cell anemia are sickle-shaped. These cells are shown magnified about 2200 times.

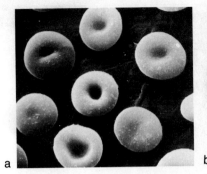

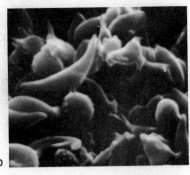

a

b

with sickle-cell anemia, hemoglobin is abnormal and the red cells are sickle-shaped (Figure 8–15). Normal red cells are disc-shaped. Analysis of hemoglobin shows that normal and abnormal hemoglobin differ in only one of the nearly six hundred amino acids that make up the molecule.

This slight difference prevents the abnormal hemoglobin from carrying enough oxygen to cells. The red cells change to a sickled shape within the blood vessels. Because capillaries are small in diameter, the sickled cells do not pass through easily and can clog the vessels. Thus, body cells do not get needed oxygen. As a result, the person might die.

Sickled cells carry less oxygen and can block small blood vessels.

Several promising treatments for sickle-cell anemia are being used. Different chemicals are being used that increase the oxygen-carrying capacity of hemoglobin. Many sickle-shaped red blood cells will return to their normal disc shape if the amount of oxygen they are carrying is increased.

A blood test can determine whether parents are carriers of the sickled-cell trait. Carriers are heterozygotes; they have one gene for the sickled-cell trait. If both individuals are carriers, they might choose not to have children.

FIGURE 8–16. (a) Sickle-cell hemoglobin differs from normal hemoglobin by only one amino acid — valine instead of glutamic acid. A representation of hemoglobin is shown with the stars marking the locations of the sickle mutations. (b) A new machine, similar to a kidney dialysis machine, is being used successfully to treat sickle-cell patients.

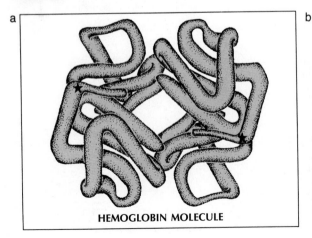

a

HEMOGLOBIN MOLECULE

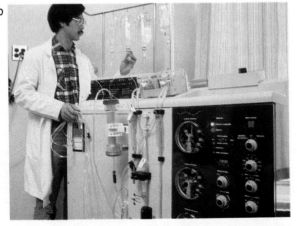

b

• **Galactosemia** (guh lak tuh SEE mee uh). People who have this disease have two recessive genes for the trait. They cannot manufacture the enzyme necessary for metabolism of a simple sugar, galactose, into glucose. Galactose is part of lactose, a sugar found in milk. Instead of being converted to glucose, the galactose is converted to another compound. As that compound builds up, damage to the nervous system results. The damage can lead to an early death. Fortunately, these problems can be avoided by early detection and a diet containing no galactose.

• **PKU.** The dangerous recessive allele that causes this rare disease is carried by about one in every 100 000 persons. PKU results from a missing enzyme and leads to severe mental retardation. The disease can be detected by analyzing urine for the presence of an abnormal compound. Symptoms of PKU are prevented by a special diet early in life.

• **Tay-Sachs** (TAY saks). This disease occurs most often in Jewish people and is caused by a recessive gene. A child with this disease begins life normally, but the nervous system fails to develop properly. An important enzyme is missing that breaks down a kind of fat. The child loses ability to move normally, eventually becoming inactive. Death usually occurs by the age of two or three.

There is no cure for Tay-Sachs. However, parents in the high-risk group can have a blood test to determine whether they carry the allele. It is then their decision whether to have children. Blood tests can also be done on fetuses to see if they have Tay-Sachs.

• **Huntington's Disease.** This disease, unlike others discussed so far, is due to a dominant gene. Thus, a heterozygous person can have the disease. Symptoms first appear at about age 40. Certain brain cells begin to deteriorate. The person may become clumsy, have memory problems, and be irritable. Later symptoms include uncontrollable jerking of the arms and legs, loss of muscle coordination, and increased loss of memory and speaking ability. Death usually occurs within twenty years after the appearance of the first symptoms.

It is estimated that about 25 000 Americans have this disease, and another 125 000 may develop it. The causes for the symptoms are not understood, and there is no treatment. Huntington's disease is always lethal. It is now known, however, that the gene for this disease is on chromosome number 4. Analysis of that gene could soon lead to knowledge of why it causes the disease and, perhaps, how to treat or prevent the disease. Furthermore, people with a family history of the disease can be tested to see whether or not they carry the gene. Knowing that one does not carry the gene would provide relief from the fear of developing the disease. Such news would also allow a person to plan on having children without fear of passing along the disease to them. On the other hand, fear of having the gene has led many people who are eligible for testing to decide against it.

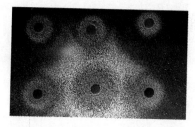

FIGURE 8–17. Most states have procedures for screening newborn babies for PKU as well as other diseases. Using the Guthrie test, a drop of blood is placed on a paper disc, which is then placed in a bacterial culture that is grown on a special medium. A chemical found in large quantities in the blood of PKU babies promotes growth of the bacterium, as seen around only one of the discs below.

Unlike many other genetic diseases, Huntington's disease is caused by a dominant gene.

FIGURE 8–18. Information about Tay-Sachs is available from the National Tay-Sachs & Allied Disease Association.

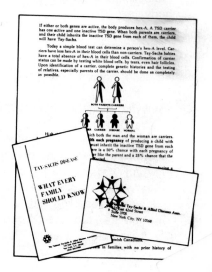

INVESTIGATION

Problem: How does a genetic disease affect red blood cells?

Materials:

microscope
prepared slide of normal human blood
prepared slide of sickled human blood

TABLE 8-1. APPEARANCE OF RED BLOOD CELLS		
Genotype	Phenotype	Term used to describe phenotype
SS	all red cells are round	normal
SS'	about half of all red cells are round and half are sickled	sickle-cell trait
S'S'	all red cells are sickled	sickle-cell anemia

Procedure:

Part A. Background

1. Examine Table 8-1. It describes how the red blood cells of a person will appear if certain genes are inherited from their parents.
2. Sickle-cell trait and sickle-cell anemia are related genetic diseases. Both affect the oxygen-carrying ability of hemoglobin. Thus, the necessary amount of oxygen may not be supplied to body cells.
3. When all blood cells are sickled, very little oxygen reaches body cells. When only half the cells are sickled, more oxygen can reach body cells.
   Answer these questions using Table 8-1 as a reference:
   (a) Which gene combination results in very little oxygen being able to reach body cells?
   (b) What name is given to the disorder described in 3(a)?
   (c) Which gene combination results in the maximum amount of oxygen reaching body cells?
   (d) Which gene combination results in more oxygen reaching body cells than in question 3(a), but less oxygen reaching body cells than in question 3(c)?

   (e) What name is given to the disorder in question 3(d)?

Part B. Observing Red Blood Cells

1. Observe a prepared slide of normal blood cells under the microscope on low power and then high power.
2. Make a copy of Table 8-2. Diagram several red blood cells in Table 8-2 as they appear under high power. NOTE: On a prepared slide, red blood cells appear pale pink. Disregard any other cells you may observe.
3. Complete Table 8-2 by describing the shape of the normal red blood cells.
4. Observe a prepared slide of sickle-shaped blood cells under the microscope on low power and then high power.
5. Diagram several sickled red blood cells in Table 8-2 as they appear under high power. NOTE: On a prepared slide, sickled red blood cells appear crescent (sickled)-shaped.
6. Complete Table 8-2 by describing the shape of the sickled red blood cells.

Part C. Solving Problems Using Punnett Squares

1. Complete the following problems using the Punnett squares provided. Indicate in the blanks to the right the proportion of offspring with each expected phenotype.

    (a) Mother's genotype is *SS'* and father's genotype is *SS'*.

    ___ normal

    ___ sickle-cell trait

    ___ sickle-cell anemia

    (b) Mother's genotype is *SS'* and father's genotype is *SS*.

    ___ normal

    ___ sickle-cell trait

    ___ sickle-cell anemia

    (c) Mother's genotype is *SS* and father's genotype is *S'S'*.

    ___ normal

    ___ sickle-cell trait

    ___ sickle-cell anemia

Data and Observations:

TABLE 8-2. OBSERVATION OF RED BLOOD CELLS		
	Appearance	Description
Normal red blood cells		
Sickled red blood cells		

Questions and Conclusion:

1. Explain why sickle-cell trait and sickle-cell anemia illustrate the inheritance pattern known as incomplete dominance.
2. (a) Where is hemoglobin found?
   (b) What is the job of hemoglobin?
   (c) Explain why abnormal, sickled hemoglobin is a problem.
3. Explain why sickle-cell trait is less severe than sickle-cell anemia.
4. How might it be possible to check one's phenotype for normal red cells, sickle-cell trait, or sickle-cell anemia?
5. Once the phenotype is determined for one's hemoglobin type, is the genotype also determined? Explain.
6. Assume that you are a genetic counselor as you complete the following problems. Show the parents how you have arrived at your answers by using Punnett squares.
   (a) A mother has normal red blood cells and the father has all sickled red blood cells. What are the chances of their children being born with sickle-cell trait? Sickle-cell anemia?
   (b) A mother has sickle-cell trait. The father has sickle-cell anemia. What are the chances of their children being born with sickle-cell trait? Sickle-cell anemia?

Conclusion: How does a genetic disease affect red blood cells?

FIGURE 8–19. A cross of parents with normal blood clotting shows the probability of hemophiliac offspring if the mother carries the hemophilia gene. A female hemophiliac is rare because she must receive a hemophilia gene from each parent. Why is the probability of this occurrence low?

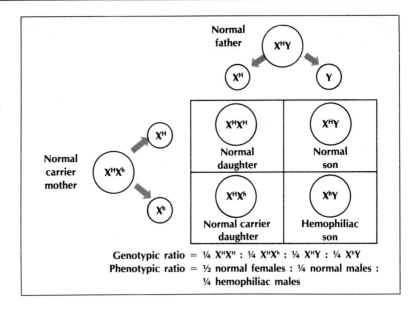

Genotypic ratio = ¼ X^H X^H : ¼ X^H X^h : ¼ X^H Y : ¼ X^h Y
Phenotypic ratio = ½ normal females : ¼ normal males :
¼ hemophiliac males

FIGURE 8–20. (a) Charts are used to detect red-green color blindness. Persons with normal vision can see a number in the top chart. What is the number? (b) To someone who is red-green colorblind, the chart appears as tones of gray.

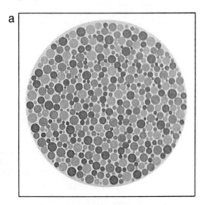

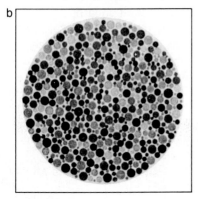

8:10 Sex-linked Diseases

Sex-linked diseases are caused by genes carried on the X chromosome. The Y chromosome carries no corresponding allele. Therefore, these diseases will appear more often in males. These diseases, just as those discussed earlier, are caused by abnormal alleles.

● **Hemophilia** (hee muh FIHL ee uh). In persons with hemophilia, the blood will not clot properly. The disease is the result of a sex-linked recessive gene.

If a male receives the recessive allele, his blood will not clot properly. Having the recessive gene results in the failure to make a certain protein. The protein, known as a clotting factor, is needed for one particular reaction in a series of reactions leading to formation of a blood clot.

Although there is no cure for hemophilia, there are treatments. One treatment consists of blood transfusions to restore lost blood. Another involves giving the hemophiliac the clotting factor. New techniques promise to make the clotting factor available more readily and more cheaply.

● **Red-green Color Blindness.** Some humans do not see red or green. Instead, anything that is red or green appears gray. These people have red-green color blindness, a sex-linked trait. The genes for color vision are located on X chromosomes. The recessive allele is responsible for the color blindness. As with other sex-linked traits, a male is more likely to have the problem. Females are often carriers of the trait.

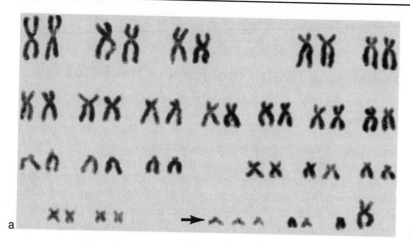

FIGURE 8–21. (a) This karyotype shows the presence of three number 21 chromosomes (arrow) that result in the Down syndrome phenotype. (b) A teen with Down syndrome helps his older brother restore a classic car.

8:11 Chromosome Problems

- **Down syndrome.** Nondisjunction (Section 8:4) can cause a variety of human defects. One cause of a disease called Down syndrome is the nondisjunction of certain autosomes. Cells of a person with Down syndrome caused in this way have one extra chromosome. Recall that when chromosomes are identified, they are numbered. A Down syndrome cell has three number 21 chromosomes.

Symptoms of this disease include mental retardation, abnormal facial traits, short arms and legs, and internal defects. At present, there is no treatment for the disorder.

Defects can occur from chromosome abnormalities other than an incorrect chromosome number. Sometimes, parts of an entire chromosome can become attached to other chromosomes. Other cases of Down syndrome result from this process. Also, a fragment of a chromosome may be missing in some cases.

Down syndrome may be caused by nondisjunction.

- **Turner syndrome** results from nondisjunction of the sex chromosomes. The person has one X chromosome and no Y chromosome. Because there is no Y chromosome, the person is female. However, she fails to develop normal female characteristics and is sterile. A person with Turner syndrome is usually short and sometimes has below average intelligence.

In Turner syndrome, a female has one X chromosome and no other sex chromosomes.

- **Klinefelter syndrome** results from an XXY chromosome pattern. Individuals with this disease are males. They fail to develop normal sex characteristics and usually are sterile. They also have very long arms and legs and below average intelligence.

A person with Klinefelter syndrome is a male with two X chromosomes and one Y chromosome.

REVIEWING YOUR IDEAS

9. In general, what two factors cause genetic diseases?
10. How can nondisjunction cause genetic diseases?

Diagnosing Health Problems in the Fetus

More than 200 000 babies each year in the United States are born with genetic disorders or congenital disabilities. Although many of these disorders are minor, others cause severe health problems. Some cause death. Several techniques allow physicians to detect dangerous problems in the fetus (unborn child).

Amniocentesis (am nee oh sen TEE sus) is a process in which a sample of fluid surrounding the fetus is withdrawn through a long, thin needle. The fluid can be analyzed for the presence or absence of certain chemicals that indicate genetic disease. Examples of diseases detected this way are hemophilia, some forms of muscular dystrophy, Tay-Sachs, and sickle-cell anemia.

Some diseases are detected by studying fetal cells found in the fluid withdrawn during amniocentesis. The chromosomes in these fetal cells can be studied. Recall that photographed and paired chromosomes make up a karyotype. Karyotypes can reveal if the fetus has certain diseases, such as Down syndrome. Other examples of diseases that show up in karyotypes are Turner and Klinefelter syndromes. (Sex of the fetus may also be determined this way.)

Amniocentesis can be used to determine if a fetus has certain genetic diseases.

FIGURE 8–22. In amniocentesis, fluid from the amnion is withdrawn. The fluid contains some of the unborn baby's cells. The cells are grown and their chromosomes are analyzed. Certain diseases can be determined from this procedure.

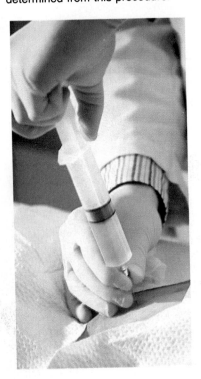

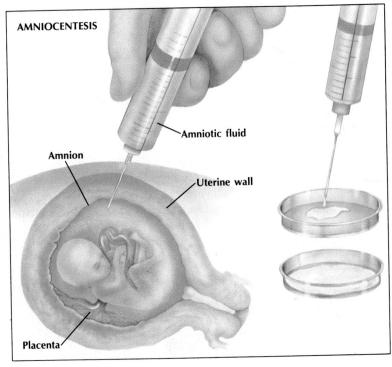

AMNIOCENTESIS

Amniotic fluid

Amnion

Uterine wall

Placenta

A technique called **ultrasonography** (ul truh suh NAHG ruh fee) is sometimes used to determine the position and anatomy of a fetus. An ultrasound probe is passed back and forth over the mother's abdomen. The probe emits high frequency sound waves that "echo" from the tissues of the fetus. Different tissues generate different wave lengths that, when put together, form an image on a screen. These images, called echograms, allow physicians to spot abnormalities, such as certain forms of heart disease. Ultrasound is also used prior to amniocentesis. Its use to determine fetal position reduces the risk of causing harm to the fetus when the needle is inserted into the mother's abdomen.

Fetoscopy (fee TAHS kuh pee) allows direct observation of the fetus and surrounding tissues. After determining the position of the fetus by means of ultrasonography, a device called an **endoscope** is used. The endoscope is inserted through a small incision in the mother's abdomen. The fetus is viewed directly through the endoscope tube. An image of the fetus is seen on a screen, too. Using other tools inserted into the endoscope, small samples of skin or blood may be withdrawn for study. These tissues can be helpful in revealing abnormalities.

For the most part, the techniques described here are used to determine whether a disease exists. They cannot cure diseases. In some cases, however, these techniques allow physicians to be prepared for treatment or surgery that can take place as soon as the infant is born.

Certain health problems, though, can be treated before birth. Fetoscopy has been used to give a transfusion to a fetus. It has also been used to remove excess fluid from around the brain of a fetus. Recently, an operation was done that temporarily solved a problem in the urinary system of a fetus. The outlook for fetal medicine is bright. It is hoped that a greater number of diseases will be detectable and that as a consequence, more individuals will benefit from early medical treatment.

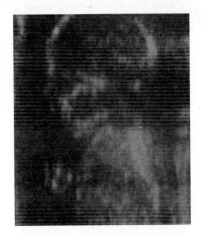

FIGURE 8–23. This echogram shows the head and shoulders of a fetus. Sound waves emitted from the probe at the mother's abdomen "echo" back from the fetus resulting in the production of the image on the screen.

In some cases fetoscopy can be used to treat a fetus.

FIGURE 8–24. (a) Fetoscopy is a procedure of inserting an endoscope into the abdomen to examine or medically treat the fetus. (b) This picture shows the hands of a 9-week-old fetus as seen through an endoscope.

a

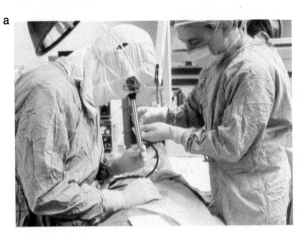

b

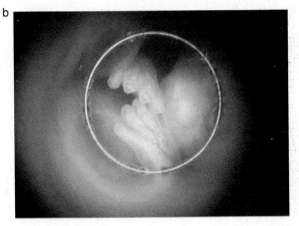

CHAPTER REVIEW

SUMMARY

1. The movements of chromosomes during the process of meiosis correspond to Mendel's Laws. Alleles are carried on homologous chromosomes. **p. 147**
2. In fruit flies and in humans, females normally have two X chromosomes. Males have an X chromosome and a Y chromosome. **8:1**
3. Thomas Hunt Morgan discovered that certain traits are sex-linked. Alleles for these traits are carried on the X chromosomes. **8:2**
4. Problems involving sex-linked traits can be solved using the same methods as for other types of genetics problems. **8:3**
5. Bridges' discovery of nondisjunction in *Drosophila* confirmed that genes are located on chromosomes—the chromosome theory of heredity. **8:4**
6. Genes located on the same chromosome are linked. Gene linkage can reduce genetic recombination and variety among offspring. **8:5**
7. Crossing-over is an exchange of corresponding segments of chromosomal material between two strands of a tetrad. Crossing-over tends to increase variety among offspring by recombining linked genes. **8:6**
8. Certain traits show continuous variation. Multiple genes probably control the expression of these traits. **8:7**
9. How genes are expressed depends on genetic makeup and the environment. **8:8**
10. Some genetic diseases are caused by the presence of certain alleles. **8:9**
11. Certain genetic diseases are sex-linked because the genes for the traits are present on the X chromosome. **8:10**
12. Nondisjunction is the cause of some human genetic diseases. **8:11**
13. Amniocentesis, ultrasonography, and fetoscopy are methods used to detect human genetic diseases in a fetus. **p. 166**

LANGUAGE OF BIOLOGY

amniocentesis
autosomes
chromosome theory of
  heredity
continuous variation
crossing-over
endoscope
fetoscopy
gene linkage
genetic map
hemophilia

lethal gene
modifier genes
multiple genes
nondisjunction
recombination
  gametes
sex chromosomes
sex-linked
  characteristic
ultrasonography

CHECKING YOUR IDEAS

On a separate paper, complete each of the following statements with the missing term(s). Do not write in this book.

1. Continuous variation is a pattern of inheritance controlled by _____.
2. _____ are located on chromosomes.
3. _____ is the failure of chromosomes to separate properly during meiosis.
4. A couple has three sons. The chance that their next child will be a boy is _____.
5. People with sickle-cell anemia have sickle-shaped red blood cells, which are not able to carry enough _____.
6. If a color-blind man marries a woman heterozygous for normal color vision, _____ of their offspring are expected to be color-blind.
7. Genes located on the same chromosome are said to be _____.
8. In order for a white-eyed female fruit fly to be produced, both parents must carry a(n) _____ gene.
9. A human with an XXXY chromosome pattern would be a _____.
10. Linked genes may be separated by _____, which takes place during meiosis.

EVALUATING YOUR IDEAS

1. Did observation of meiosis verify that genes are carried on chromosomes? Explain.
2. How many pairs of autosomes does a human cell contain?
3. How did Morgan test his hypothesis about sex-linked characteristics?
4. Did the results of Morgan's test (Question 3) verify that genes are carried on chromosomes? Explain.
5. How is nondisjunction explained?
6. State the chromosome theory of heredity.
7. In a certain organism, one chromatid contains the genes A and b. Its homologous chromatid contains a and B. What combinations of genes would be found in gametes of this organism if no crossing-over occurs? If crossing-over does occur?
8. Several plants of the same kind vary in height from 15 to 30 cm. How can this variation be explained?
9. What are modifier genes? How do such genes affect a trait such as eye color in humans?
10. Explain this statement: A particular phenotype depends upon all of an organism's genes and the organism's environment.
11. Explain how the cells of a human might contain two X chromosomes and one Y chromosome. What sex would the person be?
12. List several factors that cause human genetic diseases. Give several examples of diseases caused by each factor.
13. Discuss several techniques that are used to detect genetic diseases in human fetuses.

APPLYING YOUR IDEAS

1. In *Drosophila,* the gene for red eyes, R, is dominant to the gene for white eyes, r. This trait is sex-linked. Determine the probable genotypic and phenotypic ratios expected from a cross between (a) a heterozygous female and a red-eyed male, and (b) a heterozygous female and a white-eyed male.
2. In humans, the gene for normal blood clotting, H, is dominant to the gene for hemophilia, h. The trait is sex-linked. A woman and a man, both with normal blood clotting, have a normal son, a hemophiliac son, and two normal daughters. What is the probable genotype of each family member?
3. In a certain plant, tall, T, is dominant to short, t. Red flowers, R, are dominant to white flowers, r. A gardener crosses a $TtRr$ plant with a $ttrr$ plant. Seeds from this cross produce 52 tall, red plants and 48 short, white plants. Explain these results.
4. Alleles $A, B,$ and C are known to be linked. A crosses over with B ten percent of the time. A crosses over with C twenty percent of the time. Genes B and C cross over ten percent of the time. What is the sequence of the genes on the chromosome?

EXTENDING YOUR IDEAS

1. Determine the heights of a large number of your classmates. Draw a graph comparing the number of people versus heights. What genetic pattern does the graph suggest?
2. Conduct an experiment with *Drosophila* to show the inheritance of linked genes and the concept of crossing-over.

SUGGESTED READINGS

Anderson, W. French, "Beating Nature's Odds." *Science 85,* Nov., 1985.

Brennan, James N., *Patterns of Human Heredity: an Introduction to Human Heredity.* Englewood Cliffs, NJ, Prentice-Hall, 1985.

Grady, Denise, "The Ticking of a Time Bomb in the Genes." *Discover,* June, 1987.

THE GENETIC CODE

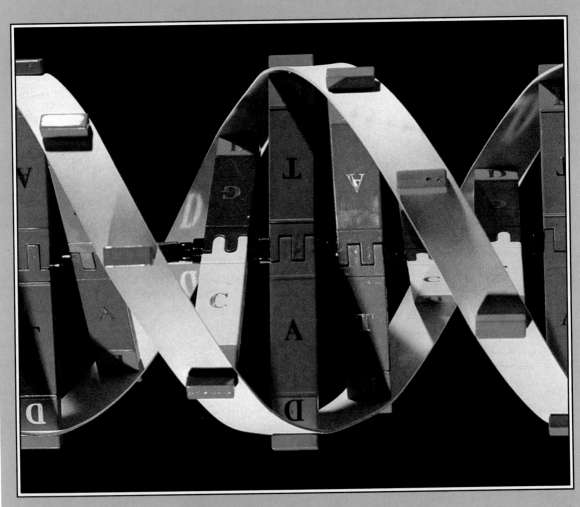

Generation after generation, genes are replicated and passed on to offspring, and the gene products result in certain traits. How is genetic information passed on in such exact form? A chemical called DNA is responsible. Shown here is a model of DNA. DNA directs replication of cells and functions as the chemical of heredity. How does DNA control heredity?

chapter 9

Handwritten margin note (top right):
Beadle + Tatum
" one gene one enzyme
hypothesis"
(polypeptide)

Mendel discovered basic laws about the passage of traits from parents to offspring. He presumed that an organism's features are determined by characters, or genes. Later work showed that genes are carried on chromosomes. Genes are replicated and distributed to new nuclei by mitosis and meiosis. Fertilization produces a new organism having a particular combination of genes. The combination of genes in an organism's cells controls that organism's characteristics.

Objectives:
You will
- trace the discovery of DNA as the genetic material.
- analyze the relationship between DNA and protein synthesis.
- discuss gene expression and its control.

STRUCTURE OF DNA

Once biologists learned that genes control heredity, they began asking other questions. What is a gene? Of what chemicals is a gene composed? Exactly how are genes replicated? How do combinations of genes (genotypes) result in particular traits (phenotypes)?

9:1 Bacterial Transformation

Experiments with the bacterium *Pneumococcus* (new muh KAHK us) helped answer some of these new and challenging questions. Several strains or varieties of this bacterium are known. (Strains differ genetically.) One strain consists of cells enclosed within a jellylike outer layer, or capsule. This strain is referred to as "smooth." Another strain has cells not enclosed by a capsule. It is referred to as "rough." Smooth cells cause the disease pneumonia, but rough cells do not.

Handwritten margin note (right):
(fungas)
Neurospora - mol
gene A → ornithene →
gene B → citrulline →
gene C → arginine

If any one gene

Smooth and rough are forms of *Pneumococcus* controlled by different genes.

Handwritten margin note (bottom):
changes you can't
get the sequence

171

FIGURE 9–1. (a) Injected smooth cells kill mice; rough cells or dead smooth cells do not. (b) When dead smooth cells and rough cells are mixed, bacterial transformation occurs. The injected mixture kills mice.

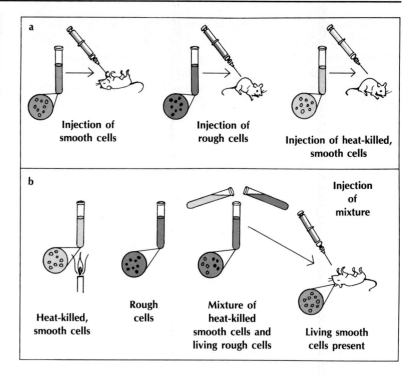

a

Injection of
smooth cells

Injection of
rough cells

Injection of heat-killed,
smooth cells

b

Injection
of
mixture

Heat-killed,
smooth cells

Rough
cells

Mixture of
heat-killed
smooth cells and
living rough cells

Living smooth
cells present

In 1928, Fred Griffith injected a mixture of heat-killed, smooth cells and living, rough cells into healthy mice. Griffith expected the mice to remain healthy because rough cells do not cause pneumonia and because the smooth cells were dead. However, some of the mice died of pneumonia. On examining the blood of the dead mice, Griffith made an unexpected observation — living, smooth *Pneumococcus* cells were present (Figure 9–1). Had the dead cells returned to life? Of course not.

There was one reasonable explanation — the dead, smooth cells must have directed the rough cells to become smooth. An inherited trait of the rough cells must have changed. This kind of change in a bacterial trait is called a **bacterial transformation.**

Bacterial transformation is one way in which a hereditary trait is changed in bacteria.

How could a trait change as it had done in these experiments? Scientists suspected that a chemical in the dead, smooth cells caused the change. They thought the chemical from the smooth cells entered the rough cells, causing them to become smooth. If a chemical had caused the change, then that chemical alone might cause a change in heredity.

It was assumed that a chemical was responsible for the bacterial transformation.

To test this hypothesis, colonies of smooth cells were grown on a nutrient medium in culture dishes. When enough colonies had formed, they were ground up to release the chemicals from the cells. The chemicals were removed, or extracted, from the cells. The solution of the removed chemicals was called an **extract.** It was thought that a chemical in the extract would cause transformation.

The extract from the smooth bacteria was added to the culture medium in each plate of another set of culture dishes. These plates were then inoculated with rough cells only. After many colonies had formed, cells were isolated for study. It was observed that many of the cells were smooth. Also, the new smooth cells produced more smooth cells. When injected into healthy mice, the new smooth cells caused pneumonia (Figure 9–2). Rough cells on a medium to which no extract was added showed no change. They did not cause pneumonia when injected into mice.

The extract had changed the inherited traits of the rough cells just as the hypothesis had predicted. It was believed that a chemical in the extract was involved in heredity. It had changed the form of one strain and had affected the cells' offspring. The chemical in the extract was called the **transforming principle.**

The transforming principle was not easy to identify. But in 1944, it was identified as deoxyribonucleic acid (DNA) (Section 3:15). DNA was discovered in 1871, but it was not thought to be involved in heredity until this work was done.

9:2 A Model

Other research supported DNA as the chemical of heredity. Biochemists found that DNA is a complex molecule composed of three smaller parts. The parts are a sugar (deoxyribose), phosphate groups, and nitrogen-containing compounds called bases. The three parts combine to form subunits called **nucleotides** (NEW klee uh tidez). Many nucleotides joined together form a DNA molecule.

In DNA, there are four different nitrogen bases. They are **adenine** (AD un een), **guanine** (GWAHN een), **thymine** (THI meen), and **cytosine** (SITE uh seen). In Figure 9–3, the chemical structure of each base is shown. Notice the colored shapes. Each shape refers to that specific base in the figures that follow. You will easily be able to tell what chemicals are being shown in the drawings if you keep in mind what the shapes represent as you examine the other figures in this chapter.

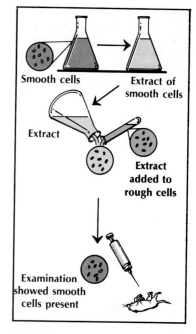

FIGURE 9–2. A chemical in the extract prepared from smooth bacteria caused some rough cells to become smooth.

DNA contains the nitrogen bases adenine, guanine, cytosine, and thymine.

FIGURE 9–3. The four bases of DNA all contain nitrogen. They are shown with colored shapes that will be used throughout the chapter to represent these chemicals.

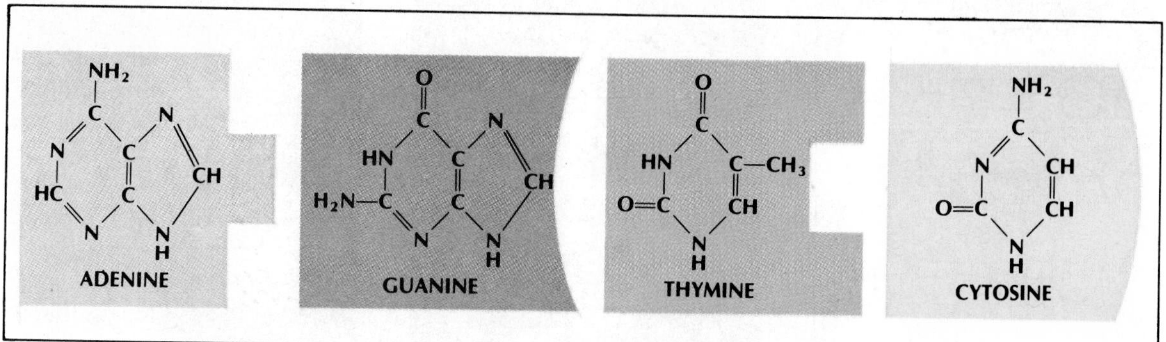

ADENINE

GUANINE

THYMINE

CYTOSINE

FIGURE 9–4. The structural formula of an adenine nucleotide shows a molecule of deoxyribose, a phosphate group, and adenine. They are shown with colored shapes that will be used throughout the chapter to represent these chemicals.

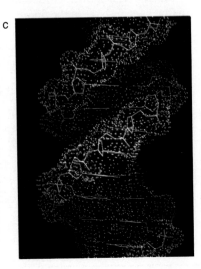

PHOSPHATE

ADENINE

DEOXYRIBOSE

A nucleotide is named for the base that it contains.

Watson and Crick's model of DNA provided an explanation of how DNA replicates and how it codes for traits.

FIGURE 9–5. (a) In the ladder model of DNA structure, the uprights are the sugar and phosphate molecules. The rungs are the matched nitrogen base pairs. (b) In a 3-D model with colored balls for atoms, the twist of the DNA ladder can be seen. (c) A computer drawing of DNA also shows the twist in the ladder.

Each nucleotide is named for the base it contains. (An adenine nucleotide contains adenine.) Figure 9–4 shows the arrangement of the three parts of a nucleotide. Note again the colored shapes that will be used in the figures to follow.

Scientists knew that it was important to learn how the nucleotides are arranged to form a complete DNA molecule. In 1953, James Watson and Francis Crick presented a model of DNA. The model was based mainly on their own work and that of Maurice Wilkins. Their model seemed to provide a basis for explaining two important properties of DNA. The model suggested how DNA replicates and how it carries the code for inherited traits.

The Watson and Crick model shows that DNA is built like a ladder (Figure 9–5) made of two chains of nucleotides joined together. The "uprights" of the ladder are composed of the sugar and phosphate parts of the nucleotides. The "rungs" are made up of the nitrogen

a

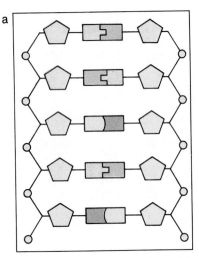

b

c

bases of the nucleotides. Each rung contains two bases joined together by weak chemical bonds. *Only certain bases can bond with one another. Adenine nucleotides always bond with thymine nucleotides and guanine nucleotides always bond with cytosine nucleotides.* Notice that the chains of nucleotides are twisted around each other to form a double spiral or double helix (HEE lihks). Watson and Crick included this twisting of DNA as part of their model because the helix is the most stable form possible.

In DNA, adenine nucleotides join with thymine nucleotides, and guanine nucleotides join with cytosine nucleotides.

DNA is in the form of a double helix.

9:3 Replication of DNA

Genes are made of DNA and are located on chromosomes. In eukaryotes, chromosomes replicate during interphase of the cell cycle (Section 6:3). As chromosomes replicate, DNA is replicated. Watson and Crick's model suggested how DNA replication (duplication) might occur. It was logical that each original chain of nucleotides in DNA act as a template (mold) for making a new chain. Thus, each new DNA molecule consists of one original chain and one new chain. Experiments later confirmed this idea.

The specific pairing of nitrogen bases is of great importance during replication. Before replication can begin, an enzyme "unzips" the double helix by breaking the weak bonds holding the two chains of nucleotides together. This "unzipping" begins at certain spots along

During DNA replication, each parent chain of nucleotides acts as a template for the synthesis of a new chain.

FIGURE 9–6. Exact duplication of a DNA molecule involves specific base pairing. The two resulting molecules are identical.

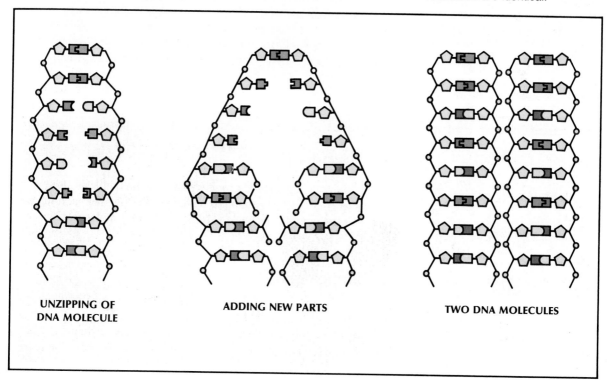

UNZIPPING OF
DNA MOLECULE

ADDING NEW PARTS

TWO DNA MOLECULES

As the DNA molecule "unzips," new nucleotides join to the nucleotides of each parent chain, A with T and G with C.

the DNA molecule. Then, a complex enzyme binds to each of the separated parent chains as well as to the four nitrogen bases, which are constantly made in the nucleus. The enzyme moves along, placing new nucleotides alongside nucleotides of the parent chain. The enzyme then bonds the nucleotides together in the growing replicate chain. The nucleotides are added in a specific order due to the fact that adenine joins only with thymine and cytosine only with guanine. As a result, each new chain is a "mirror image" of one of the two parent chains. ATP energy is needed for many of these steps, including the final one — the joining of new and old chains to form two molecules. Each new DNA molecule consists of one parent chain and one new chain and is a replica of the parent molecule.

Most errors are detected and then corrected by special enzymes after DNA replication.

Sometimes mistakes occur during DNA replication. Though rare, such copying mistakes are serious because each cell must have exact copies of DNA to function properly. The number of mistakes that are actually passed on, though, is extremely small due to the ability of the cell to repair DNA. Special repair enzymes "proofread" the newly made DNA for errors. They remove the incorrect nucleotides. Other enzymes then insert the correct ones. DNA repair reduces the chance of errors to about one in every billion nucleotides.

REVIEWING YOUR IDEAS

1. Define and identify the transforming principle.
2. What is a DNA nucleotide? Of what is it composed?
3. Explain the pairing of nucleotides in the DNA molecule.
4. Compare the DNA molecule to a ladder.
5. How is each newly made DNA molecule related to its parent molecule?

THE ROLE OF DNA

Mendel first suggested hereditary units (genes) in 1866. Almost a century later genes were found to be made of DNA. The next logical question was how do the genes, regions of DNA molecules, work? What does DNA do?

9:4 Genes and Proteins

Proteins are unique organic molecules composed of different combinations of twenty amino acids (Section 3:14). A protein consists of one or more polypeptides. Each polypeptide may contain hundreds of amino acids. The kind and sequence of amino acids make one polypeptide different from another. Each protein has a specific three-dimensional shape. The shape of a protein is related to its particular function in the cell.

Proteins play a variety of roles. They are parts of eukaryotic cell structures and some act as permeases (Section 4:7). Perhaps most important of all are enzymes (Section 5:3). Enzymes govern every cellular reaction including synthesis of other proteins.

Although all cells have much in common, each type is somewhat different. Each cell uses basically the same raw materials. What makes one type of cell different from another? The answer is the kinds of enzymes and other proteins it contains. Because cells have different sets of enzymes, they use their raw materials in different ways. Their chemical reactions vary; thus, they have unique features. The chemistry of an amoeba differs from that of a paramecium. Reactions in muscle cells differ from those in liver cells.

Just as types of cells differ, so too do organisms. Differences among organisms, even those of the same type, are due to differences in proteins. Unless you are an identical twin, you are unique. Your proteins are slightly different from those of all other organisms, including other humans.

Proteins are unique. Genes are unique. Might there be a relationship between genes and proteins? Perhaps the role of genes (DNA) is to guide the synthesis of proteins. If that is true, it would explain, for example, the difference between healthy people and those with sickle-cell anemia. Recall that they differ in terms of their hemoglobin molecules and that hemoglobin is a protein (Section 8:9). People with sickle-cell anemia have two recessive genes. Normal people have at least one dominant gene. It seems logical that different genes produce slightly different proteins, in this case hemoglobin.

Experimental evidence that genes code for proteins came in 1941, before the role of DNA was known. Research done at that time showed that genes control the production of enzymes. Since that time it has been learned that *each gene controls the synthesis of a particular polypeptide.* In the case of hemoglobin, for example, two genes are

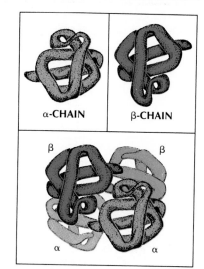

FIGURE 9–7. Hemoglobin is a blood protein made of two α chains and two β chains of amino acids. The chains are coiled and folded in a specific 3-D shape.

Each organism is unique because of its proteins.

Since both genes and proteins are unique, it is logical that genes may work by coding for proteins.

Experimental evidence shows that one gene controls the synthesis of a single polypeptide.

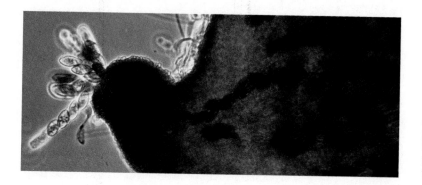

FIGURE 9–8. Initial experiments showing that genes function by producing proteins were done with the mold *Neurospora crassa*. The mold is shown here in a reproductive condition.

involved. Hemoglobin is composed of four polypeptide chains — two of one kind and two of another. Each gene controls the synthesis of one kind of polypeptide. After the polypeptides are made, they join to form hemoglobin.

9:5 DNA Code

DNA contains information for protein synthesis. Proteins are composed of amino acids. The sequence and kinds of amino acids make one protein different from another. In order for DNA (a gene) to control the synthesis of a protein, it must carry a code for putting particular amino acids together in a certain order. What part of DNA might carry the code? Only the sequence of base pairs along a DNA molecule varies. Experiments have shown that it is the sequence of bases along one of the two chains of a DNA molecule that carries the code.

The sequence of bases along one chain of DNA nucleotides carries the code for a particular protein.

There are twenty amino acids but only four kinds of bases in DNA. Thus, a single base cannot represent an amino acid. Likewise, a combination of two bases in a row cannot represent an amino acid. The four bases taken two at a time yield only sixteen possible combinations. Considering three bases at a time, however, gives more than the twenty combinations needed. In fact, it provides sixty-four possible code "words." Experiments have shown that three bases in a row do act as the code for amino acids. Each set of three bases representing an amino acid is known as a **codon** (KOH dahn). Because there are sixty-four possible codons and only twenty amino acids, some amino acids have more than one codon. Every codon along a

Three bases in a row, a codon, represent one amino acid.

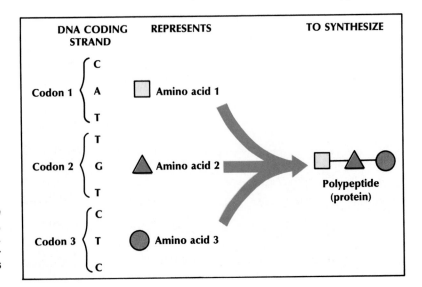

FIGURE 9–9. Each codon, three bases in a row along a DNA strand, represents a particular amino acid. The cell links the amino acids together in the same sequence as the codons to form a protein.

particular DNA chain represents one of the twenty amino acids. The cell interprets these codons, linking certain amino acids together to form a particular polypeptide (Figure 9–9).

DNA carries the code, but does not leave the nucleus during protein synthesis. Protein synthesis occurs on ribosomes (Section 4:12), which are located in the cell's cytoplasm. How, then, can the code be used to direct the making of proteins?

Another kind of nucleic acid, ribonucleic acid (RNA) (Section 3:15), works along with DNA in protein synthesis. There are different kinds of RNA in the nucleus, cytoplasm, and ribosomes. RNA is a chemical similar to DNA with a few exceptions. RNA contains ribose, rather than deoxyribose, as its sugar (Figure 9–10). Also, instead of the base thymine, T, it contains **uracil** (YOOR uh sihl), U. The other three bases — guanine, G; cytosine, C; and adenine, A — are the same as in DNA. RNA is usually a single-stranded molecule whereas DNA is double-stranded. What is the relationship among DNA, RNA, and protein synthesis? How is the DNA code passed from the nucleus to the ribosomes? How is the DNA code "read" to form proteins?

TABLE 9-1. AMINO ACIDS AND SOME DNA CODONS		
Amino Acid	Codons	
Alanine	CGA CGG CGT	
Arginine	GCA GCG GCT	
Asparagine	TTA TTG	
Aspartic acid	CTA CTG	
Cysteine	ACA ACG	
Glutamic acid	CTT CTC	
Glutamine	GTT GTC	
Glycine	CCA CCG CCT	
Histidine	GTA GTG	
Isoleucine	TAA TAG	
Leucine	AAT AAC GAA	
Lysine	TTT TTC	
Methionine	TAC	
Phenylalanine	AAA AAG	
Proline	GGA GGG GGT	
Serine	AGA AGC TCA	
Threonine	TGA TGG TGC	
Tryptophan	ACC	
Tyrosine	ATA	
Valine	CAA CAG	

9:6 Protein Synthesis: Transcription

Cells synthesize certain proteins as they are needed. When it is time to build a given protein, the gene for that protein is activated (Section 9:11). Near the gene itself is a region of DNA, the **promoter,** consisting of two sequences of bases. The promoter is recognized by a complex enzyme that binds to it. When the enzyme attaches to the promoter, a series of events much like those of DNA replication begins. The two chains of the DNA begin to "unzip" at a start signal located several bases away from the promoter. The start signal lies ahead of the codon for the first amino acid of the protein. Only one chain of the DNA molecule acts as a template for RNA nucleotides. RNA nucleotides match up with the DNA template until a stop signal

In an early step leading to protein synthesis, the two chains of DNA begin to separate. One of the chains acts as a template for linking RNA nucleotides together.

FIGURE 9–10. Uracil has a structural formula much like that of thymine. Deoxyribose is much like ribose. Uracil and ribose in RNA take the places of thymine and deoxyribose in DNA.

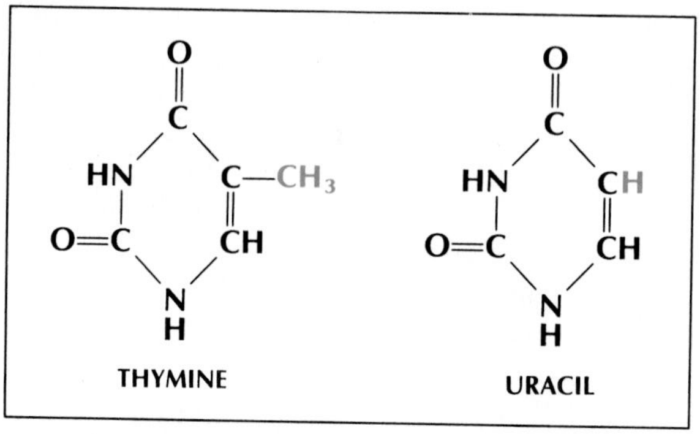

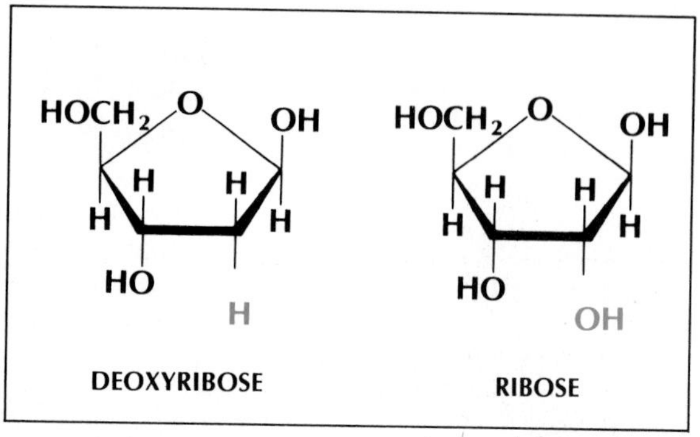

THYMINE   URACIL   DEOXYRIBOSE   RIBOSE

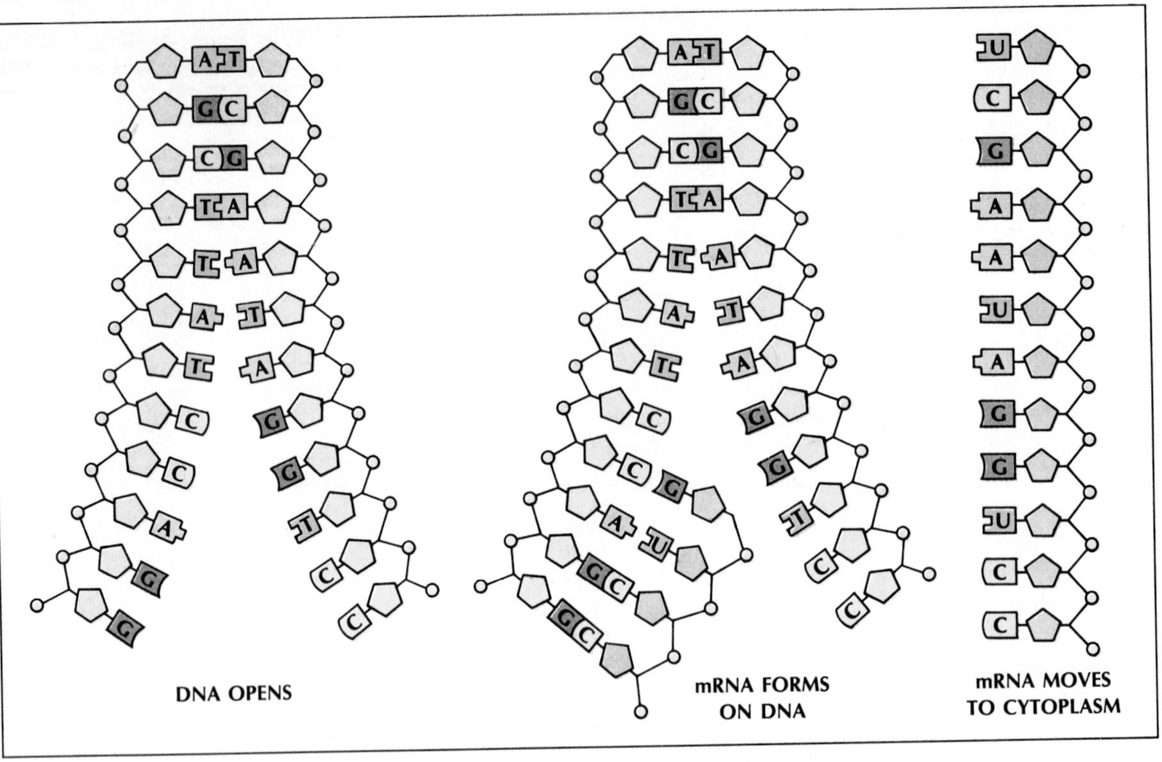

DNA OPENS

mRNA FORMS
ON DNA

mRNA MOVES
TO CYTOPLASM

FIGURE 9–11. Protein synthesis begins when a DNA molecule opens. mRNA is formed on the DNA and moves to the cell's cytoplasm. The process of forming the mRNA is transcription.

During transcription genetic information is transferred from DNA to mRNA. All forms of RNA are made by transcription.

is reached. The RNA nucleotides are bonded together, forming a single-stranded molecule of RNA that then breaks away from the DNA chain. The two DNA chains then rejoin. *The RNA molecule formed has a specific sequence of bases, which are determined by the DNA template.*

Making RNA from DNA is called **transcription.** The RNA molecule has a sequence of bases complementary to that of the DNA on which it was made. If a DNA chain has the sequence AGC TTA TCC AGG, the RNA made from it has the sequence UCG AAU AGG UCC. (Notice that in RNA, U takes the place of T.) RNA transcribed from such DNA is called **messenger RNA (mRNA).** Not all DNA, however, codes for proteins. For example, some DNA carries a code that directs the making of RNA associated with ribosomes. That RNA is called **ribosomal RNA (rRNA).** It is transcribed from DNA, just as mRNA is.

In prokaryotes, part of the mRNA molecule moves directly to the ribosomes even before the entire molecule has been completed. This event is possible because there is no nucleus. Assembly of amino acids into proteins begins right away.

In eukaryotes, however, the process is more complicated. As mRNA is produced, a "cap" and a "tail" are added at opposite ends (Figure 9–12). The cap consists of one molecule formed from guanine. The tail is made from 100-200 adenine molecules. Not only does

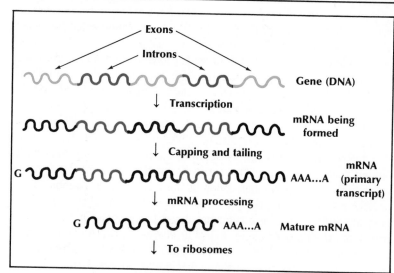

FIGURE 9–12. Splicing, capping, and tailing of mRNA are events that can be involved in gene expression in eukaryotes. These events occur in the cell nucleus.

mRNA have a cap and a tail, it also has many bases that do not code for amino acids and that must be removed. The sets of nitrogen bases to be removed are called **introns.** The other bases that code for amino acids are known as **exons.** The mRNA molecule is called the **primary transcript** at this stage.

In eukaryotes, a "cap" and "tail" are added to opposite ends of an mRNA molecule. At this point, mRNA is a primary transcript containing both exons and introns.

In order for the mRNA to become functional in protein synthesis the introns must be removed and the exons spliced together. These functions are controlled by other forms of RNA, along with special proteins and enzymes. The capping, tailing, and removal of introns from mRNA is known as **mRNA processing.** After processing is completed, the mRNA moves through pores in the nuclear membrane and travels to the ribosomes.

In mRNA processing, introns are removed and exons are spliced together. The mRNA can then take part in protein synthesis.

9:7 Protein Synthesis: Translation

In eukaryotes, processed mRNA attaches to a free ribosome in the cytoplasm and protein synthesis begins. In many cases protein synthesis is completed on free ribosomes. In other cases, however, the ribosome binds to rough ER (Section 4:12) and protein synthesis is completed there.

In eukaryotes, some proteins are synthesized entirely on free ribosomes. Other proteins are made on ribosomes that attach to rough endoplasmic reticulum.

Messenger RNA carries the code—a sequence of RNA codons—that represents a particular order of amino acids. To make the proper protein, the code must be deciphered. Synthesis of a protein from the code carried by an mRNA molecule is called **translation.** How is mRNA "language" translated to protein "language"?

Picture an mRNA molecule attached to a ribosome. In the cytoplasm is a pool of all twenty amino acids. If a protein is to be made, the amino acids must be carried to the ribosomes. Another type of RNA, **transfer RNA (tRNA),** functions to bring amino acids to the ribosomes. Like other RNAs, transfer RNA molecules are transcribed

Transfer RNA molecules transport amino acids from the cytoplasm to the ribosomes. Each tRNA is specific for a particular amino acid.

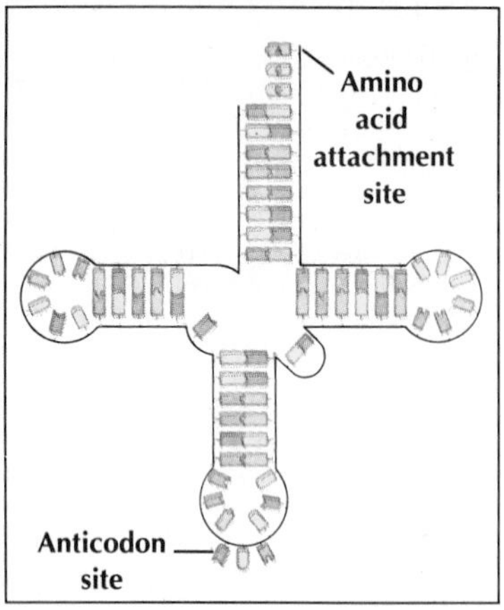

FIGURE 9–13. Transfer RNA is a cloverleaf-shaped molecule, a single strand of RNA. tRNA molecules have different anticodons. tRNA molecules join with certain amino acids and bring them to the site of protein synthesis.

Because tRNA anticodons match only certain mRNA codons, amino acids are linked together in the proper sequence.

In prokaryotes, translation may begin before transcription has been completed.

from certain regions of DNA. Each tRNA is a single chain of nucleotides folded into a clover-leaf shape (Figure 9–13). *Each tRNA combines with only one type of amino acid.* Enzymes and ATP are used to join the amino acid to one end of the tRNA molecule. At the other end of the tRNA is a set of three bases called an **anticodon.** Each type of tRNA has a different anticodon.

Correct translation depends upon proper joining of mRNA codons and tRNA anticodons. As translation begins, a ribosome moves along the mRNA strand. At the same time, a tRNA molecule carrying its amino acid approaches. The anticodon of the tRNA recognizes and joins with only a particular codon of the mRNA. For example, if the mRNA codon is AUG, the proper tRNA anticodon is UAC. The tRNA and amino acid complex remains joined to the mRNA codon. The ribosome then moves along, exposing the next mRNA codon. Again, the proper tRNA with its amino acid joins the mRNA. The new amino acid is linked to the previous amino acid by a peptide bond. The first tRNA is then freed to bring other amino acids to the ribosome. The process continues as the ribosome moves along the mRNA; thus, a polypeptide chain grows. At the end of the mRNA strand is one of three codons—UAG, UAA, or UGA—that stops the message. There are no tRNA molecules that have anticodons to match these codons. The completed polypeptide chain breaks away. It may function as is, or it may join with other polypeptides to form a more complex protein molecule.

Translation in prokaryotes is basically the same as the process described here for eukaryotes. Recall, however, that translation may begin even before transcription is completed. Another difference is that there is no ER in prokaryotes. All translation takes place on free ribosomes.

To summarize, DNA stores information for protein synthesis. The DNA code is transcribed into mRNA. Messenger RNA, tRNA, and ribosomes interact to translate the code and build proteins. The kinds of proteins synthesized make one kind of cell different from another and make each organism unique.

REVIEWING YOUR IDEAS

6. Genes (DNA) control the synthesis of what other kinds of chemicals?
7. What is a codon?
8. Compare the composition and structure of RNA to that of DNA.
9. Distinguish between transcription and translation.
10. In which parts of the cell do transcription and translation occur in eukaryotes?
11. Distinguish among mRNA, rRNA, and tRNA.

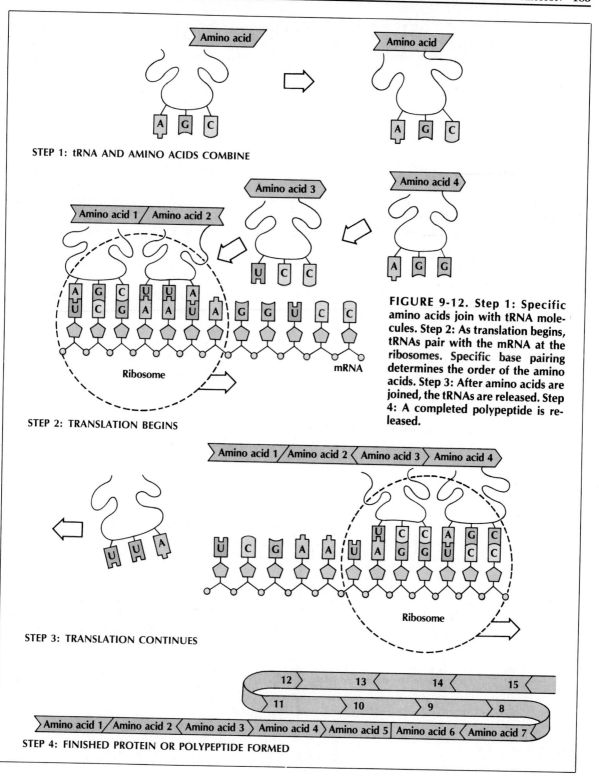

FIGURE 9-12. Step 1: Specific amino acids join with tRNA molecules. Step 2: As translation begins, tRNAs pair with the mRNA at the ribosomes. Specific base pairing determines the order of the amino acids. Step 3: After amino acids are joined, the tRNAs are released. Step 4: A completed polypeptide is released.

STEP 1: tRNA AND AMINO ACIDS COMBINE

STEP 2: TRANSLATION BEGINS

STEP 3: TRANSLATION CONTINUES

STEP 4: FINISHED PROTEIN OR POLYPEPTIDE FORMED

INVESTIGATION

Problem: What is a method for extracting DNA from cells?

Materials:

bacterial broth
graduated cylinder
250 mL flask
glass-marking pencil
lysozyme
toothpick

incubator
oven mitt
sodium lauryl sulfate
large, cold test tube
100 mL beaker
ethyl alcohol, cold

Procedure:

1. Label the flask with your name. Add 100 mL of bacterial broth to the flask.
2. Add to the flask as much lysozyme as will remain on the flat end of the toothpick.
3. Mix the bacterial broth and lysozyme by gently shaking the flask for about one minute.
4. Place the flask in a 37°C incubator for 30 minutes.
5. Using the oven mitt, remove the flask from the incubator. Add 4 mL of sodium lauryl sulfate to the flask. Gently shake the flask to mix the contents.
6. Place the flask in a 56°C incubator for 10 minutes.
7. Using an oven mitt, remove the flask from the incubator and cool by holding under cold, running tap water. See Figure 9–16 as a guide. Cool for at least 2 minutes.
8. Fill a large, cold test tube about ½ full with the bacterial broth from the cooled flask.
9. Fill a small beaker with about 20 mL of cold alcohol.
10. Hold the test tube at an angle. Slowly pour a small amount of the cold alcohol down the side of the test tube so that the alcohol forms a layer on top of the broth. Use Figure 9–15 as a guide. Continue to add alcohol until the tube is about ⅔ full.

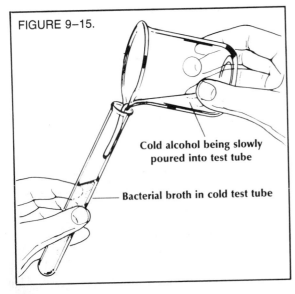

FIGURE 9–15.

Cold alcohol being slowly poured into test tube

Bacterial broth in cold test tube

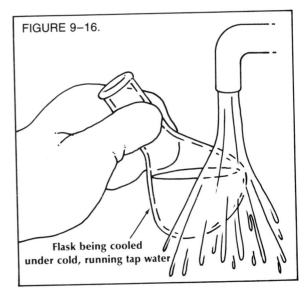

FIGURE 9–16.

Flask being cooled under cold, running tap water

11. Carefully examine the interface (point where broth and alcohol meet) in the tube. The small white strands you observe at the interface are DNA.
12. Copy Figure 9–17. Use the outline figure to diagram and label the contents of your tube.

Questions and Conclusion:

1. What is the role of DNA in living cells?
2. Describe the structural and chemical makeup of DNA.
3. What was the source of DNA in this lab?
4. What might have been the role of lysozyme (an enzyme) in this lab?
5. Would your results have been different if carrot, chicken, or mushroom cells had been used instead of bacterial cells? Explain.
6. How might the interface have appeared if you had used only frog cell cytoplasm?
7. How might the concentration of DNA have compared between two test tubes if human sperm were placed in one tube and equal numbers of human body cells were placed in another? Explain.
8. How might the concentration of DNA have compared between two test tubes if human sperm were placed in one tube and equal numbers of human eggs were placed in another? Explain.

Data and Observations:

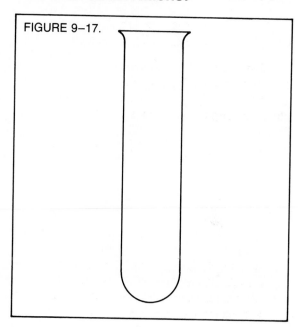

FIGURE 9–17.

9. Study the chart shown below.
   (a) Why is the percent adenine (A) always nearly equal to the percent thymine (T) in an organism?
   (b) Why is the percent cytosine (C) always nearly equal to the percent guanine (G) in an organism?

Conclusion: What is a method for extracting DNA from cells?

Organism	Percentage of Each Nitrogen Base Contained in DNA			
	A	T	G	C
human	30.2	29.8	19.9	19.8
wheat	27.3	27.1	22.7	22.8
bacteria	36.9	36.3	14.0	14.0

PEOPLE IN BIOLOGY

It was not until almost forty years after Barbara McClintock first discovered the "jumping gene" that the scientific community recognized its importance to genetic theory. McClintock began her research in the 1940s by studying the effects of crossing corn plants that had different kernel colors and textures. She observed kernel color patterns that could not be explained by the genetic principle popular at the time. This principle stated that genes were fixed structures laid out in a line. The results of McClintock's experiments led her to conclude that some genes, instead of being fixed, move around on a chromosome, affect each other's functionings, and cause massive mutations. Scientists are just now beginning to realize how important this discovery is and how "jumping genes" bring diversity to a species.

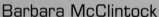

Barbara McClintock

(1902–   )

Because of the importance of her work, Dr. McClintock was awarded the Nobel Prize for Physiology and Medicine in 1983. Also, she has won the National Medal of Science and the Rosensteil Award. In 1980, she became a Prize Fellow Laureate of the MacArthur Foundation. She is still researching corn genetics at Cold Spring Harbor, Long Island, where she has been for the past forty years.

EXPRESSION OF GENES

Knowledge of DNA and protein synthesis has enabled biologists to understand more clearly how traits are passed from cell to cell and from parents to offspring. It forms a basis for explaining how genes interact to produce phenotypes and how mistakes sometimes occur in the transmission of the genetic code. It has also led to some comprehension of how genes are controlled.

9:8 Mutation

Mutations are mistakes in the genetic message resulting from changes in either chromosomes or genes.

The mechanisms that control the transfer of hereditary traits usually produce normal results. Sometimes, however, changes in the genetic code, or **mutations,** occur. Mutations can occur as a result of changes in either chromosomes or genes (DNA).

Chromosome mutations may occur during meiosis.

Chromosome mutations may occur in many ways. For example, parts of chromosomes can be dropped off or lost during crossing-over (Section 8:6) or at other times. Sometimes these "lost" parts rejoin the chromosome, but they may attach backwards or at the wrong end of the chromosome. Also, they may attach to the wrong chromosome. All these changes can result in abnormal information in the genetic code.

Changes in chromosome number also may occur. Because of non-disjunction (Section 8:4), gametes with extra or missing chromosomes can be formed. A chromosome may fail to attach to a spindle fiber, also resulting in abnormal gametes.

Many factors may cause gene mutations. Some mutations result from leaving out a nucleotide or adding an extra one during DNA replication. If a nucleotide is left out or added, the remaining codons read improperly.

The rate at which mutations occur varies. A recessive allele resulting from mutations causes hemophilia (Section 8:10). Such mutations occur about once for each 50 000 X chromosomes. It is estimated that in *Drosophila,* some gene mutations occur about once in 200 000 gametes. Some occur more often.

Radiation and high temperature are examples of external causes of mutations. Agents that cause mutations are called **mutagens** (MYEWT uh junz). Some chemicals such as formaldehyde (for MAL duh hide) are mutagens, too. Mutagens may break DNA bonds.

Regardless of their causes and rates, genetic mutations can be copied as if the code were unchanged. If they occur in a body cell of an organism, they are not important to the species as a whole. Why? But, mutations in the sex cells may affect an entire population of organisms because the information can be passed from one generation to the next. Will mutations be copied if they are very harmful? Explain.

Some mutations may have little effect or may even be helpful to the organism. Others may be harmful or lethal, resulting in diseases

Gene mutations often involve the adding or leaving out of a nucleotide during DNA replication.

FIGURE 9–18. (a) Albinism, shown by this white-tailed deer, may be due to a mutation. (b) The legs of the bassett hound are a mutant form of the normal longer legs of dogs. (c) Seedless fruit, such as grapefruit and grapes, are mutant forms of fruit.

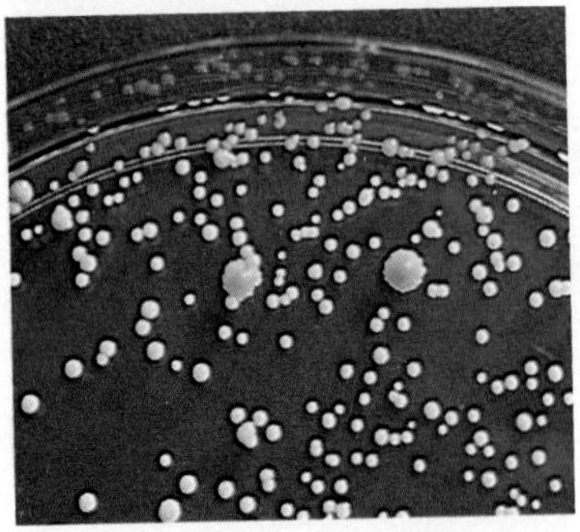

FIGURE 9–19. The petite mutation in yeast shows when the cells form small, white colonies instead of the normal, larger tan colonies. The petite mutation is a mutation in mitochondrial DNA.

such as sickle-cell anemia, PKU, and Tay-Sachs (Section 8:9). Non-lethal mutations may be important for the future by providing a "storehouse of variety" for a population.

9:9 DNA Outside the Nucleus

Not all the DNA of a eukaryotic cell is in the nucleus. Some DNA (a very small amount compared to nuclear DNA) is found in mitochondria and chloroplasts. This DNA is called **organelle DNA.** Organelle DNA is double-stranded, but is in the form of a circle and is not joined to protein as nuclear DNA is. Also, some codons of organelle DNA represent different amino acids than the codons of nuclear DNA.

Mitochondria and chloroplasts can synthesize a variety of materials. For example, DNA of a human mitochondrion makes some mRNA, tRNA, and rRNA, which become part of the mitochondrial ribosomes. It also has genes for thirteen proteins. Some of the enzymes used in cellular respiration are made within the mitochondria. Most of the proteins needed by organelles, however, are made under the direction of nuclear DNA.

9:10 DNA and Phenotypes

Knowledge of DNA and protein synthesis can explain why a certain genotype results in a given phenotype.

What you have learned about DNA and protein synthesis can explain some concepts of genetics. As an example, consider coat color in guinea pigs. Black fur *(B)* is dominant to albino *(b)*. Guinea pigs having the genotype *BB* or *Bb* will be black. Guinea pigs with the genotype *bb* will be albino. How can these patterns of inheritance be explained at the chemical level?

Genes consist of DNA that codes for specific proteins. Suppose the *B* allele (DNA) carries a code that directs the synthesis of a particular enzyme, an enzyme being a protein. That enzyme aids in the conversion of a substance in the cells of the guinea pig to melanin, a black pigment. The *b* allele (DNA) is a mutant form of the *B* allele. Its nucleotide sequence, and so its code, is slightly different. Because of the difference, the enzyme needed to make melanin is not produced properly. Thus, a guinea pig that is *BB* or *Bb* will be able to make melanin and will be black. But a *bb* guinea pig cannot make melanin. Its phenotype will be albino.

Phenotypes result from the combination of proteins produced by a given set of alleles.

Phenotypes are the results of which particular enzymes (or other proteins) are produced by certain combinations of alleles. All phenotypes have a chemical basis. An organism is a set of thousands of phenotypes, each of which is based on the synthesis of proteins. Not all phenotypes can be seen. The presence of some, such as type B blood or abnormal hemoglobin, can be determined in ways other than by *seeing* the phenotype.

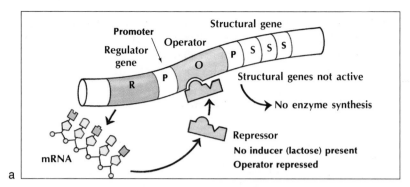

a

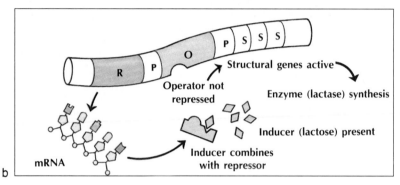

b

FIGURE 9–20. (a) When no inducer is present, a repressor molecule binds to the operator and prevents the synthesis of the enzyme.
(b) When an inducer is present,
the inducer binds with the repressor molecule. Thus, the operator is
no longer blocked, the structural
genes are functional, and enzyme
synthesis occurs.

9:11 Genetic Control in Bacteria

What is a gene? The study of protein synthesis provides one way
to define a gene — part of a DNA molecule that directs the synthesis of
a protein or polypeptide. This type of gene is called a **structural
gene.** You know that genes cannot all be making proteins all the time.
What controls when genes are active? What controls which proteins
are made within cells at certain times?

Much of the current knowledge about genes comes from research
with prokaryotic cells (Section 4:12). These cells have no true nuclei
but they do have DNA like the DNA of eukaryotic cells. Studies of
bacteria show that several types of genes exist for these cells. Some
genes control the synthesis of proteins by structural genes. Therefore, they are called **regulator genes.** Like structural genes, regulator genes synthesize mRNA. The mRNA in turn synthesizes special
proteins called **repressors.** Repressors determine whether or not a
structural gene will synthesize a protein.

Bacteria called *E. coli* thrive on a culture medium containing glucose, which is their normal energy source. Lactose, a disaccharide, is
not normally used by this species as an energy source. Therefore, *E.
coli* usually do not produce the enzyme lactase to break it down. If *E.
coli* are grown on a culture medium containing lactose instead of glucose, however, they will begin to produce lactase. Lactase enables

Structural genes direct the synthesis of proteins.

Regulator genes produce repressors, which control the activity of
structural genes.

the cells to convert lactose to monosaccharides that can be broken down and used as energy sources. This process in which a substrate (lactose) entering a cell causes the cell to begin producing an enzyme is called **enzyme induction.** The chemical that causes induction is an **inducer.** If lactose is removed, lactase production stops.

In enzyme induction, a substrate causes a cell to begin making the enzyme associated with it.

Enzyme induction depends on regulator genes and repressors. A repressor combines with a certain region of DNA called the **operator** when there is no lactose present. The operator is located next to the structural gene for lactase and two other structural genes also needed for lactose breakdown. An operator and its associated structural genes are known as an **operon.** The operator lies between the two sets of bases that make up the promoter (Section 9:6). When a repressor is bound to the operator, the enzyme that "unzips" DNA cannot bind to the promoter. Thus, transcription cannot take place. No mRNA can be formed and no lactase is produced.

In bacteria, the combination of an operator and its related structural genes is called an operon.

A repressor, when bound to the operator, blocks transcription so no enzymes can be synthesized.

When lactose is present in the medium, it enters the cell and combines with the repressor molecules. When that occurs, the operator is no longer blocked and transcription of the three structural genes can take place, thus leading to the formation of mRNA. Translation then follows, lactase is produced, and the cell can break down lactose (Figure 9–20).

When an inducer is present, the repressor joins with it rather than the operator, thus allowing transcription and translation to occur.

Enzyme induction is one means of genetic control in bacteria. What is the advantage of induction and other control systems? They result in the production of certain enzymes *only when they are necessary.* In this way, the cell conserves both materials and energy by not producing unnecessary enzymes.

9:12 Genetic Control in Eukaryotes

Discovery of gene regulation systems in bacteria led biologists to assume that similar means of control occur in eukaryotes. That has not proved to be the case. Gene control in eukaryotes is much more complex. Unlike bacterial DNA, the DNA of eukaryotes is associated with proteins on the chromosomes. The DNA is wound around spools of these proteins, called **histones,** and is highly coiled. It is the combination of histones and DNA that form chromatin (Section 4:14). Before transcription can begin in eukaryotes, the DNA must somehow be uncoiled and unwound from the histones. Not only that, but also there are many more chromosomes and genes in eukaryotes. In addition, the various steps of RNA processing must take place in eukaryotes before translation can begin (Section 9:6).

In eukaryotes, DNA is wound onto spools of histone proteins.

Eukaryotes have many more chromosomes and genes than prokaryotes do.

Regulation of gene expression may occur at several different levels in eukaryotes. (A gene is said to be expressed when the protein produced results in the phenotype.) However, the ways in which such

Control of gene expression occurs in a variety of ways and at different levels in eukaryotes.

control is carried out are not known. A variety of chemical and physical factors are thought to play a role.

One point of control is at the level of transcription, the level of control seen in bacteria. If a gene is "switched off" (made inactive), transcription cannot occur and no protein is made. If a gene is "switched on" (made active), transcription and translation then occur and the gene is expressed.

One possible means of transcriptional control in eukaryotes involves chromosomal proteins. When DNA is wound onto histone spools, it cannot be transcribed. Another type of chromosomal protein, the **nonhistones,** may function in gene regulation by causing DNA to unwind from the spools. This unwinding would serve to begin transcription, thus activating a gene. Nonhistones vary from one type of cell to another, among different organisms, and within a given cell at different times. Such variation suggests that these proteins play a functional role in gene control.

Another means of transcriptional control involves certain chemical regulators, hormones. Some kinds of hormones enter a cell, move into its nucleus, and "switch on" a gene (Section 28:8). How such a hormone activates a gene is not known.

Control of gene expression may occur after mRNA has been produced, at the post-transcriptional level. For example, some primary transcripts may be processed in a variety of ways depending on which nucleotides are removed and how the rest are spliced together. Each way of processing results in a different, final mRNA, and, thus, a different protein.

Expression of a gene also may depend upon changes in a protein after translation has occurred. For example, many proteins do not become functional until moved to a certain area within or outside the cell. Recall that carbohydrate markers play a role in moving some proteins to their destinations (Advances in Biology, Chapter 4). Other proteins must be activated to become functional. Some hormones and digestive enzymes cannot function until certain amino acids have been "snipped out."

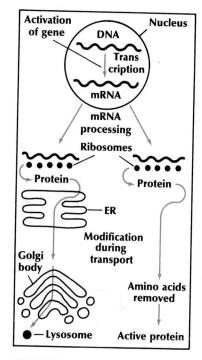

FIGURE 9–21. Genetic control in eukaryotes occurs at many levels between a gene and the expression of its product in a phenotype. These events include transport of mRNA out of the nucleus, RNA processing, protein movement to site of action, and addition of other chemicals.

Activation of proteins may be necessary before a gene is expressed.

REVIEWING YOUR IDEAS

12. What is a mutation?
13. Where is most of a cell's DNA located? Where else is it found?
14. How is DNA related to the expression of a phenotype?
15. Distinguish between a structural gene and a regulator gene in a bacterium.
16. List three points regarding genetic control in eukaryotes.

CHAPTER REVIEW

SUMMARY

1. Discovery of bacterial transformation led to knowledge that DNA is the material of which genes are composed. 9:1
2. Watson, Crick, and Wilkins developed the double helix model of DNA in 1953. They showed that DNA is composed of two chains of nucleotides joined together by weak bonds between specific nitrogen bases. 9:2
3. During DNA replication, each strand of DNA acts as a template for the formation of a new strand. 9:3
4. Genes carry codes for synthesis of proteins. In this way, genes determine features of different cells and organisms. 9:4
5. Each codon, three nucleotides of DNA in a row, represents an amino acid. 9:5
6. During transcription, one chain of DNA acts as a template for the manufacture of messenger RNA. The messenger RNA carries the DNA code to the ribosomes for protein synthesis. 9:6
7. During translation, transfer RNA molecules bring specific amino acids to the messenger RNA on the ribosomes. There, the amino acids are assembled in the proper sequence to form a polypeptide or protein. 9:7
8. Mutations, genetic changes, may occur in both genes and chromosomes. Mutations often are harmful or lethal. 9:8
9. Mitochondria and chloroplasts contain DNA that codes for several genes and for the various types of RNA. 9:9
10. Phenotypes result from the activity of specific proteins coded for by a given set of alleles. 9:10
11. One method of gene control in bacteria is enzyme induction. In induction, genes are activated only when the enzymes they code for are needed. 9:11
12. A variety of means of gene control in eukaryotes have been recognized. Expression of genes may be regulated at the transcriptional or post-transcriptional levels. *How* control takes place remains to be learned. 9:12

LANGUAGE OF BIOLOGY

anticodon
bacterial
 transformation
codon
enzyme induction
exon
inducer
intron
messenger RNA
mRNA processing
mutagen
mutation
nucleotide

operator
operon
organelle DNA
primary transcript
promoter
regulator gene
repressor
ribosomal RNA
structural gene
transcription
transfer RNA
transforming principle
translation

CHECKING YOUR IDEAS

On a separate paper, match each phrase from the left column with the proper term from the right column. Do not write in this book.

1. organelle where protein synthesis occurs
2. has an anticodon
3. genetic mistake
4. transcription occurs here in eukaryotes
5. represents an amino acid
6. contains nonnuclear DNA
7. carries DNA code to ribosomes
8. material of which genes are made
9. four kinds in DNA
10. later discovered to be DNA

a. codon
b. DNA
c. mitochondrion
d. mutation
e. mRNA
f. nucleotide
g. nucleus
h. ribosome
i. transforming principle
j. tRNA

EVALUATING YOUR IDEAS

1. How did Griffith explain that living, smooth *Pneumococcus* cells were in the blood of mice?
2. How was Griffith's hypothesis tested? What control was used?
3. What properties of DNA did Watson and Crick hope to explain by constructing a model of DNA? Did they succeed?
4. Sketch a segment of a molecule of DNA that is six base-pairs long. Use more sketches to show how this model segment of a DNA molecule replicates. When does DNA replicate?
5. How are genes related to differences among cells and organisms?
6. Why is it necessary that a codon be composed of at least three nucleotides instead of one or two?
7. Using the DNA model of question 4, assume that one strand of the segment served as a pattern for the formation of a segment of a molecule of mRNA. Sketch the segment of the mRNA molecule that would be formed. What information would be contained in this segment of mRNA?
8. How are rRNA and tRNA produced?
9. Describe the structure and function of tRNA.
10. By using diagrams, describe the translation phase of protein synthesis.
11. How are gene mutations caused? How can a mutation be passed from one generation to the next generation?
12. How can chromosome mutations occur?
13. Why are mutations that occur in a body cell of an organism not important to the entire species? How may mutations in the sex cells be important to the entire species?
14. Where is DNA located outside the nucleus? What are some functions of organelle DNA?
15. Explain enzyme induction in bacteria. How is it adaptive?
16. Explain several ways in which genes of eukaryotes are thought to be controlled.

APPLYING YOUR IDEAS

1. Suppose a gene contains the code for synthesis of an enzyme. During replication of the gene a mutation occurs so that one of the nucleotides is left out. Will the enzyme produced more likely be close to "normal" if the deleted base occurs near the beginning or the end of the gene? Explain.
2. In a certain flower, the genes red, R, and blue, R', show incomplete dominance. The heterozygous condition produces purple flowers. Using your new knowledge of DNA and protein synthesis, explain why:
   (a) the genotype RR produces red flowers.
   (b) $R'R'$ produces blue flowers.
   (c) RR' produces purple flowers.

EXTENDING YOUR IDEAS

1. Build a set of models to represent a DNA molecule, the mRNA made from it, several tRNA molecules, and amino acids. Use the model to illustrate the process of protein synthesis.
2. The experiments of Meselson and Stahl and of Taylor, Woods, and Hughes established that DNA replicates according to the Watson-Crick hypothesis. Both experiments made use of isotopes. Prepare a report on each of these experiments and explain how they verified the hypothesis.

SUGGESTED READINGS

Gerbi, Susan A., *From Genes to Proteins*. Carolina Biology Reader. Burlington, NC, Carolina Biological Supply Co., 1987.

Rensberger, Boyce, "New Genetic Codes Found." *Science Digest*, Aug., 1985.

CHANGES

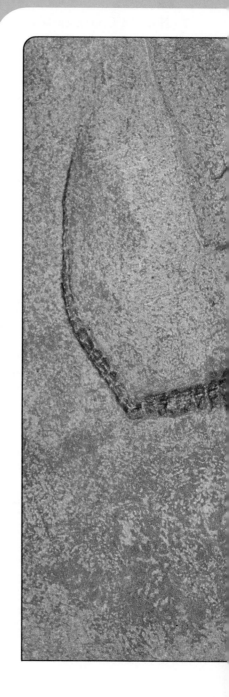

Nearly two million different kinds of organisms are known to live on Earth today. However, these same types of organisms have not always lived here. Over time environments have changed and with changing environments, the types of organisms have changed.

Organisms must adapt to their environments or they may become extinct. Shown here is a fossil of lariosaurus, an aquatic reptile that lived 225 to 280 million years ago. Lariosauruses were once well adapted to their environment but after their environment changed, they became extinct. Information about them comes from study of their fossils.

How did life originate? What has caused the changes among some organisms and the extinction of others? What information do we have about organisms no longer living? Biologists attempt to answer these and related questions using scientific methods. In Unit 3, you will study some of the evidence that has been gathered and some of the conclusions that have been made. You also will study how organisms are classified.

Fossil Lariosaurus

CHANGE WITH TIME

Fossils are remains or evidence of organisms that were once alive long ago. They can be found in almost every part of the world. Special techniques are used to dig up and preserve these remains for study. Here, workers in South Dakota are digging up the remains of mammoths. Mammoths are animals that lived several thousand years ago and resemble elephants. How is the age of a fossil determined? What kinds of information are obtained from fossils?

arth is thought to be about 4.5 billion years old. Biologists estimate that life first appeared on Earth more than three billion years ago. Both Earth and the organisms inhabiting it have changed during that time. The change in organisms over a period of time is called **evolution.**

Objectives:
You will
- describe the evidence that shows evolution occurs.
- explain how evolution is thought to occur.
- give a hypothesis that explains the origin of living things.

EVOLUTION: EVIDENCE

Different explanations have been given to explain *how* evolution occurs. Two of these explanations will be studied later in this chapter. First, it is helpful to study evidence indicating that organisms change.

Evolution is the change in organisms over time.

10:1 Fossils: Formation and Dating

Any part or trace of an organism that lived long ago is called a **fossil.** A fossil may be all, part, or an imprint of an organism. Some entire organisms have been found frozen in ice or enclosed in amber. But this type of fossil is rare. When most organisms die, they are quickly decomposed (Section 1:4), so no record of their life is left. In order for a fossil to form, something must happen to prevent the organism or part of it from decaying. A hard part of an organism, such as a bone or a tooth, may be preserved if it is surrounded and compressed by clay or sand soon after the organism dies. The surrounding sediments prevent decomposition. Then, as the sediments turn to rock over long periods of time, the part of the organism is preserved.

Fossils formed in sedimentary rock are the most common fossils. Sedimentary rocks form by the settling of particles, until layers of the particles are built up. When the layers are undisturbed, the lower

FIGURE 10–1. The impression of a fern leaf in rock is a fossil formed when the leaf is pressed into mud that hardens. The leaf decomposes, leaving the impression.

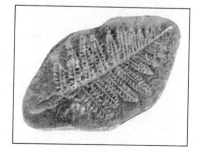

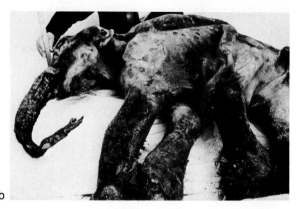

a

b

FIGURE 10–2. Parts of animals and whole animals have been found preserved. (a) Bones of dinosaurs and other animals have been uncovered in excavations at Dinosaur National Monument, Utah. (b) This baby mammoth was unearthed from frozen ground near Magadan in Russia.

Half-life is the time necessary for one half of a radioactive sample to disintegrate.

FIGURE 10–3. (a) Undisturbed sedimentary rock shows flat layers of deposits. Fossils found in lower layers of undisturbed sedimentary rock are older than fossils found in top layers. (b) When sedimentary rock is disturbed, age of fossils is sometimes harder to determine.

layers are the oldest and the upper layers are the newest. As an analogy think of this situation. If you stacked up newspapers adding the current paper each day to the top, as long as the stack was undisturbed, the oldest paper would be on the bottom and the newest paper would be on the top. Fossils found in undisturbed lower layers of rock are of organisms older than those whose fossils are found in upper layers. The relative time of appearance and disappearance of many organisms can be determined from this layering.

There are more accurate methods of dating organisms. These methods are based on the use of isotopes. Recall that isotopes (Section 3:2) are atoms that differ in the number of neutrons they have. Some isotopes have unstable nuclei and are called **radioactive isotopes.** The unstable nuclei break down and give off particles. Geiger counters are instruments that detect those particles. Energy is also given off. When radioactive isotopes decay, particles are emitted with a large amount of energy at known rates. The **half-life** of an isotope is the time needed for one half of the radioactive material to decay. Carbon 14 is a radioactive isotope with a half-life of 5730 years. After 5730 years, half of an original carbon-14 sample will have changed to energy (Figure 10–4).

a

b

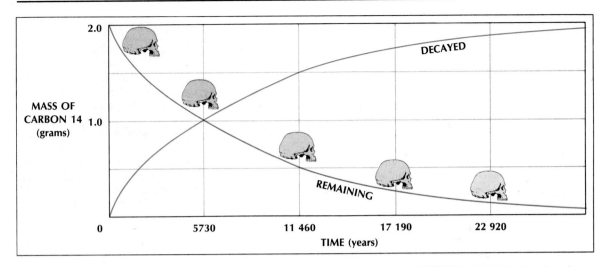

FIGURE 10–4. Radioactive decay of carbon 14 is the principle behind radiocarbon dating. The age of a fossil is determined by comparing amounts of carbon 14 present in living organisms with the amount remaining in the fossil. For example, if a fossil contained 0.5 g carbon 14 and living specimens contained 2 g, the fossil would be about 11 460 years old. In this time three-fourths of the carbon 14 decayed.

All living things contain some radioactive isotopes. During photosynthesis, plants take in radioactive carbon from the air. The carbon is in the form of $^{14}CO_2$. This carbon enters all organisms through food chains (Section 1:4). A given mass of living tissue contains a certain amount of radioactive carbon. Scientists compare the amount of radioactive carbon remaining in a fossil with the amount in the same mass of living tissue today. By using the half-life, they can approximate the age of the fossil. This method is known as radiocarbon dating and gives valid results for fossils up to about 20 000 years old.

Ages of older fossils can be estimated by studying surrounding rock samples for other radioactive elements such as uranium and potassium. These elements have a much longer half-life than carbon. Thus, they can be used for dating fossils millions of years old.

Some fossils can be dated by studying radioactive decay in rock samples.

10:2 Fossil Record: Interpretation

Earth's history can be divided into **eras** and subdivided into **periods.** Table 10–1 gives a brief account of Earth's history and the major life forms of each period. The timetable provides a time frame (in millions of years) during which the life forms originated. Much information about life in the past comes from the fossil record. In what ways does the fossil record provide evidence that evolution occurs? Many fossils are of organisms no longer present on Earth. Such organisms are said to be **extinct.** You know that organisms are adapted (Section 1:9) to certain environments. Organisms have become extinct because their environments changed over time. They could not adapt to the changes and so they died. Yet many organisms alive today are very similar to extinct forms. It seems logical that organisms living today are descendants of (have evolved from) extinct forms.

Organisms that cannot adapt to a changing environment become extinct.

Simple forms of life appear to have existed before complex forms.

Fossil studies also show that older organisms (for example, fish) are generally less complex than newer ones (such as lizards). The complexity of an organism refers to its number of specialized, interworking parts. Thus, it can be concluded that simpler forms of life existed first. Many of these simpler forms of life evolved into more complex forms. Many simple forms, though, still exist.

Looking at the fossil record as a whole, it seems clear that the number of different types of organisms was at first small. As time progressed, the diversity of life forms increased rapidly. The oldest

TABLE 10-1. GEOLOGIC TIMETABLE AND MAJOR LIFE FORMS				
Era	Period	Epoch	Millions of Years (start to present)	Life Forms
CENOZOIC	Quaternary	Recent	0.01	Modern humans; modern plants and animals
		Pleistocene	2.5	Early humans; extinction of many mammals
	Tertiary	Pliocene	7	Human ancestors
		Miocene	26	Mammals, insects, and birds; land dominated by flowering plants
		Oligocene	38	
		Eocene	54	
		Paleocene	65	
MESOZOIC	Cretaceous		136	Extinction of dinosaurs; flowering plants
	Jurassic		190	Dinosaurs dominant; primitive birds and mammals; conifers
	Triassic		225	Dinosaurs and early mammals; primitive seed plants
PALEOZOIC	Permian		280	Rise of insects; early reptiles
	Carboniferous		345	Insects and amphibians; mosses and ferns
	Devonian		395	Age of fishes; early amphibians; mosses, liverworts, and ferns
	Silurian		430	First land plants; early arthropods
	Ordovician		500	Primitive mollusks and fish; marine algae
	Cambrian		570	Sponges; jellyfish; worms; primitive algae
PRECAMBRIAN			4500	Monerans; simple protists

organisms were aquatic. Later, organisms evolved that could live on land.

The fossil record does not give a complete picture of the evolution of life. Many fossils linking one kind of organism to another have not been found. As a whole, though, the record does suggest that evolution occurs.

Fossil records indicate that life on Earth has changed.

10:3 Comparative Anatomy

Study of the structures of different organisms is called **comparative anatomy.** Figure 10–5 shows the legs of several familiar animals. Note that the bones are similar but slightly different in structure. Each animal has a leg structure that is a variation of a common pattern.

Comparative anatomy is the study of structures of different types of organisms.

The different leg structures are suited to the particular way each kind of animal walks. This similarity suggests that the animals are all related. They probably evolved from a common ancestor that had the same basic leg structure. As these animals evolved over many years to occupy different environments, their leg structures became adapted to "fit" their ways of life.

Parts of organisms that are inherited from a common ancestor are said to be **homologous.** The greater the number of homologous structures shared by different organisms, the closer the relationship among them. Do you think a cat is more closely related to a dog or to a giraffe? Why? Do you think two animals are more closely related than an animal and a plant?

Homologous structures indicate common ancestry among organisms.

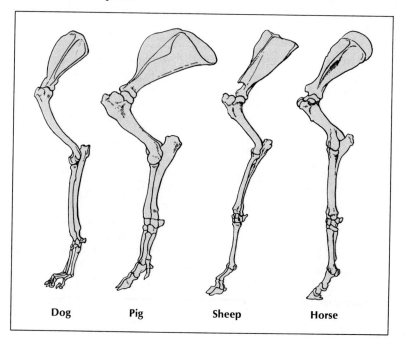

Dog **Pig** **Sheep** **Horse**

FIGURE 10–5. The leg and foot structures of many vertebrates appear similar in the number and locations of the bones. Each, however, is slightly different.

FIGURE 10–6. These limb structures are homologous. Although they function somewhat differently in locomotion, they have a common origin.

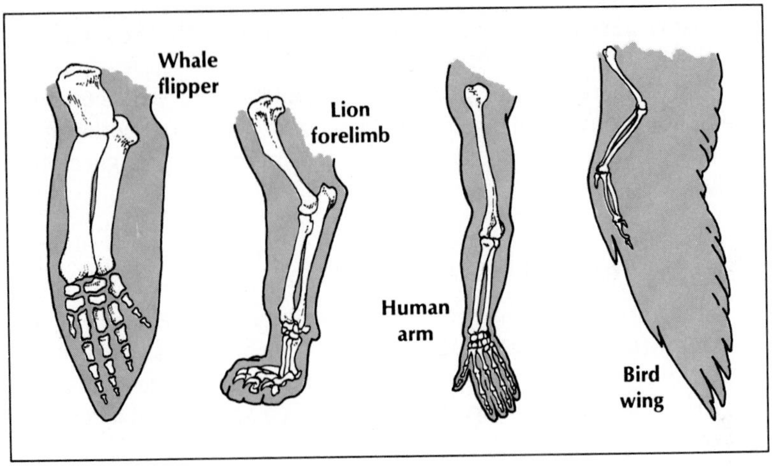

Quite often, homologous parts have similar structures and/or functions (Figures 10–5 and 10–6). But parts that appear similar are not always homologous. For example, the wings of both birds and insects have the same function, but they are not homologous. The wings develop in entirely different ways because the two organisms are not descendants of a common ancestor.

Certain structures have no functions and are called vestiges (VES tih juz) or **vestigial** (veh STIHJ ee ul) **organs.** The pig's two small toes (Figure 10–7) are vestigial because they are not used in walking. The human appendix, a small sac connected to the large intestine, is also a vestige. Why are vestiges present in organisms? At one time in the past a vestigial structure was adaptive for an organism's way of life. As time passed, the organism evolved in a changing environment. The structure, once useful, no longer served any function in the new environment. However, it continued to be inherited as part of the body plan of the organism. The fact that organisms have vestigial structures is evidence of evolution.

FIGURE 10–7. The two small toes on the pig's foot and the human appendix are examples of vestigial structures. They have no apparent function.

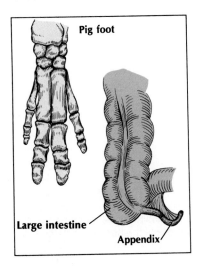

10:4 Other Comparisons

Organisms can be compared in other ways, giving additional evidence for evolution. **Comparative embryology** is the study of the development of different organisms. A developing organism is called an **embryo.** Similarities and differences can be seen in embryos of different organisms. These comparisons can show relationships.

It would be difficult for you to tell an early human embryo from an early pig embryo. Why are they so much alike? It is thought that humans and pigs are related. The more alike the development of two organisms is, the more closely related the organisms are thought to be. The human and the pig have inherited the same basic body plan from a common ancestor and develop in the same basic way. Their

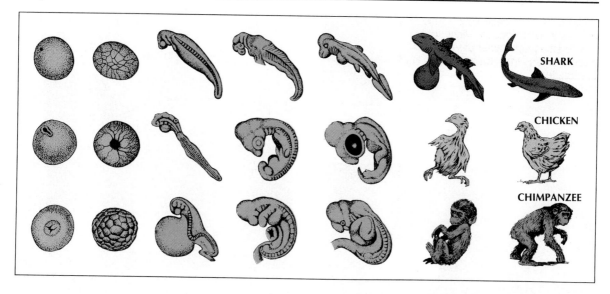

early development is quite similar. Later, however, their patterns of development change as they attain their different forms.

Often biologists compare organisms on a biochemical level. **Comparative biochemistry** also supports the idea of evolution. For example, the hemoglobin of a chimpanzee and of a human are similar. They are more similar than the hemoglobin of a human and a dog. These chemical similarities indicate common ancestry and relationships among organisms. The hemoglobin of a human is more similar to that of a chimpanzee than that of a dog because humans and chimps have a more recent common ancestor.

Notice that study of the fossil record, homologous structures, comparative anatomy, and comparative biochemistry all provide evidence for evolutionary relationships. The fact that so many different lines of evidence exist strengthens the conclusion that evolution occurs.

FIGURE 10–8. Stages of embryonic development are similar in closely related animals. The early embryos of certain animals look so much alike that it is difficult to distinguish among them. Only in later stages do they become easily recognizable.

Comparative biochemistry is a study of various organisms on a chemical level.

10:5 Genetics

Evolution is change with time. Modern knowledge of genetics adds to the evidence for evolution already presented. Each generation is linked to the next by genes (DNA). Both genes and the chromosomes on which they are carried can mutate (Section 9:8). New alleles arise through mutation of existing alleles, leading to new genotypes and phenotypes. Meiosis reshuffles alleles, producing still newer combinations and traits among offspring (Section 6:10). Thus, features of organisms can change over time. *Mutation and genetic recombination could be called the "raw materials" of evolution.* That is because they provide a basis for change among organisms.

Mutation and recombination lead to variability in populations.

FIGURE 10–9. Plants have been selectively bred (a) for better fruits and (b) as ornamentals. Some animals have been selectively bred (c) as work animals.

Selective breeding is a form of evolution controlled by humans.

Closely related organisms have very similar genes.

For thousands of years humans have made use of the fact that genetic information and traits of organisms change. They have been breeding plants and animals having traits useful to humans. They developed crops such as wheat and corn from plants having desirable traits. They also bred domestic animals such as dogs, cattle, and horses from wild ancestors of these animals. Humans selected organisms with the "best" traits and bred them so the traits would be passed on to the offspring. They did this generation after generation. Unknowingly, humans were directing the evolution of these organisms. These practices are known as **selective breeding.**

New techniques enable biologists to analyze the DNA of organisms. Specific genes can be compared by analyzing their base sequences. As in other comparative studies, close similarity among genes indicates close relationships. For example, the DNA of humans and chimps is 99 percent the same! Using known mutation rates, biologists have also been able to determine the time when certain common ancestors began to give rise to new descendants.

10:6 Evolution Observed

Evidence for evolution so far presented is indirect. "Evolution in action" has not been shown. Most evolutionary changes occur over such a long period of time that they cannot be observed. However, some changes occur rapidly. Rapid changes provide direct evidence of evolution.

Penicillin kills many kinds of bacteria. Where the bacteria are killed by the drug, a clear zone forms (Section 2:2). Usually no bacteria survive in the clear zone. Sometimes, however, a few bacteria survive and live to form colonies that are resistant to penicillin. If some of the surviving bacteria are transferred to another culture containing penicillin, they continue to produce more bacteria resistant to penicillin.

Bacteria resistant to penicillin produce offspring that also resist penicillin.

The millions of bacteria in each colony are offspring of a single penicillin-resistant bacterium. When penicillin was added (a change in the environment), resistance became very important for survival. All the bacteria that were not resistant died. Those that were resistant survived to reproduce. Thus, the population of bacteria evolved because all the organisms living after the change in the environment are adapted to the new environment.

In an environment of penicillin, only resistant bacteria survive and reproduce.

Bacteria cause many diseases in humans and animals (Chapter 19). Therefore, evolution of bacterial resistance to penicillin and other antibiotics is a serious problem. When some resistant forms of bacteria evolve, the existing drugs will not kill them. For this reason, scientists must continually search for new antibiotics to treat bacterial diseases.

Because bacteria become antibiotic-resistant, scientists search for new, effective antibiotics.

REVIEWING YOUR IDEAS

1. What is evolution?
2. What is a fossil? Give three examples of different kinds of fossils.
3. How is the relative time when an organism lived on Earth determined?
4. What is radiocarbon dating?
5. What causes extinction?
6. Within the fossil record, how does the position of a fossil provide information about organism complexity?
7. What are homologous structures?
8. What does comparative anatomy reveal about the relationships among organisms? Comparative embryology? Comparative biochemistry? Genetics?

EVOLUTION: SOME EXPLANATIONS

The preceding sections presented evidence that evolution occurs. Based on such evidence, biologists generally agree that evolution, change in organisms over time, is a fact. As you know, good science calls for facts to be interpreted, analyzed, and explained. In this section you will study and compare two explanations of *how* evolution occurs.

10:7 Lamarck

In 1809, a French naturalist, Jean Baptiste de Lamarck (luh MARK), published a book called *Zoological Philosophy*. The ideas about evolution in the book are known as Lamarckism (luh MAR kihz um). Lamarck had studied evolution and was certain of its existence and importance. He saw evolution as a "ladder of life" from the simplest to the most complex animals, with humans as the top rung of the ladder. Lamarck did little in the way of explaining the origin of this ladder, but he did offer an explanation for how organisms change over time.

Lamarck's thinking revolved around two basic assumptions. The first of these was the **law of use and disuse.** According to Lamarck, an organism could change certain body features during its lifetime. He thought that by using a certain part of its body, an animal could change the part to better fit the environment. On the other hand, by disuse, an unnecessary body part would begin to disappear. In either case, the changed feature was called an **acquired characteristic.** The change in the feature was acquired during the organism's life.

Lamarck's second assumption was the **inheritance of acquired characteristics.** He thought that acquired characteristics were passed on to (inherited by) the offspring of that individual and would occur generation after generation. Gradually a group of organisms better able to survive would evolve.

Often used to illustrate Lamarck's hypothesis is the evolution of the giraffe's long neck. Lamarck assumed that long-necked giraffes evolved from short-necked ancestors. The short-necked ancestors could graze on grasses, but the grasses began to disappear (a change in the environment) so the only remaining food source was the leaves of trees. According to Lamarck, each short-necked giraffe would try to reach the leaves of trees and thus would stretch its neck. As these individuals reproduced, the longer neck trait (an acquired characteristic) would be passed on to the offspring that would be born with slightly longer necks than those of their parents. Thus, long-necked giraffes gradually evolved.

Lamarck's hypothesis can be criticized on several points. The law of use and disuse is a poor idea because it implies that an organism can sense its needs and physically change itself to meet the needs.

Inheritance of acquired characteristics suggests that these changes (in body cells) can be passed to offspring. It is known that genetic information is passed from generation to generation in the gametes (Section 6:7). If inheritance of acquired characteristics were true, the changing body cells would need to have a way of "informing" the sex cells about the change. However, information "flows" from the gametes, to the zygote, to the body cells. There is no way for it to

Lamarck thought that characteristics could be obtained or lost as a result of use or disuse.

FIGURE 10–10. Lamarck would have explained the evolution of the giraffe's long neck by saying that individuals stretched their necks. The longer necks were then passed on to future generations.

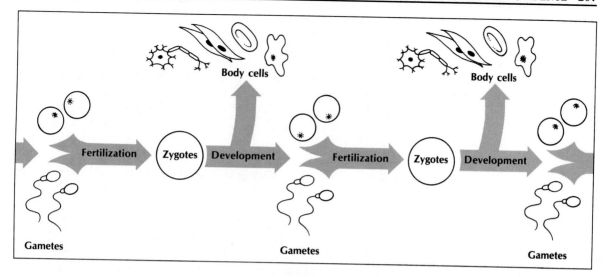

"flow" from body cells to the gametes (Figure 10–11). A mutation in a body cell does not change the genes in sex cells. Thus, even if changes could occur by use and disuse, they could not be passed to offspring. If inheritance of acquired characteristics were true, a person with well-developed muscles would produce a child with well-developed muscles, but such is not the case.

In terms of science, the major criticism of inheritance of acquired characteristics is that there has never been an experiment that supports it. To be accepted, a hypothesis must be confirmed by experiments (Section 2:5). Because all attempts to confirm Lamarck's hypothesis have failed, it must be regarded as invalid.

FIGURE 10–11. Genetic information is passed from one generation to the next in the gametes. Changes in the body cells are not transmitted to future generations.

Inheritance of acquired characteristics implies that genetic change can occur in body cells and then be transmitted to sex cells. Modern genetics shows this idea to be false.
The idea of inheritance of acquired characteristics has never been confirmed by experiments.

10:8 Darwin: Gathering Evidence

Charles Darwin (1809–1882) became ship's naturalist aboard the H.M.S. *Beagle* in 1831. The *Beagle* was destined for a five-year around-the-world cruise. Darwin had been trained in both biology and geology. During his voyage he collected many fossils and living specimens and spent a good deal of his time studying features of Earth around the world.

Darwin's observations of both organisms and geology during his trip, as well as a book by Charles Lyell, led him to begin thinking that evolution occurs. He realized that the features of Earth undergo change. He knew that organisms are adapted to their environments. If the environments change, do the organisms that live in them also change? If populations of organisms change, what causes this change?

Darwin suspected that if Earth has changed, then the organisms must also have changed.

FIGURE 10–12. (a) The scarlet crab, (b) Galapagos tortoise, (c) pigeon guillemot, and (d) marine iguana are animals found on the Galapagos Islands.

a

b

c

d

During the voyage, Darwin made discoveries similar to those discussed earlier in this chapter. From these discoveries he concluded that organisms change with time, and noted that they show close relationships to one another.

Of special interest to Darwin were the Galapagos (guh LAHP uh gus) Islands, about 970 km west of South America at the equator. There, he studied some forms of life, such as huge turtles and swimming lizards, not found anywhere else in the world. Despite their uniqueness, Darwin saw that these animals were similar to more common forms. The similarities convinced him that the Galapagos animals were related to more common turtles and lizards. By the end of the trip, Darwin believed that evolution occurs—that one form of life can evolve into another. However, he had not yet developed an explanation to account for it.

Observations of Galapagos organisms supported Darwin's conclusion that evolution occurs.

10:9 Darwin's Explanation: Natural Selection

During the next twenty years, two kinds of evidence seemed to explain some of the observations Darwin made on his trip. The first was an essay about population growth by Thomas Malthus. Malthus stated that the human population was growing faster than the food supply needed to feed it (Figure 10–13). This essay started Darwin thinking about all forms of life. He realized that in nature there is an overproduction of organisms in which many of the offspring die. Therefore, he reasoned that there must be a *struggle for existence* among organisms. Darwin envisioned many kinds of struggles such as competition for food, escape from predators, and ability to find shelter. Only some of all the organisms born can survive long enough to produce their own offspring.

The second line of evidence was selective breeding. Darwin knew that breeders can "create" desirable plants and animals by selecting parents that *already* had the desired traits. He knew that in any population, organisms have *variations that can be inherited.* Features such as color, size, mass, and number of seeds vary in plants. Cows vary in amount of milk produced. Breeders select variations in parents and produce offspring that often have the same desired features. Organisms with less desirable features are not used as parents. Darwin wondered if there was some force in nature similar to this *artificial selection* used by breeders.

Pieces of the puzzle began to fall into place. Variations exist among all organisms. Those organisms with variations that aid them in "coping" with their environment have a better chance of surviving and thus, they leave more offspring. The offspring often have the same variations and are also suited to the environment. Organisms with variations not suited to their environment leave fewer, if any, offspring. They "lose" in the struggle for existence. Darwin called this idea survival of the fittest, which is really a kind of **natural selection.** The best-adapted organisms (those most fit) survive and produce more well-adapted offspring. The least-adapted organisms may produce fewer offspring and they, along with the parents, may also die sooner.

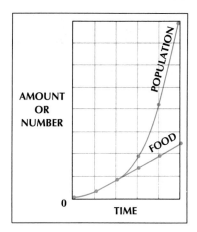

FIGURE 10–13. Malthus noted that population size increases faster than the food supply needed to support it. The result is a struggle for existence.

FIGURE 10–14. The tendency among organisms is for overreproduction. (a) Maple trees produce hundreds of seeds each year in bunches like this one. (b) Frogs lay and fertilize eggs in large masses. (c) When cut open, fruits like this melon show many seeds. Although many seeds are produced or eggs fertilized, only a few survive.

a

b

c

Evolution is a result of the interaction of organisms and their environments.

According to Darwin's theory, *evolution is an interaction of organisms and environment.* Natural selection is not the outworking of a conscious force. Chance variations either fit or do not fit the environment. Suitable variations are passed on; others are not. Darwin's concept of evolution can be summarized in the following way:

(1) In nature there is a tendency toward overproduction.
(2) Not all organisms produced survive.
(3) Variations exist in any population.
(4) Variations are inherited.
(5) Those organisms with variations that are suitable for their environment will live longer and leave more offspring on the average than organisms not having the variations. Thus, suitable variations will tend to be passed on and unsuitable ones eliminated.
(6) The resulting population as a whole will become better adapted for its environment.

Darwin constructed his theory of evolution by natural selection over a period of many years. In 1859, Darwin published a book entitled *The Origin of Species* that described his evidence supporting the theory of evolution by natural selection.

10:10 Darwin and Lamarck Compared

How would Darwin have explained the giraffe's long neck? According to Darwin, the *population* of giraffe ancestors would have included animals with many neck lengths. As long as grass was plentiful, neck length was not important, but if grass became scarce, long necks would be important. Animals with long necks would be able to graze on the leaves of the trees. Thus, taller giraffes would have a better chance of surviving and reproducing.

Because long necks were a variation, many of the giraffe offspring would also have long necks. Darwin would have said that animals with longer necks were naturally selected. Animals whose necks were not long would be less likely to survive because they would be unable to eat tree leaves when grass became scarce. They would leave fewer offspring. This pattern would repeat itself over many generations. After a long period of time, there would be no animals having necks as short as those of the original animals. The entire giraffe population would have longer necks. Would they all be the same length?

A major difference between Lamarck's and Darwin's ideas concerns *when* a variation appears. Lamarck thought that a variation (acquired characteristic) arose after environmental changes. That is, a change in the environment caused a change in the organism. Darwin believed that variations are present in a population *before* the environ-

FIGURE 10–15. Darwin would have explained the evolution of long necks in giraffes by saying that variations in neck length were originally present. Those with longer necks were better suited to the environment and survived.

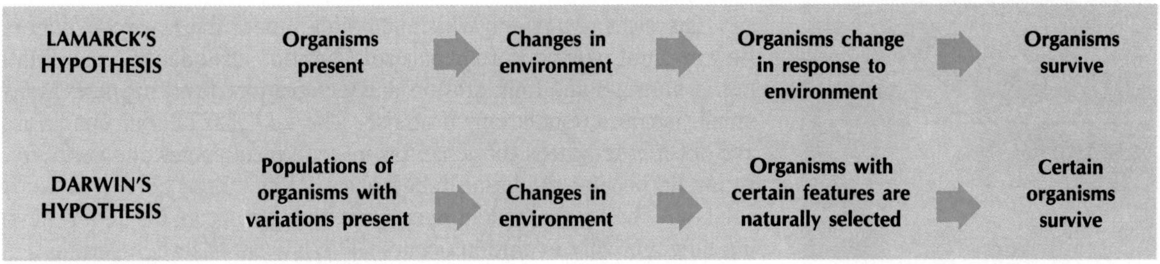

LAMARCK'S HYPOTHESIS	Organisms present	➡	Changes in environment	➡	Organisms change in response to environment	➡	Organisms survive
DARWIN'S HYPOTHESIS	Populations of organisms with variations present	➡	Changes in environment	➡	Organisms with certain features are naturally selected	➡	Certain organisms survive

ment changes. When a change occurs, some of the organisms may be able to survive. Lamarck saw evolution in terms of *individuals,* but Darwin realized that *populations* are important.

Darwin did not know that variations result from mutation and genetic recombination or that they are inherited as a result of mitosis and meiosis. It is to his credit that modern knowledge of genetics strengthens his theory of evolution by natural selection.

FIGURE 10–16. How do Lamarck's and Darwin's ideas about the origin of features differ?

Lamarck believed that variations began to occur after the environment changed.

Darwin reasoned that variations are always present in populations.

10:11 Population Genetics

Darwin recognized that a population, not an individual organism, evolves. The genes of an individual remain constant throughout its life span. However, the genes of an entire population may change with time. Mutation and genetic recombination promote such change. Geneticists picture all the genes of a population as a **gene pool.** It is from this pool of genetic information that each new organism receives its genes every generation. *The gene pool is what gradually evolves.* The study of gene pools and the changes they undergo is **population genetics.**

In 1908, Godfrey Hardy and Wilhelm Weinberg showed mathematically that, *under certain conditions,* the gene pool of a population does *not* change. That is, the frequency (percent) of alleles and of genotypes remains constant in a population. This principle is known as the **Hardy-Weinberg Law.**

The conditions of the law can be summarized as follows:

(1) Mating must be random. Random mating means that there can be no pressure to "select" a mate (due to phenotype). It also means that each member of a population mates and produces equal numbers of offspring.

(2) No genes may enter or leave the population. Within the population there can be no immigration (entrance of new organisms) or emigration (exit of organisms).

(3) There can be no occurrence of mutations.

(4) The population must be large. A large population is necessary so that changes in gene frequencies are not the result of chance alone.

The gene pool of a population evolves. The gene pool links one generation to the next.

The Hardy-Weinberg Law lists conditions under which evolution does not occur.

FIGURE 10–17. According to the Hardy-Weinberg Law, evolution does occur in this population of fur seals.

The circumstance in which all conditions of the Hardy-Weinberg Law are met *never* occurs in nature! Mutations frequently arise. Mating is nonrandom. Immigration and emigration often happen. Many small, isolated populations do exist. The fact that all four conditions are not met indicates that changes in gene frequencies and genotypic ratios do occur—evolution takes place. Why, then, bother to study the law? The law is studied because it is a useful tool for understanding how and why evolution occurs. The law can also be used mathematically to analyze gene and genotype frequencies in a population and to evaluate changes in gene pools over several generations.

REVIEWING YOUR IDEAS

9. What are the two assumptions behind Lamarck's hypothesis of the evolution of adaptations?
10. What observations caused Darwin to begin thinking that evolution occurs?
11. What is artificial selection? How has selective breeding been used?
12. What is natural selection?
13. Cite several ways in which Darwin's explanation of evolution differs from that of Lamarck.
14. What is a gene pool? Population genetics?

ORIGIN OF LIFE

Fossils have been found of photosynthetic bacteria that lived 3.5 billion years ago. It is possible that life first evolved around four billion years ago. Many scientists are conducting research to learn how life might have begun on Earth.

10:12 Formation of Organic Compounds

In 1936, a Russian biologist, Alexander Oparin, proposed a hypothesis about the origin of life. Since that time his hypothesis has been tested and revised and other explanations have been proposed. Most scientists investigating this problem agree on a major point: Conditions on early Earth were far different from what they are now. Oparin's opinion was that the early atmosphere consisted of the gases methane (CH_4), ammonia (NH_3), hydrogen (H_2), and water vapor (H_2O). Some current scientists believe the atmosphere may have contained mainly water vapor and carbon dioxide, and a small amount of hydrogen and nitrogen gas (N_2). Oxygen gas (O_2) may have also been present. Note that in either case, the main elements of organic compounds — carbon, hydrogen, oxygen, nitrogen — are thought to be present. Primitive Earth also was exposed to a great deal of energy. In addition to sunlight, it was probably bombarded with ultraviolet radiation (UV) and lightning. These energy sources may have changed some of the existing chemical bonds of the gases, forming simple *organic* compounds.

In 1953, Stanley Miller performed an experiment that simulated Oparin's ideas of the early atmosphere (Figure 10–18). He subjected the circulating gases to electricity. At the end of one week he found amino acids in the condensed water vapor at the bottom of the chamber. Similar experiments have since been performed, some based on Oparin's ideas and some on those of later scientists. Many of these experiments have shown that not only amino acids, but a variety of other organic compounds, can be formed. Sugars, fatty acids, and nitrogen bases are some of the compounds that have been produced. These experiments show that it is possible that organic compounds were produced in the early atmosphere of Earth.

Simple organic compounds are necessary for life, but more complex organic molecules are also essential. Complex molecules such as proteins, fats, and nucleotides were probably formed from simpler molecules. Some scientists suggest that these complex molecules formed after the simpler molecules were washed from the atmosphere to Earth during heavy rains. Others think they may have formed in the atmosphere, perhaps in clouds of salt water, and then reached Earth.

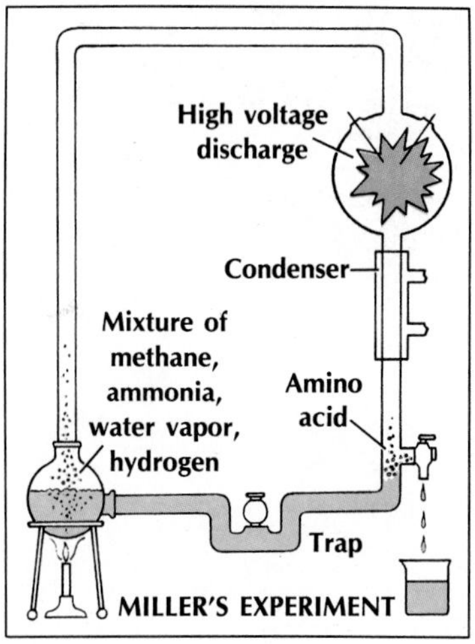

FIGURE 10–18. Miller showed that organic compounds (amino acids) can be produced when an inorganic mixture is exposed to an electric charge. He speculated that amino acids may have been formed this way in the earth's primitive atmosphere many years ago.

10:13 The First Organisms: Prokaryotes

After formation, the complex compounds probably reached the newly-formed oceans. There, life is thought to have begun. In the oceans, groups of these complex molecules could have come together. Some kind of membrane might have formed around each concentrated group (Figure 10–19). The membranes would have separated them from the surrounding water. Perhaps the chemicals surrounded by the membrane resembled a primitive cell.

According to Oparin and other scientists, other free, organic molecules in the ocean may have crossed the membrane and passed into this simple "precell." Some inorganic molecules and/or proteins may have served as enzymes. Sugars or fats might have been used as energy sources. If no free oxygen was present, early respiration would have been anaerobic. With energy from respiration came better organization. Eventually nucleic acids took over protein synthesis and reproduction. At this point, with a means of obtaining energy and producing accurate "copies," the precells would have become true cells. These cells are thought to be prokaryotes.

Complex organic compounds may have collected in the early oceans and become surrounded by a membrane.

The first organisms may have been prokaryotes with anaerobic respiration.

According to Oparin, these first cells were **heterotrophs** (consumers). Thus, his explanation is called the **heterotroph hypothesis.** How and when did **autotrophs** (producers) evolve? The hypothesis explains that there must have been competition for energy sources as the primitive population of heterotrophs grew. There would have been more organisms and fewer food molecules to satisfy energy requirements. Any organisms that could use light energy to *make* their own food (photosynthesize) would be naturally selected. These organisms might have been the first autotrophs. Photosynthesis would have led to the production of free oxygen. With oxygen present, the process of aerobic respiration could have evolved. As a result, more energy would have been available for various functions. As the amount of oxygen built up in the atmosphere, the composition of the atmosphere changed. Also, some of the oxygen was probably changed by lightning to ozone (O_3) molecules that formed a layer high in the atmosphere. This **ozone layer,** which exists today, blocks most ultraviolet radiation. When it formed, it would have prevented further synthesis of simple organic compounds in the atmosphere. However, by that time, the process of life would have been well underway.

Some scientists reason that the first organisms were autotrophs and that heterotrophs evolved later. Recall that the oldest fossils found are of *photosynthetic* organisms. Whichever explanation is more accurate, the general trends in the origin of life seem to be agreed upon. These are summarized as follows:

(1) There was production of simple organic compounds.
(2) Larger, more complex molecules were synthesized from simpler ones.
(3) Concentrations of many complex molecules became surrounded by membranes.
(4) A means of obtaining energy for life functions developed.
(5) A reliable means of reproduction evolved.

A great deal more research is necessary before it can be determined exactly how these events took place.

Competition for food is thought to have led to the evolution of the first producers.

Formation of the ozone layer may have blocked continued synthesis of simple organic compounds in the atmosphere.

FIGURE 10–19. The coming together of complex molecules that then became enclosed in a "membrane" was an important step in the development of the first organisms.

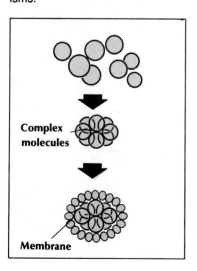

Complex molecules

Membrane

REVIEWING YOUR IDEAS

15. Describe the atmosphere that might have existed on early Earth.
16. List three of the organic compounds produced during experiments to simulate the early atmosphere.
17. What factors might have caused a struggle for existence among the first cells?
18. Explain why further synthesis of simple organic compounds in the atmosphere is not possible.

ADVANCES IN BIOLOGY

Origin of Eukaryotic Cells

Most biologists agree that the first cells were prokaryotes. Today, only the bacteria are prokaryotes. Cells of all other organisms are eukaryotes. How did eukaryotes arise?

The **symbiotic** (sihm bee AHT ihk) **theory** is one model that explains how eukaryotes evolved. **Symbiosis** (sihm bee OH sus) is a relationship in which two organisms live in close association. Often, one organism may live within the other. For example, certain bacteria living in a cow's digestive system produce an enzyme that digests grass the cow eats. Both the cow and the bacteria can obtain energy from the grass.

Perhaps certain prokaryotes "took up residence" inside other prokaryotic cells. Together, the adaptations of the "parent" organisms gave the new organism an advantage. For example, suppose a heterotrophic, aerobic prokaryote invaded a cell that carried on only anaerobic respiration. The anaerobic cell could then obtain more energy from its food. Suppose a photosynthetic prokaryote began to live within a heterotrophic one. The symbiotic arrangement would enable the heterotroph to make its own food. "Internal residents" are found in modern cells. For example, green algae are known to exist within certain hydras.

Evidence for the symbiotic theory comes from study of chloroplasts and mitochondria. It is thought that modern mitochondria and chloroplasts are descendants of prokaryotes that formed symbiotic relationships. What is the evidence? These self-replicating organelles have their own DNA, RNA, and ribosomes and carry on some protein synthesis (Section 9:9). These facts suggest that they arose from a free-living ancestor. Also, their DNA is circular and attached to very little protein, both characteristic of prokaryotes. The enzymes for making nucleic acids and proteins in these organelles are more like those found in prokaryotes. Ribosomes of the organelles resemble those of prokaryotes.

Mitochondria and chloroplasts of today's cells are regulated mainly by nuclear genes. Also, their chromosomes are much smaller than those of typical prokaryotes. How can these facts be explained? It is thought that as eukaryotes evolved, the prokaryote residents "gave up" most of their original DNA to the host cell's nucleus. The idea that genes can move is supported by a recent finding of a mitochondrial gene in the nucleus of a yeast cell.

Although evidence for the symbiotic model exists, more research is needed to determine how structures such as the nuclear membrane, ER, and Golgi bodies evolved and how the processes of mitosis and meiosis arose. Other models must also be studied.

FIGURE 10–20. Large numbers of green algae live inside this hydra, a common coelenterate, causing it to appear green.

FIGURE 10–21. One of the features that chloroplasts have in common with free-living prokaryotes is their ability to self-replicate. Shown here is a dividing chloroplast of a pea plant magnified about 18 000 times.

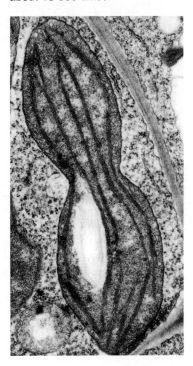

INVESTIGATION

Problem: How does comparative biochemistry support the theory of evolution?

Materials:

chart of RNA nitrogen bases

Procedure:

1. Examine the chart of RNA nitrogen base sequences of fifteen organisms provided to you by your teacher.
2. Organisms are indicated by the numbers 1–15. The series of numbers following each sequence indicates which organisms contain that particular RNA sequence.
3. Organism 1 will be used as an example of how to determine the chemical similarity between it and each of the other organisms. Look down the columns of numbers and count how many times organism 1 and organism 2 appear on the same line. The first sequence that appears in both organisms is CCUAAG (sixteen lines down). Continue the count. If you have counted correctly, you should have counted 22 sequences shared by both organisms 1 and 2.
4. In a copy of Table 10–2, record the total number of times that organisms 1 and 2 contained the same RNA sequence.
5. Complete the first row of Table 10–2 by counting how many sequences 1 and 3 have in common, then 1 and 4, 1 and 5, and so on.
6. Complete the rest of Table 10–2 by comparing only organisms 5, 10, and 15 with each of the fifteen organisms.

Data and Observations:

TABLE 10-2. COMPARISON OF RNA NITROGEN BASE SEQUENCES															
Organism	Number of RNA Sequences in Common With Each Organism														
	1	2	3	4	5	6	7	8	9	10	11	12	13	14	15
1															
5															
10															
15															

Questions and Conclusion:

1. Define:
   - (a) mRNA
   - (b) tRNA
   - (c) DNA
   - (d) nitrogen base
   - (e) comparative biochemistry
   - (f) homologous
2. What do the following letters represent in an RNA nitrogen base sequence?
   - (a) A (b) U (c) C (d) G
3. In closely related organisms, the sequences of nitrogen bases in the RNA of each organism are very similar. Thus, they are thought to have evolved from a common ancestor.
   - (a) Which two organisms show the most similar RNA sequences?
   - (b) Which two organisms show the most similar biochemistry?
   - (c) Which two organisms may have evolved from a common ancestor?
   - (d) Which two organisms may have the most similar genetic material?
   - (e) Which two organisms may have the most similar DNA?
4. Reading down column 2,
   - (a) which two organisms are the most related biochemically? Explain.
   - (b) which two organisms may show the most homologous chemical similarities?
   - (c) which two organisms may have evolved from the same ancestor?
   - (d) which two organisms are the least related biochemically? Explain.
5. Reading down column 9,
   - (a) which two organisms are the most related biochemically? Explain.
   - (b) which two organisms may show the most homologous chemical similarities?
   - (c) which two organisms may have evolved from the same ancestor?

   - (d) which two organisms are the least related biochemically? Explain.
6. Closely related organisms often show many physical similarities. Which groupings would show the most physical similarities?
   - (a) organism 2 and 15 or 2 and 1
   - (b) organism 3 and 15 or 3 and 5
   - (c) organism 7 and 5 or 7 and 15
7. How might organisms 5 and 6 be expected to appear when compared to one another? Explain.
8. How might organisms 1 and 12 be expected to appear when compared to one another? Explain.
9. An organism has the following mRNA nitrogen base sequence: ACCUCG.
   - (a) Convert the mRNA base sequence into the proper corresponding DNA sequence of nitrogen bases.
   - (b) Convert the mRNA base sequence into the proper corresponding tRNA sequence of nitrogen bases.
10. An organism has the following mRNA nitrogen base sequence: AUCCUG.
    - (a) Convert the mRNA base sequence into the proper corresponding DNA sequence of nitrogen bases.
    - (b) Convert the mRNA base sequence into the proper corresponding tRNA sequence of nitrogen bases.
11. When comparing a gorilla and a human, or a frog and a human:
    - (a) which two animals show a greater physical resemblance to each other? A lesser physical resemblance?
    - (b) which two animals would have RNA sequences that are most similar to each other? Least similar?

Conclusion: How does comparative biochemistry support the theory of evolution?

CHAPTER REVIEW

SUMMARY

1. A fossil is any remnant or trace of an organism that lived long ago. Age of fossils may be determined by using radioactive isotopes. **10:1**
2. Study of fossils shows that simple organisms appeared first and that more complex organisms appeared later. The record also suggests that many organisms are descendants of extinct forms. **10:2**
3. Comparative anatomy shows that homologous structures indicate a common ancestry and close evolutionary relationship. **10:3**
4. Comparative anatomy, embryology, and biochemistry suggest that common ancestors evolved into the present-day forms of life. **10:4**
5. Sources of change among organisms include mutation and genetic recombination. **10:5**
6. Resistance of bacteria to penicillin and other drugs is direct evidence of evolution. **10:6**
7. According to Lamarck, organisms acquire new characteristics by use or disuse. These acquired characteristics were thought to be inherited by offspring. **10:7**
8. Charles Darwin, while serving on the H.M.S. *Beagle,* gathered many lines of evidence that suggested that evolution occurs. **10:8**
9. According to Darwin, those organisms with variations best suited to their environment seem to be naturally selected while others are eliminated. Thus, with time, the entire population becomes better fit for the environment. **10:9**
10. Darwin realized that the source of change, variation, is present before the environment changes. He also recognized that it is the population that evolves, not individuals as Lamarck thought. **10:10**
11. The Hardy-Weinberg Law is useful for predicting when populations will evolve. **10:11**
12. Biologists attempt to explain the origin of life on Earth by using the methods of science. The first organisms were probably prokaryotes, but scientists disagree as to whether they were heterotrophic or autotrophic. **10:12**
13. Eukaryotes may have evolved either directly from prokaryotes or according to the symbiosis hypothesis. **10:13**

LANGUAGE OF BIOLOGY

acquired characteristic
autotroph
comparative anatomy
comparative
  biochemistry
comparative
  embryology
embryo
era
evolution
extinct
fossil
gene pool
half-life
Hardy-Weinberg Law

heterotroph
heterotroph hypothesis
homologous
inheritance of acquired
  characteristics
law of use and disuse
natural selection
ozone layer
period
population genetics
radioactive isotope
selective breeding
symbiosis
symbiotic theory
vestigial organ

CHECKING YOUR IDEAS

On a separate paper, indicate whether each of the following statements is true or false. Do not write in this book.

1. The front leg of a dog and the arm of a human are homologous.
2. More organisms are produced than can survive.
3. Domesticated animals were produced by selective breeding.
4. Closely related organisms have similar patterns of development.

5. Lamarck's conclusions are supported by knowledge of genetics.
6. The giraffe's long neck probably evolved according to the law of use.
7. Mutation and genetic recombination explain the source of variations among a population of organisms.
8. Evolution always produces organisms that can survive changes in the environment.

EVALUATING YOUR IDEAS

1. How is a fossil usually formed?
2. In general, how does what the fossil record reveals provide evidence for evolution?
3. What conclusions can be drawn from studying the leg structures of different animals?
4. How are homologous structures useful in the study of evolution?
5. How is comparison of genes useful in determining evolutionary relationships?
6. Explain how development of resistance to drugs by bacteria offers direct evidence of evolution.
7. Criticize each of Lamarck's assumptions about evolution of adaptations.
8. Summarize the evidence Darwin used to conclude that evolution occurs.
9. Outline Darwin's theory of evolution by natural selection.
10. State the Hardy-Weinberg Law. What four conditions are referred to in the law?
11. Does Miller's experiment indicate that organic compounds were produced spontaneously in the primitive atmosphere? Explain.
12. Outline the steps leading to the first organisms as proposed by the heterotroph hypothesis. According to this hypothesis, how did autotrophs evolve? How was the evolution of autotrophs important to future evolution?
13. Describe the evidence supporting the symbiosis model of eukaryote origin.

APPLYING YOUR IDEAS

1. In studying a fossil, a scientist determines that only one-quarter of the original carbon 14 remains. How old is the fossil?
2. A population of insects is characterized by a color that enables them to blend with their environment. The insects evolved this color over a long period of time. How would Lamarck have explained this? How would Darwin have explained it?
3. What possible harm could come from people taking antibiotic drugs when they are not absolutely needed?
4. Darwin recognized the existence of variations and the fact that they are inherited. However, he knew nothing about genetics. How do (a) Mendel's work, (b) events of mitosis, meiosis, and fertilization, (c) mutation and crossing-over, and (d) the role of DNA support his theory?

EXTENDING YOUR IDEAS

1. Prepare a report on the extinct flying organism, *Archaeopteryx*.
2. Prepare a report on the results of selective breeding in various plants and animals.
3. Prepare a report on life on the Galapagos Islands.
4. Investigate some "living fossils" such as the *Ginkgo* tree (Section 14:9). Find out what they are and their histories.

SUGGESTED READINGS

Gould, Stephen Jay, "Darwinism Defined: the Difference between Fact and Fancy." *Discover,* Jan., 1987.
Rensberger, Boyce, "Death of the Dinosaurs — the True Story?" *Science Digest,* May, 1986.

ADAPTATION AND SPECIATION

Organisms are adapted to their particular environments. An adaptation is a trait that improves chances for survival and reproduction. This polar bear has thick fur to protect it from the cold. The white fur also provides excellent camouflage as it hunts in its snowy environment. What special adaptations do you have that aid your survival in different environments?

chapter 11

D arwin arrived at the idea of natural selection even though he had no knowledge of genetics. Today, biologists use both Darwin's ideas and modern concepts of genetics to explain change among organisms. Natural selection and genetics explain two aspects of evolution: (1) how a population of organisms adapts to changes in the environment, and (2) how completely new types of organisms evolve.

Objectives:
You will
- give examples of adaptations and describe their evolution.
- describe a model that explains the origin of new species of organisms.
- list the stages in the evolution of humans.

ADAPTATION

An adaptation is an *inherited* trait or set of traits that aids the chances of survival and reproduction of an organism in a particular environment (Section 1:9). The source of new adaptations is variation. Variations that aid survival will be selected and will spread through a population. Remember, organisms with suitable variations have the best chance of reproducing (Section 10:8). After many generations, most members of the population will have the variation. At that point, it can be called an adaptation.

Adaptations promote survival and reproduction.

11:1 Origin of Adaptations

Adaptations rarely occur "overnight" by means of a single change in the genetic message. Natural selection operates with what is available; it cannot "invent." *It can only affect what is already there.*

As clay can be molded into many forms, the features of a population can be changed in form through natural selection. But with natural selection, there is no final mold. If changes occur that improve

FIGURE 11–1. A tiger's stripes camouflage this animal by breaking up the outline of the body, making the tiger more difficult to see. This adaptation aids the tiger in food getting.

A complex adaptation evolves over a long period of time.

the chances of survival, then by natural selection the features will be continued. The population will change as new genes in new combinations spread throughout the population.

As an example, complex eye structure may be the result of millions of years of slow evolution. Each step probably gave organisms slightly better vision. Perhaps this organ began only as an area that could distinguish light and dark. This feature would improve the chances of survival and reproduction. Organs that could detect light and dark were adaptive. What advantages could they give? The organ may have later evolved into an eye with a primitive lens that gave blurred vision. This change also gave an advantage. Seeing even a blurred image is better than seeing no image at all. Later changes may have led to a sharpening of focus. Eventually, eyes that could clearly see things evolved. Each of the steps along the way was adaptive. It improved the chances of survival and reproduction of the organisms. Natural selection has slowly changed these organs by selection of structures already present.

Is it difficult to understand how such complex adaptations and so many different life forms have evolved? If it is, think about the length of time evolution has been going on. According to the theory of evolution, natural selection has been at work for more than three billion years. How big is three billion? Suppose you had three billion dollars and you spent a dollar a second, day and night. It would take you over 95 years to spend all the money.

11:2 Evolution of Adaptations Observed

Evolution of adaptations does not always take long periods of time. In Chapter 10, you studied how bacterial resistance to penicillin can occur quickly. The few bacteria that can live in the presence of penicillin reproduce. As a result, many bacteria that are resistant to penicillin are produced.

This example is important for two reasons. First, it shows that evolution is an interaction of organism and environment. Second, it points out that evolution and natural selection work on the basis of genetic change within a population. What was the source of the change? Did the resistance to penicillin originate in response to the change in the environment? Lamarck would have said yes, but according to Darwin's ideas, the answer is no. The allele for resistance to penicillin already was present in the bacterial gene pool, but it was quite rare. It was not until the environment changed that bacteria resistant to penicillin became more numerous. Natural selection removed the bacteria that were not resistant. The number of alleles in the gene pool for resistance increased.

FIGURE 11–2. The complex structure of the human eye may be the product of millions of years of evolution.

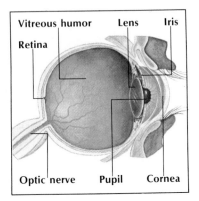

FIGURE 11–3. Before industrialization, tree bark was light in color. Light-colored peppered moths were well camouflaged. Thus, they had an adaptive advantage. The dark-colored peppered moths were easier prey for birds. The adaptive advantage of the dark color was not apparent until after the industrial revolution.

Another example of observable evolution occurred in England around 1850 when almost all the peppered moths were very light in color. Black moths with a pigment called melanin were rare. During the day, the moths rest on tree trunks. At one time the trunks were very light in color, so the light-colored moths blended with their environment. Birds that fed on peppered moths easily saw and ate the dark-colored moths. Thus, the light-colored moths had an adaptive advantage and produced more offspring.

Coloration of peppered moths may protect them from predators.

As industry in England increased, the tree trunks became black with soot and smoke. Then, the black forms of the peppered moth had an adaptive advantage. On the dark tree trunks, birds could not see them, but light-colored moths became easier prey. During the last 100 years, the gene pool has changed so that today, over 90 percent of the peppered moths are black. This evolution in peppered moth populations is called **industrial melanism** (MEL uh nihz um). Industrial melanism shows again that evolution is an interaction of organisms and environment. It also shows that evolution has a genetic basis. The allele for black color may have originated long before the environment changed. When it first appeared, the allele for melanin was neither useful nor lethal. Thus, some black moths survived and passed on the allele. Only after the environment changed did the allele for melanin give the black moths an advantage. They began living longer and leaving more offspring. Thus, the percentage of alleles for melanin in the population increased. The frequency of the allele for light color decreased.

Industrial melanism illustrates the relationship between organisms and environment as well as the role of natural selection in the process of evolution.

Neutral and nonlethal mutations may accumulate in a population and may become important if the environment changes.

Many mutations that appear in a population are neutral. They neither help nor hinder organisms when they first appear. Many other mutations, such as those in peppered moths, are not neutral, but they are not lethal. Although they are not useful when they first arise, they increase variation within a population. This variation is important later when the environment changes.

FIGURE 11–4. Many adaptations are related to food getting. (a) The anteater's long snout and sticky tongue are adaptations for eating ants. (b) The filament protruding from the angler fish's skull is an adaptation that serves as a lure in attracting prey.

Morphological adaptations involve body structures.

11:3 Types of Adaptations

Organisms as they are today are adapted to their environments. If they were not, they could not survive and reproduce. Adaptations can be grouped into three major types — morphological (mor fuh LAHJ ih kul), physiological (fihz ee uh LAHJ ih kul), and behavioral (bih HAY vyuh rul).

Morphological adaptations are those that involve the structures of organisms (their anatomy). These adaptations are the most obvious of the three types, such as the structure of the bones of the hand, the beaks of birds, or the hooves of horses. How is each of these structures adapted for its function?

Many interesting morphological adaptations in animals are for obtaining food. A woodpecker's tongue is narrow and very long. It can probe small openings in trees that have been pecked away by the bird's sharp beak. Insect larvae living beneath the bark are picked out by the tongue and eaten by the woodpecker. An anteater's sticky tongue inserted into an anthill attracts ants as it moves. If the ants touch the tongue, they stick to it. When the anteater pulls its tongue in, the ants are trapped easily.

Organisms have other interesting adaptations for food getting. Protruding from the top of the angler fish's skull is a long filament (Figure 11–4b) that is broad and flat at its tip. It hangs in front of the fish's mouth and attracts prey much like a fishing lure. When the prey "bites" at the lure, the fish sucks the prey into its mouth. Some deepwater angler fish of the North Atlantic have modified fins. Part of a fin glows in the dark waters and attracts prey. The prey then can be easily trapped and swallowed.

Physiological adaptations are involved with the functions of organisms. The enzymes needed for digestion, clotting of blood, or muscular contraction in animals all have a physiological basis. Secretion of a poison venom by a snake is another example. Protein materials in a spider's web are made chemically. An enzyme released by sperm cells enables them to break down the outer wall of an egg. These adaptations are physiological.

Physiological adaptations include biochemical features.

Organisms have many **behavioral adaptations,** those involving reactions to the environment. Migration of birds, hunting and storing of nuts by squirrels, and tracking abilities of hunting dogs are all behavioral adaptations. They all improve the chances of survival and reproduction. Plants, too, have certain behavior even though they have no nerves. Their behavior is controlled by hormones (Section 22:9). The growing of a plant toward light is an example of plant behavior.

Behavioral adaptations govern an organism's reactions to the environment.

Of course, these adaptations are artificially grouped. Any one type of adaptation depends on other types. For example, the behavioral adaptation of bird migration depends on morphological adaptations such as feathers, lightweight bones, and strong muscles. Also, nerve and muscle coordination and energy production are among the physiological adaptations involved. In the final analysis, an organism is a bundle of adaptations!

11:4 Camouflage and Other "Tricks"

Adaptations involving deception and camouflage are widespread in nature. Camouflage is a kind of "disguise" that blends with the environment. Often, animals' colors can match their surroundings almost perfectly. The color of peppered moths is a type of camouflage in which the moths blend with their environment. Thus, predators do not easily see them. This type of protection is called **cryptic** (KRIHP tihk) **coloration.**

Camouflage not only serves the hunted but can also help the hunter. A tiger's stripes break up the general body outline of the animal so that another animal may not see a tiger approaching.

Many animals can change colors in different environments (Figure 11–5b). They can change colors because they have special cells called **chromatophores** (kroh MAT uh forz), which are cells filled with one or more pigments. When the pigment spreads throughout the entire cell, one color pattern is seen. When the pigment shrinks to one part of the cell, another color is seen. These changes are probably triggered by the action of nerves and chemicals. Vision and other senses may be involved. A chameleon (kuh MEEL yun) is a reptile that can change colors. The coloration pattern is determined by the interaction of three pigments—black, yellow, and red.

FIGURE 11–5. Cryptic coloration makes animals more difficult to see. (a) A toad blends in well with the sand. (b) A Carolina anole changes colors and matches its background.

a

b

FIGURE 11–6. (a) The katydid resembles the leaf on which it is feeding. Therefore, the katydid is less likely to be preyed upon. (b) Pipefish can swim in an upright position. Thus, they are easily hidden among the coral.

a   b

Shape and behavior, as well as color, may be important in camouflage.

In addition to camouflage by color, some animals "hide" by means of their shape and behavior. The katydid and many other insects have wings shaped like leaves (Figure 11–6a). The veins within the wings have a central vein from which many smaller veins branch like the veins of leaves. Also, the insect may remain very still. Pipefish (Figure 11–6b) can resemble the coral among which they live by swimming in an upright position. Many insects look like the twigs of plants upon which they rest. In all these cases, the combination of color, shape, and behavior provides protection. How would Lamarck have explained the origin of these adaptations? How can they be explained using Darwin's theory of evolution?

Warning coloration involves a display of bright colors and patterns that announce rather than hide animals. Some of these animals may taste bad, or smell bad, or sting, or secrete poisons. Their obvious colors or features enable predators to learn to recognize and avoid them in the future. The yellow and black stripes of a bumblebee warn predators that the bee is distasteful and can sting. A predator that once ate a bumblebee or got stung would learn to avoid bumblebees later. The milkweed beetle is poisonous and the velvet ant stings. The skunk emits a foul odor. Each of these animals has a distinct coloration easily recognized by potential predators. Often, behavioral adaptations are related to warning coloration. Some animals show their presence by sudden movements or changes in position.

Some distasteful or harmful organisms advertise their presence by warning coloration.

a

b

FIGURE 11–7. Warning coloration announces the presence of an organism. (a) Certain moths and (b) larvae have spots that have the appearance of eyes.

MODEL: Monarch butterfly

MIMIC: *Gryllacris* beetle

MODEL: Bombardier beetle

MIMIC: Bee fly

MODEL: Honey bee

MODEL:

MIMIC: Viceroy butterfly

MIMIC: Hornet moth

MODEL: Hornet

MIMIC: Robber fly

MODEL: Bumble bee

Mimicry (MIHM ih kree) is a type of deceptive adaptation. In one form of mimicry, organisms that are neither harmful nor distasteful evolve a color pattern similar to that of organisms with warning coloration. These harmless, tasty organisms are known as *mimics*. The organisms with warning coloration are *models*. Predators quickly learn to avoid both models and mimics. Thus, both populations are protected. Figure 11–8 shows a variety of models and mimics. The robber fly is a mimic of the bumblebee, the model. The two insects are not closely related, but their likeness in color deceives predators. As a result, predators ignore the mimic just as they do the model. The two populations involved in mimicry need not be closely related. In fact, in the canopy of a tropical rain forest, a parrot snake (reptile) is mimicked by a caterpillar (insect).

How do these adaptations involving color evolve? As with all adaptations, variations occur that increase the chances of organisms' survival. The variations then spread through the population. Complex adaptations involving coloration do not usually result from one variation. Rather, slight variations, each adaptive, arise gradually, producing the pattern seen today.

FIGURE 11–8. Examples of mimicry are prevalent among the insects. Each mimic gains some protection because it resembles a model.

A model population is distasteful and has warning coloration. A mimic population is not distasteful but resembles the model population.

REVIEWING YOUR IDEAS

1. What are the three types of adaptations? Give an example of each type in a human.
2. What is cryptic coloration?
3. Distinguish between warning coloration and mimicry.
4. Give an example of warning coloration and an example of mimicry.

ORIGIN OF NEW GROUPS

Darwin noticed many kinds of birds on his visit to the Galapagos Islands. Of special interest to him was a group now known as "Darwin's finches." There were thirteen forms of finches on the islands. Although similar in many ways, the different forms were distinct enough to make each a separate species. Such observations set Darwin thinking about how new species evolve.

11:5 Evolution of Species

A species is a group of organisms that normally interbreed in nature to produce fertile offspring.

What is a species? A **species** is a group of organisms that normally interbreed *in nature* to produce *fertile offspring*. Many organisms can be crossbred when removed from their natural environment. A lion and a tiger can be crossed in captivity to produce a "liger" (or tiglon). This probably would not occur in nature. A female horse and a male donkey can mate and produce a mule. Because mules are not fertile, two mules cannot mate to produce more mules. Thus, horses and donkeys are not the same species. Species also can be defined as a closed gene pool into which "foreign" genes cannot enter by normal mating.

Because the Galapagos finches share so many features, Darwin thought they must have had a common ancestry. It is thought that the ancestor of the island finches was from the South American mainland. It was probably a ground-living, seed-eating species. Today's species of Galapagos finches differ mainly in their beak structures and their sizes. Although their beaks may not seem very different, the beaks' structures and sizes determine what the birds can eat. Looking at

FIGURE 11–9. (a) A horse and (b) a donkey are each members of distinct species that can mate. (c) A hinny is the offspring of a male horse and a female donkey. A hinny, except for rare cases, is infertile.

a

b

c

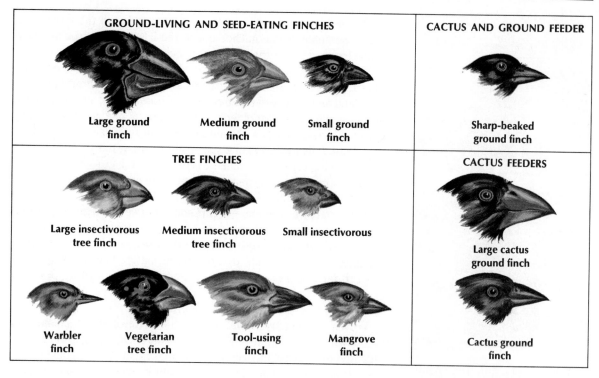

GROUND-LIVING AND SEED-EATING FINCHES

Large ground finch

Medium ground finch

Small ground finch

CACTUS AND GROUND FEEDER

Sharp-beaked ground finch

TREE FINCHES

Large insectivorous tree finch

Medium insectivorous tree finch

Small insectivorous

CACTUS FEEDERS

Large cactus ground finch

Warbler finch

Vegetarian tree finch

Tool-using finch

Mangrove finch

Cactus ground finch

beak characteristics seems to be the main way these birds recognize other members of their own species. The Galapagos finches consist of three species that live on the ground like the ancestor and eat seeds. Also, there are two cactus feeders, one cactus and ground feeder, and seven tree finches (Figure 11–10). How might these species have evolved?

In order for most new species to evolve, biologists feel that there must be a division that isolates one part of a population from another. The division is often caused by a large barrier such as a river, canyon, or mountain. This type of separation is known as **geographic isolation.** Because the groups are isolated by a barrier, organisms of one group cannot interbreed with those of another group. Since mutations arise by chance, each group begins to have slightly different alleles, and genes are recombined in different ways. The environment also differs from group to group. Different variations are naturally selected, and the gene pools gradually become more and more distinct. As time passes, each isolated group evolves different traits in response to its own changing environment. If the groups remain isolated, they may reach a point where they can no longer interbreed, even if the barrier were removed. At that point, the groups are separate species. Their gene pools are closed. Evolution of a new species is called **speciation** (spee shee AY shun).

FIGURE 11–10. Each of these Galapagos finches is a distinct species evolved from a common ancestor. They have adapted to several different environments.

Geographic isolation is the starting point for evolution of most new species.

Isolated gene pools are subject to differences in mutations, genetic recombination, and environment. Thus, each gene pool evolves differently.

Speciation is completed when populations cannot interbreed even if not isolated geographically.

FIGURE 11–11. (a) The Abert squirrel and (b) the Kaibab squirrel probably evolved from a common ancestor. As a result of geographical isolation by the Grand Canyon, the original population evolved into separate species.

This sequence of events may have occurred with the Galapagos finches. Somehow some finches got to the islands. As a result, they became isolated from others of their species by 900 km of ocean. The two environments were different, so island birds would have evolved differently from mainland birds. Further isolation occurred between each island and also between birds on each island. As a result, many gene pools would have evolved independently of one another. In this way, the thirteen species presently living on the islands could have come about. The evolution of a species into two or more species with different characteristics is called **divergence** (di VUR junts).

Recall that different species cannot mate and produce fertile offspring in nature. Their gene pools are too different for this to occur. A number of factors prevent interbreeding between distinct species. With no interbreeding, there is no gene exchange among species, a condition known as **reproductive isolation.** The factors that prevent interbreeding include differences in mating habits, inability of sperm to fertilize eggs, and seasonal differences in mating. Even if eggs are fertilized, gene and chromosome differences almost always prevent an embryo from developing properly.

Why is reproductive isolation important? Recall that each species has its own set of chromosomes with specific alleles, and each species is adapted to a certain environment. Suppose that species could interbreed freely and randomly. The resulting zygotes would almost always receive "nonsense" genetic messages. Even if they could develop, they would probably not be suited to their environment.

FIGURE 11–12. Speciation can result when a geographic barrier such as a river isolates organisms. The groups of organisms then evolve separately.

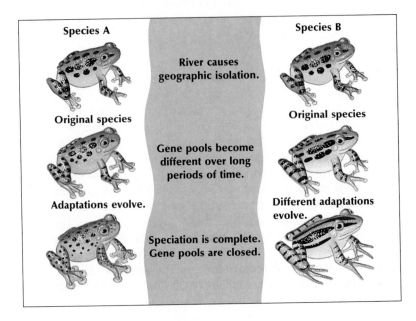

Species A

River causes geographic isolation.

Species B

Original species

Original species

Gene pools become different over long periods of time.

Adaptations evolve.

Different adaptations evolve.

Speciation is complete. Gene pools are closed.

11:6 The Tempo of Speciation

Darwin thought that speciation occurs over very long periods of time, millions of years, during which small changes build up in a population. That idea, still shared by many biologists today, is known as **gradualism.**

Recently, though, some biologists have begun to think that speciation occurs much more rapidly. They believe that a species remains in equilibrium — basically the same — for a long period of time. Then, that equilibrium is punctuated (interrupted) as a result of some crisis such as disease, drought, or flood. Those that survive the crisis rapidly evolve into a new species. This idea, first proposed by Niles Eldridge and Stephen Jay Gould, is known as **punctuated equilibrium.** The term "rapid" must be thought of in terms of geological time. A rapid change may be as many as 100 000 years.

Gradualism suggests that many forms linking one species to another should be found in the fossil record. Recall, though, that such "in-between" forms are not always present. Often the fossil record shows that a given species remains the same over long periods of time and then, quite suddenly, a new species appears. Punctuated equilibrium would explain this pattern. However, a major problem with interpreting the fossil record is that only the hard parts of organisms are preserved. There may have been changes in the soft tissues of evolving organisms that cannot be studied. This is true, too, of physiological and behavioral features.

Mosaic evolution explains speciation in terms of both gradualism and punctuated equilibrium. According to this hypothesis, basic internal functions may remain constant or change very slowly. Structures and processes related to crises in the environment may evolve quite quickly.

The debate over the *rate* at which speciation occurs is far from over. However, most biologists still use Darwin's ideas to explain *how* such change occurs. Natural selection remains the cornerstone of evolutionary theory.

FIGURE 11–13. Punctuated equilibrium is the idea that when a crisis such as disease, drought or flood occurs, organisms that survive the crisis rapidly evolve into new species.

The idea of punctuated equilibrium is that organisms remain basically the same for long periods of time and then evolve into new species quite rapidly.

Mosaic evolution suggests that some features remain the same for long periods of time while others change rapidly.

11:7 Adaptive Radiation

Evolution of the Galapagos finches is an example of **adaptive radiation,** the evolution of many new species from a common ancestor in a new environment. For adaptive radiation to occur, several conditions must be met. (1) There must be a way for the ancestral form to reach the new environment. (2) The ancestral form must have some basic adaptations that are suited to its new habitat. (3) The new environment must be free from competition with similar, better-adapted species.

Adaptive radiation is the evolution of several new forms from a common ancestor.

FIGURE 11–14. One type of Galapagos finch is adapted to fill the ecological niche of a woodpecker. The woodpecker finch picks insects from trees with a cactus thorn.

For adaptive radiation to occur, a variety of ecological niches must be available.

The ancestor of the island finches is thought to have met these conditions. Because they could fly, finches could get to the new environment. They had basic adaptations because the islands' environment was not too different from that of the mainland. Also, there was no competition. Very few species of birds were already on the islands, and they occupied different ecological niches (NIH chuz). An **ecological niche** is the specific role an organism plays in a community.

Once on the islands, the ancestral species could evolve rapidly. Because similar types of bird life were rare, the birds were free from environmental pressures on the islands. Food was plentiful, competition for food was slight, and the islands formed a chain in which each island was isolated. Thus, there were many ecological niches to be filled.

One type of finch filled the ecological niche of a woodpecker. A normal woodpecker has strong neck muscles, a hard, sharp beak, and a narrow, very long tongue. Although the "woodpecker finch" has a short tongue (a long tongue never evolved), the finch adapted in another way. Once it pecks a hole, it picks up a cactus thorn or twig in its beak and probes the hole with it. Insects caught on the thorn or twig provide food for the "woodpecker finch" (Figure 11–14).

This adaptation shows that natural selection may modify what is already present and that evolution occurs by chance. Variations for beaks that could be used to peck holes were present in the population. These variations were selected. However, there were no variations leading to evolution of narrow, long tongues. Therefore, tongue structure did not change. But a behavioral variation—using a twig— arose. This behavioral variation was also selected.

11:8 Convergence

In convergence, distantly related species produce descendants that have similar features.

Convergence may occur when different organisms evolve in similar environments.

Most often, evolution results in new species that are different from one another (divergence). Sometimes, though, convergence (kun VUR junts) occurs. **Convergence** is a process in which species that are not closely related produce descendants with similar traits.

Think about fish and dolphins. The ancestor of fish is thought to be a simpler marine (sea living) organism. Dolphins are descendants of a land mammal. Thus, ancestors of fish and dolphins are not closely related. But, in some ways, fish and dolphins are similar. Both have a streamlined shape for swimming. Fish have fins; the limbs of dolphins have been modified to form flipperlike structures.

The similarities between fish and dolphins can be explained in terms of their environment. Both organisms evolved in water. In each organism, natural selection favored variations suited for an aquatic environment. But the organisms did not undergo convergence in all

a

b

c

d

FIGURE 11–15. (a) The ring-tail lemur and (b) the raccoon show convergence in their coat markings. (c) The dolphin's flipperlike structures and (d) the fish's fins also show convergence.

their traits. For example, the fish has gills, but the dolphin has lungs. Each can live in water with these organs, but the dolphin must come to the surface for oxygen. Can you think of other differences in these sea animals?

REVIEWING YOUR IDEAS

5. What is a species? Why are horses and donkeys two species?
6. What is divergence? What is convergence?
7. What is reproductive isolation?
8. Distinguish between gradualism and punctuated equilibrium.
9. What is adaptive radiation?

HUMAN ORIGINS

What place do humans have in nature? They gather food and plant crops. They breed animals to use for work, food, and protection. Humans build shelters, cities, roads, and factories. They create masterpieces of art and write poetry. When you think of human activities, you probably picture the types of things mentioned here. Many of these are unique features of human life.

Humans are unique organisms.

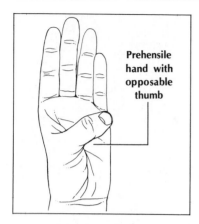

FIGURE 11-16. The human has an opposable thumb and is prehensile (able to grasp).

Prehensile hands permit humans to grasp objects.

FIGURE 11-17. (a) The eyes of some animals, such as cows, are located on the sides of their skulls. They perceive width and height. (b) In primates, such as lemurs, the eyes are usually large and located in the front of the skull. Primates have three-dimensional vision that enables them to perceive depth as well as width and height.

11:9 Important Human Traits

What traits have been important in making humans unique? The first ones may seem obvious to you. Humans have upright posture and **bipedal** (bi PED ul) **locomotion,** locomotion on two feet. Most other animals do not walk around on their back legs. Humans are called bipeds because of this locomotion pattern.

A second important feature of humans is **prehensile** (pree HEN sul) hands. Prehensile hands are hands that can grasp or wrap around objects. Humans can grasp mainly because the thumb on the human hand opposes (moves in a direction opposite) the fingers. Each finger of the hand can be touched to the thumb.

The eyes of many animals are located on the sides of their skulls. They see things only in two dimensions, width and height. The eyes of humans are located in the front of their skulls. The three-dimensional vision of humans enables them to perceive depth as well as width and height.

How did these traits originate? Humans are primates, animals that include monkeys, chimpanzees, and gorillas (Section 16:14). Prehensile hands and three-dimensional vision are adaptations that first arose in primates for life in trees. Together, they make possible **hand-eye coordination,** the ability to use sight and touch together for delicate movements. Having these traits, primates are better equipped to grasp and move about on branches or to swing rapidly from one branch to another. Changes in limb structure also occurred as an adaptation to tree life. These changes may have made possible later evolution of bipedalism.

Evolution of these adaptations for tree life are thought to be important in later evolution of humans. When primitive **hominids** (early humans) moved from trees to the ground, their adaptations for tree life could be modified for other uses. Evolution of bipedalism probably freed the forelimbs for other functions. Hand-eye coordina-

a

b

tion permitted hominids to grasp objects and use them as tools. Use of tools was especially important. It allowed early humans, though not well-adapted biologically for life on the ground, to hunt and defend themselves from other predators. They began to live by their wits as well as by their "biological equipment." Hominids having a greater intelligence were doubtless selected. As time passed, the size of the brain nearly tripled. With high intelligence came other important human traits—language and speech, emotions, and personality. These adaptations form the basis for complex human societies.

11:10 African Origins

Humans are indeed unique, but because they are organisms, they are thought to be a product of biological evolution. They are a branch on the evolutionary tree of primates. Another branch is the modern great apes. It is believed that some common primate ancestor, extinct for millions of years, diverged to form these two modern groups. *Note that humans are not thought to be descended from the great apes. The two groups are contemporary organisms.* As evolution progressed, many species evolved and became extinct. They form the fossil record giving insight into the evolution of humans.

Fossils of human ancestors are rare. Early hominids lived in tropical or subtropical regions that are not well-suited for fossil formation (Section 10:1). In warm areas, the rate of decay and decomposition of dead organisms is high. Thus, the chance of sediment covering an organism before it decays is small. It would be good to have more evidence, but some fossils have been found that offer an exciting, yet somewhat limited, glimpse into the past.

Most primitive of all hominids is a group represented by fossils found in various parts of Africa. First discovered in 1924, these early humans are classified in a group known as australopithecines (aw stray loh PITH uh seenz). Several species are thought to have existed.

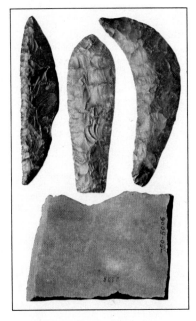

FIGURE 11–18. Early humans made many types of primitive tools from stone, wood, and bones. Some of these tools were used for hunting. Other tools were used for jobs such as preparing animal hides.

Some common primate ancestor gave rise to the modern great apes as well as modern humans.

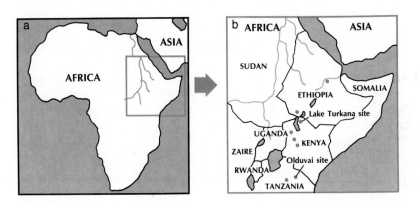

FIGURE 11–19. (a) Humans probably originated in East Africa. (b) Eight spots on the map mark sites of major archaeological finds.

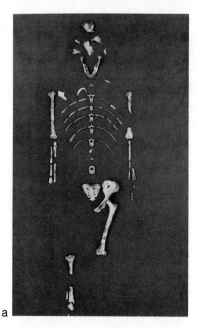

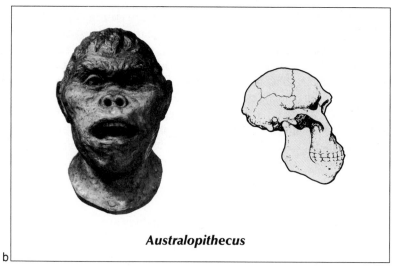

Australopithecus

FIGURE 11–20. (a) In 1974, the fossil skeleton of a twenty-year-old australopithecine was discovered. (b) This cast made from *Australopithecus* fossil skulls shows a smaller brain area and more prominent brow ridges than modern humans.

Australopithecines were the first humanlike organisms.

Later australopithecine species were larger and had a bigger brain than earlier forms.

Oldest of the species is *Australopithecus afarensis*. The first fossil discovered, in 1974, was of a fairly complete skeleton of a twenty-year-old female that scientists named Lucy. The skeleton was 3.5 million years old and caused great excitement in the scientific world. Lucy was about 1.1 m tall and weighed close to 30 kg. Like all australopithecines, she had both apelike and humanlike features. They all were bipedal and had teeth more like those of humans than of apes. The hip and leg bones joined together the same as in a human. Their skull structures, though, were more like those of apes. Other members of *A. afarensis* have since been discovered. (Sometimes, when the first part of a scientific name is understood, just the first letter of that name is used.) They had a brain capacity of about 375 to 500 cm^3. (Modern humans average about 1350 cm^3.) This species first appeared close to four million years ago and died out about 2.8 million years ago.

The first australopithecine was discovered in 1924. It was named *Australopithecus africanus*. This form lived between two and three million years ago. It was about the same size as *A. afarensis* but had a slightly greater average brain capacity.

Yet another fossil of australopithecine was unearthed in 1959. The fossil is now classified as another species, *Australopithecus robustus*. It was so named because this hominid was heavier and larger (more robust) than the other species. It was about 0.4 m taller and had teeth suited for grinding. Its brain size averaged 530 cm^3. This species seems to have lived between 1.5 and 2 million years ago.

Living at about the same time as *A. robustus* was *Australopithecus boisei*. This fourth species had about the same brain size as *A. robustus* but differed in facial traits and teeth.

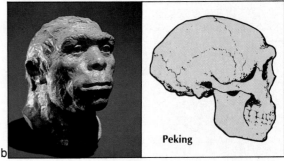

Java

Peking

FIGURE 11–21. Because of similarities in the fossils found and the similar ages of these fossils, both (a) Java people and (b) Peking people are considered in the group *Homo erectus.*

11:11 Earliest Humans

Other discoveries have revealed another important fact — other groups of hominids were living at the same time as *Australopithecus.* In 1961, a fossil that was 1.8 million years old was found. It was more humanlike than *Australopithecus* and had a larger brain (700 cm^3.) The size and structure of its brain suggest that it may have had a primitive form of speech. Because this organism was more like a true human, it is classified by some as *Homo.* It was not a modern human but was a separate form, *Homo habilis.* It is thought that *H. habilis* was a tool-maker because many pebble tools were found in the area where the fossil was discovered. *H. habilis* is thought to have died out about 1.5 million years ago.

During the late nineteenth and early twentieth centuries, fossils of two other forms of early humans were discovered in Asia. One was named Java Man; the other, Peking Man. These fossils represent a more modern form of human now classified as *Homo erectus.* Later discoveries indicate that this form of human first appeared in Africa and later migrated to Asia.

H. erectus lived between 1.6 million and 300 000 years ago. This species still had an apelike face with heavy, large brow ridges, a receding chin, and protruding jaws and teeth. The average brain capacity was 1000 cm^3. Discovery of hearths in caves indicates that these people used fire. In 1984, a new *H. erectus* fossil was found. It is the most complete hominid fossil that has ever been discovered. Analysis of its teeth and bone structure indicate that it was a male about twelve years old. Especially interesting is the fact that this young person was nearly 1.7 m tall. It is estimated that he would have grown to be 1.8 m tall as an adult. If that is true, *H. erectus* may have been larger on the average than modern humans. Fossil evidence shows that *H. erectus* was the first of the humanlike ancestors to have left Africa. In addition to those fossils found in Africa, fossils have been found in India, China, and Southeast Asia.

The earliest human may have been *Homo habilis* who lived two million years ago.

FIGURE 11–22. *Homo erectus* has an apelike face with heavy, large brow ridges, a receding chin, and protruding jaws and teeth.

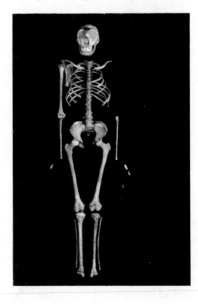

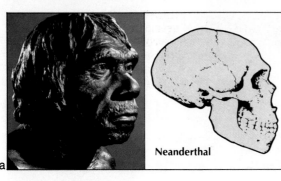
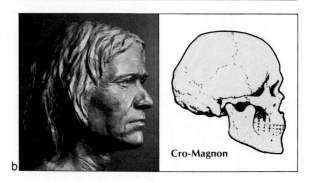

a | Neanderthal

b | Cro-Magnon

FIGURE 11–23. (a) Neanderthals and (b) Cro-Magnons are the most recent of fossil hominids. These are classified, along with modern humans, as *Homo sapiens.*

Neanderthals were an early, more primitive form of modern humans, *Homo sapiens.*

Neanderthals made excellent tools and had a primitive form of religion.

Cro-Magnons were humans very much like those of today. They are distinguished by their artwork.

11:12 Modern Humans

The earliest forms of modern humans, *Homo sapiens,* are thought to have first appeared in Europe about 300 000 years ago. The best known of these was discovered in 1856 in the Neander Valley near Dusseldorf, Germany. This form, called **Neanderthal** (nee AN dur thawl), lived from about 35 000 to 125 000 years ago. These people were short, stocky, and very powerfully built, having thick bones and large muscles. Although their faces had some primitive characteristics, they had a brain capacity of over 1400 cm^3 — larger than that of today's humans.

Evidence indicates that Neanderthals made excellent tools and were good hunters. Sometimes they buried their dead with sacrifices. Note that despite their primitive appearance, Neanderthals are classified as modern forms. If they were alive now, they would be considered a race of today's humans.

Modern humans essentially like those of today lived from 10 000 to 50 000 years ago. Most famous among them are the **Cro-Magnons,** discovered in France. Not only were they advanced toolmakers, but they were also excellent artists. Artwork is found on the walls of the caves where these people lived.

FIGURE 11–24. Artwork has been found on walls of caves where early humans lived. Artwork, done mostly by Cro-Magnons, is the earliest known step in the development of writing.

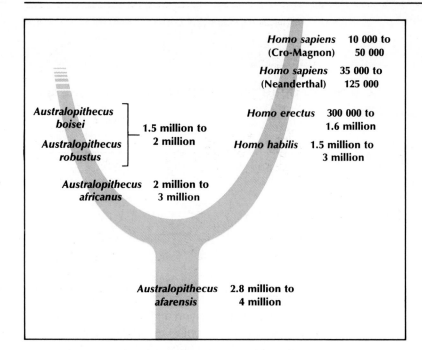

FIGURE 11–25. This evolutionary tree shows one possible set of relationships among early hominids. Other trees have been proposed.

Because the fossil evidence for human evolution is limited, the exact relationships among the various species is not yet known. One possible evolutionary pathway suggests that *Australopithecus afarensis* is a common ancestor of the other australopithecines as well as of the various species of *Homo* (Figure 11–25). Other scientists feel that both australopithecines and humans evolved from a yet undiscovered common ancestor. Either way, recall that *H. habilis* and *H. erectus* also inhabited Africa. It is possible that *H. habilis* replaced the australopithecines and that it, in turn, was eliminated by *H. erectus*. Perhaps *H. erectus* migrated from Africa and gave rise to the Neanderthals. Finally, the Neanderthals may have been eliminated by the most modern *H. sapiens,* Cro-Magnon. A great deal more evidence is needed before the puzzle of human evolution can be solved.

It is possible that *Australopithecus afarensis* is the common ancestor of both other australopithecines and of humans.

REVIEWING YOUR IDEAS

10. Why are human fossils rare?
11. Why have biologists concluded that *Australopithecus* was more like humans than apes?
12. What do *Homo habilis, Homo erectus,* and *Homo sapiens* all have in common?
13. Compare the classification of modern humans and Neanderthals. Explain this classification.

INVESTIGATION

Problem: What evidence can be used to determine if an animal is or was a biped or a quadruped?

Materials:

metric ruler
tape
scissors
cardboard

Diagram A
Diagram B
Diagram C

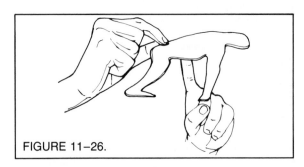

FIGURE 11–26.

Procedure:

Part A. Comparing Characteristics of Bipeds and Quadrupeds

1. Make a copy of Table 11–1.
2. Examine Diagram A. The organisms shown are drawn to scale. The figure on the left is that of an ape. Its pattern of locomotion is called quadrupedalism (walking on all four legs).

   The figure on the right is that of a human. Its pattern of locomotion is called bipedalism (walking on only two legs).
3. Determine the length, in millimeters, of the arm (not including the hand) and the leg (not including the foot) for both figures. Record each measurement in Table 11–1.
4. Note and record in Table 11–1 if the normal position of the hind legs of each organism are straight or bent.
5. Observe the lumbar (lower back) region indicated on each organism. Record in Table 11–1 if this region appears to be straight or has a slight inward (S) curve to it.

Part B. Locating the Center of Gravity

1. Examine Diagram B of an ape and a human. Cut out each figure.
2. Tape each figure onto the cardboard. Cut each figure out of the cardboard.

3. Trace the position and outline of the pelvis for each animal onto the cardboard figure.
4. Attempt to balance each figure on the tip of your finger. Use Figure 11–26 as a guide.
5. Mark the spot on each figure where your fingertip touches when the figure is in perfect balance. This spot marks each animal's center of gravity.
6. Record in Table 11–1 whether each animal's center of gravity is in front of or in the center of the pelvis.

Part C. Evidence for Bipedalism in Early Humans

1. Examine Diagram C. Diagram C shows the shape and normal position of the upper thigh bone (femur) of three animals as seen from a front view. NOTE: *Australopithecus* is considered to be an early human ancestor.
2. Draw a straight, vertical line from the X to the bottom of each figure.
3. Compare the line drawn on each figure to the others. This line indicates where the weight of the upper body is borne on the femur.
4. Record in Table 11–1 if the weight-bearing axis passes along the inside or the outside edge of the lower portion of each femur.

Data and Observations:

TABLE 11-1. COMPARISON OF QUADRUPED AND BIPED CHARACTERISTICS			
	Ape	Human	*Australopithecus*
Arm length			
Leg length			
Position of hind leg			
Shape of lumbar region			
Position of center of gravity			
Location of weight-bearing axis			

Questions and Conclusion:

1. Define each of the following terms:
   (a) *Australopithecus*
   (b) natural selection
   (c) bipedalism
   (d) quadrupedalism
2. Compare arm and leg lengths in a biped.
3. Compare arm and leg lengths in a quadruped.
4. Explain how the position of the hind legs in bipeds differs from the position of the hind legs in quadrupeds.
5. If the center of gravity is located above and in front of the pelvis of an animal, the animal has a tendency to fall over when standing on only two legs. Which animal that you studied in this investigation might have this problem?
6. An animal having more body weight in the upper trunk shows a center of gravity above the pelvis.

   (a) Describe how body weight is distributed in humans.
   (b) Explain the affect of weight distribution on an animal's ability to stand upright.
7. Explain how the shape of the lumbar region of a biped differs from the shape of the lumbar region of a quadruped.
8. Does the evidence indicate that *Australopithecus* was a biped or quadruped? Explain.
9. Explain if each of the following fossil animals was bipedal or quadrupedal:
   (a) arm length = 452 mm; leg length = 520 mm
   (b) outside edge of knee joint more worn than inside edge
   (c) inside edge of knee joint more worn than outside edge

Conclusion: What evidence can be used to determine if an animal is or was a biped or a quadruped?

CHAPTER REVIEW

SUMMARY

1. Adaptations arise from the natural selection of suitable variations present. **11:1**
2. Industrial melanism in peppered moths illustrates evolution of adaptation and shows evolution as an interaction of organisms and environment. **11:2**
3. Adaptations can be grouped as morphological, physiological, or behavioral, but adaptations may fit in more than one category. **11:3**
4. Cryptic coloration, warning coloration, and mimicry are special adaptations that "confuse" predators or prey. **11:4**
5. Speciation occurs as a result of geographic isolation and the evolution of separate gene pools. **11:5**
6. New species are groups of organisms with closed gene pools. **11:5**
7. Gradualism, punctuated equilibrium, and mosaic evolution are hypotheses that attempt to explain the rate at which speciation occurs. **11:6**
8. Adaptive radiation occurs when many new species evolve from a common ancestor in a new environment. **11:7**
9. Convergence is a type of evolution in which different forms of life come to resemble one another. **11:8**
10. Among important human traits are upright posture, bipedal locomotion, prehensile hands, and three-dimensional vision. These traits originated among primates as adaptations to life in trees. **11:9**
11. Earliest of humanlike animals were the australopithecines who lived in Africa between 1.5 and 4 million years ago. **11:10**
12. The earliest species to be classified as humans were *Homo habilis* and *Homo erectus*. Both evolved in Africa. They were larger, less apelike, and had a larger brain than the australopithecines. **11:11**
13. Modern humans, *Homo sapiens,* first appeared about 300 000 years ago. **11:12**

LANGUAGE OF BIOLOGY

adaptive radiation
behavioral adaptation
bipedal locomotion
convergence
cryptic coloration
divergence
ecological niche
gene pool
geographic isolation
gradualism
hominid

industrial melanism
mimicry
morphological adaptation
mosaic evolution
physiological adaptation
prehensile
punctuated equilibrium
reproductive isolation
speciation
species
warning coloration

CHECKING YOUR IDEAS

On a separate paper, complete each of the following statements with the missing term(s). Do not write in this book.

1. Similarities in the leg of a grasshopper and the leg of a frog show the pattern of evolution known as _____.
2. Speciation begins as a result of _____ isolation.
3. _____ is the general term given to early humans.
4. Migration of birds is an example of a(n) _____ adaptation.
5. Every organism occupies a special _____ in its environment.
6. _____ is any barrier to interbreeding.
7. _____ may occur when many species evolve from a common ancestor in a new environment.
8. Java, Neanderthal, and modern humans are all classified as _____.
9. In mimicry, the _____ is distasteful or harmful.
10. Protection by camouflage is called _____.

11

EVALUATING YOUR IDEAS

1. Explain: The measure of a population's success is its ability to reproduce.
2. Do complex adaptations such as the eye appear all at once? Explain.
3. Explain industrial melanism as it occurred in peppered moths. Would it be easily explained in Lamarckian terms? Explain.
4. Give three examples for each of the three kinds of adaptation. How is each adaptive?
5. Why do isolated gene pools evolve differently from one another?
6. When does one species become two?
7. Which hypothesis, gradualism or punctuated equilibrium, seems best supported by the fossil record? Why? How does the idea of mosaic evolution fit with the fossil record?
8. Explain the conditions necessary for adaptive radiation to occur. How were they met by Darwin's finches?
9. Explain the origin and usefulness of prehensile hands, three-dimensional vision, and bipedal locomotion. How are each of these traits thought to have been useful later to the earliest human ancestors?
10. Why would there have been pressure for evolution of larger brains among hominids? How does greater intelligence make humans unique?
11. Describe the features of *Australopithecus*. When and where did this prehuman live?
12. Why is *Homo habilis* considered more humanlike than *Australopithecus?* When and where did it live?
13. Describe the features of *Homo erectus*. What effect might early *Homo erectus* have had on *Homo habilis* and *Australopithecus?*
14. How is geographic isolation important?
15. When did Neanderthals live? Where? How are Neanderthals different from *Homo erectus?*
16. In what ways were Cro-Magnons unique?

APPLYING YOUR IDEAS

1. Learning is a type of behavior in complex animals. For example, a dog can learn a trick. Should learning be considered a behavioral adaptation? Would it be better to say that the basis for learning is an adaptation? Explain.
2. Human eye color and human blood types are examples of inherited traits. Do you think that these traits are adaptations? Can you offer a hypothesis to explain why all humans have these traits?
3. Recent evidence suggests that pollution in England is decreasing. How do you think decreasing pollution is affecting the peppered moth populations?
4. Consider the four conditions of the Hardy-Weinberg Law (Chapter 10). Which condition applies to the case of industrial melanism?

EXTENDING YOUR IDEAS

1. Find out about the Irish elk and the possible causes of its extinction.
2. Prepare a report on the evolution of the small bones of the mammalian ear.
3. Carefully study the external and internal anatomy of an animal, such as a frog or a grasshopper. Make a list of its morphological adaptations. Explain how the organism's structures suit their functions.

SUGGESTED READINGS

Inouye, David W., "The Ant and the Sunflower." *Natural History,* June, 1984.

Leakey, Richard, and Walker, Alan, "Homo Erectus Unearthed." *National Geographic,* Nov., 1985.

Weaver, Kenneth F., "The Search for Our Ancestors." *National Geographic,* Nov., 1985.

CLASSIFICATION

If you were asked to classify the animals shown here, how would you begin? First you would be sure to notice the features that identify them as birds. What features distinguish birds from other animals? What features distinguish each of these birds from the others? How might you name each animal so other people would be able to classify birds with the same characteristics? Observation of physical differences is the first step in classification.

chapter 12

Humans often organize sets of objects by classifying them. For example, musical instruments are placed in several groups — brass, percussion, woodwind, and string. Motor vehicles can be categorized as cars, trucks, and vans. Each broad group is then subdivided into several more specific groups. Woodwinds consist of clarinets, saxophones, and oboes. Cars consist of sedans, coupes, and convertibles. As new instruments or motor vehicles are developed, they can be placed in the proper category in the classification system. A classification system brings order and logic to a set of related objects.

Recall that there are nearly two million known species of organisms. Biologists have long been interested in classifying them. Today, organisms are classified in categories called **taxa** (singular, taxon). The science of classifying organisms is called **taxonomy** (tak SAHN uh mee). A classification system for organisms brings order to the great diversity of life forms. It also serves as a basis for identification of newly discovered organisms.

Objectives:
You will
- describe the need and basis for classifying organisms.
- discuss the development of the classification system.
- explain the system used in modern classification.

Taxonomy is the science of classifying organisms.

THEORY OF CLASSIFICATION

Rather than dealing individually with millions of different organisms, biologists place them in major groups. Each group has a certain set of features. When a new organism is discovered, its characteris-

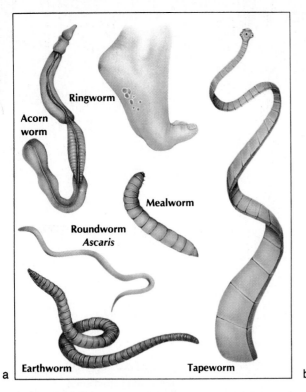

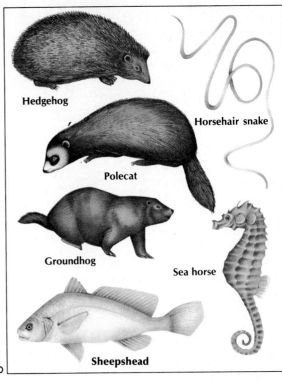

FIGURE 12–1. Common names are confusing for scientific work. (a) A common name, such as worm, may have many different meanings. (b) Some organisms have common names that include common names of other, totally different organisms.

tics are studied, and it is then added to the proper group. If its features are unique, it may lead to the formation of an entirely new group. New organisms are being discovered continually. For example, recent exploration of marine caves has led to discovery of more than twenty different species of crustaceans (the group that includes shrimp, lobsters, and crabs).

12:1 The Need for Classification

What are some of the reasons for classifying organisms? One reason is the need for order and organization. You know it would be difficult to find a certain book in a disorganized library. The same is true for organisms. It would be difficult to find information about a certain organism if organisms were not in some order. Because they are put in a certain order, information about similar organisms is also easier to find.

Taxonomy enables biologists to study and identify organisms more easily.

Another reason to classify organisms is that a logical means for naming organisms is needed. Common names are inadequate for use in a uniform classification system. The word *frog*, for instance, suggests a certain mental image to you, but it is inaccurate as a scientific label. What kind of frog is it? Is it a grass frog, a tree frog, or a bullfrog? Consider also the common word *worm*. To you, this name

Classification results in precise names of organisms.

Common names for organisms can be confusing.

probably suggests an animal that is slimy and soft; but biologists are familiar with many worms such as roundworms, flatworms, and segmented worms. Also, there are organisms such as ringworms, mealworms, and acorn worms that are not worms. A ringworm is a fungus, a mealworm is an insect in the larval stage, and an acorn worm is a simple relative of the vertebrates.

Also, common names vary country to country and language to language. They even vary region to region. Consider that puma, cougar, and mountain lion refer to the same organism.

12:2 Binomial Nomenclature

Many early biologists devised classification schemes. Aristotle divided organisms into two groups—plants and animals. Plants were classified on the basis of structure and size—herbs, shrubs, and trees. Animals were subdivided on the basis of habitat—air, land, and sea. But a classification system should have the same basis for all groupings. In the eighteenth century, Carolus Linnaeus (luh NAY us) developed a classification system having just one basis—structural features. According to Linnaeus, each type of organism was a distinct species. If organisms had the same set of features, they were the same species. Linnaeus' decision to group organisms on the basis of structure was important. Many of his groupings are still used today. Different species having similar features were classified together in broader groups.

Linnaeus introduced a two-term naming system, **binomial nomenclature** (bi NOH mee ul ● NOH mun klay chur), for classifying organisms. Each organism is given a two-word Latin name. The first word, a noun, is the **genus** (JEE nus; plural genera, JEN ur uh) to which an organism belongs. Its first letter is capitalized. The second word, an adjective, represents the **species** (the specific name). Its first letter is not capitalized. Notice that the name of a species consists of both the genus and the specific name.

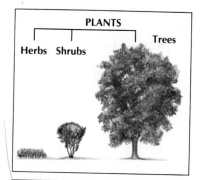

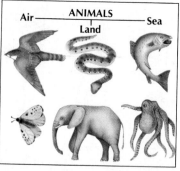

FIGURE 12–2. Aristotle divided organisms into plants and animals. He classified plants on the basis of structure and animals on the basis of habitat.

In binomial nomenclature, each organism is named by genus and species.

FIGURE 12–3. Scientific names list the genus and species of organisms. Both animals are considered frogs. However, (a) the bullfrog, *Rana catesbeiana*, and (b) the tree frog, *Hyla versicolor*, are classified in different genera. What is the genus of each?

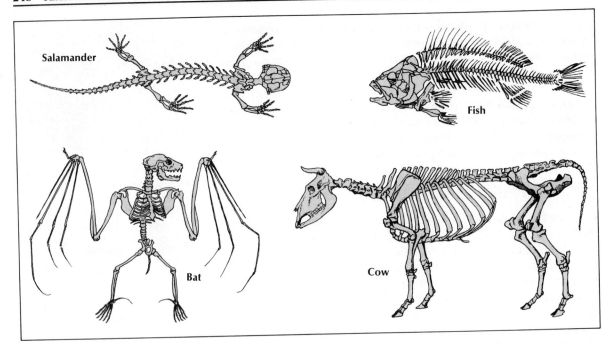

FIGURE 12–4. Similarity in structure is a basis for classification. These animal skeletons show many homologous structures.

Genus and species are different taxa. Genus is a broader category than species. Thus, a single genus may contain many different species. For example, many cats belong to the genus *Felis* (FEE lus), but there are many species of cats. A wildcat is *Felis sylvestris* (sihl VES trus), an ocelot is *Felis pardalis* (PAR duh lus), a cougar is *Felis concolor,* and a house cat is *Felis domesticus* (doh MES tih kus). Oak trees belong to the genus *Quercus* (KWUR kus). A red oak is *Quercus rubra* and a white oak is *Quercus alba*. Note that in print these names are usually italicized. In handwriting the names should be underlined. Sometimes the genus name is abbreviated by using only the first letter. *Felis sylvestris* is sometimes written as *F. sylvestris*.

12:3 Bases for Classification

Knowledge of evolution is used today in classifying organisms.

Understanding of evolution provides the modern basis for classification. Biologists group organisms based on their evolutionary relationships. Lines of evidence that support the concept of evolution are used today as guides to classification. Often, several types of evidence are used together. Organisms that have a common ancestor are closely related, and they are grouped together. The various species of cats are thought to have evolved from a common ancestor. They are all members of the same genus, *Felis*. Organisms not descended from a common ancestor are placed in different groupings. For example, snakes and worms are in the same kingdom, but their classification differs from that point on.

A major line of evidence for classification is the presence of homologous structures (Section 10:3). Although Linnaeus did not realize it, by using structure as his basis, he was grouping organisms based on evolutionary relationships. That is why much of his classification remains valid today.

Often, problems in classification can be solved using other lines of evidence. Comparative biochemistry (Section 10:4) is often useful. For example, the horseshoe crab (Figure 12–5) was thought to be a true crab as the common name implies. However, blood studies showed that this animal is more like a spider. As a result, its classification was changed, and it is now included as a relative of spiders. Comparative embryology (Section 10:4) has shown that the major group to which vertebrates belong is most closely related to the major group that includes the starfish!

Modern genetics has become a very important tool of taxonomy. For example, organisms that have similar numbers and kinds of chromosomes may be closely related. Recall that DNA (genes) of different species can be compared (Section 10:5). How is this done? Double-stranded DNA from each of two different organisms is separated into single strands. The DNA of one of the organisms is made radioactive. The DNA of the two organisms is then mixed. Due to base-pairing, the single strands bond with one another to form double-stranded molecules. Bonding will only occur where the genes (sequences of bases) are similar. Thus, the more bonding, the more closely related the organisms. This technique has shown that closely related organisms have many genes in common.

Recently, mitochondrial DNA (Section 9:9) has been used to study evolutionary relationships. DNA of mitochondria from different species can be analyzed and compared. This DNA is known to mutate at a certain rate. Using these known mutation rates and comparing the DNA, biologists can approximate the time when two species

FIGURE 12–5. The horseshoe crab, *Limulus polyphemus* (bottom view shown), is not a true crab, but a relative of the spider. Comparative studies of the blood of these organisms have caused the reclassification of horseshoe crabs with spiders.

Organisms having similar DNA are classified in similar ways.

FIGURE 12–6. Research comparing the mitochondrial DNA of the quagga, zebra, and horse may be useful in understanding their evolution.

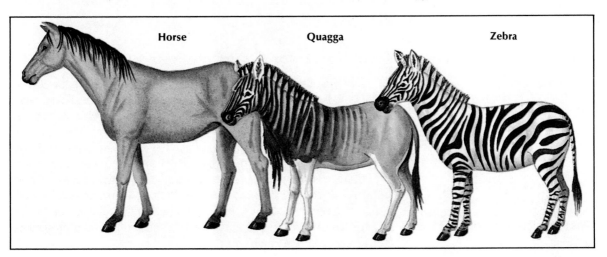

Horse   Quagga   Zebra

began to diverge from a common ancestor. Species having mitochondrial DNA with very few differences in base sequence are closely related. Such studies can be a basis for classifying organisms. Recent studies have even analyzed mitochondrial DNA of the quagga, a zebra extinct since 1883! (The DNA was extracted from some preserved muscle tissue.) Two segments of quagga DNA have been analyzed and compared to DNA of other zebras. This research may provide new insight into the evolution and classification of various zebra species and the relationship between zebras and horses.

REVIEWING YOUR IDEAS

1. What is taxonomy? What are taxa?
2. How did Aristotle classify organisms?
3. What is binomial nomenclature?
4. Why is Linnaeus' system better than Aristotle's system?
5. What is the basis of modern taxonomy?

SYSTEM OF CLASSIFICATION

Using the evidence discussed in Section 12:3, a system of classifying living things has been devised. Each organism is given a two-word name (binomial nomenclature) that is its genus and species. This name is used because it is the most specific. Each species is a distinct form of life.

FIGURE 12–7. (a) The paper birch, *Betula papyrifera,* and (b) the yellow birch, *Betula alleghaniensis,* are members of the same genus.

12:4 Classification Groups

Genus and species are not the only classification groups. Before organisms are placed in these quite specific categories they are first grouped more broadly. Placement in each taxon is based on similar features. In the complete classification of an organism the groups are **kingdom, phylum** (FI lum), **class, order, family,** genus, and species.

Each group from kingdom to species becomes more specific as each step narrows the number of organisms of the previous group. A kingdom is the broadest of all taxa. Recall that there are five kingdoms in today's classification scheme (Section 2:8). All the organisms in each kingdom share the same basic features. Each kingdom is subdivided into phyla. Phyla in turn are subdivided into classes, classes into orders, orders into families, families into genera, and genera into species. Each species is one certain type of organism.

As presented, the sequence is in its simplest form. Classification is often more complex because each of these major groups may be further subdivided. For example, a single species may be made up of several subspecies. Subspecies are sometimes called varieties or races. The various breeds of dogs are all subspecies.

FIGURE 12–8. The Brahman cattle and cattle egrets shown here are in separate classes of the same phylum, Chordata. The birds are in the class Aves, and the cattle are in the class Mammalia. The grasses are classified in a separate kingdom, the plants.

12:5 Some Examples

Consider the complete classification of a common house cat. Because it is an animal, it is placed in Kingdom Animalia (an uh MAY lee uh). A cat is classified as an animal based on the very broad features of all animals. A cat eats rather than makes its own food. A cat also moves around and has a nervous system.

TABLE 12-1. CLASSIFICATIONS OF SOME ANIMALS				
Division	House Cat	Dog	Human	Grasshopper
KINGDOM	Animalia	Animalia	Animalia	Animalia
PHYLUM	Chordata	Chordata	Chordata	Arthropoda (ar THRAHP uh duh)
SUBPHYLUM	Vertebrata	Vertebrata	Vertebrata	
CLASS	Mammalia (muh MAY lee uh)	Mammalia	Mammalia	Insecta (ihn SEK tuh)
ORDER	Carnivora (kar NIHV uh ruh)	Carnivora	Primates (PRI may teez)	Orthoptera (or THAHP tuh ruh)
FAMILY	Felidae (FEE luh dee)	Canidae (KAN uh dee)	Hominidae (hoh MIHN uh dee)	Locustidae (loh KUS tuh dee)
GENUS	Felis	Canis	Homo (HOH moh)	Schistocerca (shis tuh SUR kuh)
SPECIES	Felis domesticus	Canis familiaris	Homo sapiens (SAY pee unz)	Schistocerca americana

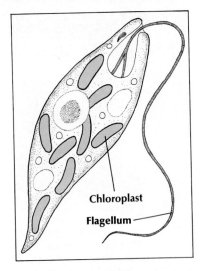

Chloroplast

Flagellum

FIGURE 12–9. Classification of organisms such as *Euglena* poses special problems. *Euglena* moves like an animal but can make its own food like a plant.

Many organisms cannot be easily grouped as either plants or animals.

Because of certain other features, a cat is then put into Phylum Chordata (kor DAHT uh). An important subphylum of Chordata is Vertebrata (vurt uh BRAHT uh), animals with backbones. The complete classification of a common house cat appears in Table 12–1. Each group represents more specific features than the previous group. Notice that in writing the species name, the genus is repeated.

Compare the classification of a cat to that of a dog (Table 12–1). Because cats and dogs share the same broad features, they are in the same kingdom, phylum, subphylum, class, and order. At the family level, the classification differs, because a dog's features are distinct from those of a cat. Each of the remaining groups is, of course, also different from the others.

Now compare the cat and dog to humans (Table 12–1). How closely do humans seem to be related to dogs and cats? How many groups do humans have in common with dogs and cats? At what level do the classifications differ?

Finally, compare these three animals to the common American grasshopper (Table 12–1). The only common group among the grasshopper, cat, dog, and human is kingdom — all of these organisms are animals. What are some of the features of the grasshopper that cause biologists to classify it in a phylum different from that of the cat, dog, and human?

12:6 The Kingdom Problem

You know that most biologists today use a five-kingdom system of classification. For many years, though, all organisms were classified as either plants or animals. This two-kingdom scheme was reasonable for most familiar, multicellular organisms. However, the later discovery of microscopic organisms posed a problem. Some unicellular forms were easily placed in one kingdom or the other. However, many did not fit neatly with either group. For example, *Euglena* (yoo GLEE nuh) has features of both kingdoms. *Euglena* is mobile like an animal and autotrophic like a plant. Also, at certain times it may lose its chlorophyll and become heterotrophic.

There are other problems with a two-kingdom system. For example, how should fungi such as mushrooms and molds be classified? They are like plants in structure and do not move, but they are heterotrophic like animals. Sponges are heterotrophic, but they do not move around and they show little response to changes in their environment.

Development of the five-kingdom system has solved some of these problems. However, the new system was not invented just to solve those problems; it is based on current knowledge of evolutionary relationships among organisms. Problems have been solved because the new system makes a great deal more "scientific sense."

PEOPLE IN BIOLOGY

William Montague Cobb has spent most of his life studying the structural features of humans. He has been an anatomist for fifty-one years at the Howard University Medical School. Dr. Cobb founded a comparative anatomy museum to assist his students in studying the relationships between many different organisms and their structures. He has also made a collection of over 600 human skeletons to research the effects of aging on the skeletal system.

William Montague Cobb

(1904–   )

Besides his work in anatomy, Dr. Cobb has also done research in physical anthropology, public health, and medical education. He has published papers and articles in all these fields and has become especially well known for his work on the growth and development of American blacks.

Dr. Cobb has received many honors for his work. He was the first black president of the American Association of Physical Anthropologists, and he served as editor of the *Journal of the National Medical Association* for 28 years. He has received the Distinguished Service Medal of the National Medical Association and the U.S. Navy Distinguished Public Service Award. Dr. Cobb was married in 1929, has two daughters, and now lives in Washington, D.C.

Other classification systems have also been proposed. Some systems are based on four kingdoms; others, on six or even seven. A classification system based on five kingdoms is presented in Appendix A. Use this system as a reference.

In the next unit you will study the major phyla and some classes of organisms in detail. You will learn about the features that determine each group's classification. In Chapters 17 through 30, you will compare how life functions are carried out by the various forms of life. The similarities in life processes among organisms will become apparent.

Although organisms differ in physical characteristics, they have many functions in common.

REVIEWING YOUR IDEAS

6. What are the major classification groups from broadest to most specific? Which of the groups are used in the scientific name of an organism?

7. Why do many biologists use a five-kingdom system of classification?

INVESTIGATION

Problem: What information can be gained from classifying living organisms?

Procedure:

Part A. Determining Major Kingdom Traits

1. Make a copy of the outline shown as Figure 12–10.
2. Complete the brackets on Figure 12–10 so that the diagram correctly names the traits of the kingdoms listed at the end of the outline. Choose from the following list of terms:

   usually unicellular   multicellular
   autotrophic   heterotrophic
   prokaryotic   ingest food
   eukaryotic   absorbs food

3. Complete Figure 12–10 by listing three examples of organisms for each kingdom. Choose your examples from those shown on page 33 in your text.

Part B. Binomial Nomenclature

1. Copy the two columns marked Table 12–2.

TABLE 12-2. MATCHING SCIENTIFIC AND COMMON NAMES	
I	II
___ Beta vulgaris	a. black pepper
___ Salmo gairdneni	b. perch
___ Citrus limon	c. rat
___ Piper nigrum	d. canary
___ Ipomoea batatus	e. penicillin
___ Tarpor atlanticus	f. herring
___ Crocodylus americanus	g. beet
___ Rattus norvegicus	h. carrot
___ Perca flavescens	i. crocodile
___ Daucus carota	j. lemon
___ Serinus canarius	k. sweet potato
___ Clupea harengus	l. salmon
___ Penicillium chrysogenum	m. Atlantic tarpon

2. Many of the common names of organisms are derived from scientific names. For example, the scientific name of the elephant is *Elephas maximum*.
3. Match each of the scientific names in column I with the correct common name in column II of that organism.
4. Copy the two columns marked Table 12–3.

TABLE 12-3. MATCHING SCIENTIFIC AND COMMON NAMES	
I	II
___ Aralia quinquefolia	a. dog
___ Crotaphytus collaris	b. seventeen-year locust
___ Helianthus annus	c. cat
___ Eurycea bilineata	d. paper birch
___ Alytes obstetricans	e. sea cucumber
___ Ursus horibilis	f. sugarcane
___ Magicicada septendecem	g. collared lizard
___ Betula papyrifera	h. earthworm
___ Secale cereale	i. two-lined salamander
___ Felis domesticus	j. sunflower
___ Saccharum officinarum	k. midwife toad
___ Lumbricus terrestris	l. grizzly bear
___ Cucumaria frondosa	m. rye
___ Canis familiaris	n. five leaflet ginseng

5. All scientific names are derived from Greek or Latin. These Greek or Latin terms may sound like or are synonyms for English words. For example, the scientific name for the rubber plant is *Ficus elastica*. Elastica sounds like elastic, which describes rubber.
6. Analyze each of the scientific names in column I. Match each of the scientific names with the correct common name of that organism in column II.

Data and Observations:

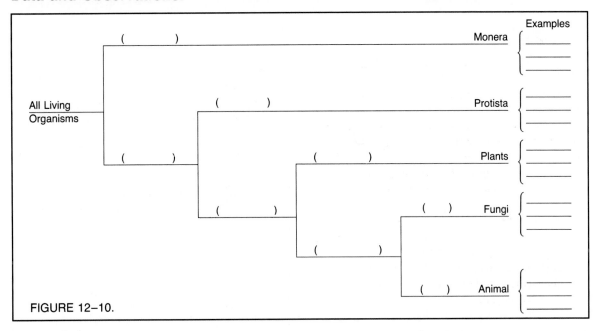

FIGURE 12–10.

Questions and Conclusion:

1. Define:
   (a) binomial nomenclature
   (b) taxonomy
   (c) eukaryotic
   (d) prokaryotic
   (e) autotrophic
   (f) heterotrophic
2. (a) What category of classification is used for the first term in a scientific name?
   (b) What category of classification is used for the second term in a scientific name?
3. Describe three rules that are used in writing a scientific name. Do not use the information in the question above.
4. An organism is found to be eukaryotic and to ingest food. Using your completed Figure 12–10:

   (a) name the kingdom to which this organism belongs.
   (b) list two other traits of this organism.
5. An organism is found to be prokaryotic. Using completed Figure 12–10, name the kingdom that this organism belongs to. Explain.
6. An organism belongs to the plant kingdom. Using completed Figure 12–10, list three traits that this organism will have.
7. An organism belongs to the fungi kingdom. Using completed Figure 12–10, list three traits that this organism will have.
8. List two reasons why there is a need for classifying organisms.
9. Explain why the language of classification is based mainly on Latin.

Conclusion: What information can be gained from classifying living organisms?

255

CHAPTER REVIEW

SUMMARY

1. Classification brings order to the great variety of organisms alive today. Precise classification of organisms is necessary so that each living thing has a name recognizable to biologists everywhere. **12:1**
2. Linnaeus devised a system of classification based on structural similarities. His system of binomial nomenclature is still used today. **12:2**
3. Using the system of binomial nomenclature, organisms are given a two-word Latin name. The first identifies the genus and the second, the species. **12:2**
4. Modern taxonomy depends on knowledge of evolution. Classification is based on factors such as homologous structures, biochemical similarities, and genetics. **12:3**
5. The classification categories, from general to specific, are kingdom, phylum, class, order, family, genus, and species. Each of these major categories may be subdivided. **12:4**
6. Comparison of classification of organisms reveals that closely related forms are classified very similarly. **12:5**
7. Many systems of classification employ a five-kingdom plan: monerans, protists, fungi, plants, and animals. **12:6**
8. The five-kingdom system of classification is based on current knowledge of evolutionary relationships and solves many of the problems associated with the older systems. **12:6**

LANGUAGE OF BIOLOGY

binomial	kingdom
nomenclature	order
class	phylum
family	taxa
genus	taxonomy

CHECKING YOUR IDEAS

On a separate paper, complete each of the following statements with the missing term(s). Do not write in this book.

1. _____ is the broadest classification category.
2. The two-term name for the system of naming organisms is _____.
3. _____ is the first taxon after class.
4. Another term for race is _____.
5. The first taxon above genus is _____.
6. _____ is the most specific taxon.
7. In Aristotle's system of classification, animals were classified on the basis of _____.
8. Analysis of mutation rates of _____ may be a modern basis for classification.
9. Without knowing it, Linnaeus used _____ as a basis for classification.
10. Organisms classified in the same phylum *must* also be classified in the same _____.

EVALUATING YOUR IDEAS

1. Why is it necessary to classify and clearly name living things?
2. In modern terms, why was Aristotle's basis for classifying plants better than his basis for classifying animals?
3. Upon what did Linnaeus base his system of classification? Why is his system better than Aristotle's? What later information might have been helpful to him?
4. Is there any relationship between the words species and specific? Explain.
5. Why are the many lines of evidence of evolution also used in classifying organisms?
6. Explain several ways in which analysis of DNA can be useful in taxonomy.
7. What structural features of humans put them in the same kingdom, phylum, subphylum,

and class as cats and dogs? Why are humans in a different order from cats and dogs?

8. Give several reasons why a five-kingdom system of classification works better than a two-kingdom system.

APPLYING YOUR IDEAS

1. Both cats and dogs are grouped into the same kingdom, phylum, subphylum, class, and order. List their common structural features. Cats and dogs belong to different families, genera, and species. Therefore, they must have some different structural features. List some of the ways in which they differ.

2. Based on your own knowledge of their characteristics, set up a scheme to place the animals listed below in groups and subgroups. No scientific terms are needed.

(a) baboon
(b) bear
(c) cheetah
(d) clam
(e) earthworm
(f) frog
(g) fruit fly
(h) horse
(i) human
(j) jellyfish
(k) lobster
(l) mosquito
(m) ostrich
(n) panther
(o) planaria
(p) robin
(q) turtle
(r) leech

3. How can the similar features of two different species (such as tiger and cheetah) be explained in terms of evolution?

4. Two groups of salamanders resemble each other very closely. One group lives in northeastern United States. The other group lives in southeastern United States. A biologist discovers that the southeastern group mates in April, and the northeastern group mates in June. How would you classify the two groups? Why? Would the two groups interbreed if brought together? Explain.

5. Many common names of organisms have come from their scientific names. What is the com-

mon name of each of these organisms?

Rattus norvegicus
Equus zebra
Elephus maximus
Gorilla gorilla
Pinus ponderosa
Camelus bactrianus

6. Would similarity in proteins be an indication of close relationship between two different kinds of organisms? Explain.

EXTENDING YOUR IDEAS

1. Obtain a classification key to woody plants. Use the key to identify woody plants near your home.

2. Use a classification key to identify protists from a pond, lake, or mud puddle.

3. Subdivide a culture of *Euglena* among 4 containers. Leave one culture exposed to light at room temperature as a control. Place another culture in a dark area at room temperature. Place a third in a warm, lighted environment. If available, add streptomycin (an antibiotic) to another culture (in light, room temperature). Note the effect of darkness, heat, and streptomycin on the color of *Euglena*.

SUGGESTED READINGS

Barnes, R.S.K., ed., *A Synoptic Classification of Living Organisms*, Sunderland, MA, Sinauer Associates, Inc., 1984.

Cole, Charles J., "Taxonomy: What's in a Name?" *Natural History*, Sept., 1984.

Gilbert, Susan, "Taking Inventory of Life." *Science Digest*, April, 1986.

Margulis, Lynn, and Schwartz, Karlene, *Five Kingdoms: An Illustrated Guide to the Phyla of Life on Earth*. San Francisco, W. H. Freeman and Co., 1982.

DIVERSITY

Natural selection and adaptation have produced the great diversity of organisms on Earth today. Many forms of life are now extinct; still others have yet to be discovered. Each life form has certain physical traits, some of which show relationships among organisms. By studying traits of organisms, biologists can place organisms in various groups.

In this unit you will study the major groups of organisms, as well as some of the traits used to divide these major groups into subgroups. You will survey the features of the life forms in the five kingdoms of organisms—monerans, protists, fungi, plants, and animals.

This tidepool shows both plant and animal life forms. They have different traits that are used to classify them into their respective kingdoms. Diversity can be seen not only in the contrasting colors and shapes of all the different organisms shown, but also in the different colors of just the starfish. When you finish these chapters, you will have not only an appreciation for the diversity of life, but also a better appreciation for the ways in which organisms are classified.

Tidepool

MONERANS, PROTISTS, FUNGI, AND VIRUSES

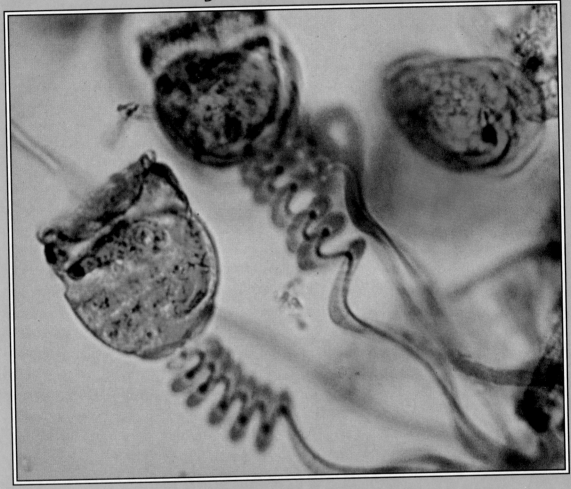

Some organisms cannot be classified in either the plant or the animal kingdom. They fall into three other kingdoms—monerans, protists, and fungi. Vorticella is a microscopic protist, shown here magnified about 1400 times. Vorticella, like many other simple organisms, lives in water. What are the main traits of some of the simple organisms? How are simple organisms economically and ecologically important?

chapter 13

In the two-kingdom system of classification, many organisms were considered as either plants or animals even though their features did not match those groups. One major problem in classifying these organisms was that they were quite simple in structure. Many are unicellular. All lack the complex structure of true plants and animals. Many have features of both kingdoms. For these reasons, relationships could not be easily determined.

The gradual change to a five-kingdom system has solved many of these problems. These simple organisms are now classified as members of the three new kingdoms — monerans, protists, and fungi. Although simple in structure and less familiar than plants and animals, these organisms play important roles in nature. Their activities affect all natural communities and the lives of humans and other organisms.

Objectives:
You will
- compare the features of monerans, protists, and fungi.
- list characteristics and examples of organisms in these kingdoms.
- discuss viruses and their structure.

Monerans, protists, and fungi play important roles in nature and affect the lives of many organisms.

KINGDOM MONERA

Kingdom Monera includes two groups of bacteria — the **Archaebacteria** (AR kee bak TIR ee uh), ancient bacteria, and the **Eubacteria,** (YEW bak TIR ee uh), true bacteria. Each group includes several phyla. All monerans (bacteria) are prokaryotes (Section 4:12). Some may be descendants of the first types of cells to have evolved on Earth.

Eubacteria and Archaebacteria are the two major groups of monerans.

13:1 Characteristics of Prokaryotes

A prokaryote has no true nucleus or nuclear membrane. However, its single circular chromosome is located in a certain area called a **nucleoid.** The chromosome consists of DNA, which, unlike that of

Prokaryotes have neither a true nucleus nor membrane-bound organelles.

261

TABLE 13-1. DIFFERENCES BETWEEN PROKARYOTES AND EUKARYOTES	
Prokaryotes	Eukaryotes
no true nucleus or nuclear membrane	true nucleus and nuclear membrane
single, circular chromosome of DNA	several, linear chromosomes of DNA and protein
small ribosomes	large ribosomes
no mitochondria, ER, Golgi bodies, lysosomes	mitochondria, ER, Golgi bodies, lysosomes
chlorophyll, if present, not in chloroplasts	chlorophyll, if present, in chloroplasts
cell wall, if present, containing murein or other substances	cell wall, if present, made of different substances
no microtubules in flagella of most forms	microtubules in flagella of all forms

eukaryotes, has no histone proteins. Small ribosomes are present, but cell parts such as ER, Golgi bodies, and lysosomes are absent. Prokaryotes have no mitochondria; aerobic respiration, if it occurs, takes place on the inner surface of the cell membrane. Many prokaryotes are photosynthetic. However, they have no chloroplasts. Cell walls are often present and are usually composed of a substance called **murein** (MYOOR ee un). This substance differs from any of those found in cell walls of fungi, certain protists, and plants. In most forms, flagella, if present, lack microtubules.

Prokaryotic cells tend to be much smaller than eukaryotic cells. In general, they are about one-tenth the size of cells in more complex organisms. Table 13–1 summarizes the main differences between prokaryotes and eukaryotes.

FIGURE 13–1. The three basic shapes of bacteria, (a) coccus (b) bacillus and (c) spirillum, are shown in scanning electron micrographs. The cocci are magnified 6000 times. The bacilli are magnified 4000 times. The spirillum is magnified 8000 times.

13:2 Structure of Bacteria

Bacterial cells have one of three shapes. Spherical forms are called **cocci** (KAHK si), rod-shaped forms are **bacilli** (buh SIHL i), and spiral-shaped cells are called **spirilla** (spi RUHL uh). Many bacteria are unicellular.

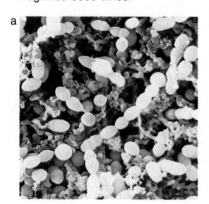

a

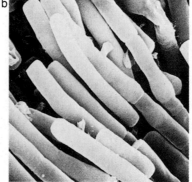

b

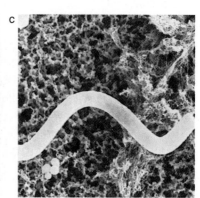

c

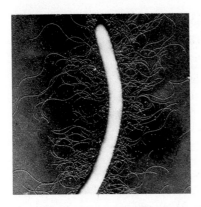

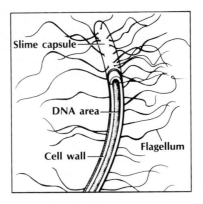

FIGURE 13–2. The flagellated bacterium, *Proteus vulgaris,* shows many bacterial cell parts. This cell is shown magnified 21 000 times.

Reproduction in bacteria is always asexual (ay SEK shul). **Asexual reproduction** is the production of one or more offspring from a single parent. After reproduction occurs in some species, the cells remain together to form pairs, chains, or clusters with each cell within the group being an independent organism. In certain photosynthetic species of bacteria, the cells form filamentous structures. These filaments are considered to be primitive multicellular organisms.

Most bacterial cells are enclosed by a cell wall. This wall protects the cell and helps maintain an osmotic balance (Section 4:6) between the bacterium and its environment. Penicillin is effective against many bacteria because it hinders the production of murein needed to make new cell walls. Thus, the bacteria cannot reproduce. Often the cell wall is surrounded by an outer slime capsule, which may protect the cell. Many bacteria can move, most often by means of flagella.

In addition to the main chromosome, many bacteria have smaller, circular segments of DNA called plasmids. Replication of plasmids occurs independently from that of the main chromosome. Sometimes part of a plasmid may join the chromosome and replicate along with it. Other plasmids never join the chromosome. Some plasmids carry genes that enable the bacterium to resist antibiotics. When the bacterium comes into contact with an antibiotic, these plasmids may replicate many times. This replication aids the survival of the bacterium because many copies of the gene that confers resistance are made.

Many bacteria, especially bacilli, are adapted for withstanding harsh conditions. When living conditions become unfavorable, a tough, protective wall begins to form around the bacterial DNA and a small bit of cytoplasm. The DNA, cytoplasm, and wall form a highly resistant, dormant structure called an **endospore.** Once the endospore has formed, the rest of the cell may die. The endospore, however, can resist periods of freezing, boiling, or drought. When conditions become favorable again, the endospore develops into an active cell. Note that endospore formation is not considered a means of bacterial reproduction.

The bacterial cell wall is important in maintaining an osmotic balance between the cell and the environment.

FIGURE 13–3. Certain bacteria, such as *Clostridium,* shown here magnified about 25 000 times, form endospores that enable them to withstand conditions that would kill an active cell. When conditions are right, the endospore germinates, giving rise to an active bacterial cell.

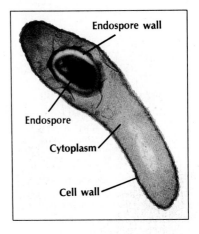

13:3 Heterotrophic Eubacteria

Two phyla of Eubacteria are heterotrophic. These phyla are classified on the basis of whether or not chemicals in the cell wall react with Gram's solution, a solution used to stain cells for microscopic observation. Members of the phylum in which there is a reaction with Gram's solution are referred to as **Gram-positive bacteria.** The members of the other phylum, in which no reaction occurs with Gram's solution, are known as **Gram-negative bacteria.** These bacteria include both gram-positive and gram-negative rods and cocci.

These heterotrophic bacteria are everywhere — in air, food, soil, and on almost everything you touch. Many are parasites, living in or on other organisms, and some of these forms cause diseases in humans and other organisms. Other species are saprophytes (Section 1:2) and play an important role in recycling materials. Most species respire aerobically, but some species carry out anaerobic respiration (Chapter 5).

13:4 Autotrophic Eubacteria

Several phyla of Eubacteria are autotrophic. A major group is Phylum **Cyanobacteria,** formerly known as blue-green algae. These bacteria undergo aerobic respiration. In addition to chlorophyll, they contain a blue pigment called **phycocyanin** (fi koh SI uh nun). Together, these two pigments give many cyanobacteria their blue-green color. Yellowish pigments called **carotenes** (KER uh teenz) are also present, and some forms also contain a red pigment. Depending on the combination of pigments present, Cyanobacteria may be a variety of colors including black, yellow, green, and red.

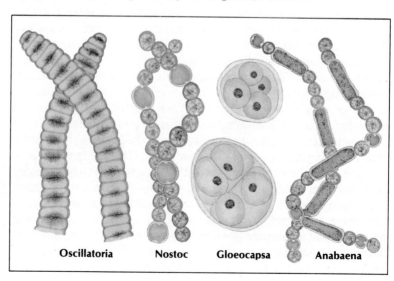

Oscillatoria   Nostoc   Gloeocapsa   Anabaena

FIGURE 13–4. Cyanobacteria can be single cells or cells grouped in colonies or filaments. The cells are prokaryotic, lacking nuclei and membrane-bound organelles.

Pigments of Cyanobacteria are located in flattened membranes within the cytoplasm. These membranes appear similar to those found in chloroplasts of eukaryotes (Section 4:13). The cytoplasm sometimes contains storage granules of a starchlike substance.

Although some species of Cyanobacteria exist as multicellular organisms, there is little, if any, division of labor in such forms. Each cell is basically the same. Some forms can move, although they have no flagella. How they move is not known.

Cyanobacteria are important producers of aquatic communities. They are common in ponds, lakes, puddles, streams, and even moist places on land. Many forms are important in recycling nitrogen (Section 33:5).

Other autotrophic Eubacteria include the **green-sulfur bacteria** and the **purple bacteria.** These phyla have a unique form of chlorophyll. Photosynthesis occurs in the absence of oxygen, and no oxygen is given off. Also, water is not used as a reactant. Phylum **Prochlorophyta,** discovered in 1976, consists of organisms with pigments very much like those of eukaryotic autotrophs. Their photosynthetic reactions are also like those of eukaryotes.

Some true bacteria are autotrophic, but not photosynthetic. Rather, they derive energy from inorganic materials such as sulfur and nitrogen compounds. The energy is used to make organic compounds in a process called **chemosynthesis** (kee moh SIHN thus sus). The various means by which the Eubacteria obtain energy is a major reason for the success of these primitive organisms.

Cyanobacteria have no chloroplasts.

Cyanobacteria are producers in aquatic communities.

Chemosynthetic bacteria obtain energy from inorganic substances.

13:5 Archaebacteria

The ancient bacteria are prokaryotes, but they differ from the true bacteria in several ways. Their cell wall is not made of murein, and the lipids in their cell membranes have a different structure. The sequence of bases in their rRNA is quite different from that of the Eubacteria. They also show differences in tRNA structure and reactions to antibiotics.

As their name implies, Archaebacteria are very primitive. All are anaerobic and live in places that may be very similar to environments found when life was first evolving on Earth. There are three phyla, each inhabiting a unique environment.

One phylum includes organisms found in areas such as bogs and marshes, environments where there is a great deal of decaying plant material. These organisms also are found in the digestive tracts of many grazing mammals, humans, and in certain sewage treatment devices. They obtain energy by chemosynthesis, converting carbon dioxide and hydrogen to methane (marsh gas). For this reason, they are known as **methane-producing bacteria.**

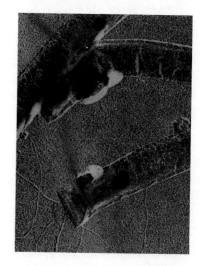

FIGURE 13–5. The Archaebacteria are very primitive bacteria that inhabit bogs, salt lakes, and hot sulfur springs.

The salt-loving and heat-and-acid-loving bacteria inhabit Earth's harshest environments.

Other phyla inhabit even harsher habitats. One group is known as the **salt-loving bacteria.** They live in salt lakes and are the only organisms able to inhabit the Dead Sea. They require a high concentration of salt in order to carry out a special kind of photosynthesis. They have no chlorophyll. The other phylum of ancient bacteria occupies areas that are very hot and acidic. They are known as the **heat-and-acid-loving bacteria.** Members of this phylum include producers that live deep in the ocean near underwater volcanoes (Chapter 34). They produce organic compounds by chemosynthesis. Other members of this phylum live in hot sulfur springs having temperatures near 80° C and a pH of 2!

You can see why Archaebacteria have been classified separately from the Eubacteria. Some biologists feel that the ancient bacteria are so different that they should be placed in a kingdom all their own. Perhaps the three groups — Archaebacteria, Eubacteria, and eukaryotes — have a common ancestor that was a very early form of life on Earth.

13:6 Beneficial Bacteria

Bacteria have important roles in nature and in human economy as well. Most decomposers are bacteria that break down the organic material of dead organisms and organic wastes into its basic inorganic parts. These materials are formed into organic material once again in other organisms. Such cycling of materials could not occur in the absence of certain bacteria.

Bacteria are important in a variety of industrial processes.

FIGURE 13–6. Bacteria are used in many processes in the dairy industry. *Lactobacillus bulgaricus* is a bacterium commonly used in the production of yogurt.

Bacteria are used in many processes in the dairy industry. Products such as yogurt and buttermilk are made by adding certain bacteria to milk. The bacteria thrive and multiply rapidly in the milk and secrete substances that flavor the milk. Production of cheese depends on bacteria that cause the milk products to turn to solids. Cream is converted to butter as a result of bacterial action. Cream is soured by a type of bacterium, then churned into a solid form, butter. Bacteria are involved indirectly in the production of milk in a cow. These bacteria grow in the stomach of the cow and aid in breaking down plant material from which the cow gets its nutrient requirements.

Bacteria also are used in other industries. Bacteria aid in converting alcohol to vinegar. Alcohol is changed to acetic acid, which gives vinegar its odor. Some bacteria are used to break down the material holding together cellulose fibers in plants such as flax and hemp. The fibers then can be used in making linen or rope. Preparing skins for making leather also involves bacteria. Many of today's antibiotics are produced by bacteria.

While many bacteria are very helpful, still others are harmful and dangerous. The role of bacteria in disease is discussed in Chapter 19.

KINGDOM PROTISTA

Kingdom Protista includes simple eukaryotic organisms. Most forms are unicellular, but some are very simple multicellular organisms with little division of labor. Some protists are autotrophic, some are heterotrophic, and some may obtain food either way. The protists include three main groups: simple algae, protozoa, and a group called slime molds.

Protists are simple eukaryotes that may be unicellular or simple, multicellular organisms.

13:7 Euglenoids

The algal protists are mostly unicellular producers. Their chlorophyll and other pigments are located in chloroplasts. They are mostly aquatic, living in fresh water, salt water, and moist places on land. They are classified into phyla on the basis of color and structure.

Algal protists are aquatic producers and are classified on the basis of color and structure.

Euglenoids (yew GLEE noydz), phylum **Euglenophyta** (yoo gluh NAHF uh tuh), are unicellular algae. Many forms are autotrophic, but some are heterotrophic. Some may live as animal parasites.

Euglenoids are motile, having a long flagellum (or two) to propel them. They have no cell walls. They are mostly aquatic, usually freshwater forms. All forms reproduce asexually by mitosis. No means of sexual reproduction is known.

Euglena is an interesting genus of organisms. These organisms contain chlorophyll, and when they are present in high concentrations, the water appears green. They swim swiftly through water, occasionally changing from cigar shape to round shape. A *Euglena* (Figure 13–7) has a reddish-orange eyespot or **stigma** near the base of its flagellum. The stigma is sensitive to light. Thus, the *Euglena* can respond and move to a source of light. How is this behavior adaptive? A contractile vacuole (Section 4:12) pumps out water, maintaining osmotic balance in the cell. Excess food is stored in a structure called the **pyrenoid** (pi REE noyd). Pyrenoids are located on the chloroplasts. *Euglena* is an example of a euglenoid that may lose its chloroplasts and in doing so become heterotrophic. The loss of chloroplasts results from exposure to prolonged darkness.

FIGURE 13–7. The euglenoids include *Euglena*, a single-cell organism with certain plant and animal traits.

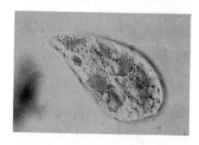

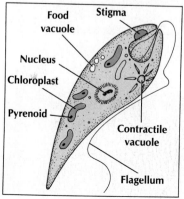

Food vacuole
Stigma
Nucleus
Chloroplast
Pyrenoid
Contractile vacuole
Flagellum

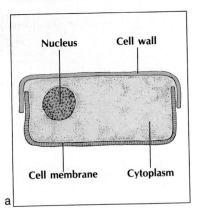

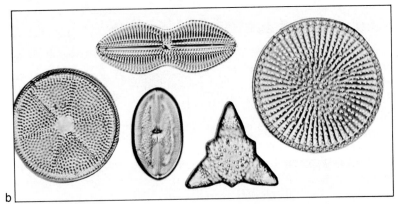

FIGURE 13–8. (a) The shell of a diatom, a golden alga, consists of one part that overlaps the other and is made of a glasslike material. (b) The beautiful forms diatoms take are many and varied. The diatoms shown are magnified about 2000 times.

FIGURE 13–9. (a) Representative dinoflagellates show that each has a pair of flagella for locomotion. (b) A "red tide" is the result of rapid reproduction of certain dinoflagellates.

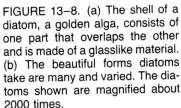

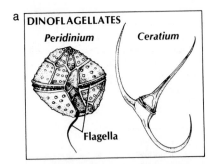

13:8 Golden Algae and Dinoflagellates

The **golden algae**, phylum **Chrysophyta** (kruh SAHF uh tuh), are mostly unicellular organisms. Their colors may range from yellow-green to golden-brown depending on the carotenes they contain. The most common and best known members of this phylum are the **diatoms** (DI uh tahmz). Diatoms inhabit both fresh water and salt water. They are the most numerous of the many small producers that float on the surface of the sea. Their numbers make them one of the most important marine producers.

Diatoms are unicellular and have many shapes and colors, making them some of nature's most beautiful organisms (Figure 13–8). Each diatom is composed of a two-part outer shell in which one part of the shell overlaps the other like the lid of a box. The shell of a diatom consists of the cell wall covered with a glasslike material. When diatoms die, the cell walls collect producing huge deposits known as diatomaceous (di ut uh MAY shus) earth, used as a filtering agent for juices and other liquids and as an ingredient in scouring powders, cosmetics, and toothpaste.

Dinoflagellates are members of the smallest phylum of algae, **Pyrrophyta** (puh RAHF uh tuh). These unicellular forms are found in both fresh water and salt water and are food for aquatic heterotrophs. Each cell has two flagella for locomotion (Figure 13–9a). Some species of dinoflagellates have the ability to produce light, a process referred to as bioluminescence (bi oh lew muh NES uns). When large numbers of these organisms are emitting light, the ocean may appear to sparkle or glow. Certain species produce a red pigment that colors the water to produce "red tides," especially when these organisms multiply rapidly. During red tides, substances produced by the algae can poison large numbers of fish. In summer, red tides cause the deaths of thousands of organisms along the coastal United States. Red tides do not happen every summer, but they kill or damage seafood when they do occur.

13:9 Sarcodines

Protozoa are unicellular, animallike organisms. Most of them can move and are classified into phyla based on how they move. They are heterotrophs, and many have special features for obtaining food. Some actively trap food, while others are parasites. They undergo both sexual and asexual reproduction.

One of the simplest and most often studied protozoa is the amoeba (Figure 13–10). Amoebae are **sarcodines** (SAR kuh dinez), phylum **Sarcodina** (sar kuh DI nuh), which move by means of extensions of the cytoplasm called **pseudopodia** (sewd uh POHD ee uh), or false feet. Such movement is called **amoeboid** (uh MEE boyd) **motion.** An amoeba can form pseudopodia because of its flexible plasma membrane and its constantly moving and changing cytoplasm. The cell membrane extends as the cytoplasm flows toward it. The shape of the amoeba changes as new pseudopodia form and others disappear.

Pseudopodia are used both in movement and food-getting. Pseudopodia engulf particles of food by phagocytosis (Section 4:9). The food particle enters the cytoplasm within a vacuole and is digested there. Food vacuoles are often visible in an amoeba.

Two marine forms of sarcodines are well known. One type, the *Foraminifera* (fuh ram uh NIHF ra), or forams, has an outer protective skeleton composed of calcium carbonate. The second group, the *Radiolaria* (rayd ee oh LER ee uh), has an internal skeleton that contains a glasslike material. In some radiolarians the skeleton may protrude outward from the cytoplasm forming spines that can be moved or withdrawn. In many species the pseudopods are sticky, which enables them to trap food. This type of pseudopod is not used in locomotion.

Skeletons of forams form much of the limestone and chalk on Earth, including the white cliffs of Dover, England. Radiolarian skeletons form a rock called chert.

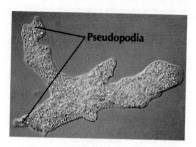

FIGURE 13–10. The amoeba, *Chaos carolinense*, is a sarcodine, shown magnified about 185 times. It moves with pseudopodia, extensions of the cytoplasm.

Protozoa are unicellular heterotrophs, classified on the basis of how they move.

Pseudopodia (singular, pseudopodium) function in locomotion and food-getting.

Forams and radiolarians are eaten by larger ocean dwellers.

FIGURE 13–11. Two marine sarcodines, (a) *Radiolarians* and (b) *Foraminiferans*, have skeletons. (c) Foram skeletons have formed much of the white cliffs of Dover, England.

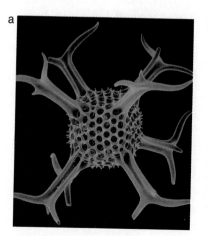

a

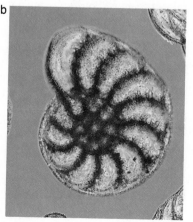

b

c

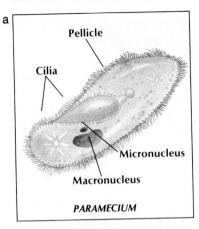

PARAMECIUM

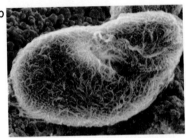

FIGURE 13–12. (a) *Paramecium* is a ciliate commonly studied by biology students. (b) The cilia, the hairlike structures on the surface, show clearly in this photo magnified about 775 times.

Most flagellates are parasites.

13:10 Ciliates

Ciliates (SIHL ee ayts), phylum **Ciliophora** (sihl ee AHF uh ruh), have many hairlike cilia on their surfaces. Cilia are used for obtaining food and for locomotion. Like sarcodines and most other protozoa, ciliates are unicellular. Unlike amoebae, they have a definite shape due to a stiff covering, the **pellicle** (PEL ih kul).

Genus *Paramecium* (Figure 13–12) is the group of ciliates most often studied by biology students. A paramecium is a complex organism with much division of labor. The single cell contains special organelles for a variety of functions. These functions include intake and digestion of food, elimination of undigestible materials, and pumping out of excess water.

Movement is accomplished by the coordinated beating of the cilia that extend from the cell. Usually a paramecium rotates as it glides rapidly through water. If disturbed, a paramecium may curl up and suddenly reverse its direction.

Ciliates have two types of nuclei within a single cell. A large **macronucleus** controls the basic activities of the cell. The smaller **micronucleus** is involved in reproduction. Some ciliates have many micronuclei. Some biologists think that two kinds of nuclei are needed because the cells are very complex.

13:11 Flagellates and Sporozoans

Protozoa that move by means of flagella are called **flagellates** and belong to the phylum **Mastigophora** (mas tuh GAHF uh ruh). Some flagellates are free-living organisms in both fresh and salt water. Most forms, though, live within other organisms.

One species is a parasite that causes African sleeping sickness. This organism, a member of the genus *Trypanasoma,* is transmitted to humans by the tsetse (SEET see) fly. It lives in the blood and releases a poisonous substance that attacks the nervous system causing weakness and then death.

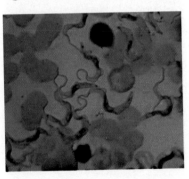

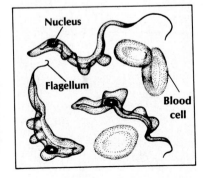

FIGURE 13–13. *Trypanosoma* is a flagellate that causes African sleeping sickness and is transmitted by the tsetse fly.

Another flagellate lives in the gut of termites. It secretes enzymes that digest the wood termites eat. Thus, both the flagellate and the termite benefit. Such a relationship, in which two organisms live in close association and benefit each other, is called mutualism (MYEW chuh lihz um) (Section 33:6).

Sporozoans (spor uh ZOH unz), phylum **Sporozoa** (spor uh ZOH uh), include protozoa that have no means of moving and that reproduce by means of sporelike structures. Since they have no way to move, they cannot actively obtain food. They are parasitic and depend on their hosts for food. As parasites, some cause diseases in humans and many animals.

Most dreaded of the sporozoans are members of the genus *Plasmodium*. They are responsible for malaria, a disease that has caused millions of deaths. Even today in tropical regions, at least a million persons per year die of malaria. This disease is discussed in Section 19:6.

13:12 Slime Molds

Slime molds, phylum **Gymnomycota** (jim no mi KOH tuh), are one of the most fascinating groups of protists. They have a unique life cycle with features of both protozoa and fungi. Some stages are unicellular, while others are multicellular. In some species the life cycle begins as a **plasmodium** (plaz MOHD ee um), a yellowish, slimy mass of material composed of many nuclei but no cell walls. Note that the plasmodium stage of a slime mold is not the same as the name of the organism that causes malaria. A slime mold plasmodium very slowly oozes along the forest floor over rotting logs and other decaying materials. The plasmodium takes in organic materials by phagocytosis. Thus, slime molds act as decomposers.

As a plasmodium creeps along the forest floor, it may slow down and undergo an amazing change. The plasmodium develops into many fruiting bodies that produce spores by mitosis and then release them. The spores develop into flagellated swarm cells that join if moisture

The plasmodium of a slime mold resembles a giant amoeba.

During its life cycle, the slime mold changes form several times.

FIGURE 13–14. (a) The plasmodium, a slimy mass of protoplasm, is a stage in the life cycle of some slime molds. (b) Fruiting bodies develop from the plasmodium. The fruiting bodies are spore-producing structures.

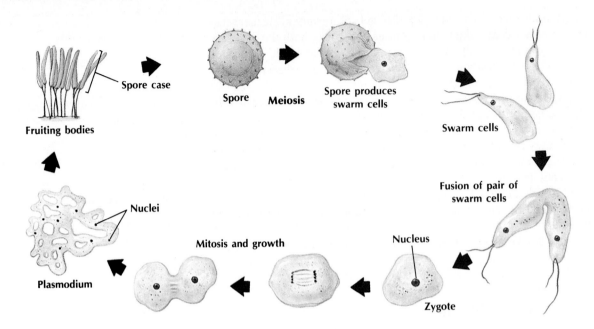

FIGURE 13–15. The life cycle of a slime mold includes many changes in form. Notice that swarm cells resemble flagellates, and the plasmodium resembles amoeboid sarcodines.

and food are available. These cells are attracted to each other by a chemical they secrete. Swarm cells fuse to form a zygote that loses its flagella and begins to move like an amoeba. The production of a new plasmodium involves mitosis and growth without cell division. Sometimes several plasmodia fuse to form a large plasmodium.

Features of slime molds are interesting to biologists because the life cycle involves many changes in form. These different forms resemble other forms of life. Spore production is like that of fungi. Individual, flagellated swarm cells are like protozoa. Biologists study slime mold development to try to better understand the development of more complex organisms.

REVIEWING YOUR IDEAS

5. On what basis are the algal protists subdivided?
6. Describe the structure of *Euglena*. What features of this organism make it difficult to classify?
7. What are the three phyla of algal protists?
8. How may certain dinoflagellates be harmful?
9. On what basis are the protozoa subdivided? List the name of each phylum of protozoa.
10. Describe the uses of pseudopodia and cilia.
11. How do sporozoans obtain food? Can they obtain food in other ways? Why?
12. How are slime molds similar to other simple forms of life?

KINGDOM FUNGI

Fungi are plantlike in that many are stationary. However, they are heterotrophic; they do not have chlorophyll. Because they cannot move to capture food, fungi are either parasites or saprophytes. They absorb small molecules of food from a host or the environment.

Most true fungi have filamentous stalks called **hyphae** (HI fee). In some fungi, each hypha is a mass of cytoplasm containing many nuclei and no cell walls. In others, the hyphae are composed of definite cells. Often, a mass of hyphae are tangled together forming a **mycelium** (mi SEE lee um). The cell walls of most fungi are composed of **chitin** (KITE un), a carbohydrate material. Some forms are unicellular and lack hyphae.

Fungi reproduce by forming spores and by other means. Thick cell walls of both hyphae and spores are adaptations that permit fungi to live on land. *Fungi are classified on the basis of their spore-producing structures.*

fungi (singular, fungus)

Fungi are heterotrophs that obtain nourishment by absorbing small food molecules.

hyphae (singular, hypha)

mycelium (plural, mycelia)

13:13 Sporangium Fungi

The common bread mold, *Rhizopus* (RI zuh pus), is a member of the phylum **Zygomycota** (zi goh mi KOH tuh), or sporangium fungi (Figure 13–16). In this phylum, spores are produced in **sporangia.** Sporangia are located at the tips of certain hyphae called **sporangiophores** (spuh RAN jee uh forz). They extend from the food source giving the fungus a fuzzy appearance. Other hyphae called **stolons** (STOH lunz) spread along the surface of the food supply, or substrate (SUB strayt). In addition to asexual reproduction by spores, bread mold can reproduce sexually (Section 17:6).

The common bread mold, *Rhizopus,* is an example of a sporangium fungus.

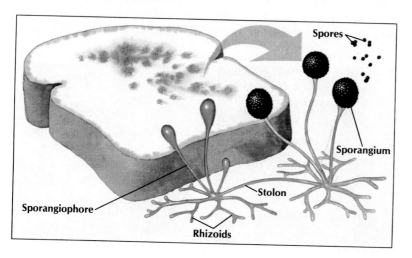

FIGURE 13–16. The black dots of common bread mold, *Rhizopus nigricans,* are sporangia. The mold can reproduce asexually by releasing spores from the sporangia.

FIGURE 13–17. (a) Downy mildew, *Peronospora manshurica,* shown here growing on a soybean plant, is a parasitic fungus. (b) Its rhizoids penetrate and digest the tissues of the host plant.

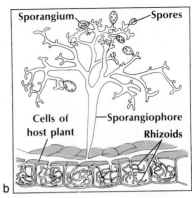

a

b

Rhizoids anchor fungi and digest and absorb food.

Still other hyphae, **rhizoids** (RI zoydz), anchor the fungus to the food source. Thus, rhizoids resemble roots in function. Enzymes secreted by rhizoids break food molecules into simpler molecules. The simple molecules diffuse into the rhizoids and throughout the mold.

Many sporangium fungi are important saprophytes, but some are dangerous parasites.

Most relatives of bread mold are saprophytes, which decompose dead organisms for food. As a result, valuable materials are returned to the soil and atmosphere. However, some species are parasites and cause diseases by feeding on plants, such as potatoes and cereal grains. They produce rhizoids that penetrate the tissues of the host plant, thus robbing it of nutrients.

13:14 Club Fungi

Club fungi produce spores on basidia (singular, basidium).

Mushrooms are a class of fungi with club-shaped, spore-producing structures called **basidia** (buh SIHD ee uh). They are members of the phylum **Basidiomycota,** or **club fungi.** Also in this phylum are shelf fungi, rusts, smuts, and puffballs (Figure 13–18).

FIGURE 13–18. Club fungi include (a) mushrooms, (b) shelf fungi, and (c) puffballs. Club fungi have spore-producing structures called basidia.

Mushrooms are important because many of them are edible. The edible part of a mushroom is only the reproductive part of the organism. Underground is a branching network of hyphae that obtain nourishment in much the same way as bread mold does. The button por-

a

b

c

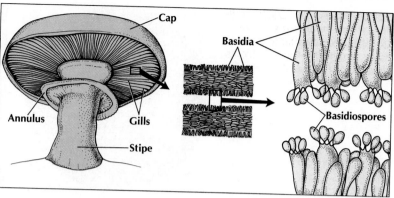

tion of the mushroom develops as a small outgrowth of the mycelium in the soil. This portion pushes through the soil and develops into a stalklike section, the **stipe,** and an umbrella-like **cap.** Beneath the cap are spokelike **gills** on which basidia are located. Each basidium produces four spores that are dispersed by the wind.

Poisonous mushrooms are commonly called toadstools. Toadstools can cause illness, and some, such as a few *Amanita* (am uh NITE uh) species, can cause death if eaten (Figure 13–19). *Therefore, an untrained person should not eat mushrooms found growing wild.*

Whereas mushrooms are valuable to humans, many other club fungi are harmful. Mycelia of rusts, for example, invade the internal parts of wheat, oats, and rye. In this way, they can destroy entire crops of these and other plants. Geneticists have developed some strains of plants that are resistant to these parasites.

FIGURE 13–19. (a) *Amanita verosa,* the "destroying angel," is a poisonous mushroom. (b) Mushroom basidiospores are produced on the gills.

13:15 Sac Fungi

Yeasts, cup fungi, and powdery mildews are all members of phylum **Ascomycota,** the **sac fungi.** In each of these organisms, the zygote develops into a saclike structure, the **ascus.** Spores are produced in the ascus.

In sac fungi, spores are produced in the ascus (plural, asci).

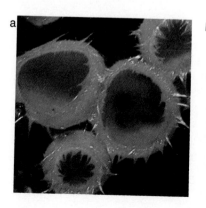

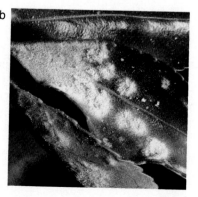

FIGURE 13–20. (a) Cup fungi and (b) powdery mildew (shown growing on a leaf) are examples of sac fungi. They produce spores in structures called asci.

FIGURE 13–21. (a) *Penicillium* mold, shown here growing in a Petri dish, is a sac fungus. One species produces the antibiotic penicillin. (b) A morel is an edible sac fungus that resembles a mushroom.

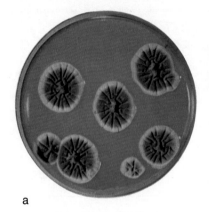

a   b

Yeasts are versatile organisms, used to conduct research in a variety of fields.

FIGURE 13–22. Lichens take various forms and can live in barren places.

Yeasts are unusual fungi in that they are unicellular. Recall that they are important in both the brewing and baking industries (Section 5:9). They are very important in research because they are simple eukaryotes. Yeast have proven to be of value in studying diseases such as cancer and AIDS. Cellular and genetic processes also have been described by using yeasts as research organisms.

Other types of sac fungi are also important in industry and medicine. One species of *Penicillium* mold is used in the production of penicillin (Section 2:5). Other species of *Penicillium* are used in making cheeses such as Roquefort (ROHK furt) and Camembert (KAM um ber) in which enzymes produced by the molds give the cheeses their flavors. Morels (Figure 13–21) are sac fungi that resemble mushrooms and are edible.

While some sac fungi are useful, many others cause diseases in a variety of plants. For example, Dutch elm disease results from invasion by a sac fungus that destroys the tree's conductive tissues. The fungus is transmitted by bark beetles. It has been responsible for the deaths of millions of elm trees. Biologists are working to control the disease by using bacteria that inhibit growth of the fungus. Another approach involves the use of natural chemicals to attract the beetles to traps.

13:16 Lichens

Most noticeable in desolate regions, but found throughout the world, are organisms called lichens. A **lichen** is a "dual organism" consisting of a fungus and either a Cyanobacterium (such as *Nostoc*) or a green alga (Figue 13–23). The mycelium of the fungus surrounds the other organism. Classification of lichens is based on the fungus part of the organism. Lichens grow on soil, rocks, and trees. Some grow flat and close to the surface, while others grow upward and may appear shrublike.

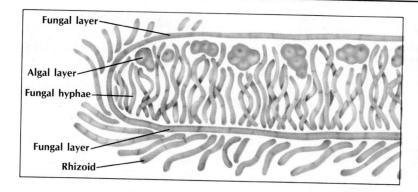

Fungal layer

Algal layer

Fungal hyphae

Fungal layer

Rhizoid

FIGURE 13–23. A lichen is an association of a Cyanobacterium or green alga and a fungus. The green alga or Cyanobacterium produces food; the fungus may provide protection and moisture.

Lichens can live in barren places such as on bare rock or on Arctic ice. Neither the fungus nor the Cyanobacterium or alga could survive in such harsh environments alone. What kind of living arrangement do the two organisms have? The alga or Cyanobacterium is autotrophic; it provides food for itself and the fungus by means of photosynthesis. Exactly how the fungus aids the other organism is not clear. Perhaps the fungus provides it with protection and moisture. The fungus also may be a source of inorganic materials. A lichen is another example of a living arrangement based on mutualism.

The most efficient means of reproduction in lichens is asexual. Some new lichens develop from pieces of old lichens. Others produce powdery grains containing a few autotrophic cells surrounded by fungal hyphae. In either case, a new lichen can grow because both kinds of organisms are present and dispersed together. Sexual reproduction of lichens is not as well understood as asexual reproduction.

Lichens may serve as food for animals. An example is reindeer "moss," an Arctic species that is eaten by reindeer (caribou). Lichens serve yet another important ecological function. Fungi secrete acids that begin the breakdown of rock into particles of soil (Section 34:1). When lichens die, they decompose, thus enriching the soil. Eventually, more complex plants can grow in regions that at one time were habitable mainly by lichens.

In a lichen, the alga or Cyanobacterium produces food for the fungus.

The fungus may provide protection and moisture for its partner organism.

Lichens may serve as food sources for animals. They are also important in the process of soil formation.

REVIEWING YOUR IDEAS

13. How are fungi classified?
14. What is the spore-producing structure of a mushroom? Of *Rhizopus*? Of yeast?
15. Describe the structure of bread mold.
16. Describe the structure of a mushroom.
17. What is a lichen? Why can lichens live in barren places?

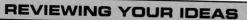

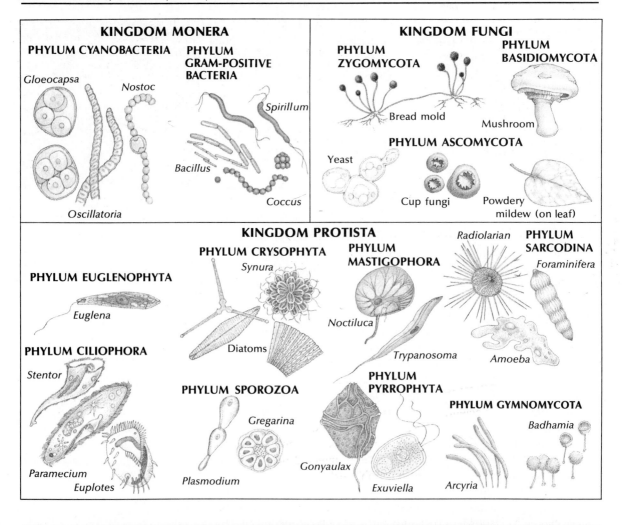

KINGDOM MONERA

PHYLUM CYANOBACTERIA

PHYLUM GRAM-POSITIVE BACTERIA

Gloeocapsa
Nostoc
Spirillum
Bacillus
Coccus
Oscillatoria

KINGDOM FUNGI

PHYLUM ZYGOMYCOTA

PHYLUM BASIDIOMYCOTA

Bread mold
Mushroom

PHYLUM ASCOMYCOTA

Yeast
Cup fungi
Powdery mildew (on leaf)

KINGDOM PROTISTA

PHYLUM EUGLENOPHYTA
Euglena

PHYLUM CILIOPHORA
Stentor
Paramecium
Euplotes

PHYLUM CRYSOPHYTA
Synura
Diatoms

PHYLUM SPOROZOA
Gregarina
Plasmodium

PHYLUM MASTIGOPHORA
Noctiluca
Trypanosoma

PHYLUM PYRROPHYTA
Gonyaulax
Exuviella

Radiolarian

PHYLUM SARCODINA
Foraminifera
Amoeba

PHYLUM GYMNOMYCOTA
Badhamia
Arcyria

TABLE 13-2. SUMMARY OF MAJOR CHARACTERISTICS OF MONERANS, PROTISTS, AND FUNGI				
Kingdom	Common Name	Structure	Nutrition	Importance
Monera	true bacteria	prokaryotic; mainly unicellular	heterotrophic, autotrophic	producers; disease; dairy; recycling; recombinant DNA
	ancient bacteria		heterotrophic, autotrophic, chemosynthetic	producers
Protista	euglenoids	mostly unicellular	autotrophic	producers and consumers in aquatic habitats; disease; recycling; industry
	golden algae		autotrophic	
	dinoflagellates		autotrophic	
	sarcodines		heterotrophic	
	ciliates		heterotrophic	
	flagellates		heterotrophic	
	sporozoans		heterotrophic	
	slime molds	multicellular stages in slime molds	heterotrophic	
Fungi	sporangium fungi, club fungi, sac fungi	multicellular	heterotrophic	foods; decomposers; plant and animal diseases

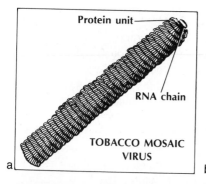

FIGURE 13–24. (a) Tobacco mosaic viruses (TMV) are rod-shaped and have an RNA center surrounded by a protein coat. (b) When TMV infects a tobacco plant, the plant leaves become mottled and wilted.

VIRUSES

Monerans, protists, and fungi are simple organisms compared to plants and animals. However, there are even simpler "organisms" in nature. They are the viruses. Because their characteristics are unique, they are not classified in a kingdom. As you study their features, try to decide whether viruses are organisms or particles of chemicals.

13:17 Characteristics

For many years scientists knew of diseases that were not caused by any known organism. Such an agent of disease was called a **virus,** which means poison. One of the many organisms affected by a virus is the tobacco plant. Because the leaves of an infected plant become mottled and take on a mosaic (moh ZAY ihk) pattern (Figure 13–24), the disease was named tobacco mosaic disease.

A virus is composed of a nucleic acid surrounded by protein.

In 1935, the tobacco mosaic virus was isolated in crystal form. Since then, many other viruses have been isolated. Although viruses cannot be seen with a light microscope, much has been learned about them with electron microscopes and biochemical analysis.

Viruses have several shapes including spherical, needlelike, cubical, and many-sided (Figure 13–25). Each virus particle consists of an outer coat of protein and an inner part of nucleic acid. The nucleic acid

FIGURE 13–25. (a) T_4 bacteriophage, shown magnified 150 000 times, is a DNA virus that infects bacterial cells. (b) Adenovirus, shown magnified about 500 000 times, is a DNA virus that infects animal cells.

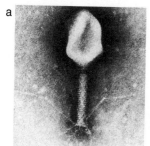

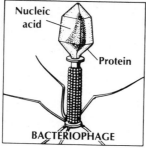

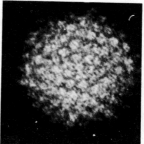

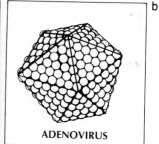

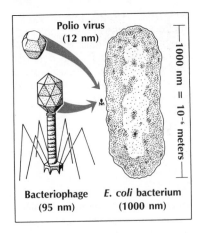

Polio virus
(12 nm)

1000 nm = 10⁻⁶ meters

Bacteriophage
(95 nm)

E. coli bacterium
(1000 nm)

FIGURE 13–26. Viruses, though they vary in size and shape, are all much smaller than bacteria.

A virus requires a specific host for reproduction.

Viruses may be descendants of more complex organisms.

can be either DNA or RNA depending on the type of virus. A virus has no organelles and no cellular level of organization. Viruses are so small that a bacterium seems huge in comparison (Figure 13–26). Some viruses have enzymes, but viruses do not undergo respiration or other common cell processes. The only life function a virus carries out is reproduction (Section 17:7), which it cannot do by itself. To reproduce, a virus requires a specific host cell.

Viruses are considered parasites, and they are responsible for a variety of diseases in plants and animals. Viruses are responsible for human diseases such as AIDS, the common cold, and polio. They also are involved in certain forms of cancer (Chapter 19).

13:18 Some Questions

Is a virus alive? This question is debatable. In terms of the basic features of life discussed in Chapter 1, a virus does not seem to qualify as a living system. However, viruses can reproduce, which is the most universal function of life.

A question that may be more important than whether or not viruses are alive is that of the origin of viruses. Many biologists today think that viruses are products of what is called **degenerative** (dih JEN uh ruh tihv) **evolution.** That is, they probably are descendants of a more complex, nonparasitic living system. In becoming parasitic, viruses may have lost certain features.

Another explanation for the origin of viruses is that some viruses may have been the result of DNA escaping from a cellular organism. In either case, what is left is mainly the chemical basis for reproduction, the genes.

REVIEWING YOUR IDEAS

18. Describe the structure of a virus.
19. Why is it difficult to decide whether or not viruses are alive?

ADVANCES IN BIOLOGY

Recombinant DNA

Bacteria have long been used in practical ways. A new technique now enables biologists to make use of bacteria as "chemical factories." The bacteria can be engineered genetically to produce a wanted protein. In this procedure, DNA (one gene) is removed from a eukaryote and recombined with the DNA of a bacterium. The bacterium is then said to have **recombinant DNA.** The bacterium behaves as if the gene were its own and produces the protein for which the foreign DNA codes.

Many recombinant DNA experiments make use of certain *E. coli* bacteria that have plasmids (Section 13:2). The plasmids are the structures with which the genes of other organisms are recombined. The foreign gene is obtained from a donor cell. For insertion to occur, the circular plasmid must first be broken at a specific place. That place is where the nitrogen bases of the plasmid's DNA match those of the gene to be inserted. The two DNAs join and the plasmid regains its circular shape. Thus, the two DNA segments are "fused." Several enzymes are needed for these steps. The plasmids, now with recombinant DNA, are then placed into a culture of *E. coli* cells that have no plasmids. Some of the *E. coli* cells absorb the plasmids bearing the foreign gene.

Plasmids containing genes for resistance to an antibiotic are used in this process. The bacteria are exposed to the antibiotic. Only those bacteria that have picked up the plasmids will survive, because they have obtained the gene for resistance. These surviving bacteria are then used to produce the protein coded for by the foreign gene. The protein is extracted from the *E. coli* growth medium and purified. Once its identity is verified, the protein can be mass-produced.

Many proteins have been made by bacteria using recombinant DNA techniques. Human insulin is a hormone being produced in this way. Insulin is used in the treatment of diabetes mellitus (Section 28:1). Bacteria have been used to produce human growth hormone and interferon, a protein that fights viruses (Section 19:10).

Techniques have been developed recently for introducing foreign genes into a variety of eukaryotic cells. For example, genes of donor animals have been successfully inserted into cells of different animals. One use of this technique has been in the production of a substance called clotting factor, which causes blood to clot. The gene for clotting factor, missing in hemophiliacs, has been isolated. The gene was introduced into cultures of hamster cells, which then produced the factor.

A promising use for recombinant DNA technology is gene therapy. In **gene therapy,** defective genes are replaced by normal ones. In a recent test, the gene for normal hemoglobin was spliced into plasmids. The plasmids were then placed in solution with fused mouse-human cells. In some of the cells, the plasmid gene became part of the chromosome at its naturally occurring location. Such experiments offer hope that many human genetic diseases can be treated or eliminated by gene therapy.

FIGURE 13–27. Human growth hormone produced by recombinant DNA techniques is packaged for shipment.

Recombinant DNA contains genes from two different organisms.

Various human (and other) proteins have been produced by bacteria using recombinant DNA techniques.

FIGURE 13–28. Using recombinant DNA techniques, *E. coli* bacteria can be engineered to produce a desired protein.

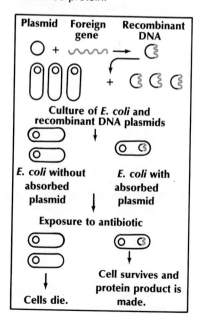

INVESTIGATION

Problem: **What are the traits of organisms in the Kingdoms Monera, Protista, and Fungi?**

Materials:

microscope
glass slide
coverslip
dropper
water
hand lens

methylene blue stain
diatoms
yogurt
razor blade
lilac leaf with
  Microsphaera
3 samples of
  unknown organisms

Procedure:

Part A. Kingdom Monera

1. Copy Figure 13–29 onto a sheet of paper. You may wish to enlarge the chart so that you can draw in it.
2. Place one drop of yogurt onto a glass slide. Yogurt is a source of bacterial cells. Add one drop of methylene blue stain to your slide.
3. Add a coverslip.
4. Observe the slide under low power, then high power. Bacteria will appear as very small, deep blue rods. Reducing the amount of light aids observation.
5. Diagram several bacterial cells in your chart as seen under high power. Make your drawing to scale. Beneath your diagram, indicate the magnification.

Part B. Kingdom Protista

1. Place one drop of diatom culture onto a glass slide. Add a coverslip.
2. Observe the slide under low power, then high power.
3. Diagram several diatoms in your chart as seen under high power. Make your drawing to scale. Beneath your diagram, indicate the magnification.

Part C. Kingdom Fungi

1. Examine the underside of a lilac leaf with a hand lens. You will observe tiny black dots that are a fungus infection called *Microsphaera*. A common name for this fungus is mildew.
2. Use a razor to carefully scrape off some of the black dots. Place them on a glass slide. Add one drop of water and a coverslip to the slide.
3. Observe the slide under low power. Diagram the mildew in your chart as seen under low power. Make your drawing to scale. Indicate the magnification of your diagram in the space beneath your drawing.

Part D. Classifying Unknown Organisms

1. Note the unknown samples provided by your teacher. They are marked A, B, and C. Also note any special instructions next to the unknowns telling you whether or not a wet mount is to be prepared.
2. Follow the instructions next to each unknown and observe any slide you make under low power. Observe the unknown at high power if the instructions next to it tell you to do so.
3. Observe and diagram the first unknown in your chart. Make your drawing to scale. Indicate the magnification of your diagram in the space beneath your drawing. NOTE: If a microscope was not used, label the magnification as 1X.
4. Complete your chart by indicating the kingdom to which you believe this unknown organism belongs.
5. Repeat steps 1 through 4 for the other two unknown samples.

Data and Observations:

FIGURE 13-29.

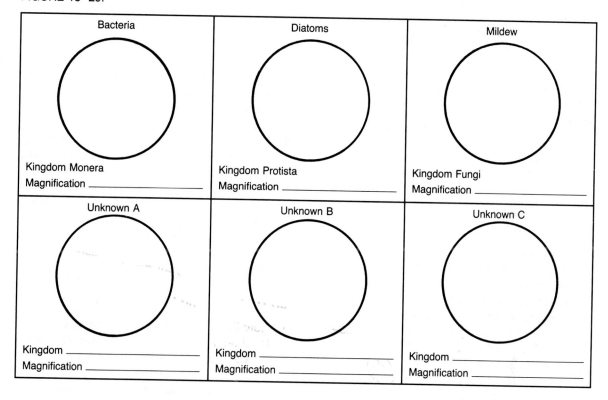

Bacteria

Kingdom Monera
Magnification _____

Diatoms

Kingdom Protista
Magnification _____

Mildew

Kingdom Fungi
Magnification _____

Unknown A

Kingdom _____
Magnification _____

Unknown B

Kingdom _____
Magnification _____

Unknown C

Kingdom _____
Magnification _____

Questions and Conclusion:

1. List three traits of Kingdom Monera.
2. (a) Distinguish between the two main groups of Eubacteria in Kingdom Monera.
   (b) Based on your answer to 2(a), to what group does the organism observed in Part A belong?
3. List three traits of Kingdom Protista.
4. (a) Distinguish among the seven phyla in Kingdom Protista. (Omit slime molds.)
   (b) Based on your answer to 4(a), to what phylum does the organism observed in Part B belong?
5. List three traits of Kingdom Fungi.

6. (a) Distinguish among the three phyla in Kingdom Fungi. (Omit lichens.)
   (b) Based on your answer to 6(a), to what phylum does the organism observed in Part C belong? (HINT: This fungus produces asci.)
7. List several traits that aided you in determining to which kingdom:
   (a) unknown organism A belonged.
   (b) unknown organism B belonged.
   (c) unknown organism C belonged.

Conclusion: What are the traits of organisms in Kingdoms Monera, Protista, and Fungi?

CHAPTER REVIEW

SUMMARY

1. Kingdom Monera includes all prokaryotic organisms. It is composed of two main divisions. They are the Archaebacteria and the Eubacteria. **p. 263**
2. Prokaryotes are primitive cells lacking true nuclei as well as membrane-bound organelles. **13:1**
3. Bacterial cells are either spherical, rod-shaped, or spiral-shaped. Most are enclosed by a cell wall and contain plasmids. **13:2**
4. Gram-positive and Gram-negative bacteria are heterotrophic phyla of true bacteria. They are either parasites or saprophytes. **13:3**
5. Four phyla of true bacteria are autotrophic. Most are photosynthetic, but some are chemosynthetic. **13:4**
6. Ancient bacteria consist of three phyla, each of which occupies a unique environment. **13:5**
7. Bacteria are important in the recycling of materials and are useful for various industrial processes. **13:6**
8. Euglenoids are unicellular algal protists, most of which are autotrophic. **13:7**
9. The golden algae are mostly unicellular protists. Most common among them are the diatoms. Dinoflagellates are important producers in aquatic communities. Some species are responsible for causing "red tides." **13:8**
10. Sarcodines are protozoa that move and trap food by means of pseudopodia. **13:9**
11. Ciliates move and trap food by means of hair-like cilia. **13:10**
12. Flagellates are protozoans that move by means of flagella. Sporozoans, unlike other protozoa, cannot move. All Sporozoans are parasitic. **13:11**
13. Slime molds have a life cycle that includes unicellular and multicellular stages. **13:12**
14. Sporangium fungi, such as the common bread mold, produce spores in sporangia. **13:13**
15. In club fungi, such as mushrooms, spores are produced in basidia. **13:14**
16. Yeasts, cup fungi, and powdery mildews are sac fungi, which are organisms that produce spores in an ascus. **13:15**
17. Lichens consist of two organisms, a fungus and either a Cyanobacterium or green alga. They are mutualistic. **13:16**
18. Viruses are composed of a protein coat and a nucleic acid core. They are not cellular and reproduce only in living host cells. **13:17**
19. Viruses may be products of degenerative evolution or might have evolved from DNA that escaped a cellular organism. **13:18**
20. Bacteria and some eukaryotes are used in recombinant DNA research. Foreign genes are spliced into these organisms to produce useful proteins. **p. 282**

LANGUAGE OF BIOLOGY

asexual reproduction	gene therapy	pseudopodium
ascus	hypha	pyrenoid
bacillus	lichen	recombinant DNA
basidium	murein	rhizoid
carotene	mycelium	spirillum
chemosynthesis	nucleoid	sporangium
chitin	pellicle	stigma
coccus	phycocyanin	stipe
diatom	plasmid	stolon
endospore	plasmodium	virus
	protozoan	

CHECKING YOUR IDEAS

On a separate paper, complete each of the following statements with the missing term(s). Do not write in this book.

1. _____ are cells that have no true nuclei.
2. Fungi have filaments called _____.
3. Protozoa are classified on the basis of their means of _____.

4. Kingdom _____ includes both autotrophic and heterotrophic eukaryotes.
5. Each virus has an inner part that is made up of _____.
6. Spore-producing structures are a feature of Kingdom _____.
7. In bacteria, _____ is carried out on the inner surface of the cell membrane.
8. Presence of two types of nuclei in a cell is a characteristic of the _____.
9. The plasmodium is a stage in the life cycle of the _____.
10. Red tides may be caused by certain _____.

EVALUATING YOUR IDEAS

1. Compare prokaryotic and eukaryotic cells in terms of:
   (a) nucleus and nuclear membrane
   (b) kinds of organelles
   (c) location of chlorophyll
   (d) chromosome number and structure.
2. Describe the formation and importance of endospores.
3. What are plasmids? How may certain plasmids be important to bacteria?
4. Where does photosynthesis take place in Cyanobacteria?
5. Identify the means of nutrition among Gram-positive, Gram-negative, green-sulfur, purple bacteria, and Prochlorophyta.
6. Identify the three phyla of ancient bacteria and explain how each obtains energy.
7. Describe the features of the algal protists.
8. How are diatoms important to humans?
9. List the main characteristics of protozoa.
10. Describe food-getting and locomotion in an amoeba and a paramecium.
11. Describe the life cycle of a slime mold.
12. List the characteristics of fungi.
13. How are fungi adapted for life on land?
14. In general, how do fungi reproduce?

15. How are the two organisms of lichens mutualistic?
16. Explain two ideas concerning the possible origin of viruses.
17. Describe the technique used in recombinant DNA work with *E. coli* bacteria.

APPLYING YOUR IDEAS

1. Do you think heterotrophic euglenoids have a stigma? Explain.
2. How do you think protozoa and fungi were classified in the two-kingdom system of classification? Why?
3. Consider the life cycle of a slime mold (Figure 13–15). Which stages are animallike? Plantlike? Explain.
4. Could staining be a means of identifying certain bacteria? Explain.
5. Could something like today's virus have been the first living thing? Explain.

EXTENDING YOUR IDEAS

1. Prepare a report on the uses of bacteria in the dairy industry.
2. Prepare a report on the life cycle of a parasitic fungus such as wheat rust or corn smut.
3. Test the effect of various disinfectants and "germ killers" on bacteria.
4. Prepare a report on the factors governing changes in form in slime molds.

SUGGESTED READINGS

Allman, William F., "Drugs in Feed: Fatter Cattle, Fitter Bacteria." *Science 84,* Dec., 1984.

Chinnici, Madeline, "The Promise of Gene Therapy." *Science Digest,* May, 1985.

Weaver, Robert F., "Beyond Supermouse: Changing Life's Genetic Blueprint." *National Geographic,* Dec., 1984.

PLANTS

Plants are organisms that manufacture their own food. They can be found in almost every environment on Earth. However, different types of plants are found in different environments. What are some adaptations of these plants that enable them to live in their environment? How do humans depend on plants? What new uses are being found for plants?

chapter 14

In the five-kingdom classification system used in this text, plants are nonmotile, autotrophic organisms. Almost all are multicellular, and they have cell walls made of cellulose. Most plants have a life cycle involving diploid and haploid phases. The plant kingdom is composed of five phyla, three of which are algae. These algae are included here because they have a more complex structure than the algae of the protist kingdom (Chapter 13).

Plants occupy a wide variety of habitats. Many forms live on land, but others live totally or partly in water. Regardless of their environment, plants are the producers of their communities.

Objectives:
You will
- discuss the features of three phyla of algal plants.
- compare the characteristics of the major groups of land plants.
- describe the external structures of roots, stems, and leaves of flowering plants.

ALGAL PLANTS

Those algae classified as plants are mostly multicellular organisms. Most of them have no true tissues or organs, but they do exhibit some division of labor. The entire body of an alga is called a **thallus.** In some complex algae, the thallus is thickened, but in most forms it is a flattened structure. Development in these algae occurs in the water. The developing organisms are not protected by the parent plant. As with the algal protists, algal plants are subdivided on the basis of color and structure.

A thallus is an entire plant body that lacks roots, stems, and leaves.

Algal plants are classified on the basis of color and structure.

14:1 Green Algae

Cells of **green algae,** phylum **Chlorophyta** (kloh RAHF uh tuh), closely resemble cells of true plants. It is believed that complex plants evolved from these algae. Chlorophyll is the major pigment in green algae, but yellow carotenes (Section 13:4) add to the color of the cells. When the amount of carotene is high, algae have a light, yellowish-green color.

Green algae contain chlorophyll and carotenes.

287

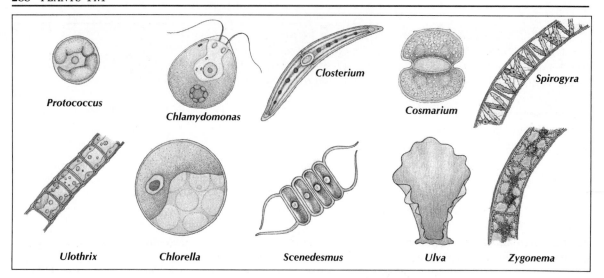

Some green algae are unicellular.

Green algae are usually found in fresh water, but some live in the sea. *Protococcus* (proht uh KAHK us) can grow on the bark of trees when enough moisture is present.

Many green algae are filamentous, colonial, or multicellular. Thus, they seem more like other plants. Some forms, though, are unicellular. *Protococcus* is unicellular. *Chlamydomonas* (klam ud uh MOH nus) is also a one-celled green alga.

Chlamydomonas has a single large chloroplast and moves with two flagella (Figure 14-1). It may seem to you that classification systems do not make much sense because *Chlamydomonas* in many ways is similar to *Euglena* (Section 13:7), a protist. Both carry on photosynthesis and move with flagella. *Chlamydomonas,* however, has other traits that lead biologists to classify it as a plant. No system can solve all problems of classification.

Desmids (DEZ mids), such as *Closterium* (klah STER ee um) and *Cosmarium* (kahz MAR ee um), are one-celled green algae that have interesting and often beautiful forms. A desmid is composed of two halves connected by a narrow section in which the nucleus is located. Desmids float freely in water. Desmid cells can join to form colonies or filaments.

Spirogyra and Ulothrix are filamentous green algae in which most cells are alike.

Examples of filamentous algae are *Spirogyra* (spi ruh JI ruh) and *Ulothrix* (YEW loh thrihks). *Spirogyra* is commonly found in ponds where many filaments may grow together forming a dense mat. Individual filaments look like very thin threads. Each filament contains many identical cells. The chloroplasts are ribbon-shaped structures that form spirals throughout each cell. *Ulothrix* usually grows in fresh water where it is anchored by a special cell called a holdfast. The other cells of the filament are all alike. Any one of them may produce spores that may develop into a new filament.

Volvox (VAHL vahks) is a colonial green alga. *Volvox* cells, which resemble *Chlamydomonas* cells, form a hollow sphere. The colony has a slight amount of specialization. For example, some cells are sensitive to light. Reproductive cells divide to form daughter colonies, which can be seen in the hollow cavity of the parent colony. The parent colony will burst to free the daughter colonies. A colonial form such as *Volvox* may represent a stage in evolution between unicellular and multicellular organisms.

14:2 Brown Algae

Brown algae, phylum **Phaeophyta** (fee AHF uh tuh), are complex multicellular algae, almost all of which are found in salt water. They contain chlorophyll and a special carotene that gives them their brown color. Many forms grow in cool water along rocky coasts where they are anchored by rootlike holdfasts. These forms often become visible during low tide. Some grow unattached in warmer waters. One such form is *Sargassum,* which grows so densely that it covers much of the surface water of the Atlantic Ocean near Bermuda. This area, called the Sargasso Sea, covers millions of square kilometers.

Brown algae are commonly called seaweeds or kelps. They may be as long as 50 m. Many forms have a thickened thallus with specialized parts. Sometimes the thallus is branched and "treelike." *Fucus* (FYEW kus), or rockweed, is a brown alga with air-filled bladders that act as floats. Some branches of the thallus have receptacles that contain sex organs. The thallus of *Laminaria* (lam uh NER ee uh), a kelp, has a holdfast, a stemlike portion, and a leaflike region. It also has a variety of primitive tissues for protection, conduction of food,

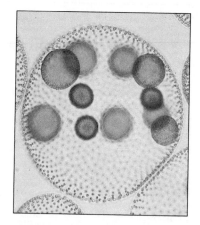

FIGURE 14–2. The colonial green alga *Volvox* is a group of many cells forming a sphere. Eleven daughter colonies can be seen in the center. The *Volvox* shown here is magnified about 400 times.

Brown algae are among the most complex algae with some forms having specialized tissues.

FIGURE 14–3. Brown algae are commonly called seaweeds and may grow up to 50 m long.

FIGURE 14–4. The color of red algae is due to the presence of pigments such as phycoerythrin.

and storage. These forms are among the most complex algae. Brown algae are important as a source of iodine. Certain kelps can be used as a fertilizer. Others are a source of algin, a substance used in making a variety of foods including ice cream.

14:3 Red Algae

Red algae belong to the phylum **Rhodophyta** (roh DAHF uh tuh). Most are multicellular and more complex than the other phyla of algae. The red color is due to the pigment **phycoerythrin** (fi koh ER ih thrun). Their bright color and feathery structure make many red algae quite beautiful.

Due to their special pigments (for example, phycoerythrin), red algae can trap light at great depths in the ocean. Thus, many forms can live deeper in the ocean than other groups of algae. A species recently discovered in the Bahamas grows at a depth of nearly 270 m! Other red algae live anchored in mud along coasts.

Red algae are used in a wide variety of ways. *Chondrus crispus* (KAHN drus • KRIHS pus), commonly called Irish moss, is used by some people for food. Agar is a substance made from various species of red algae and is used to make gelatinous culture media. Substances from red algae are also used in making ice cream, puddings, and icings for cakes.

REVIEWING YOUR IDEAS

1. What pigments are found in green algae?
2. Describe the various forms that may be observed among the different species of green algae.
3. Compare the structure of brown algae to the structure of green algae.
4. What are some uses of red algae?

LAND PLANTS

Land plants are adapted for life in a dry environment.

Most algae are simple in structure. They have adaptations that allow them to produce food and reproduce in an aquatic environment or, in some cases, a moist habitat on land. Most habitats on land, however, are too dry for algae to occupy. The land is occupied by more complex plants. The complex adaptations of land plants enable them to live in a wide variety of environments.

14:4 From Water to Land

Many lines of evidence suggest that land plants evolved from some form of green alga. During this process, land plants became well adapted for life in a dry environment. Their adaptations involve ways of:

(1) obtaining water and dissolved minerals,
(2) distributing water and dissolved minerals to all tissues,
(3) distributing food to cells that cannot make their own,
(4) preventing excess evaporation of water,
(5) obtaining carbon dioxide from air,
(6) supporting tissues that grow upright,
(7) increasing the chances of fertilization, and
(8) protecting the developing plant.

You will study such adaptations among the major groups of land plants alive today.

On the basis of the presence or absence of vascular (VAS kyuh luhr) tissue, land plants are classified into two major phyla — Bryophyta (bri AHF uh tuh) and Tracheophyta (tray kee AHF uh tuh). **Vascular tissue** transports food, water, and minerals throughout the plant. Bryophytes (BRI uh fites) do not have this tissue and are called **nonvascular plants.** Tracheophytes (TRAY kee uh fites) have vascular tissue and are called **vascular plants.**

14:5 Bryophytes

Phylum Bryophyta includes two major plant groups — **liverworts** and **mosses.** The major features of this phylum are (1) small size, (2) lack of specialized tissues for transport of materials, and (3) lack of true stems, roots, or leaves. Both groups are widely distributed. Although many bryophytes live in moist environments, some have adaptations that permit them to inhabit dry areas.

Liverworts have "leathery" photosynthetic structures that lie flat upon the water or soil in which they grow. Liverworts are so named because the thin "leaves" are often liver-shaped (Figure 14–5a). Mosses are somewhat more complex than liverworts in that their "leaflets" are attached to upright stalklike structures (Figure 14–5b). In both groups of bryophytes, the familiar, predominant stage of the life cycle is the haploid phase. The diploid phase is small and grows from the haploid parent. It is dependent upon the parent and lasts only a short time (Section 20:3).

Some bryophytes are useful to people. *Sphagnum* is a moss that grows in lakes and bogs in the form of floating mats. As *Sphagnum* decomposes, peat moss is formed, which is commonly used as a fertilizer to improve soil quality in gardens. Mosses are often early invaders of new environments. Their decay results in the formation of rich soil suitable for other plants.

Land plants are classified according to the presence or absence of vascular tissue.

FIGURE 14–5. (a) A liverwort has a flat leaflike photosynthetic structure. (b) Mosses have leaflike structures attached to stalks. Liverworts and mosses are bryophytes.

a

b

FIGURE 14–6. A liverwort shows several adaptations for life on land. This section of the leaflike structure shows pores, rhizoids, and cutin.

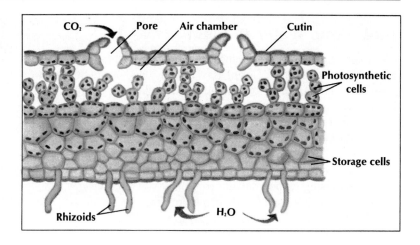

Adaptations for land life among the bryophytes can be seen by studying the structure of a liverwort. Figure 14–6 shows a cross-section of such a plant.

The source of water for most land plants is the soil. Bryophytes have no roots, but they do have rootlike **rhizoids** (RI zoydz) that anchor the plants. Rhizoids extend into the moist soil where they obtain water by osmosis and dissolved minerals by diffusion.

Because bryophytes have no vascular (conductive) tissues, water moves from cell to cell by osmosis. This process is slow but adaptive because the plant is small. Also, food made by photosynthesis is transported by diffusion.

A plant on dry land loses water by evaporation. **Cutin,** a waxy covering found on the outer surfaces of many bryophytes and most other land plants, helps prevent excess water loss by forming a watertight seal. The size of the single, thin "leaf" of liverworts also promotes water conservation. Because its surface area is small, the amount of water loss from the leaf is small.

Aquatic plants take in carbon dioxide dissolved in the water around them. Land plants get this gas from air. The atmosphere contains only 0.03 percent carbon dioxide; therefore, land plants must have adaptations for getting enough carbon dioxide. Some bryophytes have **pores,** small openings in "leaves." Air enters through these pores. In order to enter cells of plants, gases must first be dissolved in water. Land plants have evolved in such a way that the cells are surrounded by a thin film of water. Thus, carbon dioxide entering a plant through the pores dissolves in the water and then diffuses into the cells.

Part of the life cycle of liverworts and mosses involves sexual reproduction. Sperm produced by the male plant swim by means of flagella to the eggs of the female plant. This reproduction can occur only if moisture is present.

Bryophytes obtain water through rhizoids.

In bryophytes, water is distributed by osmosis and food is distributed by diffusion.

Water loss from leaves is reduced by the presence of cutin.

Carbon dioxide enters bryophytes through pores. Inside the plant it dissolves in water before diffusing into the cells.

In sexual reproduction of bryophytes, sperm swim to the eggs of the female.

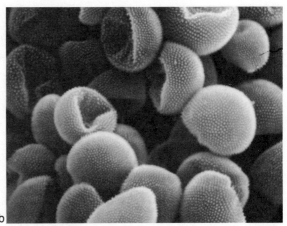

FIGURE 14–7. (a) Moss plants produce spore capsules (the light-colored structures) from which (b) spores are released. The moss spores shown are magnified 3150 times.

The bryophyte life cycle also includes spore production. Spores are an adaptation for land life in that their walls are tough and resist evaporation. When spores settle upon moist land, they may germinate into a new haploid plant.

In both liverworts and mosses, the zygote is formed within the female reproductive organ. This structure provides a sterile, moist environment for the zygote and early embryo. Therefore, there is less chance of the embryo's being destroyed.

Adaptations for support are not found in bryophytes. In more complex plants, vascular tissue provides support, but bryophytes do not have vascular tissue. Thus, they grow very close to or directly upon the ground.

Bryophytes have no adaptations for support.

Although adapted for life on land, bryophytes are not as common as some other plants. They need moisture for sexual reproduction (as well as for photosynthesis and other life functions). Because adequate water is not available everywhere, most bryophytes are limited to a moist habitat. They have no true vascular tissue for transport or support, which limits their size.

Bryophytes, because of their need for water, are limited in size and distribution.

14:6 Characteristics of Tracheophytes

Plants of the phylum **Tracheophyta** have vascular tissue and (in most forms) true roots, stems, and leaves. Unlike the bryophytes, most forms are not restricted to a moist habitat. The diploid stage of tracheophytes is predominant. Included in this phylum are early forms found mostly in warm, tropical climates. Modern groups are well distributed over Earth. The most widespread of these plants are those that reproduce by seeds.

Tracheophytes have vascular tissue.

Most tracheophytes have true roots, stems, and leaves. The diploid stage of the life cycle is predominant.

Ancestors of today's tracheophytes were the major plant forms during the Carboniferous (kar buh NIHF rus) Period about 280 million years ago (Table 10–1). This period is so named because the plants living then formed the vast coal (carbon) deposits.

FIGURE 14–8. (a) A club moss and (b) horsetail are primitive tracheophytes. Club mosses and horsetails reproduce by spores.

FIGURE 14–8. (a) A club moss and (b) horsetail are primitive tracheophytes. Club mosses and horsetails reproduce by spores.

14:7 Club Mosses and Horsetails

Club mosses are small, primitive tracheophytes that produce spores in conelike structures.

Club mosses (Figure 14–8a) are primitive tracheophytes. They are so named because of their resemblance to true mosses. Club mosses have true roots, stems, and leaves. They grow close to the ground. Most forms live in damp woods, but some grow in deserts and on mountains. These plants produce spores in conelike structures. Club mosses were very prominent at one time, some species being as large as trees.

Horsetails (Figure 14–8b) are also simple tracheophytes. They grow in moist, sandy areas. *Equisetum* (ek wuh SEET um) is the only genus that exists today. It includes 25 species. Most forms are small with a slender stem. Groups of small, wedge-shaped leaves are arranged at several points along the stem. Some stems are reproductive and release spores. Other stems possess no reproductive organs. Horsetails are commonly called scouring rushes because they have been used to scour utensils. At one time these tracheophytes were widespread.

FIGURE 14–9. Ferns have true leaves, stems, and roots. The leaves, sometimes called fronds, often are branched and appear lacy.

14:8 Ferns

Modern **ferns** range in size from quite small to as large as trees. Large forms in the tropics reach a height of ten to fifteen meters. In temperate areas, ferns often grow as low shrubs on the damp forest floor. Some forms float in water or are rooted in mud, and some live in very dry areas such as on cliffs. Species living in dry regions are dormant when moisture is not available but resume life activities when water is present.

The diploid phase is the major form in the life cycle of ferns. The diploid fern has a **rhizome** (RI zohm), an underground stem from which roots and leaves develop. In tree ferns, an upright stem develops as a separate, aboveground structure. But in most ferns, only the rhizome exists. Leaves, sometimes called **fronds,** are usually highly branched and often have a lacy appearance (Figure 14–9).

During the fern's life cycle (Section 20:4), spores are produced in specialized structures of the diploid plants. The spores are released and travel through the air. If a spore encounters favorable conditions, it develops into a small, nonvascular, haploid form. The haploid form produces gametes. Fertilization results in the development of a diploid plant.

Ferns have some adaptations not found in the bryophytes. For example, the diploid forms have vascular tissues that transport water and food between leaves and roots. The presence of vascular tissue permits a larger size. Specialized cells in the stem provide support for these plants, and water loss is reduced by cutin on the leaves and the presence of special cells that control the size of the pores (Section 21:1).

Like many of the mosses and liverworts, most ferns are limited to a moist environment that provides the sperm with a thin film of moisture in which to swim to the eggs. Because the haploid form has no vascular tissue, water moves by osmosis from rhizoids to other cells. Most ferns must live in a moist area to obtain enough water.

As in bryophytes, fertilization occurs within the female sex organ and development of the embryo begins there. However, the young plant is soon "on its own" because the parent dies.

FIGURE 14–10. (a) A tree-sized fern, found in tropical regions, and (b) a maidenhair fern, found in temperate forest regions, show the great variation found in members of this group of tracheophytes.

In most ferns, roots and leaves develop from an underground stem, the rhizome.

A fern's life cycle includes both asexual and sexual reproduction.

Compared to bryophytes, ferns are larger, have a means of support, and conserve more water.

Like many bryophytes, most ferns are limited to a moist environment.

a   b   c

FIGURE 14–11. (a) Pine trees, (b) redwoods, and (c) *Ginkgo biloba* are examples of gymnosperms. Pines and redwoods are conifers, whereas the *Gingko* is in a separate order in which no cones are produced.

Gymnosperms and angiosperms reproduce by seeds.

In most gymnosperms, gametes are produced in cones.

Unlike bryophytes and ferns, conifers do not require water for reproduction.

Conifers are supported by xylem.

14:9 Gymnosperms

There are two major classes of seed plants — **Gymnospermae** (jihm nuh SPUR mee) and **Angiospermae** (an jee uh SPUR mee). Most **gymnosperms** living today have seeds that develop partly covered by the scales of cones. These cone-producing gymnosperms are commonly called **conifers.**

In gymnosperms the diploid phase is dominant; there is no independent haploid phase (Section 20:5). Separate male and female cones each produce gametes. Male gametes are contained in pollen grains. After the gametes fuse, the seeds form.

Most gymnosperms are trees, but shrubs and even vinelike forms exist. Familiar forms include the evergreens, such as pines, firs, and spruces, which have needlelike leaves. *Sequoia* (sih KWOY yuh) is a conifer that may reach a height of more than 100 meters and an age of more than 4000 years (Figure 14–11b). Conifers are distributed over most of Earth, and large forests once were common in temperate regions. Today, they are found mainly in northern latitudes. These forests provide most of the lumber for commercial use.

Conifers are well adapted for living on land. Unlike bryophytes and ferns, a watery habitat is not required for fusion of gametes. Pollen grains resist evaporation and provide protection. Seeds protect the delicate embryos during early development.

Conifers may grow very large because of the large amounts of xylem (ZI lum) (Section 21:1), commonly called wood, that supports the plants. Gymnosperms have adaptations, such as bark on the stems and cutin and special cells on the leaves, that reduce water loss.

An unusual relative of the conifers is the *Ginkgo* (GING koh) tree of China (Figure 14–11c). Because *Ginkgo* is the only living member of an order of gymnosperms that was common at one time, it is called a "living fossil." Only one species now exists, *Ginkgo biloba* (BI loh buh). It is grown in the United States mainly as an ornamental tree.

The species name, *biloba,* was given to *Ginkgo* because the leaves have two lobes. Leaves grow in clusters along the branches and conelike structures form on short branches of the tree. Each tree has either male or female structures. Pollen is transported from male trees to female trees by wind. Ginkgo seeds are spherical and soft on the outside. Although they look like fruits such as cherries, they are not. Plants like *Ginkgo* were probably the earliest seed producers.

14:10 Angiosperms

More than 250 000 species of **angiosperms,** flowering plants, are known. They grow in almost all types of climates. Angiosperms are adapted for life at the equator as well as the arctic regions. Most of the plants you see every day are flowering plants.

Flowering plants have some of the same basic features as conifers. However, the organ of reproduction in angiosperms is the flower. The angiosperms are a more successful group than the gymnosperms due to differences in reproduction. Angiosperms have better means of pollination and well-protected seeds. Also, the seeds are dispersed in more ways.

Like gymnosperms, angiosperms produce pollen. In gymnosperms, pollination occurs mainly by wind. Pollination in angiosperms also may involve wind, but it often involves animals. Thus, the chance of pollination is greater than in gymnosperms and as a result more offspring can be produced.

The flower is the reproductive organ of angiosperms.

Angiosperms produce pollen.

FIGURE 14–12. (a) Angiosperms, flowering plants, may have obvious flowers, as this water lily shows. Other types of angiosperms have inconspicuous or modified flowers as shown by (b) the red oak tree and (c) corn.

a

b

c

FIGURE 14–13. Some plants have complex flower-insect relationships to help ensure pollination. (a) The European orchid resembles a certain female wasp. (b) Male wasps attracted by the orchid's appearance transfer pollen from one flower to another.

a

b

Both flowers and insects have various adaptations that increase the chances for pollination.

Some intricate insect-flower relationships have evolved for pollinating flowers. One relationship involves a species of wasp and the European orchid, in which the flower parts look like the female wasp. In attempting to mate with the "female," a male wasp picks up pollen and transfers it to the next flower it visits. As a result, pollination is quite likely to occur.

After pollination and fertilization, an embryo forms within the flower (Section 20:8). As in gymnosperms, the zygote develops into an embryo within a seed that protects it from the dry environment. The seeds of angiosperms are enclosed within a fruit that first protects the seed and later aids in its dispersal (Section 20:9). Thus, the chance for survival of the new angiosperm plant is good. In gymnosperms, the seeds are not so well protected and dispersal is mainly by the wind; therefore, embryos are less likely to find a suitable place to carry out their development.

FIGURE 14–14. The great variation among angiosperms is shown by (a) albino Indian pipe and (b) mistletoe.

a

b

Angiosperms have many sizes, shapes, and forms. Some have no chlorophyll and are, therefore, saprophytes. Indian pipe is an albino plant that grows in damp forest soil. Mistletoe is a plant that lives as a parasite on other plants although it may also produce food by photosynthesis.

14:11 Monocots and Dicots

Angiosperms are classified as either monocots (MAHN uh kahts) or dicots (DI kahts). **Monocots** are plants whose seeds have one **cotyledon** (kaht ul EED un), a food-storing structure. Seeds of **dicots** have two cotyledons.

In addition to the number of cotyledons per seed, monocots and dicots differ in other ways. In monocots, **vascular bundles,** groups of conductive tissue, are scattered throughout the stem, but in dicots, they are arranged in an outer circle. Monocots have long, narrow

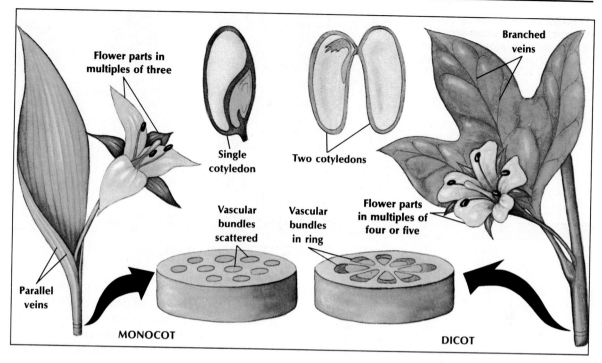

FIGURE 14–15. Monocots and dicots differ in seed structure, stem structure, vein pattern, and number of flower parts.

leaves with parallel veins. Dicot leaves are broad with branched veins. Flower parts of monocots are arranged in threes or multiples of three, but flower parts of dicots are arranged in fours or fives or multiples of four or five. Common monocots include plants of great value to humans, such as bananas, and cereals such as corn, wheat, rice, and barley. Ornamental flowers that are monocots include tulips, orchids, and lilies. Beans, carrots, peas, and potatoes are dicots. The number of species of dicots is much greater than the number of species of monocots. Internal structures of monocots and dicots will be studied in Unit 6.

REVIEWING YOUR IDEAS

5. What features can be used to distinguish between vascular and nonvascular plants?
6. For what functions do bryophytes need water?
7. Name the three main groups of tracheophytes.
8. Which phase is dominant in each of the three main groups of tracheophytes?
9. What are the reproductive structures of gymnosperms and angiosperms?
10. Compare monocots and dicots in terms of the number of cotyledons, the arrangement of vascular bundles and leaf veins, and the number of flower parts.

FIGURE 14–16. (a) A fibrous root system has many secondary roots. (b) These various kinds of taproots show the characteristic large primary root with a few secondary roots.

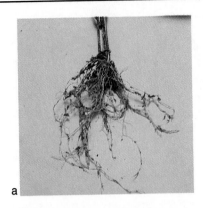

a

b

FEATURES OF ANGIOSPERMS

Most of the plants that you see around you are angiosperms, the major plant form on Earth today. Because of their wide distribution and importance, it is useful to learn more about them. Other than flowers, the three major organs of angiosperms are the root, stem, and leaf. In this chapter, you will study the external features of these organs. In later chapters, you will learn about their internal structures and functions.

14:12 Root Types

Roots anchor plants, store food, and absorb water and minerals.

Major functions of roots are anchorage, food storage, and absorption of water and minerals from the soil. The role of roots in absorption of water and minerals will be studied in Chapter 22. Most woody plants (and many nonwoody species) have a **fibrous** (FI brus) **root system** that consists of many secondary roots and root hairs (Figure 14–16a). The secondary roots are branches of the primary root, which forms during early development.

Some plants have a **taproot system** that has one large, long primary root with a few secondary roots (Figure 14–16b). Carrot and beet plants are examples of plants with a taproot system. In these examples, the taproot is a food storage organ and is an edible portion of the plant.

14:13 The Stem

A stem supports a plant and is a transport link between roots and leaves.

A plant has no skeletal system but is supported by the stem. Also, the stem is a transport link between the roots and the leaves, and some stem cells store food.

Stems are either herbaceous (hur BAY shus) or woody. **Herbaceous stems** are soft, green, and often juicy, with very little or no woody tissue. Plants that live for a single growing season are usually herbaceous. A plant with a herbaceous stem is supported by the pressure of water in the cells of the stem. This pressure causes **turgidity** (tur JIHD ut ee), or stiffness, which results in the plant standing upright. In the event of a water shortage, the stem loses turgidity (wilts) and bends down. **Woody stems** are composed mainly of certain tough, dead, xylem cells. These cells support the plant. Figure 14–17 illustrates the external structures of woody stems.

Buds are protected, dormant tissues that may develop into new stems, flowers, or leaves. During winter, buds are protected by modified leaves called **bud scales. Terminal buds** are buds at the tips of branches and are responsible for lengthening stems during each growing season. **Lateral buds** lie along the sides of branches and give rise to new branches. Lateral buds are often smaller than terminal buds.

Nodes are places on a stem where leaves develop. The space between two nodes is called the **internode** (IHNT ur nohd). **Leaf scars** are the marks left on a stem after leaves fall off. The leaf scar pattern on a twig indicates the arrangement of leaves. If there is one leaf scar (or leaf) per node, leaves are arranged in an *alternate* pattern on the twig. If there are two leaf scars (or leaves) per node, leaves are arranged in an *opposite* pattern. Presence of three or more leaf scars (or leaves) at the node indicates a *whorled* arrangement of leaves.

Flower scars and **fruit scars** can be found on the stems of flowering plants. **Bud scale scars** also may be found on stems. Bud scale scars mark the places where bud scales surrounded a bud. Because most bud scales form around terminal buds, the age of a twig can be determined by counting the bud scale scars of previous terminal buds. Amount of growth can be determined by measuring the distance between the bud scale scars.

On young stems, porous regions, or **lenticels** (LENT uh selz), are easily seen. Exchange of gases between the stem and the atmosphere occurs through the lenticels.

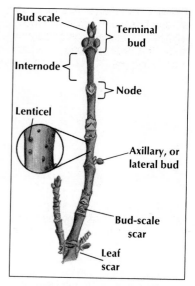

FIGURE 14–17. The external structure of a woody stem shows buds and scars that can be used to identify different kinds of stems.

FIGURE 14–18. The leaf is the main photosynthetic organ of a plant. Externally, the leaf is composed of a blade (where most photosynthesis occurs) and a petiole, or stalk, which attaches the blade to the stem.

14:14 The Leaf

Flowering plants exhibit diversity in leaf types and forms. Although the details of leaf structure vary, the main function of leaves is the same in nearly all plants. The leaf is the major organ of photosynthesis (Section 21:1).

Most leaves are attached to the stem by a slender stalk, the **petiole** (PET ee ohl). The expanded part of a leaf is the **blade.**

In leaves with branching veins, several vein patterns, called **venation** (ve NAY shun), can be observed. A feathery pattern in which veins branch from a central vein, or midrib, is called **pinnate**

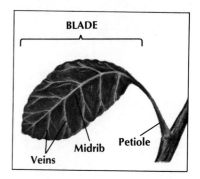

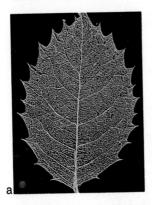

a b c d

FIGURE 14–19. Leaf shape and vein patterns vary among the many species of plants and can be used in plant identification. (a) A simple leaf with pinnate venation, (b) a simple leaf with palmate venation, (c) a palmately compound leaf, and (d) a pinnately compound leaf are shown.

Pinnate and palmate patterns apply to the arrangement of both veins of leaves and leaflets of compound leaves.

(PIHN ayt) **venation** (Figure 14–19a). This pattern is found in oaks and elms. In trees such as maples, many large veins originate at the point where the petiole is attached to the blade. This pattern is called **palmate** (PAL mayt) **venation** (Figure 14–19b).

If a leaf is composed of one blade, it is a **simple leaf.** However, when the blade is subdivided into distinct parts, called leaflets, it is a **compound leaf.** Patterns in compound leaves correspond to venation patterns. Leaves are **palmately compound** when the leaflets are all attached to the petiole at a central point (Figure 14–19c). Clover is a common example of a palmately compound leaf. When leaflets branch out along different points of the petiole, they are said to be **pinnately compound** (Figure 14–19d). Leaves of the rosebush and walnut tree have this pattern.

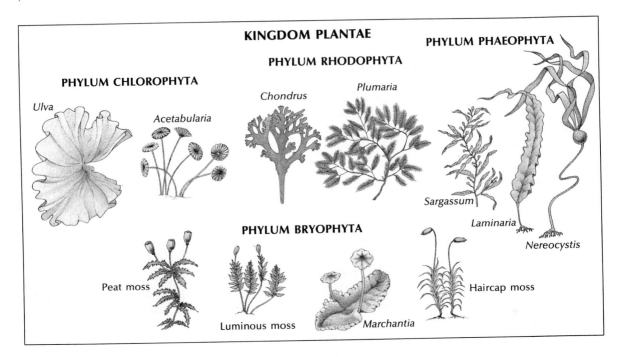

KINGDOM PLANTAE

PHYLUM RHODOPHYTA

PHYLUM PHAEOPHYTA

PHYLUM CHLOROPHYTA

Ulva

Acetabularia

Chondrus

Plumaria

Sargassum

Laminaria

Nereocystis

PHYLUM BRYOPHYTA

Peat moss

Luminous moss

Marchantia

Haircap moss

REVIEWING YOUR IDEAS

11. What are the major functions of a root? A leaf?
12. What are buds? What are terminal and lateral buds?

TABLE 14-1. SUMMARY OF MAJOR CHARACTERISTICS OF PLANTS

Classification	Common Name	Vascular Tissue	Structure	Habitat
Phylum Chlorophyta	Green algae	Absent	Unicellular and multicellular; little division of labor	Water
Phylum Phaeophyta	Brown algae	Absent	Multicellular; tissues in some forms	Water
Phylum Rhodophyta	Red algae	Absent	Mostly multicellular	Water
Phylum Bryophyta	Mosses/Liverworts	Absent	Multicellular; no true roots, stems, or leaves	Moist areas on land
Phylum Tracheophyta	Club mosses/Horsetails	Present	Multicellular; true roots, stems, and leaves	Moist areas on land
Phylum Tracheophyta Class Filicineae	Ferns	Present	Multicellular; true roots, stems and leaves, develop from rhizome	Moist areas on land
Phylum Tracheophyta Class Gymnospermae	Conifers	Present	Multicellular; true roots, stems and leaves	Land
Phylum Tracheophyta Class Angiospermae	Flowering plants (Monocots and dicots)	Present	Multicellular; true roots, stems and leaves	Land

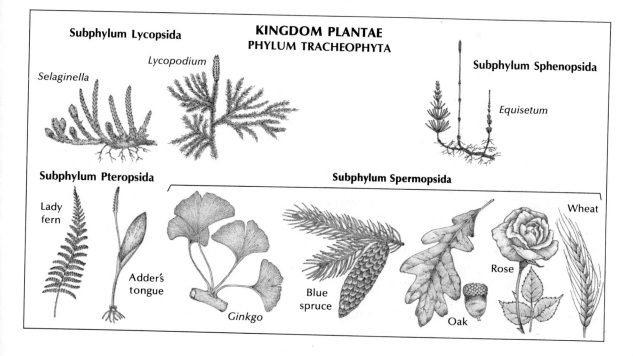

KINGDOM PLANTAE
PHYLUM TRACHEOPHYTA

Subphylum Lycopsida

Selaginella

Lycopodium

Subphylum Sphenopsida

Equisetum

Subphylum Pteropsida

Lady fern

Adder's tongue

Subphylum Spermopsida

Ginkgo

Blue spruce

Oak

Rose

Wheat

INVESTIGATION

Problem: What are the characteristics of gymnosperms?

Materials:

microscope
glass slide
colored pencils
forceps
water
coverslip

gymnosperm branch
male cone
female cone
prepared slide of
   gymnosperm leaf
   cross section

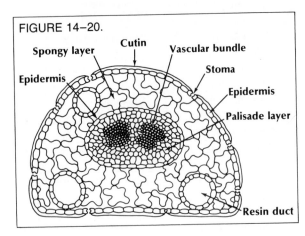

FIGURE 14–20.

Procedure:

Part A. External Anatomy of a Gymnosperm Leaf

1. Copy the data chart onto a sheet of paper. Include the outline diagram.
2. Examine the leaves on the gymnosperm branch. Each leaf is a single, needle-shaped structure.
3. Record the following observations in your chart:
   (a) general shape of the leaf
   (b) number of leaves per bundle
   (c) color of the leaf
   (d) presence or absence of cutin.

Part B. Internal Anatomy of a Gymnosperm Leaf

1. Examine the prepared slide of the leaf cross section under low-power magnification.
2. Locate and identify the following leaf structures using Figure 14–20 as a guide:
   (a) cutin–thin, nonliving, outer waxy layer that protects the leaf and is most easily seen at the corners of the leaf.
   (b) epidermis–single layer of protective cells directly below cutin.
   (c) stomata–small spaces along the edge of the epidermis; allow gas exchange.
   (d) spongy layer–thickest layer, which is green in the living leaf and the site of food production.
   (e) resin ducts–round structures located in the spongy layer, which are filled with resin.
   (f) endodermis–single layer of protective cells forming the inner ring around the palisade layer.
   (g) palisade layer–layer surrounding the vascular bundle and the site of some food production.
   (h) vascular bundle–two round structures in the center of the leaf, which conduct food and water through leaf. The vascular bundle is composed of xylem and phloem.
3. Color the diagram in your chart as follows:
   (a) red for all protective tissues.
   (b) green for all photosynthetic tissues.
   (c) blue for all transporting tissues.

Part C. Gymnosperm Cones

1. In gymnosperms, the male and female reproductive structures, cones, are usually found

on the same plant. Examine both types of cones and record the answers to the questions in Part C of your chart.

2. Use forceps to remove one small scale from the male cone. Place the scale onto a glass slide and add two drops of water. Gently mash the scale with the end of the forceps. Add a coverslip and observe under low and then high power.

3. Look for pollen cells, each of which resembles a cartoon of a mouse head. The structures that resemble the "mouse's ears" are air bladders. Diagram one or two pollen cells as seen under high power in the space provided in your chart. Label the air bladders and the pollen grain and fill in the magnification.

4. Estimate and record in your data chart the number of pollen cells present in one cone scale. Base your estimate on the number of pollen cells seen under low power.

5. Remove one scale from an open female cone. Note that each seed was one egg.

6. Count the number of egg cells present on one cone scale and record your answer in your data chart.

Questions and Conclusion:

1. Define the following terms: gymnosperm, pollen grain, and cone.

2. Refer to Appendix A and determine the kingdom, phylum, subphylum, and class for a pine tree.

3. Compare the number of pollen grains (male sex cells) to the number of egg cells (female sex cells) on one scale.

4. Pollen cells are carried by the wind to female cones where fertilization of egg cells occurs. Explain the adaptive advantage of air bladders on the pollen grains.

Conclusion: What are the characteristics of gymnosperms?

Data and Observations:

Part A.

(a) general leaf shape _____

(b) number of leaves per bundle _____

_____

(c) color of leaf _____

(d) presence or absence of cutin _____

Part B.

Part C.

(a) Which cone type is larger? _____

(b) Which cone type is smaller? _____

(c) Which cone type appears singly? _____

(d) Which cone type appears in groups? _____

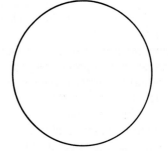

(e) estimate of number of pollen cells _____

_____

(f) number of egg cells _____

CHAPTER REVIEW

SUMMARY

1. Plants are nonmotile, mostly multicellular, autotrophic organisms. They have cell walls composed of cellulose. Most plants have a life cycle involving both diploid and haploid phases. **p. 289**
2. The algal plants are mostly multicellular with some division of labor. They are subdivided on the basis of color and structure. **p. 289**
3. Green algae exist in a variety of forms: unicellular, filamentous, and colonial. **14:1**
4. Brown algae, known as seaweeds or kelps, have a complex structure. Almost all forms are marine. **14:2**
5. Red algae are complex multicellular forms, some of which can live deeper in water than any other plants. **14:3**
6. The complex structure of land plants is related to their life out of water. Such plants are classified based on the presence or absence of vascular tissue. **14:4**
7. Bryophytes include mosses and liverworts, which are small plants with no true roots, stems, or leaves. They are limited in size because of lack of vascular tissue and in habitat because of their need for moisture for fertilization. **14:5**
8. Tracheophytes include plants with true roots, stems, and leaves. They have vascular tissue and are better adapted for life on land than the bryophytes. **14:6**
9. Club mosses and horsetails are primitive tracheophytes that reproduce by spores. **14:7**
10. Most ferns consist of an underground rhizome from which roots and leaves develop. They are larger than bryophytes but, like them, require moisture for fertilization. **14:8**
11. Gymnosperms are seed plants, mostly conifers, in which seeds form on the undersides of cone scales and are dispersed by wind. Because they require no moisture for fertilization, they are widely distributed. **14:9**

12. Angiosperms, flowering plants, are best adapted for life on land. Animals often are involved in pollination, making the chances of fertilization greater than in gymnosperms. Also, the fruit better protects the seeds and aids in their dispersal. These factors make flowering plants the most abundant forms. **14:10**
13. Angiosperms are divided into monocots and dicots. These groups differ in the following ways: the number of cotyledons per seed, the number of flower parts, the shapes of leaves, vein patterns, and the arrangement of vascular bundles. **14:11**
14. Roots anchor a plant, absorb water and minerals, and store excess food. Both fibrous and taproot systems are found among angiosperms. Fibrous roots consist of a series of secondary roots that are branches of the primary root. Taproots have one large, long primary root. **14:12**
15. The stem is an organ of support and transport. Stems are either woody or herbaceous. The external structure of a stem is characterized by the presence of buds, nodes, and scars. **14:13**
16. Leaves are the organs where most photosynthesis occurs. Leaves can be characterized by their vein patterns, by whether they are simple or compound, and by their arrangement. **14:14**

LANGUAGE OF BIOLOGY

angiosperm	fibrous root	rhizoid
blade	system	rhizome
bryophyte	gymnosperm	taproot system
bud	herbaceous	thallus
conifer	stem	tracheophyte
cotyledon	lenticel	turgidity
cutin	monocot	vascular tissue
dicot	node	venation
fern	petiole	woody stem

CHECKING YOUR IDEAS

On a separate paper, indicate whether each of the following statements is true or false. Do not write in this book.

1. All land plants have vascular tissue.
2. The body of an alga is called a thallus.
3. Vascular bundles have conductive tissue.
4. The diploid stage is predominant in all plants that live on land.
5. Land plants have adaptations that enable them to conserve water.
6. Food storage is the major function of most plant stems.
7. Pollen is produced by all land plants.
8. A fibrous root system has one large root often used for food storage.
9. Most algae have no true tissues.
10. Most bryophytes need moist habitats.

EVALUATING YOUR IDEAS

1. Describe the traits of algal plants.
2. Why do some biologists think that an organism like *Volvox* might represent an intermediate form of life between unicellular and multicellular organisms?
3. Explain ways in which some brown algae resemble land plants.
4. Why can red algae live deeper in the ocean than other algae?
5. List the functions that are carried out by land plants.
6. How do bryophytes carry out the jobs in answer 5?
7. How are bryophytes limited? Explain.
8. Describe the structure of club mosses and horsetails.
9. How are ferns better adapted for life on land than mosses? How are ferns limited?
10. Compare conifers to ferns in terms of support and sexual reproduction.

11. How are angiosperms better suited for life on land than gymnosperms?
12. Compare herbaceous and woody stems.
13. Distinguish between a fibrous root system and a taproot system.
14. Describe venation and leaf forms.

APPLYING YOUR IDEAS

1. Compare algal plants to algal protists.
2. How do plants differ from fungi?
3. Embryos of land plants are protected by the parent plant. Algal embryos develop in water. What is the significance of this difference?
4. How would you explain the origin of Indian pipe, a heterotrophic angiosperm?
5. What are some advantages of a fibrous root system? A taproot system?
6. Why do you think there are usually longer branches low on a tree?
7. How might venation and leaf patterns be useful to a biologist?

EXTENDING YOUR IDEAS

1. Make a collection of leaves of many woody plants near your home. Use a classification key to identify each plant. Also record the venation and leaf pattern of each plant. Put your specimens in a plant press and then mount each one on colored paper. Record the information for each plant on the sheet.
2. Find out what you can about Spanish moss. Is it really a moss?

SUGGESTED READINGS

Beller, Joel, *Experimenting with Plants.* New York, Simon and Schuster, 1985.

McFarlane, R. B., *Collecting and Preserving Plants for Science and Pleasure.* New York, Arco Publishing, Inc., 1985.

ANIMALS: SPONGES THROUGH MOLLUSKS

E ven simpler animals have adaptations that help them survive in their environments. Here, feather-duster worms, segmented aquatic worms, produce water currents with their ciliated filaments. The currents bring food particles into their mouths. What kinds of simple animals are there? What kinds of adaptations do these less complex animals have? Are any of their adaptations similar to adaptations you have?

chapter 15

Animals must trap and ingest food. Many characteristics of animals are related to their heterotrophic way of life. Features such as movement, support systems, sensory equipment, and nervous control work together to aid an animal in obtaining food. These features depend upon specialized tissues, organs, and systems not found in other organisms.

Animals can be divided into two groups — the **invertebrates** (ihn VURT uh brayts), which do not have backbones, and the **vertebrates** (VURT uh brayts), which do have backbones. Invertebrates are classified into about thirty distinct phyla; vertebrates make up only one *subphylum* of animals. In this chapter, you will study the major features of the less complex invertebrates. In Chapter 16, you will learn about the complex invertebrates and the vertebrates.

Objectives:
You will
- list the major features of several phyla of invertebrate animals.
- compare the traits that make one phylum more or less complex than another.
- discuss the biological and economic importance of animals in these phyla.

Vertebrates have backbones; invertebrates do not.

THE SIMPLEST ANIMALS

Unlike animals with which you are familiar, many of the simplest animals do not move or actively search for food. They lack the higher levels of organization seen in more complex animals. However, they do have adaptations for the heterotrophic way of life.

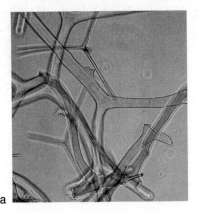

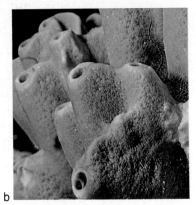

a   b   c

FIGURE 15–1. (a) A support system of spicules is found in sponges. (b) Incurrent and excurrent pores and (c) varied colors and shapes are also characteristics of sponges.

oscula (singular, osculum)

Sponges have two layers of cells separated by a jellylike layer.

Mature sponges have no locomotion.

Sponges are supported by structures called spicules.

15:1 Sponges

Simplest of all animals now living are the sponges, phylum **Porifera** (puh RIHF uh ruh). They are so named because they have many openings, called **incurrent pores,** through which water enters. The body of a sponge also has one or a few large openings called **oscula** (AHS kyuh luh), or **excurrent pores.** Water leaves a sponge through the oscula.

Most sponges are marine. Often they are part of a colony and individual organisms cannot be seen easily. Each sponge has two layers of cells. The outer layer is protective. Cells of the inner layer have flagella that aid in drawing in water. Between these cell layers is a jellylike layer containing amoebalike cells. The amoebalike cells are involved in digestion (Section 25:1).

Newly developing sponges can move, but mature sponges remain attached in one spot. An attached organism is said to be sessile (SES ul). Most sponges are attached to rocks or shells on the ocean floor. In some sponges, there are cells that can contract, or shorten. Some movement can be detected in these cells. However, there is no evidence of any true locomotion in adult sponges.

Sponges feed on microscopic organisms they filter from the water. Gases are exchanged with the water, and wastes are carried away by water.

Sponges do not have bony skeletons. However, they are supported by structures called **spicules** (SPIHK yewlz). Spicules are produced by the amoeboid cells and stick out from the jellylike layer between the ectoderm and endoderm.

Sponges are grouped into three classes on the basis of the minerals in their spicules. One class has calcium carbonate spicules. Another class, known as glass sponges, has silica spicules. Their skeletons, when dry, have a beautiful, glasslike appearance. A third class has members with no spicules but with an elastic substance. These sponges are dried and sold as bath sponges. Are most commercial sponges sold now really animal remnants?

Sexual reproduction occurs in sponges. Some of the cells in the jellylike layer become modified as eggs or sperm. In some species, individuals are hermaphrodites (hur MAF ruh dites). **Hermaphrodites** are animals that produce both sperm and eggs. In other species, separate sexes exist. Fusion of sperm and eggs produces zygotes. Zygotes develop into free swimming, immature organisms that eventually attach to the ocean bottom and develop into new sponges.

Asexual reproduction also occurs in sponges. Sometimes a new sponge begins to grow from the parent and then breaks away. Sometimes a piece of a parent sponge may break away and then develop into a new sponge. Some sponges produce branches that develop into new sponges. Others asexually produce flagellated embryos that swim away and become attached to form new sponges. These processes give rise to colonies of sponges. At times, gemmules (JEM yewlz) form inside the sponge. **Gemmules** are groups of cells that have become enclosed by a tough outer covering. They are resistant to both dryness and cold temperatures. When conditions become favorable, such as after a dry season or winter, the cells leave the gemmule and form a new sponge.

Sponges are the least complex animals. They do have specialized cells, but the cells are not organized into tissues (Section 1:8). Because they have no nerve cells, there is no coordination between parts. If one dissociates (separates) the cells of a sponge, they move back together and form a sponge again. In many respects, a single sponge is not much more complex than a colonial protozoan. Because sponges are so different from most other animals, some biologists feel that they represent a special subkingdom.

Sponges reproduce both sexually and asexually.

Sponge cells are specialized but are not organized into tissues.

A sponge does not have a nervous system.

FIGURE 15–2. (a) In commercial sponge-fishing, sponges brought up by divers are hung up to allow the living part of the sponges to decay. (b) Sponge culturing (growing) has replenished some areas that have been depleted by too much sponge-fishing. Fragments of sponges, attached to tiles, are lowered to the bottom. Under good conditions, they grow quickly into sponges like the one shown.

a

b
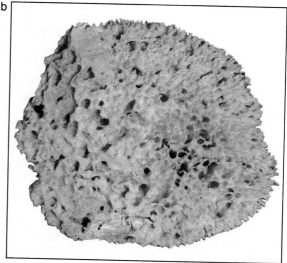

a

b

FIGURE 15–3. (a) The sea anemone and (b) the white coral are examples of coelenterates. Coelenterates have two tissue layers— ectoderm and endoderm.

Coelenterates have two tissue layers: ectoderm and endoderm.

Nerves and other tissues are found in coelenterates.

Tentacles are used in trapping food.

15:2 Coelenterates

More complex than Porifera are organisms of phylum **Coelenterata** (sih len tuh RAH tuh). This phylum includes organisms such as jellyfish, *Hydra,* corals, and sea anemones (uh NEM uh neez). **Coelenterates** (sih LENT uh raytz) have two tissue layers—**ectoderm** (outer layer) and **endoderm** (inner layer). Between them is a jellylike **mesoglea** (mez uh GLEE uh), which may contain some unspecialized cells.

Coelenterates, unlike sponges, have true tissues. Another difference is the presence of nerve cells that form a network throughout the organism. This network coordinates movements, but there is no brain as found in more complex animals. No specialized tissues for gas exchange or excretion are present.

Coelenterates have a hollow body with one opening, the mouth. Around the mouth is a ring of **tentacles** (TENT ih kulz). **Nematocysts** (nih MAT uh sihsts), sometimes called "stinging cells," are special capsules on the tentacles that "shoot out" poisonous filaments, paralyzing prey. The tentacles then surround the food and transfer it to the mouth.

Coelenterates have two body forms. The **polyp** (PAHL up) form has a cylindrical body and is usually sessile. The mouth faces upward. Hydras are examples of polyps. The other form of coelenterate is called a **medusa** (mih DEW suh). A medusa is a free-swimming form with the characteristic jellyfish shape that has the mouth facing downward.

Polyps are sessile forms of coelenterates. Medusae are motile.

Some coelenterates are colonial and have specialized polyps and medusae. For example, *Stephallia* has both of these body forms at the

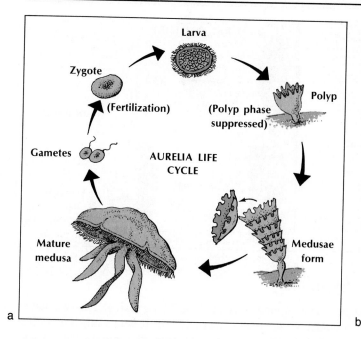

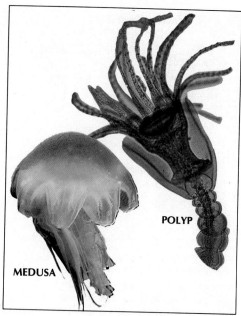

a

b

FIGURE 15-4. (a) The life cycle of *Aurelia* shows an alternation of two body forms. (b) The body forms of a coelenterate are the polyp and medusa.

same time. The Portuguese man-of-war is a floating, colonial coelenterate composed of a float and trailing specialized polyps. Its tentacles can paralyze large animals such as fish. Its sting is quite painful and can be fatal.

Many coelenterates have a life cycle that alternates between polyps and medusae. *Aurelia* is an example of such an organism. As a polyp, it is sessile and reproduces asexually. Parts of the polyp are transformed into medusae. The medusae reproduce sexually. The fertilized eggs develop into attached polyps (Figure 15-4).

Many coelenterates have a life cycle including both polyp and medusa forms.

a

b

FIGURE 15-5. (a) The Portuguese man-of-war, *Physalia physalis*, is a colony of polyps that floats on the water surface due to the gas-filled chamber. (b) The cigar jellyfish is a medusa that moves by contraction of muscles.

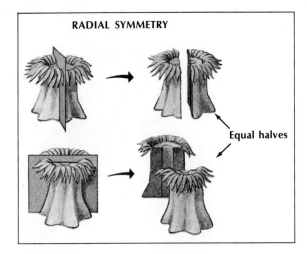

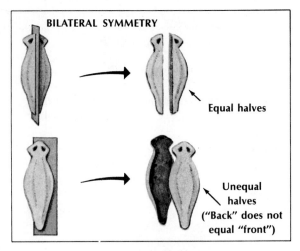

FIGURE 15–6. A radially symmetrical animal can be cut lengthwise through the center in any direction, and the result is two equal halves. A bilaterally symmetrical organism can only be cut lengthwise through the center in only one place to give equal halves.

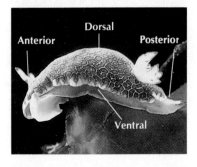

FIGURE 15–7. Names of areas of a bilaterally symmetrical animal, a nudibranch, are shown. These names are used to accurately give locations of body parts.

15:3 Symmetry

Suppose an anemone were cut in half lengthwise. The two halves produced would be identical (Figure 15–6). It would make no difference in which direction across the diameter the lengthwise cut was made. An anemone has the same features all around its cylindrical body. This body plan is called **radial symmetry** (RAYD ee ul • SIH muh tree). It is a feature of coelenterates, some sponges, and several other animal groups.

Animals with a specific direction of locomotion have a different body pattern. It is called **bilateral** (bi LAT uh rul) **symmetry.** One side of such an organism is a mirror image of the other. One cut down the midline of an animal from head to tail would result in identical halves (Figure 15–6). Other lengthwise cuts would not produce identical halves.

Animals with bilateral symmetry have definite body regions. The names of these areas are related to the animal's direction of movement. There is a front, or **anterior** (an TIHR ee ur) end. There is a hind, or **posterior** (pah STIHR ee ur) end. There is a top, or **dorsal** (DOR sul) side, and a bottom, or **ventral** (VEN trul) side. There are definite sides (edges) or lateral areas.

REVIEWING YOUR IDEAS

1. Name the two major divisions of animals.
2. Why are sponges called Porifera?
3. Describe the coelenterate nervous system.
4. Distinguish between polyp and medusa.
5. How are sponges and coelenterates different?
6. Compare radial and bilateral symmetry. Distinguish between dorsal and ventral, anterior and posterior.

WORMS AND MOLLUSKS

Although most worms and mollusks may not be very familiar to you, they are ecologically important. Certain worms play an important role in decomposition and recycling of materials. Some land mollusks cause damage to plants. Mollusks such as clams, oysters, and squid are human foods.

Worms and mollusks are more complex than the animal groups you have studied so far. They have characteristics such as true organs, simple muscles for locomotion, specialized systems, and primitive brains.

15:4 Flatworms

Flatworms, phylum **Platyhelminthes** (plat ih hel MIHN theez), are aquatic or semiaquatic animals, the simplest animals with bilateral symmetry. Figure 15–8 shows a type of free-living flatworm called a planaria (pluh NER ee uh). Flatworms have three tissue layers. Between the ectoderm and endoderm is a third layer called **mesoderm.** With bilateral symmetry and mesoderm, there is also a greater degree of differentiation.

Unlike sponges and coelenterates, flatworms have definite organs. Nervous tissue (brain) and sensory organs are concentrated at the anterior end of these organisms as seen, for example, in planarians. A planarian has a digestive system with a mouth and intestine. As in coelenterates, the flatworm mouth is also the anus. Specialized parts for maintaining osmotic balance are also present. Muscles aid locomotion. There are no specialized tissues or organs for gas exchange, waste removal, or transport of materials.

Flatworms have three tissue layers: ectoderm, endoderm, and mesoderm.

Flatworms have specialized tissues and organs for some, but not all, life functions.

FIGURE 15–8. *Planaria* is a nonparasitic flatworm that has nervous and sensory tissue concentrated at its anterior end.

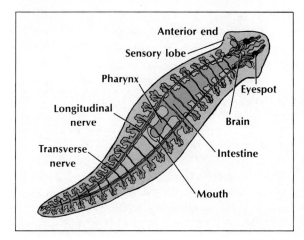

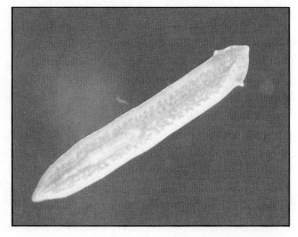

Planarians may reproduce sexually or asexually.

Planarians are hermaphrodites though they do not self-fertilize. Instead, one planarian exchanges sperm with another. Asexual reproduction also may occur as a result of part of the planarian breaking off. Both parts may then regrow missing tissues, thus producing two planarians.

In addition to free-living forms such as planarians, this phylum includes many parasitic animals. The parasitic forms include the flukes and tapeworms.

Often, the parasites have a complex life cycle involving two or three different hosts. The Chinese liver fluke, for example, spends part of its life in a snail, a fish, and in the human liver. The dog tapeworm spends part of its life in dog fleas and lice, and part of its life in the intestine of dogs. These parasites do not exhibit the same features as their free-living relatives. They have adaptations related to the kinds of hosts they invade.

PEOPLE IN BIOLOGY

Libbie Henrietta Hyman

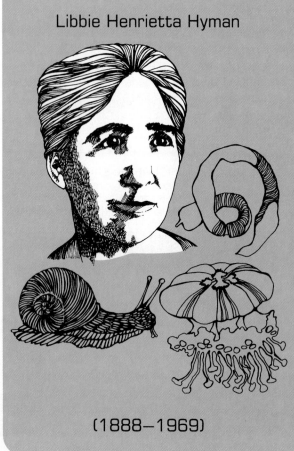

(1888–1969)

Growing up in a poor family in Fort Dodge, Iowa, Libbie Hyman never imagined that she would one day be the foremost expert on invertebrates in the Western Hemisphere. Libbie Hyman developed her interest in zoology while earning her B.A. and Ph.D. degrees at the University of Chicago. After graduating, Dr. Hyman stayed at the university as a research assistant and wrote two textbooks still in use today: *A Laboratory Manual for Elementary Zoology* and *A Laboratory Manual for Comparative Vertebrate Anatomy*.

In 1931 Dr. Hyman moved to New York City and began work on a treatise on invertebrates. At first Dr. Hyman thought the book would take two volumes. She soon discovered that to cover the one million classified invertebrates, she would need to write ten. Because she had no money to hire an artist, Hyman drew all her own pen-and-ink illustrations for each volume.

After only three volumes had been published, Dr. Hyman was awarded the Daniel Giraud Medal of the National Academy of Science. She also received the Gold Medal of the Linnaen Society of London. Hyman went on to write three more volumes before her death.

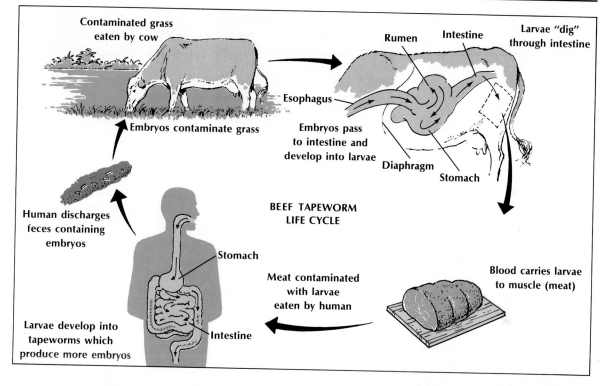

Contaminated grass eaten by cow

Rumen Intestine Larvae "dig" through intestine

Esophagus

Embryos contaminate grass

Embryos pass to intestine and develop into larvae

Diaphragm

Stomach

Human discharges feces containing embryos

BEEF TAPEWORM LIFE CYCLE

Stomach

Meat contaminated with larvae eaten by human

Blood carries larvae to muscle (meat)

Larvae develop into tapeworms which produce more embryos

Intestine

FIGURE 15–9. The life cycle of the beef tapeworm, *Taenia saginata*, alternates between cattle and human hosts.

15:5 The Tapeworm: A Parasitic Flatworm

A mature tapeworm lacks a mouth and a digestive system. It gets food by diffusion. Figure 15–9 shows the life cycle of a beef tapeworm. A cow grazing on grass contaminated with tapeworm embryos may eat some of them. The capsules around the embryos are digested in the cow's intestine. The embryos grow, change form, "dig" through the wall of the intestine, and enter the bloodstream. Then, they can enter the cow's muscles.

Humans get the tapeworms by eating poorly cooked beef. The tapeworms reach the human intestine where they become adults. The adult attaches to the intestinal wall by suckers and/or hooks. It lives on the host's digested food.

Tapeworms are highly specialized for sexual reproduction. The body of the tapeworm is composed of a series of flat sections produced from the neck region. Each section contains reproductive organs. Sperm fertilize eggs (often from a different section) and a capsule forms around each zygote. The sections of the worm then fall off, full of these fertilized eggs. The zygotes develop into embryos that pass out of the host's intestine along with wastes. Human wastes with these embryos can then contaminate grass. What conditions promote the life cycle of the beef tapeworm?

FIGURE 15–10. Tapeworms have adaptations for attachment on their heads. The hooks and suckers of a pork tapeworm are shown here.

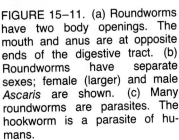

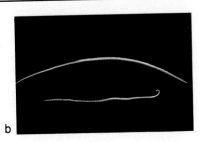

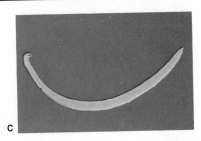

FIGURE 15–11. (a) Roundworms have two body openings. The mouth and anus are at opposite ends of the digestive tract. (b) Roundworms have separate sexes; female (larger) and male *Ascaris* are shown. (c) Many roundworms are parasites. The hookworm is a parasite of humans.

Many roundworms are parasites.

15:6 Roundworms

Members of phylum **Nematoda** (nem uh TOH duh), the **roundworms** (Figure 15–11), are found almost everywhere. This group includes free-living forms that live in soil and water. However, most nematodes are parasites. At least fifty species are parasites of humans alone!

Ascaris (AS kuh rus) is a parasite that can live in the intestine of humans and most other mammals, robbing the host of digested food. If enough worms are present, they may block the intestine, which could kill the host.

The hookworm is a parasite of humans that attaches to the wall of the intestine and sucks blood. Outside the bodies of their hosts, hookworms need warm soil temperatures. Therefore, they are common in southern parts of the United States. They can enter the human body through bare feet.

Filaria (fuh LER ee uh) worms cause the disease elephantiasis (el uh fun TI uh sus). In this disease, enormous swelling of tissues occurs. Filaria nematodes are transmitted by a certain species of mosquito. Their occurrence is rare in the United States.

Trichinella worms spend part of their life cycle in muscles of hogs. Failure to cook infected pork thoroughly can result in human infection. Worms enter the human's intestine, reproduce, and eventually spread to the muscles. The result is a painful, sometimes fatal disease called trichinosis. Trichinosis can be prevented by thorough cooking of pork.

Unlike sponges, coelenterates, and flatworms, nematodes (NEM uh tohdz) have two body openings. Food is taken in through the mouth. Undigested materials are egested (ih JEST ud), or removed, through the anus (AY nus). The mouth and anus are openings at opposite ends of the digestive tract.

Roundworms, unlike flatworms, have two body openings.

The round body of roundworms is tapered at each end. These animals have muscles that run only along the length of the body, causing their movements to appear jerky and random. Unlike flatworms, which are hermaphrodites, roundworms are either male or female. Sex organs are located in the space between the outer muscle layer and the digestive tube. Fertilization is internal; it occurs within the female. Development is external.

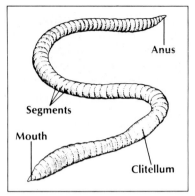

Anus

Segments

Mouth

Clitellum

FIGURE 15–12. The earthworm is a common annelid that plays a role in soil ecology. Annelids have easily visible segments.

15:7 Segmented Worms

Members of phylum **Annelida** (uh NEL ud uh) are the **segmented worms.** The common earthworm, *Lumbricus* (LUM bruh kus), is an example of an annelid. The cylindrical bodies of these worms are segmented. Each segment, or unit, is distinct and visible on the outside of the worm. Most segments contain the same internal structures as one another. The common earthworm usually has around one hundred segments (Figure 15–12). Giant earthworms of Australia can be three meters long.

Earthworms play an important role in soil ecology. As they tunnel through soil, earthworms make a network of spaces that aid in aerating the soil and improving water drainage. They consume a lot of dirt and decaying plant matter in the process and then undigested materials, mostly soil, pass through the worms and are deposited on the surface as castings. In this way, earthworms are constantly turning over the soil and enriching it. All these factors improve growing conditions for plants. For this reason, earthworms have been called "nature's plows."

Although the earthworm is the annelid most familiar to you, most annelids are aquatic. *Nereis* (NIHR ee us) (Figure 15–13) is an ocean dweller that lives near the coastline, sometimes in burrows in the mud. On each segment (except the first and last) are a pair of paddlelike **parapodia** (PER uh poh dee uh) that aid in locomotion. Parapodia also aid gas exchange. *Nereis* has separate sexes and external fertilization; fertilization occurs outside the organism. The young develop in the water. In earthworms, there are no parapodia and eggs and sperm are deposited into a capsule secreted by the animal's body. Fertilization takes place in the capsule, which is then deposited in the soil. Earthworms develop without a larval stage and are then released from the capsule.

Leeches, most of which are aquatic parasites (Figure 15–13), are another class of annelids. They have suckers at both ends of their bodies. The posterior sucker attaches to a host. Sometimes the host

Earthworms enrich and aerate soil.

parapodia (singular, parapodium)

FIGURE 15–13. (a) This sandworm is a segmented worm with parapodia. It is shown here on a coral reef along with a nudibranch. (b) A leech is a parasitic segmented worm.

a

b

is a human; often the hosts are fish. Sharp teeth in the anterior sucker puncture the skin. Then, the leech sucks the blood of its host. A leech can store blood in special pouches. Thus, a single feeding can last a leech for several months.

An important adaptation in segmented worms is a circulatory system. A series of hearts pumps blood throughout the organism in a system of closed tubes. Blood carries food molecules, gases, and certain wastes.

Another feature not present in less complex phyla is a body cavity or "true" **coelom** (SEE lum). (Nematodes have a similar cavity, but it forms in a different way and is known as a "false" coelom.) The coelom is the space between the outside body wall and the digestive tract. Within this cavity are the major organs.

A more complex digestive tract may be found in segmented worms, although as in roundworms, there are two body openings. Most annelids have a specialized system for gas exchange. However, the earthworm has no such system and gases are exchanged across its moist skin. Annelids have two types of muscles. Thus, annelids have more efficient locomotion than do the nematodes. Also, segmented worms have a simple excretory system that removes nitrogen wastes.

Figure 15–14 compares in cross section the general body plan of a hydra, a planarian, a nematode, and an earthworm. How many tissue layers does each have? Which have digestive tubes rather than just digestive cavities? Which has a circulatory system? Which have some kind of a body cavity, or coelom?

An annelid has a circulatory system.

Annelids have a digestive system and excretory system, and most have a gas exchange system.

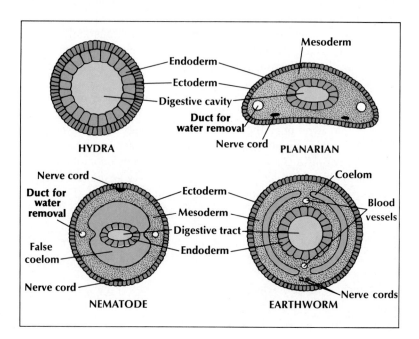

FIGURE 15–14. Cross sections of the bodies of a hydra, planarian, nematode, and earthworm show increasing complexity.

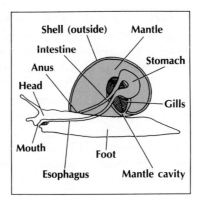

Shell (outside)   Mantle
Intestine
Anus   Stomach
Head
Gills
Mouth
Foot
Esophagus   Mantle cavity

FIGURE 15–15. A snail is a univalve, a mollusk with one shell. Mollusks use the thick, muscular foot for locomotion.

15:8 Mollusks

Mollusks (MAHL usks), members of phylum **Mollusca** (mah LUS kuh), are animals with soft bodies. They are usually enclosed within a hard, protective outer shell. Mollusks commonly have a **mantle,** a folded tissue, covering the internal organs. They move by means of a thick, muscular organ called a **foot.** Most species are marine. However, they also live in fresh water and on land. They have highly specialized systems for circulation, digestion, gas exchange, and excretion. Reproduction is sexual.

Like worms, mollusks have three tissue layers. A coelom is also present. However, the anatomy of members of this group of animals is quite different. On the basis of both internal and external features, adult mollusks do not seem to be closely related to annelids. However, comparative embryology indicates a close similarity between these phyla.

There are three main classes of mollusks. **Univalves** (YEW nih valvz) represent one class of mollusks, most of which have one shell. The shells of mollusks are called **valves.** Univalves include snails (Figure 15–15) and slugs. Although most univalves are aquatic, many snails and slugs are land forms. They have tentacles that are used as sensory organs. Snails have a coiled shell. Although classified as univalves, slugs have no shell at all. Both forms have a mouth structure called the **radula** (RAJ uh luh). The radula looks like a tongue covered with toothlike structures. Univalves feed by rubbing the radula against the surface of algae-encrusted rocks. Snails often are added to an aquarium where they scrape algae from the glass.

Bivalves (BI valvz) are mollusks with two shells. Members of this group include clams, oysters, scallops, and mussels (Figure 15–16). Oysters and mussels are sessile — they remain attached to rocks and other surfaces during their lifetimes. They filter food from the water (Section 25:1).

Mollusks have soft bodies, usually enclosed in a hard shell.

Mollusks and annelids are closely related.

FIGURE 15–16. Mussels are bivalves, mollusks with two shells. Mussels are sessile (nonmoving) and are filter feeders.

Some head-foot mollusks are active predators. Their shells may be absent or greatly reduced.

Most mollusks are slow-moving organisms. But the squid and octopus, called **head-foot mollusks,** are active predators (Figure 15–17). The foot has been modified into several arms, or tentacles, originating at the head region. Also included in this group are the cuttlefish (KUT ul fihsh) and chambered nautilus (NAHT ul us). Head-foot mollusks have eyes that may aid them in being more active than other mollusks. The similarities between vertebrate eyes and mollusk eyes are the result of convergence (Section 11:8). A squid has a reduced internal shell and an octopus has no shell. Squids measuring over seventeen meters and octopuses of about ten meters in length have been found. These are the largest invertebrates.

REVIEWING YOUR IDEAS

7. What are two hosts in a beef tapeworm's life cycle?
8. How are flatworms and roundworms different?
9. What is a coelom?
10. What tissue layers do annelids and mollusks have?
11. What are the distinguishing features of mollusks?

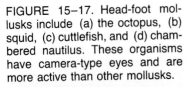

FIGURE 15–17. Head-foot mollusks include (a) the octopus, (b) squid, (c) cuttlefish, and (d) chambered nautilus. These organisms have camera-type eyes and are more active than other mollusks.

a
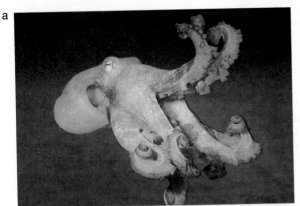

b

c

d

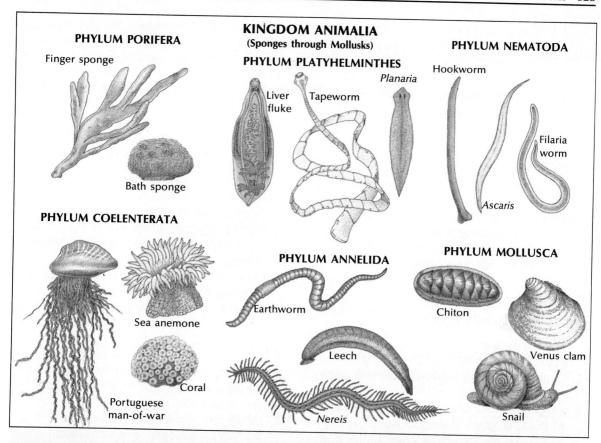

KINGDOM ANIMALIA
(Sponges through Mollusks)

PHYLUM PORIFERA

Finger sponge

Bath sponge

PHYLUM PLATYHELMINTHES

Planaria

Liver fluke

Tapeworm

PHYLUM NEMATODA

Hookworm

Filaria worm

Ascaris

PHYLUM COELENTERATA

Sea anemone

Coral

Portuguese man-of-war

PHYLUM ANNELIDA

Earthworm

Leech

Nereis

PHYLUM MOLLUSCA

Chiton

Venus clam

Snail

TABLE 15-1. SUMMARY OF MAJOR CHARACTERISTICS OF SIMPLE INVERTEBRATES

PHYLUM	Porifera	Coelenterata	Platyhelminthes	Nematoda	Annelida	Mollusca
COMMON NAME	sponges	coelenterates	flatworms	roundworms	segmented worms	mollusks
LOCOMOTION	none	sessile; free-floating	muscles; cilia	muscles	muscles	muscles
SYMMETRY	none or radial	radial	bilateral	bilateral	bilateral	bilateral
NUMBER OF BODY OPENINGS	pores and canals	one	one	two	two	two
NUMBER OF TISSUE LAYERS	none	two	three	three	three	three
NERVOUS SYSTEM	none	present	present	present	present	present
DIGESTIVE SYSTEM	none	none	present	present	present	present
EXCRETORY SYSTEM	none	none	none	present	present	present
CIRCULATORY SYSTEM	none	none	none	none	present	present
RESPIRATORY SYSTEM	none	none	none	none	present	present
SKELETAL SYSTEM	none	none	none	none	none	none

INVESTIGATION

Problem: What does the pork worm *Trichinella spiralis* look like and what is its life cycle?

Materials:

stereomicroscope   microscope
metric ruler   life cycle diagram
prepared slides of:
  Trichinella spiralis adult female
  Trichinella spiralis adult male
  Trichinella spiralis larval stage in muscle

Procedure:

Part A. Adult Female Pork Worm

1. Make a copy of Table 15–2 in which to draw your diagrams and record your data.
2. Examine the slide marked *Trichinella spiralis*-adult female.
3. Diagram the female pork worm in the space marked "diagram of life-size worm" in Table 15–2. Indicate the actual length of the worm in millimeters.
4. Observe the slide under a stereomicroscope. Diagram the female pork worm in the space marked "diagram of magnified worm" in Table 15–2. Record the magnification.

Part B. Adult Male Pork Worm

1. Examine the slide marked *Trichinella spiralis*-adult male.
2. Diagram the male pork worm in the space marked "diagram of life-size worm" in Table 15–2. Indicate the actual length of the worm in millimeters.
3. Observe the slide under a stereomicroscope. Diagram the male pork worm in the space marked "diagram of magnified worm" in Table 15–2. Record the magnification.

Part C. Larval Pork Worm

1. Use low power of your microscope to observe the prepared slide of *Trichinella spiralis* larvae in muscle.
2. Locate the following structures:
   (a) larva—appears as a small worm coiled into a ball
   (b) cyst—egg-shaped covering surrounding the larva
   (c) muscle—long, banded cells surrounding the cyst

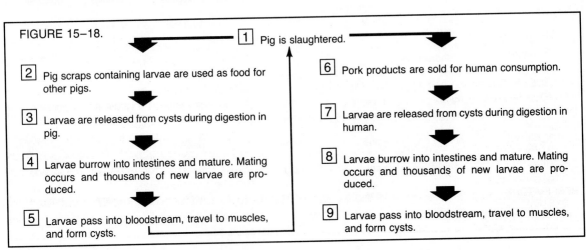

FIGURE 15–18.

1 Pig is slaughtered.

2 Pig scraps containing larvae are used as food for other pigs.

3 Larvae are released from cysts during digestion in pig.

4 Larvae burrow into intestines and mature. Mating occurs and thousands of new larvae are produced.

5 Larvae pass into bloodstream, travel to muscles, and form cysts.

6 Pork products are sold for human consumption.

7 Larvae are released from cysts during digestion in human.

8 Larvae burrow into intestines and mature. Mating occurs and thousands of new larvae are produced.

9 Larvae pass into bloodstream, travel to muscles, and form cysts.

3. Diagram and label a cyst and larva in the space marked "diagram of magnified worms" in Table 15–2. Record the magnification.

4. Assume that your low-power field of view has a total diameter of 1.5 mm. Estimate the actual size of one larva in millimeters. NOTE: Remember that the larva is coiled up. Its length should reflect how it would appear if it were uncoiled. Record this number in Table 15–2.

Part D. Pork Worm Life Cycle

1. Examine Figure 15–18, which shows the stages in the life cycle of *Trichinella spiralis*.

2. Each stage in the life cycle has been numbered. Stages 2–5 occur within the body of a pig host. Stages 6–9 occur within the body of a human host. The arrows indicate the order of events. Note that the pig is an essential part of the life cycle of *Trichinella spiralis*.

Data and Observations:

TABLE 15-2. APPEARANCE OF *TRICHINELLA SPIRALIS*		
Trichinella spiralis	Diagram of life-size worms	Diagram of magnified worms
Female	_____ mm	_____ X
Male	_____ mm	_____ X
Larva and Cyst		_____ X Estimated length: _____ mm

Questions and Conclusion:

1. Define the following terms:
   (a) parasite
   (b) host
2. Classify *Trichinella spiralis* according to kingdom, phylum, genus, and species.
3. List several characteristics for the phylum to which this animal belongs.
4. What evidence do you have to support the observation that these worms are not hermaphrodites?
5. List the stage in the life cycle of the worm in which infection of other pigs could be prevented. Explain your answer.

6. List the stage in which infection of humans could be prevented. Explain your answer.
7. List the stage at which each of these symptoms may occur:
   (a) weakness and muscle soreness
   (b) internal bleeding and infections
   (c) diarrhea, abdominal pain, and nausea
8. To find evidence of pork worms in a human, would you look for larval or adult worms? Explain your answer.

Conclusion: What does the pork worm *Trichinella spiralis* look like and what is its life cycle?

CHAPTER REVIEW

SUMMARY

1. Most of the features of animals can be related to their heterotrophic way of life. Most animals have mobility, a nervous system, and sense organs. They must digest food and excrete wastes. Animals can be subdivided into two groups—invertebrates and vertebrates. **p. 311**
2. Sponges are the least complex animals. They lack true tissues and are sessile. **15:1**
3. The coelenterates have a nerve net and other tissues. The body has one opening and may be either a polyp or medusa. **15:2**
4. Except for some sponges, animals have either radial or bilateral symmetry. **15:3**
5. Flatworms include free-living and parasitic forms. They have one body opening, three tissue layers, organs, and a primitive central nervous system. **15:4**
6. The beef tapeworm is an example of a parasitic flatworm. It is specialized for completing its life cycle in two host species, cow and human. Tapeworms are specialized for sexual reproduction. **15:5**
7. The roundworms have two features not seen in less complex animals—two body openings and separate sexes. Many types of roundworms are parasites. **15:6**
8. Segmentation, a coelom, and systems for transport and respiration are traits of annelids. **15:7**
9. Mollusks have soft bodies and often hard, mineral shells. The major mollusk groups include bivalves, univalves, and head-foot mollusks. **15:8**

LANGUAGE OF BIOLOGY

anterior	incurrent pore	posterior
bilateral symmetry	invertebrate	radial symmetry
	mantle	

coelom
dorsal
ectoderm
endoderm
excurrent pore
gemmule
hermaphrodite

medusa
mesoderm
mesoglea
nematocyst
osculum
parapodium
polyp

radula
spicule
tentacle
valves
ventral
vertebrate

CHECKING YOUR IDEAS

On a separate paper, match the phrase in the left column with the proper term in the right column. Do not write in this book.

1. simplest phylum of animals with two body openings
2. simplest phylum with three tissue layers
3. edge of an animal
4. form of coelenterate
5. characterized by many pores
6. same features on all sides
7. have two tissue layers and a mesoglea
8. front of an animal
9. soft body and hard shell
10. simplest phylum with a circulatory system

a. anterior
b. coelenterates
c. flatworms
d. lateral
e. medusa
f. mollusks
g. radial symmetry
h. roundworms
i. segmented worms
j. sponges

EVALUATING YOUR IDEAS

1. List several features of animals that are related to the animal's heterotrophic way of life. Justify your answers.
2. Why do you think biologists classify sponges as animals, rather than as another group?
3. What is the highest level of organization found in sponges?
4. Describe the various ways in which sponges are able to reproduce.

5. How do coelenterates obtain food?
6. Describe the life cycle of *Aurelia*.
7. How many tissue layers do flatworms have? What kind of symmetry and levels of organization do flatworms have? Give examples of flatworms.
8. Describe the life cycle of a beef tapeworm.
9. What physical features first appeared among the roundworms?
10. Compare reproduction in flatworms with that of roundworms.
11. List some diseases caused by roundworms.
12. What systems are present in most segmented worms? What systems do they lack?
13. How are annelids important in soil ecology?
14. Is an earthworm a land animal? Explain.
15. How do aquatic annelids reproduce? How does the earthworm reproduce? What might be an explanation for the differences?
16. List some of the unique physical characteristics of mollusks.
17. What is the basis for dividing mollusks into three groups?
18. Compare methods of food getting among the mollusk groups.
19. The animal phyla in this chapter have been presented from simple to complex. Using Table 15–1, what trends do you see as animals evolved from simple to complex with regard to each of the following: (a) symmetry (b) number of body openings (c) number of tissue layers (d) variety of systems present?

APPLYING YOUR IDEAS

1. In what ways are animals similar to protozoa? How are they different?
2. Why do you think that some biologists suggest that sponges be classified as protists?
3. How is radial symmetry an advantage to sessile animals?
4. What adaptations must tapeworms have to be able to live in the intestine?

5. Most hermaphrodites mate with other hermaphrodites. Can you explain why tapeworms fertilize themselves?
6. Explain why, long ago, sick people were treated by attaching leeches to their skin.
7. Consider the three major classes of mollusks. Relate their locomotion to differences in their structure.
8. In what ways are bivalves similar to sponges?
9. How is food getting different between animals and fungi?
10. What ecological approach could be used to eliminate tapeworms?

EXTENDING YOUR IDEAS

1. Prepare a report on the development of mollusks and annelids. Why is it commonly assumed that mollusks and annelids are closely related?
2. Report on several of the animal parasites in this chapter. Include the life cycles.
3. Compare and contrast flukes and leeches.
4. Find out about comb jellies and brachiopods. What other animals do they resemble?

SUGGESTED READINGS

Angier, Natalie, "Time Machines in the Sea." *Discover,* Aug., 1982.

Hansen, James, "Leeches, Lancets, and Fleams." *Science 81,* Oct., 1981.

Levine, Joseph, "Life in Search of Light." *Science Digest,* July, 1983.

Roper, Clyde F.E. and Boss, Kenneth J., "The Giant Squid." *Scientific American,* April, 1982.

Rose, Kenneth Jon, "Living Lights." *Science Digest,* Jan., 1984.

Sebens, Kenneth P., "The Anemone Below." *Natural History,* Nov., 1986.

Sumich, James L., *An Introduction to the Biology of Marine Life,* 2nd ed. Dubuque, IA, William C. Brown Co., 1980.

ANIMALS: ARTHROPODS THROUGH VERTEBRATES

like simple animals, more complex animals have adaptations that aid survival and reproduction in a variety of environments. What types of adaptations aid the survival of this Siberian tiger? How are these adaptations different from those needed by a tiger living in a more temperate region? You are classified in the same class with this tiger. What traits do you have in common with it?

chapter 16

Study of the characteristics of simple invertebrates (Chapter 15) reveals a variety of animal adaptations. Each phylum of organisms has particular features that make it unique. Yet, the animals have much in common as they carry out the same basic life functions. Their adaptations enable them to survive and reproduce in a variety of habitats. In this chapter, you will study three other phyla of animals—arthropods, echinoderms, and chordates. They, too, have a variety of adaptations and particular features that make each group unique. The arthropods and vertebrates have complex adaptations that make them the most successful of the animal phyla.

Objectives:

You will
- list the major features of several phyla of complex animals.
- compare the traits that make one phylum more or less complex than another.
- discuss the biological and economic importance of animals in these phyla.

ARTHROPODS AND ECHINODERMS

Arthropods (AR thruh pahdz) are the most advanced invertebrates. This phylum has more living members than all other phyla combined. In this sense, these animals are the most successful. They are found almost everywhere. The many species illustrate the diversity of adaptations that have evolved in living systems.

Echinoderms (ih KI nuh durmz) are less complex and diverse than arthropods. They are slow-moving organisms with a body plan different from other animals. However, they seem to be more closely related to the chordates than to any other animal phylum.

Arthropods are the most abundant of all animals.

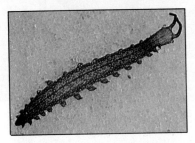

FIGURE 16–1. *Peripatus* has features common to both annelids and arthropods.

16:1 Arthropods

Arthropods, phylum **Arthropoda** (ar THRAHP uh duh), have segmented bodies, which is also a feature of annelids (Section 15:7). Arthropods and annelids may have evolved from a common ancestor. *Peripatus* (puh RIHP ut us) is an example of an animal that has features common to both annelids and arthropods (Figure 16–1). For example, it is segmented and has an outer skin like that of annelids, but it has claws and a respiratory system like that of some arthropods. It belongs to a phylum that may have branched off from the evolutionary line leading to the arthropods.

Modern arthropods have the following features:
(1) segmented body covered by an **exoskeleton** (ek soh SKEL ut un), or external skeleton
(2) pairs of jointed **appendages** (uh PEN dihj uz), such as limbs and antennae
(3) well-organized muscles with definite points of attachment to the exoskeleton
(4) body segments in most forms fused into distinct body regions — **head; thorax,** or chest; and **abdomen**
(5) specialized mouthparts
(6) complex **nervous system** and specialized sensory organs for detecting incoming stimuli.

Arthropods have well-organized muscles.

Studying these and other features shows why arthropods can survive in many environments. An exoskeleton, jointed appendages, and a developed muscular system allow arthropods greater mobility than worms have. For example, consider the jumping of a grasshopper or the scurrying of a millipede. The exoskeleton gives body support. Muscles are arranged in special ways and are attached to definite points on the exoskeleton (Section 30:1). This arrangement permits movement of certain parts as well as movement of the entire organism. Without a support system and appendages, animals are limited to either crawling or floating. In addition, the exoskeleton provides some protection for the arthropod against predators and reduces water loss.

The exoskeleton provides support and protection and improves locomotion.

Because they live in many environments, arthropods have various diets. Several types of mouthparts have evolved. Some mouthparts are toothed and are used for chewing. Others help hold and transfer food to the mouth.

Fusion of segments into distinct body sections (compartments) is another kind of specialization that is displayed by members of the arthropod phylum. The arrangement of the body into compartments provides definite areas for certain organs and functions. This body plan is more advanced than that found in less complex invertebrates, which do not have specific body compartments.

In most arthropods, body segments are fused into distinct regions.

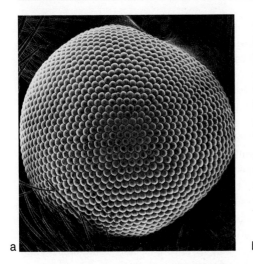

a

b

Like annelids, arthropods have a ventral nerve cord. However, their nervous system is more advanced with greater control at the anterior end of the body. The complex nervous system receives information from a variety of sensory organs. Antennae function for hearing, tasting, feeling, and smelling in different species. Arthropods also have complex eyes.

Arthropods are bilaterally symmetrical. Reproduction among these animals is mostly sexual, and sexes are usually separate.

FIGURE 16–2. Arthropods have highly developed, specialized sensory organs. (a) The compound eye and (b) antennae (here, of a moth) are adaptations that aid survival of arthropods in many environments.

16:2 Centipedes and Millipedes

Five important classes of arthropods presently live on Earth. Two classes are **Chilopoda** (ki LAHP uh duh), the centipedes or "hundred leggers," and **Diplopoda** (duh PLAHP uh duh), the millipedes or "thousand leggers." In both forms, the body consists of a head region and a trunk made up of a series of segments.

a

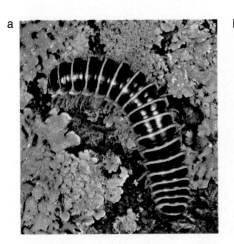

b

FIGURE 16–3. (a) A millipede, class Diplopoda, has two pairs of legs per segment. (b) A centipede, class Chilopoda, has one pair of legs per segment.

Centipedes are carnivorous land dwellers.

A centipede has a pair of antennae on the head and mouthparts called mandibles. **Mandibles** are adapted for biting and chewing. Centipedes are land dwellers and carnivores. The first pair of body appendages contains poison claws through which poison can be injected into prey. The claws are not jaws; they are modified appendages. Centipedes are active at night, during which time they hunt for prey. Their diet includes insects, earthworms, and slugs. Sometimes they eat small reptiles or mammals.

Most segments of a centipede bear one pair of walking legs.

Each of the body segments of a centipede, except for the first one and last two, has one pair of walking legs. The total number of legs ranges from as few as thirty to more than three hundred. Centipedes exchange gases with the atmosphere by means of special tubes called **tracheae** (TRAY kee ee), which open through the exoskeleton to the outside of the body (Section 27:3).

Millipedes have two pairs of walking legs on most segments.

Millipedes also have one pair of antennae and mandibles. They are not carnivorous but feed on decaying plant material. They have no poison claws. Each of the first four body segments has one pair of walking legs. The rest of the body has two pairs of walking legs per segment. Like centipedes, millipedes exchange gases by means of tracheae.

16:3 Crustaceans

Most crustaceans breathe by gills and live in aquatic or moist environments.

Class **Crustacea** (krus TAY shee uh) is a diverse group of arthropods with a variety of adaptations. They have mandibles, two pairs of antennae, and most breathe by means of gills. Their body forms vary a great deal. Most crustaceans live in the sea, but others are found in fresh water and moist places on land.

Most familiar to you are large crustaceans such as the lobster, shrimp, and crab. Commonly called shellfish, these animals are a major part of the seafood industry. Red tides (Section 13:8) may cause severe economic loss by poisoning or killing these animals.

FIGURE 16–4. (a) Shrimp and (b) crayfish are common crustaceans that, like most animals in this class, breathe by means of gills. (c) Sow bugs are among the few crustaceans that live on land. They have simple tracheal systems.

a

b

c
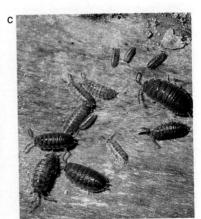

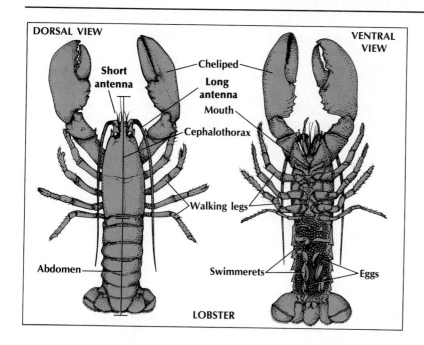

FIGURE 16–5. The lobster has two body sections, the cephalothorax and the abdomen. Swimmerets can be seen on the ventral side.

The lobster, like many other forms, has two body sections, the **cephalothorax** (sef uh luh THOR aks), a fusion of head and thorax, and the abdomen. It also has five pairs of legs on the cephalothorax. The first pair is modified as claws called **chelipeds** (KEE luh pedz), which are used in trapping food and for defense. One cheliped is larger than the other and is used for crushing prey. The smaller cheliped is used for holding and tearing food. The other four pairs of legs are walking legs.

The lobster's abdomen has one pair of appendages on every segment except the last. These appendages, called **swimmerets,** aid in gas exchange and are functional in reproduction. The first pair of swimmerets in males is specialized for transferring sperm to the female. Fertilized eggs are attached to some of the female's swimmerets during early development. The last pair of abdominal appendages is fused with the last segment to form a flipper. You can see in Figure 16–5 how different appendages are specialized for different functions — sensing, feeding, locomotion, and reproduction.

An example of a land crustacean is the sowbug. This animal, like most crustaceans, is quite small. It has tracheae for gas exchange and lives under logs or stones where it is moist.

Although most crustaceans are motile, adult barnacles (Figure 16–6) are sessile. They are marine forms that may be attached to other animals, to rocks, or to bottoms of ships. The pattern of the barnacles is sometimes used to identify sea animals, such as whales, to which they attach. Barnacles secrete mineral shells that surround their bodies. Their jointed appendages reach out of the shell and trap small animals for food.

Most crustaceans have two body sections, cephalothorax and abdomen.

FIGURE 16–6. Barnacles secrete shells in which they live, remaining sessile. Barnacles are shown here attached to a rock.

FIGURE 16–7. Insects are not only the most numerous type of arthropod, but with over 800 000 species, they are also the most numerous type of animal.

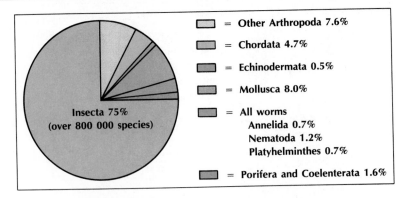

= Other Arthropoda 7.6%

= Chordata 4.7%

= Echinodermata 0.5%

= Mollusca 8.0%

= All worms
  Annelida 0.7%
  Nematoda 1.2%
  Platyhelminthes 0.7%

= Porifera and Coelenterata 1.6%

Insecta 75%
(over 800 000 species)

16:4 Insects

Insects are adapted for a great variety of environments.

The largest group of arthropods is class **Insecta** (ihn SEK tuh). Insects are widely distributed. With the exception of salt water, insects have successfully invaded almost every habitat and ecological niche. Their large numbers, high rate of reproduction, and diversity of adaptations have enabled many of them to become pests to humans. Beetles, earwigs, and locusts are insects that eat crops and garden plants. Thus, they compete with humans for food. Termites destroy wood both in living trees and in houses. Gypsy moth caterpillars can strip trees bare as they devour the leaves. Other insects are beneficial to humans and are important in nature. They may prey on harmful insects, pollinate flowers, or make products such as honey or silk.

Insects have three body sections, three pairs of legs, and, in most forms, two pairs of wings.

Insects have three main body regions. These regions are the head, the thorax, and the abdomen. The head has one pair of antennae and a pair of mandibles. The mandibles and other mouthparts of insects are specialized for many different ways of feeding. Consider a caterpillar chewing a leaf or a mosquito piercing your skin and sucking blood. Insects have three pairs of walking legs, one pair on each segment of the thorax. In most insect forms, the second and third segments of the thorax each have a pair of wings. The abdomens of some insects have small, vestigial appendages or appendages used in reproduction. Insects breathe by means of a network of tracheae. Oxygen enters the tracheae through small openings located between abdominal segments.

Many insects have complex behaviors that aid their survival (Section 31:11). Some insect forms, including ants, termites, and bees, are social and live in colonies. The colonies show division of labor with different insects carrying out special roles.

Their diversity of adaptations is the primary reason for the success of insects in various environments. Some of the major orders of insects are described in Appendix A. Representative examples from the orders are pictured in Figure 16–8.

16:5 Arachnids

Class **Arachnida** (uh RAK nuh duh) is a group that includes spiders, ticks, mites, and scorpions. **Arachnids** (uh RAK nuhdz) have two body sections—cephalothorax and abdomen. Unlike other classes of arthropods, they have no antennae. Also, they have chelicerae (kih LIHS uh ree) instead of mandibles. **Chelicerae** are either fanglike or pincerlike mouthparts. Posterior to the chelicerae are a pair of **pedipalps** that hold and tear apart food. Behind the pedipalps are four pairs of walking legs. These six pairs of appendages are all located on the cephalothorax. The abdomen usually has no appendages.

Most arachnids exchange gases by means of tracheae. Others have special structures called **book lungs** for exchanging gases. These structures are so named because they resemble an opened book in appearance.

FIGURE 16–8. These insects, (a) stinkbug, (b) aphid, (c) wasp, (d) moth, (e) housefly, (f) dragonfly, (g) stag beetle, and (h) grasshopper, are representatives of different orders of the class Insecta.

Arachnids have two body sections and no antennae. Unlike other arthropods, they have chelicerae rather than mandibles.

Arachnids have four pairs of walking legs.

a

b

c

FIGURE 16–9. Class Arachnida includes (a) the giant hairy scorpion, (b) tarantula, and (c) dog tick. Arachnids have four pairs of legs.

Spiders' webs and poisons are adaptations for trapping prey.

Scorpions are active predators.

Many spiders, as you know, spin webs to trap prey. The webs are made from threads of silk that leave the spider's body through structures called **spinnerets,** which are located at the tip of the abdomen. Insects become trapped in the sticky threads of the web and thus become food for the spider. Spiders have fanglike chelicerae that they use to transfer poison and paralyze their prey. The poisons of a few spiders, such as the black widow and the brown recluse, may be fatal to humans. However, more spiders are beneficial to humans than are harmful. Spiders are an important means of natural control of numerous insect pests.

Scorpions are common in the southwest United States. They are inactive during the day but search for insects and spiders at night. A scorpion has a poisonous stinger at the end of the abdomen. Prey is held by pincerlike chelicerae while the poison from the stinger is injected into it. The sting of a scorpion may be painful to a human and cause illness. However, it rarely causes death.

TABLE 16–1. SUMMARY OF CHARACTERISTICS OF ARTHROPODS						
Class	Examples	Body Sections	Antennae	Mouthparts	Number of Walking Legs	Gas Exchange
Chilopoda	centipedes	head and body segments	1 pair	mandibles	1 pair per segment	tracheae
Diplopoda	millipedes	head and body segments	1 pair	mandibles	2 pair per segment	tracheae
Crustacea	lobster, crab, shrimp	cephalothorax and abdomen	2 pair	mandibles	5 pair in most forms	gills in most
Insecta	wasp, beetle, mosquito	head, thorax, and abdomen	1 pair	mandibles	3 pair	tracheae
Arachnida	spider, tick, scorpion	cephalothorax and abdomen	none	chelicerae	4 pair	tracheae; book lungs

a
b

FIGURE 16–10. (a) The sand dollar and (b) brittle star are representative echinoderms. Echinoderms have radial symmetry as adults.

16:6 Echinoderms

Most animals in phylum **Echinodermata** (ih ki nuh dur MAH tuh) are spiny skinned. Plates of minerals with projecting spines are embedded in the soft tissues of the body. These minerals form an **endoskeleton** (en doh SKEL ut in), or internal skeleton. Adult echinoderms are radially symmetrical. However, larval (developing) echinoderms have bilateral symmetry. The starfish, sea urchin (UR chun), and sand dollar are members of this phylum. All echinoderms live in salt water. Most are bottom-dwellers that are either sessile or quite slow-moving.

Starfish are the most familiar echinoderms. The starfish's body is made up of a central disc from which "arms" (usually five) stick out. These arms provide a means of movement and food-getting. The slow movement is controlled by a special water-pumping system (Section 30:7). Starfish often feed on clams (Section 25:5), competing with humans for this food source.

In a starfish, materials are distributed within the fluid of the coelom. Gases are exchanged through gill-like structures. Simple nervous and sensory systems are present. There is no excretory system. Reproduction involves external fertilization.

A starfish has a spiny skin and an internal skeleton.

16:7 The Acorn Worms

Acorn worms (not really worms) are quite primitive. They are so named because their body is wormlike and has a structure resembling an acorn at one end. These marine animals burrow and take sand into their bodies. Organic particles in the sand are used as a food source by the animals.

Acorn worms have some features common to chordates (Section 16:8) and were once classified in that phylum. Now they are classified in a separate phylum, **Hemichordata**. These animals are also similar to echinoderms in a very important way. The larval stages of both

FIGURE 16–11. Acorn worms, shown as they live burrowed in sand, were once thought to be invertebrate chordates. Now, they are classified in a separate phylum, the Hemichordata.

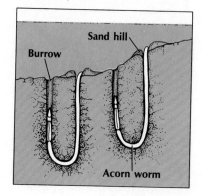

Sand hill

Burrow

Acorn worm

Echinoderms, hemichordates, and chordates are thought to have evolved from a common ancestor.

echinoderms and hemichordates are bilaterally symmetrical and develop similarly. The larvae also are similar in terms of biochemistry. Because of the similarities among echinoderms, hemichordates, and chordates, biologists believe that the three phyla are closely related. It is assumed that all three phyla evolved from a common ancestor.

REVIEWING YOUR IDEAS

1. What is an exoskeleton? What are appendages?
2. List the five major classes of arthropods. Give an example of each class.
3. What are the three body regions of an arthropod?
4. Why do you think echinoderms are sometimes called "spiny-skinned" animals?
5. Where do acorn worms live? How do they obtain food?

CHORDATES

You are a **chordate** (KOR dayt), a member of phylum **Chordata** (kor DAHT uh). All chordates, at some stage of their lives, have a stiff dorsal rod of cartilage called a **notochord** (NOHT uh kord). Chordates also have gill slits or pouches. Gill slits or pouches, like the notochord, do not remain in all chordates throughout their entire lives. Often, they are present only in the early embryo. Other traits of chordates are the presence of a dorsal nerve cord and a ventral heart.

16:8 Characteristics of Chordates

All vertebrates (animals with backbones) are chordates, but not all chordates are vertebrates. Chordates that are not vertebrates are the tunicates (TOO nuh kayts) and lancelets (LANS luts). They are marine dwellers. Most tunicates are sessile and grow singly or as colonies attached to rocks. They have a tough body wall in which there are two openings. When the body contracts suddenly, water is forced out of the openings. For this reason, tunicates are known as sea squirts. Tunicates obtain food from water that passes into the body through numerous gill slits.

Lancelets more closely resemble vertebrates. *Amphioxus* (am fee AHK sus) (Figure 16–12) is a small lancelet that you might mistake for a fish. Its body is pointed at each end and is streamlined. Although it can swim, it usually lives partly buried in sand. It feeds on microorganisms in water drawn into its mouth.

FIGURE 16–12. (a) Tunicates and (b) a lancelet are chordates that are not vertebrates.

a

b

Vertebrates are members of the subphylum **Vertebrata** (vurt uh BRAHT uh). In vertebrates, the embryonic notochord is replaced in the adult by bony structures called **vertebrae** (VUR tuh bray). It is from these structures that the group receives its name. Vertebrae connect to form the spinal column, or backbone. In addition to those features shared by all chordates, the following features are common to almost all vertebrates:

(1) internal skeleton of bone and/or cartilage
(2) specialized muscle system
(3) advanced centralized nervous system with a true brain and a spinal cord
(4) complex sensory equipment
(5) complex **integumentary** (ihn teg yuh MENT uh ree) **system** (outer covering)
(6) paired appendages for locomotion (although never more than two pairs).

An internal skeleton provides excellent protection and support. The different parts of the skeleton have shapes that suit their functions. Muscles are large and provide vertebrates with greater speed and agility than is found in many invertebrates. Because vertebrates have large, specialized brains, they often have more complex behavioral patterns. A well-developed nervous system controls the smoothly coordinated movements of vertebrates. In general, sensory organs are much more complex and effective in this group of animals than in any other.

Vertebrates have a wide variety of outside coverings such as scales, thin skin, "armor," feathers, and hair. As a result of these and other adaptations, vertebrates occupy a wide range of habitats. Seven classes of vertebrates exist today. Each of those classes will be considered in turn in Sections 16:9–16:15.

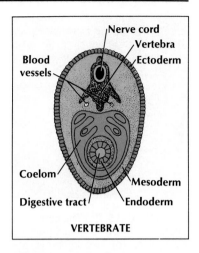

FIGURE 16–13. The cross section of the body of a generalized vertebrate shows the vertebra, bone that replaces the notochord.

Vertebrates have a variety of complex adaptations.

FIGURE 16–14. (a) Feathers, (b) fur or hair, and (c) scales are three types of integumentary adaptations.

a

b

c

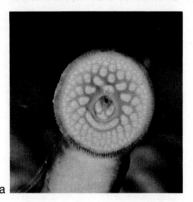

FIGURE 16–15. (a) The lamprey's sharp teeth aid its attachment to other fish. (b) Two lampreys are shown attached to a carp, feeding on its tissues and blood.

16:9 Jawless Fish

Three classes of vertebrates are fish. Fish exchange gases by means of gills, and most have scales. Their limbs are in the form of fins. Their body temperature is affected by that of the environment. For example, when the environment becomes cold, the body temperature decreases somewhat. Such animals are called **ectotherms** (EK tuh thurmz).

Fish have gills and are ectotherms. Most forms have paired fins and scales.

The most primitive fish belong to the class **Agnatha** (AG nuh thuh). At one time, these fish belonged to a very diverse group, the members of which were covered with armorlike plates. Today, this small class is composed of two groups, lampreys and hagfish. Lampreys inhabit fresh and salt water. Hagfish live in salt water.

These fish have a slimy skin with no scales. They lack paired fins, and their gill openings are not covered. They have a skeleton of cartilage and retain a notochord as adults. The notochord is not replaced by vertebrae, but these animals are still considered vertebrates. You can again see difficulties in making a classification scheme.

An agnathan has slimy skin without scales, a skeleton of cartilage, and a permanent notochord. It lacks paired fins.

Instead of hinged jaws, an agnathan has a sucker-shaped mouth. Thus, agnathans are called **jawless fish.** The fish's mouth is lined with sharp teeth. The lamprey hooks onto prey such as trout or other fish. Then, with its toothed tongue, it rasps a hole and feeds on the blood or tissues of its host. Sometimes the attack is fatal to the host. However, many fish survive and have scars that indicate previous lamprey attacks. Lampreys can cause great economic loss by destroying fish. In the Great Lakes, for example, lampreys have destroyed many trout.

Agnathans are jawless fish that prey on other fish.

Hagfish are eel-like and live in deep parts of cold water. A hagfish often lies buried in the sand except for its head. Most hagfish are not parasites. They feed upon invertebrates or dead animals. They often attack fish caught in nets. They may bore into the bodies of these fish, eating the fish from inside. In this way, hagfish are responsible for economic loss in the fishing industry.

16:10 Cartilage Fish

Like the agnathans, members of class **Chondrichthyes** (kahn DRIHK thee eez) have skeletons composed of cartilage. These **cartilage fish** include sharks, skates, and rays (Figure 16–16). They have skin covered with toothlike scales, two pairs of lateral fins, a pair of dorsal fins, a tail fin for locomotion, and hinged jaws. Their gill openings are not covered. These fish are unusual in that most have internal fertilization.

Sharks range in size from pygmies, which may be only 0.6 m, to whale sharks, which may be 15 m long with a mass of several thousand kilograms. The dogfish shark is common along both the Atlantic and Pacific coasts. It is about 1.0 m long and has a mass of 3.5 kg.

Almost all sharks are meat eaters. Their diet includes fish, squid, crustaceans, jellyfish, sea turtles, and even sea lions. Most sharks have wide mouths lined with rows of sharp teeth slanted backwards to hold their prey. Often, these teeth are replaced when lost. Their sense of smell, swimming ability, and size make sharks excellent predators. Although the shark is feared by many humans, shark attacks are quite rare. The shark species most dangerous to humans are the white, hammerhead, tiger, and sand shark. The whale shark, largest of all, is harmless to humans.

Sharks are commercially important in several ways. Shark liver oil is a source of Vitamin A and is used for tanning leather, in preserving wood, and as a lubricant. Shark skin can be made into leather. The meat of some sharks is eaten by humans and is a good source of protein. Shark fin soup also is eaten.

Rays and skates have flattened bodies. Some, like the stingray, are dangerous. They have a barbed tail that secretes a venom. When stepped on, the ray lashes its tail inflicting a very painful wound. Both rays and skates are bottom dwellers in coastal waters. They usually feed on mollusks.

A cartilage fish has a skeleton of cartilage, scaly skin, paired fins for locomotion, and hinged jaws. Most have internal fertilization.

Sharks are well adapted for a predatory way of life.

FIGURE 16–16. (a) A sand tiger shark and (b) a spotted eagle ray are fish with skeletons made of cartilage.

a

b

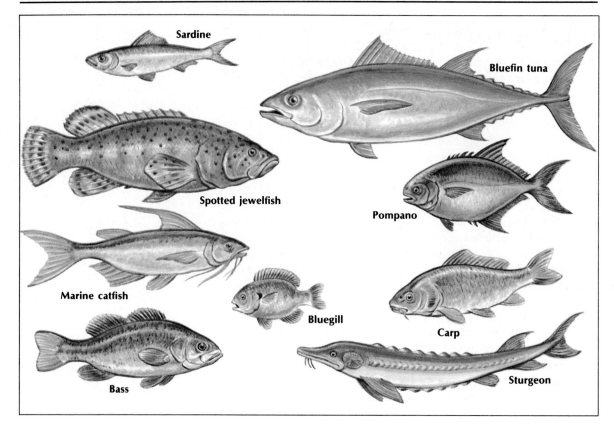

FIGURE 16–17. Bone replaces cartilage in the skeletons of bony fish, members of the class Osteichthyes.

Bony fish exhibit a variety of unique adaptations.

16:11 Bony Fish

Modern **bony fish** belong to the class **Osteichthyes** (ahs tee IHK thee eez). They are the most diverse group of fish and live in both fresh water and salt water. A few can even live on land for short periods of time.

Although many bony fish have a similar appearance, there are a variety of unique adaptations and forms. Eels, for example, are snakelike in appearance, often lacking one or both of the pairs of fins found in most bony fishes. The electric eel and some other fish have a specialized organ that can generate up to 400 volts of electricity for several milliseconds. This adaptation enables the animal to stun or kill small fish.

In certain angler fish (Section 11:3) and other fish, a glowing structure that sticks out from the fish's head lures prey into the fish's mouth. Luminescence may aid some fish in finding mates in dark waters. Often, the light is actually produced by luminescent bacteria that live under the skin of the fish. In some cases, light is produced during chemical reactions within special organs of the fish.

As their name implies, bony fish have a skeleton made of bone. They have hinged jaws. Also, they have two pairs of lateral fins as well as a tail fin, two dorsal fins, and a ventral or anal fin. These fins are used for both locomotion and balance. Some may also aid in defense. These fish have scales made of flat discs of bone that grow as other bones do. The body is covered with slime, which aids in swimming and offers protection. The gills are covered and protected by a hard plate, the **operculum** (oh PUR kyuh lum). You can see the operculum open and close as a bony fish exchanges gases with the water. As in all vertebrates, reproduction is sexual. In most cases, fertilization is external.

A bony fish has a skeleton of bone, hinged jaws, paired fins for locomotion, and scaly skin. The gills are covered by opercula.

REVIEWING YOUR IDEAS

6. List two kinds of chordates that are not vertebrates.
7. What structure is replaced by vertebrae in vertebrates?
8. What is an ectotherm?
9. List distinguishing features of jawless, cartilage, and bony fish.

16:12 Amphibians

Amphibians, class **Amphibia** (am FIHB ee uh), may have evolved from an extinct group of fish with lungs. Amphibians were the first class of vertebrates to live on land. Modern amphibians include frogs, toads, newts, salamanders, and apodes. Newts are small, salamanderlike animals that live in water. Apodes are limbless, blind, burrowing animals.

In most amphibians, fertilization and development occur in water. Thus, these animals are "chained" to the water of streams and ponds. They have adaptations for life on land, but most return to water to reproduce. Some species live exclusively in water.

Although external fertilization is the rule, some species have internal fertilization. For example, some male salamanders deposit sperm clumps called spermatophores on the ground near or on twigs

Amphibians are adapted for life on land.

FIGURE 16–18. Amphibians include (a) toads, (b) frogs, and (c) salamanders. Amphibians were the first class of vertebrates to live on land.

a

b

c

a

External gills

b

FIGURE 16–19. Unusual developmental characteristics are found in class Amphibia. (a) The pygmy marsupial frog carries the fertilized eggs in a dorsal pouch until development is complete. (b) Under certain conditions, salamanders may retain external gills as adults. This is an example of neoteny shown here in a tiger salamander.

in water. The sperm clumps are taken into the female's reproductive system where internal fertilization occurs. Although external development occurs in most amphibians, some have internal development. Some frogs are mouth breeders. The male stuffs fertilized eggs into his vocal sacs where they develop into young. In a certain type of Australian frog, fertilized eggs are kept in a dorsal brood pouch on the female. Young frogs are "born" from the pouch. These behaviors are adaptive in areas where water is in short supply.

Before they reach maturity, amphibians have gills that obtain oxygen. In most forms, gills are replaced by lungs as the organism becomes an adult. As an adult, an amphibian has thin, moist skin that contains a rich supply of blood vessels. Oxygen is taken in through the skin as well as by the lungs. In one kind of aquatic frog, the male has hairlike filaments on its legs. These "hairs" contain blood vessels and aid in getting oxygen.

Some salamanders, such as the mud puppy, retain gills as adults, even though they also develop lungs. The gills of a mud puppy are external, projecting out in front of the front legs. Axolotls are salamanders that retain some characteristics of immature salamanders throughout their lives and reproduce without ever becoming "mature" adults. Retaining immature traits indefinitely in the "adult" form is called **neoteny** (nee AHT un ee), and several species exhibit this trait.

Amphibians are adapted for swimming and locomotion on land. When moving quickly on land, salamanders and newts slither along the ground with their legs hardly touching the ground. This movement is a sort of "swimming on land." When moving more slowly, salamanders and newts walk on their four feet. Frogs and toads are adapted for jumping, which they can do quickly and to large distances. Certain "flying" frogs can jump 12 to 15 meters. Tree frogs are adapted for climbing in trees. In these frogs, toe discs, or digital pads, aid in climbing. These pads are covered with a sticky mucus enabling the frog to cling to the tree.

16:13 Reptiles

Dinosaurs, once the largest and most dominant reptiles, lived about 200 million years ago. Today's **reptiles,** class **Reptilia** (rep TIHL ee uh), include snakes, turtles, lizards, and alligators. All have a means of internal fertilization and most of the young develop outside the body in eggs. The shelled egg of reptiles is a moist, protected environment for development. Therefore, reptiles do not need to reproduce in water as amphibians do. This adaptation has allowed reptiles to occupy a variety of niches on land, but some species, such as the sea turtles, do live in water.

Reptiles have internal fertilization and can inhabit a variety of ecological niches.

a   b   c

FIGURE 16–20. (a) The snake, (b) turtle, and (c) alligator are representative reptiles. Reptile skin is dry and scale-covered, an adaptation that aids survival on land.

Except for snakes, reptiles have limbs that are located toward the ventral side of the body. Thus, the body is lifted off the ground, an adaptation for locomotion on land. Claws are present on the feet.

Throughout life, reptiles breathe by means of large lungs. They are protected by dry, scaly skin that prevents water loss by evaporation. Lungs and scales are among the important adaptations for land life. Reptiles are ectotherms.

Reptiles are ectotherms, breathe by lungs, and have dry, scaly skin.

Lizards are a diverse group of reptiles that live in a variety of habitats including the Arctic, the tropics, and the desert. Geckos are small lizards only 3 cm long with a mass of a few grams. Geckos have toes with special pads for climbing; they can even walk on ceilings. The largest lizards are the monitors, which may be three meters long with a mass of 150 kg. Most lizards move on four legs though some have no limbs and are snakelike.

Most lizards eat insects and small rodents. In some parts of the world, small lizards, such as geckos, are welcome in and around homes because they prey on these pests. Most lizards are harmless to humans. But a few forms, such as Gila (HEE luh) monsters, are poisonous. Gilas release venom from poison glands in the mouth. The poison affects breathing. Gilas will attack humans if provoked.

A variety of adaptations for defense and escape are found among lizards. Sometimes a predator may catch a lizard by the tail. There are certain planes within the tail at which it will break. The lizard's tail breaks off and the lizard escapes. The missing part of the tail is usually regrown. Some lizards have spiny tails, which they use as defense weapons. Others have tissues around the throat that enlarge and scare off a predator. Many lizards can change color to blend with the environment (Section 11:4).

Crocodiles and alligators live in shallow waters of tropical environments. Both types of these ancient-looking reptiles are meat eaters. The eyes, ears, and nostrils are located on the top of the head and remain above water, while the rest of the body floats beneath the surface hiding the animal from its prey. Crocodiles and alligators feed on fish, small mammals, and birds. Occasionally, they have been

FIGURE 16–21. An adaptation for quick escape of lizards is a tail that breaks off at certain caudal (tail) planes. When a lizard tail regrows, the place where it broke off can be seen because the new growth looks different from the old part of the tail.

Caudal plane

FIGURE 16–22. Snakes molt, or shed their skins, several times a year. This red rat snake is crawling from its old skin.

Poisonous snakes inject venom through their fangs.

FIGURE 16–23. The fossil bird *Archaeopteryx* has features of both reptiles and birds. Reptile features include teeth, solid bones, and a long tail.

known to attack deer, cattle, and humans. Their strong jaws and sharp teeth make them good predators.

Turtles live on land, in fresh water, and in the sea. These reptiles are protected by a shell of plates and bones. The shell also slows their movement. Most turtles are small animals, but the Atlantic leatherback may grow to nearly four meters in length and have a mass of 60 kg. Turtles live longer than any other vertebrates. Some have lived for 150 years.

The diet of turtles includes worms, insects, and other small animals. Large aquatic turtles eat fish, birds, and small mammals. Some turtles can go weeks without eating. The alligator snapping turtle of the southern United States has a lure on the floor of its mouth that attracts prey.

Both turtle meat and eggs are eaten by humans. The shells of many species are prized as ornaments. Because turtles have been killed in large numbers by predators, some forms are now protected by law.

Snakes are legless reptiles that occupy both aquatic and land habitats. In a process called **molting,** snakes shed their outer skin, which is a layer of scales. Molting may occur several times each year. During molting, the old layer of skin loosens and the snake then crawls out of it.

Very few snakes are poisonous. Those that are have fangs with which they inject venom. Venom of snakes such as cobras and rattlesnakes can be fatal to humans. The poisons in snake venom may act on either the nervous system or circulatory system of the prey. Poisons that affect the nervous system interfere with nerves that control breathing and heartbeat. Poisons affecting the circulatory system may destroy red blood cells and the walls of blood vessels. Antidotes to snake venom are prepared from venom "milked" from the fangs. Because of the danger of the few poisonous species, many people fear snakes.

16:14 Birds

Birds, class **Aves** (AY vayz), represent animals that probably arose as an early branch of the reptile group. *Archaeopteryx* (ar kee AHP tuh rihks) is a fossil bird with several features of reptiles. Today's birds have many features similar to those of reptiles. If you examine the legs of a bird, you will see that they are covered by scales and have claws. Like some reptiles, birds have a tough beak. Birds also develop in shelled eggs similar to those of reptiles.

Modern birds have a constant body temperature and are called **endotherms** (EN duh thurmz). They are covered with feathers that help insulate the bird and maintain a constant body temperature. Most

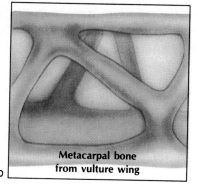

Metacarpal bone
from vulture wing

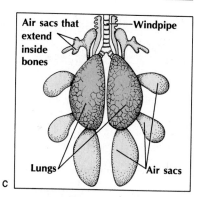

Air sacs that extend inside bones — Windpipe

Lungs

Air sacs

FIGURE 16–24. (a) Birds are well adapted for flight, as this ring-billed gull shows. (b) Bones within bird wings are hollow with braces crossing them. The bones are therefore light and strong. (c) The lung system has extra air sacs that fill spaces in the body and even within some bones. The sacs are part of a highly specialized respiratory system and they help make the animal lighter.

The limbs and bones of birds are adapted for flight.

birds gradually molt (replace old feathers) once a year. Many birds molt seasonally.

Although some birds, such as penguins and ostriches, do not fly, most birds are well adapted for flight. The forelimbs are modified as wings, the shape of which is important to flight. The bird's skeleton is composed of thin, lightweight bones that contain hollow cavities. Some of these hollow bones contain small cross braces of bone, which add strength. Many of the skeletal bones are fused and reduced in size. The breastbone is large, providing an attachment site for flight muscles. Downward movement of the wings is accomplished by contraction of the large pectoralis muscles that can make up as much as twenty percent of the bird's total body weight.

In addition to the skeletal and muscular systems being adapted for flight, other body parts are as well. Feathers play an important role in balance and flight. Those few birds that lose all their feathers at one time during molting remain flightless until new feathers grow. The respiratory system is also adapted for flight. A system of **air sacs** connects with the lungs. The air sacs extend into the body cavity and even into some of the bones. The air sacs, as well as the lungs, fill with air. They also empty more completely when the bird exhales. The flow of air through the air sacs and lungs enables a bird to obtain more oxygen than it could with lungs alone. With more oxygen, cellular respiration can take place more efficiently. Thus, more energy is available to the flight muscles.

The bird's skeleton is also well adapted for walking. The entire body weight of a bird is on the legs when the bird is on the ground. The pelvis and associated muscles are adapted to this form of locomotion.

Bill (beak) structure is one of the many features of birds that exhibits a great degree of variation. For example, cardinals are seed eaters and have crushing bills; pelicans have bills adapted for "fishing." Bill structure, along with other traits such as plumage color, foot structure, and behavior, is used to classify birds.

A bird's respiratory system delivers a plentiful supply of oxygen to the flight muscles.

			Body	Limb		
Class	**Examples**	**Integument**	**Temperature**	**Structure**	**Gas Exchange**	**Fertilization**
Agnatha	lamprey, hagfish	slimy skin	ectotherm	no paired limbs	gills	external
Chondrichthyes	shark, skate, ray	scales	ectotherm	2 pair of fins	gills	internal
Osteichthyes	perch, bass, trout	scales and slimy skin	ectotherm	2 pair of fins	gills	external
Amphibia	frog, toad, salamander	slimy skin in most forms	ectotherm	2 pair of legs; no claws	gills; lungs	external
Reptilia	turtle, lizard, snake, alligator	dry; scaly	ectotherm	2 pair of legs; claws	lungs	internal
Aves	robin, eagle, pelican	feathers; scales on legs	endotherm	1 pair of wings; 1 pair of legs with claws	lungs	internal
Mammalia	bear, whale, kangaroo	hair	endotherm	2 pair of legs; claws in most forms	lungs	internal

(Table 16-2. Summary of Characteristics of Vertebrates)

16:15 Mammals

Mammals are the most complex vertebrates.

Mammals, class **Mammalia** (muh MAY lee uh), are the most complex group of vertebrates living today. They are adapted for life in a variety of ecological niches from land to water. What features make mammals so successful?

In mammals, development may occur in shelled eggs or may be partially or completely internal.

Most of the mammals with which you are familiar bear their young alive. The young develop completely within the female and then are born. These mammals are called **placentals** (pluh SENT ulz). In mammals called **monotremes,** young develop in eggs similar to those of birds and reptiles. The spiny anteater and duckbill platypus are monotremes. Pouched mammals, called **marsupials** (mar SOO pee ulz), have partial development within the female. In these animals, such as the kangaroo and opossum, the young leave the mother's body before development is complete. They complete development within a pouch on the mother's abdomen.

FIGURE 16–25. The three main groups of mammals are represented by (a) the duckbill platypus, a monotreme, (b) the Tasmanian devil, a marsupial, and (c) the field mouse, a placental.

 a
 b
 c

Whatever the means of development, mammals care for their young until they are able to survive on their own. This caring gives added protection for the newborn animal and ultimately for the species. Young are fed milk from mammary (MAM uh ree) glands. Development of mammals will be studied in Chapter 24.

Mammals are endotherms. Their bodies are covered with hair or fur, and they breathe by means of lungs. The nervous system of mammals is the most complex of all animals. They have more complex behavior (Chapter 31) and more highly developed brains than other animal groups. There are many orders of mammals. Appendix A summarizes the features of some of them.

Mammals are named for the presence of mammary glands.

Mammals are covered with hair, breathe with lungs, and are endotherms.

FIGURE 16–26. Mammals are a diversified class of animals adapted to many environments. Included are the (a) camel, (b) raccoon, (c) armadillo, (d) orangutans, (e) dolphin, (f) bat, (g) walrus, and (h) opossums.

The order to which you belong is the primates. Primates also include the lemurs, monkeys, and great apes. The great apes are the orangutan, gibbon, chimpanzee, and gorilla. Most primates are adapted to life in trees.

The great apes, family **Pongidae** (PAHN juh dee), have no tails. Many spend at least part of their lives on land where they walk, sometimes on their back feet. The gorilla is almost entirely a land-dweller, returning to the trees only to sleep. The great apes have large brains and appear to be the most intelligent of the nonhuman primates.

Humans are the only living members of the family of primates called Hominidae.

Humans are members of a family of primates called **Hominidae** (hoh MIHN uh dee). Modern humans are the only living members of this family. They belong to the genus and species *Homo sapiens* (HOH moh • SAY pee unz).

A high degree of mental development enables humans to learn and apply their knowledge.

Humans are distinctive in certain respects. For example, their offspring have a long childhood in which they develop both physically and mentally. This period of time allows for training and learning. In most human societies pair-bonds are formed; that is, a male mates with a single female. Thus, the family is the basic unit of most human societies. Mating can occur at any time instead of during a specific breeding season as in other animal species. Humans have a high degree of mental development. This ability permits humans to learn and to apply their knowledge to the world around them (Section 32:10).

REVIEWING YOUR IDEAS

10. What are the major groups of amphibians? Of reptiles?
11. List several adaptations of birds for flight.
12. Distinguish among the mammal groups: placentals, monotremes, and marsupials.
13. To which family of mammals do you belong?
14. What is the genus and species of humans?
15. List some distinctive human features.

TABLE 16–3. SUMMARY OF MAJOR CHARACTERISTICS OF COMPLEX ANIMALS

PHYLUM	Arthropoda	Echinodermata	Chordata
Common name	arthropods	echinoderms	chordates
Locomotion	muscles, appendages	water pumping system	muscles, limbs
Symmetry	bilateral	radial	bilateral
Number of body openings	two	two	two
Number of tissue layers	three	three	three
Nervous system	present	present	present
Digestive system	present	present	present
Excretory system	present	absent	present
Circulatory system	present	present	present
Respiratory system	present	present	present
Skeletal system	external	mineral deposits	internal

KINGDOM ANIMALIA
(Arthropods through Vertebrates)

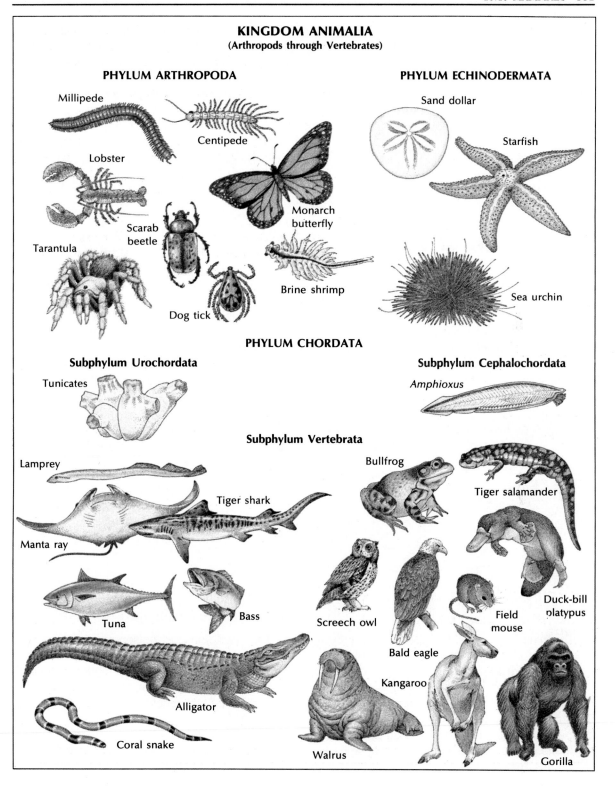

PHYLUM ARTHROPODA

Millipede

Centipede

Lobster

Monarch butterfly

Scarab beetle

Tarantula

Brine shrimp

Dog tick

PHYLUM ECHINODERMATA

Sand dollar

Starfish

Sea urchin

PHYLUM CHORDATA

Subphylum Urochordata

Tunicates

Subphylum Cephalochordata

Amphioxus

Subphylum Vertebrata

Lamprey

Bullfrog

Tiger salamander

Tiger shark

Manta ray

Tuna

Bass

Screech owl

Field mouse

Duck-bill platypus

Alligator

Bald eagle

Kangaroo

Coral snake

Walrus

Gorilla

INVESTIGATION

Problem: How do poisonous and nonpoisonous snakes compare?

Materials:

nonpoisonous snake, preserved
colored pencils
map of poisonous snake ranges

Procedure:

Part A. Comparing Features of Poisonous and Nonpoisonous Snakes

1. Make copies of Figures 16–27 through 16–29 and Figure 16–31 for recording your observations.
2. Examine Figure 16–27a. It shows the head of a typical poisonous snake. Note the fangs and the shape of the pupil.
3. Examine the head of a nonpoisonous preserved snake. Complete Figure 16–27b by drawing in the teeth and the pupil as they appear on this snake.

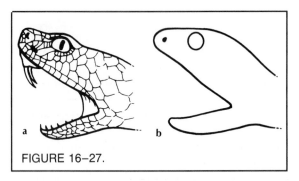

FIGURE 16–27.

4. Examine Figure 16–28a. It shows the underside of a poisonous snake tail. The vent is the area where the anus opens to the outside. Note the scale pattern.
5. Examine the underside of the nonpoisonous snake tail. Locate the vent and note the scale pattern. Complete Figure 16–28b by drawing in the vent and scale pattern as they appear on this snake.

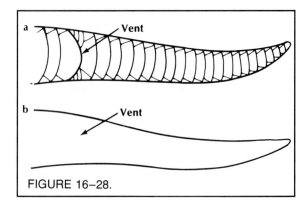

FIGURE 16–28.

6. Examine Figure 16–29a. It shows the bite pattern of a poisonous snake.
7. The bite pattern of a nonpoisonous snake is similar to Figure 16–29a except that no fang marks are present. Complete Figure 16–29b by drawing the expected bite pattern of a nonpoisonous snake.

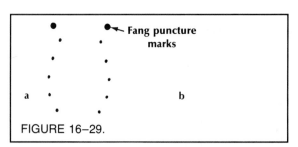

FIGURE 16–29.

Part B. An Exception to the Rule

1. The coral snake is poisonous but does not show the traits of other poisonous snakes. The pupils are round and the underside of the tail shows a double row of scales. However, the bite pattern, shown in Figure 16–30, resembles an hour glass.
2. The coral snake does show warning coloration. In Figure 16–31, color the labeled areas as follows: R-red, B-black, and Y-yellow.

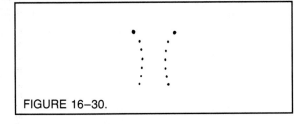

FIGURE 16–30.

Note how the color bands differ in the three snakes represented in Figure 16–31.

Part C. Distribution of Poisonous Snakes

1. Obtain range maps for poisonous snake species from your teacher. Examine the maps, which show the ranges of seven different species.
2. Make a copy of Table 16–4. Using check marks, complete the table to show where these snakes are located.

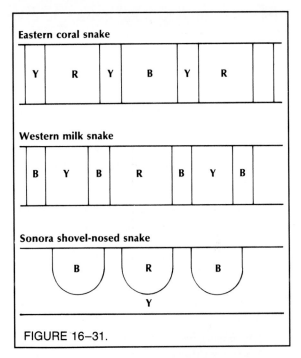

FIGURE 16–31.

Data and Observations:

TABLE 16-4. LOCATIONS OF SOME POISONOUS SNAKES					
	Your State	California	Florida	Maine	Texas
Eastern coral					
Southern copperhead					
Northern copperhead					
Eastern cottonmouth					
Western cottonmouth					
Eastern massasauga					
Western rattlesnake					

Questions and Conclusion:

1. (a) List five traits of subphylum Vertebrata.
   (b) List three traits of all reptiles.
2. List three ways that most poisonous snakes may be distinguished from nonpoisonous snakes.
3. Explain how the coral snake may be distin-guished from nonpoisonous snakes.
4. Which states listed in Table 16–4 have the:
   (a) fewest poisonous species?
   (b) most poisonous species?

Conclusion: How do poisonous and nonpoisonous snakes compare?

CHAPTER REVIEW

SUMMARY

1. Arthropods, the most numerous animals, have exoskeletons and jointed appendages. Their segmented bodies are organized into sections. **16:1**
2. Both centipedes and millipedes have body plans consisting of heads and trunks. Centipedes are carnivores; millipedes feed on plant material. Members of both classes have mandibles and a pair of antennae, and breathe by tracheae. **16:2**
3. Crustaceans are arthropods that are mainly marine. Most have a cephalothorax and abdomen and breathe by gills. They have two pairs of antennae and mandibles. **16:3**
4. Insects have bodies composed of head, thorax, and abdomen. They have three pairs of legs, a pair of antennae, mandibles, and most have wings. **16:4**
5. Arachnids have chelicerae, rather than mandibles. They have a cephalothorax and abdomen, four pairs of legs, and a pair of pedipalps. They have no antennae. **16:5**
6. All echinoderms are marine and most have an endoskeleton of mineral spines. They are bottom-dwellers. **16:6**
7. Hemichordates, acorn worms, are marine forms similar to echinoderms and chordates. The three phyla are closely related and probably have a common ancestor. **16:7**
8. Chordate characteristics are a notochord and gill slits, or pouches, at some stage of life. Chordates have dorsal nerve cords and ventral hearts. Tunicates and lancelets are invertebrate chordates. The vertebrates represent a subphylum of chordates in which the notochord is replaced by vertebrae. They have internal skeletons of cartilage and/or bone. **16:8**
9. Jawless fish have slimy, scaleless skin and no fins. They retain the notochord and have skeletons of cartilage. Like all fish, they breathe by gills. **16:9**
10. Cartilage fish have skeletons of cartilage, two pairs of fins, and toothlike scales. They also have internal fertilization. **16:10**
11. Bony fish have skeletons and scales of bone and fins for locomotion and balance. **16:11**
12. Amphibians are adapted for life on land, but most require water for reproduction, which involves external fertilization and development. They have slimy skin and lack claws. Most adults breathe by lungs. **16:12**
13. Reptiles have internal fertilization and most develop in shelled eggs. They have dry, scaly skin, breathe by lungs, and have claws. **16:13**
14. Birds have adaptations for flight, including wings, feathers, lightweight bones, and powerful muscles. Their legs have scales and claws. Unlike fish, amphibians, and reptiles, they are endotherms. They have internal fertilization, develop in shelled eggs, and breathe by lungs. **16:14**
15. Mammals are the most complex vertebrates. Most undergo complete internal development. Young are fed milk and are cared for by parents. They have hair, breathe by lungs, and are endotherms. **16:15**

LANGUAGE OF BIOLOGY

air sac	marsupial
appendage	molting
cephalothorax	monotreme
chelicera	neoteny
cheliped	notochord
ectotherm	operculum
endoskeleton	pedipalp
endotherm	placental
exoskeleton	spinneret
integumentary system	swimmeret
mandible	trachea

16

CHECKING YOUR IDEAS

On a separate paper, indicate whether each of the following statements is true or false. Do not write in this book.

1. Reptiles are well adapted for life on land.
2. All chordates are vertebrates.
3. Arthropods have jointed appendages and an external skeleton.
4. Salamanders develop in eggs.
5. Spiders and ticks are in the class Insecta.
6. Echinoderms are the closest living relatives of chordates.
7. Reptiles live only on land.
8. Insects and spiders are arthropods with the same kinds of mouthparts.
9. Most primates are adapted for life in trees.
10. Lizards and snakes are amphibians.

EVALUATING YOUR IDEAS

1. Why are arthropods so successful?
2. Compare and contrast the body plans of annelids and arthropods.
3. Compare arthropod classes in terms of number of body sections and walking legs, kinds of mouthparts, and means of gas exchange.
4. What are the functions of chelipeds? What are the functions of swimmerets?
5. In what ways are insects pests to humans? How are they beneficial?
6. Why are echinoderms, hemichordates, and chordates considered related groups?
7. What features are common to all chordates?
8. What kinds of animals are chordates but not vertebrates? Where do they live? How do they obtain food?
9. What are integumentary systems? Give examples from different vertebrate classes.
10. Distinguish between agnathans and other fish.
11. Distinguish between cartilage and bony fish.

12. Why are most amphibians "chained" to the water and not completely terrestrial?
13. Describe the various ways by which amphibians obtain oxygen.
14. List some ways in which reptiles are adapted for life on land.
15. How is beak structure important to birds? Give some examples.
16. Distinguish between molting in reptiles and molting in birds.

APPLYING YOUR IDEAS

1. Why are most barnacles hermaphrodites?
2. How would you explain the fact that many birds eat quite frequently?
3. Why do ectotherms migrate or hibernate?
4. Why are arthropods and vertebrates the most successful animals? How are they similar?
5. Placentals and marsupials live in different parts of the world, yet many forms resemble one another. How can this be explained?

EXTENDING YOUR IDEAS

1. Collect sowbugs from a damp area such as the forest floor. Examine their structure with a hand lens or binocular microscope. Describe their responses to light and touch.
2. With what kinds of animals is each of the following branches of biology concerned: ichthyology, entomology, herpetology?

SUGGESTED READINGS

Diamond, Jared, "In Quest of the Wild and Weird." *Discover,* March, 1985.

Jacques, Richard J., Jr., "Insects, Man, and Disease." *Biology Digest,* March, 1985.

Wilson, E. O., "In Praise of Sharks." *Discover,* July, 1985.

SIMPLE ORGANISMS

You have had a glimpse of the great diversity of life on Earth. In spite of their physical differences, organisms carry out certain common processes. They reproduce, grow, and develop. They make or obtain food and use the food energy for cellular work. They also use food and other materials to build important compounds. They exchange materials with the environment, transport materials, and rid themselves of wastes. They have a means of support and/or locomotion, and respond to stimuli in their environments.

Simple living organisms—monerans, protists, and fungi—are no exceptions in carrying out these life functions. Many of these organisms are adapted in special ways. These birds'-nest fungi are adapted to living on decaying logs. In this unit, you will study other adaptations of simple organisms and how these organisms carry out life functions.

Birds'-nest Fungi

SIMPLE ORGANISMS: REPRODUCTION

Simple organisms reproduce by various methods. Some of these methods involve only one parent. The parent organism may produce special reproductive cells that develop into offspring genetically identical to the parent. The special reproductive cells of a puffball, a type of fungus, are shown here. Of what are these reproductive cells composed? How are they formed? What other methods of reproduction are found among the various kinds of simple organisms?

chapter 17

Genetic information is present in the DNA of a cell. You learned that it is transcribed to form RNA and then translated in the making of proteins in a cell. Transmission of the code occurs when the cells divide, each new cell receiving an exact copy of the code. Chromosomes are the means by which DNA is moved from cell to cell and from one generation to the next.

The processes just described are similar in many organisms and ensure that the genetic code is passed to offspring when living things reproduce. The ways that organisms reproduce, however, vary. In this chapter, you will learn about the kinds of reproduction found in the simplest organisms — monerans, protists, and fungi — and in viruses.

Objectives:
You will
- describe the various methods of asexual reproduction.
- discuss the types of sexual reproduction in monerans, protists, fungi, and viruses.
- compare the advantages and disadvantages of asexual and sexual reproduction.

ASEXUAL REPRODUCTION

Recall that asexual reproduction is the production of one or more offspring from *one parent*. It occurs in simple organisms in several ways, but each time it occurs, an identical set of genes is formed from one set. Therefore, each offspring has the same genotype and phenotype as its parent. Asexual reproduction is adaptive because it passes the *same* genes to the offspring. Because the offspring gets the same set of genes as its parent, the offspring is suited to its environment, just as the parent was. The only source of variation is mutation. Is variation helpful or harmful to a group of organisms? How helpful is asexual reproduction in a *changing* environment?

Offspring produced by asexual reproduction are genetically identical.

FIGURE 17–1. Bacteria can reproduce asexually by fission. First the chromosome duplicates and then the cell wall lengthens and pinches in, forming two cells.

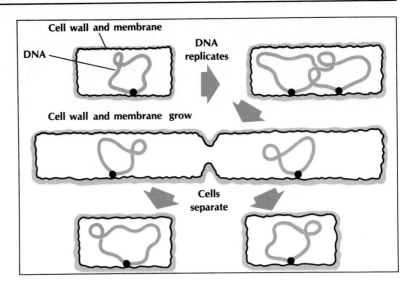

17:1 Fission in Monerans

Fission (FIHSH un) is the splitting of a cell into two smaller cells (offspring) that are equal in size. Monerans reproduce by fission. These organisms are prokaryotes (have no true nuclei), and so they do not undergo mitosis. Recall that mitosis is a process that occurs only in cells with nuclei. A moneran cell has one circular chromosome made of DNA. As cell fission begins, the chromosome attaches to the inside of the cell membrane and a duplicate is formed (Figure 17–1). The new chromosome also attaches to the cell membrane at a point near the first chromosome. While this attachment occurs, a new cell membrane and cell wall form, and the cell gets longer. At the center of the cell, more new membrane and cell wall form and begin to push inward. As the new membrane and wall grow inward, the two chromosomes become separated. Two separate daughter cells are formed, each with the same genetic material.

Fission is the most important form of reproduction in cyanobacteria. Other bacteria also reproduce asexually by fission. Under ideal conditions, some bacteria can reproduce by fission every twenty minutes.

In fission, one unicellular organism is split into two offspring of equal size.

In prokaryotes, fission does not involve mitosis.

Most reproduction in prokaryotes occurs by fission.

17:2 Fission in Eukaryotes

Some simple eukaryotic organisms reproduce by fission. Because they have true nuclei, fission in eukaryotes involves mitosis and cell division. Unicellular forms such as *Amoeba, Euglena,* and *Paramecium,* some fungi, and many one-celled algae undergo fission as a means of reproducing.

Fission in eukaryotes involves mitosis and cell division.

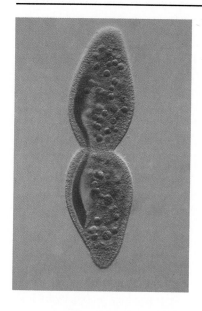

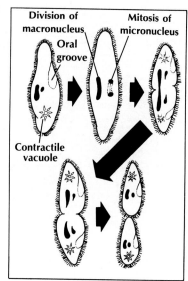

Division of macronucleus

Oral groove

Mitosis of micronucleus

Contractile vacuole

FIGURE 17–2. When *Paramecium* reproduces asexually by fission, both macronucleus and micronucleus divide. The micronucleus divides by mitosis.

In organisms such as *Amoeba* and *Euglena,* fission may occur as often as once each day. Chromosomes replicate, become attached to spindle fibers, and identical sets are pulled to each end of the cell. The cell membrane then pinches in to form two distinct offspring.

Fission in *Paramecium* is more complex than fission in *Amoeba* or *Euglena.* This organism contains a macronucleus and one or more micronuclei (Section 13:10). As reproduction occurs, the chromosomes of both nuclei replicate. The division of the macronucleus does not involve mitosis. It appears to "stretch out" and pull apart into two equal-sized parts. The micronucleus divides by mitosis. Its chromosomes are pulled to opposite poles of the cell. The cell membrane pinches in midway along the length of the cell, and two daughter cells are produced. Each new *Paramecium* has one of each type of nucleus (Figure 17–2).

In Paramecium, the macronucleus divides without mitosis, but mitosis occurs in the division of the micronucleus.

Paramecium is a complex cell with specialized parts such as an oral groove and contractile vacuoles. As fission occurs, the oral groove disappears. New oral grooves begin forming at each end of *Paramecium* before the original cell has finished its division. At the same time, new contractile vacuoles begin to form in each half of the parent cell. Both the oral grooves and contractile vacuoles are fully formed by the time the daughter cells separate.

New specialized cell parts of Paramecium form as the cell divides.

17:3 Budding and Fragmentation

Budding is a type of asexual reproduction in which an outgrowth forms on the parent organism. This outgrowth eventually breaks off, and the "bud" then grows into a complete organism.

In budding, an outgrowth develops on and then breaks away from the parent organism.

FIGURE 17–3. (a) Yeast cells reproduce asexually by budding, a process in which outgrowths of the cell, buds, grow and eventually break away. This budding yeast cell is shown magnified about 3600 times. (b) *Nostoc,* a Cyanobacterium, may reproduce asexually by fragmentation at a heterocyst. *Nostoc* is shown here magnified about 1200 times.

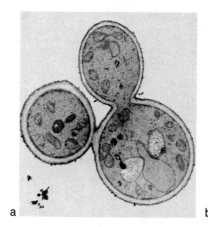 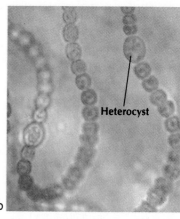

a   b

Budding can occur in both unicellular and multicellular organisms. Because it involves mitosis, offspring have the same traits as the parent. Yeast, a unicellular fungus, may reproduce asexually by budding (Figure 17–3). In a yeast cell, a portion of the cell wall grows out to form a bud. When the nucleus replicates, one of the nuclei remains in the parent cell and the other one moves to the bud. The new cell may give rise to a chain of buds or may be pinched off. Yeasts can also reproduce in other ways (Sections 17:2 and 17:6).

In fragmentation, a piece of an organism breaks off and becomes a new organism.

Breaking off of large parts of an organism can lead to another form of asexual reproduction called **fragmentation.** For example, certain cyanobacteria such as *Nostoc* may break apart into two or more smaller filaments. The breaks usually occur at a heterocyst. **Heterocysts** (HET uh roh sihsts) are special cells in the filament that have thick walls, no DNA, and that appear empty. After fragmentation, each filament adds new cells by fission.

17:4 Spore Formation

Spores are specialized reproductive cells.

Some organisms develop asexually from spores. Each spore is a specialized cell that contains DNA and a small mass of cytoplasm. A tough, outer wall protects many spores until conditions are favorable for development.

Anabaena is a filamentous cyanobacterium. Along the filament are special oval cells with thick walls. These cells are spores. Because of their thick walls, these spores can resist harsh winter conditions. The may remain dormant in a pond during the winter and begin their development the next spring.

One phylum of protozoans, Sporozoa (Section 13:11), reproduces by spores. The nucleus of the parent cell undergoes mitosis many times to form the spores. The parent cell later bursts, releasing the spores, which develop in a particular host. *Plasmodium,* which causes malaria (Section 19:6), is an example of a sporozoan.

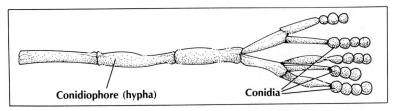

Conidiophore (hypha)   Conidia

FIGURE 17–4. *Penicillium* mold produces conidia at the ends of the hyphae. The brush shape of the hyphae with conidiospores gave rise to the mold's name. The Latin word for small brush is *penicillus.*

Asexual reproduction by spores occurs in different ways in many types of fungi. The common bread mold, *Rhizopus* (Section 13:13), produces thousands of spores by mitosis in each sporangium. The sporangium breaks open, releasing spores that may be carried by wind or animals. If a spore lands in a suitable environment, it will develop into new hyphae.

Penicillium molds may also reproduce asexually by spores. Chains of spores, called **conidia** (kuh NIHD ee uh), form at the tips of the hyphae by mitosis (Figure 17–4). Thus, the hyphae are called **conidiophores** (kuh NIHD ee uh forz). Each spore produced by the conidiophore may develop into a new mold.

Fungi reproduce asexually by spores.

REVIEWING YOUR IDEAS

1. What is asexual reproduction?
2. How do asexually reproduced offspring compare genetically with their parent and one another?
3. Distinguish between fission in prokaryotes and fission in eukaryotes.
4. Distinguish between budding and fragmentation.
5. What is a spore?

SEXUAL REPRODUCTION

Asexual reproduction is a *division* process that involves one parent. However, the opposite is true of sexual reproduction. **Sexual reproduction** is the *fusion* (union) of two sets of DNA that usually come from different parents.

Sexual reproduction is the union of two sets of DNA.

Fusion of different sets of DNA results in genetic recombination (Section 6:10) and variety among offspring. Variety in offspring shows as new features. Because of this variety, *sexually reproducing organisms can adapt to changing environments.* This advantage explains the fact that almost all types of organisms have some means of sexual reproduction in their life cycles.

Sexual reproduction provides variety among offspring.

In simple organisms, conjugation is a common form of sexual reproduction. **Conjugation** (kahn juh GAY shun) is the fusion of nuclear material of two cells. Unlike sexual reproduction in more complex organisms, simple conjugation does not involve gametes.

Conjugation involves fusion of nuclear material but does not involve gametes.

FIGURE 17–5. Only one of these bacteria was "labeled" with a radioactive substance (indicated by the black grains) prior to conjugation. The presence of the black grains in both cells indicates that DNA from the F⁺ cell moved into the F⁻ cell.

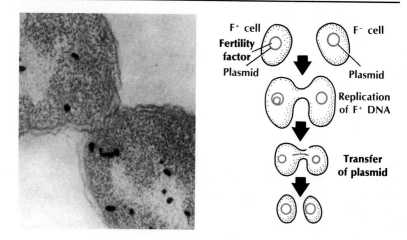

17:5 Genetic Recombination in Bacteria

Bacteria produce asexually by fission. Even though they are limited to asexual reproduction, several means of genetic recombination are known to occur. One such means is conjugation. Although there are no sexes among bacteria, different mating types exist. One is called F^+. F stands for **fertility factor**. F^+ bacteria have this factor, which is located on a plasmid (Section 13:2). The other type, F^-, has no fertility factor.

Bacterial conjugation leads to genetic recombination among bacteria.

Only F^+ and F^- bacteria can conjugate. During conjugation, bridges of cytoplasm form between bacteria. The plasmid of the F^+ type moves through the bridge and into the F^- cell. The F^- cell then contains its own DNA plus DNA from the other cell. Although genetic recombination has occurred, there has been no reproduction. After conjugation, each bacterium reproduces asexually by fission.

Not only is the fertility factor transferred, but also other plasmid genes. Recall that some of these genes make bacteria resistant to antibiotics. An F^- bacterium with no resistance to antibiotics could become resistant as a result of conjugation and genetic recombination. Offspring then produced by fission would also be resistant.

In bacteria, recombination can result from transformation.

Another form of genetic recombination occurs in bacteria. **Transformation** (Section 9:1) occurs when one bacterium breaks open and part of its DNA enters another bacterium. Recall that transformation occurred in *Pneumococcus*. It caused a change in the genotype and phenotype of rough cells. The rough cells were changed to smooth cells that caused the disease pneumonia.

17:6 Conjugation in Fungi

In addition to asexual reproduction by means of spores, bread mold may reproduce sexually by conjugation. This process involves different types of hyphae, called *plus* and *minus*. Branches grow from

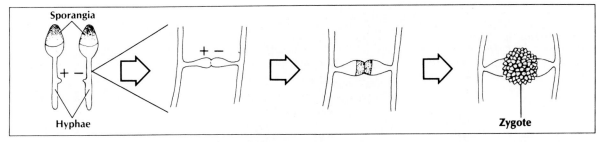

each of these hyphae and meet. The tip of each branch becomes a separate cell and acts as a "gamete." The "gametes" fuse to form a zygote that is resistant to drying (Figure 17–6). The zygote first goes through a dormant period. Later, the zygote grows, meiosis occurs, and a new hypha develops. The hypha later develops a sporangium and reproduces asexually by spores.

Other fungi also undergo sexual reproduction. In yeasts, individual cells can fuse to form a zygote (Figure 17–7). The zygote can bud or it can undergo meiosis to form four spores. The zygote is modified into an ascus (Section 13:15), which holds the spores. Spores are then freed and develop into separate cells.

In club fungi such as mushrooms, different types of underground hyphae fuse and produce the rest of the mycelium. Cells of the basidia (Section 13:14) contain two nuclei. These nuclei fuse to form a zygote. The zygote then undergoes meiosis to produce four spores. Each spore may develop into a new underground mycelium.

FIGURE 17–6. Bread mold reproduces sexually by conjugation. A resistant zygote results.

Conjugation is common in many fungi.

REVIEWING YOUR IDEAS

6. What is sexual reproduction?
7. How do sexually reproduced offspring compare genetically with their parents and one another?
8. What is conjugation? In which organisms does it occur?
9. What is transformation?

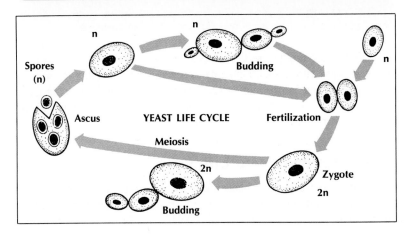

FIGURE 17–7. Both haploid and diploid cells exist in the life cycle of yeasts. The life cycle shows asexual and sexual reproduction.

PEOPLE IN BIOLOGY

It was not until after he studied astronomy and physics in college that Max Delbrück found a scientific "home" in biology. In 1937, Delbrück left his German homeland and moved to the California Institute of Technology (Caltech). There, Delbrück became fascinated with viruses and their methods and rates of reproduction. Very little was known at the time about viral reproduction, except that a virus invades a bacterium, reproduces at a fast rate, and then the offspring explode from the bacterial cell. Dr. Delbrück wanted to see if different types of viruses could reproduce inside a single bacterial cell. He injected two different types of viruses into one bacterium, and then he examined the offspring. Delbrück found viruses like the two he had injected. He also found new viruses that had been produced by recombination of DNA from the two viruses that had been injected into the bacterial cell. Delbrück had shown that two viruses, if present in the same bacterial cell, could exchange DNA when reproducing. For this discovery, and for his other work on viruses, Max Delbrück was awarded the Nobel Prize in 1969.

As physics instructor at Vanderbilt University and biology professor at Caltech, he was a model for many scientists doing research. Besides the Nobel Prize, Dr. Delbrück was elected to the National Academy of Sciences in 1949 and received the Kimber Genetics Award in 1965.

Max Delbrück

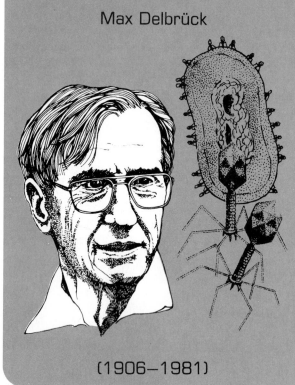

(1906—1981)

REPRODUCTION IN VIRUSES

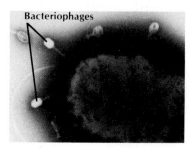

FIGURE 17–8. Viruses can reproduce only by infecting living cells. Bacteriophages, magnified about 31 000 times, are shown infecting a bacterium.

Viruses are unique particles with no cellular structure or metabolism. In fact, their only lifelike characteristic is their ability to reproduce. Even their reproduction can occur only within a *specific* host cell. As "reproductive parasites," viruses may cause illness in or death to their hosts.

17:7 Bacteriophage Life Cycles

A virus that attacks a bacterium is a **bacteriophage** (bak TIHR ee uh fayj), or **phage.** Like all viruses, phages have an outer coat of protein. Inside the phage is a core of DNA. The phage has a tail-like portion by which it attaches to a bacterial cell. It "recognizes" a specific host because the host has special receptors in its wall. It is thought that a protein in the tail reacts with the receptor. Part of the tail pierces the cell wall, and the phage's DNA is injected into the bacterium. The viral DNA takes over the bacterial cell and uses the bacterium's materials, enzymes, ATP, and cell parts to reproduce.

Within a very short time, the bacterium splits open releasing many new virus particles (Figure 17–9). This breaking open of bacteria or other cells by viruses is called **lysis** (LI sus). The viral life cycle that results in lysis is called the **lytic cycle.** The new phages continue the cycle by attacking other bacteria.

Sometimes a phage infects a host cell but does not destroy it. The DNA of the phage becomes part of the bacterium's DNA and is replicated when the bacterium reproduces (Figure 17–9). This viral life cycle is called the **lysogenic** (li suh JEN ihk) **cycle.** A virus that has become part of its host's DNA is called a **provirus.** The provirus may cause some phenotypic changes among the resulting bacteria. Also, the viral genes prevent other viruses from infecting the bacteria. Eventually, certain changes occur and the bacterial cells begin making new phages. The bacteria then lyse and release the phages. It is thought that many organisms, including humans, have proviruses in their cells. Genes of the proviruses may affect the phenotypes of the host organism. In time, changes within the body of the host or from outside sources may affect proviruses and trigger them to become active again. Some diseases may occur as proviruses change to active forms and destroy their host cells.

Phage DNA makes use of bacterial materials to direct the production of more phages.

In the lytic cycle, viruses reproduce, destroy host cells, and then attack other host cells.

In the lysogenic cycle, the viral DNA becomes part of and replicates along with the host cell's DNA.

FIGURE 17–9. Phage DNA enters a host cell. It may take over the host's metabolic "machinery" to produce new phages (shown in the lytic cycle on the left). Or, it may attach to the host's DNA, divide when the host cell divides (shown in the lysogenic cycle on the right), and then take over the "machinery." Many phages are released when the host cell lyses.

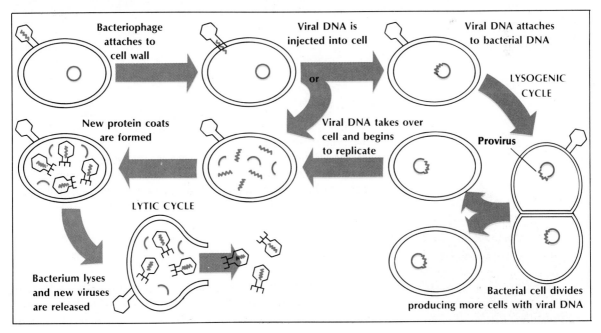

Bacteriophage attaches to cell wall

Viral DNA is injected into cell

Viral DNA attaches to bacterial DNA

LYSOGENIC CYCLE

or

Viral DNA takes over cell and begins to replicate

Provirus

New protein coats are formed

LYTIC CYCLE

Bacterium lyses and new viruses are released

Bacterial cell divides producing more cells with viral DNA

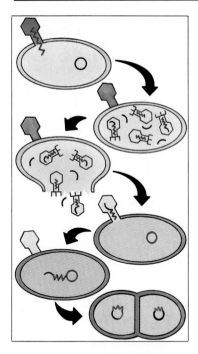

FIGURE 17–10. In transduction, a newly formed virus picks up part of the bacterial chromosome. Part of the first bacterial chromosome is then transferred to another bacterium when the virus infects it.

The life cycles of other viruses that attack other kinds of cells are generally similar to those of bacteriophages. Nonbacterial viruses may enter their host cells whole. But, like phages, only their nucleic acid is important for reproduction.

17:8 Transduction

Sometimes, during the lytic cycle, a new phage's DNA may combine with a small bit of the DNA of the bacterium. If this phage then attacks another bacterium, it will inject the portion of the first host's DNA into the second host. The DNA of the two different bacteria may then combine. This process is known as **transduction** (trans DUK shun) (Figure 17–10). Transduction results in genetic recombination. It is a source of variety among bacteria.

Transduction can be used to learn about the genes of bacterial cells. A practical method based on transduction may prove important in gene therapy (Chapter 13). In this procedure, a normal human gene would be added to the nucleic acid of a "synthetic" virus. The virus would "attack" a patient's host cells where it would become a provirus. The normal human gene would then be expressed. Thus, the genetic disease could be cured or avoided.

REVIEWING YOUR IDEAS

10. What is required for a virus to reproduce?
11. What is a bacteriophage?
12. What part of a virus is involved in its reproduction?

ADVANCES IN BIOLOGY

RNA Viruses

So far you have studied reproduction in viruses containing DNA as the primary genetic code. Reproduction in such viruses can be understood in terms of the usual interaction of DNA, RNA, and protein synthesis. However, many plant and animal viruses contain RNA as the primary genetic code. How do these RNA viruses make proteins and reproduce?

In order to synthesize proteins, an RNA virus must make mRNA. To reproduce, the virus must make copies of all the RNA that will become part of the "offspring" viruses. Both these processes require special enzymes for the transcribing of RNA molecules.

RNA viruses have special enzymes for copying their RNA.

One group of RNA viruses has enzymes called RNA transcriptases. These enzymes are needed to make new RNA molecules using the RNA of the virus as a "guide." Some of the new RNA strands are

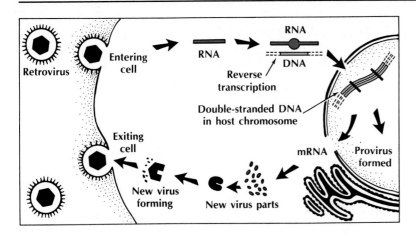

used as mRNA, directing the synthesis of proteins for the coats of the new viruses. Other new RNA strands become the RNA of the offspring. Note that in this group of RNA viruses, reproduction does not involve the host cell's nucleus but takes place in the cytoplasm of the host cell.

RNA transcriptases are used to produce both mRNA and new viral RNA.

Other RNA viruses follow a different sequence of events in their life cycle. The RNA of **retroviruses** is not used directly to make more RNA. Instead, in a unique process, the RNA is used to make DNA! In a first step, a viral enzyme, *reverse* transcriptase, produces a DNA copy of the viral RNA. The two strands, RNA and DNA, are joined to form a double-stranded molecule. This process is called **reverse transcription** because an RNA master is used to make a DNA copy. In a later step, the RNA strand is replaced by DNA. Thus, a double-stranded DNA molecule is produced. This DNA molecule contains all the genetic information of the infecting virus.

In retroviruses, DNA is made from RNA by reverse transcription.

Once formed, the DNA becomes a provirus, part of the DNA of the host cell (Section 17:7). Some retroviruses remain in the provirus stage and do not reproduce. From the nucleus of the host cell, the viral genes are expressed along with those of the host cell. Many retroviruses are known to cause cancer in animals. It is thought that the provirus stage of these retroviruses transforms a healthy host cell into a cancerous one by introducing a cancer gene into the host cell's DNA (Section 19:12).

Some retroviruses remain in the provirus stage and do not reproduce.

Other retroviruses do reproduce within the host cell. The viral DNA is transcribed to make mRNA for synthesizing new viral protein coats and new viral RNA. Viruses are assembled in the host cell's cytoplasm and move to the inner side of the cell membrane. They are then surrounded by the cell membrane and "pinched off" in a process similar to budding. The viruses may then infect other host cells and continue their life cycles. AIDS is caused by a retrovirus that attacks cells of the immune system (Section 19:11). The virus may remain dormant for years and then begin to reproduce, invading more cells. Unlike other retroviruses, the AIDS virus often kills its host cells.

Many retroviruses have a life cycle in which they do reproduce and pinch off from the host cell. The host cell usually is not killed.

INVESTIGATION

Problem: How do different organisms reproduce asexually?

Materials:

microscope glass slide
coverslip dropper
yeast culture
prepared slides:
   paramecium undergoing fission
   Rhizopus nigricans
preserved sample of *Oscillatoria*

Procedure:

Part A. Fission

1. Make a copy of Table 17–1 in which to record your observations.
2. Observe a prepared slide of paramecium fission under low power.
3. While viewing the slide, slowly move it until you locate a single paramecium (shaped like a slipper).
4. Continue to move the slide until you locate two paramecia that look attached at the tips. These cells are the result of fission. One cell has divided to form the two cells. The cells will soon separate to form two independent paramecia.
5. Diagram a single paramecium and a paramecium undergoing fission in your table.

Part B. Budding

1. Place two drops of yeast culture on a glass slide. Add a coverslip and observe under low power and then high power.
2. While viewing the slide under high power, you will see many yeast cells that appear almost perfectly round.
3. Slowly move the slide until you locate a yeast cell that appears to have a smaller yeast cell attached to it. The small, attached yeast cell is a bud.

4. The bud will soon detach from the parent cell to form an independent yeast cell.
5. Diagram and label several single yeast cells and several yeast cells with buds attached in your table.

Part C. Fragmentation

1. Using a dropper, place a small sample of *Oscillatoria* on a glass slide. Add a coverslip and observe under low power and then high power.
2. Locate one strand, or filament. Move the slide slowly until you find a cell that looks like the one shown in Figure 17–12. This is a heterocyst cell. *Oscillatoria* will break, or fragment, at this cell to form two independent filaments.
3. Diagram and label parts of a single filament and one with a heterocyst in your table.

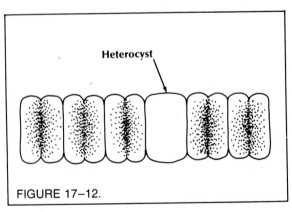

FIGURE 17–12.

Part D. Spore Formation

1. Observe a prepared slide of *Rhizopus nigricans* (bread mold) under low power.
2. Slowly move the slide until you locate a structure that resembles a lollipop on a stick.

3. The lollipop-like part is a sporangium. It produces thousands of tiny round cells called spores, which can be seen best under high power. The spores may be seen clinging to the sporangium or as individual cells that have broken loose.

4. Diagram and label a sporangium with spores in your table. Use either low or high power.

Data and Observations:

TABLE 17-1. OBSERVATIONS OF ASEXUALLY REPRODUCING ORGANISMS		
Organism	Single Organism	Organism Reproducing
Paramecium		
Yeast		
Oscillatoria		
Rhizopus nigricans		

Questions and Conclusion:

1. Define: asexual reproduction, fission, budding, fragmentation, spore.
2. Explain why the four types of reproduction studied are asexual and not sexual. Include features such as parent number and type of cell reproduction.
3. To which kingdom does each of the following organisms belong: paramecium, yeast, *Oscillatoria, Rhizopus nigricans?*
4. Explain how the cells formed during budding differ from those formed during fission.
5. How do the number of cells formed by budding or fission compare to the number of spores formed by *Rhizopus nigricans?*
6. How may the number of spores formed be important to the survival of the bread mold?
7. (a) How do offspring formed during asexual reproduction compare to the parent?
   (b) How may this be helpful to the offspring?

Conclusion: How do different organisms reproduce asexually?

371

CHAPTER REVIEW

SUMMARY

1. Asexual reproduction is common among simple organisms. All offspring produced asexually are genetically identical; therefore, asexual reproduction does not increase variety. **p. 359**
2. Fission is a common means of asexual reproduction in monerans. **17:1**
3. Many unicellular eukaryotes reproduce asexually by a fission process involving mitosis and cell division. **17:2**
4. Budding is a form of asexual reproduction in which an outgrowth forms and then breaks away from the parent. In fragmentation, part of an organism breaks off from its parent and becomes a new organism. **17:3**
5. Spore formation is a means of asexual reproduction involving the production of special reproductive cells. **17:4**
6. Sexual reproduction is the fusion of two sets of DNA. It results in genetic recombination and variety among offspring. Most sexual reproduction in simple organisms occurs by conjugation, the fusion of nuclear materials of two cells. **p. 363**
7. Although bacteria do not reproduce sexually, they undergo conjugation and transformation. Both processes lead to genetic recombination. **17:5**
8. Fungi such as bread mold, yeast, and mushrooms reproduce sexually by conjugation. **17:6**
9. Viruses reproduce only at the expense of a specific host cell. The life cycle of a phage may be lytic or lysogenic. **17:7**
10. During transduction, a portion of one bacterial chromosome may be injected into another bacterium. This process is a source of variety among bacteria. **17:8**
11. RNA viruses have special enzymes for the transcription of their RNA. In retroviruses, RNA is transcribed to form DNA. **p. 368**

LANGUAGE OF BIOLOGY

bacteriophage
budding
conidia
conidiophore
conjugation
fertility factor
fission
fragmentation
heterocyst

lysogenic cycle
lytic cycle
provirus
retrovirus
reverse transcription
sexual reproduction
transduction
transformation

CHECKING YOUR IDEAS

On a separate paper, complete each statement by filling in the blank with the best term. Do not write in this book.

1. Sexual reproduction involves the fusion of two sets of _____.
2. Most sexual reproduction in simple organisms occurs by _____.
3. _____ are chains of spores that form in certain kinds of fungi.
4. Monerans have only _____ reproduction.
5. During _____, DNA of one bacterium moves into another bacterium.
6. _____ reproduction results in genetic recombination among organisms.
7. _____ are cells that are specialized for asexual reproduction.
8. In a certain form of asexual reproduction called _____, an outgrowth forms on the parent organism.
9. Fission in prokaryotes does not involve the process of _____.
10. A virus that attacks a bacterium is a _____.
11. During the _____ cycle of phages, the host cell is destroyed.
12. A _____ is a virus that becomes part of the host's DNA.
13. In the process of _____, an RNA master is used to make DNA.

EVALUATING YOUR IDEAS

1. Describe fission among monerans.
2. How is fission in eukaryotes different from that in prokaryotes?
3. Describe asexual reproduction in yeasts.
4. Compare asexual reproduction by spores in *Rhizopus* and *Penicillium*.
5. How are the tough outer walls of many spores adaptive?
6. How is sexual reproduction more advantageous than asexual reproduction?
7. Describe conjugation in bacteria.
8. Discuss two processes other than conjugation that may lead to genetic recombination in bacteria.
9. Describe sexual reproduction in *Rhizopus*.
10. Compare sexual reproduction in yeasts to that of mushrooms.
11. Describe how a bacteriophage reproduces.
12. Describe the lysogenic cycle.
13. What is unique about protein synthesis and reproduction in RNA viruses?
14. Describe the process of reverse transcription in retroviruses. What happens to the DNA produced?
15. How are new retroviruses produced in and then released from a host cell?

3. Why were the mutants in question 2 above not able to synthesize certain substances?
4. After conjugation, what is the mating type of the original F⁻ bacterium? Explain.
5. Why can asexual reproduction be called "multiplication by division"?
6. Explain why viruses usually cause diseases in their hosts.
7. How is it adaptive that most retroviruses do not destroy their host's cells?

EXTENDING YOUR IDEAS

1. Start a yeast culture by adding either powdered or cake yeast to solutions of sugar water of various concentrations. Take samples from the culture and prepare wet mount slides. Observe the yeast cells for signs of budding. Does the amount of budding vary in different sugar solutions?
2. Using library sources, find out about "high frequency recombination" (Hfr) bacteria. Compare their means of conjugation with that of F⁺ and F⁻ bacteria.
3. You have learned that asexual reproduction in *Paramecium* is a complex process. Find out how *Paramecium* reproduces sexually and describe this process.

APPLYING YOUR IDEAS

1. Consider the complete life cycle of *Rhizopus* and yeast. Which structure(s) is (are) haploid? Which is (are) diploid?
2. Normal bacteria can synthesize substances A, B, C, and D from a basic growth medium. These substances are essential for life. Two mutants are found. One cannot synthesize A and B. The other cannot synthesize C and D. The two mutants are transferred to a basic medium. Soon a few colonies of bacteria appear. How might this result be explained?

SUGGESTED READINGS

Blackmore, Richard P. and Frankel, Richard B., "Magnetic Navigation in Bacteria." *Scientific American*, Dec., 1981.

Fincham, J. R. S., *Genetic Recombination in Fungi*, 2nd ed. Carolina Biology Reader. Burlington, NC, Carolina Biological Supply Co., 1983.

Gibbons, Don L., "Closing in on the Virus-Cancer Link." *Sciquest*, April, 1981.

Sanders, F. Kingsley, *Viruses*, 2nd ed. Carolina Biology Reader. Burlington, NC, Carolina Biological Supply Co., 1981.

SIMPLE ORGANISMS: OTHER LIFE FUNCTIONS

Simple organisms carry out life functions other than reproduction. They must make or obtain food, digest it, eliminate wastes, and distribute needed materials within and among cells. How is this slime mold obtaining food? How is the food distributed from cell to cell? How do other simple organisms carry out other life functions?

chapter 18

In Chapter 17 you learned how the simplest organisms reproduce. In this chapter you will study how these simple living things carry out other vital functions. To survive, they must carry out the same functions that more complex organisms do. Yet, most of these organisms are one-celled, and all lack true tissues and organs. Each function they have involves interaction with the environment in some way.

Objectives:
You will

- compare nutrient requirements of autotrophs and heterotrophs.
- explain digestion in protists and fungi.
- discuss transport, excretion, gas exchange, movement, and response in simple organisms.

NUTRITION

Cells require a variety of chemicals for life. Some chemicals are used as energy sources; others are needed for synthesis of new living material. The materials required for these purposes are called **nutrients** (NEW tree unts).

Nutrients enter cells from the environment. Some nutrients are simple compounds or mineral ions. These simple nutrients may be changed by cells to more complex compounds. Other nutrients enter cells already in complex form. In either case, the complex compounds can be used either as energy sources or for building protoplasm.

Nutrients are chemicals required for energy sources or making protoplasm.

Nutrients, which come from the environment, include both simple and complex chemicals.

18:1 Nutrient Requirements

You learned in Chapter 10 that autotrophs are organisms that make their own food, and heterotrophs are organisms that feed on other living or once-living organisms. Autotrophs and heterotrophs require different nutrients.

Carbon dioxide and water are important nutrients for autotrophs. By photosynthesis, these nutrients are changed to glucose. Glucose can be used as an energy source or changed to other sugars, starches, fats, or other compounds needed for growth. Other than

TABLE 18-1. MAJOR FOOD SOURCES AND THE PRODUCTS OF DIGESTION	
Food Type	Digestive Products
Carbohydrates	Simple sugars
Fats	Fatty acids, glycerol
Proteins	Amino acids

carbon dioxide and water, autotrophs require mineral nutrients. For example, nitrogen is needed to make amino acids, and magnesium is used in the production of chlorophyll. *In general, autotrophs take in simple nutrients and change them to complex organic substances.*

Because they cannot make their own food, *heterotrophs must obtain organic nutrients that are already in complex form.* These nutrients must enter a cell's cytoplasm to be useful. However, most complex molecules are too large to pass across membranes of cells (or of food vacuoles). Thus, they must first be broken into smaller, simpler molecules. Once the complex molecules have been broken down, they can be taken into the cell. Then the small, simple molecules can be resynthesized into complex molecules for the cell's use.

Chemical digestion involves hydrolysis reactions.

The process by which an organism breaks large molecules into smaller ones is called **digestion.** Digestion is a chemical change; it requires enzymes. Digestion is accomplished through hydrolysis reactions. Table 18–1 lists the major food types (organic nutrients) and the products formed from their digestion.

Heterotrophs require water, minerals, and vitamins as well as complex, organic compounds.

In addition to complex organic nutrients, heterotrophs require water, small amounts of minerals, and vitamins. Minerals are taken in as ions in food or water. Vitamins are present in food. Although they may be complex, most vitamins are easily absorbed and need not be digested. Most vitamins are coenzymes (Section 5:6).

Although they take in different nutrients, cells of both autotrophs and heterotrophs must have organic compounds.

In summary, autotrophs take in simple, inorganic nutrients. These nutrients easily enter cells where they are converted to more complex, organic compounds. Heterotrophs take in some simple nutrients, but also need complex, organic compounds that cannot enter cells. The complex molecules are digested to smaller ones that can pass through membranes. In either case, the cell builds or obtains organic molecules that can be used in respiration or converted to other substances.

18:2 Digestion Within Simple Autotrophs

Autotrophs, such as euglenoids, diatoms, and some bacteria, take in *small, inorganic* molecules and mineral ions by passive or active transport. By photosynthesis, they make organic molecules such as glucose, amino acids, and ATP.

Although autotrophs take in small inorganic molecules, they do perform some digestive functions. For example, starch is a complex molecule made from glucose. When the supply of glucose decreases, enzymes in the cells hydrolyze stored starch to glucose molecules. Then the glucose can be "burned" for energy. The breakdown of molecules in these simple organisms occurs inside the cells and is called **intracellular** (in truh SEL yuh lur) **digestion.** To some extent, intracellular digestion occurs in all organisms.

Digestion within cells is called intracellular digestion.

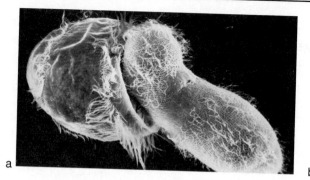
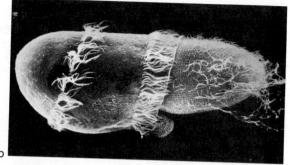

a

b

18:3 Digestion Within Simple Heterotrophs

Digestion is intracellular in protozoa. Because these organisms live in water, their food supply often surrounds them. They have many adaptations that aid their chances of obtaining food. Amoebae trap and ingest food by phagocytosis (Section 4:9), and the food is enclosed within a **food vacuole** (Figure 18–2). The food, which may be other small protozoa or algae, is digested within the vacuole. Enzymes for digestion are within lysosomes (Section 4:12) that fuse with the food vacuole. The food that is digested into small molecules can diffuse across the vacuole membrane and into the amoeba's cytoplasm. Undigested food remains in the vacuole. This food leaves the cell by exocytosis (Section 4:9). In general, elimination of undigested food is called **egestion.**

In *Paramecium* (Figure 18–2), currents caused by the beating of cilia draw food into the **oral groove.** The food is swept down the groove where it collects. A food vacuole forms around it. As in *Amoeba,* food vacuoles fuse with lysosomes and digestion occurs. As a vacuole moves slowly through the cytoplasm of the *Paramecium,* small molecules diffuse out across the vacuole membrane. Movement of the vacuole allows distribution of digested food to all parts of this cell. Undigested food remains in the food vacuole until the vacuole fuses with the **anal pore** where egestion occurs.

FIGURE 18–1. *Didinium,* a ciliate (a) captures and (b) ingests whole another ciliate, *Paramecium.* The organisms, shown magnified about 500 times, are both simple heterotrophs.

Many protozoa digest food in special food vacuoles.

Lysosomes fuse with food vacuoles and provide the enzymes necessary for digestion.

Egestion is the elimination of undigested food.

Food is drawn into the oral groove of a *Paramecium* by the action of cilia.

FIGURE 18–2. (a) *Amoeba* traps and ingests food by phagocytosis. (b) *Paramecium* draws food into the oral groove by beating cilia.

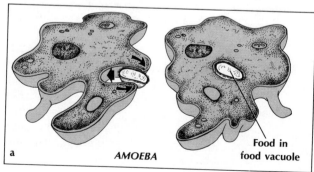

a   AMOEBA

Food in food vacuole

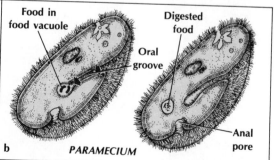

Food in food vacuole

Digested food

Oral groove

b   PARAMECIUM

Anal pore

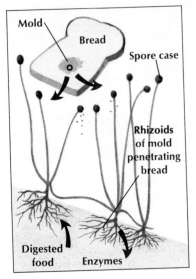

FIGURE 18–3. Enzymes secreted from rhizoids of bread mold digest starch in the bread. Glucose molecules then are absorbed.

FIGURE 18–4. Some fungi have adaptations allowing them to capture nematodes. This fungus, *Arthrobotrys oligospora*, forms hyphal rings in which worms, in this case *Ditylenchus dipsaci*, become trapped. The fungus then digests the worm and absorbs the food.

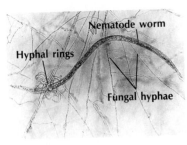

18:4 Digestion by Fungi

Fungi must obtain food from their hosts or substances upon which they live. Unlike protozoa (or animals), fungi do not ingest chunks of food and then digest it. Instead, they first digest food outside their cells, a process called **extracellular** (ek struh SEL yuh lur) **digestion.** Then the food is absorbed by the fungus.

As an example of extracellular digestion in fungi, consider *Rhizopus* (Figure 18–3), common bread mold. *Rhizopus* usually obtains nourishment from bread. Because carbohydrates in bread are in starch form, they must be digested. Enzymes that change the starch to glucose are secreted from the rhizoids of the mold into the bread. Rhizoids absorb the glucose, and it is then moved to the rest of the plant.

Other saprophytic fungi, such as mushrooms or cup fungi (Section 13:15), digest and absorb food as bread mold does. Recall that these fungi have a large underground mycelium. These hyphae secrete enzymes into the soil or decaying material. Digestion of organic materials is extracellular. Small molecules are absorbed by the hyphae and then pass to other cells by diffusion. Like saprophytic fungi, many bacteria also undergo extracellular digestion. Together, these bacteria and fungi play a key role in recycling materials.

Many fungi are parasitic, obtaining nutrients from plant or animal hosts. In so doing, they may cause diseases and economic loss. For example, fungi cause damage to crops such as wheat.

Some small fungi live within individual cells of their host and may absorb food directly without digestion. Other forms send hyphae into or between cells of the host. The hyphae secrete enzymes, digest food, and then absorb the digested food, which passes to other cells by diffusion.

Some fungi have adaptations for trapping small animals, such as nematode worms (Section 15:6). The hyphae of some of these fungi are sticky. Nematodes stick to the hyphae and become trapped. Other "predator" fungi have hyphae that form a ring of cells. If a nematode enters the ring, the cells swell and trap the worm. In either case, other hyphae penetrate the worms. Digestion and absorption of digested food then occur as in other fungi.

REVIEWING YOUR IDEAS

1. What are nutrients?
2. What is digestion? Distinguish beween intracellular and extracellular digestion.
3. Which type of digestion occurs in autotrophs? In protozoa? Fungi? Saprophytic bacteria?

OTHER LIFE PROCESSES

In addition to nutrients, organisms must obtain other materials from their environment and distribute them within cells or from cell to cell. Waste materials are deposited in the environment. Also, many organisms must have a way of moving in their environment. Finally, they must be able to respond to changes that take place in their environment.

Besides obtaining nutrients, organisms interact in a variety of other ways with their environments.

18:5 Transport

Organisms transport or move materials within a cell or between cells. The transported substances include gases, such as oxygen and carbon dioxide. They also include food molecules and certain wastes. Because monerans, protists, and fungi are unicellular, or composed of relatively few cells, substances within them do not have to be transported very far.

In unicellular forms, such as bacteria, simple algae, and protozoa, materials are often distributed by diffusion. Also, the cytoplasm of most cells flows in a pattern called **cytoplasmic streaming.** Streaming aids in distributing larger substances such as food. It moves materials more quickly than diffusion.

Diffusion and cytoplasmic streaming help to move materials within cells.

Although some simple organisms have more than one cell, the organisms still are usually small. Also, in many forms, such as filamentous cyanobacteria, each cell is independent. That is, it acts almost like a separate "organism." Therefore, transport in these forms occurs as it does in unicellular organisms. Fungi are larger organisms, but they have no special tissues for transport. They, too, distribute materials by diffusion and streaming.

No specialized transport structures exist in simple organisms.

18:6 Gas Exchange

Oxygen and carbon dioxide are gases important to almost all living systems. In autotrophs, carbon dioxide is a raw material of photosynthesis, and oxygen is given off as a by-product. Most heterotrophs and autotrophs require oxygen to "burn" food for energy by means of cellular respiration (Section 5:8). All organisms obtain needed gases from and release others to the environment. In general, this movement of oxygen and carbon dioxide between organisms and their environment is called **gas exchange.**

Gas exchange is the movement of oxygen and carbon dioxide between organisms and their environments.

In order for gases to pass into or out of cells, they must be dissolved in water or another liquid. Thus, moisture is needed for gas exchange to occur. Most simple organisms live in water or a moist environment in a host. Fungi live on land, but primarily in areas that remain damp.

Gas exchange requires moisture.

FIGURE 18–5. Simple organisms that exchange gases solely by diffusion live in wet environments. The gas exchange in organisms such as these fungi depends on moisture.

Cells of simple organisms exchange gases directly with the environment.

In unicellular organisms, gases are exchanged by diffusion across the cell membrane. Diffusion accounts for gas exchange in simple multicellular organisms, too. In filamentous cyanobacteria, each cell is in direct contact with the environment. Thus, each cell can exchange gases directly. Sometimes an organism is several cells thick, and diffusion can still serve as the means of gas exchange. Each cell is close enough to the external environment for diffusion to take place.

18:7 Excretion

Because metabolism occurs constantly within all living systems, cells produce many needed products and by-products. Often, the by-products are harmful and must be removed from the organism. The release and removal of harmful by-products, or waste materials, is called **excretion** (ek SKREE shun).

Excretion is the removal of metabolic by-products.

Many chemicals are included as waste material, but what may be a waste material in one organism may not be a waste in another. Carbon dioxide is such a substance. In protozoa, carbon dioxide is a dangerous waste material, but in algae, it is necessary for the photosynthetic reaction.

Ammonia is a nitrogenous waste produced by protozoa.

Protozoa produce ammonia (NH_3) as a waste product. It is formed during the breakdown of proteins. Because it contains nitrogen, ammonia is called a **nitrogenous** (ni TRAHJ uh nus) **waste.** Ammonia is poisonous and must be removed. Protozoa are in direct contact with water. Thus, ammonia can diffuse directly into the water. It is removed without a special system. Excretion of nitrogenous wastes occurs only in heterotrophs.

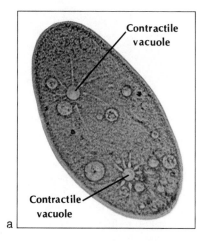

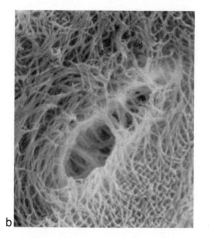

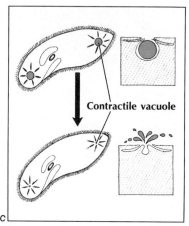

a

b

c

FIGURE 18–6. (a) Protozoa such as *Paramecium* have contractile vacuoles that remove excess water. (b) The opening of a contractile vacuole of a ciliate at the cell surface is shown magnified about 3000 times. (c) When contractile vacuoles fill, they contract, forcing water out the opening.

Although water is not a waste product, its level must be carefully controlled by all organisms. In some cases, water must be stored, while in others, it must be removed.

Simple marine organisms, such as diatoms and dinoflagellates, have little trouble keeping a proper water balance with their environment. The salt water in which they live is like the contents of their cells. Therefore, water enters and leaves the organism in equal amounts.

In simple freshwater organisms, the fluid of the cells and the surrounding water are not balanced because there are more dissolved substances inside the cells. As a result, water molecules enter the cells by osmosis. Constant osmosis of water into these organisms could cause the cells to swell and burst (Section 4:6).

In simple freshwater organisms, water enters the cells by osmosis.

Paramecia and many other protozoans have contractile vacuoles (Section 4:12) that remove excess water. Water entering a paramecium cell fills canals that surround these vacuoles. The water slowly passes from the canals into the vacuole. When a vacuole is full, its membrane contracts forcing the water from the vacuole and through the plasma membrane. This process requires energy.

Many protozoans have contractile vacuoles that are used to remove excess water.

18:8 Locomotion: Amoeboid Motion

Many protozoans actively obtain food. To trap food, they must be able to move. Movement is also involved as protozoans respond to their environment.

An amoeba moves by forming extensions of cytoplasm called pseudopodia (Section 13:9). This motion, amoeboid motion, involves cytoplasmic streaming. If you examined a living amoeba under a microscope, you would see that some of its cytoplasm flows. As it flows, it causes the membrane ahead of it to bulge outward.

Amoeboid motion involves cytoplasmic streaming.

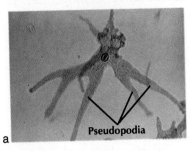

a

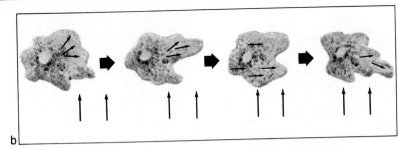

b

FIGURE 18–7. (a) *Amoeba proteus* moves by (b) formation of pseudopodia. Sol-gel transformations occur as the pseudopodia form.

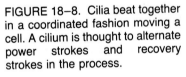

Pseudopodia are formed by sol-gel changes in the cytoplasm.

Not all the cytoplasm in an amoeba flows freely. At certain times, part of the cytoplasm is in a more fixed, jellylike state known as a **gel.** At other times, the cytoplasm is more like a fluid and is known as a **sol** (SAHL). Amoeboid motion depends upon changes between the sol and gel states of cytoplasm. Cytoplasm in the sol state flows toward a part of the membrane, which then begins to bulge out and form the pseudopodium. The cytoplasm then moves to the other side of the amoeba where it changes to a gel. The gel again changes to sol that flows toward the pseudopodium. The flow of cytoplasm is controlled by contraction of microfilaments (Section 4:12). As they contract, the microfilaments are thought to push the cytoplasm out and toward the pseudopodium.

18:9 Locomotion: Ciliary and Flagellar Motion

Many protists have either cilia or flagella (Section 4:13) for locomotion. *Paramecium* (Section 13:10) has hundreds of tiny cilia, while *Euglena* (Section 13:7) moves by means of a single flagellum. Locomotion by cilia is called **ciliary** (SIHL ee er ee) **motion,** and by a flagellum it is called **flagellar motion.** ATP is a necessary energy source for this work.

Exactly how cilia and flagella operate is not known. However, the hundreds of oarlike cilia on a cell work together. Like coordinated movements of the oars of a rowboat, the beating of cilia causes motion in a certain direction.

FIGURE 18–8. Cilia beat together in a coordinated fashion moving a cell. A cilium is thought to alternate power strokes and recovery strokes in the process.

Cilia and flagella have microtubules (Section 4:12), which have the ability to contract. According to one hypothesis of ciliary and flagellar motion, contraction of certain microtubules results in their sliding over other microtubules. This action causes the cilium or flagellum to bend forward in a power stroke. A recovery stroke follows the power stroke. While some microtubules contract, others relax. Alternation of power strokes and recovery strokes results in locomotion (Figure 18–8).

Many bacteria also move by means of flagella. However, the flagella do not contain microtubules. Movement is not so smooth as in protozoa. The means of flagellar movement in bacteria is not completely understood.

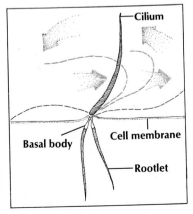

18:10 Response to Environment

All cells can respond to stimuli. For example, many bacteria can sense the presence of chemicals in their environment. They move toward useful substances and away from harmful ones. The movement of an organism toward or away from a stimulus is called a **taxis.** It is thought that the bacteria detect stimuli by means of receptor proteins in their cell membranes and walls. Stimulation of receptors leads to production of chemicals within the cell. These chemicals are then "analyzed" and an electrical signal is sent to the flagella. The flagella respond by beating in a direction that propels the cell toward or away from the stimulus.

Some simple eukaryotes have specialized parts for responding to the environment. *Euglena* has a stigma (Section 13:7) that detects light. Some ciliates have **neurofibrils** (noor oh FIBE rulz) that transmit impulses. In *Euplotes* (yoo PLOHD eez), the neurofibrils are connected to a control center (Figure 18–9). The control center and neurofibrils coordinate the movement of large bundles of cilia called **cirri** (SIHR i).

Response to the environment may be related to protection. *Paramecium* has special structures called **trichocysts** just inside the cell membrane. Certain chemical changes in the environment cause the trichocysts to discharge long threads of cytoplasm. *Paramecium* then has a "fuzzy" appearance. It is thought that these threads work something like spears and protect the organism.

Paramecium can also respond to objects in its way. If *Paramecium* comes into contact with an object, it will reverse the beating of its cilia and back away from the object. Then it turns at an angle of about 30° and moves forward again. *Paramecium* continues to respond in this way until it passes the object (Figure 18–10). *Paramecium,* however, does not "choose" to turn 30°. A "goal" is not involved in this behavior. Rather, the protist merely "carries out" its genetic instructions for dealing with objects in its path. Responses in *Paramecium* seem mainly to be under electrical control.

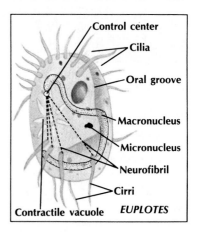

FIGURE 18–9. In the ciliate *Euplotes* neurofibrils transmit impulses from the control center to the cirri.

Paramecium can respond to objects in its path by reversing its motion and turning.

FIGURE 18–10. *Paramecium* exhibits specific turning behavior when it runs into objects in its path.

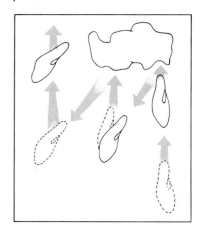

REVIEWING YOUR IDEAS

4. What are some materials that must be transported by organisms?
5. What is gas exchange?
6. What is excretion? What chemical substance is excreted by protozoa?
7. List two types of locomotion in protozoa.
8. What is a taxis? Give an example.

Problem: How can you measure respiration rate in yeast?

Materials:

4 stoppers with glass tube
4 test tubes   metric ruler
graduated cylinder   4 flasks
water   yeast cake
5% sucrose solution   clock
25% sucrose solution   glass-marking pencil
single-edged razor blade

Procedure:

1. Make a copy of Table 18–2 in which to record your data.
2. Obtain four stoppers with glass tubing inserted in each. Do not attempt to reposition the glass tubes.
3. Using the glass-marking pencil, label the tubes 1–4.
4. Using the razor blade, cut the yeast cake into three 0.5 cm cubes.
5. Fill the tubes as follows:
   Tube 1 — 10 mL water and one yeast cube
   Tube 2 — 10 mL 25% sucrose solution
   Tube 3 — 10 mL 25% sucrose and one yeast cube
   Tube 4 — 10 mL 5% sucrose and one yeast cube.
6. Mix the contents of each tube by gently shaking until the yeast cube has dissolved.
7. Insert a stopper with glass tubing into each test tube. NOTE: If the end of the glass tube is not below the surface of the liquid, notify your teacher. Use Figure 18–11 as a guide.
8. Place each tube into a flask. Use Figure 18–12 as a guide.
9. Place a ruler next to each flask. Position the ruler so that the 0 mm mark is at the bottom of the flask. Measure the height of the liquid in

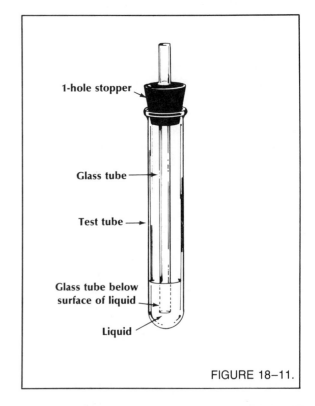

1-hole stopper

Glass tube

Test tube

Glass tube below surface of liquid

Liquid

FIGURE 18–11.

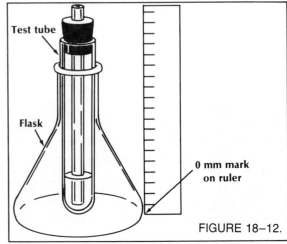

Test tube

Flask

0 mm mark on ruler

FIGURE 18–12.

each tube in millimeters. Record these numbers in Table 18–2 under the heading "starting height." Use Figure 18–12 as a guide for measuring.

10. Wait five minutes. After waiting, measure and record the height of the liquid in each tube. Record these numbers in the column marked "height after 5 minutes" in Table 18–2.

11. Repeat step 10 three more times. Record your numbers in the appropriate columns in Table 18–2.

12. Calculate and record the total distance that the liquid moved in each glass tube. If the height of the liquid on the last reading is lower than the starting height, record it as a negative number.

Data and Observations:

TABLE 18-2. READINGS OF HEIGHT OF LIQUID IN TUBES						
	Starting Height	Height After 5 Minutes	Height After 10 Minutes	Height After 15 Minutes	Height After 20 Minutes	Total Distance Liquid Moved
Tube 1						
Tube 2						
Tube 3						
Tube 4						

Questions and Conclusion:

1. Define cellular respiration.
2. (a) What gas is released by yeast cells as they carry out cellular respiration?
   (b) What is the role of sucrose in the experiment?
3. As the yeast cells released gas, pressure within the test tube increased. This pressure forced the liquid within the test tube to rise into the glass tube.
   (a) Which tubes showed evidence of gas release?
   (b) How were the contents of these tubes alike?
4. What experimental evidence do you have that:
   (a) yeast without food does not carry out cellular respiration?
   (b) food itself is not responsible for gas release?

5. Yeast is used in several baking processes. Dough rises due to the trapping of gas within the dough.
   (a) What gas is trapped within the dough?
   (b) Where does this gas come from?
6. Unleavened bread is made without yeast and is usually very flat. Explain why.
7. (a) What chemical produced by yeast is used in the production of beer and wine?
   (b) Why do yeast cells produce this chemical?
8. Design an experiment to test the following hypotheses:
   (a) Temperature influences the respiration rate of yeast cells.
   (b) The pH of a liquid influences the respiration rate of yeast cells.

Conclusion: How can you measure respiration rate in yeast?

CHAPTER REVIEW

SUMMARY

1. Each cell gets nutrients from its environment. Autotrophs require simple substances that are then built into complex organic compounds. Heterotrophs take in some simple materials and large, complex organic molecules. **18:1**

2. Large molecules must be digested, or broken down, so that they can pass through membranes and into cells. **18:1**

3. Autotrophs carry on intracellular digestion, digestion within the cells. **18:2**

4. Many protozoans carry out intracellular digestion in food vacuoles. **18:3**

5. Fungi carry on extracellular digestion, digestion outside the cells. They absorb small molecules from their host or environment. **18:4**

6. Organisms must transport materials within and between cells. In simple organisms, such as *Amoeba* and *Paramecium,* transport is aided by the processes of diffusion and cytoplasmic streaming. **18:5**

7. Gas exchange occurs between organisms and their environment. For gas exchange to occur, moisture is needed. In simple organisms, most cells are in direct or close contact with moisture. Simple organisms exchange gases directly with the environment by means of diffusion. **18:6**

8. Excretion is a process that removes nitrogenous wastes. The amount of water in living organisms also must be controlled. Excretion in protozoa occurs by diffusion. Many protozoa have contractile vacuoles that maintain water balance by pumping out excess water. **18:7**

9. Amoeboid motion is a form of locomotion that depends upon changes between sol and gel states of the cytoplasm. **18:8**

10. Many protists move by means of ciliary or flagellar motion. **18:9**

11. Simple organisms respond to their environment in a variety of ways. Among the things to which they react are light, touch, food, and chemicals. **18:10**

LANGUAGE OF BIOLOGY

anal pore	gas exchange
ciliary motion	gel
cytoplasmic streaming	intracellular digestion
digestion	nitrogenous waste
egestion	nutrient
excretion	oral groove
extracellular digestion	sol
flagellar motion	taxis
food vacuole	trichocyst

CHECKING YOUR IDEAS

On a separate paper, match the phrase from the left column with the proper term from the right column. Do not write in this book.

1. process that transports substances more quickly than diffusion does
2. elimination of dangerous products
3. involves pseudopodia
4. movement of oxygen and carbon dioxide
5. movement toward or away from a stimulus
6. digestion in protozoa
7. hydrolysis reaction
8. involves microtubules
9. fuses with lysosomes during digestion
10. digestion in fungi

a. amoeboid motion
b. ciliary motion
c. cytoplasmic streaming
d. digestion
e. excretion
f. extracellular digestion
g. food vacuole
h. gas exchange
i. intracellular digestion
j. taxis

EVALUATING YOUR IDEAS

1. What kinds of nutrients do autotrophs require? Heterotrophs? Explain the differences.
2. Why is digestion a necessary life process for organisms?
3. Why is digestion in autotrophs different from that in heterotrophs? Do autotrophs digest molecules? Explain.
4. By what chemical process are molecules digested?
5. Describe ingestion, digestion, and egestion in *Amoeba* and *Paramecium*. How are they similar? How are they different?
6. Discuss several ways in which fungi obtain and digest food.
7. Explain transport in simple organisms.
8. Why must gas exchange occur in a moist environment? How do simple organisms exchange gases?
9. How does excretion occur in protozoa?
10. Why do many saltwater protists have little trouble maintaining proper water balance?
11. By what means does *Paramecium* rid itself of excess water?
12. Explain how amoeboid motion occurs.
13. How are ciliary and flagellar motion thought to occur?
14. Explain how a bacterium is thought to detect and respond to stimuli.
15. How does *Paramecium* protect itself? How does it react to objects in its path?

APPLYING YOUR IDEAS

1. How is extracellular digestion in fungi an adaptive trait?
2. How is digestion in fungi important to community life?
3. What organelle functions in intracellular transport?

4. Explain several ways in which protozoa are more complex than other simple organisms. Can you suggest a reason for the complexity of protozoa?
5. Certain bacteria are able to synthesize a great variety of substances. How is this capability of producing substances adaptive?
6. Explain some different responses to the environment made by *Euglena* and *Paramecium*. How is each type of response an adaptation to the organism's way of life?

EXTENDING YOUR IDEAS

1. Observe a drop of *Paramecium* culture under high power on a microscope. Add a drop of iodine or vinegar to the slide. Observe the discharge of trichocysts.
2. Place another drop from a *Paramecium* culture on a clean slide. Add a few threads of cotton and a coverslip. Describe the response of *Paramecium* when it bumps into a thread.
3. Remove a drop from an *Amoeba* culture, make a wet mount, and observe it under high power on a microscope. Describe the movement of *Amoeba* and the streaming motion of its cytoplasm.

SUGGESTED READINGS

Baker, John R., *The Biology of Parasitic Protozoa*. Baltimore, MD, Arnold, Edward, Publishers Ltd., 1982.

Burchard, R.P., "Gliding Motility of Bacteria." *Bioscience*, March, 1980.

Dixon, Bernard, *Magnificent Microbes*. New York, Athenum, 1979.

Jahn, Theodore L. and Jahn, Francis F., *How to Know the Protozoa*, 2nd ed. Dubuque, IA, William C. Brown Co., 1979.

Pietsch, Paul, "The Mind of a Microbe." *Science Digest*, Oct., 1983.

SIMPLE ORGANISMS AND DISEASE

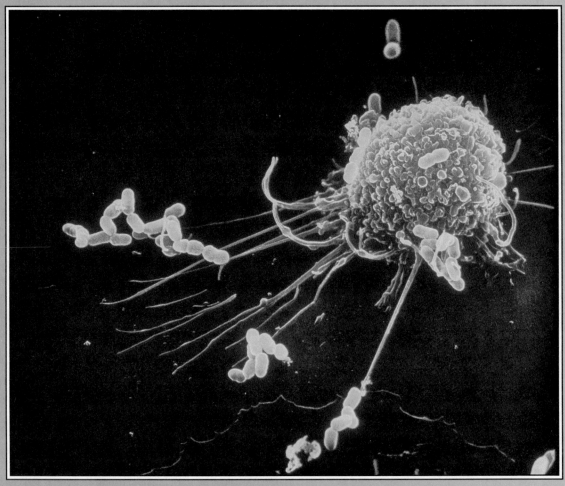

Although many simple organisms are beneficial to humans and the environment, others are harmful and cause disease. The human body has barriers designed to prevent entry of disease-causing organisms. However, should the organism enter, the body has other mechanisms to defend itself. Here, a macrophage traps bacteria that have entered the body. What are some of the other mechanisms that the body has for defense against disease?

chapter 19

Many simple organisms play important, beneficial roles in nature and in human activities. Algae are producers in their aquatic communities. Certain fungi and bacteria decompose organic material and help in the recycling of materials. A variety of forms are used in producing foods and other substances for human use.

Although many simple organisms are helpful or useful, others are quite harmful. They are **pathogens** (PATH uh junz), or disease-causing organisms. Pathogens cause disease in plants, animals (including humans), and in other organisms. Their activities may result in illness, death, and destruction of crops. Many diseases caused by pathogens are called infectious diseases. **Infectious diseases** can be spread from one organism to another.

Objectives:
You will
- identify examples of bacterial and viral diseases and explain how they are spread.
- name diseases caused by fungi and protozoa.
- discuss how the body defends itself against diseases.

Pathogens are agents of infectious diseases.

BACTERIAL AND VIRAL DISEASES

Many of the infectious diseases with which you are familiar are caused by bacteria. However, the role of bacteria as agents of diseases was not well understood until the end of the 19th century. At that time, Louis Pasteur proposed the **germ theory of disease.** That theory states that bacteria can cause diseases.

Bacteria cause many infectious diseases.

19:1 Koch's Postulates

At about the time that Pasteur proposed the germ theory, a German physician, Robert Koch, identified two pathogenic bacillus bacteria. One bacillus causes the disease anthrax (AN thraks), which affects sheep, cattle, horses, and humans. The other causes tuberculosis

Koch's postulates provide a method for identifying pathogens.

in humans. The methods by which Koch determined that a bacterium causes a specific disease are known as **Koch's postulates:**

(1) The organism suspected of causing the disease must be present in the diseased host and isolated from it.

(2) The organism must then be grown in a pure culture.

(3) Organisms taken from that pure culture and injected into a healthy host must cause the disease in the host.

(4) The organism must be isolated from the new host, grown in pure culture, and compared to the original culture.

Since Koch's time, these rules have been and still are used by many others. Use of the rules has led to the discovery of many bacterial pathogens. Use of Koch's postulates also has led to the discovery of nonbacterial pathogens. The procedure of isolation and reinfection has become standard in determining the cause of an infectious disease regardless of the kind of organism involved.

19:2 Bacterial Diseases and How They Are Spread

Many bacteria have been found to cause infectious diseases in humans and other animals. Among the diseases caused are bacterial pneumonia, bubonic plague, pertussis (whooping cough), strep throat, and meningitis. Several sexually transmitted diseases, such as syphilis and gonorrhea, are also caused by bacteria.

Bacteria may cause disease in plants as well as in humans and other animals.

Bacteria cause some plant diseases, such as pear blight and fire blight. Bacteria may cause rotting or wilting in a variety of plants. Some cause galls, or swellings, in plant tissues.

Infectious organisms may be airborne, waterborne, or spread by direct contact.

Infectious diseases may spread from one organism to another in several ways. Some diseases are airborne. Bacteria are carried through the air on small droplets produced when an infected organism

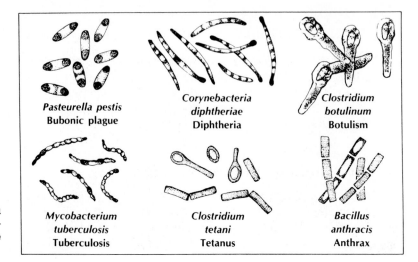

FIGURE 19-1. Many bacteria cause diseases. Six such organisms are shown with the disease each causes listed below it.

Pasteurella pestis
Bubonic plague

Corynebacteria diphtheriae
Diphtheria

Clostridium botulinum
Botulism

Mycobacterium tuberculosis
Tuberculosis

Clostridium tetani
Tetanus

Bacillus anthracis
Anthrax

a   b

FIGURE 19–2. Some types of bacteria cause plant diseases. (a) The large tumor, called a Crown gall tumor, on this *Euonymous* plant, is the result of the presence of the bacterium *Agrobacterium tumefaciens.* (b) Wilts are common diseases of plants caused by bacterial infections. The cucumber bacterial wilt shown is caused by *Erwinia tracheiphila.*

sneezes or coughs. Other diseases are waterborne. They are spread by drinking water that is contaminated by sewage containing human wastes. Still other diseases are spread by direct contact with an object or living thing in or on which bacteria are located. Sexually transmitted diseases are spread by sexual intercourse. Some bacteria are transmitted by arthropods. For example, bubonic plague is transmitted by fleas. Epidemic typhus fever is transmitted by lice, and Rocky Mountain spotted fever is transmitted by ticks. Often, the bacteria enter the body when the arthropod bites or pierces the skin of an animal.

19:3 How Bacterial Diseases Occur

When pathogens enter the body, they may cause diseases in several ways. Many bacteria produce poisonous chemicals called **toxins** (TAHK sunz). **Exotoxins** are released by living bacteria. They travel through the bloodstream and affect particular tissues. For example,

FIGURE 19–3. Transmission of disease organisms occurs in several ways. (a) *Pneumococcus* bacteria responsible for pneumonia are airborne organisms. (b) *Neisseria gonorrhea,* bacteria that cause the disease gonorrhea, are transmitted by sexual contact. (c) Human plague organisms, *Yersinia pestis,* are transmitted by arthropods.

a

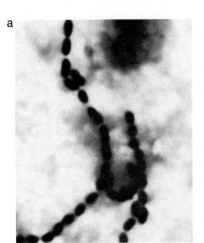

b

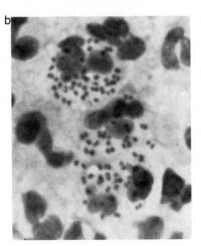

c

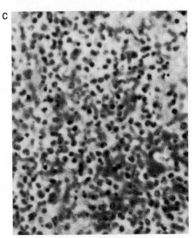

one species of genus *Clostridium* (klah STRIHD ee um) causes tetanus (lockjaw). It secretes a toxin that stimulates certain nerves, causing some muscles to remain contracted. As a result, the condition can be very painful and death may result. The bacterium that causes diphtheria secretes a toxin that may damage the heart, kidneys, and nerves. Although it is not known how toxins produce harmful effects, most of these diseases can now be either prevented or cured.

Endotoxins are released when bacteria die.

Endotoxins remain in the bacterial cytoplasm. When bacteria die and break open, the endotoxins are released. Endotoxins are the causes of a variety of diseases such as cholera, tuberculosis, and bubonic plague.

As bacteria reproduce, they may affect the host's metabolism.

Bacteria may cause diseases in other ways. As they invade tissues, they may reproduce in such great numbers that they may alter the host's metabolism. Sometimes the reaction of a host, as it responds to the foreign invader, causes changes in the host's cells. The cells may become less active as chemical changes occur within them. Chemical changes can interfere with respiration or upset the osmotic balance between cells and their fluid environment. These abnormalities may lead to the death of the cells.

19:4 Viruses and Disease

Viruses were not isolated until 1935. Since that time, a large number of viruses have been discovered. Although some are harmless to humans, many are pathogenic. Viral diseases are transmitted in the same ways as bacterial diseases. Human diseases caused by viruses include polio, smallpox, influenza, measles, mumps, AIDS, rabies, and viral pneumonia. One form of herpes virus causes a sexually transmitted disease. Other herpes viruses cause infectious mononucleosis and cold sores. Animal diseases caused by viruses include distemper in dogs and cats, foot-and-mouth disease in cattle, and swine influenza in pigs.

Usually there is only one kind of host for a given viral disease. Some viruses that affect humans, however, also occur in other animals. Encephalitis, an inflammation of the brain, is a disease of humans and horses. Viruses sometimes spread from one kind of host to another. Rabies can be spread among animals such as raccoons, dogs, and humans.

FIGURE 19–4. Rabies vaccination of pets is required by law in many areas. The virus cannot be transferred to vaccinated animals.

The symptoms of viral diseases vary with the virus and the tissues "under attack." Viruses reproduce in specific tissues or organs. For example, polio virus attacks nerves in the brain and spinal cord. As polio viruses reproduce, they destroy nerve cells. Destruction of nerve cells may lead to paralysis and, in some cases, even death. The effects of other viruses are also related to the part of the host that is invaded by the virus.

Recently, unique structures called viroids have been discovered. **Viroids** are small segments of RNA now known to cause diseases in a variety of plants, including potatoes and tomatoes. They may cause diseases by regulating the host's genes in an abnormal manner. The origin, transmission, and effect of viroids must be studied further. Their name suggests that they are related to viruses, but that idea is not an established fact.

Viroids are known to cause some plant diseases.

The effect of a virus depends on the kind of tissue it invades.

REVIEWING YOUR IDEAS

1. What is a pathogen?
2. What is an infectious disease?
3. List some diseases caused by bacteria.
4. What is a toxin? What are the two types of toxins?
5. List some diseases caused by viruses.

OTHER AGENTS OF DISEASE

Fungi and protozoa, which are heterotrophs, may also be parasites. As some forms of these organisms obtain food, they cause diseases in their hosts. Thus, they are pathogens.

19:5 Fungal Diseases

Some fungi cause diseases in animals and humans. Often, these fungi invade the skin. Ringworm (a disease of the scalp, skin, or nails) and athlete's foot are human fungal infections. Both of these diseases involve the cracking and scaling of the skin and the formation of small blisters. These symptoms occur as the hyphae (p. 273) invade the skin.

Sometimes fungi attack internal organs, such as the lungs, where they may cause serious diseases. Histoplasmosis is a fungal disease which affects the mouth, throat, and lungs of humans. Found mostly in the United States, it results from spores that enter the body from the air. These spores are often found in birds' nests.

Some fungi cause diseases in animals as they invade the skin.

Fungi cause several respiratory diseases in humans.

a 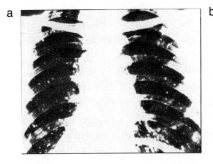   b

FIGURE 19–5. Histoplasmosis is a disease caused by the fungus *Histoplasma capsulatum.* Airborne spores cause infections primarily in the lungs. (a) White spots on lung X rays are used to diagnose the disease. These spots are sores hardened with calcium deposits. (b) *Histoplasma* spores often are found in abundance in nests of certain birds and in the ground beneath the nests.

FIGURE 19–6. Rusts and smuts are plant diseases caused by parasitic fungi. (a) Blackberry rust and (b) corn smut cause extensive damage to crops.

a

b

Rusts and smuts are club fungi (Section 13:14) that cause diseases in a variety of plants. Many parasitize grains such as wheat, barley, oats, and corn. They spread rapidly and destroy millions of dollars in crops each year.

Wheat rust has a life cycle involving two hosts.

In order to deal with diseases caused by some fungi, it is necessary to study the life cycles of the fungi. Wheat rust has a complex life cycle involving two hosts — wheat and the barberry bush. The fungus attacks and damages the stem and leaf cells of young wheat plants (Figure 19–7). Spores produced at this time invade other, healthy wheat plants. Different spores are produced later. After a dormant period during the winter, the spores invade the barberry bush. In the bush, more spores are produced. They attack young wheat plants to complete the cycle.

FIGURE 19–7. Life cycle of wheat rust fungus *Puccinia graminis* involves both barberry and wheat hosts.

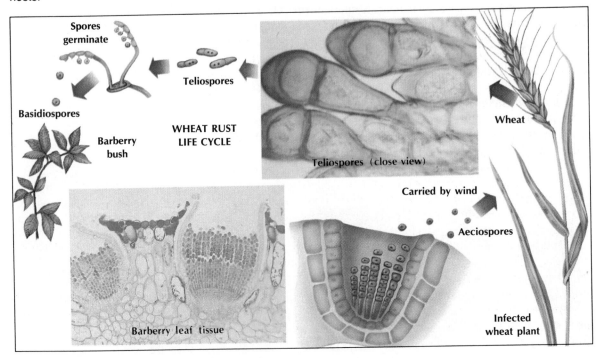

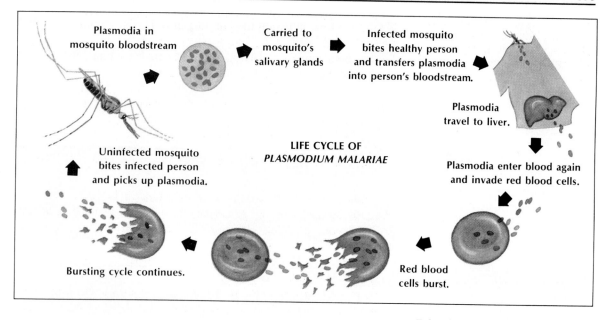

Plasmodia in mosquito bloodstream

Carried to mosquito's salivary glands

Infected mosquito bites healthy person and transfers plasmodia into person's bloodstream.

Plasmodia travel to liver.

LIFE CYCLE OF PLASMODIUM MALARIAE

Uninfected mosquito bites infected person and picks up plasmodia.

Plasmodia enter blood again and invade red blood cells.

Bursting cycle continues.

Red blood cells burst.

FIGURE 19–8. Malaria is caused by the protist *Plasmodium malariae,* and is transmitted by female *Anopheles* mosquitoes.

19:6 Protozoan Diseases

Some human diseases are caused by protozoa. African sleeping sickness (Section 13:11) is caused by a flagellate. Amoebic dysentery is a disease caused by a certain species of amoeba. It is a serious and often long-lasting infection of the large intestine.

One of the most lethal diseases in human history is malaria, caused by *Plasmodium,* a sporozoan. *Plasmodium* is transmitted to humans by female *Anopheles* (uh NAHF uh leez) mosquitoes. If this insect bites a person with malaria, it sucks some of the parasites along with the blood cells. The sporozoans then pass along with the blood into the mosquito's stomach where they grow and reproduce. Some of the parasites pass into the mosquito's bloodstream and eventually reach its salivary glands.

Malaria is caused by the protozoan *Plasmodium.*

When the mosquito bites a healthy person, some of the parasites enter the human's blood through the skin. The parasites travel to the liver where they develop for about eight days. Then they reenter the bloodstream and invade the red blood cells. The red cells provide an oxygen-rich environment and a food source for the parasites. In each blood cell, many parasite spores form. Eventually the red cells burst and release the spores, which then enter other red cells. These cells later burst and a regular cycle of invasion and bursting develops. Cells usually burst every twenty-four to forty-eight hours.

After developing in the liver, malarial parasites travel to the red blood cells where they produce and release spores.

At the height of each bursting cycle, the host experiences chills, headache, high fever, and heavy sweating. These symptoms result from the parasites' waste products being emptied into the blood.

Symptoms of malaria occur when the red blood cells burst.

During the time between the bursting of red cells, the host may feel well, although weak.

In some kinds of malaria, the cycle may continue for several weeks and then seem to disappear. But, the symptoms may suddenly reappear months later. In another form of malaria, death may result within several hours after the onset of symptoms. The type of malaria depends on the species of *Plasmodium* involved.

REVIEWING YOUR IDEAS

6. What human diseases are caused by fungi?
7. How is agriculture affected by fungi?
8. What organism is responsible for malaria? How is it transmitted?

DEFENDING AGAINST DISEASES

Even though pathogenic microbes may be present, they often do not cause diseases. For diseases to occur, pathogens must first gain entry to the cells and tissues of the body. In many cases, even if disease does occur, the host survives. The following sections describe adaptations that either prevent the entry or fight off the effects of pathogens.

Adaptations that destroy or prevent entry of microbes into the body include the skin, filtering of pathogens in air, trapping of pathogens by mucus, and action of acids and enzymes.

19:7 Barriers to Pathogens

A major barrier to the entry of pathogens is the skin. The outer layer of skin is made of dead cells that microbes cannot penetrate. When the skin is broken, pathogens can enter the host.

Openings in the body, such as the nostrils and mouth, are points at which microbes may enter. However, other adaptations usually prevent further entry of the microbes. Hairs in the nose filter air. Mucus in the lining of the nose traps and destroys microbes. The lining of the windpipe and smaller tubes leading to the lungs contains cells with many cilia. The cilia move mucus from the lungs to the mouth. As air moves down toward the lungs, microbes are trapped in the mucus. Action of the cilia and mucus prevents respiratory infection. You can see how damage to these cilia by means of cigarette smoke hurts one of the body's natural defenses.

Acids and enzymes in the body kill organisms that may enter through body openings. The digestive organs have substances that kill microbes. An enzyme in tears kills bacteria that may enter the tear ducts of the eyes. That enzyme breaks down the cell walls of the bacteria.

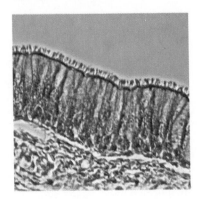

FIGURE 19–9. Cilia lining the windpipe form a barrier against entering pathogens. These cilia, from a cat windpipe, are shown magnified about 330 times.

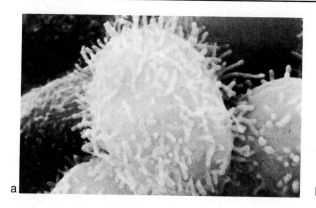
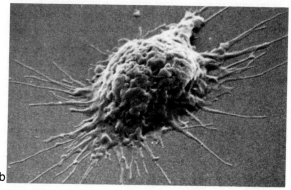

FIGURE 19–10. (a) A lymphocyte (shown magnified about 30 000 times) and (b) a macrophage, are two important cell types involved in the immune response.

19:8 The Immune System: Structures

Attacks by pathogens are often serious and sometimes fatal. Yet, in many cases, the host survives and regains its health. Fighting off pathogens is the function of the **immune system,** a network of cells and tissues located throughout the body. Knowledge of the immune system and its roles explains how the body defends itself against foreign chemicals and particles. It also explains why transplanted organs are often rejected by the patient's body.

The immune system defends against pathogens.

Early in development, an embryo has the ability to detect a difference between its own chemicals and those that are foreign to it. In general, any foreign chemical is called an **antigen.** An antigen may be a protein in the membrane of a bacterium or virus or a chemical secreted by another organism.

An antigen is any foreign chemical or substance.

Before ridding the body of pathogens, the immune system must detect the presence of antigens. These responses involve certain white blood cells and certain tissues. **Lymphocytes** (LIHM fuh sites) are white blood cells produced by bone marrow, the tissue in the hollow parts of some bones. Some lymphocytes pass from the bone marrow and mature in the **thymus gland,** located beneath the breastbone. Such lymphocytes are called **T cells.** Others that bypass the thymus gland are called **B cells. Macrophages** (MAK ruh fayj uz) are another kind of white cell involved in immunity. These various white cells enter the blood, the spleen, and the lymph nodes. The **spleen** is an abdominal organ. **Lymph nodes** are small structures that filter a fluid known as lymph. White blood cells are also present in lymph.

The immune system is composed of lymphocytes, macrophages, thymus, spleen, and lymph nodes.

19:9 The Immune System: Functions

B cells play a role in defending against bacteria, viruses, and toxins that may be present in the blood. The number of possible antigens is very large. How can B cells recognize all of them? B cells have protein molecules called **antibodies** (ANT ih bahd eez) on their sur-

An antibody is a specific protein that matches a certain antigen.

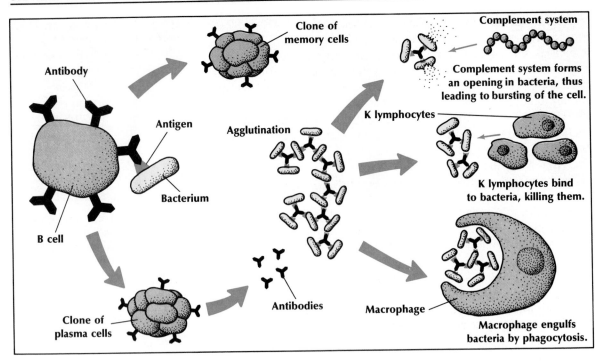

FIGURE 19–11. B cells defend against bacteria, viruses, and toxins in the bloodstream.

FIGURE 19–12. Macrophages attack foreign matter other than bacterial cells. Here a macrophage attacks a tumor cell. Holes have appeared in the tumor cell's membrane due to the macrophage's attack.

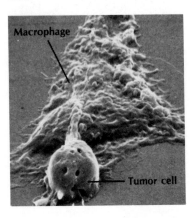

faces. Different B cells have different antibodies, each unique because of the order of amino acids it contains. Thus, millions of different antibodies can be made. Each different antibody has a shape that matches a specific antigen molecule. Antigens and antibodies match in much the same way that enzymes and substrates do (Section 5:4).

Suppose a particular type of bacterium enters the body. The bacteria have antigen molecules on their surfaces. Certain B cells have antibodies that match the bacterial antigens. Each antibody molecule has a Y-shape. The two branches of the Y are the same, and each branch has the ability to join with one bacterial antigen molecule.

Once the antibodies of a B cell are linked to antigens, the B cell becomes activated and begins to enlarge. This large B cell divides many times to form a cluster of cells called **plasma cells,** each of which is genetically identical to the other. A group of genetically identical cells is called a **clone.** The plasma cells produce thousands of antibody molecules. These antibodies, the same as on the original B cell, are released into the bloodstream. There, the free antibodies bind to antigens on the invading bacteria. This binding results in **agglutination,** or clumping together, of the bacteria.

Once clumped together, the bacteria may be destroyed in one of several ways. Macrophages may engulf the bacteria by phagocytosis and digest them. Often, this process occurs in lymph nodes, which become swollen and tender during an infection. Sometimes, special **K**

lymphocytes, or killer cells, bind to the bacteria and kill them. How they do this is not known. In some cases, a set of blood proteins known as the **complement system** may play a role. Through a series of reactions, the complement system forms an opening in the bacterium. Water enters the opening by osmosis and causes the bacterium to burst.

Certain T cells affect the functions of B cells and macrophages. **Helper T cells** have membrane receptors similar to the antibodies on B cells. These receptors recognize and bind to activated B cells. The helper T cells release chemicals that stimulate the B cells to divide and form plasma cells. The chemicals also activate macrophages and other kinds of T cells, the suppressor T cells. **Suppressor T cells** counteract the effect of helper T cells. They release chemicals that inhibit activity of B cells and macrophages. The number of suppressor T cells increases as an infection is brought under control.

Helper T cells also stimulate other T cells, the **cytotoxic T cells.** Unlike B cells, cytotoxic T cells do not attack free antigens in the blood. Rather, they recognize antigens on certain host cells. Such cells include cancer cells, cells infected by viruses, and transplanted cells. The cytotoxic T cells attach to the antigens on these cells and act on the cell membranes. Death of the cells results from lysis.

Destruction of bacteria may result from the action of macrophages, K lymphocytes, or the complement system.

Helper T cells release chemicals that stimulate cloning of activated B cells. Suppressor T cells secrete chemicals that inhibit activity of B cells.

Cytotoxic T cells attack and destroy foreign, infected, or transplanted cells.

19:10 Immunity and Treatment

Not all B cells and T cells produced during an infection are active in fighting off pathogens. Some of the clones formed become **memory cells** and remain in the body long after an infection is over. The presence of memory cells allows the body to react very quickly if the same pathogen invades again later. The early steps of the immune response are bypassed. The antigen is recognized very quickly and the pathogen is destroyed before it can cause disease symptoms. The presence of memory cells is the basis of active immunity (ihm YEW nut ee). **Active immunity** is immunity or disease resistance resulting from production of antibodies by the host. It explains why a person who has had measles, for example, will not get the disease again.

The principles of active immunity may be used to prevent certain diseases from occurring. Solutions of antigens prepared from weakened or dead microorganisms or from toxins are called **vaccines** (vak SEENZ). Vaccines injected into a healthy person do not cause the disease but do cause antibody formation. The person may suffer mild discomfort from the vaccine, but the antibodies give protection from the disease organisms or toxins when the body is later exposed.

Many vaccines have been developed. Among them are vaccines for polio, measles, mumps, some forms of influenza, tetanus, diphtheria, and bubonic plague. When someone receives a vaccination,

FIGURE 19–13. Chickens are vaccinated for Marek's disease before they hatch. Often they are exposed to the disease virus immediately after hatching. Therefore, vaccinations given after hatching are too late to give immunity.

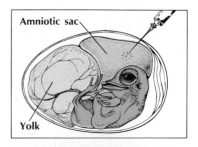

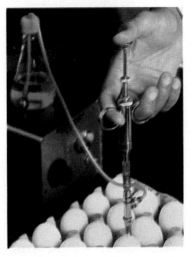

the person is said to become immunized against the disease. Without vaccines, exposure to many of these diseases could be fatal. Children are required to receive certain vaccinations before entering school. Because some diseases, such as polio, are less common now, some people are neglecting to have their children vaccinated. Why is this practice dangerous?

Often when a person becomes infected, the body cannot produce antibodies fast enough to fight the disease. A vaccination of weakened or dead organisms also would not result in production of antibodies rapidly enough to fight the disease. In these cases (such as tetanus), a person may receive antibodies made by an animal. The antibodies are extracted from the animal's blood. The solution of antibodies and blood serum is called an **antiserum.** Injection of an antiserum provides "borrowed" antibodies to fight off a disease. Because the person is not making antibodies, this process results in what is called **passive immunity.** Passive immunity works more quickly than active immunity, but its effects last for a shorter time.

Vaccines and antisera are two ways to prevent a variety of infectious diseases. People can recover from some infectious diseases naturally as a result of their immune responses. But the immune system may not always work quickly enough to fight off a disease. If just the immune system were at work, many diseases would be fatal.

Modern medicines are used to *treat* many diseases. Bacterial diseases are treated with antibiotics (Section 2:5). These drugs work along with the immune system to destroy bacteria and restore health. Each kind of antibiotic has a specific kind of effect. Some work on many kinds of bacteria, others work only on certain types. Because some people are allergic to certain antibiotics, these drugs should never be taken unless they are necessary and prescribed by a physician. Many people have developed allergies because of overuse of antibiotics.

Bacterial diseases can be treated with antibiotics.

Because viruses are not organisms, viral diseases must be treated differently than bacterial diseases. Drugs that disrupt a virus's life cycle without destroying the host cell are now being tested. Several points of attack are possible. Some drugs prevent replication of the viral nucleic acid. Others interfere with transcription or translation. Still others might work by preventing entry of viruses into host cells.

To be effective, antiviral drugs must interfere with the virus's life cycle without harming the host cell.

In 1957, it was discovered that protection from a first attack by viruses is not the result of antibody production. Instead, a certain protein is released by the host cell that has been attacked by a virus. The protein travels to and enters a healthy cell where it prevents viral reproduction. Because the protein interferes with the virus, it is called **interferon.** Interferon may work in several ways. It may activate an enzyme that breaks down viral mRNA. It may also prevent translation of viral mRNA or stop the packaging and release of new viruses.

Interferon protects healthy cells by preventing reproduction of viruses.

Can interferon be used as a drug? Using recombinant DNA techniques, scientists now have large quantities of interferon available for research. They are testing to see whether or not interferon can be used to prevent or treat a variety of viral diseases or certain forms of cancer. Results of early tests, however, have not been promising. A great deal more research is needed to determine the true potential of interferon as a drug.

Tests are underway to determine whether interferon might be used to treat or prevent viral diseases.

19:11 AIDS

Acquired immune deficiency syndrome (AIDS) was first recognized in 1981. AIDS is a fatal disease. It is caused by a retrovirus (p. 368) that attacks the immune system. Because the virus affects the immune system, it is named **human immunodeficiency virus (HIV).** In infected persons, HIV is present in body fluids such as blood and semen, a fluid that contains sperm. Infection occurs when the virus enters a person's bloodstream.

Transmission of the virus occurs in four known ways, all of which involve exchange of body fluids.

HIV may be transmitted by intimate sexual contact, on contaminated needles, by transfusion of infected blood, or from pregnant woman to fetus.

(1) The virus can be transmitted from one person to another during intimate sexual contact via the semen or vaginal secretions.

(2) Another means of transmission is by the sharing of needles during intravenous drug use. A small amount of blood can remain on a needle used by a person infected with the AIDS virus. If this needle is used by a second person, the virus present in the blood can be injected into the second person's blood.

(3) Many cases of infection with the virus have resulted from transfusions of blood or blood products that contained the AIDS virus. Since 1985, this means of transmission has declined as a result of a blood test that screens for the presence of AIDS antibodies. Currently, there is an extremely small chance that the AIDS virus can be spread through transfusions.

(4) The virus can be transmitted from the blood of an infected female to the fetus during pregnancy. The virus can pass to the fetus through the placenta.

FIGURE 19–14. HIV, or the AIDS virus, is a retrovirus that attacks the immune system, causing it to break down. The host then becomes prone to a variety of diseases that normally would be fought off by the immune system.

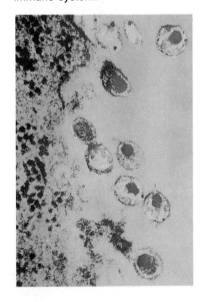

After gaining entry into the blood, the HIV virus attacks and destroys helper T cells. Because helper T cells trigger the production of antibodies by B cells, the immune system begins to break down. In addition, the activities of macrophages, cytotoxic T cells, and suppressor T cells are affected. Antibodies are not produced fast enough

HIV infects helper T cells, thus setting off a chain of events that weakens the immune system.

and the immune system becomes overwhelmed. The infected person is then prone to a variety of opportunistic diseases, or diseases that normally would be fought off by the immune system. The two major opportunistic diseases associated with AIDS are a rare type of pneumonia and a rare type of cancer that appears as spots on the skin or in the mouth.

Persons with AIDS often die from a rare type of pneumonia or a rare form of cancer that appears as spots on the skin or in the mouth.

Early in the course of AIDS, a patient may have symptoms similar to those of a cold. Weight loss and fatigue are common. Later, many persons contract the pneumonia associated with AIDS. Some develop the cancer associated with AIDS. These secondary diseases are the leading causes of death among AIDS patients.

In some people, AIDS may develop several months or years after HIV infection occurs.

When HIV infects helper T cells, reverse transcription takes place and the viral DNA becomes a provirus (Section 17:7). In many persons, the proviral DNA may remain dormant for months or years, during which time the person has no symptoms. Once the virus becomes active, however, it reproduces and kills the host's T cells, causing the symptoms of AIDS to appear. Some scientists believe that anyone infected with the virus will develop AIDS eventually. As of fall 1987, more than 42 000 cases of AIDS had been diagnosed in the United States, and more than half of the diagnosed individuals had died. The number of reported cases of AIDS continues to increase at a rapid pace.

Avoidance of risk behaviors associated with transmission of HIV can prevent the spread of AIDS.

Clearly, AIDS is a major health problem that must be taken seriously. To stop the spread of infection with HIV, risk behaviors associated with transmission of the virus from one person to another must be avoided. Needles used to inject drugs must not be shared. People must exercise good judgment about their sexual activities and choice of partners. A person who is infected with HIV may have no symptoms and can unknowingly infect others. It is estimated that in the United States, more than two million people are infected with HIV but have no symptoms.

Because there are many strains of HIV, no effective vaccine has been produced yet.

Ideally, AIDS could be prevented if a vaccine against HIV could be produced. A major problem in developing such a vaccine is the fact that the structure of HIV varies greatly. There are many strains of HIV, each with slight differences in RNA and proteins. Also, the virus mutates frequently, creating further differences. For those reasons, a single AIDS vaccine has not yet been produced. Scientists are working to find an antigen common to all strains of HIV. Discovery of such a common antigen, if it exists, might lead to development of an effective vaccine.

Drugs that disrupt reproduction of HIV may prove to be effective in treating AIDS.

Another direction of research is development of drugs to relieve symptoms or destroy the virus. AZT, a drug that blocks reverse transcription, was approved recently by the FDA. Although the drug has been successful in slowing progress of AIDS symptoms, its possibly toxic side effects must be studied. Meanwhile, other drugs are being tested.

PEOPLE IN BIOLOGY

Medicine was a family tradition in Jane Wright's family — both her father and grandfather were doctors — so it surprised no one when she decided to go to medical school after graduating from Smith College. She received her M.D. in 1945 from the New York Medical College and began to specialize in cancer research, especially chemotherapy. Chemotherapy is the use of chemicals, often injected into the bloodstream, to treat diseases. She succeeded her father as head of the Harlem Hospital Cancer Research Foundation in 1949. In 1955 she joined the New York University Medical School as an instructor in research surgery, and later became director of cancer chemotherapy there. In 1967 Dr. Wright was appointed dean and professor of surgery at the medical school — the highest post ever held at that time by a black woman in an American medical school.

Dr. Wright is most noted for her work in researching drugs and surgical procedures effective against cancer. In 1975 she was honored by the American Association for Cancer Research for her contributions to research in clinical cancer. The mother of two daughters — a doctor and a clinical psychologist — Jane Wright also enjoys swimming, mystery novels, and painting watercolors.

Jane C. Wright

(1919–   )

19:12 Cancer

Cancer is a disease that affects various organs of the body. Although each form of cancer is different, the disease always involves uncontrolled growth and reproduction of cells. These abnormal cells differ in shape from normal cells and have unique antigens on their surfaces. They may also spread to other parts of the body.

Cancer cells differ in several ways from normal cells.

A large mass of cancer cells is a tumor. Growth of such a tumor leads to the destruction of healthy cells. Each tumor develops from one original cell that becomes cancerous. For many years, scientists have been working to find the cause(s) of this disease. Recent evidence suggests that many cancers result from a gene mutation. Such a mutation may be triggered by radiation or a carcinogen (kar SIHN uh jun). A **carcinogen** is a cancer-causing agent in the environment. Apparently, the gene that mutates is one that codes for a protein involved in controlling normal cell growth and reproduction. When the

Carcinogens are cancer-causing agents.

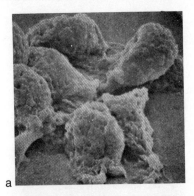

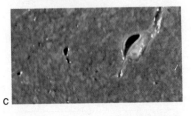

FIGURE 19–15. (a) Cancer cells have surface properties different from other cells. (b) Cancerous liver tissue has areas of uncontrolled cell production (yellow). (c) Healthy liver tissue is pink and uniform.

gene mutates, it becomes a cancer-causing gene or **oncogene** (AHN koh jeen). Because mutations are passed on, descendant cells inherit the oncogene and a tumor forms. Some scientists think that several oncogenes are necessary to change a normal cell to a cancerous one. Other factors also may be involved in this process.

Some retroviruses are known to cause cancer in animals (p. 368). The virus picks up a normal gene from an animal and that gene is changed to an oncogene while in the virus. When the virus infects another animal, it causes cancer in that host. Transmission of a gene from one animal to another by means of a virus is an example of transduction (Section 17:8). Two rare forms of human leukemia, cancer of the blood-forming tissues, are caused by retroviruses similar to HIV. These viruses do not produce oncogenes, but they do cause cancer by a different means.

Discovery of oncogenes is an exciting and promising advance in cancer research. Perhaps further research will provide better means of detection, prevention, and treatment of the disease, which is currently a leading cause of death among humans.

REVIEWING YOUR IDEAS

9. What is the relationship between antigens and antibodies?
10. Distinguish between B cells and T cells.
11. Name and describe each of the two forms of immunity.
12. What is interferon?
13. What causes AIDS? How is AIDS transmitted?
14. What is cancer? How do cancer cells differ from normal cells?

ADVANCES IN BIOLOGY

New Techniques in Immunology

Better understanding of the immune system and improved technology have led to the development of new approaches in the prevention and treatment of certain diseases. One recent advance is in the production of synthetic vaccines for certain viral diseases. Most viral vaccines are produced using whole, inactive or weakened viruses. However, the entire virus is not required. A specific antigen, a peptide in the virus's coat, is the portion recognized by the host. Such recognizable peptides are called **epitopes** (EP uh tohps). Synthetic vaccines contain only epitopes.

A suspected epitope is tested by injecting it into a rabbit. If it is an epitope, it will cause antibodies to form. Such antibody formation is verified by analyzing blood serum from the rabbit. The serum is then injected into mice that have also been injected with whole viruses. The serum should neutralize the viruses. The epitope then is introduced into a guinea pig. The pig is later injected with whole viruses. If the pig remains healthy, the epitope may be an effective synthetic vaccine. A synthetic vaccine against viral hepatitis recently has been approved for use.

Synthetic vaccines have several advantages over vaccines produced from whole cells or viruses. They are less expensive to make. Also, they are pure, containing no unwanted chemicals or live organisms. Other vaccines sometimes have side effects, some of which can be fatal. Another advantage is the possibility of producing a single synthetic vaccine that would provide immunity against several diseases.

A second technique provides large amounts of pure antibodies for a variety of uses. To produce the antibody, a mouse is injected with a bacterium, virus, or chemical. Lymphocytes in the mouse's spleen respond by making antibodies against the antigens present in the injection. Each lymphocyte makes a specific antibody.

The lymphocytes are then removed from the spleen and fused with certain cancer cells, which divide repeatedly. Each fused cell, or hybridoma (hi brih DOH muh), is cultured and forms a clone. The clone is made of many cells that make only one kind of antibody. The antibodies produced are called **monoclonal antibodies.**

After the antibody produced by each clone is identified, the clones are separated and grown in culture. As they grow, they produce one kind of antibody, which is very pure. Even greater quantities of pure antibody can be obtained by injecting a mouse with the hybridomas. The hybridomas form a tumor that produces a large amount of the antibody. The antibody is later filtered from the body fluids of the mouse.

Because monoclonal antibodies are specific, they can "seek out" abnormal or unusual conditions in the body. For example, they might be used to detect diseases that could not be diagnosed as early by other means. Some cancers may be detected in this manner. Also promising is a technique that uses monoclonal antibodies to treat cancer. Antibodies that detect antigens on the surface of cancer cells can be synthesized. Antibodies alone, however, will not destroy the cancer cells. To destroy the cells, a natural poison is linked to the monoclonal antibody. The antibody locates the cancer cell and the poison enters and kills it. So far this technique has proved successful in killing certain cancer cells in culture. Much work remains to be done to determine whether or not it can be used to treat the disease in humans.

Synthetic vaccines contain only the particular peptide that causes the immune response in a host.

Because they are pure, synthetic vaccines have no side effects.

FIGURE 19–16. Antibody-producing lymphocytes are fused with cancer cells so they can be grown rapidly in culture. After screening, selected hybridomas are injected into different mice and specific antibodies are formed in quantity.

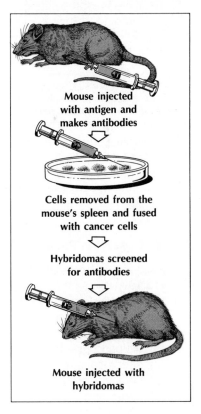

Mouse injected
with antigen and
makes antibodies

Cells removed from the
mouse's spleen and fused
with cancer cells

Hybridomas screened
for antibodies

Mouse injected with
hybridomas

INVESTIGATION

Problem: What are certain traits of bacteria?

Materials:

microscope   coverslip
depression slide   toothpick
prepared slides of   3 droppers
  three bacterial shapes   3 bacterial samples
petroleum jelly

Procedure:

Part A. Observing Shapes of Bacteria

1. Make a copy of Table 19–1 in which to record your data.
2. Examine the three shapes of bacteria shown in Figure 19–17.

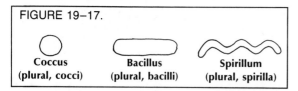

FIGURE 19–17.

Coccus   Bacillus   Spirillum
(plural, cocci)   (plural, bacilli)   (plural, spirilla)

3. Observe a prepared slide of a bacterial shape under low and high power. Record the species name in Table 19–1.
4. Describe the shape in Table 19–1.
5. Repeat steps 3 and 4 for the other two prepared slides.

Part B. Observing Patterns of Bacteria

1. Analyze the bacterial patterns shown in Figure 19–18.
2. In Table 19–1, write the letter of the term that best describes each pattern.
   (a) coccus—single cell
   (b) diplococcus—two cells
   (c) streptococcus—chain of cells
   (d) staphylococcus—cluster of cells
   (e) tetrad or gaffkya—four cells
   (f) cube or sarcina—cube of eight cells

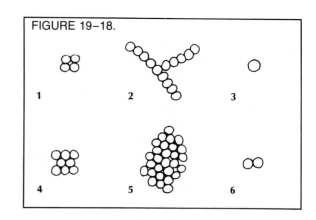

FIGURE 19–18.

3. Analyze the bacterial patterns shown in Figure 19–19.
4. In Table 19–1, write the letter of the term that best describes each pattern.
   (g) bacillus—single cell
   (h) diplobacillus—two cells
   (i) streptobacillus—chain of cells
   (j) palisade—series of stacked cells
5. Note that spirilla always occur singly.

Part C. Observing Motility of Bacteria

1. Some bacteria have motility, the ability to move about. Motile bacteria have flagella.

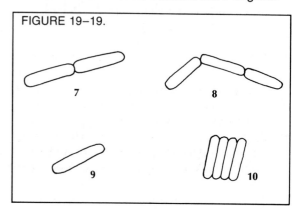

FIGURE 19–19.

FIGURE 19–20.

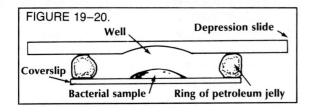

Well — Depression slide

Coverslip

Bacterial sample — Ring of petroleum jelly

2. Use the following method to determine if the bacteria in sample A are motile. Use Figure 19–20 as a guide.
   (a) Use a toothpick to place a thin ring of petroleum jelly about the size of a penny on a coverslip.
   (b) Add one drop of sample A to the center of the petroleum jelly ring.
   (c) Place a depression slide over the coverslip so that the well is directly over the drop of bacteria.
   (d) Lightly press down on the slide to form a seal around the bacteria.
   (e) Quickly and carefully invert the slide. The sample should form a drop that hangs from the center of the coverslip.
   (f) Observe the slide under high power. Motility is observed when the bacteria move in only one direction. NOTE: It may be necessary to reduce the amount of light.
   (g) Using "yes" or "no," record your observations in Table 19–1.
3. Repeat step 2 for samples B and C.

Data and Observations:

TABLE 19–1. TRAITS OF BACTERIA		
Part	Observations	
A. Shape	Species name	Shape
	_____	_____
	_____	_____
B. Pattern	Pattern 1 _____ Pattern 2 _____ Pattern 3 _____ Pattern 4 _____ Pattern 5 _____	Pattern 6 _____ Pattern 7 _____ Pattern 8 _____ Pattern 9 _____ Pattern 10 _____
C. Motility	Sample	Motile (yes or no)
	_____	_____
	_____	_____
	_____	_____

Questions and Conclusion:

1. Define bacteria, flagellum, motility.
2. Analyze the scientific names and descriptions of the following bacteria. Diagram and label these species of bacteria on a separate sheet of paper:
   (a) *Staphylococcus aureus* — nonmotile, causes infection of any body organ
   (b) *Streptococcus pyogenes* — nonmotile, causes sore throat, scarlet fever
   (c) *Clostridium tetani* — motile, bacillus, occurs in pairs and chains, causes tetanus

Conclusion: What are certain traits of bacteria?

CHAPTER REVIEW

SUMMARY

1. Many simple organisms are pathogens and are sources of infectious diseases. The idea that microorganisms can cause diseases was proposed by Louis Pasteur in his germ theory of disease. **p. 389**
2. Many pathogens have been discovered using Koch's postulates. **19:1**
3. Bacterial pathogens may be airborne or waterborne, or they may be spread by direct contact. **19:2**
4. Many pathogenic bacteria produce toxins that affect certain parts of the body. Both exotoxins and endotoxins are produced. Other pathogenic bacteria affect metabolism or destroy tissues. **19:3**
5. Many viruses cause diseases such as AIDS, flu, the common cold, measles, and mumps. **19:4**
6. Fungi may cause diseases in humans, plants, and animals. Great economic losses can occur as a result of fungal diseases that infect crops. **19:5**
7. Malaria, a disease responsible for millions of human deaths, is caused by the protozoan *Plasmodium*. **19:6**
8. Barriers of the body to pathogen entry include skin, hair, mucus, cilia, digestive juices, and enzymes in liquids such as tears. **19:7**
9. The immune system includes the blood, lymph, and other tissues. Antigens are recognized by two types of lymphocytes — T cells and B cells. **19:8**
10. B cells defend against bacteria, viruses, and toxins in the blood. Activity of B cells is controlled by T cells. Certain T cells destroy infected, cancerous, or transplanted cells. **19:9**
11. Protection against pathogens may result from active or passive immunity. **19:10**
12. Many bacterial diseases are treated with antibiotics that kill the bacteria. Drugs that disrupt a virus's life cycle are being developed. **19:10**
13. Interferon is a protein produced by cells when they are attacked by viruses. It protects healthy cells from being attacked by the viruses. **19:10**
14. AIDS is a fatal disease caused by a retrovirus that destroys the immune system. Drugs to treat the disease and vaccinations to prevent the disease are being developed. **19:11**
15. Cancer is a disease that involves abnormal cell growth. Oncogenes are thought to play a role in the onset of certain types of cancers. **19:12**
16. Synthetic vaccines contain only the recognizable antigen of a virus and are purer than vaccines made from whole viruses. Monoclonal antibodies may be used in the detection or treatment of diseases. **p. 404**

LANGUAGE OF BIOLOGY

active immunity
agglutination
AIDS
antibody
antigen
antiserum
B cell
carcinogen
clone
complement system
cytotoxic T cell
endotoxin
epitope
exotoxin
germ theory of disease
helper T cell
HIV
immune system
infectious disease

interferon
K lymphocyte
Koch's postulates
lymph node
lymphocyte
macrophage
memory cell
monoclonal antibody
oncogene
passive immunity
pathogen
plasma cell
spleen
suppressor T cell
T cell
thymus gland
toxin
vaccine
viroid

CHECKING YOUR IDEAS

On a separate paper, indicate whether each statement is true or false. Do not write in this book.

1. The skin is a barrier to pathogen entry.
2. An antigen is matched by a specific antibody.
3. Exotoxins remain in bacterial cells until the cells die.
4. Viruses can be treated successfully by means of antibiotics.
5. Malaria is transmitted by a mosquito.
6. The germ theory of disease was first stated by Koch.
7. Interferon is a natural means of protection against viruses.
8. Memory cells are important in long-lasting immunity.
9. Infectious diseases may be spread from one organism to another.
10. The common cold is caused by a bacterium.

EVALUATING YOUR IDEAS

1. What is the germ theory of disease?
2. Explain the procedure for determining if a given microorganism causes a certain disease.
3. Explain three ways in which infectious diseases may be spread.
4. Compare exotoxins and endotoxins.
5. What are viroids? How do they cause disease?
6. Give the stages in the life cycle of wheat rust.
7. Relate the stages in the life cycle of *Plasmodium* to the symptoms of malaria.
8. Explain how pathogens are prevented from entering a host's cells and tissues.
9. How do B cells recognize antigens? Describe how the pathogen is then destroyed.
10. Which T cells play a role in the activity of B cells? How do they function?

11. What is the role of cytotoxic T cells?
12. What is active immunity? What is passive immunity? How does each develop?
13. How are bacterial diseases often treated? How can viral diseases be treated?
14. Explain how interferon aids in fighting viruses. How might interferon be useful?
15. Explain how HIV results in AIDS. Why has it been difficult to develop an AIDS vaccine?
16. What role do oncogenes play in cancer?
17. How are synthetic vaccines made? What are the advantages of synthetic vaccines?
18. What are monoclonal antibodies? How might they be used against cancer?

APPLYING YOUR IDEAS

1. The common cold is caused by more than 100 different viruses. Why is there no vaccine?
2. Many vaccinations require later "booster" shots. Why are they necessary?
3. What problem usually follows an organ transplant? Why?
4. Why does a physician usually examine the lymph nodes during a physical examination?

EXTENDING YOUR IDEAS

1. How important are tonsils? Study the pros and cons of tonsil removal.
2. Prepare a report on the Salk polio vaccine.
3. Find out what Jenner and Lister contributed to knowledge of infectious disease.

SUGGESTED READINGS

Ellis, Ronald W., "Disease Busters." *Science 85,* Nov., 1985.

Janet, Peter, "Our Immune System—the Wars Within." *National Geographic,* June, 1986.

Langone, John, "Cancer—Cautious Optimism." *Discover,* March, 1986.

PLANTS

Plants may seem like inactive organisms to you because they have no adaptations for locomotion. However, plants carry out the same basic functions as other forms of life. They reproduce and develop in cycles involving both haploid and diploid forms. They make their own food, and a few plants also trap insects that become food. Plants transport water, minerals, and food, and exchange gases. Although plants do not move, some have parts that move. These movements are a plant's response to stimuli in the environment. In addition, many plants have a means of support.

Evergreens can flourish in areas where soils are poor and the climate harsh. Evergreens are adapted to extract nutrients efficiently from poor soil. Because of the needle structure, fog can condense on the needles and later drip down to water the tree. The famous Old Jeffrey Pine growing from the top of Sentinel Dome in Yosemite National Park shows the signs of its struggle with high winds and barren soil. Yet this pine carries out the same basic life functions as a cassava plant that lives in a tropical rain forest. In this unit you will learn about plant life functions and how plants are adapted to carry out these functions.

Old Jeffrey Pine

PLANT REPRODUCTION AND DEVELOPMENT

Like other organisms, plants reproduce and develop. Both asexual and sexual reproductive patterns are found among plants. The flower is the structure of sexual reproduction in many plants. Here, male and female parts of a columbine are visible. What are the parts? How does fertilization in a flower occur? How do nonflowering plants reproduce? What is involved in plant development?

chapter 20

Although you may not realize it, evidence of plant reproduction and development is common. Seeds, cones, flowers, and fruits are plant reproductive structures with which you are familiar. If you have hay fever, you know that your symptoms may be due to a reaction to certain types of pollen in the air. You may have seen seeds or bulbs develop into plants. What is involved in plant reproduction and development?

Plant life cycles may involve some stages in which the cells of the plant are haploid and other stages in which the cells are diploid. Although plant life cycles differ, each cycle includes both meiosis and fertilization. These processes ensure variety within a population of plants.

Objectives:
You will
- describe the life cycles of several algal plants and land plants.
- explain various means of asexual reproduction among plants.
- discuss development in flowering plants.

LIFE CYCLES OF ALGAL PLANTS

Most multicellular plants produce both gametes and spores during their life cycles. Plant gametes are produced by *mitosis* from certain cells of a haploid plant. Fertilization, fusion of gametes, results in a diploid stage of the life cycle. Certain cells of the diploid plant produce spores by *meiosis*. In these plants, the haploid and diploid stages follow one another in an **alternation of generations.** In true alternation of generations, each generation is multicellular.

Alternation of generations is a life cycle in which the diploid and haploid generations follow one another.

20:1 *Spirogyra* and *Ulothrix*

Spirogyra is a filamentous green alga (Section 14:4). A filament is composed of identical, haploid cells. Asexual reproduction in *Spirogyra's* life cycle occurs by fragmentation. New cells are formed along a new filament as a result of mitosis and cell division.

Spirogyra reproduces asexually by fragmentation.

413

FIGURE 20–1. *Spirogyra* reproduces sexually by conjugation. The series of events that occur in this process are shown from left to right.

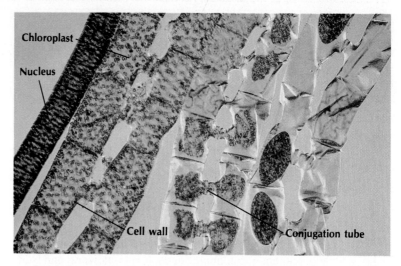

FIGURE 20–2. The life cycle of *Ulothrix* involves both asexual and sexual reproduction. Asexual reproduction occurs by fragmentation and zoospore formation. Sexual reproduction involves fusion of gametes.

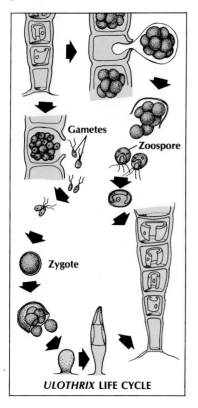

ULOTHRIX LIFE CYCLE

Sexual reproduction in *Spirogyra* occurs by conjugation. Bridges of cytoplasm form between cells of two haploid filaments lying side by side (Figure 20–1). The cell contents from one filament pass through the bridges into the cells of the other filament. There the contents of the cells fuse to form diploid zygotes.

Many zygotes can be seen along a filament after conjugation. Each zygote develops a tough outer "shell" and is called a **zygospore**. A zygospore is a *diploid* spore resulting from fertilization. When conditions are favorable, the zygospore divides by meiosis to form four haploid cells. Three of the cells die, but the surviving cell divides by mitosis to form a new filament.

The life cycle of *Ulothrix,* a green alga, also includes asexual and sexual reproduction. But the means of reproduction are different from those in *Spirogyra*. Sometimes, asexual reproduction occurs by fragmentation if a portion of the haploid filament breaks off. More often, asexual reproduction involves production of special cells, **zoospores.** Zoospores are produced by mitosis from certain cells located along the filament. The zoospores have four flagella. They actively swim through the water for a while, settle down, divide by mitosis, and develop into a new filament.

Sexual reproduction can occur when gametes with two flagella fuse to form a diploid zygote. The zygote later undergoes meiosis to produce four haploid spores. Each spore then can develop into a new haploid filament by mitosis (Figure 20–2).

20:2 Ulva

Both *Spirogyra* and *Ulothrix* have life cycles dominated by the haploid stage. Only the zygote is diploid. Because there is no *multicellular* diploid stage, these algal plants do not have true alternation of generations.

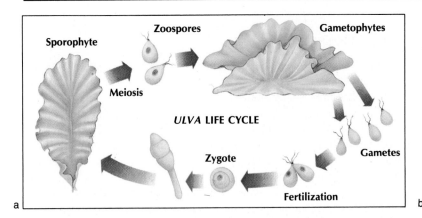

a   b

Ulva, or sea lettuce, is a green alga that undergoes true alternation of generations. Multicellular stages exist in each generation. Haploid and diploid stages look the same in *Ulva*. Cells in sporangia of the diploid stage undergo meiosis to produce spores that develop into multicellular haploid plants. Certain cells of a haploid *Ulva* produce gametes by mitosis. These gametes fuse to form a zygote that develops into a multicellular diploid plant (Figure 20–3).

Like *Ulva*, the life cycles of complex plants have true alternation of generations. Multicellular stages are present in each generation. The multicellular haploid stage is called the **gametophyte** (guh MEET uh fite) because it produces gametes. The multicellular diploid stage is called the **sporophyte** (SPOR uh fite) because it produces spores.

FIGURE 20–3. (a) The life cycle of *Ulva* involves both zoospore and zygote formation. (b) *Ulva* has the common name sea lettuce.

FIGURE 20–4. A generalized plant life cycle shows alternation of generations.

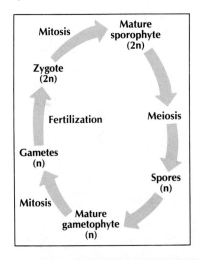

REVIEWING YOUR IDEAS

1. What is alternation of generations?
2. What kind of reproductive cells are made by haploid plants? By what process?
3. What kind of reproductive cells are made by diploid plants? By what process?
4. Which stage is predominant in *Spirogyra*? *Ulothrix*? *Ulva*?
5. Distinguish between sporophyte and gametophyte.

LIFE CYCLES OF HIGHER PLANTS

The relative importance of gametophyte and sporophyte generations varies from one type of plant to another. An examination of the major groups of land plants illustrates the differences in life cycles. Compare their life cycles to the diagram of a generalized life cycle shown in Figure 20–4.

20:3 Mosses

Gametophyte moss plants produce sperm and eggs.

Mosses are among the simplest true land plants. In the gametophyte generation, a moss plant has multicellular sex organs that protect the gametes—sperm and eggs.

Each gametophyte, male and female, makes many haploid gametes by mitosis. Sperm that are made by the **antheridium** (an thuh RIHD ee um), the male sex organ, must swim through a film of moisture to the egg. The egg stays in the female sex organ, the **archegonium** (ar kih GOH nee um), where fertilization and zygote formation occur. The diploid zygote, still embedded, then begins to divide by mitosis to become a sporophyte. The sporophyte carries on some photosynthesis but also obtains nourishment from the female gametophyte. When mature, the sporophyte produces many spores by meiosis. If conditions are favorable, each spore develops into a small structure, the **protonema** (proht uh NEE muh), that grows into a mature gametophyte.

Sperm swim to the eggs, which are in the archegonia.

FIGURE 20–5. The gametophyte generation is predominant in the life cycle of a moss. The sporophyte (which cannot live alone) grows from the female gametophyte.

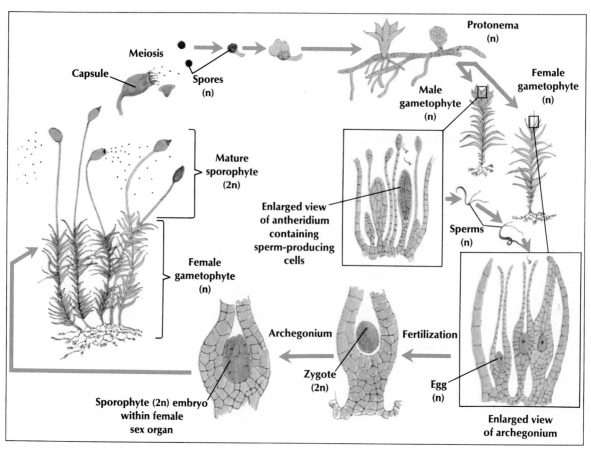

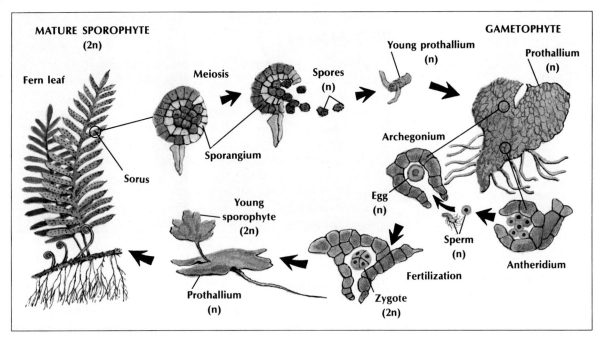

MATURE SPOROPHYTE (2n)

Fern leaf

Meiosis

Sorus

Sporangium

Spores (n)

GAMETOPHYTE

Young prothallium (n)

Prothallium (n)

Archegonium

Egg (n)

Sperm (n)

Antheridium

Young sporophyte (2n)

Prothallium (n)

Zygote (2n)

Fertilization

20:4 Ferns

Unlike mosses, ferns have a predominant sporophyte generation. The gametophyte is very small and lives for only a short time. It is photosynthetic. The plant that you would recognize as a fern is the sporophyte. By meiosis, the sporophyte produces many haploid spores. Spore cases can be seen on the underside of the fern fronds. These structures are clustered in groups called **sori** and contain the haploid spores.

Each spore may develop into a tiny, heart-shaped gametophyte called the **prothallium** (proh THAL ee um). This tiny structure is about 2 cm in diameter and contains both archegonia and antheridia (Figure 20–6). It has no vascular tissue. The archegonia produce eggs and the antheridia produce sperm. The sperm swim to the eggs in the archegonia and fertilize them. The zygote begins its development within the archegonium, but it soon grows systems of its own and separates from the gametophyte to become an independent plant, which is the sporophyte. The gametophyte then dies. The sporophyte has become independent of the gametophyte in this group of plants, and the haploid generation is reduced greatly.

Because mosses and ferns require water for sexual reproduction, they are most often found in moist environments. But many live elsewhere. Some can occupy constant dry environments by reproducing asexually. Others survive temporary dry conditions. They remain dormant until water is available for sexual reproduction.

FIGURE 20–6. The sporophyte generation is predominant in the life cycle of a fern. However, a distinct gametophyte generation, the prothallium, also exists. The young sporophyte begins its development on the prothallium.

In ferns, the sporophyte is dominant.

A spore develops into a distinct gametophyte, the prothallium.

Most ferns, like mosses, inhabit a moist environment.

FIGURE 20–7. Diploid cells in the male cone (a) produce pollen grains. Eggs develop in a female cone. (b) A sperm fertilizes an egg beginning a new sporophyte generation.

20:5 Conifers

Plants that reproduce by seeds are the most abundant on Earth today. In one group of these plants, the conifers, seeds are produced in structures called **cones** (Section 14:8). Conifers have two kinds of cones, male and female (Figure 20–7). Female cones are larger than male cones. In most conifers, male and female cones are found on separate branches of the same tree. *The gametophyte is multicellular but much reduced.* Spores of seed plants develop into male and female gametophytes. The gametophytes, though, do not develop into independent plants. Instead, they give rise to sperm and eggs. Fertilization occurs, producing new sporophytes.

In the spring, certain diploid cells in the male pine cones produce haploid spores by meiosis. By mitosis the spores are changed to **pollen grains,** the male gametophyte. Pollen grains are carried by the wind to female cones on the same or a different tree. Pollen grains have tough outer walls that resist water loss. The process by which pollen reaches the female reproductive organ is called **pollination.** After releasing pollen grains, the male cone falls off the tree.

Special diploid cells in the female cones undergo meiosis to produce one surviving spore. That spore, by many mitoses, produces the female gametophyte. Later, eggs develop in the gametophyte. When pollination occurs, a haploid sperm cell is produced within the pollen grain. The sperm fertilizes an egg, producing a diploid zygote, the beginning of a new sporophyte generation.

The zygote develops into an embryo. A hard coat forms around the embryo and the rest of the female gametophyte. The gametophyte becomes a food source for early development. Together, the outer coat, embryo, and food source are a **seed.** A single female cone may bear many seeds, each partly protected by the overlapping scales

Spores produced by female cones give rise to female gametophytes.

The seed of a conifer contains an embryo sporophyte and a food source.

of the cone. Each seed develops a "wing" that aids in the dispersal of the seeds by wind. The time between pollination and seed formation is often longer than a year.

The life cycle of the conifers shows several adaptations not seen in mosses and ferns. Water is not needed for fertilization. Sperm cells are carried to eggs in pollen grains. Thus, the sperm does not dry out, and conifers are not restricted to a moist habitat. Also, the embryo is protected by the outer seed coat and by the cone in which the seed is formed. Similar adaptations are seen in the flowering plants.

20:6 Flowering Plants: Reproductive Structures

Most numerous of the seed plants are the flowering plants. The reproductive structures of flowering plants are the flowers. A flower and its parts are diploid. Both male and female organs may be located within a single flower, or they may be in separate flowers. Flowers with both male and female organs are said to be *perfect* while those with only male or female organs are said to be *imperfect*.

The female organ of a flower is a long, vase-shaped structure called the **pistil** (PIHS tul). A pistil is divided into three parts. The top is a sticky structure called the **stigma** (STIHG muh). The stalklike portion is the **style,** and the swollen, lower region of the style is the **ovary.** Within the ovary are one or more **ovules** (OHV yewlz), female sporangia.

The male organ of a flower is the **stamen** (STAY mun). The stalk-like portion of a stamen is called the **filament.** At the tip of the filaments are **anthers** (AN thurz), which contain **pollen sacs,** male sporangia.

FIGURE 20–8. The reproductive structure of flowering plants is the flower. (a) Pollen is transferred from the male part, the stamen, to the female part, the pistil. (b) Pollen is adapted for adhering to the stigma. Notice the edges on the Lamb's quarter pollen shown here magnified about 1450 times.

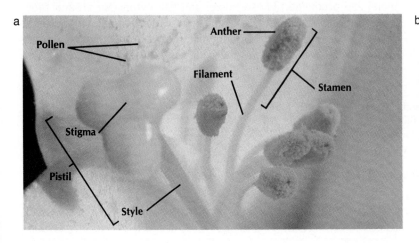

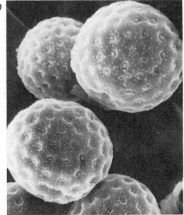

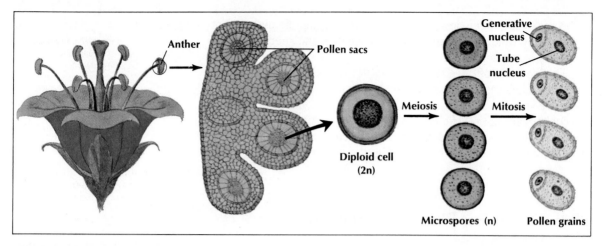

FIGURE 20–9. Microspores are produced by meiosis within the pollen sacs of the anther. Microspores are converted to pollen grains.

20:7 Flowering Plants: Sporophyte to Gametophyte

Sexual activities begin within the ovules and anthers of a flower. Diploid cells in the pollen sacs undergo meiosis (Figure 20–9) to produce four haploid cells known as **microspores** (small spores). Each microspore nucleus divides by mitosis to form two haploid nuclei, the **tube nucleus** and the **generative nucleus.** When the outer wall of the microspore hardens, the structure becomes a pollen grain.

Within diploid pollen sacs, meiosis occurs producing haploid microspores that develop into pollen grains.

A similar type of activity takes place within each ovule (Figure 20–10). A single diploid cell of the ovule undergoes meiosis to produce four haploid cells of which only one survives. This surviving cell is called a **megaspore** (MEG uh spor), which means large spore. In some flowers the megaspore nucleus divides by mitosis to form two haploid nuclei. Mitosis continues until a total of eight haploid nuclei are produced. These nuclei and the cytoplasm around them represent the entire female gametophyte generation. Of the eight nuclei produced, only three are important in reproduction: two **polar nuclei** in the center of the ovule and the true egg at one end. The remaining five haploid nuclei die.

A diploid cell within the ovule undergoes meiosis producing a haploid megaspore that gives rise to the egg and polar nuclei.

FIGURE 20–10. After meiosis and several mitoses, an egg and two polar nuclei are formed within an ovule. The other haploid cells that result from this process do not survive.

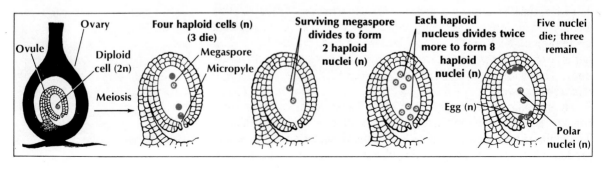

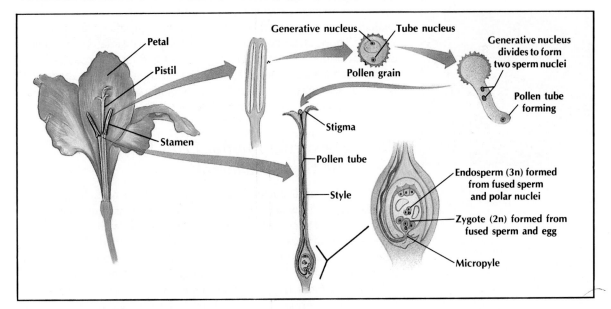

20:8 Flowering Plants: Fertilization and Seed Formation

When and if a pollen grain reaches the pistil (pollination), it sticks to the surface of the stigma (Figure 20–11). There the tube nucleus causes a **pollen tube** to form and "dig" its way down through the style. It eventually reaches a small opening in the ovule called the **micropyle** (MI kruh pile). Meanwhile, the generative nucleus divides (by mitosis) to produce two haploid sperm nuclei that are carried down the pollen tube into the ovule. (The nucleus of the pollen tube and the two sperm nuclei make up the male gametophyte generation.)

Two haploid sperm nuclei are produced by mitosis from the generative nucleus.

Within the ovule, one of the sperm nuclei joins with the two polar nuclei to form a triploid (TRIHP loyd), or 3n, structure called the **endosperm** (EN duh spurm) **nucleus.** The other sperm nucleus fertilizes the egg to form the diploid zygote. These two fusions are known as **double fertilization.** Zygote formation marks the beginning of a new sporophyte generation.

In double fertilization, one sperm nucleus fuses with the two polar nuclei to form the triploid endosperm nucleus. The other sperm nucleus fertilizes the true egg to form the diploid zygote.

After fertilization, the zygote develops into an embryo. The endosperm nucleus divides many times to form a mass of tissue called **endosperm.** Endosperm is the food source for the embryo. The outer covering of the ovule hardens into a protective device, the **seed coat.** The ovule is now a seed. *Within each seed is the plant embryo and a food source.*

After fertilization, the ovule becomes the seed containing the developing plant and a food source.

In summary, a flowering plant has a life cycle in which the sporophyte generation is dominant. Within a seed is a diploid embryo. The embryo grows and develops into a multicellular sporophyte and produces flowers. By meiosis, diploid cells within the ovules produce

FIGURE 20–12. Fruits have a variety of forms. These walnuts are one example of a fruit.

The ovary becomes the fruit of the plant.

Fruit protects developing seeds. Many fruits aid in seed dispersal.

FIGURE 20–13. Seeds may be dispersed in many ways, including wind, water, and animals. Some plants have pods that burst open, dispersing seeds.

haploid megaspores, and cells within anthers produce haploid microspores. These structures are the beginning of a gametophyte generation. Cells derived from the megaspores and microspores undergo mitosis to become eggs and sperm. Fertilization results in the formation of a diploid zygote, marking the beginning of a sporophyte generation and completing the life cycle.

20:9 Flowering Plants: Fruit Formation and Seed Dispersal

Sometime after pollination, most of the flower parts begin to die, but the ovary (in which the seeds are located) enlarges rapidly. An enlarged ovary becomes a **fruit.** You are familiar with fleshy, sweet fruits such as oranges, grapes, and peaches. Also, many "vegetables" such as tomatoes, green beans, pea pods, and kernels of corn are really fruits. Many kinds of nuts and grains are fruits, too.

Once formed, fruit plays an important role in the reproductive process of flowering plants. Each fruit contains one or more seeds, and each seed contains an embryo. In addition to protecting the seeds, many fruits have another important function: they aid in **seed dispersal** (dihs PUR sul), the scattering of seeds. If seeds are to **germinate** (JUR muh nayt), or begin development, they must land where there is a suitable environment. Because of competition for light and water, all seeds produced by a plant cannot germinate in the immediate area of the parent plant. Adaptations for seed dispersal combat this competition.

Because many fruits are sources of food, they are eaten by animals. The seeds of the fruit pass through the digestive systems of the animals and are usually deposited at a distance from the parent plant. If conditions are favorable, they will germinate.

Not all fruits can be eaten. Cockleburs are inedible fruits, but they stick easily to the fur of animals and may be transported in that manner. Other fruits, such as those of maples and elms, are "winged" and may be carried great distances by the wind. Coconuts and other seeds float and may be dispersed by water.

Seeds that pop from plants

Seeds carried by wind

Seeds carried by water

Seeds carried by humans and other animals

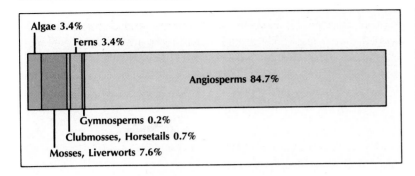

Algae 3.4%

Ferns 3.4%

Angiosperms 84.7%

Gymnosperms 0.2%

Clubmosses, Horsetails 0.7%

Mosses, Liverworts 7.6%

FIGURE 20–14. Flowering plants make up the majority of all plants. Their success is due in part to the fact that they do not need water for sexual reproduction and to the evolution of various methods of pollination.

20:10 Success Story

There are many reasons why seed plants are so plentiful on Earth. Methods of reproduction are more specialized than in mosses and ferns. Because sperm are carried in pollen grains, water is not necessary for fertilization. Also, there are various ways pollination occurs.

A plant that has flowers with both male and female sex organs can be pollinated within itself. This process, called **self-pollination,** can result from physical contact between the stamens and the pistil to which pollen is transferred. Self-pollination can also occur between different flowers of the same plant. When pollination occurs between flowers of separate plants, the process is called **cross-pollination.** In both self- and cross-pollination, pollen may be transferred by the wind, insects, or small birds. Many of the elaborate shapes, colors, and scents of flowers are adaptations that attract these pollen-transferring animals, and thus, increase the chance of fertilization. Pollination in the conifers occurs mostly by wind transfer of pollen.

The seed plants have other adaptations not found in mosses and ferns. The embryos they produce are protected within seeds and the seeds are protected by fruits or cones. Also, seeds are adapted in many ways for a variety of methods of dispersal.

Specialization increases from mosses, to ferns, to conifers, to flowering plants. As plants become more specialized, the diploid phase of the life cycle becomes more dominant. Seed plants, almost totally diploid, are the dominant group of land plants. All the plants studied have the capability for sexual reproduction and the variety it produces. Why, then, is there this trend in more complex plants to have more predominant sporophytes (diploid phases)? What is adaptive about the predominant sporophyte stage?

The haploid organism has only one possible allele for each trait while diploid organisms have two. The presence of two alleles offers a better chance of survival because the chances of harmful traits or mutations appearing in an organism are reduced. A harmful allele may be masked by another allele in a diploid organism.

Pollination in flowering plants is aided by wind and animals.

FIGURE 20–15. One reason for the success of flowering plants is that pollination is aided by insects and birds. Notice the yellow pollen covering this bee.

FIGURE 20–16. (a) Many plants can be grown vegetatively from bulbs they form. (b) Strawberry plants reproduce from runners. (c) Potato "eyes" are buds, each of which can produce a new plant.

In vegetative reproduction a new organism is produced from a nonsexual part of the parent.

Runners, buds, and bulbs are means of vegetative reproduction in plants.

Vegetative reproduction produces offspring that are genetically identical to the parent organism.

20:11 Vegetative Reproduction

In addition to a life cycle involving alternation of generations, many plants also reproduce asexually. They do so by **vegetative** (VEJ uh tayt ihv) **reproduction,** a process involving reproduction from a nonsexual (vegetative) part of the parent plant. This form of reproduction may occur in either the gametophyte or sporophyte generation, depending on the plant. It is not a regular part of the life cycle, although it may occur in nature or be artificially carried out by humans.

Examples of vegetative reproduction in nature are numerous. Strawberry plants develop outgrowths of stems called runners. These grow outward along the ground. At points along the runners, new strawberry plants develop. Bud formation is a form of vegetative reproduction in potatoes. The part of the potato plant that you eat is a swollen underground stem called a tuber. The "eyes" of the potato are buds that can develop into a complete plant. A short underground stem surrounded by many scales (modified leaves) is called a **bulb.** Daffodils, tulips, onions, and garlic are examples of plants grown vegetatively from bulbs.

Artificial vegetative reproduction is carried out in several ways. One way involves cuttings from a plant stem. The cuttings, or cut pieces of stem, are removed from a plant and placed in moist sand. Special **adventitious** (ad ven TIHSH us) **roots** grow from the stems, and the stems grow into mature plants. Other vegetative parts, such as roots and leaves, can be used for cuttings. Because mitosis is involved, artificial vegetative propagation preserves the genetic traits of plants. As a result, gardeners and nursery people can preserve suitable types of plants for generations.

Fruit crops are artificially propagated in another way. In this process, buds are removed from one plant and grafted to the stem of another plant. The grafted buds later produce new shoots on which fruit develop. Apples, pears, and peaches are examples of fruits produced this way. Why would orchard keepers prefer to propagate crops by artificial vegetative propagation?

Grafting is a method of artificial vegetative reproduction.

REVIEWING YOUR IDEAS

6. Which stage of the life cycle is predominant in mosses? In ferns? In conifers? In flowering plants?
7. What are the reproductive structures of a conifer?
8. What are the sex organs of a flower?
9. Distinguish between self- and cross-pollination.
10. Name several means of vegetative reproduction in plants. Give an example of each.

FLOWERING PLANTS: DEVELOPMENT

The zygote of a flowering plant develops into an embryo within the seed. Cells of the embryo have sets of DNA that guide its development into a mature sporophyte plant. Recall that development (Section 1:5) involves a series of changes leading to a mature form. How does development of a flowering plant occur?

20:12 Germination

A seed enters a period of inactivity, or **dormancy** (DOR mun see), that is necessary before development can continue. During dormancy, chemical changes occur that prepare the embryo for further change. When a seed reaches a suitable environment, the embryo within the seed germinates. Germination occurs if conditions such as amounts of oxygen and moisture and a suitable temperature are right. Each type of seed has unique growth requirements. Oxygen is needed for respiration to provide the rapidly growing embryo with energy.

Seed plant embryos have regions that develop into definite structures during germination. The **radicle** (RAD ih kul) is the first part of the embryo to emerge from the seed. It becomes the primary root. The radicle is the tip of the **hypocotyl** (HI puh kaht ul). In some seeds, such as a bean, the hypocotyl forms an arch, pushes above the soil, and becomes a small part of the base of the stem. In other seeds, such as corn and peas, the hypocotyl remains in the soil and becomes the upper part of the primary root (Figure 20–17). The **epicotyl** (EP ih kaht ul) is a part of the embryo that gives rise to the shoot (stems and leaves).

FIGURE 20–17. The epicotyl (upper structure) and hypocotyl (lower structure) can be seen in this germinating corn seed.

FIGURE 20–18. As germination occurs, food stored in the cotyledons of a bean seed is used until the first leaves are formed.

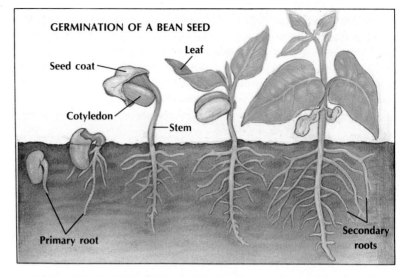

GERMINATION OF A BEAN SEED

Seed coat

Cotyledon

Leaf

Stem

Primary root

Secondary roots

The embryo is attached to one or two cotyledons (Section 14:11), or seed leaves. Plants such as beans are dicots and have two cotyledons per seed. Corn, a monocot, has one per seed. In monocot seeds, food is stored mainly in endosperm tissue. In many dicot seeds, the cotyledons absorb food from the endosperm as the seed matures. In either case, food stored in seeds provides a source of enegry during early development of the sporophyte.

20:13 Further Development

As an example of plant development, consider the growth of a young tree root (Figure 20–19). The tip of the growing root has an outer **root cap** composed of protective cells. Cells in the root cap are scraped off as the root tip pushes downward through the soil. However, new cap cells are produced by tissue located just above the cap. This tissue, found at the tips of roots and stems, is called **apical meristem** (MER uh stem) tissue. A **meristem** is a region of plant growth. Cells of meristems actively divide at certain times. As the root grows, new cap cells are being produced. New cells are also being formed on the opposite side of the apical meristem. These cells form the **elongation region.** Cells in this area grow only in length. This growth pattern adds to the total length of the young root and aids in pushing it deeper into the soil. Above the region of elongation is the **maturation region.** Cells here are even larger and develop into different types of specialized tissue. Changes in cells that result in the formation of specialized parts are called **differentiation** (dihf uh ren chee AY shun). In a young root, early differentiation results in the formation of different types of tissues for conduction of water and food, storage of food, protection, and uptake of water and minerals. Later differentiation produces a meristem called the **vascular cam-**

FIGURE 20–19. The growth pattern of a root results in three definite regions. Cell division occurs in the meristematic region. Lengthening of the root occurs in the region of elongation. Cell specialization occurs in the maturation region.

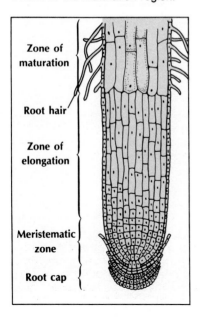

Zone of maturation

Root hair

Zone of elongation

Meristematic zone

Root cap

bium located near the center of the root. All of these types of cells develop from cells produced by the meristem. You will study the details of root structure in Chapter 22.

The events described so far lead to an increase in the length of the root. Once the vascular cambium is formed, an increase in diameter also occurs. This increase occurs as the vascular cambium undergoes mitosis to provide many cells that form conductive tissues. Cells on the inside of the vascular cambium differentiate into tissue that transports water. On the outside, new cells develop into tissue that transports food. Each year the vascular cambium produces new conductive tissues and moves farther away from the center of the root. This pattern results in the increase in thickness of the root.

The root's vascular cambium is a meristem that gives rise to cells that become conductive tissues.

The pattern of growth and differentiation in a root is similar to that in other plant parts. Meristems in stems and buds produce cells that become specialized for particular functions. Some seed plants live for many years. During their lives, they continue to produce new tissues and organs year after year, for example, cones, leaves, and flowers. How does the development of a plant such as an oak tree compare to your development?

Meristems in stems and buds produce cells that develop into specialized tissues.

REVIEWING YOUR IDEAS

11. What is germination? What conditions are necessary for it to occur?
12. List the various parts of a seed.
13. What are the regions of a developing root?

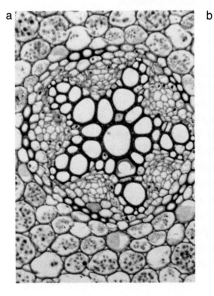

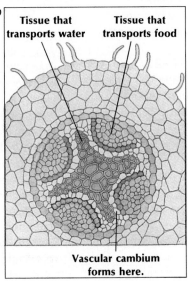

Tissue that transports water

Tissue that transports food

Vascular cambium forms here.

FIGURE 20–20. (a) A cross section of the center of a root shows the vascular cambium. (b) The vascular cambium will develop into tissue that transports water and tissue that transports food.

INVESTIGATION

Problem: How can you tell if a seed is still alive?

Materials:

single-edged razor
  blade
soaked corn seeds
iodine stain
dropper
tetrazolium solution
2 petri dishes

glass marking pencil
graduated cylinder
hot plate
beaker
water
oven mitt
2 paper towels

Procedure:

Part A. Seed Anatomy and Function

1. Make a copy of Figure 20–21. It will be used for labeling and for recording data.
2. Observe a soaked corn seed. Lay the seed flat on a protected surface. Carefully cut the seed in half with the razor blade. Use Figure 20–22 as a guide for the proper direction in which to cut the seed.
3. Identify the following seed structures:
   (a) seed coat—hardened, outer layer
   (b) endosperm—large, internal, white area
   (c) embryo—small, yellow area near pointed end
4. Label the structures on Figure 20–21.
5. Using your fingernail, gently scrape the surface of the endosperm and embryo. Add one

drop of iodine to the scraped surface of each structure. **CAUTION:** *If iodine spillage occurs, rinse with water immediately.*

6. After two minutes, observe any color change of the iodine. A blue color change indicates that starch or stored food is present.
7. Shade that area on your diagram of the seed in which there was a color change of the iodine. This will indicate the area of stored food.

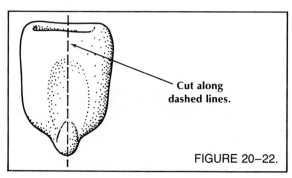

Cut along
dashed lines.

FIGURE 20–22.

Part B. Testing for Presence of Living Tissue

1. Make a copy of Table 20–1. Use it to record your data.
2. Place 20 soaked corn seeds in a small beaker. Cover the seeds with water. Using a hot plate, boil the seeds for 10 minutes.
3. Label one petri dish "Boiled." Pour 10 mL of tetrazolium solution into the dish.
4. Remove the beaker from the hot plate. *Be sure to wear an oven mitt.* Drain the water from the seeds.
5. Lay each seed on a protected surface. Using the razor blade, carefully cut each seed in half as in step 2 of Part A.
6. Discard one-half of each seed. Place the remaining half, cut surface down, into the petri dish.

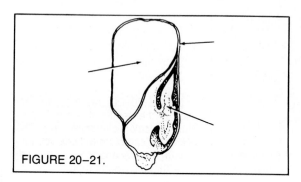

FIGURE 20–21.

7. Label the second petri dish "Unboiled" and add 10 mL of tetrazolium solution to it.
8. Obtain another 20 soaked corn seeds. Cut these seeds in half as in step 2 of Part A.
9. Discard one-half of each seed. Place the remaining half, cut surface down, into the dish.
10. Allow the boiled and unboiled seed halves to remain in the tetrazolium solution for 20 minutes.
11. Remove the seeds from one dish and place them on a paper towel. Mark the towel as to the type of seeds present, boiled or unboiled. Count and record in Table 20–1 the number of seeds that have a concentrated pink area

on the cut surface. Tetrazolium solution causes tissue that is alive and undergoing cellular respiration to turn pink. If a pink area is present, note if it is in the endosperm or embryo area.
12. Repeat step 11 using the seeds from the remaining dish.
13. Complete Table 20–1 by calculating the percentage of boiled and unboiled seeds that contain living tissue.
14. Gather the class data for number of seeds containing living tissue and record the data in Table 20–1. Complete it by calculating the percentage of boiled seeds and unboiled seeds that contain living tissue.

Data and Observations:

TABLE 20-1. PERCENTAGE OF SEEDS WITH LIVING TISSUE				
	Your Results		Class Results	
	Boiled	Unboiled	Boiled	Unboiled
Total number of seeds				
Number of seeds with living tissue				
Number of seeds with no living tissue				
Percentage of seeds with living tissue				

Questions and Conclusion:

1. Define:
   (a) seed embryo
   (b) endosperm
   (c) cellular respiration
2. Which area of the seed turned blue when iodine was added? What does this tell you about that area within the seed?
3. In which area of the seed did the pink color of tetrazolium concentrate? What does this tell you about that area within the seed?

4. How do the class percentages of living tissue for boiled and unboiled seeds compare? What process was halted as a result of the boiling?
5. Explain why the presence of living tissue is an indication of which seeds will germinate.
6. How might farmers use the tetrazolium test before they plant large batches of seeds?

Conclusion: How can you tell if a seed is still alive?

CHAPTER REVIEW

SUMMARY

1. Plants have a life cycle in which there is an alternation of generations—a diploid, spore-producing stage and a haploid, gamete-producing stage. **p. 413**
2. *Spirogyra* and *Ulothrix* have life cycles with no multicellular diploid stage. **20:1**
3. In *Ulva* and more complex plants, the life cycle includes a multicellular haploid stage, the gametophyte, and a multicellular diploid stage, the sporophyte. **20:2**
4. Mosses have a life cycle with a predominant gametophyte generation. Moisture is required for sexual reproduction. **20:3**
5. In the life cycle of ferns, the sporophyte generation is predominant. Like mosses, ferns need water for sexual reproduction. **20:4**
6. Neither conifers nor flowering plants have an independent gametophyte stage. In conifers, pollination results in production of seeds. Water is not needed for fertilization. **20:5**
7. The flower is the reproductive organ of flowering plants. **20:6**
8. Microspores, produced by male flower structures, develop into pollen sacs. Megaspores, produced by female flower structures, give rise to the egg and polar nuclei. **20:7**
9. Seed formation follows double fertilization in flowering plants. The seed consists of embryo, endosperm, and seed coat. **20:8**
10. Developing seeds are protected within fruits of flowering plants. Many fruits aid in dispersal of seeds. **20:9**
11. Adaptations such as pollen, protection of embryos and developing seeds, and a predominant sporophyte generation make seed plants more successful than other plants. **20:10**
12. Many plants reproduce asexually by vegetative reproduction. **20:11**
13. During germination, special regions of the plant embryo develop into roots, stem, and leaves. **20:12**
14. A developing plant root consists of a root cap and zones of cell reproduction, elongation, and maturation. **20:13**

LANGUAGE OF BIOLOGY

alternation of generations	germinate
antheridium	hypocotyl
apical meristem	maturation region
archegonium	pistil
cone	pollen grain
differentiation	pollination
double fertilization	prothallium
elongation region	radicle
endosperm	sporophyte
epicotyl	stamen
fruit	vascular cambium
gametophyte	vegetative reproduction

CHECKING YOUR IDEAS

On a separate paper, complete each of the following statements with the missing term(s). Do not write in this book.

1. In plants, gametes are produced by the process of _____.
2. Sporophytes contain the _____ number of chromosomes.
3. The _____ of a seed develops into the primary root.
4. _____ plants have no independent gametophyte stage.
5. Fruits aid in the dispersal of _____.
6. _____ is another name for asexual reproduction in plants.
7. Describing chromosome number, each pollen grain contains two _____ nuclei.
8. In mosses, the _____ generation is predominant.

EVALUATING YOUR IDEAS

1. Compare sexual reproduction in *Spirogyra* and *Ulothrix*. How are they different?
2. What stage of *Ulva's* life cycle is not part of the life cycle of *Spirogyra* and *Ulothrix?*
3. Describe a generalized plant life cycle.
4. Describe the life cycle of a moss.
5. In terms of reproduction, how have mosses adapted to life on land?
6. Describe the life cycle of a fern. How are ferns limited?
7. Describe the conifer life cycle. What adaptations in it are not present in the life cycle of mosses and ferns?
8. How is a pollen grain formed in flowering plants?
9. Describe the activities within an ovule.
10. Describe pollination and double fertilization in flowering plants.
11. Describe how dispersal is accomplished in flowering plants. Why is it necessary?
12. Why are seed plants the most common plants? Why are flowering plants even more successful than conifers?
13. Describe the structures of the embryo within a seed. List the structures' functions.
14. What is the energy source for germination? Where is it located?
15. How does a root increase in length? What tissues develop during this period?
16. How does an increase in the thickness of a root occur?
17. How is artificial vegetative reproduction advantageous?

APPLYING YOUR IDEAS

1. How might a tree be "produced" that has both plums and peaches on separate branches?
2. Why would you not expect to see a moss growing in an open field?

3. Find out if the following are flowering plants: (a) saguaro cactus (b) maple tree (c) pine tree (d) cotton plant (e) redwood tree (f) blueberry bush (g) palm tree (h) yucca tree (i) bamboo (j) wheat (k) mustard (l) grass
4. In what way is a plant produced by vegetative reproduction a clone?
5. Many of the events in the development of a stem are like those of a root. What differences are there in the development of these two organs?
6. What kind of tissue allows plants to develop new structures over and over again?
7. Compare vegetative reproduction in plants to asexual reproduction in simple organisms.

EXTENDING YOUR IDEAS

1. Using library resources, report on methods of artificial vegetative reproduction.
2. Use reference materials to study the life cycles of *Chlamydomonas, Oedogonium,* and *Fucus.* How do these life cycles compare to those of land plants (mosses, ferns, and seed plants)?
3. Trace the events that occur when a sweet potato is placed in water. What kind of reproduction is this plant formation?
4. Determine the rate of growth of seedlings during germination. Record the height of each seedling daily. At the end of the experiment, prepare graphs of (a) total height each day and (b) increase in height each day.

SUGGESTED READINGS

Galen, Candace, "The Smell of Success." *Natural History,* July, 1985.
Newman, Cathy, "Pollen—Breath of Life and Sneezes." *National Geographic,* Oct., 1984.
Stiles, Edmund W., "Fruit for All Seasons." *Natural History,* Aug., 1984.

PLANT NUTRITION

Plants are the producers in almost every community on Earth. Through the process of photosynthesis, plants convert the energy of sunlight to the chemical energy of glucose. Spanish moss and water lilies are uniquely adapted for food production in the conditions found in the Okefenokee Swamp in southeastern Georgia. How does photosynthesis occur? What raw materials are required? What properties of light are important in the process? In what different ways are plants adapted for obtaining nutrients?

chapter 21

Plants are essential members of every type of community. In oceans, lakes, forests, and deserts, algae or more complex plants are producers. All other forms of life depend directly or indirectly on plants for food.

In Chapter 5, you learned how all organisms obtain usable energy from food by cellular respiration. The food most often used by living systems is glucose. Glucose is important because it has chemical potential energy. Only autotrophs, mostly plants, can make glucose. They trap light energy and change it to chemical energy. In this chapter, you will study how glucose is made and how plants meet their nutrient requirements.

Objectives:
You will
- explain the role of chlorophyll and other pigments in trapping light energy.
- relate the events of the light and dark reactions of photosynthesis.
- describe types of digestion that occur in plants.

TRAPPING ENERGY

Photosynthesis can occur in all plant cells containing chlorophyll. In complex plants, most photosynthesis occurs in one special organ, the **leaf.**

Leaves are the major sites of food production in plants.

21:1 The Leaf: Internal Structure

A leaf is composed of several distinct cell layers. The upper layer of a leaf is called the **upper epidermis** (ep uh DUR mus). Its main function is to protect the other layers of the leaf. Epidermis is classified as a protective tissue in plants. Beneath the epidermis is a layer of palisade (pal uh SAYD) cells. **Palisade cells** are long cells arranged vertically. These cells, along with the **spongy layer** beneath them, contain chlorophyll. Photosynthesis occurs in these cells of a leaf.

Photosynthesis occurs in the palisade and spongy layer cells of a leaf.

433

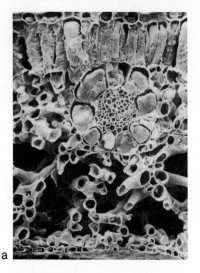

a

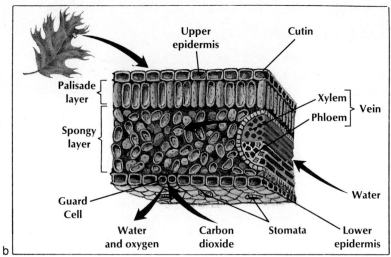

b

FIGURE 21–1. The leaf is the major organ of photosynthesis. (a) This scanning electron micrograph shows the detail of the internal leaf structure. It is shown magnified about 300 times. Compare these structures with those shown in (b) the illustration.

Guard cells regulate the opening and closing of stomata.

stomata (singular, stoma)

A general biological principle: structure is related to function.

How does a leaf illustrate the relationship between structure and function?

Shape and arrangement of leaves are adaptations for trapping light energy.

Located throughout a leaf are veins with two types of vascular (conductive) tissue, xylem and phloem. **Xylem** (ZI lum) cells carry water and minerals. **Phloem** (FLOH em) cells carry food. The veins of the leaves connect with the veins of the stem, which are continuous with those of the roots. Through the veins, water and minerals move from roots to leaves, and food is carried from leaves to other parts of the plant.

The lower layer of a leaf is the lower epidermis. It contains guard cells. **Guard cells** surround and regulate the opening and closing of pores called **stomata** (STOH mut uh).

Both the upper and lower epidermis of a leaf are covered by cutin (Section 14:5). In the leaves of some plants, such as ivy, cutin is especially thick and gives the leaf a shiny appearance.

All the materials necessary for glucose production enter the leaves. Carbon dioxide enters through stomata and passes to tiny air spaces between palisade and spongy cells. It then dissolves in the watery film around the cells and diffuses into them. Water enters the leaf through xylem vessels that originate in the roots of the plant. Thus, *the leaf structure is adapted for photosynthesis.*

A leaf also provides for the distribution of glucose. Once glucose has been produced, it is changed to sucrose and distributed by the phloem. Thus, all parts of the plant receive materials for energy or growth. Oxygen (produced during photosynthesis) and some water escape from the leaf through the stomata.

Structure of a plant as a whole is also related to function. Besides the materials needed for photosynthesis, energy is needed to put the materials together. This energy is sunlight. Plants have morphological adaptations that result in leaves being positioned on a plant where they receive the most sunlight. Also, the shape of the leaf determines the amount of sunlight it can receive.

21:2 Light

To understand how radiant energy is used to make glucose, you must learn about light. Light is a form of radiant energy. All radiant energy travels in waves. The distance between consecutive crests of the waves is called **wavelength** (Figure 21–2). The human eye is sensitive to radiant energy with wavelengths of about 400 to 700 nm. This portion of radiant energy is called white light.

When white light is passed through a prism, several different colors can be seen. As it passes through the prism, the wavelengths separate because each is bent at a different angle. As a result, white light is seen as a band of colors known as the **visible spectrum** (SPEK trum). The colors of the different wavelengths are red, orange, yellow, green, blue, and violet. Red light has the longest wavelength of the visible spectrum, and violet light has the shortest wavelength. Ultraviolet and infrared wavelengths, on either end of the visible spectrum, are invisible to humans.

Have you ever worn a dark blue or black sweater on a sunny day? If you have, you know that after a short period of time, you begin to feel quite warm. Why? Molecules of dye in the sweater absorb certain wavelengths of light energy. As the light energy is absorbed, it is changed into heat, which you feel. Whenever light energy is absorbed, it is changed to another form of energy. A solar cell traps light energy and transforms it to electric energy.

Not all light is absorbed when it strikes an object. A dark blue sweater appears blue because its molecules reflect mostly the blue portion of the spectrum. The portion of the spectrum reflected is the color of an object. A blue object reflects blue light; a red object reflects red light. Window glass and cellophane are common objects that transmit light. Reflection and absorption of light are important concepts in understanding the trapping of light in photosynthesis.

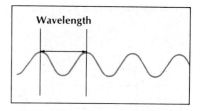

FIGURE 21–2. Light travels in waves. One wavelength is the distance between two consecutive crests. Different wavelengths of light are different colors.

White light is composed of many wavelengths.

When light energy is absorbed, it is changed to another form of energy.

Colors that you see are wavelengths reflected from an object.

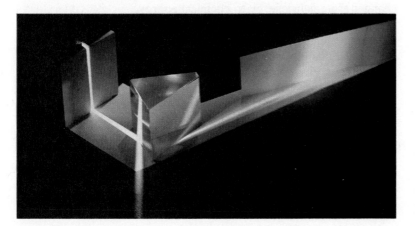

FIGURE 21–3. When white light is passed through a prism, light of different wavelengths is separated, producing a spectrum. The visible light spectrum includes wavelengths of 400 nm (violet) to 700 nm (red).

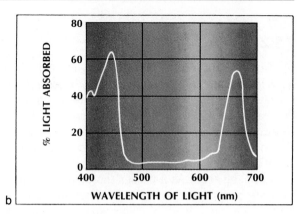

FIGURE 21–4. (a) When white light is passed through a solution of chlorophyll *a,* mostly green and yellow wavelengths are transmitted. Other colors are mainly absorbed. (b) A graph of the absorption spectrum of chlorophyll *a* shows the extent to which different wavelengths are absorbed.

Chlorophyll *a* absorbs certain wavelengths and transforms them into chemical energy.

In eukaryotes, chlorophyll is located in chloroplasts. Prokaryotes have no chloroplasts.

Various pigments interact to trap light energy needed for photosynthesis.

21:3 Chlorophyll and Other Pigments

Photosynthetic organisms contain the pigment chlorophyll (Section 1:4) that *traps light energy and changes it to chemical energy.* There are several types of chlorophyll, but chlorophyll *a* is the most important in photosynthesis. Its formula is $C_{55}H_{72}O_5N_4Mg$. Another form, chlorophyll *b,* differs by just a few atoms. In eukaryotes, chlorophyll is found in chloroplasts. Chlorophyll of cyanobacteria is attached to membranes in the cytoplasm. In other photosynthetic eubacteria, chlorophyll is located in areas called chromatophores.

Photosynthetic parts of autotrophs usually appear green because chlorophyll reflects most of the green portion of the spectrum. Certain other wavelengths are also reflected. Still others are absorbed. This pattern of absorption and reflection can be demonstrated by passing white light through a solution of chlorophyll *a* and then through a prism. Because certain wavelengths are absorbed by the chlorophyll, the spectrum produced is called an **absorption spectrum** (Figure 21–4). Violet and red wavelengths are greatly absorbed as well as some blue wavelengths. These absorbed wavelengths are changed from light energy to chemical energy.

Besides chlorophyll *a* and *b,* chloroplasts also contain yellow and orange pigments called carotenes (Section 13:4). Carotenes absorb mainly blue and green wavelengths of light and pass this energy to chlorophyll *a.* Thus, other wavelengths are also important in photosynthesis. Chlorophyll *a,* chlorophyll *b,* and carotenes probably interact to trap wavelengths necessary for photosynthesis.

Carotenes and other pigments are also present in other parts of a plant. These yellow, orange, and red colors are most often seen in flowers and fruits. Carotenes and other pigments in leaves often are masked by chlorophyll. In northern latitudes, the manufacture of chlorophyll stops in autumn. With less chlorophyll present, the carotenes and other pigments become visible so the leaves turn from green to yellow, orange, and red.

a

b

FIGURE 21–5. Leaves contain pigments other than chlorophyll. (a) In some plants, such as the coleus, these pigments are always visible. (b) In leaves of many trees, pigments other than chlorophyll become more obvious in autumn.

REVIEWING YOUR IDEAS

1. What is the chief function of leaves?
2. Why is glucose important to cells?
3. What is light? How does it travel?
4. What happens to light when it is absorbed?
5. Why is chlorophyll green? Why do green leaves turn orange or yellow in autumn?

PHOTOSYNTHESIS

Photosynthesis is a complex process by which an autotroph makes food. Like cellular respiration, photosynthesis is a set of many separate reactions. *It is the process by which light energy is absorbed and then converted to the chemical energy of glucose.* Because energy (light energy) is absorbed, photosynthesis is an endergonic process (Section 5:7).

In photosynthesis, light energy is absorbed and then converted to the chemical energy of glucose.

21:4 The Light Reactions

The many reactions of photosynthesis are summarized by the following equation:

$$6CO_2 + 6H_2O \xrightarrow[\text{light energy}]{\text{enzymes, chlorophyll}} C_6H_{12}O_6 + 6O_2$$

There are two main sets of reactions in photosynthesis. They are known as the light reactions and the dark reactions.

The **light reactions** involve a series of changes dependent on light. They occur in the grana (Section 4:13) of chloroplasts of eukaryotes. During these reactions, light is absorbed by chlorophyll and carotene molecules. These molecules trap the energy of light and pass it to a special chlorophyll molecule. Then, by a series of steps, the light energy is converted to the chemical energy of ATP molecules.

Another part of the light reactions involves the water used in photosynthesis. The water molecules are split into hydrogen ions and oxygen. The oxygen is given off as a by-product. The hydrogen ions are used in later steps of photosynthesis.

The oxygen given off during photosynthesis comes from water molecules.

Note that the light reactions do not involve carbon dioxide, and no glucose is produced. See Appendix C for more details about the light reaction. In summary, the light reaction involves two main events:

During the light reactions, light energy is changed to the energy of ATP.

1. Light energy is trapped and converted to chemical energy in the bonds of ATP.
2. Water is split into hydrogen ions and oxygen.

PEOPLE IN BIOLOGY

Charles French became fascinated with photosynthetic reactions as a college sophomore. Using simple, precise measurements of light absorption, gas exchange, and growth rate of photosynthetic cells, French has researched the process of photosynthesis. Dr. French has spent much of his life trying to concentrate the photochemically active parts of chloroplasts to learn how each part works. To aid him in his research, Dr. French invented several scientific instruments. These instruments include one that disintegrates chloroplasts and one that records the fluorescence (energy given off as light) of plant pigments. Using these and other instruments, French confirmed that the extra pigments in a plant cell, such as chlorophyll *b* and carotenoids, absorb energy from light and then transfer the energy to chlorophyll *a* for use by the cell. Since his retirement in 1973, Dr. French has been measuring the wavelengths of light used in some of the reactions that occur in photosynthesis.

For his work on photosynthesis, Dr. French was elected to the National Academy of Sciences and the American Academy of Arts and Sciences in 1963 and has received several honorary doctorates.

Charles Stacy French

(1907–   )

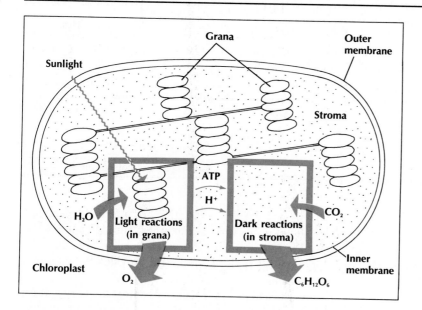

FIGURE 21–6. During photosynthesis, the light reactions provide for chemical energy and hydrogen needed for synthesis of glucose during the dark reactions.

21:5 The Dark Reactions

The events of the light reactions are preparation for the dark reactions. Synthesis of glucose occurs during the **dark reactions.** The dark reactions do not have to occur in the dark; they simply do not require light.

The dark reactions occur in the stroma (Section 4:13) of chloroplasts in eukaryotes. During these reactions, carbon dioxide and hydrogen are combined to form glucose. (Glucose is made of carbon, hydrogen, and oxygen — the same elements found in carbon dioxide and water.) This conversion requires energy. The energy comes from ATP. The hydrogen comes from water molecules. Both the ATP and hydrogen used in the dark reactions come from the light reactions. See Appendix C for more details about the dark reactions. Figure 21–6 shows the relationship between the light and dark reactions of photosynthesis.

Although glucose is the main product of photosynthesis, other important compounds are also made. Some of the glucose can be converted to fats and amino acids (Chapter 3), and some is used directly as an energy source. Glucose transported to other areas of a plant can be stored as starch or converted to cellulose, the substance contained in plant cell walls.

Synthesis of glucose occurs during the dark reactions.

The ATP and hydrogen atoms needed in the dark reactions come from the light reactions.

Glucose produced in photosynthesis may be converted to other compounds.

21:6 Energy Relationships

In some ways, photosynthesis is the opposite of respiration. The reactants (raw materials) of one reaction are the products of the other, and vice versa. Also, respiration is exergonic while photosynthesis is endergonic.

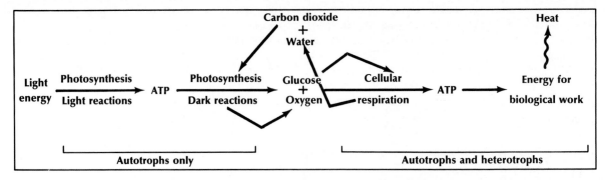

FIGURE   21–7. Photosynthesis and cellular respiration are interdependent processes. The products of each reaction are the reactants of the other. Note that materials (CO_2, H_2O, and O_2) are recycled, but energy is not recycled.

Photosynthesis and respiration are interdependent.

Glucose provides energy for the plant in which it was produced and for any animal that eats the plant or eats another animal that ate the plant.

Photosynthesis and cellular respiration are interdependent processes. One could not occur without the other. During photosynthesis, carbon dioxide and water, low-energy molecules, are converted to glucose, a high-energy molecule. Sunlight provides the energy to build glucose molecules.

Once formed, glucose is a rich source of potential energy. It can be used by the plant that produced it or be passed along a food chain (Section 1:4). Whether in the cells of producers or consumers, glucose is broken down by cellular respiration. As respiration occurs, what was originally light energy is transferred to ATP (Section 5:7). ATP is then used for biological work. Energy of ATP, once used, cannot be recycled. But the low-energy molecules, carbon dioxide and water, produced during cellular respiration can be recycled. These molecules are recycled as plants produce glucose by photosynthesis. Thus, more energy-rich glucose is made available. Can you see why glucose is referred to as the "crossroads" of energy in living systems?

REVIEWING YOUR IDEAS

6. What is photosynthesis?
7. What are the reactants of photosynthesis?
8. What are the two main sets of reactions in photosynthesis? In what organelle do they occur in eukaryotic cells? Where does each set of reactions occur within the organelle?
9. In what ways is photosynthesis the opposite of cellular respiration? Compare the two processes in terms of reactants, products, and energy.
10. In what ways is photosynthesis dependent on cellular respiration? How does cellular respiration depend on photosynthesis?
11. Compare the potential chemical energy of glucose with that of carbon dioxide and water.
12. Why is photosynthesis an endergonic process? What is the source of energy for photosynthesis?

DIGESTION IN PLANTS

Although most plants do not take in food and digest it, large molecules of starch, stored in roots or seeds, may be digested by enzymes. The digestion is intracellular. Digestion of starch results in glucose molecules that can be used as energy sources.

Intracellular digestion occurs in roots and seeds of plants.

21:7 Insectivorous Plants

Some autotrophic plants, although capable of photosynthesis, take in food in an organic form. They are called **insectivorous** (in sek TIHV uh rus) **plants** because they "eat" insects. These plants usually live in places such as bogs with soil low in nitrogen and other elements. Insects are an added source of these elements for the plants. Because insectivorous plants can synthesize food, they do not require insects as food for their survival.

Insectivorous plants trap and digest insects. The insects provide nitrogen and other elements for the plants.

One type of insectivorous plant is Venus's-flytrap. The upper part of a leaf of this plant is bordered by spines and is hinged along a midrib. There are sensory hairs on the upper surface of each leaf lobe. When these hairs are stimulated by the touch of an insect, the leaf quickly closes and traps the insect. Then, digestive enzymes are secreted into the hollow formed by the lobes, and extracellular digestion of the insect takes place. The digested molecules are absorbed and transported throughout the plant.

Other insectivorous plants include the sundew plant and the pitcher plant. Sundew plants trap insects differently from Venus's-flytraps. Long "tentacles" extending from the leaves secrete a sticky substance that traps small insects. Then, the "tentacles" bend inward toward the center of the leaf where the insect is digested. Pitcher plant leaves form pitcherlike structures. Insects are attracted to the pitchers and fall into the traps where they are digested by enzymes.

FIGURE 21–8. (a) The Venus's flytrap, (b) sundew plant, and (c) pitcher plant have adaptations that aid in catching insects. Insectivorous plants get nitrogen and other nutrients from insects.

a

b

c

21:8 Plant Nutrients

Water and carbon dioxide are materials necessary for photosynthesis. Carbon dioxide enters a plant from the surrounding environment. Water enters through roots of land plants or by osmosis in algae.

Plants require a variety of nutrients.

In addition, plants require a variety of other important nutrients (Section 18:1). For example, magnesium is needed because it is a part of chlorophyll. Nitrogen is needed for the production of ATP, the nitrogen bases of nucleic acids, and amino acids. Sulfur is also a part of several amino acids. Other important nutrients in plants include iron, phosphorus, potassium, and calcium.

These nutrients enter a plant in the form of ions (Section 3:5) dissolved in water. They pass into algae and mosses by passive or active transport. In higher plants, they are absorbed by roots and move through the stem to the leaves (Section 22:4) in vascular tissue.

REVIEWING YOUR IDEAS

13. Describe how starch is digested in plants.
14. What are insectivorous plants? Why are insects important to them?
15. List five nutrients required by plants.
16. How do these nutrients enter a plant?

ADVANCES IN BIOLOGY

FIGURE 21–9. Petrochemicals are used to make a variety of products.

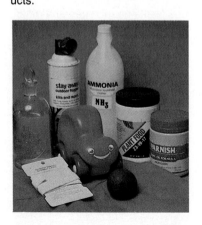

Fuel from Plants

Nearly all the energy used in the highly industrialized countries of Western Europe, the United States, and Japan comes from fossil fuels. **Fossil fuels** (coal, oil, and natural gas) are made deep within Earth from the remains of plants and animals and take millions of years to form. Fossil fuels are also called hydrocarbons because they are made up of hydrogen and carbon. These fuels not only run cars and produce electricity and heat, but extracts from the fuels, called petrochemicals (peh troh KEM ih kulz), are used to make plastics, synthetic fibers, detergents, pesticides, and many other products. In the face of increasing energy demands and decreasing supplies of fossil fuels, scientists are looking for other ways to produce fuels and petrochemicals similar to those derived from fossil fuels. Some scientists see plants as a possible solution to energy problems.

During photosynthesis, plants use sunlight, water, and carbon dioxide to make carbohydrates (Section 3:12). Some plants, such as sugarcane and corn, are very efficient in making large amounts of sugar. In order to make fuels, these sugars are fermented. In this process, microorganisms attack the sugars, take out some of the oxygen, and make alcohol. Alcohol can be mixed with gasoline to produce gasohol (GAS uh hahl) that will fuel a standard car. If the engines are slightly changed, cars will run on alcohol and water (95% alcohol, 5% water). Gasohol provides about the same engine performance and fuel mileage as gasoline. In Brazil, alcohol is being distilled further to produce substances similar to petrochemicals to be used in manufacturing. Because synthesis of these fuels is by artificial processes, they are called **synfuels** (SIHN fyewlz).

Production of alcohol for use in gasohol involves fermentation of sugars from sugarcane or corn.

Dr. Melvin Calvin of the University of California, Berkeley, is experimenting with plants of the Euphorbiaceae (yew for bee AY see ee) family in an attempt to produce synfuels. These plants, which include the rubber tree and gopher plant, take photosynthesis one step further and produce actual hydrocarbons. In the past, these hydrocarbons, in the form of latex, have been used to make rubber for tires. By using different species of *Euphorbia* (yoo FOR bee uh), Calvin has been able to produce a sap composed of about one-third hydrocarbon that may be a possible petroleum substitute.

Certain species of plants may be useful in producing hydrocarbons that may be used as a petroleum substitute.

In Brazil, Calvin has discovered a tree, the copa-iba tree, that produces an oily sap that can be used as diesel fuel without having to be processed. Calvin has germinated over 2000 trees to be cultivated in an experimental plantation to try to "grow" oil. A company in Japan is also trying to "grow" oil. By planting a *Euphorbia* in specially treated soil, they expect to produce between five and ten barrels of oil per acre per year. Some other species are expected to yield between ten and twenty barrels of oil per year.

Certain plants produce oily substances that can be used as fuels. These plants are being experimentally cultivated.

Some people are concerned about the use of plants for fuel. They feel that growing plants for fuel rather than growing plants to feed people and livestock may bring on an increase in world hunger. Also, land and oil may be depleted by overfarming of the fuel crops that take vital nutrients from the soil. If demand for fuel crops becomes high, marginal land that is not very productive may have to be used to grow food crops.

These problems have been recognized by scientists doing synfuels research, and some answers are being found. Many of the *Euphorbia* species currently being tested grow only in semiarid regions, so more productive land can be used for food crops. Dr. Calvin is experimenting with an artificial membrane that would work like plant membranes in converting the sun's energy to fuel. Instead of producing sugars, these membranes would make hydrogen and hydrocarbons. Using these artificial membranes would decrease or eliminate the need for using farmland to produce fuel.

FIGURE 21–10. *Euphorbia villosa is an oil producing plant.*

INVESTIGATION

Problem: How can you estimate the number of stomata on a leaf?

Materials:

microscope
water
dropper
glass slide
onion leaf

metric ruler
coverslip
single-edged
   razor blade

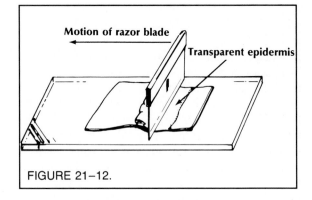

FIGURE 21-12.

Procedure:

1. Make a copy of Table 21-1. You will record your data in this table.
2. Obtain an onion leaf. Carefully cut it open lengthwise using a single-edged razor blade.
3. Measure the length and width of the leaf in millimeters. Use the widest part of the leaf when measuring its width. Record this data in Table 21-1.
4. Use a razor blade to carefully cut a small section of about 40 mm from the leaf. Position this section on a glass slide as shown in Figure 21-11. The dark green side must be facing down.

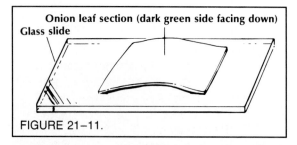

Onion leaf section (dark green side facing down)
Glass slide

FIGURE 21-11.

5. Add several drops of water to the leaf section. With a scraping motion in one direction only, use the razor blade to gently scrape away leaf tissue, leaving a small area of only the transparent epidermis. Use Figure 21-12 as a guide.

6. Cut away any remaining green tissue leaving only the epidermis. Add 2 drops of water to the transparent epidermis that remains. Add a coverslip and observe the epidermis under low power.
7. Slowly move the slide on the stage as you look for guard cells and stomata. Guard cells contain chloroplasts and look like a pair of lips. Two guard cells surround each stoma, which is simply the space between the guard cells. As the guard cells swell and relax, they change the size of the stomata.
8. Continue to search for stomata until you find an area that contains many of them.
9. Switch to high power. Count and record the number of stomata in this area. Consider this as area 1.
10. Move the slide to a new area and repeat step 9. Record this data as area 2.
11. Repeat step 9 three more times and record these counts as areas 3, 4, and 5.
12. Calculate the average number of stomata seen in the five fields of view. Record this number in Table 21-1.
13. Calculate the number of stomata on the entire onion leaf by completing Table 21-1. An example has been done for you.

Data and Observations:

TABLE 21–1. ESTIMATION OF NUMBER OF STOMATA ON A LEAF		
	Sample Data	Your Data
Length and width of leaf	length = 140 mm width = 10 mm	
Total area of leaf (length × width)	140 mm × 10 mm = 1400 mm²	
Number of stomata observed in: Area 1	4	
Area 2	6	
Area 3	3	
Area 4	5	
Area 5	2	
Total	20	
Average number of stomata observed	$\frac{20}{5} = 4$	
Number of high power fields of view on entire leaf (total area ÷ area of high power field of view –0.07 mm²)	$\frac{1400 \text{ mm}^2}{0.07 \text{ mm}^2} = 20\,000$	
Number of stomata on entire leaf (number of stomata in one high power field of view × number of high power fields of view on entire leaf)	20 000 × 4 = 80 000	

Questions and Conclusion:

1. Define:
   (a) guard cell
   (b) stoma
   (c) leaf epidermis
2. (a) Name the gases that enter and leave a leaf by way of the stomata.
   (b) Where do these gases originate or come from?
   (c) How are these gases related to photosynthesis?
3. Explain the relationship between stoma size and the action of guard cells.
4. Name two places in the procedure that would help to improve the accuracy of your final number of stomata for the entire leaf.
5. A student estimated the number of stomata on three different kinds of leaves:

Leaf	Upper epidermis	Lower epidermis
A	0 stomata/mm²	214 stomata/mm²
B	27 stomata/mm²	31 stomata/mm²
C	350 stomata/mm²	0 stomata/mm²

Use the following plant types to match with plant A, B, or C: water lily, cactus, oak tree. Explain your choices.

Conclusion: How can you estimate the number of stomata on a leaf?

CHAPTER REVIEW

SUMMARY

1. All forms of life depend directly or indirectly on plants for food. **p. 433**
2. The structure of a leaf is suited to its function of photosynthesis. **21:1**
3. When light strikes an object, certain wavelengths may be absorbed and changed to another form of energy. **21:2**
4. Chlorophyll *a*, chlorophyll *b*, and carotenes each absorb certain wavelengths of light. The energy of the wavelengths absorbed is changed to chemical energy. **21:3**
5. During the light reactions of photosynthesis, light energy is converted to chemical energy of ATP. Water is split into hydrogen and oxygen. The oxygen is released. **21:4**
6. In the dark reactions of photosynthesis, the ATP and hydrogen from the light reactions are used in the synthesis of glucose. Glucose made in photosynthesis may be used in respiration or converted to fats, amino acids, starch, or cellulose. **21:5**
7. Photosynthesis and respiration are interdependent reactions. During photosynthesis, energy is used to produce glucose from carbon dioxide and water. During respiration, energy is transferred to ATP as glucose is converted to carbon dioxide and water. ATP energy is used for work. **21:6**
8. Most plants undergo intracellular digestion as they convert starch to glucose. Insectivorous plants obtain needed nitrogen and other nutrients by the extracellular digestion of insects. **21:7**
9. In addition to carbon dioxide and water, plants require nutrients such as nitrogen, magnesium, sulfur, potassium, phosphorus, calcium, and iron. **21:8**
10. Scientists are doing experimental work to derive fuels from plants without using farmland. **p. 443**

LANGUAGE OF BIOLOGY

absorption spectrum
dark reactions
fossil fuel
guard cell
insectivorous plant
leaf
light reactions
palisade cells
phloem
photosynthesis
spongy layer
stoma
synfuel
upper epidermis
visible spectrum
wavelength

CHECKING YOUR IDEAS

On a separate paper, indicate whether each of the following statements is true or false. Do not write in this book.

1. Chlorophyll *a* absorbs mostly wavelengths of light in the green part of the spectrum.
2. Overall, the reactions of photosynthesis are endergonic.
3. Synthesis of glucose occurs in the dark reactions.
4. The products of respiration are the reactants of photosynthesis.
5. Hydrogen in a glucose molecule comes from water.
6. Light energy is changed to chemical energy of glucose during the light reactions.
7. Carbon dioxide and water are the only nutrients of plants.
8. Digestion is extracellular in most plants.
9. Carotenes are present in leaves only during autumn.
10. Oxygen given off during photosynthesis comes from carbon dioxide.

EVALUATING YOUR IDEAS

1. List the parts of a leaf and give the function of each.
2. In terms of energy, why is photosynthesis so

important to all living systems? How does photosynthesis affect you?

3. What does an absorption spectrum reveal about chlorophyll *a*? Explain the significance of the absorption spectrum of chlorophyll *a*.

4. What are carotenes? When are they visible in most leaves?

5. What are the reactants and products of photosynthesis?

6. Describe the major events of the light reactions.

7. Explain how the light reactions and dark reactions are related.

8. Once glucose is produced in photosynthesis, what are some ways in which it is used by the plant?

9. How is photosynthesis the opposite of respiration?

10. How is the chemical energy produced during photosynthesis used? What happens to some of the chemical energy?

11. Write an essay about the interdependence of photosynthesis and respiration. Your essay should include a discussion of both materials and energy.

12. How does a Venus's-flytrap trap and digest insects?

13. Explain how magnesium, nitrogen, and sulfur are used by plants.

14. Explain several ways in which photosynthesis might be used to provide new fuels.

APPLYING YOUR IDEAS

1. Relate the events of photosynthesis to the structure of the leaf.

2. What would the absorption spectrum of a red pigment be like? Explain.

3. Why are fertilizers often used in growing plants?

4. Explain how photosynthesis and respiration result in the recycling of materials.

5. Where does the oxygen necessary for aerobic respiration originate?

6. Trace the "flow of energy" from light to energy used for work. Is energy recycled? Explain.

7. Nitrogen is important to plants. What is the source of available nitrogen? How is it made available to plants?

8. What is phosphorus used for in plants?

EXTENDING YOUR IDEAS

1. Using reference materials, prepare a brief report on the contributions of van Helmont, Priestley, and Ingen-Housz to our understanding of photosynthesis.

2. Will a developing plant grown in the dark produce chlorophyll? Conduct a controlled experiment to determine the answer. As a follow-up, test other developing plants using different colored lights. Does any particular color seem to favor chlorophyll production?

3. Conduct an experiment to determine the effect of different wavelengths of light on the rate of photosynthesis.

4. Find out how Spanish moss, which grows on the branches of trees, obtains its nutrients.

5. Grow two groups of insectivorous plants during the school year. Keep all conditions for both groups the same except feed group A insects and do not feed group B. Compare the growth and appearance of the two groups.

SUGGESTED READINGS

Honda, Makoto, "Cobras of the Pacific Northwest." *Natural History,* April, 1983.

"Why We Need Tropical Forests." *Bioscience,* June, 1984.

Yulsman, Tom, "How Do Plants Harness the Sun?" *Science Digest,* Oct., 1985.

PLANTS: OTHER LIFE FUNCTIONS

In addition to photosynthesis, plants carry out a variety of other life functions. Materials such as water, dissolved minerals, and food must be transported. Oxygen and carbon dioxide must be exchanged with the environment. Pictured here is a touch-me-not. How is the plant adapted for water transport? What other life processes are carried out by plants?

chapter 22

Green plants produce glucose by photosynthesis. The energy stored in glucose is then used for a variety of life processes. One use of energy is reproduction and development (Chapter 20). In this chapter, you will learn about other activities that plants carry out—transport, gas exchange, and response to factors of the environment.

Objectives:
You will
- explain gas exchange and transport in simple and vascular plants.
- discuss the role of hormones in various plant responses.

STRUCTURE OF ANGIOSPERMS

Recall from Chapter 14 that the plant kingdom includes certain algae and the land plants. Algae have a simple structure. Most land plants are more complex, having tissues, organs, and systems.

Land plants can be grouped as nonvascular, those without special transport cells, and vascular, those with special transport cells. The nonvascular plants include liverworts and mosses. Vascular plants include club mosses, horsetails, ferns, gymnosperms, and angiosperms. Of the vascular plants, the angiosperms, flowering plants, are predominant. Therefore, it is important to your understanding of plant functions that you have a knowledge of the internal structures of angiosperms. You learned about angiosperm leaf structure in Chapter 21. Root and stem structure are presented here.

Liverworts and mosses are nonvascular. Other land plants are vascular.

22:1 Roots: Internal Structure

The outer part of an angiosperm root is the **epidermis,** which protects the root. **Root hairs** are outgrowths of the epidermis. Much of a root is **cortex,** which is used for storage. Within the cortex is a central cylinder called the **stele** (STEEL). The stele of the root is

Various tissues interact to carry out functions of roots.

FIGURE 22–1. Cross sections of dicot and monocot roots show that both have a central cylinder composed of xylem and phloem. However, the arrangements of xylem and phloem differ.

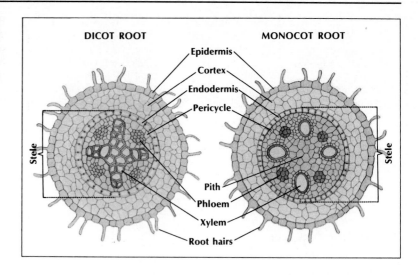

surrounded by a layer of cells called the **endodermis** (en duh DUR mus). The outer tissue of the stele is a layer of cells called the **pericycle** (PER uh si kuhl). The pericycle is a kind of meristem (Section 20:13). Branch roots originate in the pericycle. The inner part of the stele is composed of the vascular tissues — xylem and phloem.

Xylem cells carry water and minerals from the roots to the leaves of the plant. These cells are thick-walled and dead. They are hollow cylinders arranged end to end to form a tube. In cross section, they appear as empty circles. Phloem cells carry food made by the leaves to other parts of the plant. These phloem cells are also cylindrical but are not dead. They contain cytoplasm and have thinner walls and a smaller diameter than xylem cells.

Xylem cells transport water and minerals. Phloem cells transport food.

In a dicot root, xylem tissue is usually arranged in a star-shaped pattern with phloem vessels between the arms of the star (Figure 22–1). In a monocot root, the stele is somewhat larger than that of the dicot. It contains a great deal of **pith** tissue in the center. Pith aids in storage of food and water. Large xylem vessels are arranged in a circle around the pith with phloem vessels between them (Figure 22–1).

22:2 Stems: Internal Structure

A plant has no skeletal system but is supported by the **stem.** Also, the stem is a transport link between the roots and the leaves, and some stem cells store food.

Herbaceous stems are soft and green.

Stems are either herbaceous (hur BAY shus) or woody. **Herbaceous stems** are soft, green, and often juicy, with very little or no woody tissue. Plants that live for a single growing season are usually herbaceous. A plant with a herbaceous stem is supported by the pressure of water in the cells of the stem. This pressure causes turgidity,

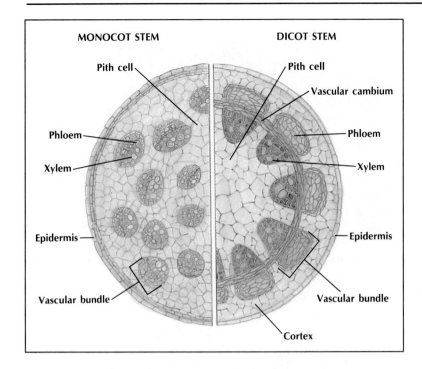

FIGURE 22-2. Herbaceous monocot and dicot stems both have vascular bundles. In monocots, the bundles are scattered throughout the stem. In herbaceous dicots, they are located close to the edge of the stem in a ring.

or stiffness, that causes the plant to stand upright. In the event of a water shortage, the stem loses turgidity (wilts), and bends down.

As in roots, there are slight differences between monocot and dicot stems. In herbaceous monocots, vascular bundles are scattered throughout the stem. In herbaceous dicots, they are located close to the edge of the stem, and there is usually a layer of vascular cambium between phloem and xylem. The vascular cambium is a meristem that produces cells that later develop into new xylem and phloem tissue. Both stem types contain pith for food storage. Also, the dicots have cortex. Some cortex tissue stores food while other cortex tissue contains chloroplasts and makes food.

A woody stem is more complex than a herbaceous stem. The outermost part of the stems of trees or woody shrubs is called **outer bark.** It is composed of special cells, cork, produced from cork cambium, the deepest layer of outer bark. Cork cells are coated with an oily material called **suberin** (SEW buh run) that protects inner cells against water loss. Inside the outer bark lies a region of **inner bark** made up of cortex and phloem tissue. This cortex is a food storage layer. As the plant ages, cortex is crowded out by other cells.

Inside the inner bark lies the vascular cambium. This cambium produces new xylem and phloem cells each growing season. Phloem cells are formed toward the bark side. Xylem cells are formed toward the center of the stem. Phloem cells flake off from the bark of a woody stem. Thus, the phloem layer never becomes very thick. However,

Vascular bundles of most dicots contain a vascular cambium, a meristem that gives rise to new xylem and phloem cells.

The outer bark protects the stem of woody plants and helps prevent water loss.

FIGURE 22–3. A cross section of a woody stem shows that much of the stem is made of xylem, the wood of a tree.

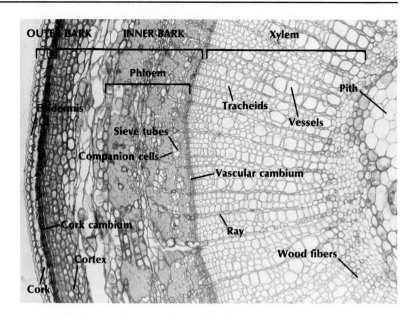

Mainly because of buildup of xylem, woody stems continue to grow in diameter.

xylem remains within the stem and builds up each year. Old xylem is the wood that forms the bulk of a woody stem.

Xylem closest to the cambium is active, or new, xylem tissue that transports water and minerals up the stem. This xylem is composed of two kinds of cells called **tracheids** (TRAY kee udz) and **vessels.** Closer to the center of the stem are old xylem cells filled with waste matter. Old xylem makes up most of the xylem. Cells closest to the center of a stem are the oldest because new cells are produced from the cambium. Wood produced in the spring, **spring wood,** is mostly large xylem vessels. **Summer wood** is composed of vessels of smaller diameter. In a cross section of a woody stem, lines are visible between layers of cells. These circular lines are the **annual rings** that represent growth during a year. Annual rings provide a means of determining the age of a tree.

Annual rings, lines of growth, are a means of determining the age of a tree.

FIGURE 22–4. (a) Annual rings of a woody stem indicate the age of the tree. How old was this locust tree when it was cut? (b) Core samples can be taken from a tree so that the age can be determined without cutting the tree down to count rings. The number of bands on the core sample indicate the tree's age.

a

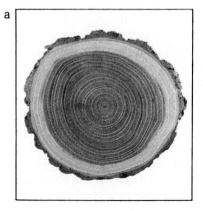

b

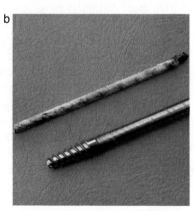

The center of a woody stem is composed of pith cells that store nutrients. The area of pith in a young stem is large, but it "shrinks" as the stem grows because no new pith is produced. As the stem grows, a series of rays form which extend from the inner bark to the pith. **Rays** transport nutrients laterally between phloem and pith.

Careful study of a cross section of a woody stem shows that structure and function complement each other. A stem transports materials and also supports a plant. Cells and tissues are adapted for these functions. For example, the "pipeline" structure of active xylem vessels transports water. Dead xylem fibers closely packed are tough materials that support the stem as it grows upright.

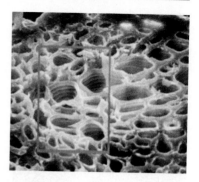

FIGURE 22–5. Xylem tissue of a woody stem magnified 250 times. Xylem transports water in plants.

REVIEWING YOUR IDEAS

1. What is the stele of a root? What kind of tissues are located in the stele?
2. Distinguish between herbaceous and woody stems.
3. Compare the location of vascular bundles in monocot and dicot stems.
4. What are annual rings?
5. How are structure and function of stem parts related?

TRANSPORT AND GAS EXCHANGE

The major materials transported within plants are water, dissolved minerals, and food. Oxygen and carbon dioxide are exchanged directly with the environment.

22:3 Simple Plants

In algae, there are no specialized tissues for transport or gas exchange. Most cells of algae act independently. Each is in direct or very close contact with the environment. Water, minerals, and gases enter each cell directly from the water. Also, most cells make food by photosynthesis. Thus, food "transport" is limited to within the cell and occurs by diffusion or cytoplasmic streaming.

Liverworts and mosses are more complex than algae. However, they are small plants that can distribute materials without specialized transport tissues. Water enters the rhizoids and is transported to other cells by osmosis. Food moves to all cells from photosynthetic cells by diffusion. Gases are exchanged through pores in liverworts.

Algae have no specialized tissues for transport or gas exchange.

In liverworts and mosses, transport occurs by diffusion or osmosis. Gases are exchanged through pores in liverworts.

FIGURE 22-6. Root hairs increase surface area for absorption of water. Water moves into root hairs by osmosis or through intercellular spaces.

FIGURE 22-7. Root pressure may contribute to the transport of water through stems. Root pressure can support a column of water as shown. However, there is not enough pressure to account for transport over long distances.

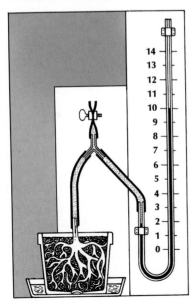

22:4 Uptake of Water and Minerals in Vascular Plants

Most vascular plants live on land and obtain water from the soil through their roots. Water present in a film of moisture around soil particles may move into root hairs by osmosis. The millions of tiny root hairs together have a very large surface area for absorbing water. Entrance of water into the epidermal cells results in a greater concentration of water molecules there than in the layer of cortex cells inside. Thus, water moves by osmosis into the cortex. From there, movement of water continues across the many cell layers of a root until the water finally enters the xylem.

Some water may enter a root without passing through cells. Instead of passage by osmosis, the soil water moves through spaces between root cells until it reaches the endodermis. The water then passes through the endodermis and into the xylem.

Mineral ions enter roots by diffusion, facilitated diffusion, or active transport. Some minerals are used in root cells or are simply stored there. Other minerals enter the xylem. The xylem of roots extends into the stem and to the leaves. Water and dissolved ions are transported upward through this "pipeline."

22:5 Transpiration-Cohesion Theory

Water entering the xylem in roots must travel through the stem to the leaves. Sometimes this movement involves a distance greater than one hundred meters. How can this process occur?

Water enters a root and passes into the xylem of the stele. The water in the xylem exerts a pushing force called **root pressure.** Perhaps you have seen the effects of root pressure. Sometimes, when root pressure is high, water is forced out of the ends of leaf veins. The water forms droplets around the edges of the leaf, a process known as guttation.

Root pressure can also be shown by an experiment. A piece of tubing can be attached to the cut stem of a plant whose roots are immersed in water. Water will rise in the tube (Figure 22-7). Root pressure may play a role in transport of water in small plants. However, it cannot explain water transport in most plants.

Most biologists favor another explanation, called the **transpiration** (trans puh RAY shun)-**cohesion** (koh HEE zhun) **theory** for water transport through a stem. According to this theory, water is pulled up the stem. In order for such a pull to occur, water must form continuous columns from roots to leaves. Two properties of water, cohesion and adhesion (ad HEE zhun), make this water movement possible. **Cohesion** is the clinging together of the same kind of molecules, and **adhesion** is the attraction of unlike molecules. Cohesion

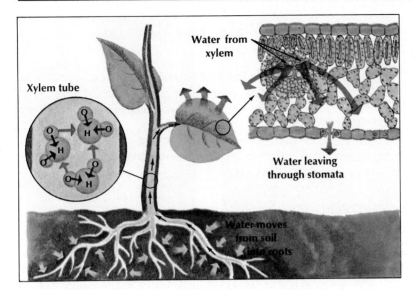

Xylem tube

Water from xylem

Water leaving through stomata

Water moves from soil into roots

FIGURE 22–8. Movement of water in tree stems is explained by the transpiration-cohesion theory. Cohesion among water molecules in a xylem vessel makes the water a continuous column. Transpiration from leaves exerts a "pull" on the columns of water in xylem tubes.

causes water molecules to be "sticky." Also, they adhere to the walls of the xylem tissues. *In these very narrow tubes,* the combination of cohesion and adhesion gives a column of water the properties of a metal wire — continuous with tightly packed molecules. This assumption is important to the theory.

What causes the pull? Water constantly is leaving the leaves of plants via stomata (Section 21:1). This water loss is called **transpiration.** As water molecules leave, other water molecules from the xylem vessels of the leaf replace them. Because of cohesion, this movement creates a *tension, or pull,* on the rest of the water column. This pull can be thought of as a "stretching" effect. The "stretching" force extends all the way to the roots. The more quickly water is lost, the more quickly it is pulled up through the plant. As the water in the root xylem is pulled up, more water moves into the xylem from the surrounding root cells. The water comes from spaces in the soil.

Transpiration pull is not the same as drawing a liquid through a straw. Even a vacuum pump could not lift water to the height it travels in tall trees. The pull occurs only because the column of water is continuous. Without cohesion and adhesion of water in the narrow xylem tubes, transport could not occur. The combination of these properties produces an unbroken column of water that can be pulled.

It is important that no air gets into a column of water. Air bubbles would break the continuous column of water. Thus, they would prevent an effective tension on the water column.

There is experimental evidence to support this theory. But, scientists think that although this explanation is the best available, it is not the complete story. Some details must still be worked out.

Because of cohesion and adhesion, water in xylem vessels forms a column similar to a thin wire.

Transpiration creates a pull on the water columns in xylem, and water is "stretched" from roots to leaves.

Air bubbles in a xylem vessel would prevent transpiration pull on the water.

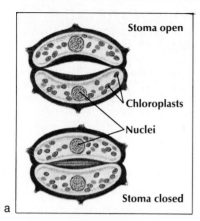
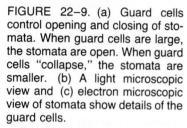

Stoma open

Chloroplasts

Nuclei

Stoma closed

a

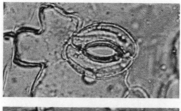
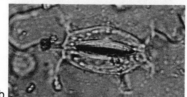

b

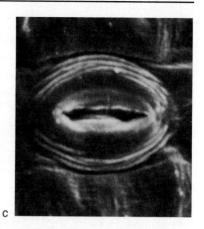

c

FIGURE 22–9. (a) Guard cells control opening and closing of stomata. When guard cells are large, the stomata are open. When guard cells "collapse," the stomata are smaller. (b) A light microscopic view and (c) electron microscopic view of stomata show details of the guard cells.

During the day, stomata are large, permitting carbon dioxide to enter and water to escape.

Decrease in the size of stomata at night reduces water loss.

In vascular plants, phloem tissue transports food.

22:6 Control of Stomata

Transpiration is important in bringing water and its dissolved mineral ions to upper parts of the plant. However, if too much water escapes, a plant will die. An adaptation involving guard cells and stomata prevents excess water loss.

During the day, carbon dioxide enters a leaf, so the stomata must be open. Guard cells have chlorophyll and undergo photosynthesis when light is available. As photosynthesis occurs, guard cells swell, their shape changes, and the size of the space (the stoma) between them is increased (Figure 22–9). Carbon dioxide can then enter the leaf. However, water also escapes from the stomata during the day. Transpiration is this loss of water from stomata.

At night, because light is not present, no photosynthesis occurs in the guard cells. The guard cells collapse, their shape changes, and they move together, reducing the size of the stomata between them. As stomata become smaller, loss of water is decreased. In summary, this adaptation prevents excess water loss. Some water escapes during the day when the stomata are fully open. However, at night the openings become smaller, so less water escapes.

22:7 Transport Through Phloem

Food is transported by phloem cells located in the vascular bundles. Like xylem tissue, phloem tissue forms a continuous "pipeline" from leaves to roots. Because phloem cells do not lose their cytoplasm as they mature, the pipeline they form is not hollow.

The end wall of each phloem cell is perforated so it resembles a sieve. Therefore, phloem tubes are called **sieve tubes.** The sidewalls of phloem cells have small holes, and strands of cytoplasm pass horizontally between phloem cells.

Although the cytoplasm remains in mature phloem cells, the nucleus dies. Smaller **companion cells,** which have nuclei, often lie next to sieve tubes. Perhaps they control the cytoplasm of sieve tubes.

As food is made within a leaf, it is dissolved in water. Dissolved food enters the phloem cells of the veins (Section 21:1). From here, food may be transported in any direction within the plant. The transport of food within a plant is called **translocation** (trans loh KAY shun). Usually, food moves downward through the phloem of the stem into the roots where it enters the cortex. In the cortex, simpler sugars are converted to starch and stored. Food also may be transported laterally through the strands of cytoplasm between cells.

Exactly what causes the transport of food within phloem is not known. One hypothesis suggests that food moves through the phloem as a result of pressure. As sugar enters cells of sieve tubes in a leaf, water enters from other leaf cells by osmosis. The entrance of water increases the pressure inside the sieve cell. Because of this pressure, the sugar and water are forced into the next sieve cell where pressure is lower. These differences in pressure in sieve cells from leaves to roots may account for the downward movement of food. At the roots, both sugar and water leave the sieve cells. Thus, pressure remains lower in the roots. This explanation is known as the **pressure-flow hypothesis.** Further testing is needed to determine whether it is correct.

In addition to food, certain mineral ions move through phloem. These ions are first transported upwards in xylem. Later, they may move up or down to places where they are needed. For example, phosphorous ions move through phloem from older to newer leaves.

22:8 Gas Exchange

In plants, carbon dioxide is a raw material for photosynthesis, and oxygen is given off as a by-product. Therefore, you may think of gas exchange in plants as an inward movement of carbon dioxide and an outward flow of oxygen. This exchange occurs only in photosynthetic tissues when light is present. However, all plant tissues, both photosynthetic and nonphotosynthetic, take in oxygen for cellular respiration and give off carbon dioxide.

Within roots, cells obtain minerals from the soil by active transport. Energy needed for active transport and other processes comes from cellular respiration. The oxygen needed for respiration is in the air spaces between soil particles. Oxygen from the air dissolves in the moisture of the soil and then diffuses into the root hairs. Then it is

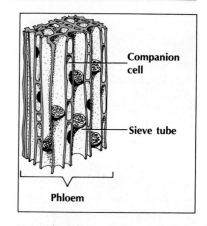

FIGURE 22–10. Sieve tubes, the conductive parts of phloem tissue, are in close association with companion cells. Food moves through plants in the phloem.

Pressure differences between sieve cells may move food through phloem.

Some mineral ions are translocated through phloem.

Root cells exchange gases with soil.

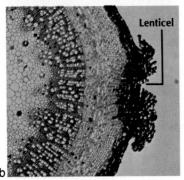

FIGURE 22–11. (a) Lenticels are visible as white dots on this twig. (b) A cross section of a stem through the lenticel shows the lenticel as an opening in the tissue.

Both oxygen and carbon dioxide may be used by leaves.

distributed by diffusion to other root cells such as the cortex. As respiration occurs within the root cells, carbon dioxide is produced as a by-product. This carbon dioxide diffuses from the root cells to the root hairs and then from the root hairs into the soil.

Most of a woody stem is composed of dead cells (Section 22:2). However, some cells of the outer and inner bark region are alive and need oxygen for respiration. The combination of dead cork cells and suberin forms a continuous protective layer around the stem. However, oxygen can enter and carbon dioxide can leave the stem through tiny openings called lenticels (Section 14:13). Lenticels also are found on large roots lacking root hairs. These roots, like stems, are covered by a layer of cork.

As photosynthesis occurs in leaves, carbon dioxide enters and oxygen is given off. The photosynthetic cells are surrounded by a thin film of moisture. Carbon dioxide in the air enters a leaf through stomata (Section 22:6). The carbon dioxide enters air spaces, dissolves in the thin film of moisture, and then diffuses into the cells. Oxygen, a by-product of photosynthesis, diffuses from the cells, moves into the air spaces, and leaves through stomata. There are a great many palisade and spongy cells within each leaf. Taken together, these cells provide an enormous amount of surface area for gas exchange. Thus, photosynthesis occurs very efficiently.

Because cells in the leaf also undergo respiration, some of the oxygen from photosynthesis may be used for respiration. Also, some of the carbon dioxide from respiration may be used in photosynthesis. At night, only cellular respiration occurs in leaves. At that time, there is an intake of oxygen, and carbon dioxide is released. Photosynthesis does not occur at night because the light reactions require energy of sunlight.

REVIEWING YOUR IDEAS

6. What are the major materials transported in plants?
7. How are materials transported in simple plants? How do the materials enter cells of simple plants?
8. What is the function of root hairs?
9. Distinguish between xylem and phloem. What does each of them carry?
10. How does water reach the stele of a root?
11. What theory accounts for the transport of water and materials in vascular plants?
12. Distinguish between cohesion and adhesion.
13. What hypothesis explains the translocation of food in vascular plants?
14. What is the function of guard cells?

RESPONSE TO ENVIRONMENT

Although their activities are usually not directly observable, plants are by no means inactive. Plants, like other organisms, respond to changes in the environment. Plants respond to certain stimuli such as light, water, and gravity. A **stimulus** is anything that causes an activity or change in an organism. Many responses of plants involve growth. Such responses usually occur slowly. Thus, they are not always obvious. Control of most plant responses involve chemical signals within the organism.

Many plant responses involve growth and occur slowly.

22:9 Discovery of Auxins

Many plants respond to stimuli by movements called tropisms (TROH pihz umz). A **tropism** is a plant growth response caused by unequal stimulation on opposite sides of a plant. Growth of a plant part toward the stimulus is a positive tropism. Growth away from the stimulus is a negative tropism. One part of a plant may have a positive tropism to a certain stimulus. But, another part of the plant may have a negative tropism to the same stimulus. For example, the shoot of a plant has negative **geotropism** (jee oh TROH pihz um), or response to gravity, but the roots have positive geotropism.

Plants exhibit positive and negative tropisms.

Biologists have long been interested in these plant responses. For example, stems show positive **phototropism** (foh toh TROH pihz um), a response of growing toward light. How is growth toward light adaptive? How is it controlled?

If a young oat seedling is exposed to a light source from one side, the leafy tip of the seedling, called the **coleoptile** (koh lee AHP tul), bends in the direction of the light. But, if the tip is covered with an opaque substance, such as foil, no bending occurs. Also, no bending occurs if the tip is removed (Figure 22–12). Thus, the tip must be involved in positive phototropism, but what role does it play in this response?

The tip of a coleoptile is involved in positive phototropism of the shoot.

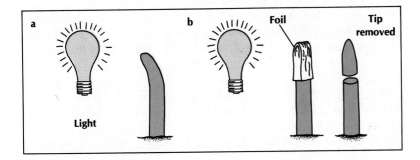

FIGURE 22–12. (a) A seedling coleoptile bends toward a light source. (b) If the tip is covered with foil or removed, no bending occurs.

FIGURE 22–13. Experiments with seedlings indicate that a chemical in the coleoptile is involved in phototropism. (a) The chemical cannot pass through mica, so no bending occurs, (b) but can pass through or (c) into and out of agar blocks. (d) The unequal distribution of the chemical causes the bending.

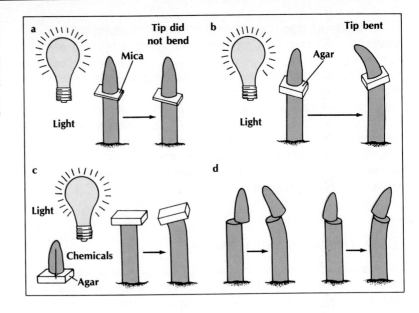

A chemical controls positive phototropism.

Perhaps a control chemical is secreted from the tip. To test this hypothesis, experiments were done in which different materials were placed between the tip and the shoot. In one experiment, a piece of mica (MI kuh) was inserted between the tip and the shoot. (Chemicals cannot diffuse through mica.) The plant was exposed to a light source from one side. If the assumption were correct, the plant should not bend toward the light. In the experiment, the plant did not bend (Figure 22–13a).

In a second experiment, a piece of agar was placed between the tip and shoot. (Chemicals can diffuse through agar.) This plant responded positively to the light (Figure 22–13b). The scientists concluded that a chemical produced by the tip plays a role in directing positive phototropism.

Scientists removed the tips and placed them upon agar blocks. They assumed that the chemical would diffuse into the blocks. These blocks alone were then placed on the rest of the shoot, and the plants bent in the direction of the light source (Figure 22–13c).

Scientists began to suspect that the chemical controls lengthwise growth and that the side of the shoot away from light received more chemical from the tip. To test this idea, tips were put on the shoot in an off-center manner (Figure 22–13d). The plants grew away from the side with the tip on it. This result occurred even in complete darkness. Thus, unequal chemical concentration on the two sides of the shoot must be responsible for the bending. Also, agar blocks containing the chemical were placed on one side of the shoot. The same effect was obtained — the plants grew away from the side with the tip on it, even in darkness.

In 1928, the chemical responsible for positive phototropism was isolated. The chemical is one of a group of plant hormones known as **auxins** (AWK sunz). A **hormone** is a chemical regulator that is produced in one part of the organism and affects other parts called target organs. The auxin that causes the lengthening of cells is called indoleacetic (IHN dohl uh seet ihk) acid (IAA).

How a stem grows toward light can now be explained. Light coming from one side causes auxin molecules to migrate away from the light. Thus, more auxin becomes concentrated on the shaded side. The auxin moves downward, causing cells on the shaded side to elongate more quickly than cells on the lighted side. Because of unequal growth rates, the stem grows (bends) toward light. In a plant exposed equally to light from all sides, the concentration of auxin is the same throughout the stem. Thus, the stem grows, but does not bend. It grows straight up.

Auxins are a type of plant hormone.

Unequal auxin concentrations on the lighted and shaded sides of a stem result in unequal cell elongation and positive phototropism.

PEOPLE IN BIOLOGY

Henrik Gunnar Lundegardh

(1889–1969)

Although most of his research concerned absorption and accumulation of salts in plant roots, Henrik Lundegardh spent more than ten years studying tropisms. Working at the laboratory of plant physiology in Lund, Sweden, Lundegardh published a series of papers on plant movements. Geotropism, the effect of gravity on plants, and phototropism, the reaction of plants to light, were studied and reports were presented to groups of plant physiologists. Dr. Lundegardh's research led to the use of film as a scientific method of documenting and recording the movements of plants under varying conditions. Even though Lundegardh did his work before the discovery of auxins (plant hormones that affect growth), his research illustrated the movements and effects of auxins within a plant.

Dr. Lundegardh was a professor of plant physiology and botany at several universities and institutions in Sweden. In 1947 Lundegardh built a private research laboratory to study the chemical reactions of photosynthesis. Dr. Lundegardh was elected a foreign member of the American Academy of Arts and Sciences in 1950 and of the U.S. National Academy of Sciences in 1964.

FIGURE 22–14. If a plant is placed on its side, (a) auxin causes negative geotropism of the stem. (b) Positive geotropism of roots is controlled by abscisic acid.

22:10 Other Effects of Auxins

Stems grow toward light, but show negative geotropism. That is, they grow away from the stimulus of gravity. A plant placed on its side, even in the dark, will make this response. The stem will bend upward. How does this happen?

Auxin causes negative geotropism in stems.

Negative geotropism of stems is also controlled by auxin. Mainly because of gravity, more auxin collects in cells of the lower side of a stem placed on its side. Because of the difference in concentration of auxin, cells on the lower side elongate more quickly than those of the upper side. As a result, the stem grows upward.

Positive geotropism of roots is controlled by abscisic acid.

Unlike stems, roots exhibit positive geotropism. This response is caused by a different hormone, abscisic acid. This hormone travels from the root tip to the elongation zone (Section 20:13). There it becomes more concentrated in cells on the lower side of the root. **Abscisic acid** *inhibits* growth of root cells. Thus, the root grows downward.

Auxin produced by the apical bud inhibits development of lateral buds.

Auxins are responsible for responses other than tropisms. One such response involves plant development. The apical (terminal) bud at the tip of a stem is a meristem tissue (Section 20:13). It produces auxin that travels downward through the stem. The high concentration of auxin stimulates the stem cells to elongate, but it inhibits much of the growth of the lateral buds. Thus, the lateral buds near the top of the stem (where auxin concentration is highest) do not develop into branches. Those farther from the apical bud develop because the auxin concentration there is lower. The responses by different parts of the plant result in the plant's having a few small branches near the top and more large branches near the bottom. This control of lateral bud development by the apical bud is known as **apical** (AY pih kul) **dominance.**

If the apical bud is removed, auxin production ceases. The lateral buds near the top can grow because they are not inhibited by high

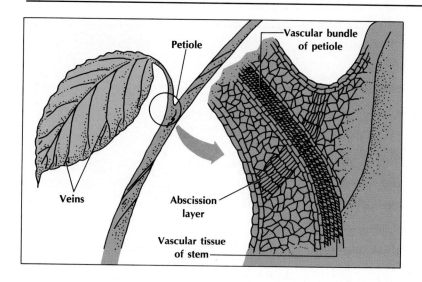

FIGURE 22–15. Decreased auxin production causes an abscission layer to form across a leaf petiole. The cells weaken in this area and the leaf eventually falls off.

concentrations of auxin. As a result, the plant becomes bushier. Many gardeners produce bushy plants by "pinching off" the apical buds.

The role of auxin in apical dominance reveals an important point about hormones. Both stem cells and lateral bud cells are stimulated by auxin. However, the responses of the two types of cells differ. *In general, it is the target organ, not the hormone, that determines the response.*

The target organ, not the hormone, determines the response.

Synthetic auxins are sometimes applied to potatoes to prevent the growth of the buds. Buds are the "eyes"; they tend to sprout with age. The high concentration of auxin inhibits the growth of the buds, and the potatoes may be stored for longer periods of time.

Auxins can be used to prevent potatoes from sprouting.

Auxin inhibits **abscission** (ab SIHZH un), the falling of leaves from trees. During the growing season, the leaf blade produces auxin constantly and leaves remain attached. In autumn, auxin production slows. Another hormone then causes the formation of a band of cells, the abscission layer, across the petiole. Cell walls in this layer become weakened, and eventually the leaf falls from the plant.

Decrease in auxin production by leaves causes the formation of an abscission layer across the petiole.

Practical applications have come from this knowledge. Spraying plants with the proper auxins can delay the falling of leaves and fruits from a tree. The falling of fruits before they are fully developed and ready for eating is prevented.

22:11 Gibberellins

Another group of plant hormones was discovered in Japan. For years, rice farmers had noticed many young, unusually tall rice plants in their crops. These tall plants did not live to maturity, and thus, were ignored as being of no practical importance. Then in 1926, it was

FIGURE 22–16. Gibberellic acid increases plant growth. The plant on the right was supplemented with gibberellic acid.

discovered that these tall plants were parasitized by a fungus called *Gibberella* (jihb uh REL uh). Later, a substance called **gibberellin** (jihb uh REL un), which causes healthy rice seedlings to grow very rapidly, was isolated from the fungus.

Since the original discovery of gibberellin, other similar compounds have been isolated from fungi and plants. They are all known as gibberellins, and some seem to play a role in cell elongation. Gibberellic acid is a gibberellin often used experimentally. When applied artificially, it causes growth in some types of plants (Figure 22–16). Such treatment often is used to produce large seedless grapes and large, tender stalks of celery. Germination of many seeds and development of buds also are speeded up by gibberellins. In nature, gibberellins work with auxins and other plant hormones to control the growth of plants.

Gibberellins cause cells of some plants to elongate.

22:12 Control of Flowering

In angiosperms, flowers develop from buds. The time of year during which flowers develop varies, depending on the plant. Some plants flower only in the spring, some only in summer, and some only in autumn or winter. Other plants have no definite flowering season.

Botanists have been studying the causes of flowering since about 1920. They first learned that flowering in many plants depends on the relative amounts of exposure to day and night in a 24-hour period. Plants such as chrysanthemums and ragweed, which flower in autumn, are called **short-day plants** because they flower when the day length is less than some critical period. Clover and beets, which flower in summer, are called **long-day plants** because they flower only when the day length exceeds some critical value. Plants such as sunflower and tomato are examples of **day-neutral plants** because their flowering does not depend on day length.

Flowering plants may be short-day, long-day, or day-neutral.

FIGURE 22–17. (a) Chrysanthemums are short-day plants, which flower when day length is less than some critical period. (b) Clover is a long-day plant. It flowers when the day length exceeds some critical value.

Although plants are grouped in terms of day length, further work has shown that it is not really the amount of light that determines time of flowering. Rather, it is the amount of continual darkness. A short-day plant is really a long-night plant, and a long-day plant is really a short-night plant. The response of flowering plants to light and dark conditions is known as **photoperiodism.**

Time of flowering is determined by the amount of continuous darkness to which a plant is exposed.

Once it was established that flowering in some plants depends on relative amounts of light and darkness, botanists then wondered *how* a plant responds to day and night length and what causes the flowering. A series of experiments has suggested that the detection of light and darkness occurs in the leaves, not in the flower buds. It is thought that the leaves contain a pigment, called phytochrome (FITE uh krohm), which is sensitive to light. **Phytochrome** detects when flowering should occur. Many lines of evidence suggest that the leaves then produce and secrete a hormone that travels through the phloem to the flower buds. There it causes the buds to flower. Although the hormone has not been isolated, it has been given the name **florigen** (FLOR uh jen). Further work is needed to determine how the amount of darkness is detected and how it is related to production of florigen. Also, florigen must be isolated. Other factors such as temperature and other hormones, especially gibberellins, seem to be involved.

Leaves contain phytochrome, a pigment sensitive to light.

Flowering may be controlled by a hormone, florigen, produced in the leaves.

REVIEWING YOUR IDEAS

15. What is a tropism?
16. Distinguish between positive and negative tropisms.
17. What controls tropisms?
18. What are some plant responses other than tropisms?
19. What effect do gibberellins have?
20. Distinguish among short-day, long-day, and day-neutral plants.

Problem: How do different environmental conditions alter transpiration rate?

Materials:

filter flask	plastic bag
glass tube, 35 cm	syringe
metric ruler	cool, boiled water
hair dryer	ringstand
transparent tape	burette clamp
clay	paper towel
Y tube	twist tie

1-hole stopper to fit flask with pine sprig
3 rubber tubes, each 5 cm

Procedure:

Part A. Water Loss Under Normal Conditions

1. Assemble the apparatus shown in Figure 22–18a. Do not secure the glass tube to the clamp. Simply rest it on top of the clamp. The syringe should be about half-filled with water before attaching it to the Y tube.
2. Tip the glass tube and ruler upward as you completely fill the flask with cool, boiled water. Continue to tip the glass tube upward to prevent water from flowing out the tube.

3. Push in on the syringe to fill the rubber tube and Y tube with water.
4. Insert the stopper with pine sprig into the flask. Return the glass tube and ruler to a horizontal position and rest it on the burette clamp. Use Figure 22–18a as a guide.
5. Check the following before you proceed. (a) Blot the area around the pine stem and stopper dry with paper towels. If leakage occurs, seal the stem and stopper with clay. (b) There must be no air trapped below the stopper. If air is present, tip the glass tube up, remove the stopper and add more water to the flask. (c) Water must fill the entire length of the glass tube. If necessary, inject more water from the syringe into the rubber collar.
6. Water in the glass tube should begin to move toward the flask. The rate that the water moves along the tube is a measure of the water lost through the pine leaves.
7. Make a copy of Table 22–1.
8. Record in the table the starting millimeter mark of the water in the glass tube.

FIGURE 22–18.

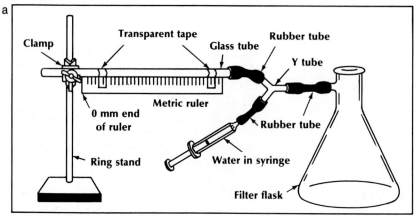

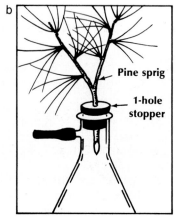

9. After 15 minutes, note and record in Table 22–1 the ending millimeter mark of the water in the glass tube.
10. Calculate the total distance that the water column moved.
11. In Table 22–1 record the total distance that the water column moved.
12. If necessary, replenish the water in the glass tube using the syringe.
13. Repeat steps 8 and 9 for your second trial.

Part B. Water Loss Under Changed Conditions

1. Cover the entire pine sprig with a plastic bag and put a twist tie around the stem.
2. Run two trials as in Part A, steps 8 and 9.
3. Remove the plastic bag. Use a hair dryer to blow cool air over the pine sprig.
4. Run two trials in steps 8 and 9 of Part A.
5. Complete Table 22–1.

Data and Observations:

		TABLE 22-1. MEASURING THE RATE OF WATER LOSS									
		NORMAL CONDITIONS			HIGH HUMIDITY (PLASTIC BAG OVER PLANT)			LOW HUMIDITY (COOL AIR BLOWING OVER PLANT)			
TRIAL		Starting mark	Ending mark	Distance moved	Starting mark	Ending mark	Distance moved	Starting mark	Ending mark	Distance moved	
1											
2											
	Total					Total				Total	
	Average					Average				Average	

Questions and Conclusion:

1. (a) Define xylem.
   (b) What is its function?
   (c) Where were xylem cells located in this experiment?
2. (a) Define stomata.
   (b) What is their function?
   (c) Where in your experiment were stomata found?
3. (a) Define transpiration.
   (b) Explain how xylem and stomata are related to the process of transpiration.
4. How was transpiration rate measured?
5. What three different conditions were used to influence transpiration rate?
6. (a) Using specific numbers from your data table, explain how each condition influenced transpiration rate.
   (b) Make a hypothesis about why transpiration rate is influenced by the changed conditions.

Conclusion: How do different environmental conditions alter transpiration rate?

CHAPTER REVIEW

SUMMARY

1. Roots are composed of tissues that function in the absorption of water and minerals and in the storage of food. **22:1**
2. Stem tissues function in the transport of water, minerals, and food as well as in support. **22:2**
3. Transport in simple plants occurs by diffusion or cytoplasmic streaming. Gas exchange occurs by diffusion or via pores. **22:3**
4. Water and mineral ions enter the roots of vascular land plants and pass to the xylem for upward transport. **22:4**
5. The transport of water through a stem is explained by the transpiration-cohesion theory. **22:5**
6. Guard cells control the size of stomata, thus regulating water loss. **22:6**
7. Translocation of food and certain inorganic materials is explained by the pressure-flow hypothesis. **22:7**
8. Exchange of oxygen and carbon dioxide in plants is needed for both photosynthesis and cellular respiration. Gas exchange in plants occurs directly with the environment. In vascular plants, gases are exchanged mainly at root hairs, lenticels, and stomata. **22:8**
9. Positive phototropism of stems is a growth response controlled by an auxin, IAA. **22:9**
10. Auxins also play a role in apical dominance, abscission of leaves and fruit, and fruit formation. An auxin controls negative geotropism of a stem. Abscisic acid controls positive geotropism of roots. **22:10**
11. Gibberellins are hormones found in some fungi and plants. They affect many processes, including cell elongation, seed germination, and bud development. **22:11**
12. Flowering in many plants depends on the relative amounts of light and darkness and seems to be controlled by florigen. **22:12**

LANGUAGE OF BIOLOGY

abscisic acid
abscission
apical dominance
auxin
companion cells
cortex
day-neutral plant
florigen
geotropism
gibberellin
herbaceous stem
hormone
long-day plant
pericycle
phototropism
phytochrome
pith
pressure-flow
  hypothesis
root pressure
short-day plant
sieve tubes
stele
stimulus
translocation
transpiration-cohesion
  theory
tropism

CHECKING YOUR IDEAS

On a separate paper, complete each of the following statements with the missing term(s). Do not write in this book.

1. Chemicals that affect certain target organs and that control plant responses are called _____.
2. Gas exchange in stems occurs through openings called _____.
3. _____ is the response of flowering plants to light and dark conditions.
4. Transport of food is explained by the _____ hypothesis.
5. In a root, vascular tissue is located in the central cylinder called the _____.
6. Roots have _____ geotropism.
7. _____ is the falling of leaves and fruits from trees.
8. Transport of water is explained by the _____ theory.
9. Vascular cambium is found in most _____ stems.
10. Transport of food in plants is called _____.

EVALUATING YOUR IDEAS

1. List the functions of the tissues of a root.
2. How do monocot and dicot roots differ?
3. Compare the structure of herbaceous monocot and dicot stems. What is the function of each structure? Is a woody plant a monocot or a dicot?
4. Distinguish between the inner and outer bark regions of a woody stem.
5. How would you distinguish spring wood from summer wood?
6. How are materials transported in simple plants? How are gases exchanged?
7. How is a transport system in vascular plants adaptive?
8. Describe how water moves to the xylem of a root.
9. How do minerals enter roots?
10. Explain the transpiration-cohesion theory.
11. How do guard cells control the size of stomata? How is this control adaptive?
12. How is food transported in plants?
13. Why do plants not need special gas exchange systems?
14. Why do root cells require oxygen? How do they obtain it?
15. What are lenticels? What is their function?
16. How are gases exchanged in leaves?
17. How did biologists discover the role of auxins in plant behavior?
18. How does auxin cause a plant stem to grow in the direction of a light source?
19. Compare and explain the responses of root and stem cells to gravity.
20. Describe the role of auxins in apical dominance, abscission, and fruit development.
21. What are the functions of gibberellins? What practical use is made of gibberellins?
22. Explain how some plants are thought to detect the relative amounts of light and darkness and to control flowering.

APPLYING YOUR IDEAS

1. How is the shape of a plant, caused by apical dominance, adaptive?
2. How could you show that flowering depends on darkness rather than on light?
3. How might you conduct an experiment to cause a short-day plant to flower in July?
4. Why are the movements of plants so much slower than those of animals?
5. Compare transport and gas exchange in algal plants with those functions in monerans, protists, and fungi.
6. How might the size of stomata change if light intensity is increased? How might it change if the amount of available CO_2 is decreased? Explain.
7. What would happen to a leaf if its lower surface were covered with a film of petroleum jelly? Why?
8. Where would a lily pad's stomata be? Why?

EXTENDING YOUR IDEAS

1. Obtain a healthy coleus plant. With a razor blade, remove only a leaf blade. Apply auxin to the cut end of the petiole. Remove another leaf blade. Leave the petiole untreated. Compare the times necessary for abscission of the petioles.
2. Design an experiment to illustrate the role of auxins in apical dominance. (See *Extending Your Ideas,* 1.)

SUGGESTED READINGS

Galston, Arthur W., *The Life of the Green Plant,* 3rd ed. Englewood Cliffs, NJ, Prentice-Hall, Inc., 1980.

"How Plants Manage to Move." *Science Digest,* Aug., 1983.

ANIMALS

In general, animals are more complex than other forms of life. This complexity is related to their heterotrophic (food getting) nutrition and active way of life. They have many adaptations related to food getting. The great horned owl is adapted in several ways for capturing its food. With eyesight that is 35 times more sensitive than human eyesight and the most acute sense of hearing in the animal world, its swiveling, mobile head enables it to sight prey in any direction without moving its body. To maintain an element of surprise, the shapes of the owl's wings and feathers reduce flight noise. The owl kills its prey instantly by sinking sharp talons into the body.

Food getting, however, is only one of the many life functions of animals. Like other organisms, animals reproduce, develop, transport materials, exchange gases, and remove wastes. Most animals have skeletal systems and a means of locomotion. They have chemical and nervous control systems that coordinate their activities and responses. In Unit 7, you will learn about the life processes of animals and how animals are adapted to carry on these activities.

Great Horned Owl with Prey

ANIMAL REPRODUCTION

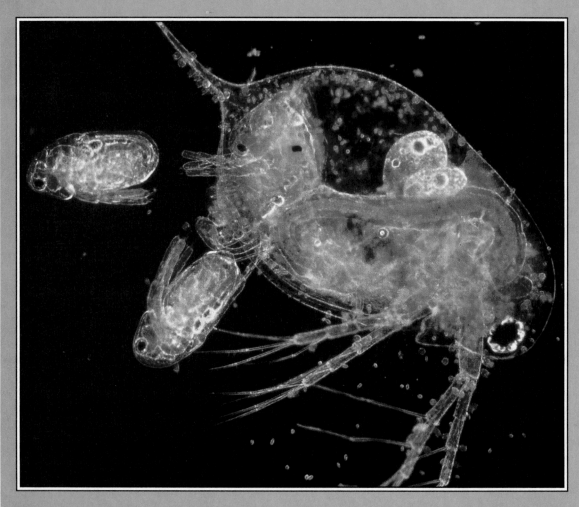

Certain animals, such as the water flea, can reproduce both sexually and asexually. Here, a water flea gives birth to young that developed inside her from eggs that were not fertilized. This development is the result of asexual reproduction. At other times, water fleas develop from fertilized eggs — from sexual reproduction. What conditions must be met for animals to reproduce successfully? How do reproductive patterns vary among animals?

n simple organisms such as monerans, protists, and fungi, asexual reproduction is common. Sexual reproduction, if it occurs, is accomplished by simple processes such as conjugation. Plant life cycles usually include alternation of generations, and plants have specialized reproductive tissues or organs. In this chapter, you will study the processes of reproduction found among animals.

Objectives:
You will
- explain patterns of reproduction among animals.
- compare mechanisms of external and internal fertilization among animals.
- discuss the functions of human reproductive structures in females and males.

PATTERNS OF REPRODUCTION

In animals, sexual reproduction is predominant and most animals have complex reproductive systems. Asexual means of reproduction do exist among some animals, however.

23:1 Animal Cycles

Life cycles of animals differ in several ways from those of plants. In plants, the life cycle most often includes alternating multicellular haploid and diploid generations. Animals, though, do not undergo alternation of generations. Except for rare cases (Section 23:2), the multicellular stage in an animal's life cycle is always diploid. Only the gametes are haploid. Recall that plant gametes are produced by *mitosis* of haploid cells (Chapter 20). However, animal gametes are produced by *meiosis* of diploid cells. After animal gametes are produced, fertilization may occur. The resulting offspring are again diploid.

In most animal life cycles, there is no multicellular haploid stage. Only gametes are haploid.

FIGURE 23–1. Earthworms are hermaphrodites that undergo internal fertilization. Sperm are exchanged by a mating pair and stored in receptacles. Later, sperm leave the receptacles and fertilize eggs.

Sexual reproduction is the most common form of reproduction among animals.

Sex of animals is determined by the presence of ovaries (female) or testes (male).

Hermaphrodites have both ovaries and testes but usually mate with other animals of the same type.

Most animals have a life cycle involving only sexual reproduction. Although asexual reproduction may occur in some animals, its occurrence is not usually necessary for the life cycle to be completed. In some animals, though, the life cycle does involve both an asexual and a sexual stage. For example, in *Aurelia* (Section 15:2) the polyp stage reproduces asexually and the medusa stage reproduces sexually. The fact that one stage follows the other makes this life cycle seem similar to that of plants. However, both stages are diploid.

Almost all sexual reproduction in animals involves separate males and females that have sex organs, or **gonads** (GOH nadz). Sperm are produced in the male gonads, the **testes** (TES teez), and eggs are produced in the female gonads, the **ovaries** (OHV reez).

A few kinds of animals have both ovaries and testes. Such animals are called hermaphrodites (Section 15:1). Most hermaphrodites do not fertilize themselves, but exchange sperm with another animal of the same species.

The common earthworm is a hermaphrodite. During mating, two worms lie parallel to one another. Sperm of one are passed to special storage receptacles of the other, and sperm of the other worm are passed to its partner. Later, both worms' eggs are fertilized.

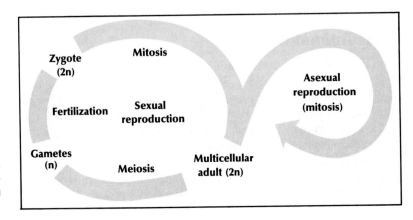

FIGURE 23–2. The normal life cycle of animals involves sexual reproduction. Note that when asexual reproduction occurs, it is not an integral part of the life cycle.

23:2 Asexual Reproduction

The usual life cycle of animals involves sexual reproduction. Asexual reproduction, if it occurs, often takes place in addition to (or in place of) the normal life cycle (Figure 23–2). Asexual reproduction involves reproduction from nonsexual structures and often occurs under special or unusual conditions. Recall that asexual reproduction results in genetically identical offspring. However, the "main" life cycle of most animals involves sexual reproduction. Thus, there is a great deal of variation possible in the population.

For example, a hydra is an animal that reproduces sexually but may also undergo budding (Figure 23–3). The bud forms as a small group of cells on the side of the parent animal. After all the specialized cells and tissues have developed, the young hydra breaks away, settles in a new location, and leads a separate life.

Fragmentation is another form of asexual reproduction in animals. In fragmentation, pieces split off from the parent and then grow into new organisms. This type of reproduction often occurs in sponges and coelenterates.

Fragmentation may also occur in the flatworm planaria. If the worm is cut in half (fragmented), each half will regrow the missing parts. The process of regrowing missing parts is called **regeneration** (rih jen uh RAY shun). Even if a planarian is cut into four sections, each section is capable of regeneration. In nature, planaria are not cut into pieces. However, they may twist around and break into a front and a tail end.

In animals, asexual reproduction is more common in less complex forms. Sponges, coelenterates, and flatworms commonly reproduce asexually. In more complex animals, though, asexual reproduction is rare (or, as in vertebrates, entirely absent). Certain insects are examples of complex animals having asexual reproduction. For example, aphids and honeybees reproduce asexually by means of parthenogenesis. **Parthenogenesis** is the development of an unfertilized egg into an adult. In honeybees, the queen lays both fertilized and unfertilized eggs. The fertilized eggs develop into females, mostly workers. The unfertilized, haploid eggs become males called drones.

FIGURE 23–3. *Hydra* produces asexually by budding. The bud grows and breaks away as a small hydra.

FIGURE 23–4. Parthenogenesis occurs in honeybees. The queen lays some unfertilized, haploid eggs that develop into drones.

REVIEWING YOUR IDEAS

1. Which form of reproduction, asexual or sexual, is more common in animals?
2. What organs produce gametes in male animals? Females?
3. Name two means of asexual reproduction in animals.

FERTILIZATION

Successful reproduction depends on fertilization of the egg and development of the zygote into a mature organism. Fertilization among animals may be either external or internal.

23:3 Conditions for Fertilization

Animals undergo either internal or external fertilization.

In **external fertilization,** eggs are fertilized outside the female. This kind of fertilization occurs in sponges, jellyfish, most worms, many fish, and frogs. In **internal fertilization,** eggs are fertilized within the body of the female. Internal fertilization occurs in insects, reptiles, birds, mammals, some fish, and some relatives of the frog. In other words, external fertilization is more common in animals that live in water, and internal fertilization is the more common form in animals that live on land.

Conditions of fertilization are the same for all animals.

In both types of fertilization, the following four conditions must always be met:

(1) The release of gametes must be properly timed so both sperm and eggs are present at about the same time.
(2) The gametes must be protected.
(3) A pathway must be present for the sperm to reach eggs.
(4) A liquid medium in which the sperm can swim to the eggs must be present.

FIGURE 23–5. Amplexus of frogs is an adaptation ensuring fertilization of eggs. The clasping behavior results in proper timing in the release of sperm and eggs.

23:4 External Fertilization

Frogs are animals with external fertilization. In females (Figure 23–6), eggs develop in ovaries. As the eggs grow, they burst from the ovaries into the coelom (body cavity). The eggs then enter two long, coiled tubes, the **oviducts** (OH vuh dukts). The opening of each oviduct is funnel-shaped and located toward the anterior end of the abdomen. Action of cilia that line the oviducts pushes the eggs to the posterior of the frog where the eggs are covered by a jellylike coat and stored in the **ovisac** (OH vuh sak). In males, sperm are produced in the testes and travel, via tubes, to the kidneys.

During the breeding season, a female frog is stimulated by bodily contact with a male. At this time, her eggs leave the ovisac, pass into the **cloaca,** a body chamber that receives wastes as well as gametes, and then enter the water. In the male, sperm travel through ducts to the cloaca and are deposited into the water at the same time as the eggs leave the female's body.

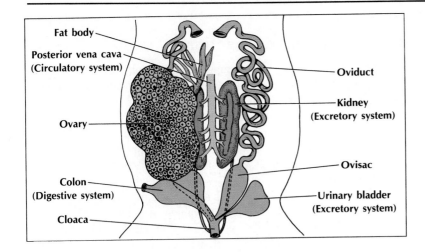

FIGURE 23–6. Eggs formed in frog ovaries are stored in the ovisacs after passing through the oviducts. How are conditions for fertilization met in frogs?

In frogs, the timing condition is met by behavioral patterns. The clasping of the female by the male results in the gametes being released at the same time. This clasping is called **amplexus** (am PLEX us) (Figure 23–5). Amplexus is an example of **courtship behavior.** Courtship involves complex behavior leading to mating. In animals with external fertilization, it stimulates both sexes to release gametes at about the same time. Proper timing is important because gametes have short life spans.

The condition of protection of the gametes is not as easily met. The environment in which the eggs have been deposited (water) is hazardous. The eggs may be eaten by other animals. Also, physical factors, such as water temperature and the amount of oxygen, are important. Frogs (and many other aquatic animals) deposit thousands of eggs at one time. There is "safety in numbers" in that at least some of the eggs will be fertilized and some of the fertilized eggs will survive. The jellylike layer around the eggs also protects them from predators and harsh conditions.

External fertilization requires the production of large numbers of gametes.

Producing many eggs and many sperm at once also aids in meeting another condition. Because both eggs and sperm are deposited in water, there is no direct pathway to the eggs for sperm to follow. A lake or a pond is a huge volume of water, and the life span of the gametes is short. Since gametes are released at about the same time and in the same place, the chances for sperm and eggs uniting are increased further by their large numbers.

Because fertilization occurs in water, the water serves as a liquid medium in which sperm can swim. The water also provides moisture that keeps the gametes from drying out. Thus, animals with external fertilization are "chained" to the water. Any attempt at fertilization on dry land would fail.

In external fertilization, water is required for sperm to reach eggs.

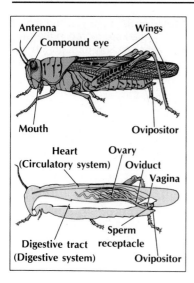

FIGURE 23–7. Sperm are deposited inside the female grasshopper and are stored in the sperm receptacle. Fertilization occurs as eggs pass through the vagina. Fertilized eggs are deposited in the soil.

Animals that have internal fertilization are not restricted to a wet habitat.

Estrous cycles ensure that mating occurs when mature eggs are ready for fertilization.

23:5 Internal Fertilization

The grasshopper is an example of an animal with internal fertilization. Eggs are produced in the female's ovaries and enter the oviducts. Sperm are produced in the male's testes, then travel to the **sperm duct** where fluids are added. During mating, the sperm and fluids pass out of the **penis** and enter the **vagina** (vuh JI nuh). The vagina is the organ in the female that receives sperm from the male. The sperm are stored in a special sac, the **sperm receptacle,** attached to the vagina (Figure 23–7). During late summer, the eggs move along the oviduct where yolk and a shell are added. Fertilization occurs as the eggs pass through the vagina. A small pore in the shell allows sperm to enter. The posterior abdominal segments of the female form two pairs of pointed structures. These structures, called **ovipositors,** are used to tunnel into the soil. Fertilized eggs leave the vagina and are deposited in the soil. The eggs remain there until the following spring.

Internal fertilization removes the dangers of predators and a harsh environment for the gametes. Sperm and eggs pass through specific reproductive structures that provide a direct pathway for gametes and increase the chances of fertilization. Protection and a pathway are provided by the internal organs. Large numbers of eggs are not produced. Also, a liquid accompanies the sperm, meeting the condition of a liquid medium. As a result, these animals are not "chained" to water as are frogs. The final condition, timing, is met by storage of sperm by the female grasshopper. The sperm are kept alive until the eggs are ready for fertilization.

Proper timing may occur in other ways in animals with internal fertilization. In many cases, the female "signals" the male that she is ready to mate. Certain hormones cause the start of the estrous (ES trus) cycle in the female. The **estrous cycle** is a series of chemical and physical changes leading up to the production of mature eggs. It occurs in mammals such as dogs and cats. When the female is in estrus, or "heat," she undergoes physical and behavioral changes that coincide with the maturation of eggs within the ovary. Because the male recognizes and interprets these changes as her readiness for mating, the female only mates when mature eggs are present.

In some animals with internal fertilization, birds, for example, proper timing is ensured by courtship behavior. In these animals, courtship leads to mating at a time when eggs are mature and ready to be fertilized. The behavior often consists of signals including sounds, dances, and other displays. Because each species has a unique behavior pattern, courtship also plays a role in identification of a mate. Proper identification of a mate is important whether animals have external or internal fertilization. Mating with a member of the same species maintains reproductive isolation (Section 11:5).

REVIEWING YOUR IDEAS

4. Distinguish between external and internal fertilization.
5. Where does fertilization occur in frogs and grasshoppers?
6. What is courtship behavior?
7. How is storage of sperm by some female animals important?
8. What is an estrous cycle?

HUMAN REPRODUCTION

In mammals, fertilization is internal. Study of human reproduction shows the basic pattern found among mammals in general.

23:6 Structures and Functions

Figure 23–9 shows the human female reproductive system. On each side of a female's abdominal cavity is one ovary. Within the ovary are groups of cells called **follicles** (FAHL ih kulz). Eggs begin to form from cells that divide by meiosis within the follicles of the ovary.

Very close, but not attached, to each ovary is a tube called the oviduct, or **Fallopian** (fuh LOH pee un) **tube.** Because the oviduct is not attached to the ovary, the end can move around the ovary somewhat. When an immature egg is released from a follicle, it is usually pulled into the oviduct by the action of tiny cilia that line the tube and its opening.

Eggs are produced within follicles of the ovary.

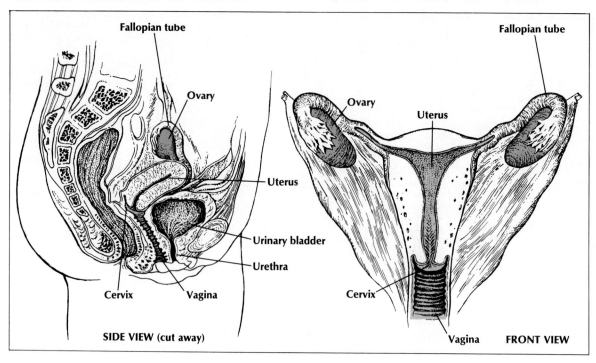

Fallopian tube

Fallopian tube

Ovary

Ovary

Uterus

Uterus

Urinary bladder

Urethra

Cervix

Cervix

Vagina

Vagina

SIDE VIEW (cut away)

FRONT VIEW

FIGURE 23–9. In humans, usually one egg is released from an ovary and passes down the Fallopian tube to the uterus.

The two oviducts join to a thick-walled, muscular organ called the **uterus** or womb. The uterus is the site of development. The hollow uterus narrows to a small "neck" called the **cervix** (SUR vihks) that leads to the vagina. The vagina is an external opening separate from the **urethra** (yoo REE thruh), the opening through which urine is excreted.

The testes of the male are two oval structures within an external sac called the **scrotum** (SKROHT um). In many animals, the testes are within the abdomen. But, in humans and some other mammals, sperm require a temperature slightly lower than body temperature. Thus, the suspension of the testes in an "external" structure is adaptive (Figure 23–10).

Sperm are produced in the testes.

Sperm are produced by meiosis from special cells of the testes and transferred along a series of highly coiled tubes. As sperm leave these tubes, they pass several structures that add fluids. Sperm have long tails, or flagella, an adaptation that aids sperm in swimming in the fluids. The combination of sperm and the fluids is called **semen** (SEE mun). Semen leaves the body through the urethra of the penis.

Semen is a combination of sperm and fluids.

During sexual intercourse (mating), semen is ejected from the male's urethra into the female's vagina. Although millions of sperm may enter the vagina, only a few thousand of them will pass through the cervix into the uterus. Many of these enter the oviduct where fertilization usually occurs. However, only one sperm can fertilize an egg.

Fertilization usually occurs in the oviduct.

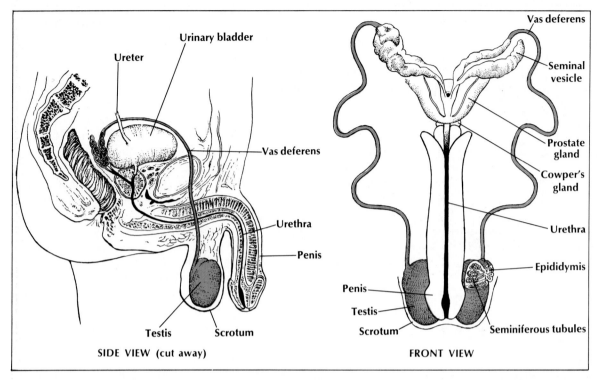

SIDE VIEW (cut away)   FRONT VIEW

In order for fertilization to occur, the sperm must penetrate the egg. The sperm secretes an enzyme that aids in forming an entryway in the outer zone of the egg. Once this secretion occurs, the outer zone becomes changed so that no other sperm can enter the egg. Why is this adaptation of one sperm per egg important? Penetration of the egg by the sperm triggers the immature egg to complete meiosis II, and the second polar body is discarded (Section 6:9). Then the haploid nuclei of sperm *(n)* and egg *(n)* combine to form the zygote *(2n)*.

If a zygote is formed, it moves downward toward the uterus where development occurs. The embryo becomes embedded in the uterine wall, which has been thickened and enriched with blood vessels in preparation for implantation.

FIGURE 23–10. Sperm are produced in the testes. Fluid from the seminal vesicle, prostate gland, and Cowper's gland is added to the sperm, forming semen.

FIGURE 23–11. Although there are many sperm present on the surface of this sea urchin egg, only one will fertilize it. Once this sperm penetrates, the outer zone of the egg will change so that no other sperm can enter.

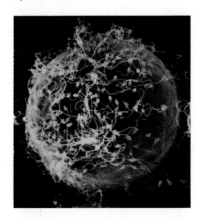

23:7 Menstrual Cycle

Human females do not undergo a seasonal estrous cycle as dogs or cats do. Instead, an egg (secondary oocyte) is released from an ovary about once every twenty-eight days. At the same time, certain uterine changes occur to prepare for a possible pregnancy. This monthly egg maturation and uterine preparation, called the **menstrual** (MEN strul) **cycle,** occurs in four distinct phases. Menstrual cycles begin around age twelve and continue until around age fifty. Ending of the menstrual cycle in the later years of life is called **menopause.**

Activities of the ovary and the uterus are controlled by an interaction of hormones. **Follicle-stimulating hormone** (FSH) is produced by the **pituitary** (puh TEW uh ter ee) **gland** located at the base of the brain. The hormone is secreted into the blood and travels throughout the body but affects only the ovaries. FSH causes primary oocytes in some follicles to complete meiosis I and form secondary oocytes. Only one of these usually survives per menstrual cycle. FSH, along with another pituitary secretion, **luteinizing** (LEW teen i zing) **hormone** (LH), causes the ovaries to produce a hormone called **estrogen** (ES truh jun). Estrogen causes the tissues of the uterus to thicken and the blood supply to increase. This series of changes lasts for about ten days and is called the **follicle stage.**

As the level of estrogen in the blood increases, it is detected by a structure near the pituitary gland called the **hypothalamus** (hi poh THAL uh mus). The hypothalamus directs the pituitary gland to decrease production of FSH and to increase production of LH. The increase in LH causes **ovulation** (ahv yuh LAY shun), bursting of a follicle and release of an egg. This is a very short stage in the menstrual cycle.

LH also causes the ruptured follicle to change to a yellowish body called the **corpus luteum** (KOR pus • LEWT ee um). Under the influence of LH, the corpus luteum produces some estrogen and a hormone called **progesterone** (proh JES tuh rohn). Progesterone further prepares the uterus for pregnancy. This stage of the cycle lasts about two weeks and is called the **corpus luteum stage.**

If an egg has been fertilized, it begins to undergo cell division. By the end of the first week of life, the zygote has developed into a ball of cells and has become embedded in the uterine lining (Section 24:6). Some of these cells produce a hormone similar to LH that "informs" the corpus luteum to keep producing progesterone. Thus, the uterine conditions necessary for pregnancy are maintained. After about five weeks, the embryo itself will produce progesterone to maintain the uterus throughout the duration of the pregnancy.

If a fertilized egg does not become implanted in the uterus, the corpus luteum begins to break down and progesterone production ceases. As the progesterone level decreases, the uterine lining begins to break down and bleeding results. Blood, some uterine tissue, and the unfertilized egg pass from the vagina. This tissue loss is called the menstrual flow. This stage, called **menstruation** (men STRAY shun), lasts three to five days. It marks the beginning of a new menstrual cycle. As menstruation is proceeding, a new egg is already beginning to mature within the ovary. Figure 23–12 summarizes the events of the menstrual cycle and the hormones that control these events.

The menstrual cycle is controlled by several hormones.

During the follicle stage, an egg matures and the uterine lining becomes thicker.

LH causes ovulation and the conversion of the follicle to the corpus luteum.

LH causes the corpus luteum to produce progesterone, the hormone that maintains the uterus for pregnancy.

If fertilization does not occur, the corpus luteum disintegrates, the uterine lining breaks down, and menstruation begins.

FIGURE 23–12. The menstrual cycle is a series of changes during which an egg matures and the uterus is prepared for possible pregnancy. This cycle is controlled by hormones secreted by the pituitary gland, ovary, and corpus luteum.

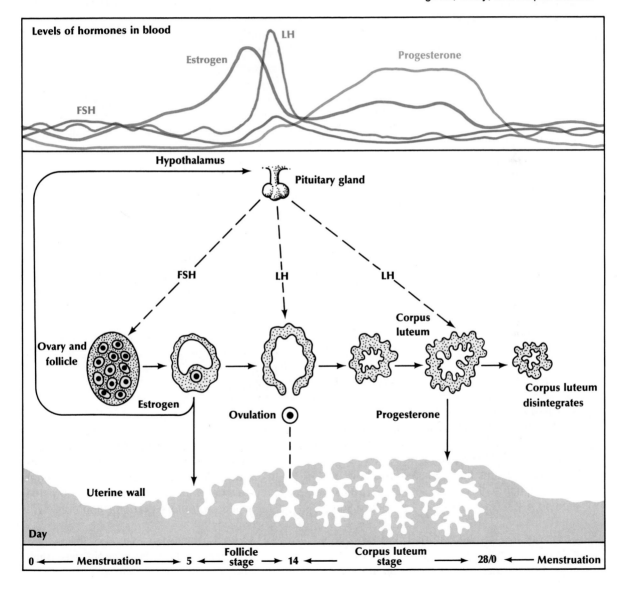

Levels of hormones in blood

LH

Estrogen

Progesterone

FSH

Hypothalamus   Pituitary gland

FSH   LH   LH

Corpus luteum

Ovary and follicle

Estrogen

Ovulation

Progesterone

Corpus luteum disintegrates

Uterine wall

Day

0 ← Menstruation → 5 ← Follicle stage → 14 ← Corpus luteum stage → 28/0 ← Menstruation

INVESTIGATION

Problem: How does *Obelia* show sexual and asexual reproduction?

Materials:

microscope
prepared slides of:
  Obelia colony, *Obelia* medusae
diagram of *Obelia* life cycle

Procedure:

Part A. Observing *Obelia* — Asexual Phase

1. Use the slide marked "*Obelia* colony" to observe the asexual phase of *Obelia* under low
power. The animal is described as being a
colony. Compare what you see under the
microscope to Figure 23–13.
2. Locate each of the following structures while
viewing the colony through the microscope:
  (a) colony — entire animal structure
  (b) feeding polyp — section of animal having
many tentacles or armlike parts
  (c) tentacles — the armlike parts on feeding
polyp
  (d) mouth — center of feeding polyp, surrounded by tentacles
  (e) gastrovascular cavity — hollow center of
animal
  (f) reproductive polyp — vaselike in shape,
separate from feeding polyp
  (g) medusa bud — small round structure with
toothed edge within reproductive polyp
3. Make a copy of Table 23–1 in which to record
your data and observations.
4. Use the structure names listed in step 2 to
correctly name the parts in Figure 23–13
labeled A–G. Use Table 23–1 to record your
answers.
5. Complete the first section of Table 23–1 by
indicating if each part in Figure 23–13 is the
sexual or asexual phase of the animal's life
cycle.

Part B. Observing *Obelia* — Sexual Phase

1. Use the slide marked "*Obelia* medusae" to
observe the sexual phase of *Obelia* under
low power. The animal is described as appearing like a miniature jellyfish. Compare
what you see under the microscope to what is
shown in Figure 23–14.
2. Locate each of the following structures while
viewing the medusae through the microscope:
  (a) medusa (plural, medusae) — bell-shaped
structure forming entire animal, looks like
a mini-jellyfish
  (b) tentacles — small, armlike parts along
outer edge of medusa
  (c) mouth — very center of medusa on underside
  (d) gonads — four dense areas, may be ovaries or testes
3. Use the structure names listed in step 2 to
correctly name the parts labeled H–K on Figure 23–14. Use Table 23–1 to record your
answers.
4. Complete the second section of Table 23–1
by indicating if each part in Figure 23–14 is
the sexual or asexual phase of the animal's
life cycle.

Part C. *Obelia* Life Cycle

1. Observe Figure 23–15 while reading the following summary of the *Obelia*'s life cycle.
  (a) Free-swimming medusae may be either
male or female. A *male medusa* contains
testes. A *female medusa* contains *ovaries*.
  (b) Testes release sperm into the water.
Sperm appear as microscopic cells, each
with a tail. Ovaries release eggs. *Eggs*
appear as small, round cells.

(c) Fertilization occurs resulting in a *zy-gote.*

(d) The zygote develops into a *larva,* a many-celled structure surrounded with cilia. The larva swims about for a while and then comes to rest on some object and grows into a *colony.*

(e) *Feeding polyps* form, which aid the colony in getting food.

(f) *Medusa buds* are formed by *reproductive polyps.* When mature, medusae break out of the reproductive polyp, starting the life cycle over again.

2. Use the structure names italicized in step 1 to correctly name the parts labeled L–W on Figure 23–15. Use Table 23–1 to record your answers.

3. Complete Table 23–1 by indicating if each structure in Figure 23–15 is part of the sexual or asexual phase of the animal's life cycle. NOTE: Consider the zygote and larva as part of the asexual phase.

Data and Observations:

TABLE 23-1. COMPARISON OF SEXUAL AND ASEXUAL PHASES IN *OBELIA*		
Letter	Structure	Sexual or Asexual Phase?
Figure 23–13 Parts		
A		
B		
C		
D		
E		
F		
G		
Figure 23–14 Parts		
H		
I		
J		
K		
Figure 23–15 Parts		
L		
M		
V		
W		

Questions and Conclusion:

1. Define:
   (a) asexual reproduction
   (b) sexual reproduction
2. Explain why
   (a) the colony form of *Obelia* is the asexual phase of the life cycle.
   (b) the medusae are the sexual phase.
3. (a) To what phylum does *Obelia* belong?
   (b) List several characteristics for this phylum that are shown by *Obelia*.

Conclusion: How does *Obelia* show sexual and asexual reproduction?

CHAPTER REVIEW

SUMMARY

1. Animal life cycles do not include alternation of generations. In most animals, only gametes are haploid. Sexual reproduction is the common mode of reproduction. **23:1**
2. In addition to sexual reproduction, some animals reproduce asexually by budding, fragmentation, or parthenogenesis. **23:2**
3. Fertilization in animals can be external or internal. Conditions for fertilization are proper timing of release of gametes, protection of gametes, a pathway for sperm to reach eggs, and a liquid medium in which sperm can swim to eggs. **23:3**
4. Frogs have a reproductive system adapted for external fertilization. **23:4**
5. The reproductive system of grasshoppers is adapted for internal fertilization. In general, internal fertilization is more efficient than external fertilization. **23:5**
6. Human fertilization occurs internally. During sexual intercourse, semen is released into the vagina. Fertilization usually occurs within the oviduct. **23:6**
7. Human females undergo a monthly cycle of events called the menstrual cycle. The cycle is controlled by hormones produced by the pituitary gland, ovaries, and the corpus luteum. **23:7**

LANGUAGE OF BIOLOGY

amplexus
cervix
cloaca
corpus luteum
courtship behavior
estrous cycle
Fallopian tube
follicle
gonad

ovipositor
ovulation
parthenogenesis
pituitary gland
regeneration
scrotum
semen
sperm duct
sperm receptacle

hypothalamus
menopause
menstrual cycle
ovary
oviduct

testes
urethra
penis
uterus
vagina

CHECKING YOUR IDEAS

On a separate paper, match each phrase from the left column with the proper term from the right column. Do not write in this book.

1. produces hormones that control menstrual cycle
2. release of egg from follicle
3. produced from ruptured follicle
4. place of sperm production
5. organ in which a human develops
6. organ where sperm are deposited
7. place of egg production
8. cycle leading to proper timing
9. sperm and fluid
10. cycle involving maturation of an egg and preparation of uterus for pregnancy
11. development of an unfertilized egg
12. regrowth of missing parts
13. behavioral pattern in frog reproduction
14. means of asexual reproduction in a hydra
15. male grasshopper reproductive structure

a. amplexus
b. budding
c. corpus luteum
d. estrous cycle
e. menstrual cycle
f. ovaries
g. ovulation
h. parthenogenesis
i. pituitary gland
j. regeneration
k. semen
l. sperm duct
m. testes
n. uterus
o. vagina

EVALUATING YOUR IDEAS

1. How does the life cycle of animals differ from that of plants?
2. Which form of reproduction, sexual or asexual, is more common among animals?
3. Describe sexual reproduction in earthworms.
4. How does asexual reproduction "fit" into the life cycle of an animal?
5. Describe asexual reproduction in a hydra, planarian, and honeybee.
6. What conditions are necessary for successful fertilization in animals?
7. List the reproductive structures of male and female frogs. What is the function of each?
8. How are the conditions of fertilization met in frog reproduction?
9. Explain why courtship behavior is necessary in some animals.
10. Describe reproduction of grasshoppers.
11. How are the conditions of fertilization met in grasshopper reproduction?
12. List the reproductive structures of the human male and female. What is the function of each structure?
13. Describe the events of the four stages of the menstrual cycle assuming
    (a) no pregnancy occurs
    (b) pregnancy occurs
14. Explain the chemical control of the menstrual cycle.

APPLYING YOUR IDEAS

1. To what plant life cycle is the animal life cycle most similar?
2. Can you suggest a reason why most animals undergo sexual reproduction?
3. Internal fertilization offers protection of and a pathway for sperm. Despite these adaptations, animals produce millions of sperm. Explain why production of so many sperm is adaptive.
4. Why do you think that animals that are parasites are often hermaphrodites?
5. Why do many hermaphrodites exchange sperm rather than fertilize their own eggs?
6. The conditions for fertilization met in animal reproduction must be met in plant reproduction. How are these conditions met in moss reproduction? In the reproduction of flowering plants?
7. Which plants undergo external fertilization? Internal fertilization?
8. In artificial insemination, semen from selected bulls is stored at a very cold temperature. Later, it is injected into cows. Why would the semen be stored at cold temperatures? Why would only certain bulls be selected as sperm donors?
9. Give examples of regeneration that do not involve the reproduction of an animal.

EXTENDING YOUR IDEAS

1. Carry out an experiment to demonstrate fragmentation and regeneration in planaria. Keep records of the appearance of each section right after cutting and each day until regeneration is complete.
2. Prepare a report on various courtship patterns in birds.
3. Use library sources to learn about fertilization of human eggs in the laboratory.
4. Find out how oral contraceptives are related to the events of the menstrual cycle.

SUGGESTED READINGS

Gold, Michael, "The Baby Makers." *Science 85,* May, 1985.
Gould, Stephen Jay, "Here Goes Nothing." *Natural History,* July, 1985.

ANIMAL DEVELOPMENT

The series of changes that occur as an organism reaches its mature form is called development. Many organisms begin their development in eggs such as these. Once hatched, they continue to change until they develop into adults. What changes have you gone through in the course of your development? Are you still going through the developmental process?

chapter 24

Most animals reproduce sexually, which involves two sets of DNA combining to form a zygote. The zygote receives a complete set of instructions that controls its development (Section 1:5). The embryo (Section 10:4) undergoes many changes before it becomes independent of the structure in which it is developing. The study of development of an embryo is called **embryology.**

Embryos are alive and have the same basic functions as any other living animal. They must secure food, exchange gases, rid themselves of wastes, and respond to their environment.

Objectives:
You will
- compare patterns of external and internal animal development.
- explain the early stages of animal development.
- discuss factors that guide the development of an animal.

PATTERNS OF DEVELOPMENT

You have learned that some animals have external fertilization and others have internal fertilization (Chapter 23). Animals also have different patterns of development. Both external and internal development occur among animals.

24:1 External Development: Adaptations for Protection

Animals that are produced by external fertilization also develop externally. Just as protection is important for an unfertilized egg (Section 23:4), it is also important for an embryo. Animals have a variety of adaptations for protecting fertilized eggs and embryos.

Fertilized frog eggs are surrounded by a jellylike layer that protects the eggs, insulates them, and helps them adhere to objects. The jellylike layer is added to the eggs as they pass through the oviducts. The layer swells when the eggs are deposited in water.

FIGURE 24–1. Frog eggs are protected by a jelly coat that swells when the eggs are deposited in water.

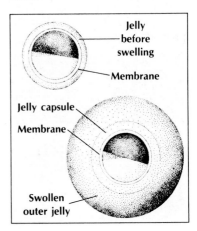

Jelly before swelling

Membrane

Jelly capsule

Membrane

Swollen outer jelly

FIGURE 24–2. (a) The color pattern of frog eggs may be a protective adaptation. (b) A protective adaptation involving the parents is the behavior of stickleback fish, in which the male guides the female into the nest for egg laying.

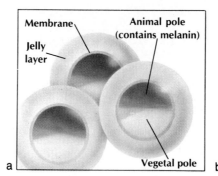

Membrane
Jelly layer
Animal pole (contains melanin)
Vegetal pole

a

b

The vegetal pole of the frog's egg contains yolk. The animal pole contains cytoplasm.

Coloration also may be protective. The bottom half of an egg, the **vegetal** (VEJ ut ul) **pole,** is light-colored because it contains yolk. The top half, the **animal pole,** is composed of living cytoplasm that has a dark pigment. Because the vegetal pole is more dense than the animal pole, it faces the bottom of a pond or stream. This pattern of coloration may blend well with the aquatic environment. Predators swimming above the eggs see a dark color that may blend with the bottom of the stream or pond. A fish swimming below an egg mass sees a light color that may blend with the sky.

Parents may help to protect eggs and embryos.

Some protective adaptations involve the parents. In stickleback fish, the male builds a nest and guides the female to it. She lays eggs and is then driven away by the male. He enters the nest and deposits sperm on the eggs. The male protects the eggs from predators and fans them with water currents to provide oxygen.

Many species of fish are mouthbreeders. Fertilized eggs are picked up by one of the parents and held in its mouth. The parent carries the eggs in its mouth until the eggs hatch and the young are able to live alone. During this period of development, which may last for many weeks, the parent does not eat.

FIGURE 24–3. (a) Male sea horses have pouches in which eggs are fertilized and incubated. (b) The cichlid is a mouthbreeding fish. Fertilized eggs are incubated in the mouth of the adult.

Sea horses are an unusual type of fish (Figure 24–3). The female transfers unfertilized eggs to a pouch on the male's abdomen. The male then releases sperm into the pouch. He carries the fertilized eggs in the pouch until they hatch and the young are released.

a

b

24:2 External Development: Metamorphosis of a Frog

Animals that develop outside the female (or male) usually go through a series of changes in structure. *These changes aid the young in surviving apart from the parents as the young mature.* After hatching from the egg, many animals develop into a **larva,** an immature animal that can live on its own. Larval forms are common in animals in which external fertilization occurs. The change of an immature animal to an adult is called **metamorphosis** (met uh MOR fuh sus). The organisms going through metamorphosis obtain food, exchange gases, and get rid of wastes independent of the parents.

A larva is immature but can survive independently.

Metamorphosis is a series of changes in the development of an immature animal into an adult.

A frog's development from egg to adult is one type of metamorphosis (Figure 24–4). Fertilized frog eggs develop for one or two weeks within their jellylike blanket, obtaining energy from the yolk and exchanging gases and getting rid of wastes by diffusion. At the end of this time, the eggs hatch and tadpoles emerge. A very young tadpole is inactive, but soon changes begin to occur. A long, tapering tail and three pairs of external gills develop. Gills pick up oxygen that is dissolved in the water. Gills also release carbon dioxide into the water. The tadpole is then a free-living, larval form of a frog. Although it has no yolk remaining as a food source, it can reach and feed on algae and plant matter. It has small horny teeth that scrape plant matter.

A tadpole is adapted for exchanging gases and obtaining food.

As a tadpole grows, further changes occur. First, the hind legs, then the front legs begin to appear. Internally a pair of lungs develop, and the circulatory system becomes coordinated with the lungs and skin. Eventually, the lungs and skin can obtain oxygen directly from the air and can get rid of carbon dioxide. Other wastes are excreted from special organs. As teeth develop within the tadpole's mouth, the tadpole's diet changes from plants to insects. The tail is resorbed by the animal; the materials of which it is composed are taken into the body. Many of the materials are used as an additional energy source for the developing frog. This entire series of events takes about three months in the common leopard frog (Figure 24–4), but the exact length of time may depend on the temperature of the water.

Changes prepare a developing frog for life on land.

FIGURE 24–4. Among the many changes that occur in frog metamorphosis, free swimming larvae develop limbs. Eventually the tail is resorbed and the tadpole becomes a frog.

FIGURE 24–5. Complete metamorphosis in insects is shown by the development of a monarch butterfly. The egg undergoes changes becoming a larva and then a pupa. The adult emerges from the pupa.

24:3 External Development: Metamorphosis in Insects

Insects undergo metamorphosis after internal fertilization has occurred. Fertilized eggs are deposited in places such as water, soil, plants, or other animals where they develop externally. The young embryo obtains food from the yolk and exchanges gases and gets rid of wastes by diffusion through the shell.

Metamorphosis may occur in either three or four stages, depending on the type of insect. **Complete metamorphosis** occurs in four stages — egg, larva, pupa (PYEW puh), and adult.

Complete metamorphosis: egg, larva, pupa, adult.

A fertilized egg first develops into a wormlike larva such as a maggot (housefly) or caterpillar (butterfly or moth). Like a tadpole, an insect larva is free-living and secures its own food. This food is needed as an energy source for the larva and is important for the changes that follow. The larva becomes inactive and changes to a **pupa.** The pupa is usually enclosed within some type of protective shell or case. During the pupal (PYEW pul) stage, the tissues of the animal are completely reorganized. The source of energy needed for this change is the food eaten during the larval stage. After a period of time, an adult emerges from the pupa case. A pupa from a maggot becomes a housefly, and a pupa from a caterpillar becomes a moth or butterfly (Figure 24–5).

During the pupal stage, a complete change of form occurs.

a

b

c

Incomplete metamorphosis occurs in three stages—egg, nymph (NIHMF), and adult. An egg hatches into a nymph, which closely resembles the adult but is smaller and often wingless. A nymph grows and usually develops wings. A reproductive system also develops. Grasshoppers undergo incomplete metamorphosis (Figure 24–6).

FIGURE 24–6. Grasshoppers are insects with incomplete metamorphosis. (a) A female deposits fertilized eggs in the ground. (b) A nymph is a young animal that resembles (c) the adult.

24:4 The Amniotic Egg

Reptiles and birds reproduce by internal fertilization. After fertilization, the zygote travels down the oviduct. As the zygote travels, a shell develops around it. In reptiles, the shell is tough and leathery; in birds, it is brittle and porous.

In addition to the shell, four membranes later develop around the embryo. One of these, the amnion (AM nee ahn), is a fluid-filled sac. Because of the presence of the amnion, eggs of birds and reptiles are known as amniotic eggs. Evolution of the amnion, along with that of the shell and internal fertilization, allowed development to occur on land. The amniotic fluid keeps the embryo moist and the waterproof shell prevents evaporation.

Shortly after fertilization, the zygote, on the surface of the yolk, divides to form a disc of cells. These cells continue to divide, producing the embryo. From the embryo, the amnion and three other membranes later develop. Beneath the porous shell lies a very thin membrane called the chorion (KOR ee ahn), which encloses the other three membranes. Attached to the developing digestive system of the embryo is the yolk sac, a membrane that encloses the yolk. The yolk (mainly fat), along with the white of the egg (protein), is a food source for the developing embryo. The yolk sac contains blood vessels through which the food passes to the embryo's cells.

Also connected to the digestive region of the embryo is a membrane sac called the allantois (uh LANT uh wus). Waste products collect in this sac. Carbon dioxide in the blood vessels of the allantois diffuses across the allantois through the shell, and into the environment. Oxygen diffuses in the other direction. Part of the allantois lies close to the chorion. Later the allantois acts with the chorion in gas exchange.

The amniotic egg is adapted for development on land.

The allantois functions in waste removal and gas exchange.

FIGURE 24–7. Reptiles lay amniotic eggs, an adaptation for life on land. These northern pine snakes are just hatching.

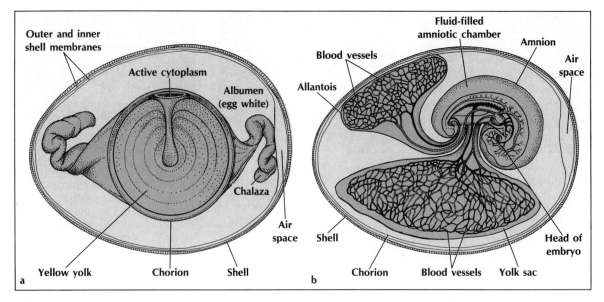

FIGURE 24–8. An amniotic egg provides protection, a food source (yolk), and a means of gas exchange and waste removal. (a) Egg parts before fertilization and (b) after development has begun are shown.

Near the end of their development period, birds obtain additional oxygen from air trapped in an area near the blunt end of the egg. Using this air, the bird's lungs start working. A short time later, the bird begins to break open its shell and breathes air from outside.

Bird and reptile eggs are cushioned by amniotic fluid and protected by the shell. Bird eggs require heat to develop. Birds are endotherms (Section 16:14). How is the heat provided? Reptile eggs are usually deposited in sand or soil and left alone. Temperature is not as important because reptiles are ectotherms.

Amniotic eggs are also produced by a group of primitive mammals called monotremes. Only two groups, the spiny anteater and duckbill platypus, are alive today. These animals, now protected in Australia, have features of both reptiles and mammals and may be very much like the earliest mammals.

FIGURE 24–9. The spiny anteater is a monotreme. Spiny anteaters live in parts of Australia, New Guinea, and Tasmania.

The duckbill platypus has fur and produces milk from glands similar to the mammary glands of other mammals, but without nipples. It also has a ducklike bill and webbed feet. The female produces several eggs that she keeps warm in a nest she builds. After the young hatch, they feed by licking milk from the mother's fur that surrounds the glands.

Amniotic eggs are adaptive. Far fewer eggs are produced by animals with these eggs than are produced by insects and aquatic animals. The amniotic egg provides an embryo with a favorable environment until it is ready to exist independently. All necessary functions are provided for within the egg. These adaptations give reptile, bird, and monotreme embryos a better chance to survive than embryos that lack amniotic eggs.

a
b

FIGURE 24–10. Amniotic eggs provide the embryo with a favorable environment until it can exist independently. (a) Gadwall duck eggs and (b) loggerhead turtle eggs give the embryos of these species a better chance to survive than embryos that must develop externally.

24:5 Partial Internal Development: The Marsupials

Although amniotic eggs provide an improved method for development, there are other patterns. Development in most mammals occurs either partially or totally within the mother's body. Partial internal development occurs in a group of pouched mammals called marsupials. These mammals, such as kangaroos, live mainly in Australia. Opossums are marsupials found in North America.

In all marsupials, eggs are fertilized internally, and the embryos begin development within the mother's uterus. The same four membranes that are in bird and reptile eggs develop from the embryo. The embryo receives most of its nourishment from stored yolk within the egg and only a small amount of nourishment from the mother while in the uterus. The embryo spends a relatively short period of time in the uterus before it is expelled. The tiny young then crawl into a pouch on the female's abdomen where mammary glands are located. The young

A marsupial begins developing within the mother's uterus and finishes developing in the mother's pouch after birth.

FIGURE 24–11. (a) After birth, small baby opossums complete their development within their mother's pouch. (b) A young red kangaroo "rides" in its mother's pouch where mammary glands are located. Opossums and kangaroos are marsupials.

a
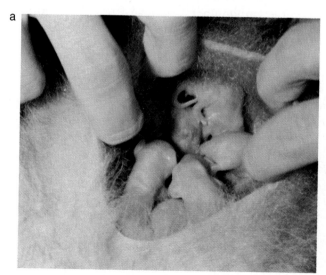

b

secure milk through the nipples of the mammary glands. When they leave the uterus, young kangaroos are only a few centimeters long. Further development and growth occur within the pouch.

Marsupial embryos are limited to partial development within the mother for several reasons. They become too large for the uterus. They may not be able to obtain enough food and oxygen or be able to get rid of wastes effectively. If complete internal development is to occur, there must be methods for a continuous food supply, gas exchange, and waste removal. These methods are found in another group of mammals.

24:6 Complete Internal Development: The Placentals

Complete internal development requires methods for food supply, gas exchange, and removal of wastes.

Complete development of embryos within the mother occurs in mammals called placentals. In these mammals, a special structure, the **placenta,** is formed in the uterus from the embryo's and the mother's tissues. Exchange of gases between the embryo and the mother occurs in the placenta. Also, the placenta is the site where food is obtained and wastes are removed. Humans, dogs, rats, horses, and deer are examples of placental mammals.

Placental mammals undergo complete internal development.

Development of a human embryo is an example of placental development. Prior to fertilization, the uterus becomes thickened, and its blood supply increases during the menstrual cycle (Section 23:7). Fertilization usually occurs within the oviduct.

The embryo develops from the inner cell mass of the blastocyst.

After fertilization, the zygote begins to divide and forms a ball of cells called a **blastocyst,** which travels down the oviduct to the uterus. Two regions of cells form as the cells divide. One is an outer layer of cells called the **trophoblast;** the other is called the **inner cell mass.** The embryo develops from the inner cell mass. During this time, the embryo obtains food from the small amount of yolk. On

FIGURE 24–12. (a) These elephant seals and (b) African elephants are examples of placental mammals. Young develop within the mother's uterus and are nourished through the placenta.

a

b

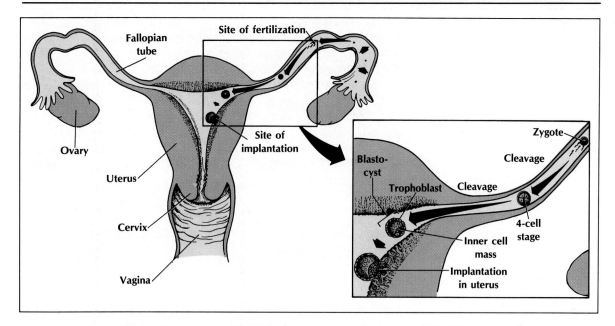

the sixth or seventh day after fertilization, the blastocyst attaches to the spongy wall of the uterus. During this process, cells of the trophoblast produce enzymes that digest some of the uterine cells. The embryo then becomes embedded in the uterine wall. The attaching and embedding of the embryo within the uterus is called **implantation.** Implantation marks the start of pregnancy.

As implantation occurs, the amnion begins to form around the embryo. Soon the trophoblast and embryo are connected to the uterine wall by a part called a **body stalk.**

In addition to the amnion, the other membranes found in bird, reptile, and marsupial embryos develop. But in placental mammals, some of the membranes have been modified for different functions, making complete internal development possible. Many branches form from the chorion and become embedded in part of the uterine wall. In these branches small blood vessels, capillaries, develop from the allantois and become part of the placenta. The rest of the placenta is made of uterine tissue. The capillaries of the placenta are part of the embryo's circulatory system. As these capillaries form, the embryo's heart and circulatory system are also developing. In humans, these changes occur during the fourth week of development.

The embryo is connected to the placenta by the **umbilical** (um BIHL ih kul) **cord,** which contains large blood vessels that transport blood between the embryo and the placenta. These blood vessels are also formed from the allantois. The cord itself develops from the body stalk. There is a separate placenta and umbilical cord for each developing embryo.

FIGURE 24–13. Usually fertilization occurs in the Fallopian tube. A zygote then begins development before implantation in the uterus.

Pregnancy begins when the fertilized egg is implanted in the spongy uterine wall.

Branches formed from the chorion become embedded in the uterine wall. Capillaries that develop in the branches become part of the placenta.

The umbilical cord contains blood vessels that link the embryo and the placenta.

FIGURE 24–14. A human fetus at sixteen weeks of development has a well developed umbilical cord and placenta.

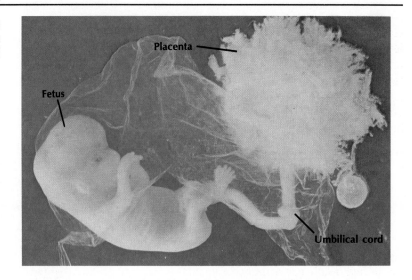

Blood pumped by the embryo's heart enters the vessels of the umbilical cord and circulates through the capillaries of the placenta. The blood comes into close contact with the blood in the mother's uterine wall. The two blood supplies normally do not mix together, but they are close enough so that certain materials diffuse between them. The mother's blood carries food molecules digested by her system and oxygen taken in by her lungs. Some of these molecules diffuse from her blood into the blood in the capillaries of the placenta. Water and vitamins also pass from mother to embryo. The blood then goes to the embryo through the umbilical cord and circulates through the embryo's blood vessels. The oxygen and food diffuse from the

Blood of the embryo does not mix with the mother's blood.

Oxygen and digested food pass from mother to embryo via placental blood.

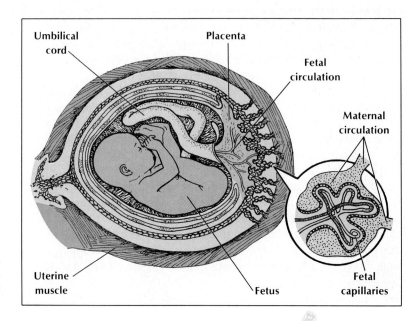

FIGURE 24–15. Needed materials and wastes are exchanged between fetus and mother through the placenta.

blood into the embryo's body cells. At the same time, carbon dioxide leaves the embryo's cells and enters its blood. At the placenta, the carbon dioxide diffuses into the mother's blood and is exhaled by her lungs. Excretory products are also transferred to the mother at the placenta and are excreted by the mother. The amniotic fluid helps cushion the embryo and keep it moist. All the needs of the developing embryo are provided. Thus, its chances of survival are very good.

24:7 Birth

As development proceeds, the embryo grows and causes the uterus to expand. After about eight weeks, the embryo is called a fetus (FEET us). Humans have a gestation (jeh STAY shun) period of about 280 days (a little over nine months). The **gestation period** is the period of internal development.

FIGURE 24–16. (a) A six-week-old human embryo shows paddles for hands and feet on the limb buds, and a large head with forming eyes and ears. (b) At eight weeks facial features become more obvious and hands and feet more developed. (c) At eleven weeks, fingers and toes are clearly visible, and the rib cage can be seen just beneath the elbow. (d) At seven months, a human fetus is well-developed and grows rapidly. If born prematurely at this time, however, it can survive only with special care.

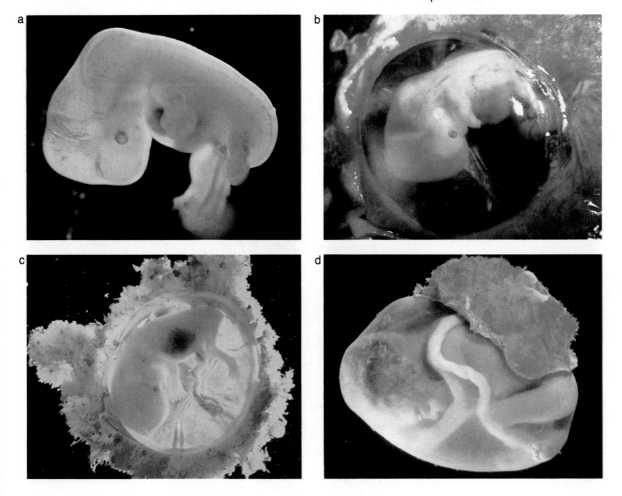

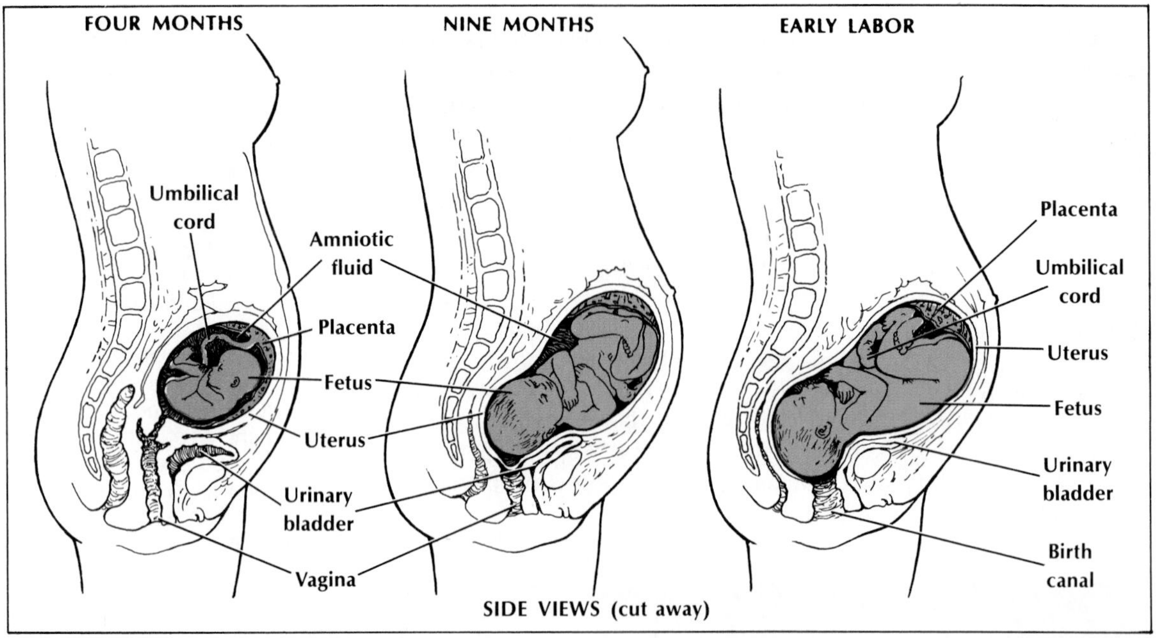

FOUR MONTHS **NINE MONTHS** **EARLY LABOR**

Umbilical
cord

Amniotic
fluid

Placenta

Fetus

Uterus

Urinary
bladder

Vagina

Placenta

Umbilical
cord

Uterus

Fetus

Urinary
bladder

Birth
canal

SIDE VIEWS (cut away)

FIGURE 24–17. The fetus is protected within the uterus during its development. Before birth, the fetus usually turns and is born head first.

Contraction of uterine muscles during labor forces the fetus out of the mother's body through the vagina.

At the end of the gestation period, certain hormones start **labor,** a series of contractions of the uterine muscles. The amniotic sac, or "bag of waters," bursts and the cervix and vaginal opening (Section 23:6) enlarge. As contractions become more frequent and intense, the fetus is pushed to the bottom of the uterus. Usually the fetus is in a headdown, facedown position. Further contractions push the fetus headfirst out of the mother's body through the vagina. This expulsion of a fetus from the uterus is called **birth.** A little later, a final series of contractions expels the placenta, or afterbirth. In humans, the umbilical cord is tied and cut. The remainder of the cord falls off the baby after several days. The **navel** (NAY vul) is the place where the umbilical cord was attached to the baby's body.

FIGURE 24–18. Parental care is important in animals other than humans, such as the mountain lion.

After a mammal is born, a period of care is needed. The young is fed milk (nursed) from the mother's mammary glands until it can eat other types of food. Besides providing nourishment, the mother's milk aids the immune response of the young mammal. The young receives some passive immunity (Section 19:10) by the passage of maternal antibodies into the milk. This immunity is soon replaced as the offspring's own immune system begins functioning.

The number of eggs produced by placental mammals at one time is relatively small. Because a small number of offspring are produced, the parent or parents can better care for the young. Along with securing food, the mother (and sometimes the father) serves as a teacher for the young mammal. Also, the parents are often actively involved in protection of the young. The nursing, training, and parental care periods vary from mammal to mammal. However, each of these factors aids in ensuring the survival of the offspring.

REVIEWING YOUR IDEAS

1. Distinguish between the vegetal pole and the animal pole of a frog egg.
2. What is metamorphosis?
3. Distinguish among larva, pupa, and nymph.
4. What is the major advantage of amniotic eggs?
5. Distinguish between a marsupial mammal and a placental mammal.
6. What factors are necessary for complete internal development?

MECHANISM OF DEVELOPMENT

Embryologists are interested in learning how a zygote develops into a complete, recognizable form. The first studies of animal development were made with animals that develop in water. Frogs, starfish, and sea urchins are favorite specimens for study because they are easy to raise and observe. Later, it became possible to study both shelled and mammalian embryos. Also, methods for fertilizing eggs in culture media and watching their development have been devised.

24:8 Cleavage

Differences in early development are found among animals, but there is a general scheme. Development begins with a series of many cell divisions called **cleavage** (KLEE vihj). Figure 24–19 shows the

Early stages of development are similar in all animals.

The zygote divides into numerous smaller cells during cleavage.

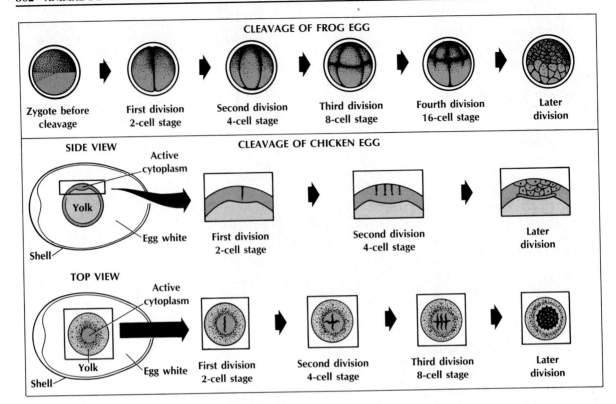

CLEAVAGE OF FROG EGG

Zygote before cleavage | First division 2-cell stage | Second division 4-cell stage | Third division 8-cell stage | Fourth division 16-cell stage | Later division

CLEAVAGE OF CHICKEN EGG

SIDE VIEW — Active cytoplasm, Yolk, Egg white, Shell

First division 2-cell stage | Second division 4-cell stage | Later division

TOP VIEW — Active cytoplasm, Yolk, Egg white, Shell

First division 2-cell stage | Second division 4-cell stage | Third division 8-cell stage | Later division

FIGURE 24–19. Cleavage in frog and chicken eggs produces blastulas. Amount of yolk in the eggs influences the pattern of cell divisions. Notice that the part of the chicken egg that divides is just a small area on top of the yolk.

stages of cleavage as they occur in a frog and a chick. Notice that the pattern differs between the two animals. The difference is due to the location and amount of yolk in each kind of egg. In a frog egg, the yolk is found throughout the cytoplasm, but there is more yolk in the vegetal pole than in the animal pole. As a result, cells of the animal pole divide more rapidly and become more numerous than those of the vegetal pole. In the chick egg, the cytoplasm is a small area on top of the yolk. Cleavage does not cut through the yolk. A disc of cells is formed as the nucleus and cytoplasm divide on top of the yolk.

In an animal embryo, cleavage produces a mass of cells that form a new, multicellular animal. Notice that there has been little growth up to this point. The large zygote has merely been divided into many smaller cells. These cells form a hollow ball called the **blastula** (BLAS chuh luh). The blastula consists of one tissue layer, the ectoderm. Inside the blastula is a cavity called the **blastocoel** (BLAS tuh seel) that is filled with fluid. The shape of the blastula allows for further rearrangement of cells in later stages of development.

Cleavage results in a hollow group of cells, the blastula.

Occasionally as a zygote is entering the two-cell stage, the cells of the embryo separate into two independent cells. Each of these cells has the potential for developing into a new and complete animal. Thus, two animals are produced instead of one. The animals have the

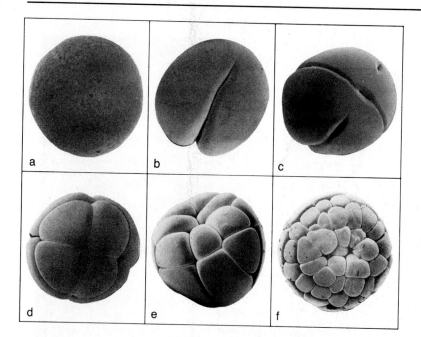

FIGURE 24–20. Frog egg cleavage results in smaller cells at one end. More yolk is present in the end with the larger cells. (a) Zygote, (b) 2-cell, (c) 4-cell, (d) 8-cell, (e) 16-cell, and (f) many-celled stages are shown. From zygote to 16 cells usually takes about five hours.

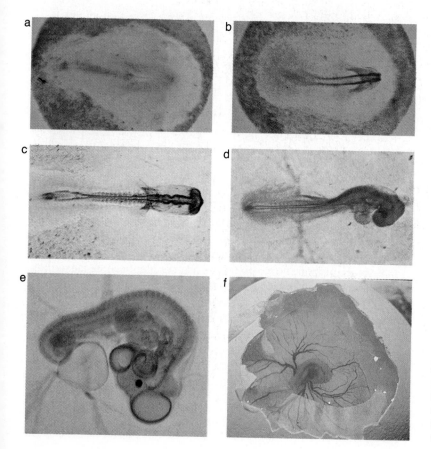

FIGURE 24–21. After cleavage in a chicken egg, continued divisions and cell movements result in the formation of a recognizable embryo. Developing chickens at (a) 18 hours, (b) 24 hours, (c) 33 hours, (d) 56 hours, and (e) 96 hours are shown. (f) The developing chicken can be seen inside the egg.

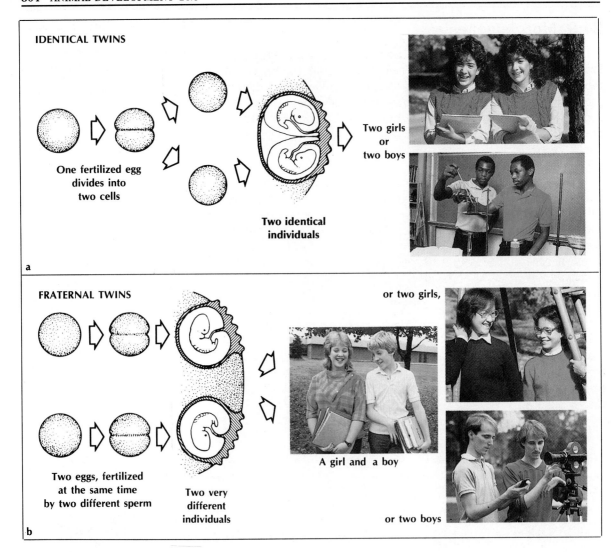

FIGURE 24–22. (a) Identical twins develop from a single fertilized egg that divides into two cells. (b) Fraternal twins develop from two fertilized eggs. Identical twins must be the same sex. Fraternal twins do not have to be the same sex. Why?

Identical twins have the same genotypes.

Fraternal twins have different genotypes.

same genotype and are physically identical because they have the same sets of DNA. These two animals are **identical twins.**

Fraternal (fruh TURN ul) **twins** are not identical, because they develop when each of two different eggs is fertilized by a different sperm. Fertilization occurs at about the same time, and development of the embryos takes place simultaneously. Each zygote begins life with a different set of DNA instructions. Therefore, fraternal twins are no more alike than any other brothers or sisters. Fraternal twins are not as common in humans as single births are because females normally produce only one egg each menstrual cycle. Must fraternal twins be the same sex? Must identical twins be the same sex? Why?

24:9 Morphogenesis

Following formation of the blastula in a frog, a small indentation forms near the vegetal pole. Some cells from the outside surface of the blastula move inward through this indentation and a curved groove, the **blastopore** (BLAS tuh por), is formed in this region. The upper part of the groove is called the **dorsal lip.** This rearrangement of cells is called **gastrulation** (gas truh LAY shun) and is shown in Figure 24–23.

Certain cells migrate through the blastopore to form a second tissue layer, the endoderm. This stage of the developing embryo is called the **gastrula.** You can picture a gastrula by thinking how an inflated balloon (the blastula) would look with your thumb stuck into one end. When pushed in, the outer "skin" of the balloon presses against itself to form two "skin" layers. The endoderm cells continue to move inward and upward to form an internal cavity called the **archenteron** (ar KENT uh rahn). This cavity will later develop into the digestive tract of the frog.

Later in gastrulation, a third tissue layer, the mesoderm, forms between the ectoderm and the endoderm. Mesoderm forms in different ways in different animals. In frogs, mesoderm forms from cells that migrate inward through the blastopore. After gastrulation is completed, the blastopore is reduced in size and will become the anus (AY nus) of the frog. The anus is the opening through which undigested food is expelled.

During gastrulation, cells of the ectoderm migrate inward through the blastopore.

The archenteron gives rise to the digestive tract.

FIGURE 24–23. Gastrulation of a developing frog embryo involves morphogenetic movement of cells. Notice the formation of the archenteron as cells push in.

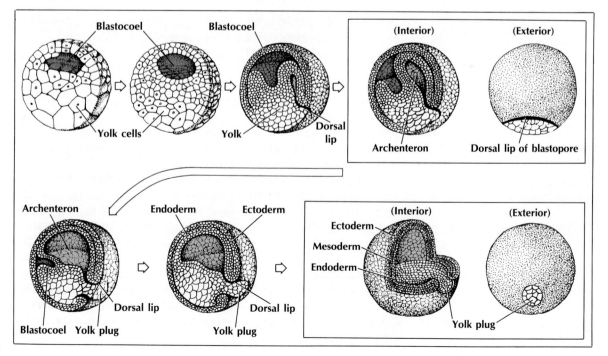

TABLE 24–1. MAIN STRUCTURES PRODUCED BY THE THREE TISSUE LAYERS		
Ectoderm	Mesoderm	Endoderm
Brain	Skeleton	Pancreas
Spinal cord	Muscles	Liver
Nerves	Gonads	Lungs
Skin (outer layer)	Excretory system	Lining of digestive
Eye cup and lens	Skin (inner layer)	system
Nose and ears		

Morphogenesis is the creation of shape or form.

The series of changes of the frog blastula from a hollow, one-layered sphere to a three-layered ball is the beginning of **morphogenesis** (mor fuh JEN uh sus). The term morphogenesis means the creation of shape or form. Formation of the three tissue layers is necessary for later development.

At this point, there is no visible evidence of any differences in the cells of the embryo. The cells are merely rearranged into three layers. Formation of ectoderm, mesoderm, and endoderm occurs in all the multicellular animals except the sponge and jellyfish groups. Each of these tissue layers gives rise to certain structures during further development. Table 24–1 lists some structures that develop from each of these layers.

24:10 Further Development

Nervous tissue develops from the neural tube.

Differentiation leads to the formation of specialized parts.

Growth adds living matter to the embryo.

After gastrulation in a frog (and other vertebrates), visible signs of further development can be seen along the dorsal surface of the animal. Morphogenesis and cell division continue. Ectoderm cells along the top of the embryo divide rapidly to form a flat **neural** (NOOR ul) **plate** (Figure 24–24). Then the sides of this plate rise up (buckle) into two folds along the entire length of the embryo. The raised cells form **neural folds** that gradually fuse together to form a hollow groove beneath them called the **neural tube.** The brain and spinal cord will form from the neural tube. Figure 24–24 shows the development of the neural tube in the embryo.

Formation of the brain and spinal cord is an example of differentiation (Section 20:13). During further development, all the specialized parts of the embryo develop. At the same time, the embryo is also growing. Growth is an increase in the amount of living material (Section 1:5). Earlier stages of development involved dividing of the large zygote into a number of smaller cells. These cells then were arranged in a specific pattern, but no growth had occurred. During gastrulation, cell division becomes less rapid and the cells grow. In the case of the frog, the embryo soon becomes recognizable as a tadpole (Section 24:2). Even at this point, though, further differentiation is necessary.

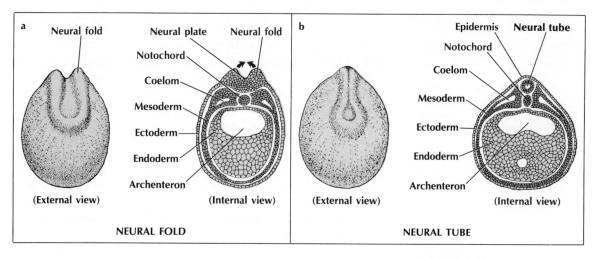

FIGURE 24–24. Formation of the neural tube begins (a) as neural folds form which (b) meet to form a tube. The brain and spinal cord form from the neural tube.

In reptiles, birds, and mammals, much differentiation is completed before the young animal is hatched or born. After that, mostly growth occurs. The period of growth varies from animal to animal. In humans, it lasts for about eighteen to twenty-one years.

24:11 Control of Differentiation

All cells of a developing embryo contain the same DNA. How, then, does differentiation occur? If all cells have the same DNA, how do different areas of the embryo become specialized as muscle or nerve or skin?

The explanation is that genes are "switched on or off" during development. Only a certain set of genes are active in a given cell at a given time. Cells produce certain enzymes depending on which genes are active. The enzymes guide the chemistry, and thus the development, of the cells. Different sets of chemical reactions result in different traits. As development proceeds, different sets of genes are active, and cells become more and more specialized.

Experiments conducted with frogs support this idea. Nuclei from blastula cells, gastrula cells, and intestinal cells of an adult frog were transferred to unfertilized eggs from which the nuclei were removed. Because the transferred nuclei were already diploid, the eggs did not have to be fertilized. Many of these eggs developed into normal tadpoles and adults. These offspring were clones (Section 19:9) because they were genetically identical to their parents. This experiment showed that *the nuclei of cells in various stages of development still have the full DNA code.* The same DNA was present in the nuclei when they were parts of blastula, gastrula, or intestinal cells, so some of it must not have been operating. But, when the nuclei were placed in the egg cells, the DNA was "switched on" again.

Cells develop differently because they have different active genes.

Experiments suggest that the nucleus of a specialized cell retains the genetic code necessary for the development of a new organism.

Tympanic membrane

FIGURE 24–25. The development of many structures is induced by other tissues near the structures. Formation of the frog tympanic membrane is induced by the cartilage beneath it.

FIGURE 24–26. Transplantation of diploid blastula nuclei into eggs from which the nuclei have been removed results in development of the tadpole. The blastula nuclei have the information necessary for total development.

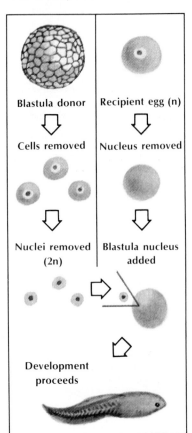

Blastula donor

Recipient egg (n)

Cells removed

Nucleus removed

Nuclei removed (2n)

Blastula nucleus added

Development proceeds

Other experiments involving cloning support this hypothesis. For example, a single, fully differentiated cell from a carrot or tobacco plant can be placed in a special culture medium. At first, the cell divides by mitosis to form an unorganized mass of cells. Later, the mass of cells becomes organized and develops into an entire new plant. This process shows that a mature cell has all the DNA necessary to produce a new organism.

Scientists are working to understand what causes genes to be switched on and off (Section 9:11). This gene regulation is probably involved in many developmental processes. How is a salamander limb regenerated? How do cells become changed in cancer? These processes are thought to be the result of different genes being switched on and off. Thus, these problems are important challenges of modern biological research.

24:12 Embryonic Induction

For genes to be switched on and off during development, some kind of "signal" is needed. Somehow a gene responds to the signal, which is chemical in nature, by becoming either activated or inactivated.

As a result of a discovery made in the 1920s, it is now known that *position of neighboring parts of an embryo is one important factor related to signals that regulate genes.* Several experiments, done by Hans Spemann and Hilde Mangold, led to the discovery. Spemann and Mangold worked with salamander embryos. They wanted to determine what causes the ectoderm on the dorsal side to develop into the neural tube (nervous tissue). As a result of two early experiments (Figure 24–27 a and b), they reasoned that the mesoderm beneath the ectoderm plays a role in nervous system differentiation. In a third experiment, a piece of mesoderm from the embryo's dorsal side was removed and transplanted to a second embryo but at a point exactly opposite from where the neural tube normally develops. (The original mesoderm from this site had been removed.) The second gastrula developed a normal nervous system in the expected manner and location. It also developed a second one where the mesoderm had been transplanted (Figure 24–27 c).

These results led to the development of the concept of embryonic induction. **Embryonic induction** occurs when one part of an embryo somehow influences, or induces, the development of another part of the embryo. During induction, one tissue, called the **organizer,** releases a chemical that passes to another tissue. That chemical, called the **inducer,** acts as the signal for differentiation. In the development of nervous tissue, the mesoderm is the organizer. It is thought to release an inducer that activates a gene (or genes) in the ectoderm. Ectoderm cells then develop into nervous tissue. Although

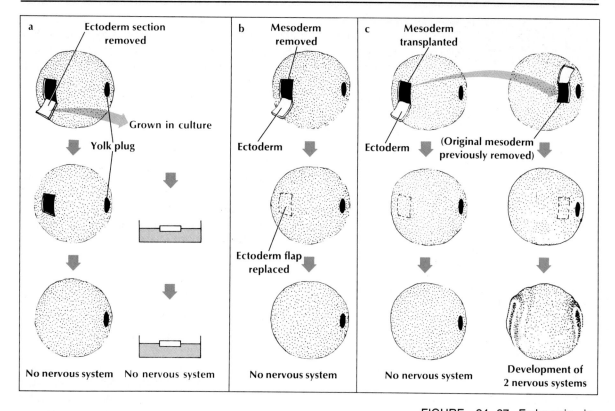

FIGURE 24–27. Embryonic induction experiments, (a) removing a portion of ectoderm, (b) removing a portion of mesoderm, and (c) transplanting a portion of mesoderm, show that mesoderm influences development of the ectoderm.

the inducer involved in nervous system development has not been identified, other inducers are known. Some inducers are hormones. For example, in humans differentiation of gonads depends upon presence of male sex hormones. The hormones induce the tissue to become testes. If the male hormones are absent, the tissue becomes ovaries.

Factors other than induction play a role in guiding differentiation. During cleavage, division of cells may occur in such a way that the chemicals in cytoplasm begin to vary from cell to cell. For example, some cells may contain more yolk than others. Also, different areas of the embryo become surrounded by slightly different chemical and physical environments. Such differences may also act as signals that regulate genes.

Different chemical and physical factors may act as signals to regulate genes.

REVIEWING YOUR IDEAS

7. What is cleavage? Describe an embryo after cleavage.
8. How is a blastula transformed to a gastrula?
9. Name the tissue layers formed during morphogenesis.
10. What is the neural tube and what does it form?
11. Explain the term embryonic induction.

Problem: How does sea urchin egg development compare to human egg development?

Materials:

microscope
prepared slide of developmental stages
  of sea urchin

Procedure:

Part A. Sea Urchin Development

1. Make a copy of Table 24–2 in which to record your observations.
2. Observe the prepared slide of sea urchin embryo development under low power of your microscope. NOTE: The slide shows all the embryo stages, but they are arranged in random order.
3. Diagram each of the following stages in the appropriate space in your table.
   (a) unfertilized egg—appears as a single cell with no clear covering, distinct nucleolus visible inside the nucleus
   (b) fertilized egg—appears as a single cell with a thick, clear covering called a fertilization membrane; nucleus is no longer visible
   (c) 2-cell stage—appears as two distinct cells that are side-by-side; each cell smaller than the fertilized egg
   (d) 4-cell stage—appears as four distinct cells that are stuck together
   (e) 16-cell stage—appears as sixteen cells that are stuck together; counting may be difficult because some cells overlap other cells
   (f) morula—appears as a solid ball of very small cells; structure is the same color throughout
   (g) blastula—appears as a ball of very small cells; cells toward center of structure appear lighter in color than cells toward outermost edge because structure is hollow
4. Label your diagrams with the names of the stages listed in Step 3. Also label the nucleolus and nucleus on diagram a and the fertilization membrane on diagram b.

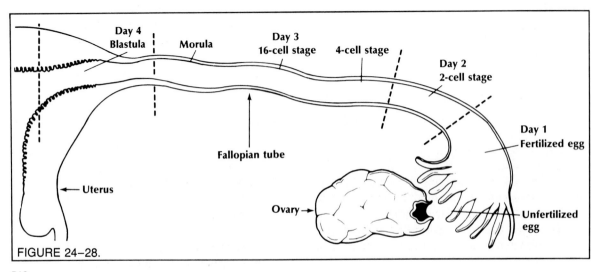

FIGURE 24–28.

Part B. Human Development

1. Make a copy of Figure 24–28. Include all the labels in your copy.
2. A fertilized human egg undergoes early developmental stages similar to those of a sea urchin. In humans, however, development occurs in the reproductive organs of the female.

3. Figure 24–28 shows half the female reproductive system. Complete your diagram by drawing in the appearance of the egg as it undergoes development. Use your seven diagrams from step 4 of Part A. Note that Figure 24–28 is divided into days. Each day indicates the time elapsed since fertilization on day 1.

Data and Observations:

TABLE 24–2. DEVELOPMENTAL STAGES OF THE SEA URCHIN	
Stage a	Stage b
Stage c	Stage d
Stage e	Stage f
Stage g	

Questions and Conclusion:

1. Define:
   (a) development
   (b) external development
   (c) internal development
2. Based on your definitions in question 1, explain how development of a sea urchin differs from that of a human.
3. Each time a cell divides by mitosis, two identical cells are formed. Describe the stage(s) that occur(s) between:
   (a) the 4-cell stage and the 16-cell stage
   (b) the 16-cell stage and the 64-cell stage
4. (a) Does cell size increase or decrease with increasing number of cells?
   (b) What evidence do you have from your observations in Part A to support your answer to 4(a)?

5. (a) In humans, how long does it take for a fertilized egg to reach the uterus?
   (b) Through what structure must the fertilized egg pass?
   (c) Describe the changes that take place in the fertilized egg between the ovary and the uterus.
6. (a) In humans, what happens to the blastula after day 4 of development?
   (b) How does this differ from what happens in the sea urchin at the same time?
7. Why can diagrams for early sea urchin development be used to represent early stages of human development?

Conclusion: How does sea urchin egg development compare to human egg development?

CHAPTER REVIEW

SUMMARY

1. Camouflage and parental behavior are means of protecting embryos. **24:1**
2. Many animals that develop externally undergo metamorphosis. Often the immature form is a larva. A larva can live apart from its parents. **24:2**
3. Complete and incomplete metamorphosis are patterns of external development among insects. **24:3**
4. Some animals with internal fertilization—reptiles, birds, and monotremes—develop within an amniotic egg. **24:4**
5. Marsupials have internal fertilization and partial internal development. After birth, development is completed within the mother's pouch, where milk is available for the young. **24:5**
6. Complete internal development occurs within placental mammals. Exchange of materials between mother and embryo through the placenta and protection of the embryo by the amnion are features of complete internal development. **24:6**
7. At the end of the development period, placental mammal young are pushed from the uterus. **24:7**
8. Animal development begins with cleavage, a series of cell divisions producing a multicellular embryo. **24:8**
9. Morphogenesis converts the blastula to a gastrula with several tissue layers. **24:9**
10. Differentiation and growth begin after gastrulation. **24:10**
11. Differentiation results from different genes being active in different cells. Particular sets of active genes give rise to different features. **24:11**
12. In embryonic induction, one part of an embryo sends a chemical signal that regulates genes, which affects differentiation of another part. **24:12**

LANGUAGE OF BIOLOGY

allantois
amnion
amniotic egg
archenteron
blastocyst
blastula
body stalk
chorion
cleavage
differentiation
embryology
embryonic induction
gastrula
gastrulation
gestation period

implantation
inducer
inner cell mass
labor
larva
metamorphosis
morphogenesis
neural tube
nymph
organizer
placenta
pupa
trophoblast
umbilical cord
yolk sac

CHECKING YOUR IDEAS

On a separate paper, complete each of the following statements with the missing term(s). Do not write in this book.

1. A _____ resembles the adult form but lacks a reproductive system and wings.
2. In amniotic eggs, wastes are stored in the _____.
3. The process of changing from an immature animal to an adult is _____.
4. _____ is marked by rapid cell division.
5. Blood travels from an embryo, through the _____, to the placenta.
6. An embryo of a bird, reptile, or mammal is surrounded by a fluid-filled sac, the _____.
7. The _____ provides for complete internal development.
8. The _____ is the length of time of internal development.
9. Bird and reptile embryos obtain most nourishment from the _____.
10. _____ is a process in which one part of an embryo causes another part to differentiate.

EVALUATING YOUR IDEAS

1. How are fertilized frog eggs protected?
2. Describe the changes that take place in the metamorphosis of a frog. Relate the structures that develop to the environment in which the frog is living.
3. What are the adaptive advantages of the presence of a larval stage?
4. What are the four membranes that develop from the embryo in an amniotic egg? What are their functions?
5. Why are marsupials limited to partial internal development?
6. Describe the implantation of an embryo of a placental mammal and the formation of the placenta and umbilical cord.
7. How is the placenta important?
8. Describe the birth process in humans.
9. How is complete internal development more efficient than other patterns of development?
10. Why is the number of mammals born at one time usually relatively small?
11. Explain how the volume of material in a blastula compares with the volume of material in a zygote.
12. Explain two ways in which cleavage is important for the rest of development.
13. How is the formation of fraternal and identical twins different? Compare their genotypes and phenotypes.
14. Describe the events that produce a gastrula from a blastula in a frog embryo. Identify the structures present in the gastrula.
15. What is differentiation? When does it begin? When does growth begin?
16. Why must it be assumed that not all genes are active during development?
17. Explain the experimental evidence that supports the hypothesis of question 16.
18. What is embryonic induction? How is it related to gene regulation?

APPLYING YOUR IDEAS

1. At what stage are yolk and shell added to bird and reptile eggs? Compare this to the stage at which they are added to grasshopper eggs.
2. How might double-yolked hen's eggs be formed?
3. In human females, ovulation usually does not occur during the time a mother is nursing a baby. Can you think of adaptive advantages of the lack of ovulation at this time?
4. A human embryo could be considered as "foreign" to its mother. What system of the mother adjusts to presence of the embryo to avoid recognition of the embryo as foreign?
5. How does a mammal develop after birth?
6. In what ways are animal and plant development similar? How are they different?
7. Why do most animals with external development develop in water or moist areas?

EXTENDING YOUR IDEAS

1. Abnormal development may result from environmental factors. Report on the effects of drugs or diseases on development.
2. Investigate methods of cloning mammals. How do these methods compare to cloning of plants, such as carrots?
3. Aging is a natural part of development. What is the error catastrophe theory of aging?

SUGGESTED READINGS

Dorfman, Andrea, "How Does a Cell Become a Complex Creature?" *Science Digest*, Oct., 1985.

McKinnell, Robert G., *Cloning: of Frogs, Mice, and Other Animals*. Minneapolis, MN, University of Minnesota Press, 1985.

Townsend, Daniel S., "Fatherhood in Frogdom." *Natural History*, May, 1987.

FOOD GETTING AND DIGESTION

To maintain life processes, animals take in and digest various foods. Here a puffin is shown with its catch of fish. These fishes will provide the bird with nutrients needed for energy, growth, and maintenance. What nutrients does your body need? What foods are the best sources of these nutrients? What would happen if you were deprived of any of these nutrients?

like other heterotrophs, animals take in complex, organic molecules. These molecules are too large to pass through cell membranes. They must be digested, or broken down, in order to enter and be used by cells.

Digestion in animals involves both physical and chemical processes. Chewing and grinding are physical processes that reduce the size of pieces of food. As in other organisms, chemical digestion in animals occurs by hydrolysis reactions (Section 3:16).

Objectives:
You will
- compare patterns of digestion in animals having one body opening with those having two body openings.
- list the major structures of the human digestive system.
- explain the processes by which food is digested and absorbed in humans.

PATTERNS OF DIGESTION

Animals have the need to take in and then digest food. The specific ways in which these functions are carried out vary from animal to animal. However, a study of many different animals reveals only a few general patterns of digestion. Each pattern is common to a great diversity of animals.

25:1 Filter Feeding

Adult sponges (Section 15:1) cannot move; they must extract food from the water around them. This method of obtaining food is called filter feeding because food is filtered from surrounding water. To filter feed, a sponge draws water into its hollow body through many incurrent pores. Lining the endoderm layer are cells called **collar cells.** These cells have flagella surrounded by collars of protoplasm. Movement of the flagella directs currents of water into the sponge. A sponge's diet is mostly microorganisms. Food is trapped when the microorganisms stick to the collar as water enters the cell. The food travels to the base of the collar and is engulfed. Digestion can occur within vacuoles of the collar cells.

Sponges obtain their food by filter feeding.

Flagella of collar cells direct currents of water into the sponge.

515

FIGURE 25–1. A sponge obtains food by filtering small organisms from the water that passes through it. The collar cell flagella direct currents of water into the sponge.

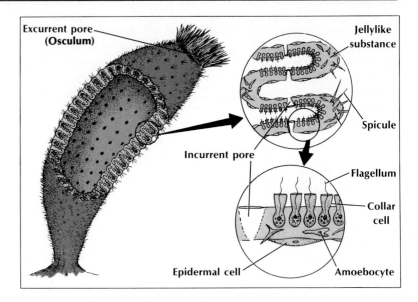

Food also enters the jellylike layer and is engulfed and digested by cells, called **amoebocytes** (uh MEE buh sites), that resemble amoebae. As amoebocytes move through the sponge, they distribute digested food to other cells by diffusion. Waste materials and excess water in the hollow body constantly are expelled from the cavity through the excurrent pore (osculum). An animal such as a sponge, which has this method of obtaining food, is called a **filter feeder.**

Filter feeding is found in other sessile animals and in some unattached animals. Oysters, for example, also filter feed. Like sponges, they do not actively seek out food. An oyster is much more complex than a sponge, but the food-getting method is similar. Cilia on the surface of the gills draw water into the oyster. Microscopic food particles stick to the gill surfaces and eventually are passed to the mouth and to the rest of the digestive system. It may surprise you to learn that some whales are filter feeders. These whales have "teeth" that strain small organisms from the water they take into their mouths.

Digested food is distributed by diffusion in sponges.

FIGURE 25–2. Cells on the tentacles of *Hydra* have nematocysts that are discharged from the stinging cells into prey.

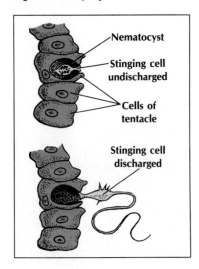

25:2 Two-Way Traffic: Hydra

A hydra (an aquatic animal of phylum Coelenterata) also obtains its food from the water in which it lives. A hydra does not swim around in search of food, but has structures for trapping food that comes close to it. When a living animal passes close to a hydra, the hydra extends its tentacles toward it. When a nematocyst (Section 15:2) is stimulated by the touch of a small organism, the filament, which is coiled inside, uncoils and is forcefully ejected. Some nematocysts have filaments that wrap around and trap prey. Others are barbed, sticking into the prey like fishhooks and preventing the prey from escaping. Poison that either paralyzes or kills the prey is then injected into it through the filament. Then the prey is drawn to the mouth of the hydra and into the body cavity. Because hydras take in large bits of food, they are called **chunk feeders.** Most animals are chunk feeders.

Nematocysts help hydra to trap food.

Hydras are chunk feeders.

There are no special regions for digestion in the cavity of a hydra. This cavity is called the **gastrovascular** (gas troh VAS kyuh lur) **cavity.** It is hollow and contains water. Lining the cavity are many gland cells that secrete digestive enzymes into the cavity. These enzymes break large molecules of food into small molecules. This process, because it occurs outside the body cells, is extracellular digestion (Section 18:4).

In hydras, digestion begins in the gastrovascular cavity.

Other endoderm cells then take in these molecules by phagocytosis. Once inside the cells, digestion continues in food vacuoles. This process, because it occurs within the body cells, is intracellular digestion (Section 18:2). Digested food passes out of the food vacuoles and into the surrounding cytoplasm. Because the animal consists of only two tissue layers, digested food can pass readily into the outer tissue layer (ectoderm) from the inner tissue layer (endoderm).

Both extracellular and intracellular digestion occur in hydra.

Hydras have only one body opening, so both incoming food and outgoing waste materials must pass through the mouth. There is a "two-way traffic" in which food and wastes move in opposite directions through the same opening. The digestive system of *Hydra* is not very specialized.

FIGURE 25–3. *Hydra* takes prey into its mouth and extracellular digestion begins in the gastrovascular cavity. Intracellular digestion follows.

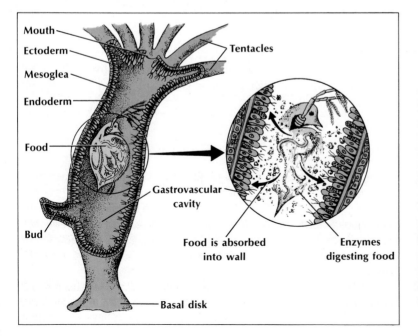

Mouth
Ectoderm
Mesoglea
Endoderm
Food
Bud
Tentacles
Gastrovascular cavity
Food is absorbed into wall
Enzymes digesting food
Basal disk

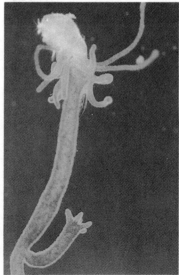

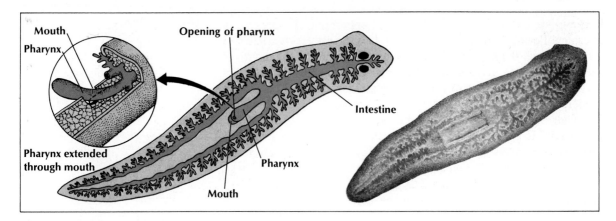

FIGURE 25–4. Extracellular digestion is carried on in the hollow of a planarian's intestine. Some intracellular digestion occurs within cells of the intestinal lining. The branching digestive system clearly is visible in the stained specimen at the right.

In planarians, food is digested in the hollow of the intestine and in cells of the intestinal lining.

Because a planarian is flat and its intestine is branched throughout the body, digested food can diffuse to all cells.

25:3 Two-Way Traffic: Planaria

A "two-way traffic" system is also found in other simple animals. A planarian has a single mouth opening on its ventral (belly) side. When an animal comes into contact with a suitable food source, a tube called the **pharynx** (FER ingks) protrudes through the mouth. Food enters the pharynx and the pharynx is withdrawn into the animal. The food moves into another tube called the **intestine** where some of the food is digested (extracellular). Digestion of the remaining food occurs in cells of the intestinal lining (intracellular).

As in *Hydra,* digested food reaches the body cells by diffusion. Although a planarian has three tissue layers, diffusion can transport food to all the cells because the body is flat and the intestine is branched throughout the animal. No cell is very far from the intestine, and digested food (Figure 25–4).

In more complex animals, direct diffusion from digestive organs to body cells is impossible because too many cells are too far from the food. In these animals, digested food is transported to body cells by blood.

25:4 One-Way Traffic: Earthworm

Organisms with two body openings have more specialized digestive systems.

In most of the other animal phyla there are two body openings, the mouth and the anus, in the digestive system. In such a system there is "one-way traffic." Food is ingested through the mouth, passed posteriorly during digestion, and waste material is egested (expelled) through the anus. *With this plan there is differentiation of digestive organs. The organs are specialized. Each organ has a certain function.*

In general, food passes through a long, hollow tube called the **alimentary** (al uh MEN tree) **canal.** Different areas of the tube have

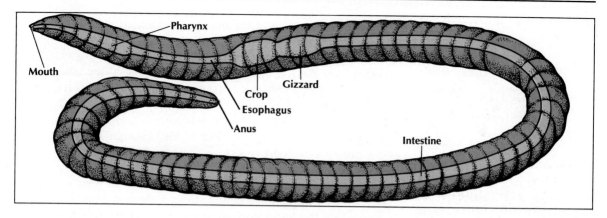

certain shapes and features. Usually one section can clearly be distinguished from another.

In earthworms, food enters the anterior end of the alimentary canal through the mouth (Figure 25–5). From the mouth, the food is moved by muscle contractions into a thickened area, the pharynx, and then to a tube called the **esophagus** (ih SAHF uh gus). The thin esophagus pushes the food onward into the first of two round organs, the **crop,** where it is stored until it can be pushed to the second round organ, the **gizzard.** In the gizzard the food is ground up with the aid of coarse soil particles ingested by the worm. After this physical digestion, food passes into a long, thin tube called the intestine. This organ is lined with gland cells that pour enzymes into the cavity. The enzymes break the food into small molecules that enter the cells of the intestinal wall. These small molecules are picked up by the blood and transferred to all parts of the body where they diffuse into the body cells. Undigested materials are pushed through the intestine and finally are expelled through the **anus** (Figure 25–5). Because digestion occurs within the cavity of the alimentary canal and not in the body cells themselves, digestion is extracellular.

25:5 One-Way Traffic: Other Invertebrates

Arthropods have a digestive system similar to the earthworm's. Arthropod mouthparts are well developed for obtaining, chewing, and grinding food. These mechanical processes prepare the food for compact storage and further digestion.

Mouthparts of arthropods are adapted for obtaining their particular foods. Butterflies have a tubelike proboscis (pruh BAHS kus), or snout, which is used to suck nectar from flowers. Female mosquitoes have mouthparts adapted for piercing skin and sucking blood. A housefly's mouthparts sponge, or suck, food through a small opening (Figure 25–6).

FIGURE 25–5. The earthworm, having both mouth and anus, has a "one-way" digestive system. Notice the digestive system is more specialized than that of less complex animals.

Food is stored in the crop and ground in the gizzard.

FIGURE 25–6. Arthropods have specialized mouth parts. A housefly sucks liquid food through an opening in the labellum.

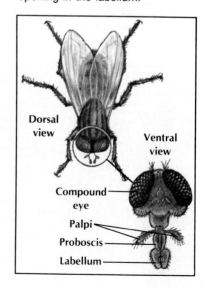

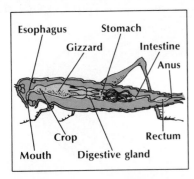

FIGURE 25–7. A "one-way" digestive system is present in a grasshopper. Digestive glands secrete enzymes into the alimentary canal.

Besides the organs of the alimentary canal, arthropods have other organs that aid in digestion. Food does not pass directly through these other organs, but these organs are important because they produce enzymes and other materials that are secreted into the alimentary canal. For example, a grasshopper has many digestive glands that surround the alimentary canal near the crop and stomach.

The starfish also has a digestive system with two body openings. The starfish moves along the floor of the ocean by using a type of water pressure system in which suction cups on the underside of each arm can attach to objects. This suction also can be used to obtain food. Using the suction cups, a starfish can pry apart the two shells of a clam. The starfish then pushes its stomach through its mouth, which is located on the lower surface of its body. The stomach secretes enzymes that partially digest the clam. Then the partially digested food passes from the stomach to digestive glands located in each ray (arm) where digestion is completed. A small amount of undigestible material is egested through the anus located on the upper body surface. Small food molecules are carried to body cells by a fluid in the coelom.

REVIEWING YOUR IDEAS

1. What is filter feeding? In what kinds of animals is it found?
2. How do hydras obtain food?
3. How is digested food distributed in hydras and planarians?
4. What is an alimentary canal?
5. How is digested food distributed in earthworms?
6. Is digestion in a starfish intracellular or extracellular?

FIGURE 25–8. The starfish turns its stomach outward into the clam and begins digestion of the clam tissues.

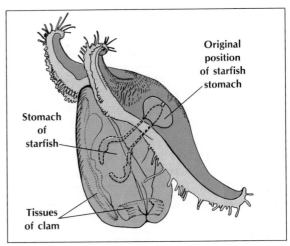

DIGESTION IN HUMANS

Humans have a complex digestive system with two body openings. The alimentary canal and other digestive organs provide for thorough digestion and absorption of food. Digestion in humans is mainly extracellular.

25:6 Nutrition

Before studying the digestive system of humans, it is important to consider human nutrition. **Nutrition** is the study of foods and how they are used by the body. Nutrients are used in two main ways — as sources of energy and as sources of molecules needed for growth and maintenance. Energy in food is commonly measured in units called Calories. Some nutrients are complex molecules — carbohydrates, fats, and proteins — that must be digested.

Water is necessary for a variety of functions and is therefore an important nutrient in the diet. It is a solvent for many cellular substances. It is also the major component of protoplasm and blood. Hydrolysis reactions and other metabolic processes use water. Water evaporating from the skin aids in cooling the body. Some of the water that is used or lost is replaced by water produced in dehydration synthesis and cellular respiration. The rest must be taken in as part of the diet.

Minerals are nutrients that are ingested as ions in food or water. They serve a variety of important metabolic functions and are parts of molecules such as hemoglobin, ATP, and DNA. Table 25–1 lists some important minerals and their functions.

Nutrients are used as energy sources or as molecules used in growth and maintenance.

Water is an important nutrient used in a variety of ways.

Minerals function in metabolism.

TABLE 25-1. SOME IMPORTANT MINERALS		
Mineral	Sources	Needed For
Calcium	Milk, cheese, meats, vegetables, whole-grain cereals	Blood clotting, muscle contraction, bone and tooth formation
Iron	Liver, leafy vegetables, meats, raisins	Production of hemoglobin, part of molecules used in respiration
Iodine	Seafood, iodized table salt	Part of hormone that controls rate of metabolism
Magnesium	Leafy vegetables, meats, potatoes	Proper functioning of some enzymes
Phosphorus	Milk, meats, eggs, vegetables	Bone and tooth formation, part of ATP and nucleic acids
Potassium	Vegetables, bananas	Conduction of nerve impulses
Sodium	Table salt, vegetables	Conduction of nerve impulses, maintaining osmotic balance

TABLE 25-2. SOME IMPORTANT VITAMINS		
Vitamin	Sources	Deficiency Symptoms
A	Green and yellow vegetables; eggs; fruits; liver	Night blindness; dry skin
B_1	Whole-grain cereals; liver; other meats; nuts; most vegetables	Weak muscles; paralysis (beriberi)
B_2	Milk; cheese; eggs; liver; whole cereals	Poor vision; sores in and around mouth
B_{12}	Liver; other meats; eggs; milk products	Anemia
C	Citrus fruits; tomatoes; some vegetables	Loose teeth; bleeding gums; swollen joints; poor muscle development (scurvy)
D	Milk; egg yolk; fish liver oil; (also made in skin)	Soft bones; deformed skeleton (rickets)
K	Green vegetables; liver; (also made by intestinal bacteria)	Slow blood clotting

Another group of vital nutrients, vitamins, are present in many foods. They may be complex molecules but are easily absorbed. Many vitamins are needed to make coenzymes (Section 5:6). Table 25–2 lists some vitamins important to humans and foods that are good sources of them. The diseases that result when the vitamins are not available in the diet in proper amounts are also given. These diseases are called vitamin-deficiency diseases. For example, if a young person's diet does not include enough vitamin D, bones and muscles will not develop properly—a disease called rickets. Lack of vitamin A results in night blindness.

Vitamin deficiencies may result in disease.

25:7 Diet and Health

A balanced diet consists of proper amounts of foods from four major groups—dairy products, breads and cereals, fruits and vegetables, and meats. A person eating a variety of foods each day receives enough of all nutrients. However, the diet should include several selections from each major group. Eating a proper, balanced diet is an important part of maintaining good health.

FIGURE 25–9. A balanced diet includes foods from each of the four basic food groups. Selections from the (a) meat group, (b) fruit-vegetable group, (c) dairy group, and (d) bread-cereal group are shown.

a

b

c

d

If one or more essential nutrients are missing from a person's diet over a period of time, **malnutrition** will occur. Malnutrition can lead to a loss of weight and a variety of health problems. For example, in children, a protein deficiency can lead to weight loss, liver damage, and anemia.

Many people today have a diet that includes too many "junk foods." These foods provide quick energy sources (carbohydrates, sugars, and fats) and are high in Calories, but provide few other nutrients. A diet made up to a large extent of these foods and few others may cause a person to suffer from malnutrition or obesity. Obesity, an overweight condition, occurs when more food is consumed than is burned up by the body. Extra weight increases the chance of developing high blood pressure by two to three times. High blood pressure may lead to heart disease or stroke. The chances of developing high blood cholesterol levels or diabetes are also greater for overweight people. The only known "cure" for obesity is to lower Calorie intake. A lower Calorie intake should be accomplished while still eating a balanced diet. Foods from all four food groups should be selected, but portions should be smaller. Removing high Calorie foods such as junk foods from the diet and increasing the amount of daily exercise are also helpful in controlling weight.

Junk foods provide sources of energy but often few other nutrients.

PEOPLE IN BIOLOGY

A Mohawk Indian raised by her Quaker grandparents, Rosa Minoka-Hill made up her mind early in life to help needy Indians. On a primitive farm on the Oneida Indian Reservation in Wisconsin, Dr. Minoka-Hill carried out that decision. Accepting everything in payment from chickens to a day's work on her farm, Rosa Minoka-Hill delivered babies, treated diseases, and ran a "kitchen clinic" stocked with herbals (plants used for medical treatment) and medicines provided

Lillie Rosa Minoka-Hill

[1876-1952]

by the doctors in Green Bay for the people of the reservation. Dr. Minoka-Hill also spent much of her time teaching the Indians about good nutrition and eating habits to try to relieve their ever present problem of malnutrition.

In addition to her medical practice, Dr. Minoka-Hill reared her six children and ran her farm alone after the death of her husband in 1916. Because of her dedication, the Oneidas adopted her into their tribe and gave her the name "you de gent" — "She who serves." Rosa Minoka-Hill was also named Outstanding American Indian of the Year in 1947 by the Indian Council Fire in Chicago, and was given an honorary membership to the State Medical Society of Wisconsin.

FIGURE 25–10. Much is learned about the nutritional value of foods, the additives they contain, and the relative amounts of ingredients present by carefully reading food labels.

Excess sugar and salt may be harmful.

An excess of some nutrients may also be harmful to the body. Many people eat an excess of sugar and salt. Besides adding these two nutrients to food during cooking and at the table, many people consume large quantities of them in packaged foods. Some "ready-to-eat" processed foods have large amounts of sugar and salt added. Excess sugar contributes to obesity, may promote tooth decay, and has been linked to hyperactivity in children. Excess salt contributes to high blood pressure and heart disease, and may cause the body to retain water, thus adding to body weight.

In recent years, consumers have become more aware of the substances added to many foods. Food companies now list product ingredients on the label in the order of the amount of each ingredient present. The ingredient present in the greatest amount is listed first, second greatest second, and so on. This type of labeling helps consumers to choose products carefully.

FIGURE 25–11. Both physical and chemical digestion begin in the mouth. The esophagus transports food from mouth to stomach.

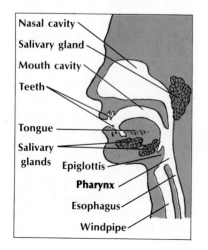

25:8 Mouth, Pharynx, and Esophagus

In humans, food is ingested through the **mouth,** or oral cavity. The major function of this organ is physical digestion; teeth grind and tear food into chunks small enough to be swallowed. The grinding action of teeth also increases the surface area of food. With more surface area available, enzymes can work more effectively.

Besides being physically broken down in the mouth, food also is moistened by **saliva** (suh LI vuh), the liquid secretion of three pairs of **salivary** (SAL uh ver ee) **glands.** Saliva contains a substance called mucin (MYEWS un) (mucus), which makes food slippery so it can be swallowed easily. Also, the water in saliva makes the food a thin paste.

Saliva also contains an enzyme called **salivary amylase** (AM uh lays), which begins the chemical digestion of food. This enzyme breaks starch into molecules of maltose, a disaccharide (Section 3:12). No digestion of fats or proteins occurs in the mouth.

From the mouth, the moistened, slightly digested food passes through the pharynx (throat) to a long, slender tube, the esophagus. No chemical digestion occurs in the esophagus; it transports the food and adds mucus to it. Food is moved by a series of alternating muscular contractions and relaxations called **peristalsis** (per uh STAHL sus). Peristalsis also occurs along the rest of the alimentary canal. The muscle action moves and churns food and controls the rate of passage of food material.

Peristalsis aids the passage of food through the esophagus and the other organs of the alimentary canal.

25:9 Stomach

Food leaves the esophagus through a muscular ring called the **cardiac sphincter** (KARD ee ak • SFINGK tur) and enters a large hollow organ, the **stomach.** Normally, the cardiac sphincter prevents food from passing back to the esophagus. When does food go from the stomach to the esophagus?

Protein digestion begins in the stomach. Lining the stomach wall are **gastric** (GAS trihk) **glands** that secrete **gastric juice** into the stomach. Release of gastric juice is stimulated by a hormone called gastrin (GAS trun). When food enters the stomach, certain stomach cells release gastrin into the bloodstream. Gastrin travels throughout the body but affects only the gastric glands by stimulating them to secrete gastric juice.

Gastrin stimulates release of gastric juice from the stomach.

Gastric juice contains pepsinogen (pep SIHN uh juhn) and hydrochloric acid. Pepsinogen is an inactive form of the enzyme pepsin. **Pepsin** breaks peptide bonds between certain amino acids in proteins, forming smaller polypeptide molecules. Hydrochloric acid gives gastric juice a pH between one and two (Section 3:8). This pH is needed for pepsin to function efficiently. The activities of these chemicals are aided by a churning motion of the stomach. By the time this phase of digestion is completed, food that is in the stomach is an acidic liquid called **chyme** (KIME).

In the stomach, proteins are hydrolyzed into polypeptides.

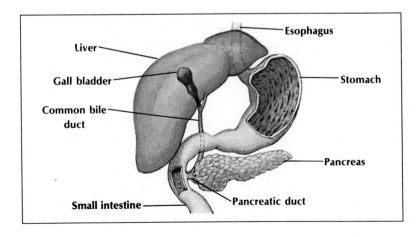

Liver

Gall bladder

Common bile duct

Esophagus

Stomach

Pancreas

Small intestine

Pancreatic duct

FIGURE 25–12. Food is liquefied in the stomach. Both the liver and pancreas produce substances necessary for digestion in the small intestine.

25:10 Small Intestine

Chyme passes in small spurts through the **pyloric** (pi LOR ihk) **sphincter** of the stomach into the **small intestine.** Thus, the stomach regulates the flow of food into the intestine. The first part of the small intestine is the duodenum (dew uh DEE num), which, in an adult human, is about 25 cm long. Next the food passes into a section called the jejunum (jih JEW num). From the jejunum, food moves into the last and longest portion of the small intestine called the ileum (IHL ee um). The total length of the coiled small intestine is about seven meters in an adult human.

Two organs that are not part of the alimentary canal are very important in digestion of food in the small intestine. The **liver** produces a greenish substance called **bile.** Bile is stored in the **gallbladder,** an organ that lies beneath the liver. From the gallbladder, bile moves through a tube called the **common bile duct** to the duodenum. Bile is not an enzyme, but it contains **bile salts** that break up fat globules in the intestine. This process is called fat emulsification (ih mul suh fuh KAY shun) and it helps speed up the rate of enzyme action on the fat by increasing fat surface area. This aspect of fat digestion is physical digestion.

The other organ, the **pancreas** (PAN kree us), lies behind and partially below the stomach. Certain cells of the pancreas secrete enzymes that are transported by the **pancreatic** (pan kree AT ihk) **duct.** This duct merges with the common bile duct from the gallbladder and empties into the duodenum.

The pancreas secretes several important enzymes. **Lipase** (LI pays) breaks fat molecules into fatty acids and glycerol. **Pancreatic amylase** changes starch to maltose. Several proteases (PROH tee ay suz), called **trypsins** (TRIHP sunz), hydrolyze polypeptides and proteins not digested in the stomach. The polypeptides are broken into smaller polypeptides. Each trypsin breaks bonds between certain amino acids.

Also released by the pancreas is sodium bicarbonate, which neutralizes the acidity of the chyme and raises its pH to about eight (slightly basic). Pancreatic enzymes do not work effectively in an acid environment. Stimulation of the pancreas to release digestive materials is controlled by a hormone.

Fingerlike projections called **villi** (VIHL i) on the intestinal lining extend into the hollow of the small intestine. Villi greatly increase the surface area for absorption of digested food. Disaccharides and small polypeptides in the intestine are actively transported across the membranes into the villi. Cells in the villi contain enzymes needed to complete digestion of these molecules. Several enzymes called **peptidases** (PEP tuh days uz) convert small polypeptides to amino acids. Other enzymes convert disaccharides to monosaccharides. **Maltase**

Bile is secreted by the liver and stored in the gallbladder.

Bile is channeled to the small intestine where it breaks up fat globules.

FIGURE 25–13. Villi contain blood vessels that pick up absorbed food molecules. The large number of villi greatly increases the surface area of the small intestine available for absorption.

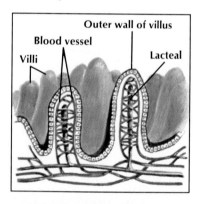

Outer wall of villus

Blood vessel

Villi

Lacteal

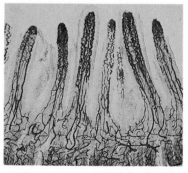

TABLE 25-3. SUMMARY OF DIGESTION IN HUMANS		
Digestive Organ	Major Physical Aspects	Major Chemical Aspects
Mouth	Chewing Grinding Moistening	Starch $\xrightarrow{\text{salivary amylase}}$ maltose
Esophagus	Moistening, Peristalsis	
Stomach	Moistening Churning Peristalsis Some absorption	Proteins $\xrightarrow{\text{HCl, pepsin}}$ polypeptides
Small intestine	Peristalsis Fat emulsification Most food absorption (villi)	Starch $\xrightarrow{\text{pancreatic amylase}}$ maltose Proteins $\xrightarrow{\text{proteases, eg. trypsins}}$ polypeptides Polypeptides $\xrightarrow{\text{peptidases}}$ amino acids Fats $\xrightarrow{\text{lipases}}$ fatty acids & glycerol Disaccharides $\xrightarrow{\text{maltase, lactase, sucrase}}$ monosaccharides
Large intestine	Most water absorption Peristalsis Waste elimination	

(MALT tays) breaks down maltose left over from previous digestion; **sucrase** (SEW krays) breaks down sucrose, or table sugar; and **lactase** (LAK tays) breaks down lactose, or milk sugar. Because these activities occur within the cells of the villi, digestion in humans is partly intracellular.

In the small intestine, all complex food molecules are converted to simple forms. Proteins are broken into amino acids, fats are split into fatty acids and glycerol, and carbohydrates are hydrolyzed into simple sugars, or monosaccharides. These small molecules can pass through cell membranes. Table 25–3 summarizes the major events of digestion.

Intestinal enzymes are responsible for the hydrolysis not only of small polypeptides to amino acids, but also of lactose, maltose, and sucrose to monosaccharides.

Amino acids, monosaccharides, fatty acids, and glycerol are small enough to pass through cell membranes.

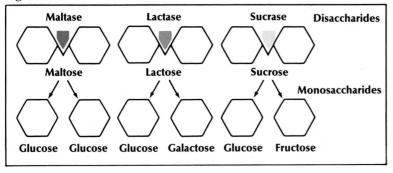

FIGURE 25–14. Maltose, lactose, and sucrose are actively transported into the cells of the villi. There they are broken down into monosaccharides by the action of the enzymes.

INVESTIGATION

Problem: How is the milk sugar lactose digested?

Materials:

2 glass slides
glass-marking pencil
toothpick
lactase enzyme

milk
glucose solution
TesTape
3 droppers

Procedure:

Part A. Models of Lactose Digestion

1. Make a copy of Table 25–4. It will be used to record diagrams and data.
2. Notice that Figure 25–15a shows a model of lactose, a carbohydrate molecule found in milk. Lactose is composed of two simpler molecules, glucose and galactose.
3. Notice that Figure 25–15b shows a model of lactase, an enzyme produced by the small intestine. Lactase is the enzyme that digests or breaks lactose molecules into glucose and galactose.
4. In the proper space in Table 25–4, diagram how the enzyme lactase fits exactly onto the lactose molecule, much as a key fits a lock.

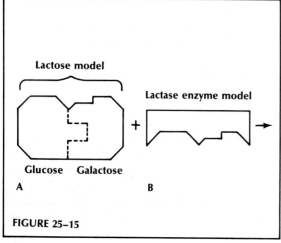

FIGURE 25–15

5. In the proper space in Table 25–4, diagram the resulting simple sugar molecules that result from the digestion of lactose by lactase. NOTE: The model in Figure 25–15a shows the two simpler molecules joined together to form lactose. You must also draw the enzyme molecule because enzymes are not used up or changed as digestion proceeds. Label each molecule.

Part B. Digestion of Lactose with Lactase

1. Use the glass-marking pencil to draw 2 circles on each of the two glass slides. Number the circles 1–4. Use Figure 25–16 as a guide.

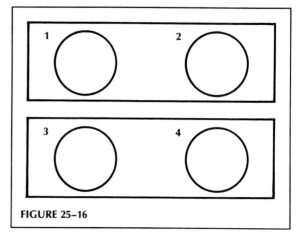

FIGURE 25–16

2. Using different droppers, add each of the following to the proper circle.
   circle 1 — 2 drops glucose solution
   circle 2 — 2 drops milk
   circle 3 — 1 drop lactase
   circle 4 — 1 drop lactase plus 2 drops milk
3. Mix circle 4 with a toothpick.
4. After waiting 5 minutes, test the liquid in each circle for the presence of glucose. This is

done by touching a different piece of Tes-Tape to each.

5. After waiting 1 minute, note any color change in the TesTape. NOTE: TesTape is normally yellow. It will turn green in the presence of glucose.

6. Record your data in Table 25-4 and complete the information requested.

Data and Observations:

TABLE 25-4. LACTOSE DIGESTION			
Lactose and lactase model joined together		Lactose model after digestion by lactase	
Circle	Contents	Resulting color of TesTape	Glucose present?
1			
2			
3			
4			

Questions and Conclusion:

1. Define:
   (a) enzyme
   (b) substrate
   (c) digestion
   (d) lactose
   (e) lactase
2. Use your models from Part A to answer these questions.
   (a) What carbohydrate molecule is present in milk before it is digested?
   (b) What two smaller molecules form a molecule of lactose?
   (c) Does the glucose in the lactose molecule exist as a separate entity before the digestion of milk?
   (d) What organ of the digestive system produces lactase?
3. Use your experimental results from Part B to answer these questions.
   (a) Was glucose present in milk?
   (b) Was glucose present in lactase?
   (c) Was glucose present in milk after the enzyme lactase was added? Explain.
4. Explain why it was helpful to test glucose with TesTape.
5. Explain the role of the enzyme lactase in the digestion of milk.
6. Some people cannot drink milk because their digestive systems do not produce lactase. If milk is drunk, it is not digested. This results in cramps, bloating, gas, and diarrhea. Suggest a treatment that would help such a person.

Conclusion: How is the milk sugar lactose digested?

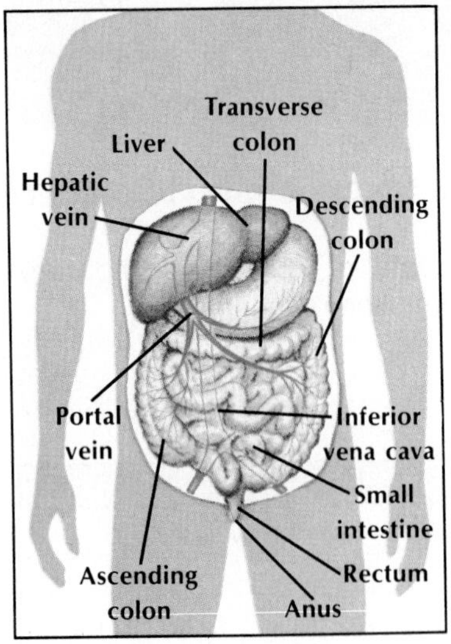

FIGURE 25–17. Blood that picks up food molecules in the small intestine passes through the liver where the sugar level is monitored and excess sugar stored as glycogen. Then the blood moves to the rest of the body.

25:11 Absorption of Food

Most absorption of the digested molecules into the blood occurs in the small intestine through the villi. The villi contain many small blood vessels called capillaries. Glycerol, amino acids, simple sugars, vitamins, and minerals enter these capillaries. The capillaries of the villi merge to form the **hepatic** (hih PAT ihk) **portal vein** that carries blood to the liver. The liver detects the level of food molecules in the blood. After circulating through the liver, blood reaches the heart and is pumped to all parts of the body. When the blood reaches capillaries, sugars, amino acids, vitamins, minerals, and other substances leave it and enter the cells.

The glucose content of blood must be within a certain range. After digestion of a meal, blood coming to the liver from the small intestine has more than the normal amount of glucose. The liver detects this excess glucose and removes it from the blood. After being converted, the glucose is stored in the liver as **glycogen,** an animal starch. Later, when the glucose content in blood begins to fall, the liver converts stored glycogen back to glucose. The glucose then enters the blood, restoring the glucose level to the desirable range.

Most digested fats are absorbed by **lacteals** (LAK tee ulz), another kind of vessel in the villi. Lacteals are part of a special transport network, the lymphatic (lihm FAT ihk) system (Section 26:8). Digested fats travel through lymph vessels and later enter the bloodstream. The fats then pass from the blood to body cells.

25:12 Large Intestine

Undigested materials and water are passed from the small intestine to the **large intestine,** or **colon** (KOH lun). The large intestine forms a loop and partly covers the small intestine. The large intestine is divided into three sections—the **ascending colon** (to which the appendix is attached), the **transverse colon,** and the **descending colon.** No digestion takes place in the large intestine.

The major function of the large intestine is to absorb the water from undigestible or undigested materials. In this way, water is conserved in the body. Absorption of water from the waste matter causes the wastes to become a solid. Solid waste, called **feces** (FEE seez), is stored in the **rectum** (REK tum) and is egested through the **anus.**

In addition to food remnants, feces contain many bacteria that live in the intestine. Some of these bacteria make vitamin B_{12}, which the body needs. A deficiency of this vitamin results in abnormal red blood cell and hemoglobin production (Table 25–2). The coexistence of

In the large intestine, undigestible materials become solid as water is absorbed from them.

these intestinal bacteria and humans is an example of mutualism (Section 13:11) because bacteria and host benefit each other. Some intestinal bacteria protect the body by inhibiting growth of pathogenic bacteria.

REVIEWING YOUR IDEAS

7. What is nutrition? What are the six kinds of nutrients required by humans?
8. What is needed for a balanced diet?
9. What is peristalsis? In what organs does it occur?
10. List the organs of the human alimentary canal. What other organs are important in human digestion?
11. What are villi? How are they important in absorption?
12. What is the function and importance of the large intestine?
13. Is human digestion mainly intracellular or extracellular?

CHAPTER REVIEW

SUMMARY

1. Some animals, mainly sessile forms, are filter feeders. **25:1**
2. A hydra has one body opening. Thus, digestion involves "two-way traffic." **25:2**
3. Planaria has one body opening for digestion. Its branched intestine and flat shape permit transport of digested food by diffusion. **25:3**
4. Earthworms have two body openings and "one-way traffic." These features provide for specialized digestive organs and more efficient digestion. **25:4**
5. Arthropods and starfish are examples of other invertebrates having two body openings and "one-way traffic." **25:5**
6. Carbohydrates, fats, proteins, water, minerals, and vitamins are important classes of human nutrients. **25:6**

7. A balanced diet includes proper amounts of foods in each of the four food groups. **25:7**
8. In humans, digestion begins in the mouth where food is chewed and moistened and starches are converted to maltose. Food passes from the mouth to esophagus, via the pharynx. **25:8**
9. In the stomach, food is churned and protein digestion begins. **25:9**
10. Bile from the liver, and enzymes from both the pancreas and the small intestine itself all function in digestion within the small intestine. **25:10**
11. Amino acids and monosaccharides are absorbed in the small intestine by the capillaries in the villi and are transported to body cells by the bloodstream. Most digested fats pass into lacteals in the villi and are transferred from the lymphatic system to the blood. **25:11**
12. The large intestine functions in the removal of water from undigested materials. Feces are

CHAPTER REVIEW

stored in the rectum and egested through the anus. **25:12**

LANGUAGE OF BIOLOGY

alimentary canal
amoebocyte
anus
bile salts
capillary
chunk feeder
collar cell
crop
esophagus
filter feeder
gallbladder
gastric juice
gastrovascular cavity

gizzard
lacteal
large intestine
malnutrition
minerals
nutrition
pancreas
peristalsis
pharynx
rectum
small intestine
stomach
villi

CHECKING YOUR IDEAS

On a separate paper, indicate whether each of the following statements is true or false. Do not write in this book.

1. Filter feeding occurs only in sessile animals.
2. Digestion in *Hydra* begins in the nematocysts.
3. Clams and sponges are chunk feeders.
4. The planaria has a digestive system with "one-way traffic."
5. In some animals, grinding of food occurs in the crop.
6. In an earthworm, digested food is transported to body cells by diffusion.
7. Substances such as iron, calcium, and sodium are nutrients.
8. Villi absorb digested food in the human intestine.
9. Most chemical digestion in humans occurs in the large intestine.
10. Digestion in humans is mainly extracellular.

EVALUATING YOUR IDEAS

1. Explain food getting and digestion in sponges.
2. How are digestion, transport of digested food, and egestion similar in a hydra and a planarian?
3. Why are the digestive systems of hydras and planarians not very specialized?
4. What is the adaptive advantage of a digestive system with two openings—a mouth and an anus?
5. How is digestion in an earthworm different from digestion in a planarian?
6. Why must digested food molecules be distributed by blood in earthworms and many other animals?
7. How does a starfish obtain and digest food?
8. What are the two main ways in which nutrients are used?
9. How is water important in the body? How is it lost? How is it replaced?
10. List several ways in which minerals are used in the body. Why are vitamins important?
11. Discuss the possible effects of obesity. How may obesity be avoided?
12. How may an excess of sugar and salt in the diet be harmful?
13. How is chewing important to human digestion? What other physical processes occur in the mouth? What might result if these processes did not occur?
14. What are the functions of the esophagus?
15. How is gastrin important to digestion in the stomach?
16. Why is hydrochloric acid necessary for digestion in the stomach?
17. Is bile an enzyme? Explain. How is bile important in digestion?
18. Why must carbohydrates, fats, and proteins be broken down? What are the products of their digestion?

19. What structures in humans secrete enzymes necessary for digestion? Does food pass through all of them? Explain.
20. In which organs of the alimentary canal does chemical digestion occur?
21. How are villi important?
22. What happens to food after it is absorbed by the villi? After it is digested?
23. How is the large intestine important?
24. How are bacteria of the large intestine important?

APPLYING YOUR IDEAS

1. In what way(s) is digestion in *Paramecium* similar to digestion in a complex animal?
2. Birds have a gizzard. How is this adaptive?
3. How might regurgitation be adaptive?
4. Why is the diet of a person whose gallbladder has been removed restricted in fat?
5. Have you ever seen commercials about "acid indigestion"? What causes acid indigestion? Why would medicines containing sodium bicarbonate be useful to relieve it?
6. The inner surface of the intestine of a shark is coiled. How is a coiled intestine adaptive?
7. How is digestion in animals different from digestion in most plants? Are there any plants with digestion similar to digestion in animals?
8. Compare food getting and digestion in *Planaria* and a tapeworm.
9. Why are many nutrients needed in greater amounts by children and teenagers than adults?
10. Vitamins and minerals are available in tablets, capsules, and liquids. Do you think most people need to take these nutrients in these forms? Explain.
11. When you swallow food, some salivary amylase enters the stomach. Does that enzyme "work" in the stomach? Explain.

12. Compare the taking in of food in fungi and animals.
13. Why may fad diets be a danger to health? Do you think people who lose weight on these diets usually keep it off? Explain.

EXTENDING YOUR IDEAS

1. Conduct a survey among classmates to determine whether each is eating a balanced diet.
2. Prepare a report on the digestive system of a cow. What is the rumen?
3. In 1822 Alexis St. Martin received a wound that resulted in a hole leading into his stomach. Use library sources to learn about the experiments that were carried out on St. Martin.
4. Find out about different kinds of food additives. Why are they used? What are the possible side effects of some of them? Are many of these additives necessary?
5. Find out about careers in nutrition. What do nutritionists do? What do dietitians do? Why are many of them hired by institutions such as schools and nursing homes?
6. Using library sources, research the different names used for sugar and salt on food labels. Use this information to survey the items in your home pantry. How much "hidden" sugar and salt does your family consume?

SUGGESTED READINGS

"How the Stomach Protects Itself." *Science Digest,* March, 1983.

Jordan, Henry A., *Finding Your Weigh to Slimming.* Carolina Biology Reader. Burlington, NC, Carolina Biological Supply Co., 1983.

"The National Nutrition Quiz." *Science Digest,* April, 1986.

TRANSPORT

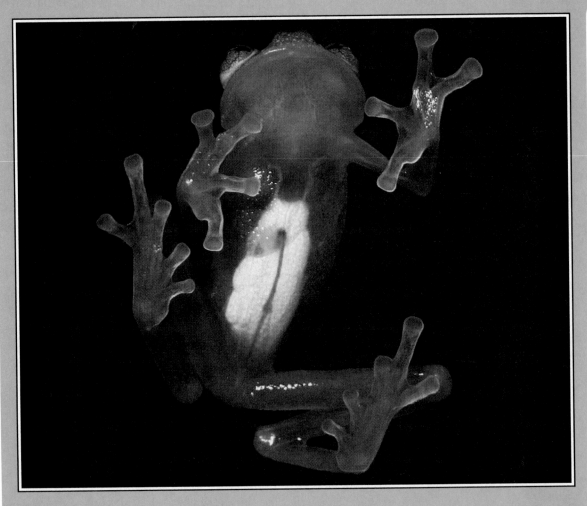

Most complex animals have systems that transport materials throughout their bodies. These systems have some kind of liquid to carry materials, and structures where the materials can be exchanged with the environment. This transport fluid is pumped by a heart, as seen through the transparent ventral surface of the bare-hearted glass frog. Through what other structures does the liquid flow? How do transport systems differ among the various kinds of animals?

chapter 26

Animals such as sponges, coelenterates, and flatworms are composed of cells only a few layers thick so each cell can exchange materials directly with the environment. In larger, more complex animals, many individual cells cannot exchange materials directly with the environment because they are too far from it. In these animals, materials enter and leave at certain points and are carried to and from these points by some kind of liquid. The liquid is part of a transport system, a link between the cells of an animal and the environment. Because exchange of materials occurs between cells and the transport fluid, the fluid often is called the "internal environment" of the animal.

Objectives:
You will
- compare the circulatory systems of a variety of animals.
- identify the structures and functions of blood in humans.
- explain various causes and forms of treatment of heart disease.

Cells of simple animals can exchange necessary materials directly with the environment.

CIRCULATORY SYSTEMS

In animals with transport systems, a pump circulates a liquid throughout the animal. The pump is a **heart** and the liquid is **blood.** The contents and functions of blood vary in different animals. In general, blood carries materials needed for metabolism and waste materials. Thus, blood is the exchange agent between internal tissues and the environment. Blood and the structures through which it passes are known as the **circulatory** (SUR kyuh luh tor ee) **system.**

Larger, more complex animals have transport systems that link cells and the environment.

In complex animals, blood is the agent of exchange between cells and the external environment.

26:1 Annelids

Segmented worms are the least complex animals to have a true circulatory system. This system is very simple in structure, but it permits rapid and efficient exchange of materials. In an earthworm, blood is transported through two major blood vessels — the **dorsal blood vessel** and the **ventral blood vessel.** These two vessels are

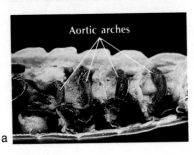

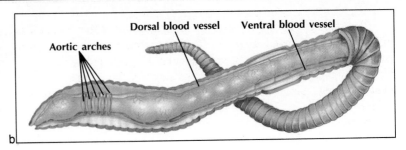

a

b

FIGURE 26–1. (a) Five pairs of aortic arches pump blood through (b) the closed circulatory system of an earthworm.

connected near the anterior end of the worm by a series of five pairs of enlarged tubes called **aortic** (ay ORT ihk) **arches** that function as hearts. As the aortic arches contract, blood is pumped through the body (Figure 26–1).

Blood flows in the dorsal blood vessel from body tissues to the aortic arches where it is pumped into the ventral blood vessel. The blood flows posteriorly in the ventral blood vessel. Other vessels in almost every segment send blood from the ventral to the dorsal blood vessel. These vessels are not the sites of exchange between blood and cells; they are merely tubes to transport the blood.

Exchange between blood and cells takes place in small branches of the major vessels, the capillaries. Capillaries have thin walls so materials pass easily into and out of them. Capillaries extend all over the body of a worm. Food molecules enter the blood in the region of the intestine and oxygen enters the capillaries near the skin. These materials are transported through the body to each cell. Capillaries are so numerous that no cell is far from the blood and so each cell is "linked" with the external environment.

Exchange of materials between cells and blood occurs through capillaries.

In a closed circulatory system, blood is always in the blood vessels.

In an earthworm, blood is always in the blood vessels. For this reason, this system is known as a **closed circulatory system.** A closed system is present in several animal phyla.

26:2 Arthropods

Arthropods have a circulatory system in which blood is not always in the blood vessels. Such a system is called an **open circulatory system.**

A grasshopper has a dorsal, segmented heart (Figure 26–2) that pumps blood anteriorly into a single blood vessel, the **aorta** (ay ORT uh). Blood empties from the aorta into a body cavity called the **hemocoel** (HEE muh seel), which is composed of spaces, or **sinuses** (SI nus uz), through which blood passes. While in the sinuses, the blood bathes the cells directly. Blood is circulated in the sinuses by muscular movements of the grasshopper. Eventually the blood collects in a sinus surrounding the heart and enters the heart through tiny openings called **ostia** (AHS tee uh).

FIGURE 26–2. A grasshopper has an open circulatory system. Blood passes through the hemocoel rather than circulating inside blood vessels all the time.

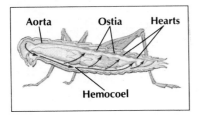

The open circulatory system of the grasshopper may seem less efficient than a closed system. In insects the circulatory system carries food to and wastes away from cells. Oxygen and carbon dioxide are transported by a series of tubes separate from the circulatory system. Thus, blood flow is not as rapid and ordered as it is in a system where blood carries these gases.

Some arthropods, such as spiders and crayfish, have open systems in which the blood does carry oxygen and carbon dioxide. In crayfish, blood bathes the gills as it moves through the sinuses. In the gills, oxygen enters the blood and carbon dioxide is removed.

In insects, gases are transported by means other than by blood.

26:3 Fish

Every vertebrate has a closed circulatory system made up of a single heart and three types of blood vessels. The heart has two kinds of chambers — atria and ventricles. (An atrium is sometimes called an auricle (OR ih kul.) An **atrium** receives blood coming to the heart and a **ventricle** pumps blood away from the heart. The three types of blood vessels are arteries, veins, and capillaries. **Arteries** carry blood away from the heart to the body. **Veins** carry blood to the heart from the body. Capillaries are thin-walled vessels that connect arteries and veins.

A vertebrate has a closed circulatory system composed of a single heart, arteries, veins, and capillaries.

atrium (plural, atria)

In fish, the heart has only two chambers — one atrium and one ventricle. Blood from the body collects in a saclike structure, the **sinus venosus** (SI nus • vih NOH sus). From there, blood enters the atrium and is forced into the ventricle. Contractions of the ventricle pump the blood into a large artery, the **ventral aorta.** The blood is under much pressure as it leaves the ventricle. Smaller arteries branch from the ventral aorta and lead to gills where the arteries are subdivided into capillaries. As blood passes through capillaries in the gills, oxygen is taken in and carbon dioxide is given off. The gill capillaries merge to form arteries leading to the **dorsal aorta.** The dorsal aorta is subdivided into smaller arteries that lead to all parts of the body. Exchange of materials occurs in the capillaries. Blood returns to the heart in veins.

Fish have two-chambered hearts. Oxygen and carbon dioxide are exchanged at the gills after the blood leaves the heart.

FIGURE 26–3. In fish, blood is oxygenated in the gills after leaving the heart. A fish has a two-chambered heart composed of one atrium and one ventricle. Within the system, the flow of oxygenated blood is indicated by red and deoxygenated blood is indicated by blue.

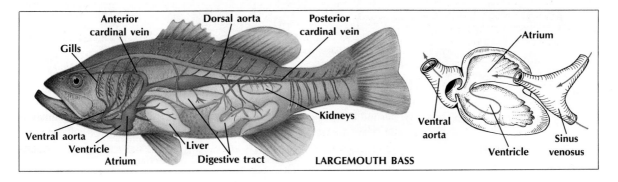

Anterior cardinal vein — Dorsal aorta — Posterior cardinal vein — Atrium — Gills — Ventral aorta — Ventricle — Atrium — Liver — Digestive tract — Kidneys — Ventral aorta — Ventricle — Sinus venosus — **LARGEMOUTH BASS**

For a fish, this type of circulatory system is adaptive. But, compared to more complex vertebrates, blood circulation is slow. As blood leaves the gill capillaries, its pressure is reduced. There is less pressure to force blood to the rest of the body.

26:4 Amphibians and Reptiles

In amphibians, the heart has three chambers — two atria and one ventricle. Deoxygenated blood (blood from the body that does not have oxygen) enters the right atrium from the sinus venosus. Oxygenated blood (blood with oxygen) from the lungs enters the left atrium through the **pulmonary** (PUL muh ner ee) **veins.** Blood from both atria enters the single ventricle so there is some mixing of blood with and without oxygen. In general, arteries from the ventricle take blood to the body, lungs, and skin. The arteries end in capillary networks where exchanges take place. The capillaries merge to form veins that return blood to the heart (Figure 26–4).

In frogs, blood pressure is reduced as blood passes through lung and skin capillaries, but oxygenated blood returns to the heart before going to the body. Therefore, there is not a great loss of blood pressure as there is in fish. By going back to the heart after being oxygenated, the blood is under the full force of the ventricle as it enters the arteries to the body. Thus, blood is pumped twice in a complete circulation of the body. It is pumped to the lungs and then again to the rest of the body.

Although blood from both the lungs and the body enters the ventricle, blood from the left atrium (oxygen-rich blood) is deflected so that it is pumped to the body. Blood from the right atrium (deoxygenated) passes into the ventricle and is pumped to the lungs and skin to pick up oxygen. Although some of the blood going to the body is not oxygenated, enough oxygen is present for use by the body cells.

FIGURE 26–4. Amphibians have three-chambered hearts. Each heart has two atria and one ventricle. A portion of the blood is oxygenated in the lungs before entering the heart. Partially mixed oxygenated and deoxygenated blood is pumped from the ventricle.

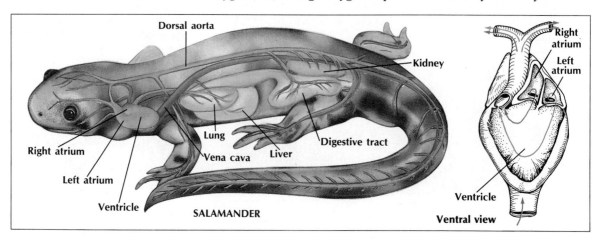

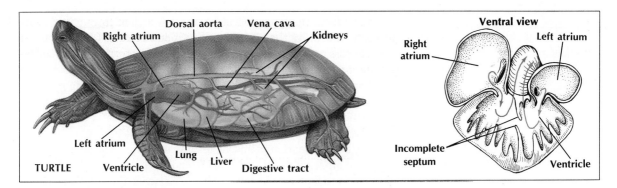

Like amphibians, most reptiles have three-chambered hearts. However, the ventricle is partly divided by a wall of muscle called a **septum** (SEP tum). This wall keeps oxygenated and deoxygenated blood almost completely separated. Therefore, deoxygenated blood is directed only to the lungs and oxygenated blood is directed to all other tissues (Figure 26–5).

FIGURE 26–5. In reptiles, oxygenated blood in the heart is directed to the body. Reptiles have three-chambered hearts. The ventricle is partly divided.

Most reptiles have three-chambered hearts.

26:5 Birds and Mammals

In birds and mammals, the heart has **right** and **left atria** and **right** and **left ventricles.** Two ventricles are present because the septum is complete (Figure 26–6). A four-chambered heart prevents deoxygenated blood from going to the body cells.

Deoxygenated blood enters the right atrium through the **superior** (from above) and **inferior** (from below) **vena cavae** (VEE nuh • KAY vee). The right atrium contracts and forces blood into the right ventricle. Contraction of the right ventricle forces blood into the **pulmonary artery.** A branch of this artery goes to each lung where oxygen is picked up and carbon dioxide is removed.

Birds and mammals have four-chambered hearts.

Deoxygenated blood enters the right side of the heart in humans.

FIGURE 26–6. In mammals, the blood moving from the heart to the body is fully oxygenated. Mammals have four-chambered hearts. Each heart has two atria and two ventricles.

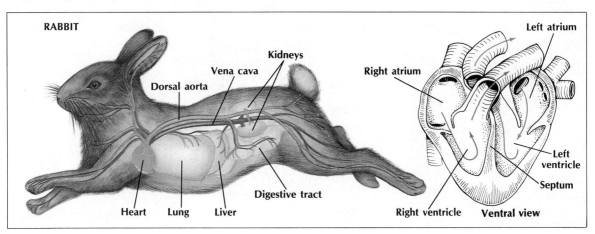

Contraction of the left ventricle forces blood out through the aorta and to the body cells.

Heart valves permit blood to flow in only one direction.

Birds and mammals have the most highly developed circulatory systems.

Blood from the lungs travels to the left atrium through the **pulmonary veins.** As the left atrium contracts, blood is forced into the left ventricle. A powerful contraction of the left ventricle forces blood into the **aorta,** the large artery from the heart. The aorta forms an arch that bends behind the heart. Smaller arteries branching from the aorta direct blood to all parts of the body. The arteries further divide into capillaries that merge to form small veins. These veins merge to form the large veins that bring blood to the right atrium.

The human heart contains valves between the atria and ventricles and at the entrances of the arteries from the ventricles. These valves allow blood to flow only in one direction. Between the right atrium and right ventricle is the **tricuspid** (tri KUS pud) **valve.** Between the left atrium and ventricle is the **bicuspid** (bi KUS pud) **valve.** At the entrances of the arteries from the ventricles are **semilunar** (sem ih LEW nur) **valves.** Why do you think blood flow in one direction is important?

The bird and mammalian circulatory systems are more complex than those of other vertebrates. In birds and mammals, all blood travels through the heart twice during each circulation around the body. By going through the heart twice and being pumped both to the lungs and the body, blood circulation is efficient. (The entire trip takes about one minute in humans.) Because there are two ventricles, a complete separation of oxygenated and deoxygenated blood is assured as well. Blood from the left ventricle for general circulation to body cells is fully oxygenated so all tissues get the needed oxygen. Blood from the right side of the heart lacking in oxygen is sent to the lungs where oxygen is picked up.

FIGURE 26–7. The human heart contains valves that keep the blood moving in one direction.

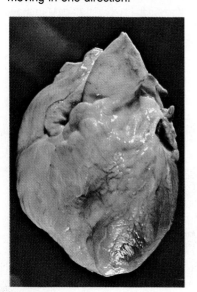

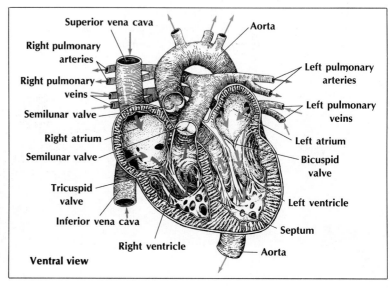

Superior vena cava

Aorta

Right pulmonary arteries

Right pulmonary veins

Semilunar valve

Right atrium

Semilunar valve

Tricuspid valve

Inferior vena cava

Right ventricle

Ventral view

Left pulmonary arteries

Left pulmonary veins

Left atrium

Bicuspid valve

Left ventricle

Septum

Aorta

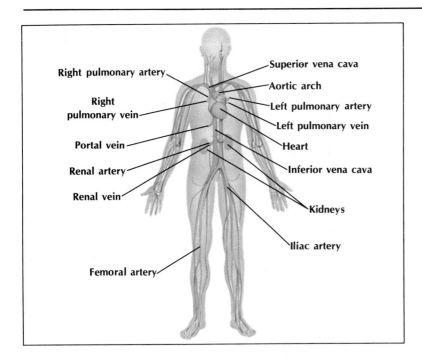

Right pulmonary artery

Right pulmonary vein

Portal vein

Renal artery

Renal vein

Femoral artery

Superior vena cava

Aortic arch

Left pulmonary artery

Left pulmonary vein

Heart

Inferior vena cava

Kidneys

Iliac artery

FIGURE 26–8. The right side of the heart pumps blood to the lungs. There it is oxygenated and returned to the heart. The left side of the heart pumps blood to the body. Oxygen moves from the blood into body cells, and carbon dioxide is picked up from the body cells by the blood.

PEOPLE IN BIOLOGY

In the late 1800s, most American medical schools admitted and trained only men. Into this male-dominated field came Florence Rena Sabin, one of the first women to be accepted at the Johns Hopkins Medical School. Sabin was well prepared to enter the medical world. While earning her B.S. in Zoology and teaching at Smith College, she had shown an exceptional ability for accurate and original observations in her laboratory work.

Florence Rena Sabin

(1871–1953)

After graduating with her M.D. in 1900, Dr. Sabin became the first woman faculty member at Johns Hopkins where she did important research on the lymphatic system. Using pig embryos, Sabin was able to describe the origins and development of the lymphatic system. Later, she did similar research on blood vessels, and finally, at the Rockefeller Institute for Medical Research, studied the body's immunity to tuberculosis.

Dr. Sabin's research brought her many awards in her lifetime including the M. Carey Thomas Prize in Science, 1935, and the Lasker Award, 1952. But Dr. Sabin's interests went beyond the medical laboratory. She enjoyed reading, cooking, and collecting Oriental art.

FIGURE 26–9. (a) Currents from the S-A node cause the atria to contract. These currents stimulate the A-V node, which causes the ventricles to contract. (b) An electrocardiogram showing two normal heartbeats has a characteristic shape.

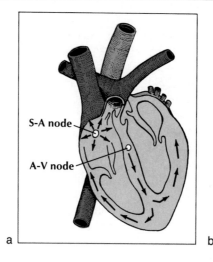

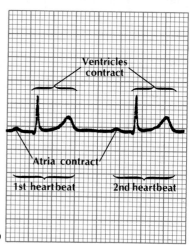

a

b

26:6 The Heartbeat

In a developing chick embryo, there is at first a slight twitching of an incomplete heart. Later, the twitching becomes a definite rhythmic beating that will continue steadily for the remainder of the animal's life. The source of each and every heartbeat is from within the heart itself.

The human heartbeat originates in a small bundle of tissue located in the right atrium. This bundle is called the **pacemaker,** or **sinoatrial** (si noh AY tree ul), or **S-A, node.** In the S-A node, a flow of ions sets up a current that travels quickly across the muscle fibers of the atria and causes the atria to contract. The current then reaches another small bundle, the **atrioventricular** (ay tree oh ven TRIHK yuh lur), or **A-V, node** located between the atria and ventricles. From here the current sweeps outward and downward along the muscle walls of the ventricles causing them to contract (Figure 26–9). The heart of a resting adult human beats about seventy times per minute.

Artificial pacemakers have been developed. An artificial pacemaker uses small batteries as its power source and sets up electric currents in much the same way as the natural pacemaker does. These devices allow people with a damaged natural pacemaker to survive. Some pacemakers can even be implanted within the body.

Control of the heartbeat is a part of maintaining homeostasis, a constant internal environment despite external change (Section 1:6). Blood is part of the internal environment of humans. Enough blood must reach all body cells under all conditions to supply needed materials and carry away wastes. At rest, a heartbeat rate of seventy beats per minute is adequate, but during heavy exercise cells need more oxygen and glucose. Also, the cells yield more wastes. Blood must bring in and take away these materials quickly in order to keep

FIGURE 26–10. An artificial pacemaker may be implanted in the chest if the natural pacemaker is failing. Much like the natural pacemaker, the artificial one sets up electric currents.

the body functioning. How is this change in heartbeat rate accomplished?

Although each heartbeat begins at the S-A node, the heartbeat *rate* is controlled by nerves and a part of the brain called the **medulla oblongata** (muh DUL uh • ahb long GAHT uh) (Section 29:10). There are two centers in the medulla oblongata—an **acceleration center** and an **inhibition** (ihn uh BIHSH un) **center.** When the acceleration center is stimulated, a message is relayed to the spinal cord. From there, impulses are sent to the S-A node along a pair of accelerator nerves. These accelerator nerves stimulate the heart to beat faster; thus, the heartbeat rate is increased. Several factors can cause increased heartbeat rate. Among them are heavy muscular activity and buildup of carbon dioxide in the blood.

When the inhibition center is stimulated, impulses from the medulla oblongata are sent to the S-A node along a pair of **vagus** (VAY gus) **nerves.** The vagus nerves slow the heartbeat. Lowered carbon dioxide content of the blood is one factor that results in stimulation of this center. Increased blood pressure is another. How are these changes involved in homeostasis?

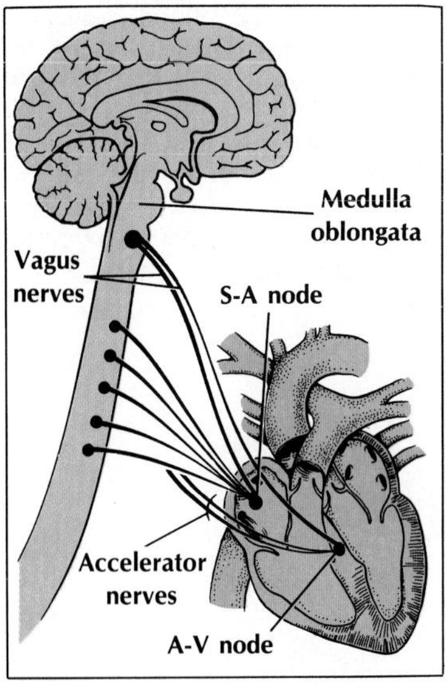

FIGURE 26–11. The medulla oblongata and accelerator and vagus nerves control heartbeat rate.

26:7 Blood Pressure

Each heartbeat consists of two phases. The forceful, muscular contraction of the ventricles is called **systole** (SIHS tuh lee). During systole, blood is pumped into the aorta. Systole is followed by a short period of "rest" called **diastole** (di AS tuh lee). During diastole, the atria and ventricles fill with blood.

As blood is pumped, it exerts a pressure on the walls of blood vessels. Because systole is a contraction of the ventricles, pressure during systole is greater than during diastole. **Blood pressure** is the ratio of systolic to diastolic pressure and is measured with a special gauge. It is usually measured in the arm. The two pressures are measured in millimeters of mercury and expressed as a fraction such as 120/80. (This fraction represents a systolic pressure of 120 and a diastolic pressure of 80, normal for a healthy adult.)

Blood pressure varies in different areas of the body. Factors that affect blood pressure include gravity, body position, and muscular activity. For example, gravity has a pronounced effect on the decreased blood pressure in the veins of the lower arms and legs. Vein structure and muscular contraction work together to overcome this. When the muscles surrounding the veins contract, the veins are squeezed, and the blood is pushed forward through "one-way" valves. Thus, the blood is channeled toward the heart.

In general, blood pressure increases gradually with age, but unusually high blood pressure, or **hypertension,** is a dangerous condition. Hypertension has been referred to as the "silent killer"

Each heartbeat consists of two phases, systole and diastole.

Blood pressure is the ratio of systolic to diastolic pressure.

Several factors affect blood pressure in different areas of the body.

FIGURE 26–12. Blood pressure, which is measured with a special gauge, should be checked periodically. High blood pressure can be reduced with proper medication.

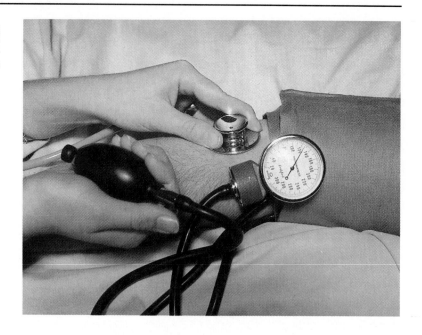

Hypertension may lead to a variety of illnesses and even death.

because it has no obvious symptoms that can be recognized by a person with the condition. Prolonged hypertension may lead to enlargement of and damage to the heart, kidney failure, heart failure, and death. Hypertension may also be related to **atherosclerosis** (ath uh roh skluh ROH sus). Atherosclerosis is a condition in which the arteries become lined with fatty deposits. These deposits make the arterial walls thicker, reduce blood flow, and further elevate blood pressure. Should hypertension exist, it can be treated with special drugs that reduce the blood pressure. Early detection and use of these drugs can prevent the harmful, long range effects of this disease.

26:8 Lymphatic System

Capillaries are the sites of exchange, but not all materials are exchanged *directly* between the body cells and the blood in the capillaries. (Carbon dioxide and oxygen probably are.) As blood passes through the capillaries, pressure is exerted on the capillary walls. The pressure is greatest at the arterial end of the capillary. This pressure forces some water and small dissolved particles out of the capillaries and into the tissues. The water and particles are called **tissue fluid.** Large molecules, such as proteins, remain in the capillaries. At the venous end of the capillary, blood pressure is lower, so some tissue fluid returns to the capillary by osmosis. Osmosis occurs because there is a larger amount of protein and a smaller amount of water in the capillaries than in the surrounding tissue fluid.

Tissue fluid is composed of water and dissolved particles derived from blood.

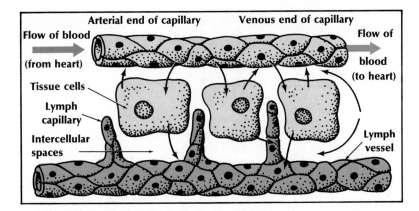

Arterial end of capillary   Venous end of capillary

Flow of blood (from heart)

Flow of blood (to heart)

Tissue cells

Lymph capillary

Intercellular spaces

Lymph vessel

FIGURE 26–13. Blood pressure forces some of the fluid portion of blood out of the capillaries. Many substances are exchanged between tissue fluid and cells. Most tissue fluid returns to the capillary but a small portion enters the lymphatic system.

FIGURE 26–14. The lymph system consists of vessels that run throughout the body and groups of lymph nodes that are located in specific places.

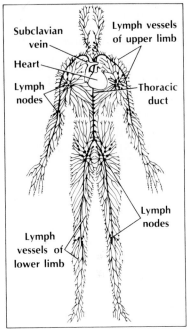

Subclavian vein

Lymph vessels of upper limb

Heart

Lymph nodes

Thoracic duct

Lymph nodes

Lymph vessels of lower limb

Tissue fluid is an important agent in the exchange of materials with cells. Materials such as nutrients and ions actually enter cells from the tissue fluid. Also, wastes and secretions such as hormones enter the tissue fluid from cells. Thus, tissue fluid is a vital part of the internal environment.

Not all tissue fluid returns to capillaries by osmosis. Some of the fluid enters a special network of **lymph vessels.** Lymph is the fluid that enters this network. Lymph vessels also carry fats that enter the lacteals of the small intestine (Section 25:11). There is no "pumping action" for lymph circulation, but random body movements force the lymph to a major collecting tube, the **thoracic duct,** located near the heart. The thoracic duct empties lymph into a large vein. In this way, the liquid portion of the blood lost at the capillaries is restored.

Scattered throughout the lymphatic system are special structures, the lymph nodes (Section 19:9). They contain cells that engulf and destroy bacteria and dead cells. Lymph nodes also remove and store small particles such as dust. Where are some of your lymph nodes? Have you ever experienced swelling in the lymph nodes? What does this swelling indicate?

REVIEWING YOUR IDEAS

1. What is a circulatory system?
2. What is the function of a heart?
3. Distinguish between a closed circulatory system and an open circulatory system.
4. Distinguish among arteries, veins, and capillaries.
5. Distinguish between atria and ventricles.
6. Compare the hearts of fish, amphibians, reptiles, birds, and mammals.
7. What is the pacemaker?
8. What is blood pressure? How is it measured?
9. How is tissue fluid important to circulation?

Problem: How does one analyze pulse and heartbeat?

Materials:

clock with second hand
ECG graphs

Procedure:

Part A. Determining Your Pulse Rate

1. Make a copy of Table 26–1 and Table 26–2. Use them to record your data.
2. Locate the pulse in your wrist using Figure 26–15a as a guide. Do not use your thumb. Use only your fingers. Exerting slight pressure with your fingers may enable you to find a stronger pulse.

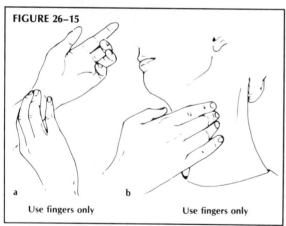

FIGURE 26–15

a Use fingers only

b Use fingers only

3. Count your pulse rate for exactly 60 seconds. Record the number you obtain in Table 26–1 as trial 1.
4. Repeat steps 2 and 3 four more times. Record these numbers as trials 2–5 in Table 26–1.
5. Calculate and record your average pulse rate in Table 26–1.
6. Locate the pulse in your neck using Figure 26–15b as a guide. Exerting slight pressure with your fingers may enable you to find a stronger pulse.
7. Count your neck pulse for exactly 60 seconds. Record the number in Table 26–1 as trial 1.
8. Repeat steps 6 and 7 four more times. Record these numbers as trials 2–5 in Table 26–1.
9. Run in place for exactly one minute. Take your neck pulse for exactly 15 seconds immediately after running. Record this number in Table 26–1 as trial 1. Multiply this number by 4 to arrive at your running pulse for one minute. Record this number in the proper column of Table 26–1.
10. Repeat step 9 four more times. Record these values as trials 2–5 in Table 26–1.
11. Calculate an average pulse rate for each column of Table 26–1.

Part B. Reading an ECG

1. Examine the graph provided by your teacher. Figure 26A shows a recording of a person's heartbeat. Placing electrodes onto a person's body allows one to detect and record this pattern of "electrical" messages. Such a recording is called an electrocardiogram or ECG.
2. Identify each of the following line segments on Figure 26A:
   (a) Segment T-P: represents no "electrical" message reaching heart muscle. Label the bracket below this segment "heart muscle relaxed."
   (b) Segment P-Q: represents "electrical" messages reaching both atria. Label the bracket above this segment "atria contracted."
   (c) Segment Q-T: represents "electrical" messages reaching both ventricles. La-

bel the bracket below this segment "ventricles contracted."
3. As "electrical" messages reach each heart area, the muscles of these areas contract and pump blood. Analyze the ECG to determine the time spent by normal atria and ventricles during the heart's normal sequence of contraction and relaxation. Complete Table 26–2. Note that each vertical line on the ECG represents 0.1 seconds.
4. Pulse rate can be calculated from an ECG. Determine the time lapse between two consecutive crests of ventricle contraction (point R). This time is 0.9 seconds on the ECG. Divide this time into 60 to arrive at the pulse per minute.

5. Use Figure 26B to draw an ECG tracing of your heartbeat. Use the average pulse rate calculated for your neck and recorded in Table 26–1. Add to your ECG tracing of your heartbeat the proper letters P-T to designate line segments. Assume the length of time for atrium and ventricle contraction to be the same as in Table 26–2.
6. Use Figure 26C to draw an ECG tracing of your heartbeat during exercise. Use the average pulse rate after running calculated for your neck and recorded in Table 26–1. Add in the proper letters P-T to designate line segments. Assume the length of time for atrium and ventricle contraction to be the same as in Table 26–2.

Data and Observations:

TABLE 26-1. PULSE RATES				
Trial	Wrist	Neck	Running pulse for 15 seconds	Running pulse for 1 minute
1				
2				
3				
4				
5				
Total				
Average				

TABLE 26-2. TIME SEQUENCE DURING NORMAL HEARTBEATS		
	Shown on ECG by which letter segment(s)	Time needed to complete event(s)
Atria relaxed		
Ventricles relaxed		
Atria contracted		
Ventricles contracted		

1. What is your pulse and what does it exactly agree with?
2. Explain why you can feel a pulse in your wrist and neck.
3. Define:
   (a) systole
   (b) diastole
4. Using the ECG of Figure 26A, determine what line segment corresponds to each of the following normal heartbeat events. NOTE: Keep in mind that when one set of chambers is contracting, the other is relaxing.
   (a) atrium systole
   (b) ventricle systole
   (c) atrium diastole
   (d) ventricle diastole

Conclusion: How does one analyze pulse and heartbeat?

THE BLOOD

Blood transports materials, repairs wounds, fights infection, and helps maintain body temperature.

As your internal environment, blood functions in the homeostasis of your body. Oxygen, digested food such as glucose and amino acids, mineral ions, and vitamins are transported by blood (Section 25:11). Waste products including urea and carbon dioxide are dumped into the blood. Blood has protective functions including repair of wounds and defense against disease. Blood also has a part in maintaining body temperature and other body functions. The materials in the blood are listed in Table 26–3.

TABLE 26-3. BLOOD CONTENTS	
Substance	Characteristics
Erythrocytes (red blood cells)	biconcave, disc-shaped cells with no nuclei; transport oxygen and some carbon dioxide
Leukocytes (white blood cells)	many types have round shape; all have nuclei; "fight off" bacteria and produce antibodies
Platelets	cell fragments that liberate materials necessary for clotting activities
Proteins	some involved with clotting activities; some are antibodies; some have other functions
Hormones	chemicals secreted by glands and transported to specific areas where they perform specific functions
Urea	waste product formed by liver; transported to kidneys where it is filtered out
Glucose, amino acids, fats, vitamins, minerals, lipids	transported to all cells and tissues

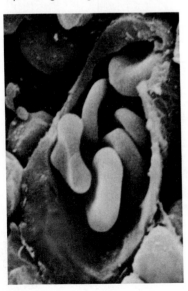

FIGURE 26–16. Gas exchange occurs in the capillaries between the red blood cells and the body cells. Red blood cells, magnified about 4800 times, are shown squeezing through a capillary.

26:9 Gas Transport

A microscope shows that a drop of human blood is packed with disc-shaped cells. These cells are indented on each side and they have no nuclei (Figure 26–16). There are about 5 million of them per cubic millimeter of blood. These cells are called **red blood cells** or **erythrocytes** (ih RIHTH ruh sites).

Red blood cells are important in the transport of gases. Red blood cells contain a pigmented molecule called **hemoglobin** (HEE muh gloh bun). Hemoglobin is a complex protein that contains iron and can carry oxygen. In the lungs, oxygen combines readily with hemoglobin to form the compound **oxyhemoglobin** (ahk sih HEE muh gloh bun), making the blood bright red. From the lungs, oxyhemoglobin goes to the heart and then to the body cells. In the capillaries, oxyhemoglobin breaks down into hemoglobin and oxygen. The oxygen diffuses from the red blood cells to the body cells.

While oxygen diffuses to body cells, carbon dioxide leaves the body cells and enters the blood. Some of the carbon dioxide ionizes in

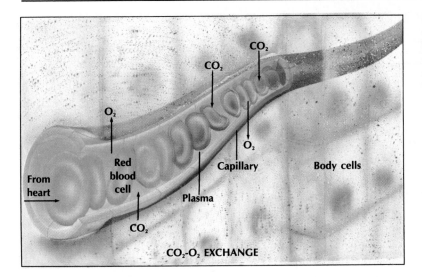

CO$_2$-O$_2$ EXCHANGE

FIGURE 26–17. As red blood cells pass through capillaries, oxygen diffuses from the red blood cells to the tissue. Carbon dioxide diffuses from tissues to blood, which transports it to the lungs.

the watery portion of the blood. Another portion of the carbon dioxide combines with hemoglobin to form a complex molecule called **carboxyhemoglobin.** The remaining carbon dioxide combines with the liquid in the red blood cells. The carbon dioxide is exchanged for oxygen in the lungs by the blood.

Red blood cells cannot reproduce. Why? Instead, they are made constantly by **bone marrow,** the tissue in the hollow part of a bone. A young red blood cell has a nucleus that disappears as it matures. Red blood cells live for about three months. Dead cells constantly are filtered from blood and destroyed by the liver and spleen (Section 19:9). The spleen is also a blood reservoir.

Production of new red blood cells sometimes does not equal the number being destroyed. The result is a type of **anemia** (uh NEE mee uh), a disease that usually results in dizziness, weakness, pale color, and lack of energy. Anemia may also be caused by lack of iron or hemoglobin in blood or by low blood volume. Foods rich in iron or injections of iron-containing drugs may help correct certain forms of anemia.

Carbon dioxide diffuses from the body cells to the blood.

Red blood cells are formed in the marrow of certain bones.

26:10 Protection Against Injury and Infection

Most people assume that a cut or bruise will heal quickly. But a cut blood vessel would be fatal if there were no way to stop the blood flowing from it. Repairs of cuts and bruises occur by the clotting of blood. Clotting of blood involves **platelets** (PLAYT luts), certain cell fragments lacking nuclei (Figure 26–18). Platelets arise from the breaking apart of very large cells produced by bone marrow. Platelets are colorless, disc-shaped, and live for only about ten days. A cubic millimeter of blood contains about 250 000 platelets.

Clotting is an essential repair function of blood.

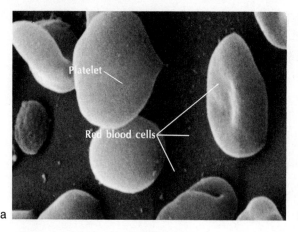

a   b

FIGURE 26–18. (a) Platelets are cell fragments that function in blood clotting. These platelets and red blood cells are shown magnified about 5600 times. (b) A red blood cell, shown magnified 8000 times, is caught in fibers that form the beginning of a blood clot.

When an injury occurs, blood platelets and the injured tissue release chemicals that trigger a set of reactions in the blood. Calcium ions, vitamin K, and other chemicals are needed for these reactions to occur. This series of reactions results in the production of a protein called **fibrin** (FI brun). A summary of these reactions follows.

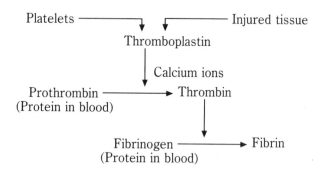

Platelets and injured tissues release chemicals that trigger reactions leading to the formation of a blood clot.

What genetic disease results in blood not clotting properly?

Fibrin is insoluble, so it settles out of the blood as long strands that form a tangled network of fibers. This network of fibers traps other blood parts and forms a **blood clot** that covers the wound and prevents excessive loss of blood.

White cells, or leukocytes, protect against infectious disease.

Protection against infectious diseases involves white blood cells or **leukocytes** (LEW kuh sites). Leukocytes are also produced by bone marrow. Although there are several types of leukocytes, they are colorless, larger than other blood cells, and they all have distinct nuclei.

Leukocytes play essential roles in the body's immune system (Chapter 19). Leukocytes are involved in recognizing antigens, producing antibodies, and destroying pathogens. Normally, a cubic millimeter of blood contains about 8000 leukocytes. However, during

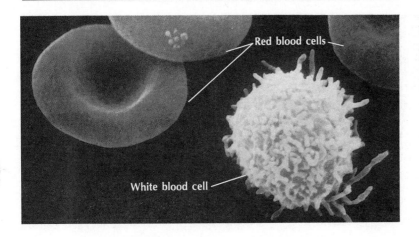

FIGURE 26–19. White blood cells function in protection against disease. They play essential roles in the immune system of the body. This leukocyte is shown magnified about 5600 times.

infection, the number may be as great as 25 000. Why? A collection of dead bacteria and white cells in an infected area is called pus. During infection, certain areas may be full of these materials.

26:11 Major Blood Types

Blood type is genetically determined. The four major blood types determined by three possible alleles (Section 7:10) are A, B, AB, and O. Differences in blood type involve antigens (Section 19:10) on the surface of red blood cells. Type A blood has A antigens and type B has B antigens. Type AB blood has both A and B antigens and type O blood has neither antigen (Table 26–4).

Antigens (proteins on the surface of red cells) are responsible for the different blood types.

		Antibodies		
Type	Antigens	(Agglutinins)	Can Receive	Can Donate to
A	A	anti-B	O, A	A, AB
B	B	anti-A	O, B	B, AB
AB	A, B	none	all	AB
O	none	anti-A anti-B	O	all

TABLE 26-4. BLOOD TYPES AND TRANSFUSION POSSIBILITIES

Blood plasma of type A blood contains an antibody (Section 19:10) called **anti-B agglutinin** (uh GLEWT uh nun). This antibody attacks type B antigens; therefore, the anti-B agglutinins link with the type B antigens. Thus, if type B blood is injected into a person with type A blood, red blood cells will clump together, or agglutinate (Section 19:10). The reverse is also true. That is, type B blood has **anti-A agglutinin** so type A blood in a type B bloodstream will clump, too. Type AB blood has neither anti-A nor anti-B agglutinins; type O has both.

Agglutination results from the action of certain antigens and agglutinins (antibodies).

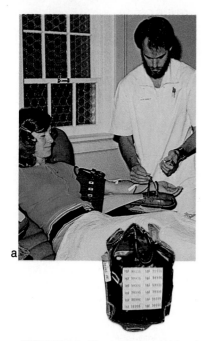

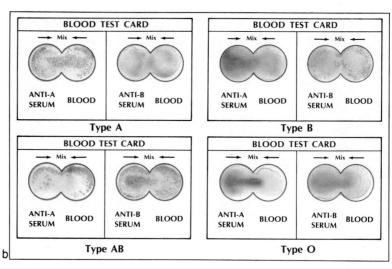

FIGURE 26–20. (a) Donated blood can be stored in bags and refrigerated for later use. (b) Before it is used, blood is carefully typed. Blood test cards for the ABO blood types are shown. Clumping shows as dotted areas, whereas blood that does not clump remains smooth looking when mixed with antisera (blood sera containing antibodies) on the cards.

Blood must be cross-matched before a transfusion is performed.

The Rh factor is another protein that may be present on red cells.

People with type AB blood could receive any blood type because no clumping will occur. For this reason, a person with type AB blood is called a **universal recipient.** A person with type O blood is called a **universal donor** because the red blood cells have no antigens. Type O blood cells will not be attacked by agglutinins.

Transfusions of incorrect blood types result in agglutination (Figure 26–20), which can be fatal. When performing transfusions, *foreign antigens* must not be introduced. Introduction of foreign agglutinins (antibodies), however, is not so serious. They are rapidly diluted by the watery part of the recipient's blood so there is little chance of agglutination. This dilution explains why type O blood is a universal donor. The anti-A and anti-B agglutinins will not cause clumping because there are so few of them. However, before any transfusion is made, the blood of the donor and the recipient should be cross-matched. Matching will indicate the presence of other antigens that might cause agglutination.

26:12 Rh Factor

Another protein called the **Rh factor** may also be present on red blood cells. The term Rh comes from "rhesus" (REE sus) because these antigens were first discovered in the blood of rhesus monkeys. About 85 percent of the people in the United States have the Rh factor and are Rh positive. The remaining 15 percent do not have the Rh factor and are negative.

Suppose Rh positive blood is given to a person with Rh negative blood. The Rh negative person does not already have antibodies against Rh positive blood, but soon begins to make them. The result will be agglutination. A person with Rh positive blood can safely receive Rh negative blood. Why?

Rh factors can harm a developing embryo if the child is Rh positive and the mother is Rh negative. A few red blood cells may pass from the child through the placenta into the mother's circulatory system near the end of pregnancy. In response to these foreign blood cells, the mother produces antibodies that may pass back across the placenta into the embryo's bloodstream. The baby's blood would be foreign protein to the antibodies. Anemia in the child or even death could result. Rh incompatibility is rarely dangerous during a mother's first pregnancy. Usually, the quantity of antibodies formed in her blood is small and they do not form until late in the pregnancy. However, the antibodies are retained in the mother's blood, so their number may increase with each pregnancy. A chemical is available to help solve this problem. It is injected into an Rh negative mother shortly after she delivers an Rh positive child. The chemical prevents the mother's immune system from recognizing that the child's red blood cells (those passed to the mother) are Rh positive, so no antibodies are built up. The foreign red blood cells that have entered her blood and the mother's own red blood cells are eventually destroyed by the liver. The injection procedure must be repeated each time the mother delivers an Rh positive child. If the chemical is not taken, later pregnancies involving Rh positive babies will cause problems.

The danger of problems from Rh incompatibility increases with each pregnancy.

FIGURE 26–21. If an Rh⁻ mother becomes pregnant with an Rh⁺ child, often some Rh⁺ red cells enter the mother during late pregnancy. A chemical can be injected into the mother that prevents her body from making antibodies that would cause incompatibility with future Rh⁺ children.

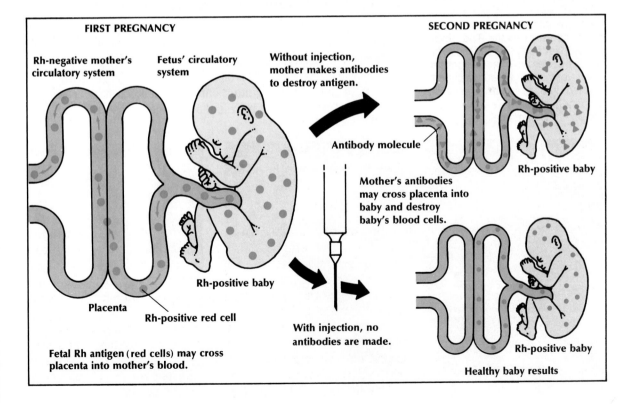

FIRST PREGNANCY

Rh-negative mother's circulatory system

Fetus' circulatory system

Without injection, mother makes antibodies to destroy antigen.

Antibody molecule

Mother's antibodies may cross placenta into baby and destroy baby's blood cells.

Rh-positive baby

Placenta

Rh-positive red cell

With injection, no antibodies are made.

Fetal Rh antigen (red cells) may cross placenta into mother's blood.

SECOND PREGNANCY

Rh-positive baby

Rh-positive baby

Healthy baby results

HEART DISEASE

Heart disease is the leading cause of death in the United States. Many coronary (heart) problems arise as heart muscle becomes damaged. Heart muscle can be weakened or even killed by a variety of factors. Prolonged hypertension (Section 26:7) can damage the lining of the coronary arteries and can also cause heart muscle to become thickened and overworked. This strain on the heart may lead to abnormal pumping action that interferes with normal blood flow and leads to death by heart failure.

In some cases, a blood clot may form or become lodged in the coronary blood vessels. The clot may block a vessel, stopping the flow of blood (and oxygen) and causing a portion of the muscle to die. This series of events is a heart attack. Heart attacks are sometimes quickly fatal because the heartbeat control system breaks down, and the rhythm of the heartbeat cannot be controlled. In persons suffering from atherosclerosis, clots are more easily trapped because the arteries are clogged with fatty deposits. Most people who have heart attacks also have atherosclerosis.

Most heart disease results from damage to heart muscle.

A clot lodged in a coronary blood vessel may lead to death of a part of the heart muscle—a heart attack.

26:13 Treatment and Prevention

Treatment of heart disease varies. Treatment may involve reduced physical activity and the use of drugs, or surgery may be performed. In one type of surgery, heart valves may be repaired or artificial valves installed. In coronary bypass surgery, veins from other parts of the body are grafted onto diseased coronary arteries in such a way that blood can bypass the damaged vessels.

FIGURE 26–22. (a) Body scanners sometimes are used to diagnose heart disease. The three scans inset on this photo show the heart muscles at work. (b) Artificial heart valves can be surgically implanted into the heart, replacing damaged valves. This valve is being used experimentally in pigs.

a
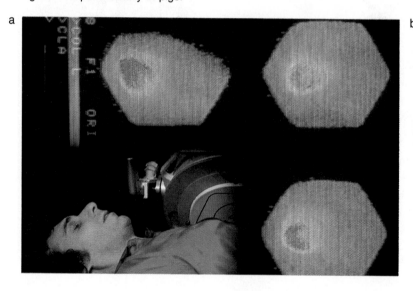

b
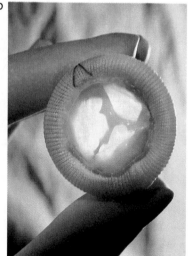

FIGURE 26–23. Regular exercise is important in preventing heart problems. Exercise programs should begin slowly and build up gradually.

Certain forms of heart disease are preventable. Eating the proper diet is one way heart disease may be prevented. Avoiding large amounts of fats, especially animal fats and those in dairy products, such as butter, cream, and whole milk, reduces the risk of heart disease. These substances add cholesterol to the blood, which some physicians think may contribute to atherosclerosis. Obesity can also lead to heart disease and put a strain on the heart. Such strain results from the heart's having to work harder in an overweight person to supply blood to all the tissues. Obesity may also lead to accumulation of fatty deposits in blood vessels. Also, fat builds up around the heart.

Other personal habits besides diet affect the chances of getting heart disease. Cigarette smokers run a greater risk than nonsmokers of getting heart disease. The nicotine present in cigarette smoke constricts blood vessels and causes the heart to beat more quickly than normal. Smoking also seems to increase the chances of developing atherosclerosis.

Exercise is also important in maintaining a healthy heart. Just as exercise may strengthen other muscles, it can also improve the efficiency of the heart. In a person who exercises regularly, the heartbeat rate is lower and more blood is pumped to the body with each beat. Thus, the heart does not work as hard in getting an adequate supply of food and oxygen to the body tissues. Swimming and bicycling are excellent forms of exercise for maintaining a healthy heart and circulatory system.

A proper diet may help to prevent heart disease.

Regular exercise strengthens the heart.

FIGURE 26–24. During heart transplants or certain other kinds of heart surgery, the patient's blood is oxygenated by a heart-lung machine, shown here in the foreground.

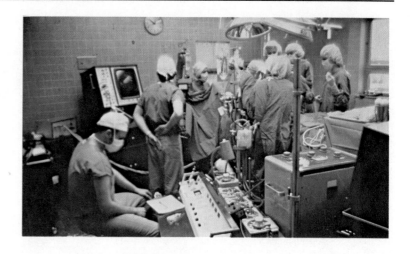

26:14 Heart Transplants

In rare cases when heart disease is severe, a heart transplant may be done. During a heart transplant, the patient's blood bypasses the heart and lungs and circulates through a heart-lung machine. This machine pumps the blood, oxygenates it, and removes carbon dioxide from it (Figure 26–24). Meanwhile, the damaged heart is removed and the new heart is attached to the major blood vessels. Once the operation is completed, blood is no longer circulated through the machine but is channeled through the lungs and the new heart of the patient.

Although this type of surgery is possible, there are problems involved. An organ contains the proteins of the body in which it developed. To the person receiving a donor organ, these proteins are antigens. The recipient's immune system begins to destroy the transplanted heart.

Use of the heart-lung machine enables surgeons to operate on the heart.

Implantation of a foreign organ results in the recipient's immune system rejecting the organ.

UNIFORM DONOR CARD

OF _____
Print or type name of donor

In the hope that I may help others, I hereby make this anatomical gift, if medically acceptable, to take effect upon my death. The words and marks below indicate my desires.

I give: (a) _____ any needed organs or parts
(b) _____ only the following organs or parts

_____
Specify the organ(s) or part(s)

for the purposes of transplantation, therapy, medical research or education;

(c) _____ my body for anatomical study if needed.

Limitations or
special wishes, if any : _____

FIGURE 26–25. Some people carry Uniform Donor Cards, which indicate their wish to donate body organs in the event of death. In some states, these cards are attached to drivers' licenses.

Without special treatment, the immune system soon rejects a foreign tissue or organ. Drugs have been developed that combat the natural tendency of the body to protect itself. One promising new drug inhibits only the rejection of the transplanted organ. It does not cause the recipient to become defenseless against diseases or infections. Often a transplant is surgically successful, but the patient dies of other causes that result from weakened body defenses. By using monoclonal antibodies (Chapter 19), physicians can learn how well the drugs are combating the immune system. The dosage of the drug can then be adjusted as needed.

Medication has been developed to help eliminate the rejection of transplanted organs.

Meanwhile, physicians compare the tissues of donors and recipient by matching tissues much like blood types are matched for a transfusion. Whenever possible they transplant organs from donors who have proteins similar to those in the recipient. The closer the tissues match, the less the chance that the organ will be rejected.

REVIEWING YOUR IDEAS

10. What are the functions of blood?
11. What are the two types of blood cells? What are platelets? List the functions of all three.
12. What is hemoglobin?
13. How are anemia and red blood cells related?
14. Why is clotting of blood important?
15. What are the four major blood types? How do they differ?
16. What is the Rh factor? What can be done to avoid problems in later pregnancies if an Rh$^-$ mother has an Rh$^+$ child?
17. What is a heart attack? In what ways can certain heart problems be prevented?
18. Why are heart transplants risky?

ADVANCES IN BIOLOGY

Artificial Hearts

Heart transplants are expensive, the number of donor hearts available is very small, and rejection of the new heart is a major problem. For many years, physicians and scientists have been working to avoid some of these problems by developing an artificial heart. One type of artificial heart was implanted in a human in 1969. It worked successfully for 65 hours, but was removed when a donor heart transplantation became available. Another kind of artificial device does not replace the heart, but takes over for the heart until it recovers. The device is then removed. This device is called a **left ventricular**

Use of an artificial heart may eliminate the problems associated with heart transplants.

Some artificial hearts consist of chambers that replace damaged ventricles.

Air from a compressor forces blood from the artificial heart to the arteries.

FIGURE 26–26. (a) This artificial heart replaces (b) the ventricles (shaded area) when implanted. Air from a compressor outside the body moves diaphragms in the heart and pushes the blood. A rhythmical pulse of air keeps the heart beating.

assist device (LVAD). LVADs have been in use since the 1970s. In 1981, a new kind of artificial heart was implanted in a human. It kept the patient alive for 54 hours before being replaced by a donor heart.

In severe heart disease, the ventricles become weakened. Today's artificial hearts consist of two chambers that replace the ventricles. The artificial heart is grafted to the patient's atria and to the pulmonary artery and aorta. In one model, a tube is attached to the base of each ventricle. The tubes pass out of the chest wall and are connected to an external air compressor. Pumping of air forces blood up and out of the ventricles and into the arteries. When compression stops a moment later, blood enters the ventricles from the atria. This "on and off" pumping of air keeps blood moving through the heart and circulating to the lungs and body.

This type of artificial heart was first implanted in a human in 1982. The patient lived for more than three months. Since that time, several other patients have received the heart. None of them has fared well, and it seems clear that many problems are associated with the heart. A major problem is the occurrence of small strokes. The strokes are thought to be caused by blood clots that may form in the artificial heart and then travel to the brain. In the brain, the clots block small blood vessels, leading to death of brain cells. This results in a stroke.

Other problems exist. Because the heart is driven by an outside source (the compressor), a person is not able to move around freely. Another problem is the possibility of infection in the areas where the air tubes enter the body. A nonmedical problem is the immense cost of patient care.

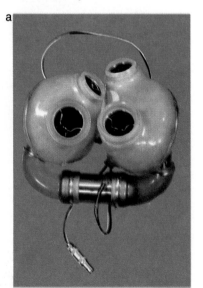

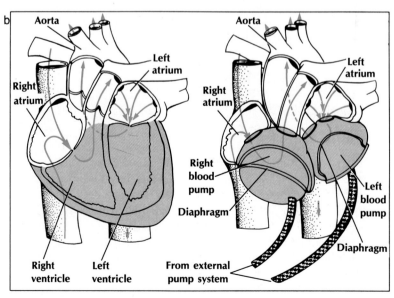

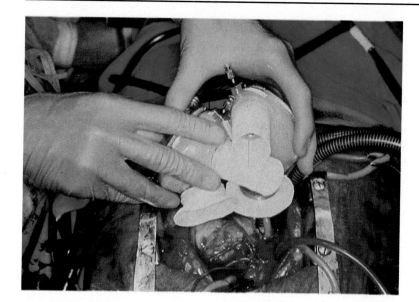

FIGURE 26–27. An artificial heart was first implanted in a human at the University of Utah.

At present the artificial heart is seen as a temporary measure at best, a device to be used until a donor heart becomes available for transplantation. But the number of donor hearts available for those who need them is small. Perhaps, with further research, a more effective type of artificial heart will be developed and available for permanent use.

CHAPTER REVIEW

SUMMARY

1. Complex animals have specialized circulatory systems that link their cells with the environment. **p. 535**
2. An earthworm has a closed circulatory system in which five pairs of aortic arches pump blood around the body. **26:1**
3. Arthropods have an open circulatory system, one in which blood does not always circulate in vessels. **26:2**
4. Fish have a two-chambered heart consisting of one atrium and one ventricle. Blood is oxygenated after it leaves the heart. **26:3**

5. Amphibian hearts have two atria and one ventricle. Although there is some mixing, oxygenated and deoxygenated blood are kept fairly separate. **26:4**
6. Birds and mammals have four-chambered hearts. Blood going to body cells is fully oxygenated and under high pressure. **26:5**
7. Each heartbeat originates within the S-A node. However, the speed at which the heart beats is influenced by the accelerator and vagus nerves. **26:6**
8. Blood pressure is the ratio of systolic pressure to diastolic pressure. Hypertension, high blood pressure, is a serious disease that can be treated with drugs. **26:7**

9. Tissue fluid directly bathes the tissues. Many substances are exchanged between the cells and the tissue fluid. Lymph reenters the circulatory system at the thoracic duct. **26:8**

10. Red blood cells are important in transport of gases. Hemoglobin combines with oxygen and transports it to body cells. Carbon dioxide also may combine with hemoglobin or ionize in the watery part of blood. **26:9**

11. Blood platelets are important in blood clotting. After an injury, they release chemicals that set off a chain of reactions resulting in the formation of a protective clot. White cells play important roles in defense against infectious diseases. **26:10**

12. Blood types differ in the types of antigens present on the surface of the red blood cells. Transfusion of incorrect blood types results in agglutination. **26:11**

13. The Rh factor can cause problems during pregnancy if the mother is Rh negative and the baby is Rh positive. **26:12**

14. Heart disease may be brought on by prolonged hypertension, blood clots, and atherosclerosis. **p. 554**

15. Proper diet and adequate exercise are factors that may prevent certain types of heart disease. **26:13**

16. Transplanting organs can have effects similar to those of incorrect blood transfusions. The recipient's immune system rejects the foreign organ. **26:14**

17. Artificial hearts have been developed. These devices may be used until a heart for transplantation becomes available or they may be permanently implanted. **p. 557**

LANGUAGE OF BIOLOGY

agglutinin	lymph vessel
aorta	open circulatory
artery	system

atherosclerosis	pacemaker
atrium	platelet
blood	red blood cell
blood clot	Rh factor
blood pressure	sinus
circulatory system	sinus venosus
closed circulatory	systole
system	thoracic duct
diastole	tissue fluid
heart	universal donor
hemocoel	universal recipient
hemoglobin	vein
leukocyte	ventricle

CHECKING YOUR IDEAS

On a separate paper, match each phrase from the left column with the proper term from the right column. Do not write in this book.

1. receives blood from left ventricle
2. liquid that exchanges materials with cells
3. heart of fish
4. combines with oxygen in blood
5. heart of bird or mammal
6. red blood cell
7. carries blood toward heart
8. source of heartbeat
9. white blood cell
10. heart of amphibian

   a. aorta
   b. erythrocyte
   c. four-chambered
   d. hemoglobin
   e. leukocyte
   f. pacemaker
   g. three-chambered
   h. tissue fluid
   i. two-chambered
   j. vein

EVALUATING YOUR IDEAS

1. Why must complex animals have a transport system?
2. How is circulation in an earthworm different from circulation in a grasshopper?

3. Compare the circulatory systems of fish and frogs.
4. Why is a fish's blood circulation slow?
5. What accounts for the high pressure of the blood going through a frog's body?
6. How is a reptile's heart different from a frog's heart? How is this difference adaptive?
7. Trace a drop of blood through the human circulatory system, beginning and ending at the right atrium.
8. How are the valves of a human heart important?
9. Name some adaptive advantages of the mammalian circulatory system. How do you think these adaptations benefit mammals?
10. What is the source of the heartbeat? Why must its rate vary? How is the rate controlled?
11. What is hypertension? What are some possible effects of hypertension? How can these effects be avoided?
12. How are tissue fluids, lymph, and the lymphatic system important?
13. How are oxygen and carbon dioxide transported by red blood cells?
14. How does blood clot?
15. Explain what can happen if an embryo with Rh positive blood is developing within a mother with Rh negative blood.
16. Explain what causes some heart attacks. What can be done to prevent them?
17. How are the dangers of organ transplants lessened?
18. Explain how an artificial heart works. How might it be used? What problems exist?

APPLYING YOUR IDEAS

1. Diameter of capillaries is controlled by nerves. How might the diameter be important in regulation of body temperature?

2. Because all capillaries in the body are not open at one time, it is not wise to go swimming after eating. Why?
3. Why are high blood pressure and lowered carbon dioxide content of blood useful signals to reduce heartbeat rate?
4. An artificial blood that carries oxygen has been developed. Could this blood replace normal blood? Explain.
5. What would happen if a person with type B blood received a transfusion of type A?
6. What would you predict about the resting heart rate of an athlete? Why?
7. Oxyhemoglobin gives off more oxygen at higher temperatures than at lower ones. How is that property important to muscle tissue?

EXTENDING YOUR IDEAS

1. Obtain a beef heart from a butcher shop. Locate the major veins and arteries leading to and from the heart. Notice the coronary arteries and veins on the surface of the heart. Make a lengthwise incision through the heart. Describe its internal structures.
2. Plan a debate on the pros and cons of heart transplants and use of artificial hearts.

SUGGESTED READINGS

Grady, Denise, "Can Heart Disease Be Reversed?" *Discover,* March, 1987.

Huyghe, Patrick, "Your Heart: A Survival Guide." *Science Digest,* April, 1985.

Langone, John, " 'The Artificial Heart Is Very Dangerous.' " *Discover,* June, 1986.

Lewis, Ricki, "Transplants — Past, Present, and Future." *Biology Digest,* Feb., 1985.

RESPIRATION AND EXCRETION

Animals exchange oxygen and carbon dioxide with the environment during respiration. Animals also rid their bodies of harmful by-products of metabolism by excretion. Complex animals have specialized systems for carrying out these processes. This whale is exhaling carbon dioxide through its blowhole. The blowhole is connected to its lungs and is part of the whale's respiratory system. What other adaptations do animals have for gas exchange and excretion?

chapter 27

G as exchange in animals involves taking in oxygen and giving off carbon dioxide. This exchange is called **respiration.** Cells need oxygen for cellular respiration and produce carbon dioxide as a product. How do cells obtain oxygen and expel carbon dioxide? In sponges, coelenterates, and flatworms, each cell exchanges gases directly with its watery environment by diffusion.

In more complex animals, though, cells cannot exchange gases directly with the outside environment, because most are too far away from it. Even if exchange of gases could occur across the membranes of outer cells, the gases could not diffuse to or from cells deep in the body quickly enough to sustain life. These animals have evolved complex systems for respiration. Gases are exchanged between certain cells of the respiratory system and the external environment. Oxygen is transported from the respiratory exchange surface to all other cells, where it then diffuses inward. Carbon dioxide diffuses from body cells, is carried to the exchange structure, and then is expelled.

Objectives:

You will

- compare respiration in aquatic and terrestrial animals.
- identify and describe several respiratory diseases and their causes.
- describe excretion in a variety of animals.

In complex animals, certain parts of the respiratory system exchange gases with the outside environment. Gases are transported between body cells and the respiratory exchange surface.

RESPIRATION IN WATER

To cross cell membranes, gases must be dissolved in water. Thus, exchange surfaces for respiration must be kept moist. This requirement is met more easily by aquatic animals than by those living on land. *The type of respiratory system an animal has is related to the environment in which the animal lives.* In addition to being kept moist, the exchange surface area must be large. A large surface area is needed so that all of the many body cells can obtain enough oxygen and expel enough carbon dioxide.

The gas exchange surface must be moist and have a large area.

FIGURE 27-1. Gas exchange in *Nereis* occurs by diffusion in the parapodia. The parapodia are the bristled paddles on the sides of the worm.

In earthworms, gases dissolve in the moisture on the skin. Oxygen diffuses inward to the blood. Carbon dioxide diffuses outward from the blood, through the skin, and into the soil.

27:1 Annelids

Most annelids live in water. *Nereis* is an example of a marine annelid. *Nereis* has rows of parapodia (Section 15:7), paired, bristled paddles that extend from the worm's segments, along the length of the body. These parapodia are rich in blood and are quite thin. Oxygen in the water diffuses across the surface of the parapodia and into the blood. The blood then transports the oxygen to body cells. Carbon dioxide diffuses from the blood in the parapodia to the water.

Earthworms are annelids that live on land. But an earthworm is "aquatic" in that it can survive only in *moist* soil. Moist soil is needed for gas exchange because an earthworm does not have a respiratory system. A thin film of moisture covering the earthworm is enough to dissolve needed gases. These gases then diffuse across the skin and into the earthworm's blood. In the blood, oxygen is transported throughout the body. Carbon dioxide diffuses into the blood from the body cells and is given off in the opposite direction. The large surface area (skin) for gas exchange provides enough oxygen for the many cells of the earthworm.

An earthworm cannot live very long outside its dark, moist environment. However, when heavy rains fall, worms must come up to the surface or drown in the excess water in their burrows. You probably have seen dead earthworms on the ground after such a rainfall. These worms, unable to get back into the soil quickly enough after the rain ended, died from lack of oxygen and exposure to ultraviolet radiation from the sun.

27:2 Animals with Gills

Gills are the organs of respiration in several groups of aquatic animals.

In mollusks, some aquatic arthropods, fish, and some other aquatic animals, gills are the respiratory organs. Gills are located close to the outside or on the outside of the body. This location is adaptive; the gills are kept moist by direct contact with the surrounding water.

FIGURE 27-2. Most mollusks have respiratory organs called gills. (a) An exception is the nudibranchs, most species of which lack gills. The extra folds of tissue on the body serve as added surface area over which diffusion occurs. (b) The gills of the scallop can be seen as rows of tissue just inside the opened shell.

a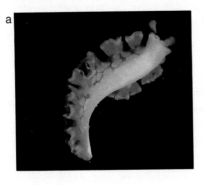

b

a
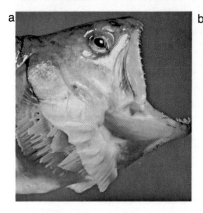

b
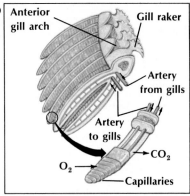

FIGURE 27-3. (a) In bony fish, the gills are covered by opercula. The operculum is raised on this fish showing gills beneath. (b) As water passes over the gills, oxygen enters blood in the capillaries and carbon dioxide diffuses into the water.

In fish, oxygen-rich water enters through the mouth and passes back to the pharynx. From the pharynx, the water goes to the **gill chamber,** an area containing the gills. As water is taken into the mouth, the operculum (Section 16:11) of the gill chamber closes. Contraction of the pharynx forces water over the gills, the operculum opens, and water is expelled.

Each gill is made up of a gill arch, a structure made of cartilage that supports the respiratory tissues. These tissues, called gill filaments, protrude from each arch and increase the surface area for gas exchange. **Gill filaments** are double rows of thin-walled tissue through which capillaries pass. The circulatory system is very closely associated with the respiratory system in these gill filaments.

Gills provide a large surface area for gas exchange.

In fish, deoxygenated blood leaves the heart and travels to the gill capillaries. As water passes over the gills, oxygen enters blood in the capillaries and carbon dioxide leaves the gills and enters the water. Oxygenated blood is then transported from the gills to all parts of the body. Gas exchange occurs between the blood and the body cells.

Gases are exchanged as water passes over gill capillaries.

REVIEWING YOUR IDEAS

1. Describe respiration in sponges, coelenterates, and flatworms.
2. Why must complex animals have a respiratory system?
3. What determines the type of respiratory system an animal has?
4. List two characteristics of the gas exchange surface in animals.
5. Compare respiration in *Nereis* and an earthworm.
6. What are gills? In what kinds of animals are they found?
7. How do gills work?

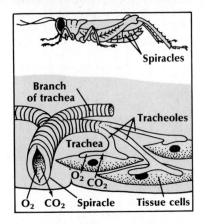

FIGURE 27-4. A grasshopper has a specialized respiratory system composed of spiracles and tracheae. The tracheae branch and rebranch until they end in tracheoles, tiny air sacs.

In insects, gases are dissolved in the moisture of the tracheoles. Diffusion of gases occurs between the tracheoles and body cells.

FIGURE 27-5. Although adult *Xenopus laevis* frogs breathe with lungs, much gas exchange occurs through the thin membranes of their hind feet. As a result, these frogs can stay underwater for long periods of time.

RESPIRATION ON LAND

Most land animals exposed to the dry air cannot undergo gas exchange on their body surfaces; there is not enough moisture to dissolve the needed gases. Instead, *dry air is brought into moistened areas within the body.*

27:3 Tracheal System

In some land arthropods (insects and their relatives), air enters the animals through small openings called **spiracles** (SPIHR ih kulz). Spiracles are located along the sides of the body and are the external openings of tubes called **tracheae** (Section 16:1). Tracheae branch and rebranch within an animal, the smallest branches ending in tiny air sacs called **tracheoles** (TRAY kee ohlz), which contain moisture (Figure 27–4). Spiracles, tracheae, and tracheoles make up the **tracheal** (TRAY kee ul) **system.**

Air is circulated through a tracheal system either by simple diffusion, or, as in large insects, by muscular contractions. Oxygen in the air dissolves inside the moist tracheoles and passes directly to the cells by diffusion. Carbon dioxide from the cells diffuses to the tracheoles and is expelled through the spiracles. The system works because no cell is very far from a tracheole. The large number of tracheoles provides a large surface area for gas exchange. In an animal with a tracheal system, the respiratory system is not related to the circulatory system. Blood does not transport gases; its main function is to transport nutrients and wastes.

27:4 Lung System

Most amphibians and all reptiles, birds, and mammals have another system for external respiration. Like insects, these animals have moist internal areas for gas exchange, but they are located in two organs, the **lungs.** Also, the respiratory system is linked to the circulatory system. Gases are transported between cells and the external environment by blood.

In many amphibians, gills are present in the larval stages but are replaced by lungs as the animals mature. Most amphibians lead a partly aquatic life; they live where moisture is plentiful. Frogs and some salamanders have lungs, but much oxygen also diffuses through the moist skin and enters the blood in the many capillaries on the inner surface of the skin. This oxygen adds to the oxygen taken in by the small lungs. The thin skin of the feet has many blood vessels. Like the

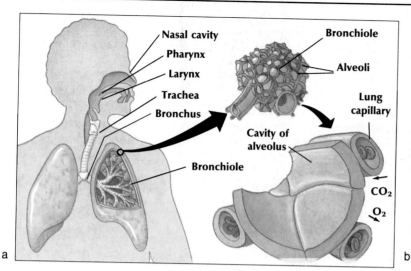

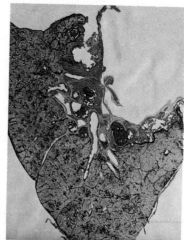

FIGURE 27-6. (a) Oxygen moves into the blood in capillaries that surround the alveoli. (b) A section of a human lung shows the spongy nature of the tissue.

outer skin, the lining of the frog's mouth is moist and rich in blood vessels. These adaptations provide another means of gas exchange. In fact, more oxygen is obtained through the mouth lining than the lungs. Why do you think the lungs are small?

In birds and mammals, exchange cannot occur through the skin. In humans, oxygen (as part of the air) enters the nose or mouth, passes to the pharynx, and then moves downward into the windpipe, or trachea. The opening of the trachea is protected by a structure called the **epiglottis** (ep uh GLAHT us). During breathing, the trachea is open. When a person swallows food, the epiglottis closes over the top of the trachea so food goes down only the esophagus. Have you ever choked from talking while eating? Choking occurs when the epiglottis fails to close properly and food starts into the trachea.

The epiglottis prevents food from entering the trachea.

Branching from the trachea are two large tubes called **bronchi** (BRAHN ki). The bronchi branch into many smaller **bronchial tubes** that further branch into **bronchioles** (BRAHN kee ohlz) within the lungs. Bronchioles end as many small, moist sacs called **alveoli** (al-VEE uh li) (Figure 27–6). Gas exchange occurs in the alveoli. Like the filaments of gills, groups of alveoli greatly increase the surface area of the lungs for gas exchange.

Gas exchange occurs in the alveoli of the lungs.

Recall that hemoglobin is the pigment in red blood cells that combines with oxygen (Section 26:9). Carbon dioxide is transported in both red cells and in the liquid portion of blood. Blood reaches the alveoli via capillaries that branch from the pulmonary arteries. This blood is rich in carbon dioxide and poor in oxygen. In the alveoli, the level of carbon dioxide is low and the level of oxygen is high. Under these conditions, the carbon dioxide diffuses from the blood into the alveoli and is exhaled. At the same time, oxygen diffuses from the alveoli into the blood and there combines with hemoglobin to form

In the lungs, carbon dioxide passes from blood to the alveoli from which it is exhaled.

oxyhemoglobin. The oxygenated blood is returned by the pulmonary vein to the left side of the heart from which it is pumped throughout the body.

At the body tissues, the relative levels of carbon dioxide and oxygen are reversed from the levels in the lungs. The tissues are rich in carbon dioxide and poor in oxygen. Under these conditions, oxyhemoglobin readily breaks down into hemoglobin and oxygen, and the oxygen diffuses from the blood to the tissues. Carbon dioxide diffuses from the tissues to the blood and is carried back toward the heart and lungs.

In the tissues, oxygen passes from blood to cells and carbon dioxide diffuses from cells to blood.

27:5 Control of Breathing

Movement of air into and out of the human respiratory system is called **breathing.** Breathing involves the action of the ribs, rib muscles, and a large muscular sheet, the **diaphragm** (DI uh fram), which separates the chest from the abdomen. During **inspiration** (taking in of air), the diaphragm moves down and the ribs move up and out. This increases the size of the chest cavity and decreases the inside pressure. As a result, air from the atmosphere rushes into the lungs. About 500 mL of air are brought into the lungs in a normal inspiration.

Breathing is controlled by the action of the diaphragm and rib muscles.

When the lungs are full, an opposite set of movements occurs. During **expiration** (expelling of air), the diaphragm moves up and the ribs move in and down. This decreases the size of the chest cavity and increases the inside pressure. As a result, air is forced out of the lungs through the trachea.

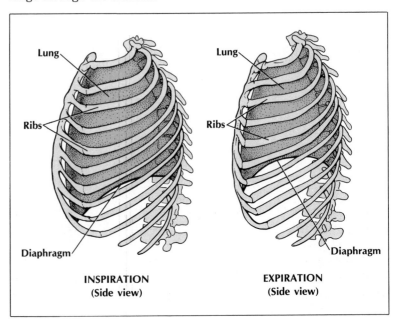

FIGURE 27-7. When the diaphragm moves down, air is taken into the lungs. When the diaphragm moves up, air is expelled. Rib muscles also play an important role in this process.

At rest, an adult breathes about fifteen times per minute. Control of breathing involves nerves on the surface of the lungs. Filling the lungs with air stimulates the nerves, sending impulses to the **breathing center** of the medulla oblongata (Section 29:10). The medulla stops sending messages to the diaphragm and rib muscles. As the muscles relax, air is forced from the lungs, and the nerves on the lungs are no longer stimulated. Thus, they do not send impulses to the medulla. The medulla once again sends messages to the rib muscles and diaphragm, which contract, causing air to enter the lungs, and the cycle to repeat again.

Continuous alternation of inspiration and expiration is controlled by nerves located on the surface of the lungs.

The breathing rate can change according to the needs of the body. Cells in the breathing center analyze the carbon dioxide content of the blood. If the carbon dioxide content is high (a sign of increased cellular respiration), nerves carry impulses from the breathing center to the muscles of the ribs and the diaphragm more quickly. The muscles are stimulated to contract and relax quickly so that the breathing rate increases. More air and oxygen are brought into the lungs so more oxygen reaches the cells. At the same time, more carbon dioxide is exhaled. This process is important because excess carbon dioxide is dangerous.

As the carbon dioxide content of the blood increases, the breathing rate increases.

When the carbon dioxide level in the blood is low, the breathing center is inhibited. The nerves transmit fewer impulses to the muscles of the ribs and diaphragm. As a result, contractions and relaxations slow down and the breathing rate decreases. Changes in the breathing rate maintain homeostasis by adjusting the rate of gas exchange to the needs of the cells. When would an increase in breathing rate occur? Why?

A low level of carbon dioxide in the blood inhibits the breathing center.

27:6 Respiratory Disease

The respiratory system is a vital part of the body that, when working properly, is taken for granted. However, if you ever have had an illness that interferes with your breathing, you know how frightening and uncomfortable having a respiratory disease may sometimes be.

Because the respiratory system is in contact with the environment, its parts may be invaded easily by microbes. Although many microbes are destroyed by the body's natural defenses (Chapter 19), some of them may grow and reproduce in respiratory tissues, thus causing disease.

Respiratory organs sometimes are easily reached by microbes.

Influenza is a viral disease that can affect various parts of the respiratory system. Influenza viruses seem to reproduce in the alveoli, but they also affect the bronchi and trachea. Many influenza deaths result from complications due to secondary bacterial infections

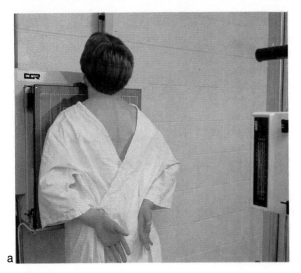

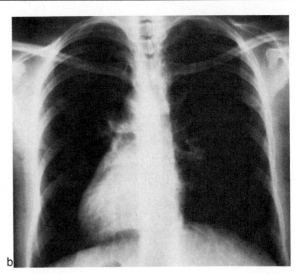

FIGURE 27-8. (a) Chest X rays can be used to diagnose certain respiratory diseases. (b) This X ray shows healthy lungs.

Many respiratory diseases are caused by bacteria.

In asthma, the diameter of the bronchial passages is reduced, making breathing difficult.

of the respiratory system. Today, there are vaccines that prevent many forms of influenza. If secondary bacterial infections occur, they usually can be treated by antibiotics. Influenza and its complications remain dangerous to young children and the elderly.

Bronchitis is another respiratory disease. Bronchitis may be caused by a virus or as a result of exposure to chemicals such as sulfur dioxide. Bronchitis is an inflammation of the bronchi and results in congestion in the bronchi and in coughing. It is usually over in several days.

A variety of respiratory diseases are caused by bacteria. Bacterial pneumonia, strep throat, and tuberculosis are examples of this type of respiratory disease. Before the discovery of antibiotics, these diseases were often fatal, because they cause damage not only to the respiratory system, but also to other parts of the body. Today, though, these diseases can usually be cured. Vaccines have also been developed for some of these diseases such as pneumonia.

Some respiratory problems result from allergies. Hay fever, caused by a reaction to certain types of pollen, affects the nose and sinuses. Hay fever results in irritation of the membranes, sneezing, and itching, watery eyes. A more serious respiratory problem is asthma. The exact causes of asthma are not known, but it is thought that it may result from an allergic reaction to some foods. Asthma attacks may also result from emotional causes. During the attack, which may come on suddenly, the muscles of the bronchial system contract, reducing the diameter of the air passages. As a result, breathing becomes more difficult. An asthma attack is usually treated by inhaling a medicine that aids in increasing the diameter of the bronchial passages, which eases breathing.

27:7 Smoking and Its Effects

The respiratory disorders so far discussed result from natural biological causes. In many cases, they are not easily avoided. However, there are certain respiratory diseases that may be avoided. These diseases are ones known to be associated with smoking.

Recall that smoking affects the heart, causing it to beat more quickly (Section 26:13), and that smoking also increases the chances of heart attacks and atherosclerosis. Data collected over many years show that smoking also increases the chances of certain respiratory diseases. Smoking may not be the only cause of these diseases (other factors such as pollution may contribute), but studies clearly show that people who smoke run a much greater risk of developing these diseases than nonsmokers.

Smokers run a higher risk of developing certain respiratory diseases than nonsmokers.

Cigarette smoke contains hundreds of different compounds. With time, these compounds build up on the linings of the brochial tubes and lungs. The cilia lining the respiratory passages stop moving and eventually are destroyed. When the cilia are not operating, mucus cannot be swept up toward the mouth. Coughing may occur, and the respiratory system is more likely to become infected because its defenses against disease are not working correctly.

Smoking may destroy cilia in the respiratory tract.

Smoking is known to contribute to a long-lasting form of bronchitis. This form of bronchitis may develop over many years and lead to blockage of the bronchioles. Shortness of breath develops, and exchange of oxygen and carbon dioxide in the lungs becomes difficult. Oxygen and carbon dioxide levels in the blood become abnormal, and body tissues become deprived of oxygen.

Emphysema is another disease linked to smoking. In emphysema, the bronchioles become inflamed. Also, the alveoli become scarred and eventually break. When the alveoli break, air can escape

In emphysema, bronchioles and alveoli are damaged.

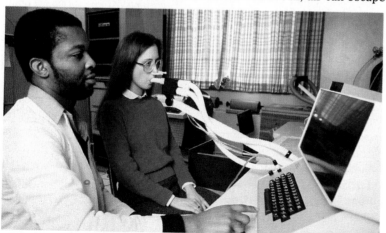

FIGURE 27-9. This patient is being given a test for emphysema. The exhaled air is tested for contents and measured for volume.

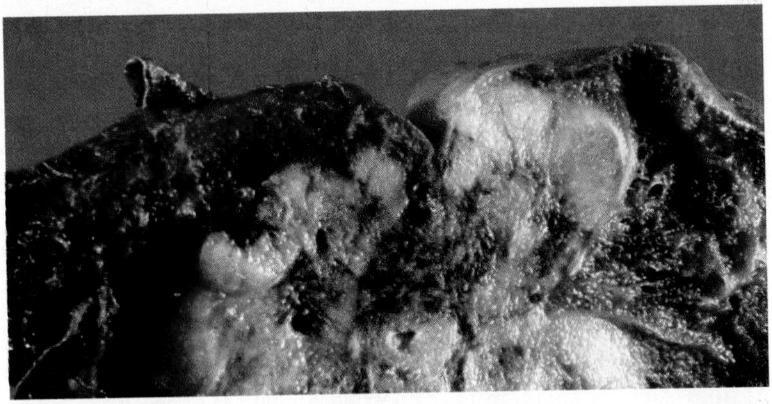

FIGURE 27-10. Lung cancer is the most serious of all smoking-related diseases. Lung cancer cells reproduce abnormally fast and destroy healthy cells.

Cigarette smoking greatly increases the chances of lung cancer.

Some damage can be repaired after a person stops smoking.

from the lungs, so less oxygen moves into the bloodstream and less carbon dioxide is removed. Breathing may become very difficult and a heavy cough may develop. The heart becomes strained as it beats faster to force more blood to the lungs. Death may result from damage to the heart and lungs.

Most serious of all smoking-related diseases is lung cancer. At least fifteen of the compounds found in cigarette smoke are known to cause cancer in laboratory animals. Studies have also shown that a person who smokes a pack of cigarettes each day is twenty times more likely to die of lung cancer than a nonsmoker. Smoking pipes or cigars does not seem as risky as smoking cigarettes, but pipe and cigar smoke are also harmful.

As in other forms of cancer, lung cancer cells reproduce abnormally fast, crowding out and destroying healthy cells. Lung cancer is especially dangerous because it is often not detected until it has spread to other body parts. Even when it is detected early, it often causes death. Fewer than twenty percent of those persons diagnosed with lung cancer survive for more than five years.

Chances of getting these respiratory diseases can be reduced by not smoking. Studies have also shown that the chances of getting these diseases are reduced if a person who has smoked quits. Some of the damage done to both the circulatory and respiratory systems by smoking can be healed over a period of time. The longer one has smoked, the longer the healing period. Think about these facts. Is smoking a reasonable habit to have?

REVIEWING YOUR IDEAS

8. What are tracheae? What kinds of animals have them?
9. What are lungs? What kinds of animals have them?
10. Distinguish between respiration and breathing. What are the two phases of breathing?
11. List some respiratory diseases caused by microbes.
12. What is asthma?

EXCRETION

Most waste products in animals are the chemicals produced from the breakdown of proteins. These substances, which are poisonous, contain nitrogen and are called nitrogenous wastes. Waste products of most animals are transported by body fluids or blood to the excretory organs where wastes are removed. Removal of wastes from these fluids affects osmotic balance between the fluids and body cells. Thus, in complex animals, the **excretory system** functions to remove waste products and to maintain osmotic balance.

27:8 Planaria

Simple marine animals are in osmotic balance with their environment. The salt water in which they live is very similar to the contents of their cells. Thus, water enters and leaves their cells in equal amounts, and no special adaptations for maintaining osmotic balance are needed.

Simple freshwater animals are not in osmotic balance with their environment. Cells of these animals contain many dissolved substances as well as water. Thus, the fluid of the cells and the surrounding water are not in balance. These animals have adaptations for maintaining osmotic balance. If they did not, their cells would burst as a result of more water moving in than out.

A planarian is a freshwater animal that has a special system to maintain osmotic balance. A planarian has no circulatory system to transport wastes and excess water to a collecting site. Instead, a system of **excretory canals** and **flame cells** is found throughout the animal. Each flame cell has several flagella, the movement of which reminded scientists of candle flames, thus the name flame cells. Excess water and some wastes are drawn into the flame cells where movement of the flagella sets up a current. This current moves the water into and along the excretory canals. At certain places, the canals branch into excretory ducts that open as pores on the surface of the planarian. Water is given off through these pores.

Most of the nitrogenous wastes of a planarian do not enter the excretory canals. They pass from cells to the digestive system and are excreted through the mouth. Some wastes may diffuse directly from cells to the water. Ammonia (NH_3) is the waste excreted by a planarian. Ammonia is poisonous and can cause death if it accumulates. It is safe for planarians because it is constantly excreted into the surrounding water. Most animals with a circulatory system do not excrete ammonia. Instead, they excrete less poisonous compounds that can be safely stored or carried by the blood until excreted.

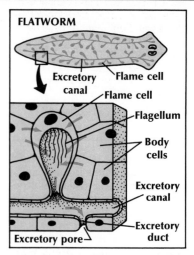

FIGURE 27-11. Flagella in flame cells of *Planaria* draw excess water from the body. The water collects in ducts and moves out of the body through pores.

Simple freshwater animals gain water by osmosis. They must have adaptations to maintain osmotic balance.

In planarians, water is collected in special channels and expelled through pores.

Planarians release ammonia through the mouth and by diffusion into the surrounding water.

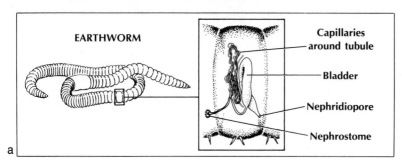

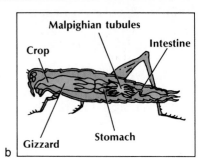

FIGURE 27-12. (a) In an earthworm, wastes are separated in the nephridia and excreted through nephridiopores. Certain materials are reabsorbed by tubules and passed to capillaries for recirculation. (b) In a grasshopper, Malpighian tubules absorb uric acid and water from blood. Water is reabsorbed but uric acid is passed to the intestine and excreted through the anus.

An earthworm has excretory units called nephridia (singular, nephridium).

In each nephridium, harmful materials are separated from necessary materials.

Insects conserve water by excreting a solid nitrogenous waste, uric acid.

Excretion of uric acid conserves water and allows safe storage of nitrogenous waste.

27:9 Earthworm and Grasshopper

Excretion in an earthworm involves the circulatory system. Pairs of excretory units called **nephridia** (nih FRIHD ee uh) are found in almost every segment of an earthworm. Each nephridium is made up of a funnel-shaped **nephrostome** (NEF ruh stohm) that narrows into a tubule. The tubule passes into the next segment and widens into a bladder. The bladder ends as a **nephridiopore** (nih FRIHD ee uh por), a small opening to the outside.

Lymph in the body cavity of an earthworm contains both useful and waste materials. Lymph enters the nephrostome and is passed along the tubule by the action of cilia and muscular contractions. Certain cells of the tubule absorb the useful materials from the lymph. These useful substances pass from the tubule cells into the many capillaries surrounding the tubule. Wastes remain in the tubule and enter the bladder. The wastes are then channeled to the outside through the nephridiopore.

This general filtering plan is found in most complex animals. The systems function in filtering a liquid to remove wastes and in returning usable substances to circulation. *The purpose of this system is separation of harmful from useful materials.*

As land animals, grasshoppers and other insects must conserve water. Thus, they cannot excrete ammonia. Instead, they convert ammonia to another waste called **uric acid.** Uric acid is in the form of insoluble crystals. Because it is insoluble, it requires no water to be excreted. Uric acid is also much less toxic than ammonia. It does not dissolve in water; therefore, it does not affect the cell's metabolism. Thus, it can be safely stored until it is excreted.

Attached to the intestine of a grasshopper (Figure 27–12) is a group of stringlike structures called **Malpighian** (mal PIHG ee un) **tubules.** The free ends of these tubules lie in the hemocoel. Nitrogenous wastes and water pass from the blood into the cells of the tubules where water is resorbed. The uric acid crystals are passed on to the intestine and later out of the anus as solid material. Thus, wastes are removed and water is conserved.

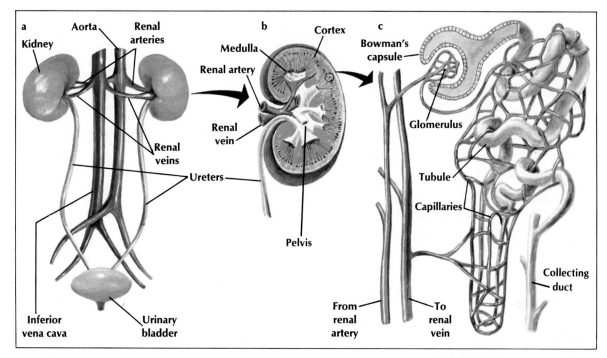

a
Kidney
Aorta
Renal arteries
Renal veins
Ureters
Inferior vena cava
Urinary bladder

b
Medulla
Renal artery
Renal vein
Cortex
Pelvis

c
Bowman's capsule
Glomerulus
Tubule
Capillaries
Collecting duct
From renal artery
To renal vein

27:10 Humans

In vertebrates, the main excretory organs are two **kidneys.** Human kidneys remove nitrogenous wastes and unnecessary materials such as salts from the blood. Kidneys also maintain proper osmotic balance in the body. Kidneys are fist-sized, bean-shaped organs (Figure 27–13) that lie near the dorsal abdominal wall. There is one kidney on each side of the body. The outer part of a kidney is called the **cortex,** the inside is the **medulla,** and the central, hollow cavity is called the **pelvis.**

Each kidney is composed of about a million tiny units called **nephrons** (NEF rahnz) that are similar to the nephridia of an earthworm. Each nephron is made up of a cup-shaped **Bowman's capsule** that narrows into a long, coiled **tubule.** In the center of each Bowman's capsule is a mass of capillaries called the **glomerulus** (gluh MER uh lus). Other capillaries from the glomerulus surround each tubule (Figure 27–14). A glomerulus begins as small arteries that branch from the **renal arteries,** the blood vessels that bring blood to the kidneys. These arteries branch to form the capillaries of the glomerulus. Another small artery leads away from the glomerulus. It divides into another set of capillaries that surround the tubule. These capillaries merge to form small veins. These small veins merge to form renal veins, which return purified blood to the general circulation.

FIGURE 27-13. (a) The human excretory system includes kidneys, ureters, and the urinary bladder. (b) The main organ, the kidney, is composed of (c) many filtering units called nephrons.

FIGURE 27-14. Inside the Bowman's capsule of each nephron is a glomerulus, a tight ball of capillaries, shown here magnified about 270 times. When blood enters the kidneys, much of the contents of the blood is forced from the glomeruli into the Bowman's capsules.

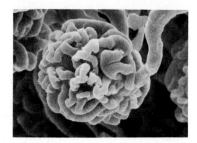

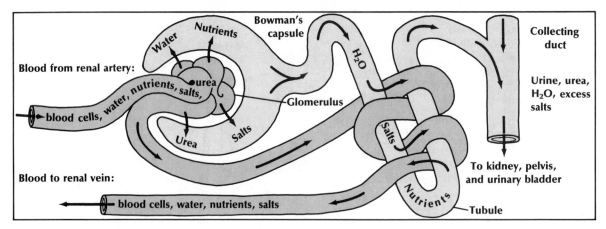

FIGURE 27-15. Blood pressure forces much of the liquid portion of the blood from the glomerulus into the Bowman's capsule. As this liquid flows through the nephron, nutrients and most salts are actively transported. Most water moves by osmosis into capillaries. Left in the nephron is urine, a combination of urea, some water, and excess salts.

In the nephrons, needed materials and most of the water are reabsorbed into the blood and tissue fluid.

Urine is a combination of urea, excess salts, and water.

Ammonia is produced from amino acids in the liver. As land animals, humans cannot excrete ammonia. They convert it to **urea** (yoo REE uh), a less dangerous nitrogenous waste. Not only is urea less toxic than ammonia, it also requires less water for its excretion. Thus, water is conserved.

Blood passing through the renal arteries contains urea and excess salts (ions). It is channeled into smaller arteries of the kidneys that lead to each glomerulus. Pressure forces most of the water and other contents of blood out of the glomerulus, and into the Bowman's capsule of the kidneys. The continuous flow of blood into the kidneys forces the liquid from the Bowman's capsule into the tubules for filtering. Cells lining the tubule actively transport needed materials such as some food molecules, hormones, and ions out of the liquid passing through. These materials then pass from the tubule cells to the surrounding capillaries. Also, almost all the water moves by osmosis from the tubules to the tissue fluid and back to the capillaries. The cleansed blood travels to the renal veins and back to the general circulation.

Urea, excess salts, and sometimes other substances, along with a small amount of water, are not resorbed but remain in the tubules. The combination of these materials is called **urine.** Table 27–1 compares the contents of urine with the contents of blood.

Urine passes from the tubules into the **collecting ducts** in the medulla and on to the kidney pelvis. Leading from the pelvis of each

TABLE 27-1. COMPARISON OF URINE AND BLOOD		
Substance	In Blood	In Urine
Urea	small amount	large amount
Water	large amount	small amount
Salts	some	more
Glucose	present	usually absent
Amino acids	present	usually absent
Proteins	present	usually absent
Blood cells	present	usually absent

kidney is a tube called the **ureter** (YOOR ut ur). The ureter transports urine to a muscular storage sac called the **urinary bladder.** When the bladder becomes filled, a muscular valve relaxes, and urine is excreted through the urethra.

Urine is collected in the kidneys and transferred to the urinary bladder for storage until it is released from the body.

27:11 Control of Osmotic Balance

Tissue fluid must be in osmotic balance with body cells. This balance is affected by the amount of water, salts, and organic compounds in the fluid and cells. If osmotic balance is not maintained, cells may dehydrate or gain too much water. Either situation can be very harmful. Besides filtering wastes from the blood, the kidneys play an important role in maintaining osmotic balance. Responding to a variety of hormones, the kidneys aid in regulating the amount of water and some salts in the body. The level of water and salts excreted in urine is adjusted to the needs of the body.

Consider the control of the water content of blood. In an average-size person, about 185 liters of blood are filtered every day. However, only about one or two liters of urine are excreted per day. This means that about 99 percent of the fluid is resorbed! The lost water is replaced by water in food and beverages.

Hormones control the level of water and certain salts resorbed by the kidneys.

Exactly how much water is lost in the urine varies under different conditions. A person may need to conserve water if much has been lost because of heavy perspiration or illness. Removal of more water than usual may be necessary for one who has consumed a large volume of liquid. The amount of water excreted in urine is controlled by **vasopressin** (vay zoh PRES un), a hormone secreted by the pituitary gland (Section 28:7).

Cells in the hypothalamus of the brain (Section 28:6) detect the amount of water in the blood. If the water level is too low, the hypothalamus stimulates the pituitary to secrete more vasopressin. Vasopressin enters the bloodstream and travels to the kidneys where it stimulates the tubules to resorb more water. The greater the need for water, the more water resorbed, and thus, urine becomes more concentrated.

The opposite occurs when too much water is present in the blood. The hypothalamus detects the excess water in the bloodstream and causes the pituitary to release less vasopressin. Therefore, the tubules resorb less water and the urine is diluted.

The kidneys control the levels of certain salts in the urine by responding to other hormones (Section 28:3). Varying the amount of water and salts excreted in urine is an important means of homeostasis. Such control is a crucial adaptation. Osmotic balance is maintained regardless of changes in diet, activity, or other factors that affect the body.

FIGURE 27-16. Vasopressin regulates osmotic balance in the body by controlling the water content of urine. The secretion of vasopressin by the pituitary gland is controlled by the hypothalamus.

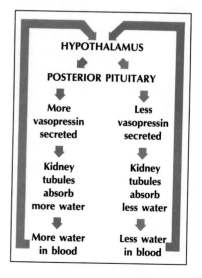

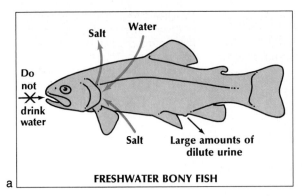

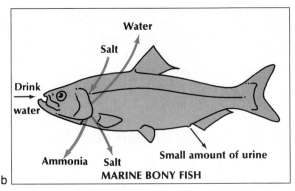

a — FRESHWATER BONY FISH

b — MARINE BONY FISH

FIGURE 27-17. (a) Freshwater and (b) marine bony fish have evolved different adaptations for maintaining osmotic balance. Blue: passive transport; red: active transport.

27:12 Other Vertebrates

Kidneys are excretory organs present in all vertebrates. They first evolved among freshwater fish. As other vertebrate forms evolved, the structure and function of their kidneys became modified for their environments.

Body fluids of bony fish are not in osmotic balance with the water in which they live. Most of the surface of a fish is covered with scales across which water and salts cannot move. However, the gills are in direct contact with the water. Therefore, water and salts may enter or leave the fish at the gills, causing an imbalance.

Body fluids of freshwater bony fish are more concentrated than the surrounding water. Thus, these fish tend to gain water and lose salts through the gill membranes. Adjustment to these conditions occurs in several ways. Their kidneys resorb most salts but very little water. As a result, the urine produced is very dilute and large amounts of urine are excreted. A freshwater fish is also adapted in another way. It drinks very little water, so no additional excess water builds up in its fluids. Together, these adaptations enable the animal to conserve salts, rid itself of excess water, and excrete wastes.

The salt content of the body fluids of marine bony fish is less concentrated than the water in which they live. As a result, they tend to lose water and take in salts. Some of the lost water is restored by drinking large amounts of seawater. However, this water contains salts, so even more salts build up in the fluids. Accumulation of salts is dangerous because it can lead to dehydration. This excess salt is removed by active transport across the gill tissue to the water. Most nitrogenous wastes in these fish are excreted in the form of ammonia from the gills. Thus, very little water enters the kidneys, and water is further conserved. The kidneys play a minor role in both excretion and osmotic balance.

Vertebrates living on land must conserve water because it readily escapes to the dry environment. Water is lost in urine and feces, is constantly exhaled, and evaporates from the body.

Saltwater fish drink large amounts of water and excrete very little urine. Excess salt is actively removed through the gills or by salt glands.

Kidneys of land vertebrates are adapted to conserve water.

FIGURE 27-18. The kangaroo rat, a desert animal, excretes highly concentrated urine and thus, conserves water. This animal obtains enough water from food and its metabolic reactions to survive without drinking water.

Amphibians produce urea as their nitrogenous waste product. As in mammals, the urea is excreted with some water in the form of urine. The water lost is replaced by drinking and eating. Birds and reptiles are more efficient at conserving water. Like insects, these vertebrates excrete uric acid. However, the excretory systems of these two groups function differently. Birds and reptiles have kidneys; insects do not. Bird and reptile kidneys resorb almost all the water from the fluid passing through the tubules. The uric acid is transported to the cloaca (Section 23:4) from which it is excreted in a pastelike form.

Excretion in mammals already has been described with reference to humans. Mammals excrete urea dissolved in water (urine). Most of the water that passes through the kidney tubules is recovered. Some desert mammals, such as the kangaroo rat (Figure 27–18), produce urine much more concentrated than that of humans. Their kidney tubules resorb most of the water that passes through them. So efficient are its kidneys that a kangaroo rat can survive without drinking water. Kangaroo rats obtain enough water from the food they eat and as by-products of metabolism.

Birds and reptiles conserve water by excreting uric acid.

Some desert mammals produce a very concentrated urine by reabsorbing a great amount of water in the kidneys.

REVIEWING YOUR IDEAS

13. What is excretion?
14. Why do planarians not swell and burst from water intake in their environment?
15. Why must nitrogenous wastes be removed? What is the source of nitrogenous wastes?
16. How is ammonia excreted from planaria?
17. What is the general function of excretory systems in the earthworm and most complex animals?
18. What nitrogenous waste product is excreted by insects?
19. What are the organs of excretion in humans? What are the functions of these organs?
20. Compare the human excretory system to excretory systems in earthworms and insects.
21. Why must water content be controlled in the body? What might happen if your kidneys removed too much water?

INVESTIGATION

Problem: How is urine used to help diagnose diseases?

Materials:

urinometer
large test tube
pH color chart
3 urine samples

TesTape
pH paper
distilled water

Procedure:

Part A. Testing Distilled Water

1. Make a copy of Table 27–2. Use your table to record all data.
2. Observe the scale on a urinometer. A urinometer is used to measure the specific gravity of urine. Specific gravity is a measurement of the weight of a given volume of urine to the weight of the same volume of water. Compare the scale with Figure 27–19a. Each line is equal to 0.001 specific gravity units. Sample readings on the scale are shown.
3. Fill a large test tube ¾ full of distilled water. Gently float the urinometer in the liquid and give the urinometer a slight twist. Note and record in Table 27–2 where the water level appears on the urinometer scale. This is the specific gravity of distilled water. Use Figure 27–19b as a guide.
4. Use the pH paper to determine the pH of distilled water. Recall that pH is an indication of whether a solution is acid, base, or neutral. Touch the pH paper to the distilled water. Compare the paper with the pH color scale.

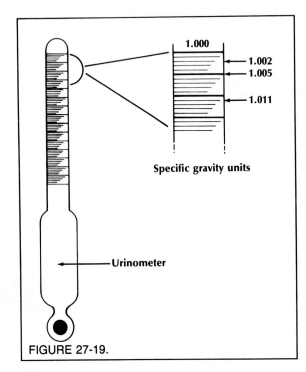

FIGURE 27-19.

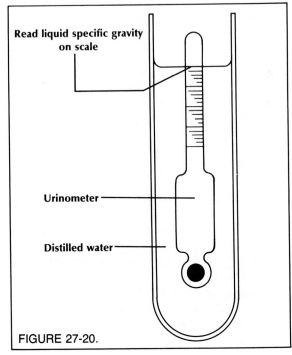

FIGURE 27-20.

Determine if the water is an acid (numbers 1–6), a base (numbers 8–12), or neutral (number 7). Record the pH of distilled water in Table 27–2.

5. Test the distilled water for the presence of glucose using a small piece of TesTape. If the TesTape turns green, glucose is present. If it remains yellow, glucose is not present. Record your data for distilled water and glucose in Table 27–2.

Part B. Testing Urine

CAUTION: Some of the samples are corrosive irritants. Avoid spilling or ingesting samples. Spillage or body contact should be flushed with water immediately.

1. Complete the specific gravity, pH, and glucose tests on samples A, B, and C.
2. Record your results in Table 27–2.

Data and Observations:

TABLE 27-2. RESULTS OF URINE SAMPLE TESTS				
Sample	Specific gravity	pH	Acid, base, or neutral	Glucose
Distilled water				
Urine A				
Urine B				
Urine C				

Questions and Conclusion:

1. Normal urine has a specific gravity of 1.010–1.025. A high specific gravity (over 1.025) means that many dissolved chemicals such as salt, glucose, and protein are present in the urine. A low specific gravity (below 1.010) indicates few dissolved chemicals.
   (a) What was the specific gravity of distilled water?
   (b) Offer an explanation for why it was so low.
   (c) Which urine samples were normal for specific gravity?
   (d) Which urine samples were abnormal for specific gravity?
2. Normal urine has a pH of about 6.
   (a) What was the pH of distilled water?
   (b) Which urine samples were normal for pH?

   (c) Which samples were abnormal for pH?
3. Normal urine has no glucose present.
   (a) Was glucose present in distilled water?
   (b) Which urine samples had glucose?
   (c) Which urine samples had no glucose?
4. These problems may show in the urine: Diabetes mellitus is indicated by glucose and a high specific gravity. Diabetes insipidus is indicated by a very low specific gravity. Basic urine indicates possible kidney infection, anemia, or kidney stones. Acid urine indicates fever or high protein diet. List the urine sample(s) that might be associated with each of these diseases or conditions: (a) diabetes insipidus (b) diabetes mellitus (c) fever (d) anemia (e) kidney infection.

Conclusion: How is urine used to help diagnose diseases?

CHAPTER REVIEW

SUMMARY

1. In simple animals, each cell exchanges gases directly with the external environment. More complex animals have specialized respiratory systems. **p. 563**
2. The respiratory exchange surface in annelids is located on or protrudes from the body where it is in contact with moisture. **27:1**
3. Many aquatic animals have gills for gas exchange. Oxygen diffuses from surrounding water across the gill filaments and into the blood. Carbon dioxide diffuses in the opposite direction. **27:2**
4. In insects, air enters the body through a tracheal system. Oxygen dissolves in moist tracheoles and diffuses directly to body cells. Carbon dioxide diffuses from cells to tracheoles. **27:3**
5. The lungs are the organs of respiration in humans. Oxygen passes from the alveoli of lungs to the bloodstream, and carbon dioxide passes in the opposite direction. **27:4**
6. Breathing is controlled by nerves that cause the diaphragm and rib muscles to contract and relax. Breathing rate is monitored by the breathing center in the medulla oblongata. **27:5**
7. Many respiratory diseases are caused by microbes. Some respiratory disorders result from allergic reactions. **27:6**
8. Cigarette smoking damages the lining of respiratory organs and has been linked to disorders including bronchitis, emphysema, and lung cancer. **27:7**
9. The flame cell system of planarians removes excess water. Nitrogenous wastes are expelled by diffusion or from the mouth. **27:8**
10. In earthworms, nephridia filter lymph and excrete nitrogenous wastes. Insects excrete uric acid, which is collected by Malpighian tubules. **27:9**
11. Kidneys are the main excretory organs of humans and other vertebrates. **27:10**
12. Human kidneys also maintain osmotic balance in the body. **27:11**
13. Kidneys of vertebrates are adapted for the animals' environments. **27:12**

LANGUAGE OF BIOLOGY

alveolus	nephron
diaphragm	respiration
excretion	spiracle
expiration	trachea
flame cell	urea
glomerulus	ureter
inspiration	urethra
kidney	uric acid
lung	urinary bladder
Malpighian tubule	urine

CHECKING YOUR IDEAS

On a separate paper, complete each of the following statements with the missing term(s). Do not write in this book.

1. _____ is the exchange of oxygen and carbon dioxide with the environment.
2. In vertebrates, nitrogenous wastes are filtered by organs called _____.
3. During inspiration the _____ moves down.
4. Insects have a _____ system for respiration.
5. Respiratory surfaces must be kept moist and have a large _____.
6. In earthworms, gas exchange with the environment occurs through the _____.
7. Most complex aquatic animals exchange gases at organs called _____.
8. _____ is the nitrogenous waste excreted by humans.

9. In planaria, excess water from the body is drawn into _____ cells.

10. Each kidney is composed of millions of filtering units called _____.

EVALUATING YOUR IDEAS

1. Describe the structure and operation of gills in fish. How is the gill system linked to the circulatory system?
2. Compare, in general, the location of the respiratory surface in aquatic and terrestrial animals. Explain the difference.
3. Describe respiration in a grasshopper. Why does the respiratory system not have to be linked with the circulatory system?
4. Describe respiration in humans.
5. How does breathing occur in humans?
6. Explain how the breathing rate in humans is controlled.
7. Explain the effect of cigarette smoking on the cilia lining respiratory organs. How does this effect interfere with normal respiratory functions?
8. What happens to respiratory tissues in emphysema and lung cancer?
9. Why do simple marine organisms have little problem with osmotic balance?
10. How do planarians rid themselves of excess water?
11. Why is ammonia not excreted by most complex animals?
12. Describe the excretory system of an earthworm.
13. How do insects excrete wastes? Why do insects excrete uric acid?
14. Describe the structure of a human kidney.
15. Explain the operation of a human kidney.
16. How is the amount of water excreted by the human kidney controlled?
17. How do freshwater bony fish maintain osmotic balance?
18. How do marine bony fish maintain osmotic balance?
19. How can a kangaroo rat live without drinking water?

APPLYING YOUR IDEAS

1. Why do most annelids live in water?
2. How is the respiratory system of humans similar to that of the grasshopper? How are the systems different?
3. Why do you suppose that the level of carbon dioxide in the blood is a signal for both breathing rate and heartbeat rate?
4. How might production of uric acid by birds and reptiles be an adaptation for development in shelled eggs?
5. Would urine be more or less concentrated in a mammal on a hot day? Explain.

EXTENDING YOUR IDEAS

1. Use library resources to learn about the operation of gills in crustaceans. Compare this system to the gills in fish.
2. Design an experiment with several classmates to test the effect of exercise on breathing rate. Also determine the time needed for breathing rate to return to normal.
3. Skin is an excretory organ in humans. Use library sources to determine the roles of skin in excretion.

SUGGESTED READINGS

Feder, Martin E., and Burgren, Warren W., "Skin Breathing in Vertebrates." *Scientific American,* Nov., 1985.

Hughes, G. M., *The Vertebrate Lung,* rev. ed. Carolina Biology Reader. Burlington, NC, Carolina Biological Supply Co., 1979.

CHEMICAL CONTROL

Many functions of animals are controlled by chemicals. The chemicals are transported throughout the body by blood or other fluids. Insect metamorphosis is a process that is regulated chemically. This cicada is shown emerging from a pupa, an intermediate stage between the larva and the adult. How is the making of chemicals controlled? What might happen if a chemical imbalance occurred?

nimals are adapted for life in certain environments. However, no environment stays the same all the time. The most successful animals are those able to make the proper responses to stimuli.

Stimuli may come from outside or inside the body. For example, outside stimuli include a sudden drop in temperature or a baseball speeding toward you. You may respond by shivering or ducking your head. Internal stimuli include hunger pangs or a high level of carbon dioxide in the blood. Responses to these stimuli include eating and a speeding up of the heartbeat and breathing rates.

Often, stimuli and responses are not obvious. You may not be aware of many changes, especially internal ones. Some responses, such as growth, occur over a long period of time.

All these responses involve the smooth coordination of the many systems of the body. All body systems working together result in homeostasis (Section 1:6), the balance of the internal operation of an organism, regardless of external changes. In this way, an organism can survive when the environment changes.

Control of responses in animals may be chemical or nervous. The effects of chemical control are discussed in this chapter. Nervous control is discussed in Chapter 29.

Objectives:
You will
- identify and state the functions of major human hormones.
- describe how hormones cause responses in target organs.
- explain the role of hormones in insect metamorphosis.

Chemical and nervous control systems regulate responses to stimuli.

ROLES OF HUMAN HORMONES

Most often, chemical control is directed by complex molecules called hormones (Section 22:9). Some hormones are proteins, polypeptides, or amino acids. Others are certain lipids called **steroids** (STIHR oydz). In animals, hormones are transported throughout the body by blood or other fluids and stimulate organs to respond in certain ways.

FIGURE 28–1. Locations of the major human endocrine glands are shown for males and females.

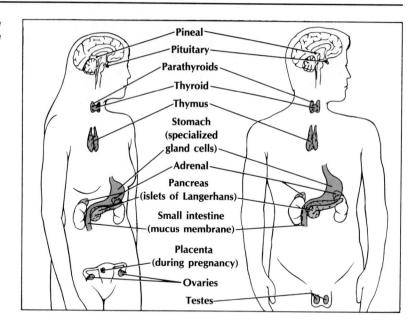

Pineal
Pituitary
Parathyroids
Thyroid
Thymus
Stomach
(specialized
gland cells)
Adrenal
Pancreas
(islets of Langerhans)
Small intestine
(mucus membrane)
Placenta
(during pregnancy)
Ovaries
Testes

Most hormones are produced by and secreted directly from endocrine glands.

In complex animals, hormones usually are produced by **endocrine** (EN duh krun) **glands.** Endocrine glands are ductless, so they pour their secretions directly into blood. (Other kinds of glands, such as salivary glands, have ducts.) The study of endocrine glands and hormones is called **endocrinology** (en duh krih NAHL uh jee). Humans have many endocrine glands that produce hormones that regulate their body functions. Information about some of the endocrine glands and the hormones they secrete follows.

28:1 Islets of Langerhans

In diabetes mellitus, there is too much sugar in the blood.

Glucose travels through the bloodstream to all the cells of the body and is used as an energy source. Proper health depends on a regulated amount of glucose in the blood. A person with too much glucose in the blood may have the disease **diabetes mellitus** (MEL ut us), or sugar diabetes. Symptoms of diabetes include excessive thirst and frequent urination. So much glucose passes through the kidneys that all the glucose cannot be returned to the blood. Thus, diabetics, when untreated, excrete a lot of glucose in the urine. Physicians can diagnose diabetes by analyzing urine for glucose.

It was learned in the late 1800s that removal of the pancreases of dogs causes diabetes. Because the digestive functions of the pancreas were already known (Section 25:10), biologists realized that the pancreas must have more than one role. Microscopic study of the pancreas showed two different types of cells. One type, greater in number, secretes digestive enzymes. The other cells had an unknown function and were named the **islets** (I lutz) **of Langerhans.**

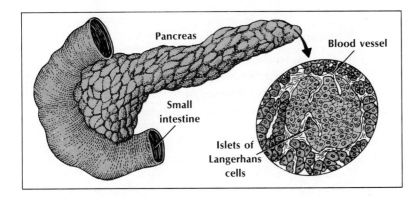

FIGURE 28–2. Islet cells in the pancreas secrete insulin, which controls blood sugar level.

Scientists thought that islet cells might affect the glucose content of the blood by secreting a hormone into the bloodstream. This belief was proven correct when a chemical made by the islet cells, a protein called insulin, was taken from animal pancreases. When insulin was injected into dogs whose pancreases had been removed, the symptoms of diabetes disappeared. Insulin is produced by only one type of islet cell, the **beta cells.**

The islet cells of the pancreas produce insulin.

Insulin controls the level of sugar in blood. After a meal, insulin causes excess blood glucose to change to glycogen in the liver and muscle cells. Insulin also causes some excess glucose to be changed to fats and increases the permeability of cell membranes to glucose. These processes reduce the blood sugar level.

Insulin regulates the level of sugar in the blood.

One form of diabetes occurs when beta cells become destroyed and not enough insulin is produced. Without enough insulin, excess glucose cannot be changed to glycogen or fats, and glucose cannot be moved out of the blood effectively. This form of diabetes usually develops in people under twenty years of age. Recent studies suggest that a virus may destroy the beta cells, especially in people with certain harmful alleles. People with this form of diabetes take regular injections of insulin. Formerly, most persons with this form of diabetes used insulin obtained from animals. Although animal insulin is slightly different in chemical structure from human insulin, it does reduce the level of glucose in human blood. The immune system of some people, though, rejects "foreign" insulin. Today, human insulin produced by recombinant DNA techniques (Chapter 13) is used and can be accepted by all diabetics.

In one form of diabetes, beta cells are destroyed and not enough insulin is produced.

Another more common form of diabetes usually occurs in people who are over forty years old and overweight. The beta cells are not destroyed and enough insulin is produced. However, glucose does not easily pass through the membranes of the muscle and liver cells, and so builds up in the blood. Diet, exercise, and certain drugs have been shown to improve uptake of glucose by muscle and liver cells and reduce the level of glucose in the blood.

In the more common form of diabetes, enough insulin is produced but glucose does not easily pass into liver and muscle cells.

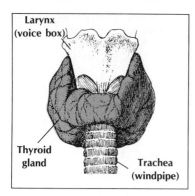

Larynx (voice box)

Thyroid gland

Trachea (windpipe)

FIGURE 28–3. The thyroid gland is located on the trachea just below the voice box. The thyroid gland secretes thyroxine.

Iodine is an essential part of thyroxine.

FIGURE 28–4. (a) Using radioactive iodine tracers, scans of the thyroid can be done to detect thyroid disease. (b) A scan of a healthy thyroid gland shows a small gland (dotted areas) whereas (c) the scan of the diseased thyroid shows a much enlarged gland.

28:2 Thyroid Gland

Located on the trachea is a small, butterfly-shaped structure called the **thyroid** (THI royd) **gland** (Figure 28–3). This gland secretes a hormone called **thyroxine** (thi RAHK sun). Thyroxine affects all body cells. In the cells, thyroxine controls the rate of metabolism and affects growth. Too little thyroxine decreases the metabolic rate. Decreased metabolism may lead to obesity (Section 25:7). Too little thyroxine can also cause several diseases. If this deficiency occurs in early childhood, a disease called **cretinism** (KREET un ihz um) results. Cretinism results in malformation, mental retardation, and dwarfism. Cretinism can be avoided by supplying thyroxine during childhood.

A disease related to the thyroid gland is goiter. **Goiter** is an enlargement of the thyroid gland itself and results in an increase in the size of the entire throat and neck. In one form of goiter, thyroid glands become enlarged because of a lack of iodine in the diet. Iodine is an important part of thyroxine. When little or no iodine is present, the thyroid gland cannot produce thyroxine. The lack of thyroxine affects cell metabolism.

In the past, certain geographic regions seemed to be "goiter belts." Many cases of goiter developed in these areas because food there was low in iodine. Persons living close to coastal areas usually did not get goiters because their diet included fresh seafood, rich in iodine. Today, iodized salt is sold. This salt provides enough iodine for normal production of thyroxine. As a result, this form of goiter is now rare in the United States.

Production of too much thyroxine results in rapid metabolism and can cause diseases also. Symptoms may include weight loss, irritability, protruding eyes, and nervousness. These symptoms can sometimes be corrected by medicines or removing or destroying part of the thyroid gland.

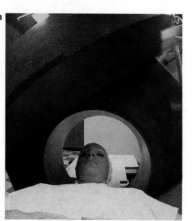

a

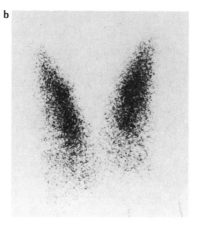

b

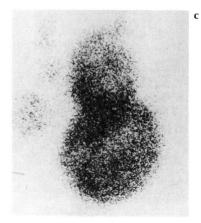

c

28:3 Adrenal Glands

On top of each kidney is a small **adrenal** (uh DREEN ul) **gland**. The outer part of this gland is the **adrenal cortex**. It differs in both structure and function from the inner part, the **adrenal medulla**.

The main hormone of the adrenal medulla is called **adrenaline** (uh DREN ul un), or **epinephrine** (ep uh NEF run). When the body is under stress, adrenaline aids responsiveness for most activities. Adrenaline increases the heartbeat rate and blood pressure. It also causes blood sugar level to rise and metabolism to occur more rapidly. More blood is sent to critical places such as the brain, heart, lungs, and skeletal muscles. How is each response adaptive?

These activities are important during stress situations, but a person can live without an adrenal medulla. These same reactions can be caused by the nervous system. Also, cells that make up the adrenal medulla are more like nerve cells than gland cells. *These facts show a close relationship between the endocrine system and the nervous system.*

The adrenal cortex is truly an endocrine structure and, unlike the medulla, is needed for life. The cortex secretes several steroid hormones. **Aldosterone** (al DAWS tuh rohn) controls the salt level in blood by affecting absorption of sodium and potassium ions in the kidney. **Cortisol** (KORT uh zawl) causes the change of fats and proteins to glucose, which helps keep the right level of sugar in the blood. Cortisol and **cortisone** (KORT uh zohn) prevent inflammation. Cortisone in synthetic form is sometimes given to reduce the pain and swelling caused by diseases and injuries. These hormones also have a role in dealing with stress. They seem to take over after adrenaline has been secreted.

The adrenal cortex of people of both sexes secretes small amounts of male and female sex hormones. **Androgens** (AN druh junz), or male sex hormones, are secreted in larger amounts than estrogens, female sex hormones.

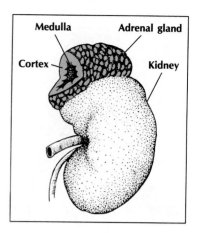

FIGURE 28–5. The adrenal glands are located on top of the kidneys. The two parts of the adrenal gland, the medulla and cortex, secrete different hormones.

The effects of adrenaline can be duplicated by the nervous system.

Hormones from the adrenal cortex control salt balance and sugar level in the blood and prevent inflammation.

Other adrenal cortex hormones are androgens and estrogens.

a DATE/TIME	
Na (136-145)	**138.** meq/L
K (3.5-5.0)	**4. 5** meq/L
Cl (96-106)	**104.** meq/L
CO_2 (24-30)	**25.** meq/l

b DATE/TIME	
Na (136-145)	**129.** meq/L
K (3.5-5.0)	**6. 1** meq/L
Cl (96-106)	**86.** meq/L
CO_2 (24-30)	**11.** meq/L

FIGURE 28–6. A kind of common adrenal gland malfunction called Addison's disease results from an insufficiency of adrenal cortex hormones. (a) A printout of tests of normal blood shows sodium (Na), potassium (K), and chloride (Cl) all within normal ranges. (b) A printout of tests on blood from an Addison's patient shows an elevated potassium and lowered sodium and chloride levels.

a

b

FIGURE 28–7. Secondary sex traits can easily be seen in the plumage of birds. (a) Immature white ibis and (b) adult white ibis have very different plumage. The black wing tips of the male birds develop as the birds mature.

Estrogens start buildup of uterine tissue during the menstrual cycle and are responsible for secondary sexual characteristics in females.

Testosterone, an androgen, causes secondary sexual characteristics in males.

28:4 Gonads

Sex hormones are produced in large amounts by the gonads, the ovaries and testes. Follicle cells of the ovary release estrogens that start the buildup of uterine tissue during the menstrual cycle (Section 23:7). Estrogens also stimulate the development of secondary sexual characteristics in the female, such as enlargement of the breasts, broadening of the pelvis, and hair growth around the external reproductive organs and in the armpits. The onset of menstruation and the development of these secondary sex traits is called **puberty.** During menopause (Section 23:7), the production of estrogens (and progesterone) decreases and menstrual cycles stop.

Androgens are secreted by the testes of a male. **Testosterone** (teh STAHS tuh rohn) and other androgens stimulate the formation of secondary sexual characteristics in males. At puberty, the external genitals enlarge and several other changes occur. The voice becomes deeper, and hair grows around the reproductive organs, in the armpits, and on the face. Fat formation decreases, and the body looks leaner. Lack of enough androgens may result in improper development of these secondary characteristics.

REVIEWING YOUR IDEAS

1. What two basic types of control systems are found among organisms?
2. What are hormones? Where are most hormones produced in complex animals? How are they transported?
3. Distinguish between an endocrine gland and other glands.
4. Give the location of the following endocrine glands and list the hormone(s) that each produces: thyroid, islets of Langerhans, adrenal cortex, adrenal medulla, ovaries, and testes.

ENDOCRINE CONTROL

Located at the base of the brain is a tiny, three-lobed structure called the pituitary gland. The two major lobes are known as the **anterior pituitary** and the **posterior pituitary.** Many of the hormones produced by the anterior pituitary control the secretion of hormones from other endocrine glands and affect other body parts.

The anterior pituitary gland secretes hormones, many of which control the activities of other endocrine glands.

28:5 Pituitary Gland: Anterior Lobe

Two of the anterior pituitary hormones control the menstrual cycle in females. They are follicle-stimulating hormone (FSH) and luteinizing hormone (LH) (Section 23:7). FSH stimulates egg ripening in the ovary follicle and causes the ovaries to secrete estrogen. LH starts ovulation and conversion of the follicle to the corpus luteum and its secretion of estrogen and progesterone. FSH and LH also are produced by the pituitaries of males. In males, LH causes the testes to produce testosterone. FSH and testosterone work together to stimulate sperm production.

FSH and LH control the menstrual cycle in females.

FSH and LH are chemically the same whether produced by males or females. However, they have different effects in males and females. These different results are due to the type of cell or tissue on which the hormone is acting, the target organ. Other evidence of the importance of the target is that the same kind of hormone may be produced by different types of animals with different effects in each of them. *Therefore, the target organ is very important in determining the response.*

In males, the testes produce testosterone in response to LH. Testosterone and FSH stimulate sperm production.

The target organ is very important in determining a response.

Anterior pituitary hormones affect other endocrine glands around the body. For example, the thyroid is controlled by a hormone called **thyroid-stimulating hormone (TSH).** The level of TSH in the

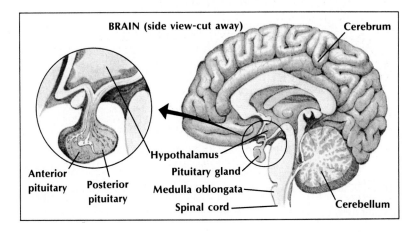

BRAIN (side view–cut away)
Cerebrum
Hypothalamus
Pituitary gland
Medulla oblongata
Spinal cord
Cerebellum
Anterior pituitary
Posterior pituitary

FIGURE 28–8. The pituitary gland is attached to the brain and is under the control of the hypothalamus. Many hormones secreted by the anterior lobe regulate functions of other endocrine glands.

FIGURE 28–9. (a) HGH deficiency in early childhood can result in dwarfism. (b) Excess HGH in childhood can result in giantism. This woman is about 228 cm tall.

a   b

TSH controls the amount of thyroxine secreted by the thyroid.

HGH influences the total growth of the body.

blood affects the amount of thyroxine secreted by the thyroid (Section 28:6). Another anterior pituitary hormone controls the adrenal cortex.

Some anterior pituitary hormones affect more than just a specific endocrine gland. For example, **human growth hormone (HGH)** influences the total growth of the body. Lack of enough HGH in early childhood can result in dwarfism or retarded growth (Figure 28–9). If too much HGH is made in childhood, giantism may result (Figure 28–9).

In 1970, scientists first made HGH in the laboratory. HGH is now being produced by using recombinant DNA techniques. It is being used to treat children who do not produce enough of it, allowing these children to grow normally. HGH has other effects on the body and might also be used in other ways, including treatment of burns.

28:6 Feedback

The hypothalamus secretes hormones that control the anterior pituitary.

Because many anterior pituitary hormones control other endocrine glands and body parts, the pituitary was once called the "master gland." Now it is known that the pituitary itself is under the control of the hypothalamus, a part of the brain near the pituitary. *This fact shows another close relationship between the endocrine system and the nervous system.*

Interaction of pituitary hormones with other endocrine glands usually involves a **feedback mechanism.** Control over other glands depends on information received from them. Control of the menstrual cycle (Section 23:7) is based on such a feedback mechanism.

Control of thyroxine production is another example of a feedback mechanism. Thyroxine circulates throughout the body in the bloodstream and reaches many tissues, including the hypothalamus. The hypothalamus contains cells sensitive to the amount of thyroxine in the blood. If the thyroxine level falls below normal, the hypothalamus secretes a hormone called a **releasing factor.** The releasing factor travels in the blood to the anterior pituitary, stimulating it to secrete TSH. TSH is secreted into the blood and reaches the thyroid gland where it stimulates thyroxine production. If there is too much thyroxine in the blood, the hypothalamus stops secreting the releasing factor and the anterior pituitary decreases production of TSH (Figure 28:10). Thus, thyroxine production is decreased. Output of other anterior pituitary hormones is controlled by different releasing factors. Some releasing factors inhibit, rather than stimulate, release of hormones from the anterior pituitary.

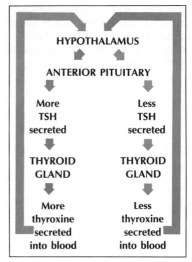

FIGURE 28–10. A feedback mechanism regulated by the hypothalamus maintains a certain amount of thyroxine in the blood.

PEOPLE IN BIOLOGY

Choh Hao Li

(1913–   )

In the 1930s, Chinese-born Choh Hao Li began the study of the chemistry and biology of anterior pituitary hormones. Throughout his years of research at the University of California, Berkeley, Li and his co-workers were either the first or among the first to isolate and identify the eight hormones secreted by the anterior pituitary.

In more recent years, Dr. Li has concentrated on researching human growth (HGH). In 1956, Li isolated HGH and later disclosed its complete structure. His research set the foundation for further research by other scientists into the use of HGH in helping children who lack the hormone to grow normally.

Li, who became an American citizen after receiving his Ph.D. at the University of California, Berkeley, is now the Director of the Hormone Research Laboratory of the University of California. He has won many awards and honors for his research including the National Award of the American Cancer Society (1971) and the Nichols Medal of the American Cancer Society (1979).

FIGURE 28–11. The posterior pituitary secretes vasopressin, which aids in controlling osmotic balance, and oxytocin, which causes labor contractions.

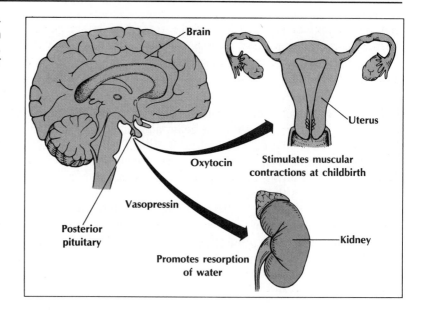

28:7 Pituitary Gland: Posterior Lobe

Oxytocin stimulates the contraction of uterine muscle during labor.

The posterior pituitary secretes two hormones, neither of which controls other endocrine glands. **Oxytocin** (ahk sih TOHS un) stimulates the contraction of uterine muscles and is secreted prior to childbirth. Oxytocin causes labor contractions (Section 24:7) and is sometimes injected to begin this process artificially.

Vasopressin regulates water balance control by the kidneys.

Vasopressin is a hormone involved in control of osmotic balance by the kidneys (Section 27:11). A lack of vasopressin prevents kidney tubules from resorbing water. Frequent urination and constant thirst are results of a lack of vasopressin. This disease is known as **diabetes insipidus** (ihn SIHP ud us). It is named diabetes because of the frequent urination and thirst symptoms that it has in common with diabetes mellitus. However, in diabetes insipidus, there is no abnormal blood sugar level. The two diseases are related to two different hormones.

The endocrine system and nervous system are closely related.

Neither oxytocin nor vasopressin is produced by the posterior pituitary. Both hormones are formed by the hypothalamus. The hormones travel to the posterior pituitary along nerve cells and are stored. Secretion of these hormones is not controlled by releasing factors. Rather, the hormones are released as a result of nervous stimulation from the hypothalamus. The production of hormones by the nervous tissue of the hypothalamus is further evidence of the *close link between the endocrine system and the nervous system.* In addition, cells of the posterior pituitary are more like nerve cells than endocrine cells. You have seen many such lines of evidence in this chapter. Perhaps chemical and nervous control should not be considered as two separate systems. It might be better to think of chemical and nervous control as two divisions of the same system. In general, responses controlled by the endocrine system occur more slowly than those controlled by the nervous system. Table 28–1 lists the activities of many human hormones.

TABLE 28-1. SOME HUMAN HORMONES AND THEIR MAIN FUNCTIONS			
Hormone	Source	Location of Source	Functions
THYROXINE	Thyroid gland	Neck	Regulates metabolic rate of body cells
PARATHORMONE	Parathyroid gland	On thyroid	Regulates calcium balance
INSULIN	Islets of Langerhans (beta cells)	Pancreas	Decreases level of sugar in blood
GLUCAGON	Islets of Langerhans (alpha cells)	Pancreas	Increases level of sugar in blood
ADRENALINE	Adrenal medulla	On kidneys	Prepares body to cope with stress
ALDOSTERONE	Adrenal cortex	On kidneys	Regulates salt level in blood
CORTISOL	Adrenal cortex	On kidneys	Increases level of sugar in blood Prevents inflammation
CORTISONE	Adrenal cortex	On kidneys	Prevents inflammation
ANDROGENS	Testes	In scrotum	Causes secondary sexual characteristics in males
ESTROGENS	Ovaries	In abdomen	Causes secondary sexual characteristics in females Prepares uterus for pregnancy
PROGESTERONE	Corpus luteum	In abdomen	Maintains uterus during pregnancy
FSH	Anterior pituitary	Base of brain	Causes maturation of egg in females Stimulates sperm production in males
LH	Anterior pituitary	Base of brain	Causes ovulation in females Causes release of androgens in males
TSH	Anterior pituitary	Base of brain	Stimulates thyroxine production
ACTH	Anterior pituitary	Base of brain	Stimulates release of adrenal cortex hormones
HGH	Anterior pituitary	Base of brain	Regulates growth of body
PROLACTIN	Anterior pituitary	Base of brain	Stimulates milk production
OXYTOCIN	Posterior pituitary	Base of brain	Stimulates contraction of uterine muscles
VASOPRESSIN	Posterior pituitary	Base of brain	Regulates water balance by kidneys
GASTRIN	Stomach cells	Stomach	Stimulates release of gastric juice
SECRETIN	Intestinal cells	Intestine	Stimulates release of pancreatic juice

28:8 Hormone Action

Hormones cause a specific response in particular target organs. How does a given hormone "recognize" cells of its target organ? Once the target organ has been "located," how does the hormone cause the response?

Cells of the target organs that respond to many protein hormones are thought to have receptors on their membranes. Cells of different target organs have different receptors. Because each hormone "recognizes" only a certain receptor, the target organ is able to be "located."

A target organ may be recognized by a specific receptor on the cell membrane.

How is a response caused in the target organ once the hormone is present? According to the **second messenger hypothesis,** the hormone does not directly cause the response. The hormone acts as

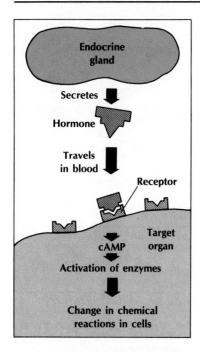

FIGURE 28–12. Some hormones act by causing the production of a second messenger, such as cyclic AMP (cAMP), which causes changes in a cell's reactions.

Steroid hormones cause responses by activating genes in the target cells.

Prostaglandins may play a role in regulating responses to other hormones.

the first messenger. When the hormone combines with the receptor, another compound known as the second messenger is formed in or enters the target cell. This second messenger actually causes the response. It does so by triggering a series of chemical changes within the target cell.

The second messenger hypothesis resulted from the study of the effects of adrenaline on glycogen stored in the liver. It was known that adrenaline causes glycogen to be converted to glucose. The glucose then can be used to provide a "burst of energy." Scientists found that adrenaline causes the formation of **cyclic AMP** (cAMP), a compound made from ATP, in the target cell. Cyclic AMP acts as a second messenger that causes reactions that lead to the activation of several enzymes. One of these enzymes begins the breakdown of glycogen.

It is now known that cAMP is a second messenger in many types of tissues and that several hormones cause cAMP to be produced. Depending on which genes are active, different types of cells have different sets of chemicals. *The effect of cAMP depends on the type of cell in which it is formed.* For example, cAMP causes synthesis of certain steroid hormones in the adrenal cortex. In the kidneys, cAMP affects water transport. It also affects the breakdown of fatty acids in the liver and muscle cells. Second messengers other than cAMP are now known to exist. Some are ions that enter a target cell, activate an enzyme, and thus cause a chemical change.

A steroid hormone passes through the membrane of its target cell and combines with a receptor inside the cell. No second messenger is produced. Instead, the steroid-receptor moves into the nucleus, where it activates a gene (Section 9:12). When the gene becomes active, it guides the formation of a protein that changes the cell's chemistry. The response depends on how the protein affects the particular target organ.

Most animal tissues produce **prostaglandins,** a special group of hormones. They are made from fatty acids (Section 3:13) and have many effects in the body. Some prostaglandins seem to play a role in how much cAMP is produced by target cells. After a hormone has combined with receptors on a target cell, the target cell releases more prostaglandins. The prostaglandins, in turn, may increase or decrease the production of cAMP. Thus, prostaglandins, as well as other hormones, may play a role in the target organ's response.

Activity of prostaglandins may occur outside the cells in which they are produced. Many travel in the blood to other tissues. Prostaglandins are involved in a variety of activities, such as controlling smooth muscle contraction, regulating pulse rate and blood pressure, causing swelling, and making the body aware of pain. One of the reasons aspirin relieves swelling and pain is that it decreases production of prostaglandins.

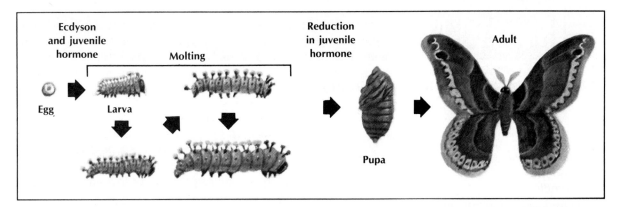

28:9 Hormones and Metamorphosis

Chemical control by means of hormones is not unique to humans. Many of the hormones discussed in this chapter are produced by other vertebrates. Invertebrates also produce hormones.

Hormones regulate metamorphosis, change in form, in both vertebrate and invertebrate animals. Recall that insects undergo metamorphosis (Section 24:3). As larvae or nymphs grow, they must shed their exoskeleton (Section 16:1) in a process called molting. The molting process is controlled by hormones.

Certain brain cells secrete a hormone that causes the secretion of another hormone, **ecdyson** (EK duh sahn). Ecdyson plays a role in causing a certain number of molts to occur. Then the nymph becomes an adult or the larva becomes a pupa. No further molting occurs after these stages are reached.

What causes the final transition to the adult form? A third hormone, **juvenile hormone,** is involved. Juvenile hormone is constantly produced by a pair of glands, the corpora allata (KOR pruh • uh LAHT uh), located in the brain. Juvenile hormone and ecdyson together cause an insect to molt and grow. As the larva or nymph goes through its series of molts, the output of juvenile hormone decreases, causing the larva to enter the pupa stage or the nymph to enter its final molt. Absence of juvenile hormone results in no further molting (Figure 29–13).

FIGURE 28–13. Complete insect metamorphosis includes a series of molts in which the larva becomes a pupa. The changes are the result of certain levels of hormones.

Molting, shedding of an insect's exoskeleton, is controlled by interaction of several hormones.

A hormone from the brain stimulates release of ecdyson, which is partly responsible for molting.

Juvenile hormone level decreases as molting proceeds.

REVIEWING YOUR IDEAS

5. Where is the pituitary gland located?
6. What is a feedback mechanism?
7. What is the hypothalamus?
8. What is the second messenger hypothesis?
9. What are prostaglandins?
10. Name a process in insects that is controlled by hormones.

INVESTIGATION

Problem: How does the body control calcium balance?

Materials:

file card
microscope
pictograph
prepared slide of thyroid and parathyroid glands
tape

Procedure:

Part A. Location of Thyroid and Parathyroid Glands

1. Make a copy of Figure 28–15 and Figure 28–16. Be sure to trace the diagram of the human trachea and larynx onto your paper as accurately as possible.
2. Two endocrine glands, the thyroid and parathyroids are located in the neck region of humans. Trace the outline diagrams of these glands (Figure 28–14) onto a file card.

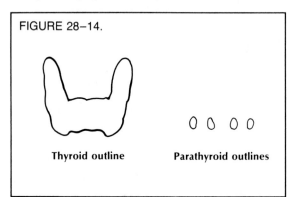

FIGURE 28–14.

Thyroid outline   Parathyroid outlines

3. Cut out the outlines of these glands and tape them in place onto the human trachea and larynx diagram of your Figure 28–15. The thyroid gland fits just below the larynx with its "wing-like" ends extending up and around the sides of the larynx. The parathyroids are located in the "wings" of the thyroid gland, two on each side with one below the other.
4. Label the following four parts: trachea (windpipe), larynx (voice box), thyroid gland, parathyroid glands.

Part B. Observing Thyroid and Parathyroid Tissue

1. Observe the prepared slide of thyroid and parathyroid glands under low power. Slowly move the slide on the stage until you find an area in which adjoining tissues appear different from each other.
2. Thyroid tissue appears more spongy with large, clear noncellular spaces (usually stained light purple). Parathyroid tissue appears as very small, compact cells.
3. Diagram each type of tissue in the space provided in your copy of Figure 28–16.

Part C. Role of the Thyroid and Parathyroids in Calcium Regulation

1. Study the pictograph provided by your teacher. It shows a series of events to be described.
2. Using the left side only of the pictograph, follow these lettered events showing the fate of excess calcium in the body.
   A. Certain foods such as *milk and cheese* contain high amounts of calcium. These foods contribute calcium to your bones and teeth.
   B. Calcium in your diet passes from the intestines into your *bloodstream.*
   C. The concentration of calcium in your blood may become too high. This *high concentration* can be detected by the *thyroid gland.* The thyroid then releases a hormone called *calcitonin.*

D. Calcitonin stimulates *bone cells* to remove the excess calcium from blood and store it in your bones.
3. Complete the pictograph by using the italicized terms in statements A–D to fill in the blanks.
4. Show the pathway of calcium by shading in the correct arrowhead in figure D. Draw five asterisks representing where calcium atoms would be deposited due to calcitonin.
5. Using the right side only of the pictograph, follow these lettered events:
   E. The diet of a person who does not consume many milk products may be low in calcium.

F. This results in a *low concentration* of calcium in the *bloodstream.*
G. This low concentration can be detected by the *parathyroid glands.*
H. The parathyroids release a hormone called *parathormone.* Parathormone stimulates *bone cells* to release their stored calcium into the blood.
6. Complete the pictograph by using the italicized terms in statements E–H to fill in the blanks on the labels.
7. Show the pathway of calcium by shading in the correct arrowhead in Figure H. Draw five asterisks representing the location of calcium atoms due to parathormone action.

Data and Observations:

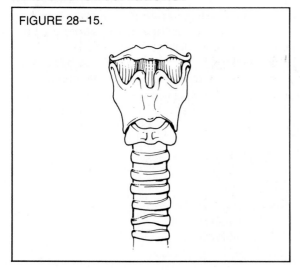

FIGURE 28–15.

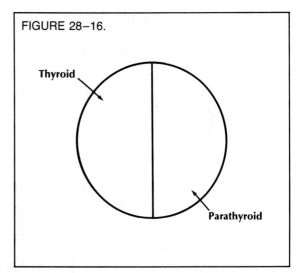

FIGURE 28–16.

Thyroid

Parathyroid

Questions and Conclusion:

1. When the calcium level in the blood increases, which hormone is released? Which is not?
2. When the calcium level in the blood decreases, which hormone is released? Which is not?

3. Is calcitonin or parathormone produced by the parathyroids? The thyroid?
4. Define endocrine system, hormone, and homeostasis.
Conclusion: How does the body control calcium balance?

CHAPTER REVIEW

SUMMARY

1. Chemical control in animals is carried out by hormones, most of which are secreted from endocrine glands. **p. 586**
2. Insulin, secreted by beta cells of the pancreas, reduces the level of glucose in blood. Diabetes mellitus is a disease characterized by an excess blood glucose level. **28:1**
3. Thyroxine, produced by the thyroid gland, regulates metabolic rate and growth. **28:2**
4. The adrenal medulla secretes adrenaline, which prepares the body for stress. The adrenal cortex produces several hormones that regulate salt and sugar levels and prevent inflammation. **28:3**
5. Estrogens and androgens control female and male secondary sex characteristics, respectively. **28:4**
6. The anterior pituitary secretes several hormones, most of which control the activities of other endocrine glands. **28:5**
7. Control of other endocrine glands by the anterior pituitary involves feedback mechanisms. **28:6**
8. The posterior pituitary secretes vasopressin and oxytocin, hormones produced by the hypothalamus. **28:7**
9. The response to many protein hormones occurs when a second messenger causes activation of enzymes in the target organ. Steroids exert their effect by activating certain genes in cells of the target organ. **28:8**
10. Metamorphosis in insects is regulated by the interaction of three hormones. **28:9**

LANGUAGE OF BIOLOGY

adrenal cortex	human growth hormone
adrenal gland	insulin
adrenaline	islets of Langerhans
adrenal medulla	juvenile hormone

androgen	oxytocin
anterior pituitary	posterior pituitary
beta cells	prostaglandin
cortisone	releasing factor
cretinism	second messenger hypothesis
cyclic AMP	steroids
diabetes mellitus	testosterone
endocrine gland	thyroid gland
feedback mechanism	thyroxine

CHECKING YOUR IDEAS

On a separate paper, match the phrase from the left column with the proper term from the right column. Do not write in this book.

1.	secretes HGH	a.	adrenaline
2.	mechanism controlling hormone production	b.	androgen
		c.	anterior pituitary
3.	involved in diabetes mellitus	d.	estrogen
4.	male sex hormone	e.	feedback
5.	located on membranes of target cells	f.	hypothalamus
		g.	insulin
6.	increases in response to stress	h.	juvenile hormone
7.	produces hormones released by posterior pituitary	i.	receptor
		j.	thyroxine
8.	involved in insect metamorphosis		
9.	contains iodine		
10.	female sex hormone		

EVALUATING YOUR IDEAS

1. Describe the causes and symptoms of diabetes mellitus.
2. What is the role of insulin?
3. What are some functions of thyroxine?
4. What is the relationship between iodine and goiter?

5. What happens to adrenaline secretion during times of stress? What are the effects of adrenaline secretion?
6. Using Table 28–1 as a reference, identify and explain the role of three hormones that regulate blood sugar level.
7. How do adrenaline and the cells of the adrenal medulla indicate a close relationship between the endocrine system and the nervous system?
8. What are the functions of the adrenal cortex hormones?
9. How are hormones important in animal reproduction?
10. FSH and LH have different effects in females and males. What principle about how hormones work does this fact support?
11. Distinguish between the effects of excess HGH and a deficiency of HGH.
12. How is thyroxine secretion controlled?
13. What are the functions of oxytocin and vasopressin? Where are these hormones made?
14. How do the structures and functions of the hypothalamus and posterior pituitary support the idea of a close relationship between the endocrine system and nervous system?
15. Explain how some protein-type hormones recognize their target organs. How is the response of the target organ caused?
16. How are steroid hormones thought to work?
17. List some effects of prostaglandins.
18. How are metamorphosis and molting controlled in insects?

APPLYING YOUR IDEAS

1. Insulin is a protein. Why must it be taken by injection rather than orally?
2. Oxytocin and vasopressin are peptides composed of eight amino acids, only two of which differ. How can their different actions be explained?

3. How would the size of an adult insect be affected if the corpora allata were removed in an early molt? Explain.
4. How would the size of an adult insect be affected if juvenile hormone were added to a larva about to undergo its final molt? Explain.
5. Before insulin can become active, an enzyme must first remove certain parts of the molecule. How might the change from inactive to active insulin be adaptive as a control process?
6. An enzyme in cells converts cAMP to a slightly different compound. How is this change a necessary reaction? (HINT: What might happen in a cell if cAMP built up?)
7. Compare chemical control in animals and plants.

EXTENDING YOUR IDEAS

1. Histamine is a chemical that works like a hormone. Find out about its effects on the body. What is an antihistamine? When is it taken? Why?
2. Use library resources to learn about insulin pumps. What are the advantages of using an insulin pump instead of taking insulin by injection?
3. Find out about a condition called hypoglycemia. How is it caused? How is it treated?

SUGGESTED READINGS

Fellman, Bruce, "A Clockwise Gland." *Science 85,* May, 1985.
Langone, John, "Is It a Disease or Isn't It?" *Discover,* Nov., 1985.
Siwolop, Sana, "Type II Diabetes: A Growing Threat." *Discover,* May, 1984.
Wechsler, Rob, "Unshackled from Diabetes." *Discover,* Sept., 1986.

NERVOUS CONTROL

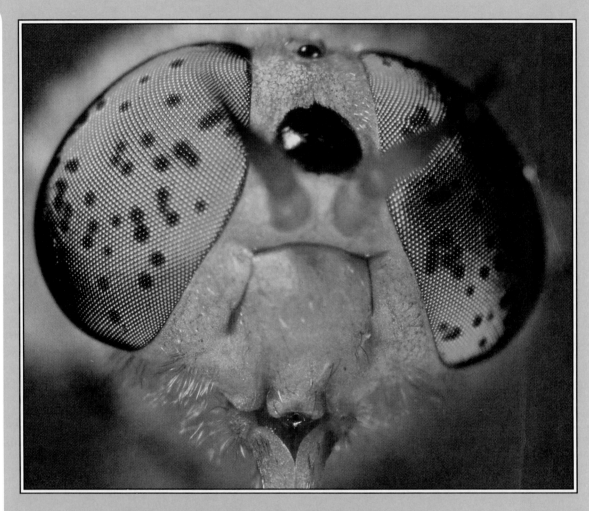

A nimals are bombarded with a broad diversity of environmental stimuli. Adaptations for receiving stimuli aid the survival of each organism. The compound eyes of this insect, which produce a mosaic image, are sensitive to very minor movements. Antennae are sensitive to touch and chemicals. The interpretation of each stimulus is controlled by the nervous system. What is involved in the movement of messages in the nervous system? What kinds of responses can be made?

chapter 29

Many animal responses are chemically controlled by hormones (Chapter 28). Chemical control is efficient and is needed for certain activities, but it is usually a slow process. Consider the responses that must be made to stimuli such as a speeding car or a hot stove. In terms of "survival," reactions to these stimuli must be made quickly. These and many other responses are controlled by the nervous system. Recall that chemical and nervous control are closely linked. You will see further evidence of this close relationship in this chapter.

Objectives:
You will
- explain the transmission of a nerve impulse and how it is passed from one cell to the next.
- describe the effects of several drugs on the nervous system.
- compare the nervous systems of several types of animals.

CONDUCTION

Three elements are needed for a nervous system response to occur. First, there must be a means of detecting a change in the environment (a stimulus). In most animals, stimuli are detected by structures called **receptors** (rih SEP turz). Can you name some receptors? Second, after the stimulus is received, it must be transmitted. A stimulus is transmitted as an impulse along a network of conductors. **Conductors** also transmit impulses to the third necessary element, the effectors (ih FEK turz). **Effectors** are the responding parts. Effectors carry out the correct responses to the stimuli. Muscles and glands are examples of effectors.

Receptors, conductors, and effectors are necessary for a nervous system response to occur.

FIGURE 29–1. Neurons are specialized cells adapted for transmitting impulses. Here, motor neurons of a cat are shown.

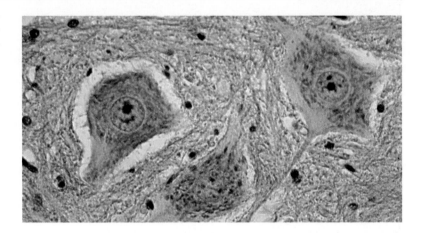

29:1 Neurons

The conductors of a nervous response are specialized cells called **neurons** (NOO rahnz). There are three types of neurons in complex animals. Neurons differ in structure and in the direction in which they carry impulses. **Sensory neurons** transmit incoming impulses from receptors to a coordinating center, usually the brain or spinal cord. **Motor neurons** transmit outgoing impulses from the brain or spinal cord to the effectors. These two types of neurons are joined in the brain and spinal cord by **interneurons.** In simple animals, interneurons may not be involved. A sensory neuron may stimulate a motor neuron directly. In some cases, a single neuron may detect a stimulus and conduct an impulse directly to an effector.

All neurons are specialized for transmitting impulses. Figure 29–2 shows the structure of a vertebrate motor neuron. Such a neuron looks like a cell that has been drawn out into a long, fine fiber. Fibers of many neurons are often grouped together to form a **nerve.** The portion of a neuron most resembling other kinds of cells is the

The three types of neurons are sensory, motor, and interneurons.

FIGURE 29–2. (a) The cell body of a neuron contains the nucleus and most of the cytoplasm. (b) A typical motor neuron is composed of a cell body, an axon, dendrites, and end brushes. Motor neurons transmit impulses to effectors. The red arrow shows the direction that an impulse travels.

a
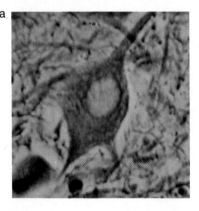

b
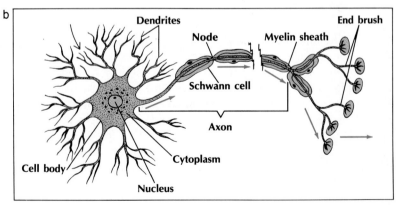

cell body, which contains a nucleus and cytoplasm. The long part of a neuron that transmits impulses *away* from the cell body is the **axon** (AK sahn). The end of the axon is subdivided into many filaments, forming the **end brush.** The other end of a motor neuron is highly branched into fibers called **dendrites** (DEN drites). Impulses are conducted *toward* the cell body from the dendrites. Thus, an impulse flows in a neuron from dendrite to cell body to axon. Then the impulse is transmitted to the dendrite of another neuron or to an effector.

Often, the axon of a neuron has an outer layer, the fatty **myelin** (MI uh lun) **sheath.** In a motor neuron, the myelin sheath comes from the cell membranes of **Schwann cells** that wrap around the axon. Schwann cells aid in the nutrition and regeneration of axons. Between Schwann cells are points where the axon is not covered, called **nodes.** Both the myelin sheath and the nodes are important in determining the speed with which impulses are conducted. Neurons lacking these features transmit impulses much more slowly than cells with a myelin sheath. Also, neurons with larger diameters conduct impulses more rapidly than those with smaller diameters.

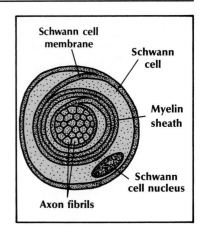

FIGURE 29-3. An axon of a motor neuron is wrapped in a myelin sheath made of Schwann cells.

PEOPLE IN BIOLOGY

Although trained as a physician, biophysicist, and cell biologist, Humberto Fernandez-Moran's first love is neurology and neuropathology—the study of the nervous system and its diseases. Growing up in Venezuela, Fernandez-Moran developed an interest in the brain and central nervous system when he saw the misery caused by tropical diseases that affect the central nervous system. Using electron microscope techniques that he developed, Dr. Fernandez-Moran discovered new brain and nerve traits. One of his main discoveries was that the myelin sheath, the fatty outer layer of some axons, acts much like a semiconductor, a material that can pass along an electrical charge. This idea helps explain why axons with myelin sheaths transmit impulses much more quickly than those without.

Humberto Fernandez-Moran

(1924–   )

Fernandez-Moran has received many awards for his work, including the John Scott Award for inventing the diamond knife, used to slice samples for microscopic study, and the Venezuelan Medalla Andres Bello, the highest award granted to a Latin American scholar. He now works at the Research Institute at the University of Chicago.

The shape, size, and arrangement of neurons vary with their functions.

Many neurons are long, thin, delicate cells, but neuron length varies considerably. A single motor neuron may extend from the spinal cord to the foot of a person, a distance of about a meter. Interneurons in the brain and spinal cord are very short and densely packed. The numbers and lengths of dendrites and axons vary with the type and function of the neurons. Axons vary also in diameter, which affects the rate at which an impulse is conducted. However, regardless of these differences, an impulse is conducted in about the same way in all neurons.

29:2 Membrane Potential

A resting neuron is polarized—there is a charge in fluids outside the membrane different from that inside the cell. What causes this difference in charge? One factor is the relative number of sodium ions (Na^+) and potassium ions (K^+). Outside the cell, the number of sodium ions is high and the number of potassium ions is low. Inside the cell, the number of potassium ions is greater than the number of sodium ions. This difference in number of ions is due to what is called the cell's **sodium-potassium pump.** ATP energy is used to actively transport ("pump") sodium ions out of and potassium ions into the cell. Three sodium ions are pumped out for every two ions that enter the cell. As a result of the pump, more positive ions exist outside the cell (sodium) than inside (potassium). Thus, the outside of the cell is *more* positive than the inside.

The sodium-potassium pump contributes to the polarization of neurons.

Other factors play a role in polarization. The cell membrane is slightly permeable to potassium, and some potassium ions diffuse outward. (The membrane is not permeable to sodium, which remains outside.) This loss of potassium ions adds to the more positive charge outside the membrane. Also, there are many ions with a negative charge inside the cell. These ions cannot cross the membrane; thus, they make the inside of the cell less positive. These many factors cause the outside of the cell to be positive with respect to the inside of the cell (Figure 29–4).

The interior of a neuron is negative with respect to the exterior.

FIGURE 29–4. Unequal distribution of ions inside and outside its membrane polarizes a neuron. Although the inside of the neuron contains positive ions, it can be considered as being negative compared to the outside.

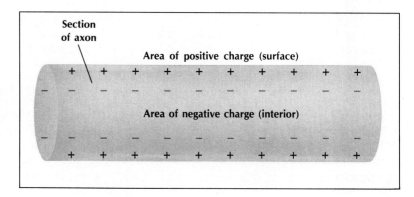

The difference in charge inside and outside a resting neuron is called **membrane potential** or **resting potential.** Membrane potential is common to all cells. Neurons are adapted to use membrane potential in a special way.

29:3 Action Potential

When an axon is excited by stimuli such as light, pressure, chemicals, or an electric current, changes in the resting potential take place. *A stimulus alters the permeability of the cell membrane.* First, the membrane becomes permeable to sodium ions. This occurs as gatelike channels (Section 4:7) open. Because there is a greater concentration of sodium ions outside the membrane than inside, sodium ions rush in, causing a reversal of polarity. The inside of the cell becomes positive with respect to the outside. After this reversal of polarity, the sodium gates close and potassium channels open. Again, because of concentration differences, potassium ions move from the inside of the cell to the outside very quickly. The membrane is once again positive on the outside with respect to the inside. The polarity changes and flow of ions caused by a stimulus results in a current and is called the **action potential.** The action potential, which can be detected by electrical equipment, shows that an impulse is being sent along a neuron.

An action potential in one region of a neuron causes a change of polarity in the next region. In this way, action potentials (currents) move along the whole length of the axon (Figure 29–6). A neuron can send 500 to 1000 action potentials per second. A series of action potentials sweeping down an axon is a **nerve impulse.** In some large neurons, impulses may be sent at the rate of one hundred meters per second.

Notice that after an action potential has passed, the original *polarity,* or charge difference, across the membrane has been restored. However, sodium and potassium ions have been exchanged across the membrane so that more sodium is on the inside and more potassium on the outside. The original *distribution* of ions is restored by the sodium-potassium pump. This pumping of ions to restore the original membrane potential of a neuron occurs in a short period of time, the **refractory** (rih FRAK tree) **period.** During this time — about 1.0 to 1.5 milliseconds — the neuron cannot carry another impulse.

Transmission of a nerve impulse differs from an electric current in a wire. An electric current travels close to the speed of light and involves the flow of electrons along a wire. A current in a nerve impulse is caused by the flow of ions across the cell membrane. Thus, a nerve impulse is an electrochemical reaction. A nerve impulse is like the sputtering of a burning fuse in that it "jumps" along the neuron.

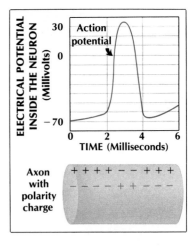

FIGURE 29–5. During an action potential, the polarity of a neuron is first reversed and then restored. Changes in polarity can be recorded and graphed. The graph shows that the inside of the neuron is first negative, then positive, and once again negative.

FIGURE 29–6. Transmission of an impulse occurs as the action potential (polarity change) moves along an axon.

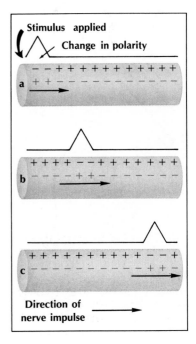

29:4 All-or-None Response

A stimulus must have a certain intensity in order for an impulse to result.

All stimuli do not cause action potentials (impulses). A weak stimulus might generate some changes in a neuron, but unless polarity is reversed, there would be no nerve impulse. A stimulus must have a certain intensity to change polarity in a neuron and start an impulse. This level of intensity is called the **threshold.** The threshold is reached when a stimulus causes a change in membrane permeability and sodium ions rush into the axon.

A stimulus stronger than threshold level will produce the same kind of impulse as a stimulus just at the threshold level. The impulse will travel at the same speed and in the same way no matter how strong the stimulus is. Thus, a nerve impulse is an **"all-or-none" response.** The impulse is either sent or not sent. If impulses are the same whether the stimulus is at or above threshold, how do you detect different strengths of stimuli? This detection depends on the number of impulses carried along a neuron in a given period of time and the number of the same kind of neurons carrying the impulse.

A nerve impulse is an "all-or-none" response.

How do you interpret *what* the stimulus is? How do you know whether you are being stimulated by light or noise or touch? The impulses are all essentially the same. The answer is that what you interpret and respond to depends on the pathway of the nerves and the *target* reached by the impulse. Once a certain pathway begins to carry an impulse, the result will be the same. What happens when you close your eyes and gently tap one eyelid? Do you see that the type of stimulus does not always determine the way you interpret it?

Responses to stimuli depend on which receptors pick up stimuli and the target to which impulses are sent, not the nature of the impulse.

FIGURE 29–7. A synapse between an axon and a muscle cell (effector) is shown. Acetylcholine is released from vesicles in the end of the axon into the synapse in transmitting the impulse.

29:5 The Synapse

An action potential can be passed from a neuron to another cell. This cell is either another neuron or an effector. There is always a small space called a **synapse** (SIHN aps) between any two such cells. An action potential must cross a synapse and can cross in one direction only. A cell that carries an impulse toward a synapse is a **presynaptic** (pree suh NAP tihk) **cell.** A cell that receives the impulse after it crosses the synapse is a **postsynaptic** (pohst suh NAP tihk) **cell.** How does an impulse cross a synapse?

Certain chemicals called **transmitters** are stored in vesicles at the ends of presynaptic axons. As an action potential sweeps toward the end of an axon, calcium ions enter it from outside the membrane. Entry of calcium causes the vesicles to move to the membrane and release the transmitter by exocytosis. The chemical diffuses toward the postsynaptic cell where it combines with receptor molecules. The combination of transmitter and receptor may cause the membrane of the postsynaptic cell to become permeable to sodium. If the mem-

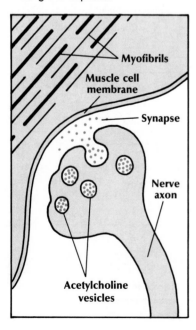

Myofibrils

Muscle cell membrane

Synapse

Nerve axon

Acetylcholine vesicles

brane becomes permeable, sodium ions rush across the postsynaptic cell's membrane, starting an action potential. In this case, the transmitter *excites* the next nerve cell or effector.

At some synapses, the transmitter causes changes that make the fluids outside the postsynaptic cell even more positive. Thus, polarity reversal cannot occur, and no nerve impulse is sent. In this case, the transmitter *inhibits* the next nerve cell or effector.

Acetylcholine (uh seet ul KOH leen) is a common transmitter. Secreted by the vagus nerve that leads to the heart, it inhibits the heart muscle cells, thereby slowing the heartbeat rate. Another common transmitter is **noradrenaline** (nor uh DREN ul un), which is very much like the hormone adrenaline (Section 28:3). The fact that passage of nerve impulses involves chemicals and the likeness between adrenaline and noradrenaline show the close relationship between endocrine and nervous control.

The effect of a given transmitter depends on the postsynaptic cell. Acetylcholine inhibits heart muscle, but it excites alimentary canal and skeletal muscles. Noradrenaline excites heart muscle, inhibits alimentary canal muscle, and has no effect on skeletal muscle. It is the target, not the chemical, that "decides" the response. Often, thousands of presynaptic axons, each with its own transmitter, lead to one postsynaptic cell. Whether the cell is excited or inhibited depends on the net effect of all the transmitters.

Transmitters are suppressed soon after they are released so that the postsynaptic fiber is not always affected. Acetylcholine is changed by an enzyme called cholinesterase (koh luh NES tuh rays) that breaks it into two smaller molecules that have no effect on postsynaptic cells. Noradrenaline is not chemically changed but is "recaptured" by the presynaptic cell that released it.

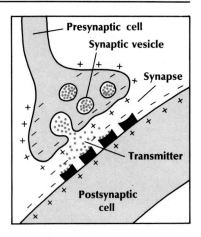

FIGURE 29–8. In transmission from nerve to nerve, a transmitter released at the synapse may cause a polarity reversal in the second neuron. Thus, the impulse is started in the second neuron.

The effect of a transmitter depends on the postsynaptic cell.

Transmitters are inactivated shortly after they are released.

29:6 Effects of Drugs and Other Chemicals

Many drugs affect the nervous system and play an important role in maintaining both physical and mental health. Used properly, these drugs may cure or treat diseases or relieve pain. If abused, though, these drugs may cause severe harm or death.

Depressants are a major group of drugs that reduce nerve transmission in the brain. Tranquilizers have this effect and are often used as calming medicines. These depressants have chemical structures very similar to those of some brain transmitters and often work by blocking impulses in the brain. Some tranquilizers decrease transmission at synapses. Others block receptors so the number of impulses transmitted is lowered. The effect of these drugs is a calming one.

Barbiturates (bar BIHCH uh ruts) are the most powerful and most often abused depressants. These drugs can be used to make a person

Depressants reduce activities of the central nervous system.

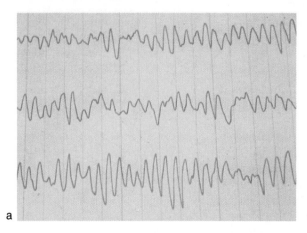
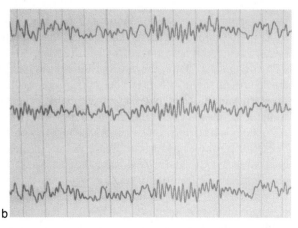

a
b

FIGURE 29–9. An electroenceph-
alogram (EEG) gives a means to
measure brain function. (a) Normal traces are much different from
(b) those of a patient taking stimulant drugs.

Cocaine is a local anaesthetic. It is
also a stimulant.

Use of cocaine can lead to depression, damage to certain tissues,
and even death.

sleep. By first blocking brain impulses and then blocking impulses throughout the body, these drugs reduce thought, memory, and awareness of light and sound.

Other depressants are morphine, codeine, heroin, and alcohol. Morphine is a painkiller that is thought to reduce the amount of transmitters released at synapses. Codeine, widely used in cough syrups, reduces the cough reflex. Heroin, a drug made from morphine, dulls pain, but because it is addictive, it is seldom used in medicine. Alcohol (C_2H_5OH) interferes with transmission at synapses.

Stimulants are drugs that increase activity in the nervous system. They make a person more alert, prevent sleep, and control appetite or depression. Caffeine, found in coffee, tea, and cola drinks, and nicotine, found in cigarette smoke, are stimulants. Amphetamines (am FET uh meenz), sometimes given to obese people for appetite control, are stimulant drugs that are often abused.

Cocaine is a particularly dangerous stimulant. Because it is a local anaesthetic, it numbs the part of the body it touches, for example the nose or throat. Its numbing effects are due to the fact that it blocks the channels through which sodium ions enter neurons. Cocaine travels in the blood to the brain and other parts of the nervous system, where it has a different effect. It prevents presynaptic cells from recapturing noradrenaline transmitters. Thus, postsynaptic cells continue to be stimulated. This constant stimulation mimics the effects of adrenaline (Section 28:3) during stress. Soon after taking cocaine, a person feels strong, confident, and mentally sharp. After the effects of cocaine wear off, a person often feels depressed. Cocaine seems to be addictive. Long-term use has been shown to damage areas of the nose and throat in those who snort cocaine and the lungs of those who smoke it. Prolonged use also induces hallucinations and mental illness. Although rarely so, cocaine can be fatal, even to first-time users. Death may result from respiratory failure, seizures, allergic reactions, and heart attacks.

Other drugs also affect the nervous system. Marijuana leaves contain drugs that change the amounts of transmitters in the brain. Nervous system enzymes and electrical activities are affected by marijuana use. LSD resembles a certain brain transmitter. LSD acts like this chemical in the brain but cannot forward signals as the real transmitter can. Thus, impulses are blocked and sensory perception is distorted when LSD is used.

REVIEWING YOUR IDEAS

1. List the three elements needed for a nervous response.
2. List the three types of neurons and their functions.
3. Distinguish between dendrite and axon.
4. How do neurons become polarized?
5. Distinguish between action potential and nerve impulse.
6. What is a synapse?
7. Distinguish between stimulants and depressants.

NERVOUS SYSTEMS

Except for sponges, all animals have a means of nervous control. In these animals neurons, and usually other nerve tissues, are organized to form definite nervous systems.

29:7 Invertebrates

Coelenterates such as *Hydra* have randomly scattered neurons that form a system called a **nerve net** (Figure 29–10). All the neurons are short and similar. Because neurons are short, an impulse must cross many synapses. In a coelenterate, impulses can travel in either direction across synapses. But, transmission is slow. Only strong stimuli cause impulses that are easily sent across many synapses. The impulses of many weak stimuli do not travel very far in a hydra. Thus, many responses involve only part of the body, such as a tentacle. However, responses are sometimes made by the whole animal. For example, a hydra can contract into a ball if disturbed.

Planarians and other flatworms have more complex systems. At the anterior end of a planarian are two bundles of neuron cell bodies called **ganglia** (GANG glee uh), which act like the brain of more complex animals. Two major nerve cords from the ganglia run the length of the body, and connecting nerves join the main cords. Smaller neurons branch from the nerve cords and act much like a nerve net. Some biologists refer to this arrangement of nerves as a **ladder-type nervous system** (Figure 29–10).

FIGURE 29–10. (a) *Hydra* has a nerve net nervous system. (b) *Planaria* has a ladder-type nervous system with the beginnings of central control in the ganglia of the head region.

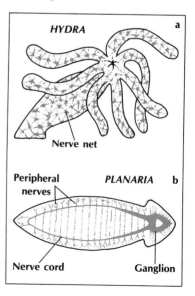

Most bilaterally symmetrical animals have a central and a peripheral nervous system.

The plan of the planarian's nervous system is similar to that of more complex animals with bilateral symmetry. Such animals have both a central and a peripheral (puh RIHF rul) nervous system. The **central nervous system** (CNS) consists of a brain and one or more longitudinal nerve cords. The **peripheral nervous system** is composed of sensory and motor neurons connected to the CNS. The peripheral system, then, consists of nerves that join the CNS with receptors and effectors.

Invertebrates such as annelids and arthropods are more complex than planarians. In these animals, the brain is located dorsally and joined to two ventral nerve cords. During the evolution of invertebrates, the brain became larger and began to exert more control.

29:8 Vertebrates

The central nervous system of a vertebrate consists of the brain and a dorsal spinal cord.

In vertebrates, there is one nerve cord located along the dorsal side of the animal. That nerve cord is called the **spinal cord** and is surrounded and protected by vertebrae. The spinal cord is an extension of the brain; together they make up the central nervous system. The peripheral nervous system of vertebrates is made up of sensory and motor neurons that carry impulses to and from the central nervous system. In humans, the peripheral nervous system is composed of twelve pairs of **cranial nerves** (attached to the brain) and thirty-one pairs of **spinal nerves** (attached to the spinal cord).

The human peripheral nervous system is composed of cranial and spinal nerves.

The spinal cord is involved in many responses, but the brain is the major coordinating center in vertebrates. Vertebrate brains differ in appearance (Figure 29–11), but each has three general areas — **forebrain, midbrain,** and **hindbrain.** Each area of the brain controls certain functions. The functioning of the brain is very complex and far from being completely understood. Much is yet to be learned about

FIGURE 29–11. Although all of these vertebrate brains have the same three regions (forebrain, midbrain, and hindbrain), the regions have different structures and functions.

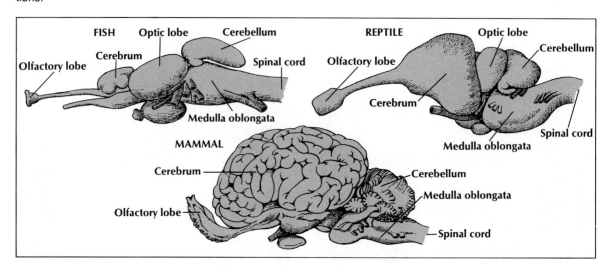

how the brain controls the body and about the exact functions of the many parts of the brain. The following sections discuss the human nervous system.

29:9 Forebrain

A human brain is protected by the cranium and three membranes called the **meninges** (muh NIHN jeez). These membranes protect and nourish both the spinal cord and brain. Between the inner two meninges is **cerebrospinal** (suh ree broh SPINE ul) **fluid.** This clear fluid cushions the brain against shock, but a severe shock can sometimes damage the brain. A concussion is a brain injury resulting from a severe blow to the head. Cerebrospinal fluid is also in the ventricles (VEN trih kulz) (cavities) of the brain. Capillaries in the walls of the ventricles exchange materials with the fluid that, in turn, exchanges materials with the brain cells. As in all vertebrates, the human brain is composed of a forebrain, midbrain, and hindbrain.

The human brain is protected by the cranium and meninges.

Cerebrospinal fluid cushions the brain. It is an exchange medium between brain cells and blood.

FIGURE 29-12. (a) Scanners such as this one are used to do brain scans to detect brain dysfunction. (b) This scan is a cross sectional view of the brain of a stroke patient. The red area of the scan is an area of the brain with poor circulation.

a

b
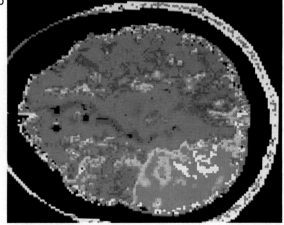

FIGURE 29–13. (a) The human brain is composed of three major regions—forebrain, midbrain, and hindbrain. (b) Each hemisphere of the cerebrum has four lobes.

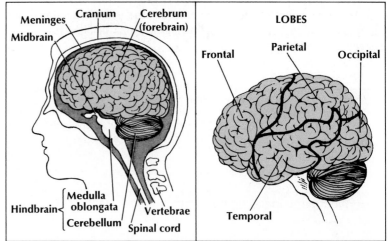

The cerebrum occupies most of the human forebrain.

The major part of the forebrain of humans is the **cerebrum** (suh REE brum), which consists of two hemispheres. Each hemisphere consists of four lobes (Figure 29–13). The surface of the cerebrum, the **cerebral cortex,** has many folds called **convolutions** (kahn vuh LEW shunz). The cortex is made up of cell bodies that together are known as **gray matter.** Convolutions provide a greater surface area for the nerve cell bodies. The great number of cell bodies is important in complex behavior and intelligence. The interior of a human brain is **white matter** made up of axons with myelin sheaths.

The cerebrum functions in speech, reasoning, emotions, and personality.

Many human abilities are related to the very large cerebrum. Speech, reasoning, emotions, and personality are all functions of this area of the brain. The cerebrum also interprets sensory impulses and starts motor impulses (Table 29–1).

The thalamus and hypothalamus are part of the forebrain.

Also in the forebrain are the thalamus and hypothalamus. The thalamus aids in sorting sensory information. The hypothalamus has many roles such as control of hunger, body temperature, and aggression. It also plays an important role in endocrine control.

FIGURE 29–14. Different areas of the cerebral cortex are activated when one reads silently. Stimulation of certain regions is indicated by increased blood flow. The yellow to red areas show above average blood flow. The green areas show average blood flow. Below average blood flow is shown by the blue areas.

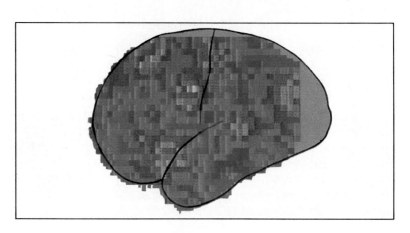

TABLE 29-1. MAJOR PARTS OF THE HUMAN BRAIN AND THEIR MAIN FUNCTIONS	
Section	Functions In
FOREBRAIN (Cerebrum)	
Temporal lobe	Taste and smell, hearing, learning and memory
Parietal lobe	Sensory input, touch
Occipital lobe	Vision, motor output, speech
Frontal lobe	Personality, learning, thought, speech
MIDBRAIN	Relay center
HINDBRAIN	
Cerebellum	Coordinating impulses, posture and balance, motor coordination, muscle tone
Medulla oblongata	Heartbeat rate and other internal control

29:10 Midbrain and Hindbrain

In less complex vertebrates, the midbrain is prominent. But in humans and other mammals, it mainly functions as a message station between the forebrain and hindbrain. The midbrain also is involved in some sight, hearing, and orientation responses.

There are two distinct areas of the hindbrain — the cerebellum (ser uh BEL um) and the medulla oblongata. The **cerebellum** is a region lying at the back of the head below the cerebrum. Like the cerebrum, the cerebellum has convolutions. Its outer surface is made up of gray matter, and the inside is white matter. The cerebellum coordinates impulses sent out from the cerebrum. It receives sensory impulses from certain receptors. If the sensory information and cerebral impulses do not agree, the cerebellum sends messages to the cerebrum and the cerebral impulses change. The cerebellum also controls posture and balance and maintains muscle tone (Section 30:8).

Extending down from the center portion of the brain is the medulla oblongata (Section 26:6). The medulla controls many involuntary responses of the internal organs. For example, breathing rate and heartbeat rate, peristalsis, and some gland secretions are controlled by this part.

29:11 The Spinal Cord

Extending down from the medulla oblongata is the long spinal cord with a fluid-filled central canal. The spinal cord is surrounded and protected by the meninges (Section 29:9) and a series of vertebrae. Like the brain, the spinal cord is composed of white and gray matter (Figure 29–15). The outer part of the spinal cord is white matter, and the inner part is gray matter, an arrangement opposite of that in the brain.

The human midbrain functions mainly as a message center.

In humans, the cerebellum coordinates impulses leaving the cerebrum and controls posture, balance, and muscle tone.

FIGURE 29–15. This section of spinal cord shows its location relative to a vertebra and spinal nerves. Spinal nerves bring sensory information to the spinal cord and carry information for motor responses to the effectors.

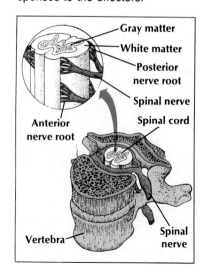

Gray matter
White matter
Posterior nerve root
Spinal nerve
Anterior nerve root
Spinal cord
Vertebra
Spinal nerve

FIGURE 29–16. The path of a reflex arc is receptor to interneuron in the spinal cord to effector.

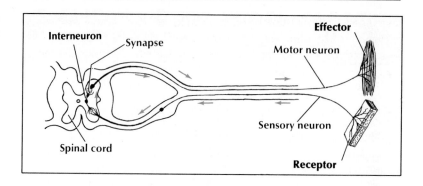

The spinal cord is a relay station between the brain and the peripheral nervous system.

A reflex involves no conscious control.

A reflex arc is the path of an impulse during a reflex.

In humans, the spinal cord is a major connecting center between the brain and the peripheral nervous system. The spinal nerves are part of the peripheral system. They connect the body parts with the spinal cord. Spinal nerves enter and leave the spinal cord through nerve roots. Spinal nerves contain both sensory and motor neurons. Impulses reach the spinal cord from receptors via sensory neurons and then travel to interneurons in the spinal cord. The impulses may be transferred to the brain for interpretation. Impulses are then sent to the internal organs or the skeletal muscles by motor neurons.

Certain responses to either internal or external stimuli do not involve a "decision" by the brain. Such a response is called a reflex. A **reflex** is a simple, automatic response that involves no conscious control. In some reflexes such as blinking, the brain acts as the control area. In other reflexes, the spinal cord acts as the control center. For example, consider the response made to the stimulus of stepping on a sharp rock while walking barefoot. Before you are aware of what has happened, certain leg muscles have contracted, lifting your foot from the rock. How is this response made?

Receptors in the skin of the foot are activated by the pressure of the rock and the pain it causes. Each receptor stimulates a sensory neuron and causes an impulse to travel to the spinal cord via a spinal nerve. The sensory neuron enters the spinal cord (Figure 29–16) and crosses a synapse to an interneuron. The impulse travels along the interneuron and then passes across a synapse to a motor neuron. The motor neuron leaves the spinal cord and branches to form synapses with several muscle cell effectors. Acetylcholine is released across the synapses, causing the muscle to contract.

The path of the impulse in a reflex is called a **reflex arc.** A reflex arc includes five parts—receptor, sensory neuron, interneuron, motor neuron, and effector.

The reflex just studied takes place within a fraction of a second and is not controlled by the brain. However, some spinal cord interneurons synapse with other neurons leading to the brain, and the brain becomes aware of what has happened. Then secondary responses occur. A loud "ouch" would be a common secondary response to the stimulus of the sharp rock. Do other animals make secondary responses? Can reflexes be modified? Consider the blinking response. How can you modify this response?

29:12 Peripheral Nervous System

The sensory and motor neurons of the spinal and cranial nerves are part of the human's peripheral nervous system. Some neurons of spinal and cranial nerves carry impulses from receptors to the CNS; others carry impulses from the CNS to the skeletal muscles. They form a part of the peripheral system called the **sensory-somatic system.** Both reflex and voluntary responses are part of this system, and the person is aware of the response.

Other neurons of spinal and cranial nerves control responses of the internal body organs. These neurons form a division of the peripheral system called the **autonomic system.** Autonomic responses are involuntary and involve smooth muscle, cardiac muscle, and glands. The person is often unaware of the response.

The autonomic system is further subdivided into two divisions — the sympathetic (sihm puh THET ihk) system and the parasympathetic (per uh sihm puh THET ihk) system. The **sympathetic system** often begins responses that prepare the body for "emergencies." These responses are similar to those resulting from secretion of adrenaline. That is, these responses speed up or strengthen body functions. The accelerator nerves to the heart are sympathetic nerves. Other sympathetic nerves cause the adrenal medulla to secrete adrenaline, the pupils to become larger, and glycogen to be changed to glucose. Sympathetic reflex arcs involve two motor neurons. The first begins in the spinal cord. It synapses with the second neuron in a ganglion outside the spinal cord. The second neuron leads to the effector. Sympathetic motor neurons secrete noradrenaline.

The sensory-somatic system controls impulses between receptors, the central nervous system, and the skeletal muscles.

The autonomic nervous system controls involuntary responses of the internal organs.

Nerves of the sympathetic system cause responses that prepare the body for emergencies.

FIGURE 29–17. Each division of the nervous system controls different parts of the body.

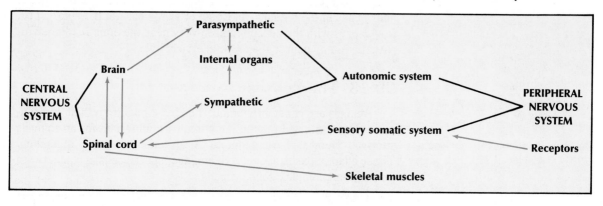

FIGURE 29–18. The autonomic nervous system controls involuntary responses. Sympathetic nerves prepare the body for emergencies whereas parasympathetic nerves counteract these effects. Those processes that sympathetic nerves speed up, parasympathetic nerves slow down.

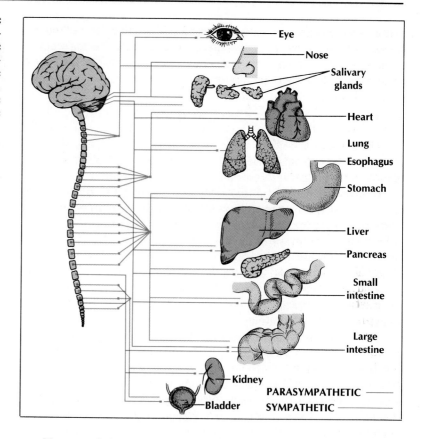

Nerves of the parasympathetic system cause impulses that balance the effects of the sympathetic system.

Nerves of the **parasympathetic system** carry impulses that counteract the effects of sympathetic nerves. This balancing is important in returning the body to normal and maintaining homeostasis. For example, one branch of the parasympathetic vagus nerve slows heartbeat rate, thus countering the effect of the accelerator nerve. Other parasympathetic nerves slow adrenaline production, reduce the size of pupils, and constrict certain blood vessels, thus slowing blood flow. Parasympathetic reflex arcs also involve two motor neurons. The first neuron lies in the medulla oblongata or at the base of the spinal cord. It synapses with the second neuron very near the effector. Parasympathetic motor neurons secrete acetylcholine.

29:13 The Senses

Humans are adapted for sensing a variety of stimuli.

Animals are constantly bombarded by stimuli from their environment. Detecting and reacting to stimuli are important for an animal's survival. Humans have adaptations to sense a variety of stimuli. These adaptations include nerve endings, special cells in contact with nerves, and complex sense organs.

There are five major human senses—vision, hearing, smell, taste, and touch. Touch is a general term that includes several related senses such as pain, pressure, heat, and cold. Each of these "touch" senses is detected by separate skin receptors. Each kind of receptor detects only one sense. The number and location of touch receptors around the body varies. For example, there are more cold receptors than heat receptors, and there are more touch receptors in the fingertips than in the legs. Whether detected by a simple nerve ending or special cells in a complex organ, sensory information causes action potentials in neurons. Each receptor or sense organ sends impulses to particular parts of the brain where they are interpreted. Correct responses can then be made.

As an example of how a sense organ works, consider the ear. The ear consists of three areas — outer ear, middle ear, and inner ear. The outer ear consists of the outer flap, the auditory (hearing) canal, and the eardrum or **tympanic membrane.** The middle ear begins on the inner side of the eardrum and consists of a chain of three small bones commonly called the hammer, anvil, and stirrup. The stirrup contacts a second membrane called the **oval window.** Also in the middle ear is the **eustachian tube,** a canal that connects the middle ear to the nasal passages and throat. The eustachian tube contains air and maintains equal air pressure between the middle and outer ear. The inner ear contains two sets of fluid-filled canals. One set controls the sense of balance and is not involved in hearing. The other, the **cochlea,** is involved in hearing. Part of the cochlea is in contact with the oval window. The cochlea contains special hair cells that make up the organ of Corti. These hair cells are each connected to fibers of the auditory nerve.

How does hearing occur? Sound waves, vibrations of air molecules, enter the outer ear, pass through the auditory canal, and cause the eardrum to begin vibrating. The vibrations are picked up by the middle ear bones, which amplify the vibrations and transfer them to the oval window. The oval window then vibrates, causing the fluid in the cochlea to move. Movement of the fluid within the cochlea causes the sensitive hairs to bend. When the hairs bend, action potentials are

FIGURE 29–19. Taste, one of the five major human senses, is perceived by taste buds located throughout the mouth and throat. This taste bud is shown magnified about 1900 times.

The cochlea is a fluid-filled chamber in the inner ear.

Sound waves cause the eardrum to vibrate. The middle ear bones and oval window then vibrate.

Movement of fluid in the cochlea causes the hair cells to bend, setting off nerve impulses that travel to the brain.

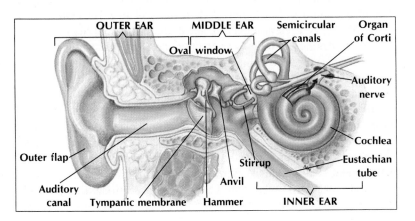

FIGURE 29–20. The human ear is a complex sense organ which receives impulses and passes them to the brain for interpretation.

created in the auditory nerve fibers. Impulses are sent along the auditory nerve to the temporal lobe of the cerebrum where they are interpreted as sound. The particular sound you hear depends on the frequency of the vibrations entering the ear, the number of vibrations, and the number and position of hairs stimulated in the cochlea.

REVIEWING YOUR IDEAS

13. What is cerebrospinal fluid? What are its functions?
14. What is the cerebrum? How is it subdivided?
15. Distinguish between white matter and gray matter.
16. Distinguish between a reflex and a reflex arc.
17. What are the two divisions of the peripheral nervous system?

ADVANCES IN BIOLOGY

Neuropeptides

In the last twenty years, great advances have been made in the study of brain chemistry. Many substances thought to exist only in tissues of the digestive system have now been found in the brain and other nervous tissues. In addition, some hormones have been found to play a role in brain functions, and newly-discovered groups of brain chemicals are being studied. All these chemicals are called neuropeptides. **Neuropeptides** are chains of amino acids made in the hypothalamus and pituitary.

Some pituitary hormones have been found to have effects in the brain as well as on their target organs.

Research on neuropeptides is leading to a new understanding of the link between the endocrine and nervous systems. For example, vasopressin and oxytocin are known to function as hormones (Section 28:7). They are produced in the hypothalamus and secreted from the posterior pituitary. In addition to their roles as hormones, these neuropeptides may influence learning and memory. Vasopressin seems to improve learning. Oxytocin has an opposite effect, reducing memory. ACTH is an anterior pituitary hormone that controls the release of adrenal cortex hormones (Section 28:3). It is now thought that ACTH can also improve learning and memory.

Certain pituitary hormones act as transmitters in the nervous system.

The newly-discovered roles of these chemicals suggest that they work not only as hormones but also as transmitters in the nervous system. For example, certain axons from the hypothalamus extend into the posterior pituitary where they secrete oxytocin to be used as a hormone. Other axons lead from the hypothalamus to other parts of the brain and spinal cord where they release oxytocin thought to be used as a transmitter.

Most transmitters, such as acetylcholine and noradrenaline, either excite or inhibit postsynaptic fibers and are then quickly inactivated. Neuropeptides seem to act differently. They do not cause a change in the resting potential of a postsynaptic fiber. However, they may affect the way the target cell responds to other transmitters. Also, some neuropeptides are not quickly inactivated, and their effects may be long lasting. There is evidence that some neurons secrete both neuropeptides and other transmitters.

Some neuropeptides seem to play many roles in the body's natural ability to perceive and cope with pain. Substance *P* is a neuropeptide found to be involved in the relaying of pain messages to the brain. Other neuropeptides seem to inhibit these pain messages. They are the **endorphins** (en DOR fihnz) and **enkephalins** (en KEF uh lihnz). The endorphins reduce sensitivity to pain and affect emotions. Enkephalins are also pain-relieving chemicals and are referred to as "natural opiates."

Recent research has suggested a link between neuropeptides and the immune system. It has been shown that endorphins and substance *P*, for example, affect activity of macrophages. If neuropeptides are related to both mood and immunity, that might explain the finding that people under stress are more apt to become sick and that people who lose hope often do not recover. People who cope well with stress and are more positive tend to fare better in defending against or fighting off disease. Perhaps the level of neuropeptides in these people keeps their immune system working more efficiently.

It seems clear that neuropeptides may prove to have a variety of practical applications. Their use as painkillers, as substitutes for addictive drugs, and in immunology and learning may all be possible as a result of future research.

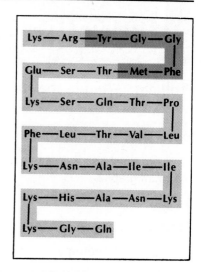

FIGURE 29–21. A model of enkephalin is shown within the larger molecule, endorphin, that contains it. Enkephalins have been found to counteract pain.

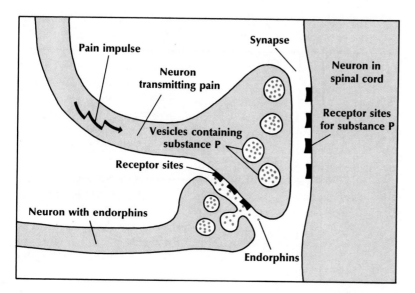

FIGURE 29–22. The neuropeptide substance *P* relays a pain message to the brain. Other neuropeptides, endorphins, latch onto receptor sites of the pain-transmitting neuron. This reduces the release of substance *P* and stops the pain.

Problem: What is the function of certain brain parts?

Materials:

11 colored pencils
cerebrum graphs
brain function chart

Procedure:

Part A. Brain Anatomy

1. Make a copy of Table 29–2. Use this table to record your data.
2. Observe Figure 29–23. It shows a side view of the human brain. Locate each of the following three regions. Each region is indicated by a bracket and a number.
   (a) cerebrum — top portion of brain, largest region
   (b) cerebellum — smaller portion toward lower back of brain
   (c) medulla oblongata — extension of central portion of brain directly above the spinal cord
3. In Table 29–2 put the name of each area beside the number it corresponds with in Figure 29–23.

4. The cerebrum, the largest area of the brain, is composed of four lobes. Each lobe is outlined with dashed lines and a number in Figure 29–23. Locate each of the following lobes of the cerebrum.
   (a) frontal lobe — area toward front of cerebrum, largest of lobes
   (b) occipital lobe — area toward back and bottom of cerebrum
   (c) parietal lobe — area toward top and back of cerebrum, between frontal and occipital lobes
   (d) temporal lobe — area of lower front cerebrum, corresponds to the area of your temple
5. In Table 29–2 put the name of each lobe beside the number it corresponds with in Figure 29–23.
6. Figure 29–24 shows a top view of the cerebrum. Note that it is divided into left and right sides. Locate each of the following parts of the cerebrum.
   (a) left cerebrum half — the half on the left side of the body

FIGURE 29–23.

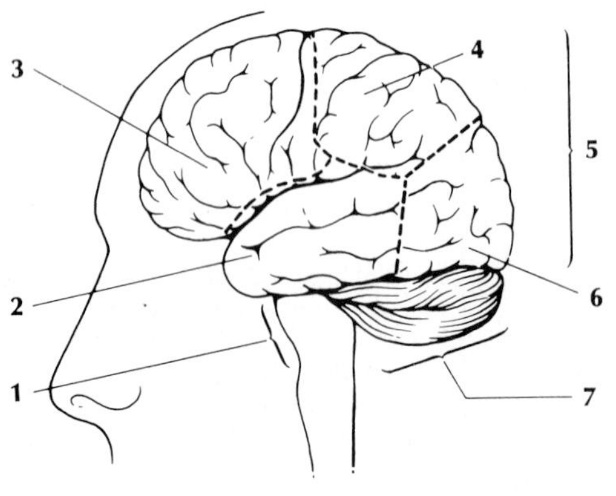

FIGURE 29–24.

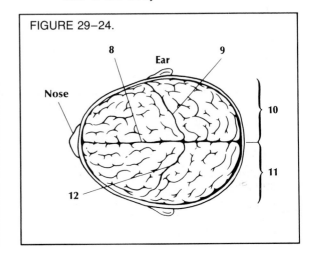

(b) right cerebrum half—the half on the right side of the body
(c) longitudinal fissure—main line or depression that divides the left and right sides
(d) left central fissure—depression that divides the left side into front and back halves
(e) right central fissure—depression that divides the right side into front and back halves

7. In Table 29–2 put the name of each cerebrum part beside the number it corresponds with in Figure 29–24.

Part B. Cerebrum Functions

1. Obtain a chart from your teacher marked Cerebrum Functions. Also obtain two graphs marked Cerebrum Graphs. The chart lists some of the jobs carried out by the cerebrum. The chart also notes the area where each function is controlled. These areas can be located on the cerebrum graphs and then colored.

2. For example, to find where the movement of the right leg and right side of the body is controlled, look at the *left* side of the cerebrum. Locate the letter M and move downward to square 2.

3. Then locate the letter N and move downward to squares 1 and 2.

4. Finally, locate the letter O and move downward to squares 1 and 2. The area covered by these five squares controls the movement of the right leg and right side of the body.

5. These five squares should now be colored solid red and the line extending into this area should be labeled "movement of right leg and right side of body."

6. Using the chart, complete the cerebrum graphs by locating, labeling, and coloring each listed function and its specific area.

Data and Observations:

TABLE 29-2. BRAIN ANATOMY	
Area	Name
1	
2	
3	
4	
5	
6	
7	
8	
9	
10	
11	
12	

Questions and Conclusion:

1. In relation to the central fissure as well as the left and right cerebrum sides, describe the location of the brain tissue that controls each of the following functions.
   (a) muscle movement on right side of body
   (b) muscle movement on left side of body
   (c) sensations on right side of body
   (d) sensations on left side of body
2. Cite evidence that:
   (a) most functions are evenly distributed on both sides of the brain.
   (b) certain functions are controlled by only one brain side.
3. Damage to each of the following areas would probably interfere with what normal functions?
   (a) right cerebrum side: L–4, 5
   (b) right cerebrum side: V–12, 13, 14
   (c) left cerebrum side: L–7, 8
   (d) left cerebrum side: H–8, 9

Conclusion: What is the function of certain brain parts?

CHAPTER REVIEW

SUMMARY

1. A nervous response involves receptors, conductors, and effectors. **p. 603**
2. There are three types of conductors—sensory neurons, interneurons, and motor neurons. **29:1**
3. During the resting potential, the outside of the membrane is positive with respect to the inside. **29:2**
4. Nerve impulses result from a reversal of polarity and a current caused by a series of action potentials. **29:3**
5. A nerve impulse is an all-or-none response. Response to and interpretation of an impulse depends on the target organ, frequency of stimulation, the number of neurons carrying the same kind of impulse, and the kind of transmitter acting. **29:4**
6. Transmitters are needed for passage of an impulse across a synapse. They may either cause or block impulses. **29:5**
7. Many drugs interfere with the normal activities at synapses. **29:6**
8. Coelenterates have a nerve net consisting of many short neurons and lacking central control. More complex invertebrates have a central and a peripheral nervous system. **29:7**
9. In vertebrates, the central nervous system consists of the brain and spinal cord. The peripheral system is composed of cranial and spinal nerves. **29:8**
10. The cerebrum, the major section of the human forebrain, controls motor output as well as speech, reasoning, and emotions. The thalamus and hypothalamus are also parts of the forebrain. **29:9**
11. The human midbrain serves as a relay station between brain parts. The human hindbrain consists of the cerebellum, which coordinates voluntary muscle activity, and the medulla oblongata, which regulates many involuntary responses of internal organs. **29:10**
12. The spinal cord is a link between the brain and the peripheral nervous system. Reflexes may be controlled by either the spinal cord or the brain. **29:11**
13. The sensory-somatic system controls voluntary responses and some reflexes. The autonomic nervous system controls reflexes involving internal organs. **29:12**
14. Stimuli are detected by receptors and specialized sense organs. **29:13**
15. The study of neuropeptides and their functions may lead to a variety of practical applications. **p. 621**

LANGUAGE OF BIOLOGY

action potential	motor neuron
all-or-none response	nerve impulse
autonomic system	neuron
axon	peripheral system
cell body	receptor
central nervous system	reflex
cerebellum	resting potential
cerebrum	sensory neuron
conductor	sensory-somatic
dendrite	system
effector	synapse
interneuron	transmitter

CHECKING YOUR IDEAS

On a separate paper, complete each of the following statements with the missing term(s). Do not write in this book.

1. Involuntary responses of internal organs are controlled by the _____ nervous system.
2. A postsynaptic cell is affected by _____ released from presynaptic cells.
3. In humans, the _____ is the part of the brain that controls intelligence.
4. Coelenterates have a nervous system called a(n) _____.

5. During an action potential, _____ ions rush into a neuron.

6. _____ neurons carry impulses toward an effector.

7. _____ are drugs that increase nervous system activity.

8. The brain and spinal cord make up the _____ nervous system.

9. Stimuli are detected by _____.

10. In humans, the _____ coordinates impulses and maintains balance and muscle tone.

EVALUATING YOUR IDEAS

1. How are the nervous system and endocrine systems different and alike?

2. Explain functions and interrelationships of receptors, conductors, and effectors.

3. How is the structure of a motor neuron suited to its function?

4. In what direction do impulses travel along a vertebrate neuron?

5. How do the myelin sheath and nodes affect nerve transmission?

6. Describe how a neuron carries an impulse.

7. What is the refractory period? Why is the refractory period necessary?

8. Compare a nerve impulse to an electric current in a wire.

9. Explain the nerve's all-or-none response.

10. Explain the factors that determine how you interpret the strength and nature of a stimulus.

11. How does an impulse cross a synapse to a postsynaptic cell?

12. Do transmitters always cause an action potential in the postsynaptic cell? Explain.

13. How does transmission across a synapse show the relationship between the endocrine system and the nervous system?

14. How are transmitters inactivated?

15. Explain several ways that drugs affect the nervous system.

16. Describe the nervous system of *Hydra*.

17. How is the human brain protected?

18. Describe the functions of the human cerebrum.

19. What is the function of the human midbrain?

20. What are the roles of the cerebellum and medulla oblongata?

21. How is the spinal cord protected?

22. What are reflexes? How are they adaptive?

23. What are the roles of the sensory-somatic system?

24. What does the autonomic nervous system control? What are the two subdivisions of this system?

25. Discuss how hearing occurs in humans.

26. Identify and state the roles of several neuropeptides.

APPLYING YOUR IDEAS

1. How is the carrying of impulses in only one direction adaptive?

2. Usually a stimulus must be repeated before transmitters are secreted across a synapse. How is this delay adaptive?

3. A person's head often feels "stuffy" as an airplane ascends or descends. What causes this feeling?

EXTENDING YOUR IDEAS

1. You can use many animals to test reflexes. Experiment with animals such as hydras and planarians and with more complex animals such as frogs. Expose them to several stimuli and note responses.

SUGGESTED READINGS

Kiester, Edwin, Jr., "Spare Parts for Damaged Brains." *Science 86,* March, 1986.

McKean, Kevin, "Pain." *Discover,* Oct., 1986.

SUPPORT AND LOCOMOTION

Most animals have some means of supporting their body weight. Often this support is in the form of an internal or external skeleton. Interactions between muscular and skeletal systems make most animals capable of locomotion. The muscles and bones of this white pelican work together to help the bird land in the river. How do your muscles and bones work together?

chapter 30

In animals, a stimulus is detected by a receptor. Once detected, the stimulus causes a nervous impulse to be conducted along one or several neurons and finally to be passed to an effector. The effector carries out the response to the stimulus.

Most responses made by animals involve movement and, in those responses, the effectors are muscles. **Muscles** are tissues that can contract. The action of muscles may result in movement of one part of the animal, or in locomotion, movement of the entire animal. Locomotion is a feature of almost all animals and is related to a heterotrophic way of life. Locomotion is needed to obtain food and to avoid becoming food.

Although some cannot move, all animals require some kind of support. Without support, an animal would collapse from its own body weight. Muscles function not only in movement, but also in support. In many animals a skeletal system also aids in support. In addition, the skeleton of an animal provides protection. In the most complex animals, the muscles and skeleton work together in locomotion.

Objectives:
You will
- compare the structures and functions of external and internal skeletons.
- explain how movement and locomotion occur in several invertebrates and in vertebrates.
- describe the structure and explain the contraction of vertebrate skeletal muscle.

In many animals, muscles and skeleton work together to provide support and locomotion. The skeleton also functions in protection.

SKELETAL SYSTEMS

Most invertebrates lack a true skeletal system. Sponges and echinoderms have mineral deposits that aid in support and protection, but not locomotion. Most mollusks have a hard outer shell for protection, but it is not a skeleton. Coelenterates and the various worm phyla also lack skeletal systems. Almost all these animals live in and are partly supported by water. Although many have muscles, most are sessile or slow moving. In contrast, the two most active groups of animals, arthropods and vertebrates, have both muscular and skeletal systems.

Animals lacking skeletal systems are most often aquatic and either sessile or slow moving.

a

b

FIGURE 30–1. Arthropods have exoskeletons and periodically molt. (a) The exoskeleton of a beetle is a hard outer skeleton. (b) The tarantula at the right has just crawled from its old skeleton at the left.

Arthropods have an exoskeleton and flexible joints.

An exoskeleton limits size and movement.

FIGURE 30–2. Flexible joints allow for movement of arthropod body parts. The joints of praying mantis legs are shown.

30:1 The Exoskeleton

Arthropods have a hard outer layer made up of a protein and carbohydrate material called chitin (Section 16:1). This external skeleton is called an exoskeleton (Figure 30–1). An exoskeleton is a tough, rigid structure that is subdivided by soft, flexible **joints**. Joints are located between body segments, at the base of appendages, and between segments of appendages. Without joints, movement of body parts and of the entire animal would be difficult. Though protected, the arthropod would be a "prisoner" in its own "armor."

An exoskeleton has certain disadvantages. The exoskeleton limits adult size and movement of an arthropod. Because a very heavy outer skeleton needed to support a large arthropod would seriously limit land movement, most adult land arthropods are small. Flying would be impossible for a large insect.

Growth is also limited in arthropods. Once the exoskeleton is made by an animal, the skeleton's size cannot be changed. Growth can take place only if molting (Section 28:9) occurs. Molting may occur as many as six times before an arthropod reaches its final adult size. After each molt, the new skeleton is soft, making an arthropod easy prey for predators. The animal has neither its armorlike protection nor the ability to move rapidly. Land arthropods also lose more water when molting. Where would you expect to find arthropods just after they have molted? Why?

Despite the limitations on size and growth, the exoskeleton is adaptive. The external "armor plate" provided by the exoskeleton shields the animal against physical and living factors in the environment. The exoskeleton also allows movement and supports the animal. More species of arthropods exist than have been discovered in all the other animal phyla combined. This great "success" of arthropods is due in part to the exoskeleton.

30:2 The Endoskeleton

In vertebrates, the skeleton lies within the soft tissues of the body rather than outside the body. Such an internal skeleton is called an endoskeleton. Although an endoskeleton and an exoskeleton function similarly, they differ in development and structure and are not homologous (Section 10:3).

An endoskeleton is subdivided into many distinct parts called **bones.** Bones provide more specialization than parts of an exoskeleton do. Each bone is adapted for certain functions and types of movements. Some bones function in protection. Skull bones, ribs, and the breastbone protect the brain, lungs, and heart. The column of vertebrae forms a protective hollow tube through which the spinal cord passes. In humans, the vertebrae also function in another way. The shape of the human vertebral column is unique. Humans have upright posture and locomotion. The double "s" curve of the vertebral column aids in distributing and supporting body weight. Thus, bipedal locomotion is possible. Other bones also function in support. The long, dense thighbones support the weight of the body and aid the speed and strength of movement. Some bones, such as the delicate finger bones, aid agility for precise, small movements. Each of the 206 bones in the human body has a specific function.

Internal skeletons protect the internal organs of the body but give little external protection. Some animals have skeletal adaptations that give more external protection. For example, turtles are covered by a shell made of bony plates. The shell includes both a dorsal part called the carapace and a ventral part called the plastron. The plates are made of bone covered by a horny material that is like the scales of reptiles. The plates are fused with the ribs of the endoskeleton but do not originate from the ribs in their development.

Vertebrates have internal skeletons, or endoskeletons.

Bones are specialized and have different roles.

Many vertebrates have adaptations for external protection.

FIGURE 30–3. The turtle has an internal skeleton. It also has a shell (carapace and plastron) that adds protection.

a

b

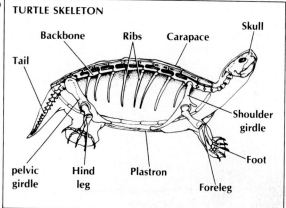

TURTLE SKELETON

Backbone Ribs Carapace Skull

Tail

Shoulder girdle

Foot

pelvic girdle   Hind leg   Plastron   Foreleg

Other animals have additional body parts that also provide external protection. Hair, quills, feathers, and skin are nonskeletal parts that protect animals from temperature variations, predators, radiation, and other factors such as dehydration and invasion by microorganisms.

Although limited size and growth are the main disadvantages of an exoskeleton, size and growth are not as limited in animals having an endoskeleton. Because bones can grow, there is no loss of protection or support as the animal grows. Figure 30–4 shows the human skeletal system.

Size and growth are not as limited in vertebrates as in arthropods.

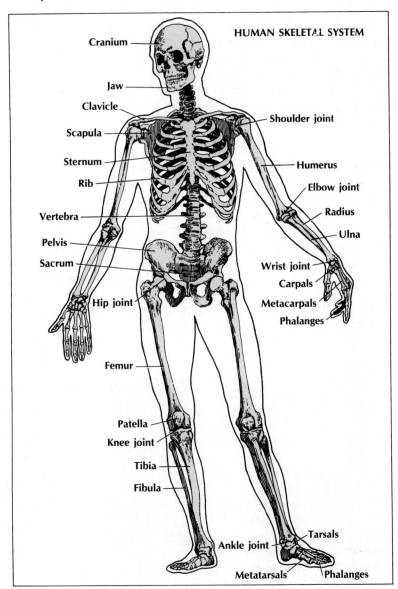

HUMAN SKELETAL SYSTEM

Cranium
Jaw
Clavicle
Scapula
Sternum
Rib
Vertebra
Pelvis
Sacrum
Hip joint
Femur
Patella
Knee joint
Tibia
Fibula
Ankle joint
Metatarsals
Shoulder joint
Humerus
Elbow joint
Radius
Ulna
Wrist joint
Carpals
Metacarpals
Phalanges
Tarsals
Phalanges

FIGURE 30–4. The human skeletal system is composed of 206 separate bones. Sixty-four bones are in the hands and arms alone.

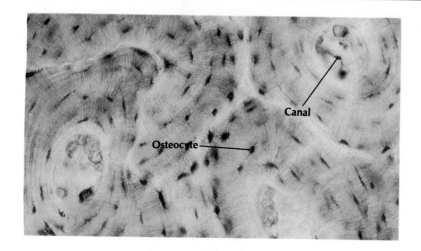

FIGURE 30–5. Bone tissue in cross section shows canals. Around each canal are the osteocytes, the bone cells.

30:3 Development of Bone

Because bone is made up of minerals and is hard, many people think that it is not living material. But a bone in a living animal is a living tissue. Within the minerals of a bone are living cells called **osteocytes** (AHS tee uh sites).

In a fetus, most of the skeleton is made up of **cartilage** (KART ul ihj), a tough, flexible tissue that has no minerals. As the fetus grows, osteocytes slowly replace cartilage cells and ossification (ahs uh fuh KAY shun) begins. **Ossification** is the formation of bone by the activity of osteocytes and addition of minerals such as calcium compounds. By the time an animal is born, many of the bones have been at least partly ossified. In humans, ossification is not completed until about age 25. During this period, the stresses of physical activity result in the strengthening of bone tissue. This process continues until about age 40, when the activity of osteocytes slows and bones become more brittle.

Calcium compounds must be present for ossification to take place. Osteocytes do not make these minerals, but must take them from the blood and deposit them in the bone. Because this process begins in the fetus, the minerals are first obtained from the mother's blood through the placenta. Thus, it is important that the mother eat a well balanced diet, taking in the proper amounts of nutrients.

Because bones grow after the birth of an animal, ossification continues and calcium compounds are still needed. Milk is a good source of calcium compounds and so it is a vital part of a teenager's diet. Vitamin D is also needed for proper bone development because it is necessary for calcium absorption from blood. Rickets (RIHK uts) is a disease that results from inadequate amounts of calcium or vitamin D and results in severe bone deformities (Section 25:6).

FIGURE 30–6. As an embryo develops, cartilage is replaced by bone. This skeleton is partly cartilage and partly bone. The bones are red in this photo. Why are the ends of the arm and leg bones still cartilage?

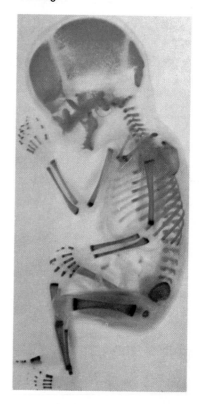

In vertebrates, most, but not all, cartilage is changed to bone.

Some bones do not develop by ossification of cartilage. The flat bones of the face and skull and a part of the collarbone form by the replacement of membrane. Ossification also occurs in these areas because of the activity of osteocytes and the addition of minerals. However, the way in which this occurs is different.

Not all cartilage is replaced by bone. In some vertebrates, such as sharks, the whole skeleton remains cartilage throughout the life of the animal. However, in most animals, only a few parts remain as cartilage. Where in the human body is cartilage found? How is that cartilage important?

30:4 Structure of Bone

Bones contain bone cells, blood vessels, and nerves.

A bone is composed of both living and nonliving substances (Figure 30–7). The minerals that make a bone hard and strong are not living. The other parts of a bone are living, and are supplied with blood vessels and nerves.

All bones are enclosed by an outer layer called the **periosteum** (per ee AHS tee um). Muscles are attached to the periostea of bones. Lying beneath this layer is the **bony layer** that contains the minerals. The bony layer is spongy near the ends of bones and hard in the midregion of a bone. Many tiny channels called **Haversian** (huh VUR zhun) **canals** run throughout the bony layer. Inside the canals are nerves and blood vessels that supply the osteocytes.

Haversian canals contain blood vessels that nourish bone tissue.

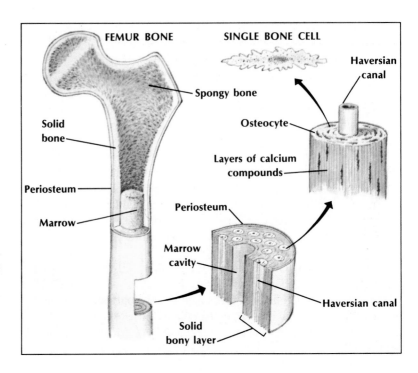

FIGURE 30–7. The human femur (upper leg bone) shows the parts of bone. All parts of the bone are living except the calcium deposits in the bony layer. Blood vessels and nerves fill the marrow cavity. Find the canals shown in Figure 30–5 in this drawing.

Some bones such as the sternum (STUR num) and the long bones of the arms and legs have central hollows filled with blood vessels and nerves. The tissue in the hollows of the bones is called **marrow.** Marrow may produce red and white blood cells and platelets or store excess fat. Marrow also functions in formation of bone cells. Bones without marrow, such as the small bones of the wrist, ankle, and toes, are solid throughout.

30:5 Joints

In vertebrates, joints are the points at which bones connect with one another. Most joints are movable and are held together by connective tissue called **ligaments** (LIHG uh munts). Ligaments are made of elastic fibers. Movable joints of vertebrates are more specialized than joints of arthropods.

One of the most flexible of all joints is the **ball and socket joint** (Figure 30–8). In humans, this type of joint is found where the upper end of the femur (FEE mur) (thighbone) joins the pelvis (hipbone). The end of the femur is rounded into a knob that fits into a depression in the pelvis. The femur can rotate as well as move from front to back or side to side. Are there other ball and socket joints in the human body? If so, where?

Hinge joints exist in many places within the human body, such as the knee. Rotation of a hinge joint is not possible; movement occurs in only one direction.

Rotation can occur in a **pivot joint.** One example of a pivot joint is located where the bones of the lower arm connect near the elbow. This joint allows the radius to rotate over the ulna when the hand is turned over. The cranium (KRAY nee um), or skull, is connected to the spine by a pivot joint. With this joint, the head can turn, as well as move up and down. Vertebrae are linked by **gliding joints.** In gliding joints, the bones move easily over one another in a back and forth manner. Gliding joints aid the flexibility of the backbone.

In adults, the individual bones of the skull are held together by **fixed joints.** In fixed joints, the bones are fused together and do not move. These joints are not connected by ligaments. In babies and children, the skull bones are not fused together but rather have soft spots between them. Why is it important that the skull bones of young children not be fused?

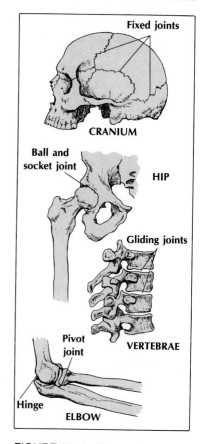

FIGURE 30–8. Bones of the skeleton are connected by different kinds of joints. Only fixed joints do not move.

Bones connected by fixed joints do not move.

REVIEWING YOUR IDEAS

1. What is muscle? What roles do muscles play?
2. List three functions of a skeletal system.
3. Distinguish between exoskeleton and endoskeleton.
4. Distinguish between ligament and joint.

LOCOMOTION

Except for sponges, all animal phyla include adult animals that move about. Animals move in many ways, each related to the animal's environment. Both locomotion and movement of body parts are important in food getting, avoiding predators, reproductive behavior, and reacting to a variety of environmental factors such as light, heat, and chemicals.

Interaction of muscles and skeleton make arthropods and vertebrates the most active animals.

Because of the interaction of their complex muscular and skeletal systems, arthropods and vertebrates are the most active animals. Their skeletons provide points of attachment for muscles. As muscles contract, they create a pull on the skeletal system. The pull results in movement. Locomotion results as many muscles and the skeleton work together in a coordinated way.

30:6 Locomotion: Simple Animals

Coelenterates have simple muscle cells that function in movement.

Coelenterates exist in two forms, polyp and medusa (Section 15:2). *Hydra* is most often a sessile polyp, but it can move by somersaulting (Figure 30–9). Medusas, such as jellyfish, float freely in the water or move by contracting their muscle cells, forcing water from the mouth to create a jet propulsion effect. The jellyfish moves in the direction opposite to that of the expelled water.

Two sets of muscles contribute to locomotion in earthworms.

Earthworms are adapted for locomotion in soil. Their movement is controlled by two sets of muscles. One set shortens the body; the other set lengthens the body. Interaction of these two sets of muscles causes movement. In addition, bristles called **setae** (SEE tee) extend from each segment of an earthworm. With setae a worm "grasps" the soil, resulting in better movement. Some aquatic annelids such as *Nereis* use their parapodia (Section 15:7) for locomotion.

FIGURE 30–9. (a) *Hydra* can "somersault," placing tentacles down and lifting the body. (b) Setae extend from the segments of the earthworm. They aid in locomotion.

FIGURE 30–10. Mollusks, in this case a Swan mussel, move on a muscular foot.

30:7 Locomotion: Mollusks and Echinoderms

Other means of locomotion occur in mollusks and echinoderms. For example, clams have a muscular foot that can be extended between the shells. To move, a clam extends the foot and anchors it in the sand. As the muscles in the foot contract, the clam is pulled forward.

Mollusks move by means of a muscular foot (or feet).

In starfish, movement is controlled by a **water-vascular system** (Figure 30–11). Water is drawn into the animal through a small opening in the central disc and is passed to canals running along each "arm." Along each canal are hollow, muscular **tube feet** that open on the underside of the starfish. The bottom part of each tube foot is a sucker. The upper part of the tube foot is a bulb-shaped **ampulla.** The ampulla contains water that comes from the canal. When the ampullae contract, water is forced down, the tube feet lengthen, and the suckers attach to an object. When muscles in the walls of the tube feet contract, the feet shorten, the starfish moves forward, and water is forced back into the ampullae. Using its hundreds of tube feet, a starfish can creep along rocks and other solid objects. The water-vascular system is also important in obtaining food (Section 25:5).

Starfish move by means of a water-vascular system.

FIGURE 30–11. (a) The water-vascular system controls movement of a starfish. Muscle contraction and water pressure cause the tube feet to act as suction cups. (b) Notice the many tube feet that cover the animal's bottom side.

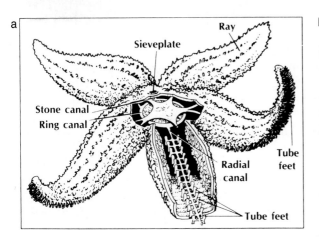

a

Ray

Sieveplate

Stone canal
Ring canal

Radial canal

Tube feet

Tube feet

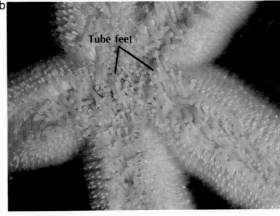

b

Tube feet

FIGURE 30–12. Movement in vertebrates and arthropods is controlled by muscles that work in pairs. Contraction of the biceps muscle bends the elbow joint. Contraction of the triceps straightens the arm. While one muscle is contracted, the other is relaxed.

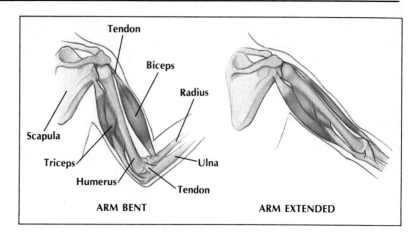

30:8 Locomotion: Vertebrates and Arthropods

In vertebrates, most muscles are attached to bones by tough connective tissues called **tendons** (TEN dunz). The muscles are attached to bones at two different sites. During contraction, one end of the muscle and the bone to which it is attached do not move. This attachment site is called the **point of origin.** The other end of the muscle and the bone to which it is attached move when the muscle contracts. This attachment site is called the **point of insertion.** Attachment of a muscle to two sites is necessary for movement.

> Most muscles attached to bones have a point of origin and a point of insertion.

Muscles that cause locomotion work in pairs in arthropods and vertebrates. For example, in humans the biceps (BI seps) and triceps (TRI seps) muscles work opposite each other to cause motion. The biceps originates on the upper end of the humerus (HYEWM uh rus) and the scapula. The other end of the biceps is attached to the radius (RAYD ee us), the point of insertion. The biceps is called a **flexor** (FLEK sor) because it causes the flexing (bending) at a joint. When the biceps contracts, a force is exerted across the hinge joint of the elbow that causes the arm to bend. In order for the arm to bend, the triceps must relax or extend.

> A flexor is a muscle that bends a joint.

To straighten the arm, the process is reversed. The triceps muscle links the back of the humerus and scapula with the ulna (UL nuh). As the triceps contracts, the ulna is pulled downward and the biceps relaxes. The arm straightens (Figure 30–12). The triceps is called an **extensor** (ihk STEN sur) because it causes the extension (straightening) of a joint. What are the points of origin and insertion of the triceps?

> An extensor is a muscle that straightens a joint.

Note that the bending of a joint is always due to contraction of one of the muscles of the pair. In other words, both flexing and extending are due to contraction. In order for one muscle of a pair to contract, the other must relax.

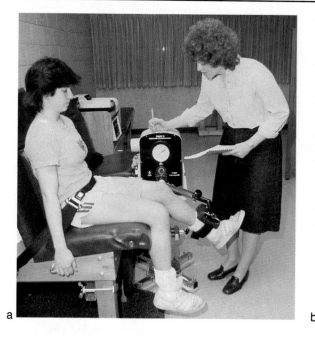

a   b

Besides bending or straightening a joint, paired muscles are important in another way. No muscle is ever completely relaxed; both muscles of a pair are always slightly contracted whether or not a movement is taking place. This condition results in **muscle tone,** which provides enough contraction of the muscles to support the body. Also, muscle tone keeps the muscles ready for quick contractions. Tetanus (TET nus) is a disease in which some body muscles are in complete and continuous contraction (Section 19:2).

The strength and size of skeletal muscles may be increased by physical training. Greater strength develops as the mass of muscles increases and nervous control of the muscles improves. Strengthening seems to depend upon how much tension is placed on muscles during exercise. Increase in size of muscles depends upon repeated contractions of a muscle. The "pumping" action involved in weight training leads to an increase in the size of muscles. Are large muscles important in all sports?

In arthropods muscles are attached to the inside surface of the exoskeleton. Contraction of one muscle exerts a pull on the exoskeleton that results in the bending of a joint and a movement. When this muscle relaxes and the opposing muscle contracts, an opposite movement occurs.

Note that the paired muscles of arthropods function as flexors and extensors. Although arthropods and vertebrates are not closely related, they have evolved a similar type of muscular control of skeletal movements. The evolution of similar functions in the two groups is an example of convergence (Section 11:8).

FIGURE 30–13. (a) Physical training can increase muscle strength. Here an athlete is having her leg strength measured. (b) Conditioning also adds to muscle tone and flexibility.

FIGURE 30–14. In arthropods, pairs of muscles are attached to the interior surface of the exoskeleton. As in vertebrates, arthropod muscles work in opposition to bend and extend joints.

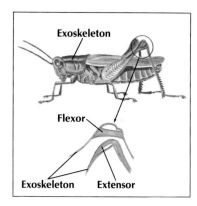

Exoskeleton

Flexor

Exoskeleton   Extensor

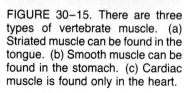
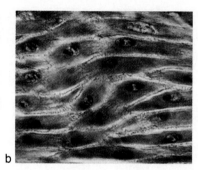

a   b   c

FIGURE 30–15. There are three types of vertebrate muscle. (a) Striated muscle can be found in the tongue. (b) Smooth muscle can be found in the stomach. (c) Cardiac muscle is found only in the heart.

Striated muscles move the bones and are voluntary.

A skeletal muscle cell is stimulated by one nerve.

Smooth muscles move many of the internal parts of the body and are involuntary.

Organs consisting of smooth muscle are controlled by a pair of nerves.

The heart is composed of cardiac muscle.

30:9 Types of Vertebrate Muscles

Of the three types of vertebrate muscles, the most important in terms of locomotion is **striated** (STRI ayt ud) **muscle** (Figure 30–15). Striated muscles are so named because of their striped appearance when viewed with a microscope. Separate striated muscle cells and their membranes are not easily visible. Each cell is called a **fiber** and contains many nuclei. Striated muscles are often called **skeletal muscles** because they are attached to the bones. Striated muscles are the muscles that move the body. Also, they are **voluntary muscles** because they can be controlled at will.

A skeletal muscle cell is stimulated by one nerve. A nerve impulse causes the muscle cell to respond (contract). When the impulse stops, the muscle cell relaxes.

As well as muscles for locomotion, vertebrates have muscles for other functions. **Smooth,** or **nonstriated muscle,** moves many of the internal parts of the body. Smooth muscle is made up of visible, tapered cells, each with a nucleus. Smooth muscles are called **involuntary muscles** because an animal cannot control the activities of these muscles at will. Smooth muscles make up the walls of the hollow organs of the body, such as those of the alimentary canal (Section 25:4). Blood vessels also have a layer of smooth muscle. Contraction and relaxation of these muscles can regulate the activity of the organs or the diameter of vessels. Peristalsis (Section 25:8) is caused by smooth muscle contractions.

Structures of which smooth muscles are a part are controlled by a pair of nerves. These nerves belong to the autonomic nervous system (Section 29:12). One nerve causes contraction; the other inhibits it. Smooth muscles contract less forcefully and somewhat more slowly than skeletal muscles and are not involved in locomotion.

Cardiac (KAHRD ee ak) **muscle** makes up the heart. Cardiac muscle is similar to striated muscle in that both have striped fibers. However, its fibers, unlike those of striated muscle, are branched. Cardiac muscle is similar to smooth muscle in that it is an involuntary muscle and it is controlled by two nerves.

30:10 Contraction of Vertebrate Skeletal Muscle

The fibers of a skeletal muscle are closely packed together. Each cylindrical fiber is made up of smaller units called **fibrils** (FIBE rulz). Fibrils consist of many microfilaments (Section 4:12), more simply called **filaments** (Figure 30–16). The arrangement of filaments gives striated muscle its striped appearance. There are two kinds of filaments, thick and thin. The thick filaments are made of the protein **myosin** (MI uh sun), and the thin ones are made of the protein **actin** (AK tun). At the ends of the thick filaments are knoblike structures that may bridge (link) the two kinds of filaments together. The thin filaments are anchored to vertical bands called **Z lines.** The part of a fibril from one Z line to the next is called a **sarcomere.**

Electron microscope studies show that Z lines of each sarcomere are closer together when a muscle is contracted than when a muscle is relaxed. Such observations as well as experiments on striated muscle have led to the **sliding filament hypothesis** (Figure 30–17), a model that explains contraction.

Nerves leading to a muscle branch many times so that individual muscle fibers are stimulated. Acetylcholine is released by a motor neuron and diffuses across the synapse to the muscle fiber (Section 29:5). Muscle fibers, like neurons, are polarized. The acetylcholine excites the fiber, and an action potential (Section 29:3) occurs in the fiber. The action potential causes a release of calcium ions that are

Striated muscle is composed of fibers that are, in turn, composed of fibrils. Fibrils are composed of filaments containing actin and myosin.

Skeletal muscle contraction is explained by the sliding filament hypothesis.

FIGURE 30–16. (a) Striated muscle is composed of (b) fibers that are made up of (c) fibrils. (d) Fibrils are composed of thick filaments of myosin and thin filaments of actin. The region bordered by two Z lines is a sarcomere.

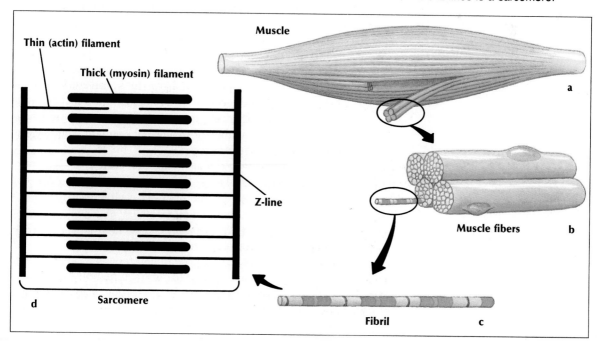

Thin (actin) filament

Thick (myosin) filament

Muscle

Z-line

Muscle fibers   b

a

Sarcomere

d

Fibril   c

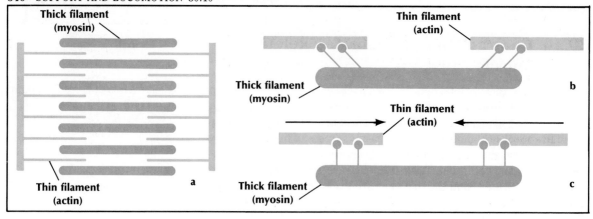

FIGURE 30–17. (a) An enlarged view shows the myosin molecules of the thick filaments. (b) When ATP is available, the knoblike parts of myosin molecules form bridges with the actin molecules of thin filaments and (c) the bridges then swivel inward, pulling the thin filaments closer together. As the thin filaments slide toward one another, contraction occurs.

After a motor neuron impulse, bridges cannot form between thick and thin filaments, and the muscle relaxes.

Strength of a muscular contraction depends on the number of fibers contracting at one time.

stored in the fiber. Calcium ions are needed for the ends of the myosin molecules of the thick filaments to form bridges with the actin molecules of the thin filaments. Bridges form as the ends of the myosin molecules "hook" into grooves along the actin molecules. Using energy of ATP, the bridges swivel inward. The bridges then leave the first groove and attach to another. Using another molecule of ATP, the bridges swivel inward again. As a result of this repeated pattern, the thin filaments slide toward and over one another. Chemical energy is transformed to mechanical energy. Note that the filaments themselves do not shorten, but the filaments' sliding over each other causes the muscle fiber to contract. After the nerve impulse is over, the calcium ions are once again stored. Bridges cannot form and the muscle relaxes.

Once an impulse reaches a muscle fiber there is an all-or-none response. That is, either the muscle fiber contracts fully or it does not contract at all; there is no partial contraction for a given fiber. Also, all contractions are of the same intensity. How, then, can a muscle contract to a greater or lesser degree? The number of fibers that contract at one time determines the strength of the contraction of the whole muscle. The greater the number of fibers that contract, the greater the contraction of the muscle as a whole.

The all-or-none response can also explain muscle tone. The nervous system constantly sends out messages of which you are not aware. These messages keep some fibers in each muscle "ready to go" (relaxed), and some fibers contracted. A muscle fiber must undergo a short recovery period after it contracts before it can contract again. The nerves constantly cause different fibers to contract so that while some fibers are contracting, others are recovering. Thus, some fibers will always be able to contract, and the muscle as a whole is always prepared for a complete contraction. Why is constant readiness of a muscle for contraction important?

30:11 Energy for Skeletal Muscle Contraction

Energy is required for the contraction of muscles. The energy needed for muscle contraction is made available in several ways, depending on muscle activity.

During rest or very light activity, ATP is produced in muscle cells by aerobic respiration. Some of the ATP produced this way is immediately used for muscle contraction. The remaining ATP is stored in the muscle cells either as ATP itself or as another compound, creatine phosphate (KREE uh teen • FAHS fayt) (CP), that can be converted quickly to ATP. CP is formed when molecules of ATP "donate" high-energy phosphate groups to molecules of a compound called creatine. As muscles become more active, the energy stored in CP can be used. CP changes to creatine and the high-energy phosphate produced combines with ADP to form ATP. The energy in the ATP can then be used for muscle contraction.

During prolonged or heavy exercise, the supplies of stored energy (both ATP and CP) are quickly used. Even though the heartbeat and breathing rates are increased during heavy exercise, not enough oxygen is delivered to muscle cells for aerobic respiration to continue. When not enough oxygen is present, the muscle cells switch to anaerobic respiration. During anaerobic respiration, glycogen stored in muscles is converted to lactic acid, and ATP is produced in the process (Section 5:9). The ATP provides energy for muscle contraction, and most of the lactic acid is transported to the liver. During this process, the muscles are said to acquire an oxygen debt. **Oxygen debt** occurs when the cell is obtaining ATP energy without going through normal, aerobic respiration. The cell is "borrowing" against future oxygen intake.

After strenuous activity, rapid breathing occurs, increasing the supply of oxygen to cells. With an increase in oxygen, aerobic respiration is again possible. In the liver, oxygen is used in the aerobic respiration of some lactic acid, producing carbon dioxide, water, and ATP. This ATP is then used to convert the rest of the lactic acid to glycogen. The oxygen debt is "paid" when all the lactic acid is converted to glycogen.

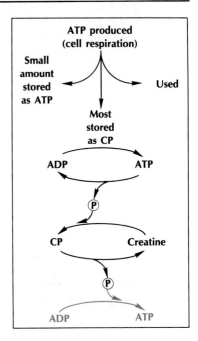

FIGURE 30–18. ATP produced in muscle cells can be used or stored as ATP itself, but most of it is stored as creatine phosphate, CP. ATP can be produced quickly from the CP when needed (shown in red).

Muscles switch to anaerobic respiration during heavy exercise.

The oxygen debt is "repaid" as lactic acid is converted to carbon dioxide and water.

REVIEWING YOUR IDEAS

5. Describe how the water-vascular system of a starfish works.
6. Distinguish between an extensor and a flexor.
7. How are muscles grouped on the basis of control?
8. Upon what does increase in muscle size depend?
9. List three ways in which energy becomes available for muscle contraction.

Problem: How does muscle shorten when it contracts?

Materials:

metric ruler
diagrams of relaxed and contracted sarcomeres

Procedure:

Part A. Anatomy of a Relaxed Sarcomere

1. Make a copy of Table 30–1. Use this table to record your data.
2. Study Figure 30–15a. It is a photograph of what a relaxed muscle fibril might look like if magnified about 2000 times. A fibril is one of many tiny units that make up a skeletal muscle fiber. Note its alternating light and dark bands.
3. Study Figure 30–19. It is a model of several sarcomeres magnified about 10 000 times.

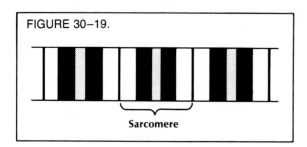

FIGURE 30–19.

Sarcomere

4. Study Figure 30–20. It is a model of one sarcomere magnified about 15 000 times. Certain areas or bands can now be identified.
   (a) Locate the thick and thin filaments that are part of the sarcomere. These filaments are protein in nature and overlap each other. They extend across each sarcomere.
   (b) Locate Z lines. They are thin dark lines extending vertically at the two ends of the

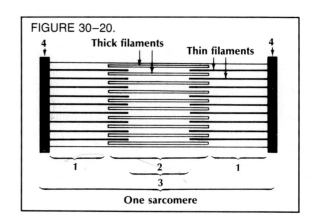

FIGURE 30–20.

Thick filaments

Thin filaments

One sarcomere

sarcomere. Thin filaments are attached to the Z lines.
   (c) Locate I bands. These areas are located at the inside of each Z line and correspond to the sarcomere area that consists only of thin filaments.
   (d) Locate the H zone. This area is located in the center of each sarcomere and corresponds to where only thick filaments are located.
   (e) Locate the A band. This area includes the middle section of the sarcomere. It corresponds to where the thick and thin filaments overlap causing what appear to be two light vertical bands on the sarcomere. It also includes the H zone.
5. In Table 30–1 list the Z lines, I and A bands, and H zone next to the number they match in Figure 30–20.

Part B. Comparing Relaxed and Contracted Sarcomeres

1. Obtain two diagrams from your teacher marked Figure 30–21 and 30–22.
2. Figure 30–21 is a model of a portion of a relaxed sarcomere magnified 20 000 times.

Label these parts: thick filament, thin filament, Z lines, I bands, A band, and H zone.
3. Measure the length of each of the structures listed in Table 30–1.
4. Record them on your data table. Record the length in millimeters in the column marked "relaxed."
5. Repeat steps 2 and 3 using Figure 30–22. This diagram is a model of a portion of a sarcomere that is contracted. NOTE: There may not be an H zone on this figure.

Data and Observations:

TABLE 30–1. PARTS AND AREAS ON FIGURE 30–20, PART A		
1		
2		
3		
4		

SARCOMERE MEASUREMENTS, PART B		
Length of	Relaxed	Contracted
one thin filament		
one thick filament		
A band		
I band (both together)		
H zone		
one sarcomere		

Questions and Conclusion:

1. Define the following terms:
   (a) skeletal muscle
   (b) muscle fibril
   (c) sarcomere
   (d) relaxed muscle
   (e) contracted muscle
2. Describe briefly what provides skeletal muscle with its striated or banded appearance.
3. As the muscle contracts, how does each of the following change?
   (a) thick filament length
   (b) thin filament length
   (c) sarcomere length
   (d) A band length
   (e) I band lengths
   (f) H zone length

4. When a muscle contracts, it shortens in length.
   (a) Which filaments are responsible for sliding when contraction occurs?
   (b) Over what do these filaments slide?
5. Explain how it is possible for a muscle to shorten during contraction when there is no actual change in the length of the thick and thin filaments. (This is called the sliding filament hypothesis.)
6. Explain how the thin filaments actually slide over the thick filaments.

Conclusion: How does muscle shorten when it contracts?

CHAPTER REVIEW

SUMMARY

1. Muscles play a role in movement and provide support. A skeletal system provides support and protection and works with muscles in locomotion. **p. 627**
2. Arthropods have a rigid exoskeleton subdivided by soft joints. Partly because they have exoskeletons, the growth and size of arthropods are limited. **30:1**
3. Vertebrates have an endoskeleton composed of cartilage and/or bone. Vertebrates are not as limited in growth and size as are arthropods. **30:2**
4. Bone is a living tissue hardened by calcium compounds. Bone replaces cartilage or other tissues. **30:3**
5. Although bone is partly composed of inorganic substances, it also consists of living parts — osteocytes, blood vessels, nerves, and, in some bones, marrow. **30:4**
6. Many bones are held together by ligaments at the joints. There are different kinds of joints, each specialized for certain types of movement. **30:5**
7. Simple animals, such as coelenterates and earthworms, have muscles for locomotion, but no skeleton. **30:6**
8. Clams move by means of a muscular foot. Movement in a starfish is accomplished by a water-vascular system. **30:7**
9. In arthropods and vertebrates, skeletal and muscle systems work together to cause locomotion. Muscles work in pairs to move the skeleton. When one muscle of a pair contracts, the other relaxes. **30:8**
10. The three kinds of vertebrate muscle are striated, smooth, and cardiac. Striated muscle is important in moving skeletal parts. Smooth muscle moves many internal body parts. **30:9**
11. The sliding filament hypothesis explains how striated muscles might contract due to the filaments' sliding over each other. **30:10**
12. ATP is needed for muscle contraction. ATP may be produced by aerobic respiration, from creatine phosphate, or during anaerobic respiration. **30:11**

LANGUAGE OF BIOLOGY

actin	myosin
ampulla	ossification
ball and socket joint	osteocyte
bone	periosteum
cardiac muscle	pivot joint
cartilage	point of insertion
extensor	point of origin
fixed joint	sarcomere
flexor	skeletal muscle
hinge joint	sliding filament hypothesis
joint	smooth muscle
ligament	striated muscle
marrow	tendon
muscle	water-vascular system
muscle tone	Z line

CHECKING YOUR IDEAS

On a separate paper, indicate whether each of the following statements is true or false. Do not write in this book.

1. Bone contains living cells.
2. An exoskeleton limits an arthropod's size.
3. Skeletal muscles are involuntary muscles.
4. Both mollusks and echinoderms have a water-vascular system.
5. The knee is a pivot joint.
6. Muscles are attached to bone by ligaments.
7. Bone tissue may replace cartilage or certain membranes.
8. A flexor bends a joint.
9. During heavy exercise, skeletal muscles undergo anaerobic respiration.
10. When skeletal muscle contracts, actin and myosin filaments get shorter.

EVALUATING YOUR IDEAS

1. How is locomotion adaptive for most animals?
2. What are the advantages and disadvantages of an exoskeleton?
3. Why are joints necessary in animals with an exoskeleton?
4. How do bones provide greater specialization than an exoskeleton?
5. Explain the two limitations of an exoskeleton. How do vertebrates overcome these?
6. How are bones formed?
7. Describe the various parts of bones and the function of each part.
8. Distinguish between the types of human joints. Why do you think the skull has several bones joined by fixed joints rather than just one bone?
9. Explain locomotion in a jellyfish, earthworm, clam, and starfish. How does locomotion in these animals compare with locomotion in insects and vertebrates?
10. How are most muscles attached to bones? Distinguish between point of origin and point of insertion.
11. Explain how the human arm is bent and straightened.
12. What is muscle tone? Why is muscle tone necessary?
13. Distinguish among the three types of vertebrate muscles. Where is each located?
14. Explain the sliding filament hypothesis.
15. What is the all-or-none response in muscle? How can muscles be contracted to a greater or lesser degree?
16. How is creatine phosphate produced in skeletal muscle? When is it used?
17. When is anaerobic respiration used to provide energy for contraction of skeletal muscle? Why?
18. When does an oxygen debt occur? How is the debt "paid"?

APPLYING YOUR IDEAS

1. Why do you think animals without skeletal systems are generally small?
2. What adaptive features do the bones of birds have? Why are these adaptive?
3. Knee injuries are common in football players. Explain why these injuries often occur.
4. How is the slow contraction of smooth muscle tissue adaptive?
5. How does locomotion in animals differ from locomotion in protozoans?
6. Why can skeletal muscles be controlled by just one nerve rather than two?
7. Why is the sternum, rather than some other bone, often used in bone marrow transplants? What illness might be treated by a marrow transplant?

EXTENDING YOUR IDEAS

1. After studying the procedure from library sources, mount the bones of a small animal to make a complete skeleton.
2. Muscles can become fatigued. Place your hand on a flat surface, fingers separated in a comfortable position. Begin tapping your index finger quite rapidly. Continue for as long as possible. What happens?
3. Investigate the processes involved in bone repair after fractures occur.
4. Find out about a molecule called myoglobin and its role in muscles.

SUGGESTED READINGS

Cameron, James N., "Molting in the Blue Crab." *Scientific American,* May, 1985.

Moore, Kenny, "Little Is Beyond His Scope." *Discover,* March, 1985.

Torrey, Lee, "Endurance Training: Pushing the Limit." *Science Digest,* Oct., 1985.

ENVIRONMENT

It is sometimes easy to forget that organisms do not live in isolation. They interact with other organisms and their physical surroundings. Here a clownfish is shown among the tentacles of a sea anemone. The tentacles of the anemone has stinging cells that can be fatal to other fishes. The clownfish develops an immunity to the sting of its host, and is therefore able to live amid the tentacles unharmed. If the fish moves to another anemone, it will need a period of adjustment before its system can combat the new host's toxins.

Living things that are not able to produce their own food depend on other living things for food. Living things must also accommodate themselves to environmental factors such as temperature, light, and water. As the environment changes, organisms respond. In Unit 8, you will study the factors involved in a population's response to and interaction with other organisms and the environment.

Clownfish in Sea Anemone

BEHAVIOR

Behavior can be classified as either innate (genetically determined) or learned. This two-toed sloth is sleeping while hanging downward from the branches of a tree. This behavior is innate. What are some other examples of innate behavior in animals? What are some examples of learned behavior in humans?

chapter 31

You have learned that organisms respond to stimuli. The pattern of activities of an organism in responding to stimuli is called **behavior.** Although some internal stimuli (such as hunger) may result in behavior (eating), internal functions such as beating of the heart are not considered to be behavior. Building a nest or playing a violin is behavior.

An organism's behavior is limited by its biological "equipment." Not just endocrine and/or nervous systems, but all parts of an organism are important in determining behavior. In general, organisms with the most complex organization have the most complex behavior. The degree of organization an organism has depends on its genetic code. *Behavior has a biological basis.*

Objectives:
You will
- explain several forms of innate behavior.
- describe different types of learned behavior.
- discuss different means of communication among animals.

An organism's behavior is related to the complexity of its organization.

INNATE BEHAVIOR

Part of every organism's behavior is innate. **Innate behavior** is inherited behavior. An organism's genes determine how it reacts to given stimuli in certain, predictable ways. Like most other inherited features, innate behavior is adaptive; it promotes survival. The curling of an amoeba when poked with a needle and the growth of a plant toward light are examples of innate behavior patterns. Each is adaptive and each is predictable. Neither the amoeba nor the plant can "choose" responses to these stimuli. Rather, the responses are "programmed."

Innate behavior is under direct genetic control.

31:1 Plant Behavior

To think about behavior in plants may seem strange, but plants do respond to stimuli. Most plant responses are controlled by hormones. For this reason, most plant behavior occurs slowly and is not easily seen. For example, a tropism (Section 22:9) may take hours or days to occur. Plant tropisms involve growth in a certain direction as a result of some external stimulus. You may notice that a plant stem is bent toward a light source or that roots bend downward, but you probably cannot see the plant stem bending or the roots growing down.

Some plant responses do occur rapidly enough to be seen. Such behavior is not controlled by hormones. The snapping shut of a Venus's-flytrap is a rapid plant response (Section 21:7). Certain species of *Mimosa* (muh MOH suh) plants also respond to touch. When the leaflets of this plant are touched, they fold together and droop (Figure 31–1).

Response to touch in *Mimosa* is controlled by changes in water pressure, or turgidity (Section 14:13). Certain areas at the bases of *Mimosa's* leaflets contain sensitive cells. When touched, these cells lose water, causing the leaflets to fold. Recent evidence suggests that stimulation of sensory hairs of the Venus's-flytrap leads to a rapid expansion of cells in the leaf, causing it to spring shut.

31:2 Animal Behavior: Reflexes

FIGURE 31–1. (a) A *Mimosa* plant before being touched has open leaflets. (b) After being touched, the leaflets close and droop.

At least some part of animal behavior is controlled by reflexes. A reflex is an innate behavior pattern resulting from the fixed pathways of the nervous system (Section 29:11). In the carrying out of a reflex

a   b

FIGURE 31–2. The response of the pupils to light is a reflex. (a) In bright light the pupils close. (b) In dim light, the pupils open allowing more light to enter the eyes.

response, a certain stimulus is first detected by a receptor. The resulting impulse is transferred by neurons to a certain effector so that the same response always occurs whenever that stimulus is detected.

Reflexes are found in all animals, including humans. They probably play a major role in the behavior of simple animals. For example, if a planarian receives an electric shock, it contracts. If it is exposed to a bright light, it glides away. The planarian always reacts the same way to these stimuli.

You already know some examples of reflexes in humans. Your hand is rapidly jerked away from a hot stove, and your foot is quickly lifted from a sharp rock. Both of these responses are reflexes. You may think about the responses later, but the responses are made before you are aware of what has happened. The rapid, automatic nature of reflex behavior in humans serves a protective function.

In a reflex, impulses normally travel over fixed pathways, but a reflex may be altered with practice and effort. A physician taps just below your kneecap, and your leg extends by reflex. However, by thinking about it, you may prevent your leg from extending. Similarly, you may force yourself not to blink. Changing of reflexes in humans involves different neurons from those involved in the reflex.

Reflexes are determined by the fixed pathways of the nervous system.

Certain human reflexes may be modified with practice and effort.

31:3 Animal Behavior: Instincts

One of the most fascinating types of innate behavior is instinct. **Instinct** is behavior that involves a set of complicated responses to a stimulus or stimuli. Instinct is common among complex invertebrates and in vertebrates. Instinctive behavior is not limited to a single response but may involve many continued activities. Some activities of an instinct depend on previous responses. In other words, one activity triggers another.

Instinct is innate behavior involving a set of complicated responses.

FIGURE 31–3. Migration of birds involves complex, instinctive behavior. Some birds, such as the Arctic tern, migrate thousands of miles. The Arctic tern spends the breeding season in the Arctic area, then "winters" in the Antarctic area.

Instinctive behavior is often triggered by a particular stimulus, the releaser.

Once instinctive behavior begins, the series of related responses continues.

Many instinctive behaviors begin when the animal recognizes a stimulus. The stimulus, because it "frees" the behavior, is called a **releaser.** An egg that has rolled from the nest is a releaser for a female goose. She leaves the nest and, using the bottom of her bill, rolls the egg back to the nest. Interestingly, once the releaser has caused the behavior to begin, it will continue regardless of later differences. For example, if the egg is removed after she has begun rolling it, the goose will continue to "go through the motions" of returning it to the nest. Also, if another object is substituted for the egg, she will treat it as an egg. This example shows two facts common to many instinctive behaviors. Once an instinct begins, the series of responses continues. Also, the releaser does not have to be the stimulus normally encountered in nature. An instinct, like all forms of innate behavior, is a fixed pattern of behavior that has been "wired" into the nervous system.

31:4 Bird Migration

Migration of birds is an example of complex behavior. The fact that some birds can migrate thousands of kilometers to places they have never been shows that instinct plays a major role in such behavior. Several stimuli may act as releasers for migration. The ratio of daylight to darkness and changes in air temperature are known to affect birds' hormone levels and metabolism. Probably a combination of changes starts the process.

Ratio of daylight to darkness may be important in starting migration in birds.

How do migrating birds know where they are going? Some birds use a sun compass, navigating by the position of the sun in the sky. Using biological "time-telling," they make adjustments for movement of the sun during the day. What about a cloudy day? When the sun is not visible, pigeons have been found to navigate by sensing Earth's magnetic field. Other birds navigate at night. When very young, they seem to learn the position of stars. Later, they rely on the stars to guide their migration.

Many birds travel to nearly the same place year after year. The American golden plover migrates from North Alaska through Nova Scotia to Argentina and back each year. How do these birds find the exact place they are going? Some birds probably learn the features of their own nesting area. Experiments show that homing pigeons that cannot see still can sense when they are near home. Again, responses to magnetism may play a role. Stimuli such as odor also may be important.

Note that the complex behavior of migration has an instinctive basis, but that learning also seems to be involved. Thus, a given behavior cannot be thought of as being purely innate or learned. The terms *innate* and *learned* are human inventions for the purpose of studying behavior.

Position of the sun, the stars, and Earth's magnetic field play a role in navigation in birds.

Migration in some birds probably involves learning.

31:5 Courtship Behavior

Courtship behavior is another example of a complex instinct in animals. Courtship is a series of ordered stimuli and responses between male and female prior to mating. The three-spined stickleback fish has a detailed courtship pattern (Section 24:1). Courtship behavior in the great crested grebe, a large water bird, involves a series of dives to the bottom of a pond or lake. During each dive, the grebe gathers weeds for nest building. The male and the female alternate dives. After each dive, the grebes face each other, stretch, shake their heads, and often touch. Time out for swimming may be taken between dives. Finally, mating occurs.

Because similar species often live together in a community, it is important that each individual mate only with a member of its own species (Section 11:5). Certain kinds of "signals" ensure that appropriate mating will happen. For instance, the males of different species of fiddler crabs wave their claws and move their bodies in slightly different ways. Females recognize and respond only to the claw waving by males of their own species.

Courtship patterns are examples of complex innate behavior.

Courtship "signals" ensure that an animal mates with a member of its own species.

FIGURE 31–4. A bower bird builds a nest to attract a mate and then decorates it with shiny or bright objects. By contortions of his body and by tossing around the prized objects, he attracts the female. This courtship behavior is instinctive.

Male building nest

Male decorating nest

Male throwing prize objects around; female watching

Male doing body contortions; female watching

31:6 Biological Clocks

Behavior of many organisms occurs in a 24-hour cycle.

Response to stimuli associated with time is probably an adaptation of all organisms. Most of these responses have about 24-hour cycles. Many flowers regularly open and close at certain times of the day. Birds migrate at certain times of the year. Mice in cages show increased activity on running wheels at certain times of the night. Even some physiological activities may occur in 24-hour cycles. Body temperature and enzyme activity vary with the time of day. The rate of photosynthesis in plants also varies.

Biological clocks control activities related to time.

Organisms are said to have **biological clocks** that control these activities. What is the clock? The answer is not known. The clock may involve some cellular or chemical mechanism. In complex animals, it may involve the pineal gland, located in the brain. Many hypotheses are being investigated.

Clocks may be "set" according to environmental stimuli.

There is strong evidence that the clock can be "reset" according to stimuli from the environment. People who have flown great distances across many time zones experience "jet lag." For a day or two, the traveler's activities are out of phase with those of other people. They are hungry or sleepy while others on local time are not. Gradually, a traveler's clock is "reset" and activities are readjusted to local times.

REVIEWING YOUR IDEAS

1. What is behavior?
2. What is innate behavior? In what kinds of animals is it found?
3. Distinguish between a reflex and an instinct and list examples of each.
4. Define courtship behavior and give an example.
5. What does a biological clock do?

LEARNED BEHAVIOR

Learned behavior is behavior that can be changed.

Innate behaviors are inherited as part of an organism's genetic message. These behaviors are carried out in a fixed and predictable way as the organism carries out its genetic instructions. By contrast, **learned behavior** is behavior that can be changed. Learning is not fixed, predictable, or inherited. Animals with the greatest capacity for learning are those with complex nervous systems. The more complex the brain, the wider the range of possible learned behavior. An animal's ability to learn is influenced by its nervous system and other physical traits as well as by its environment.

31:7 Trial and Error

At one time, scientists thought that only very complex animals could learn. Now, however, experiments suggest that simple organisms also can learn. How much simple animals can learn in nature and how they might benefit from that learning is unknown. However, simple animals can learn to change their behavior in the laboratory.

A simple method of testing learning ability is to use a maze. The simplest type of maze is called a **T-maze.** It is shaped like a T so an animal has a choice of one of two turns. If an earthworm is placed in a maze like the one shown in Figure 31-5, it will learn to avoid the chamber in which it receives an electric shock. The worm will learn to enter the dark, moist chamber in which it finds food. Many trials are necessary before the worm learns which chamber to choose. But the point is that the worm can learn. There is no innate mechanism that tells the worm to turn right or left at the end of a T-maze. The worm learns which way to turn as a result of its own experience.

Simple learning such as in the earthworm is by **trial and error.** By repeatedly trying a task, the animal learns by its mistakes. In the case of the earthworm, it learns to go into the moist chamber.

Motivation (moht uh VAY shun) is needed in the learning process for most animals. In animals, motivation usually involves satisfying a need, such as hunger or thirst. Learning in a maze can occur only if the animal is motivated. For example, the animal usually must be hungry to respond to food. Usually an animal must be thirsty to respond to water.

However, motivation alone does not ensure learning. Other factors are also important. An earthworm in a T-maze has two choices. One involves **punishment,** the shock, while the other results in a **reward,** the moist chamber containing food. The choice resulting in a reward satisfies the motivation. The punishment does not. The earthworm learns to avoid the shock and get the food reward in the moist chamber.

Human learning is also the product of motivation, reward, and punishment. In formal education, some students are motivated by the desire to do well. The reward may be a good grade, a feeling of pride, or later use of the learned material. Failure to learn and its consequences may be punishment. These same factors contribute to the learning that occurs outside the classroom throughout life. What motivates people to learn to drive cars? What motivates someone to practice playing the piano? Do you recall the experiences you had in learning to tie your shoelaces or to write your name as a young child? What was the motivation involved in those cases?

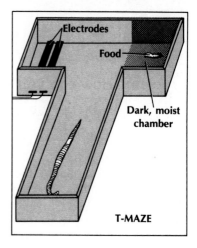

FIGURE 31-5. By trial and error, an earthworm will learn to enter the dark, moist chamber of a T-maze.

Trial and error is a simple form of learning.

Motivation, reward, and punishment are important aspects of the learning process.

INVESTIGATION

Problem: How can you study the behavior of an animal?

Materials:

3 test tubes
3 test tube stoppers
black construction paper
vinegar eel (*Turbatrix aceti*) solution
vinegar eel medium
microscope
dropper

label
lamp
tape
scissors
glass slide
coverslip
ruler
test tube rack

Procedure:

Part A. Observing Vinegar Eels

1. Make a copy of Table 31–1 in which to record your observations.
2. Place a drop of solution containing vinegar eels onto a glass slide.
3. Add a coverslip and observe the slide under low power.
4. Locate a vinegar eel and diagram it in the space provided in Table 31–1.
5. Indicate the magnification of your diagram in Table 31–1.

Part B. Testing Behavior in Response to Gravity

1. Certain animals move toward gravity, while others move against this force. Those that move toward gravity are said to exhibit positively geotropic behavior. Those that move away are said to exhiblt negatively geotropic behavior.
2. Test the response of vinegar eels to gravity by performing the following experiment:
   (a) Use a dropper to fill a tube with about 30 drops of vinegar eel solution.
   (b) Fill the rest of the tube with vinegar eel medium (vinegar).
   (c) Stopper the tube and eliminate any air trapped in the top of the tube. You may need to fill the tube with a little more vinegar eel medium.
   (d) Prepare a second tube allowing at least 3-cm air space at the top of the tube.
   (e) Place both tubes in a test tube rack. Place a label with your name on it on the rack.
3. Observe both tubes 24 hours later. Allow both tubes to remain undisturbed in the rack while observing them.
4. Note the location of vinegar eels in both tubes. They can be seen easily with the naked eye.
5. Diagram the location of vinegar eels in each tube using the outlines in Table 31–1.

Part C. Testing Behavior in Response to Light

1. Certain animals may move toward light. They are positively phototropic in their behavior. Those that move away from light are negatively phototropic in behavior.
2. Prepare a test tube containing vinegar eel solution following steps 2(a-c) of Part B.
3. Use scissors to cut a section of black paper measuring 7 cm by 11 cm.

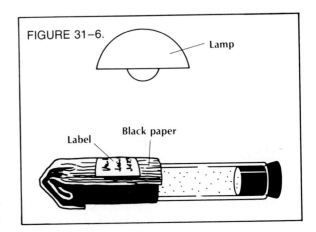

FIGURE 31–6.
Lamp
Black paper
Label

4. Cover one end of the tube with the paper by rolling up the paper and sliding the tube into it. The tube should just fit into the rolled up paper. See Figure 31–6.
5. Lay the tube flat on a table and shine a lamp onto the tube. Place a label with your name on the black paper.
6. Observe the tube 24 hours later.
7. Carefully slide the black paper off the tube making sure not to tip the tube. Tipping may cause the liquid inside to move and thus shift the locations of the eels.
8. Diagram the locations of the eels in the tube using the outline diagram in Table 31–1.

Data and Observations:

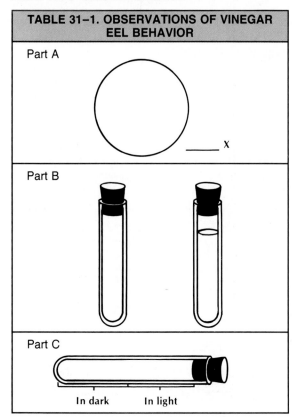

TABLE 31–1. OBSERVATIONS OF VINEGAR EEL BEHAVIOR

Part A

_____ X

Part B

Part C

In dark   In light

Questions and Conclusion:

1. Define:
   (a) behavior
   (b) positively geotropic
   (c) negatively geotropic
   (d) positively phototropic
   (e) negatively phototropic
2. Vinegar eels are roundworms. They are members of the phylum Nematoda.
   (a) List several characteristics of animals in this phylum.
   (b) Name those characteristics that you observed in vinegar eels.
3. Answer these questions based on your observations in Part B.
   (a) Are vinegar eels positively geotropic or negatively geotropic?
   (b) What evidence supports your answer to 3(a)?
   (c) What experimental evidence do you have that the animals are responding to gravity as opposed to the air at the top of the tube?
   (d) How do your observations and conclusions compare to those of your classmates?
4. Answer these questions based on your observations in Part C.
   (a) Are vinegar eels positively phototropic, negatively phototropic, or neutral with respect to light?
   (b) What evidence supports your answer to 4(a)?
   (c) How do your observations and conclusions compare to those of your classmates?
5. Based on your answers to questions 3(d) and 4(c), are experimental results obtained from studies of behavior easily interpreted? Explain.

Conclusion: How can you study the behavior of an animal?

31:8 Conditioning

In one kind of conditioning, the animal involuntarily learns to associate between a response and a stimulus that does not normally cause that response.

Another type of learning is known as conditioning. **Conditioning** is learning by association. There are two kinds of conditioning. One kind is the *conditioned response—a response to a stimulus that does not normally cause that response.*

A Russian biologist, Ivan Pavlov (1849–1936), was the first to show such a learning mechanism. Pavlov knew that if a dog smelled food, it would begin to salivate. This response seemed to be an innate one. Pavlov began to ring a bell at the same time he gave a dog its dinner. Each time, the dog's saliva would begin to flow. After a while, ringing of the bell alone caused the dog's saliva to flow. The dog was conditioned (had learned) to respond to the bell as it did when fed. The response in this case is involuntary. Under normal conditions, the bell would not cause saliva flow. This response is learning because it involves *association and change of behavior.*

Ringing a bell does not normally cause the response of salivating. Thus, salivating is a conditioned response in dogs.

Another example of this kind of conditioning is shown by planaria. The normal (innate) response of a planarian is to glide away from a bright light. Also, it curls up when electrically shocked. If a bright light is followed closely by an electric shock, the planarian curls up. Eventually the worm learns to associate the light with shock, and the light alone will cause the curling up response. Such behavior is a conditioned response. The behavior of the planarian has been involuntarily changed by its experience (Figure 31–7).

FIGURE 31–7. (a) A dog can be conditioned to salivate at the sound of a bell if the bell has been rung in the past when the dog received food. (b) A planarian can be conditioned, associating a bright light with an electric shock.

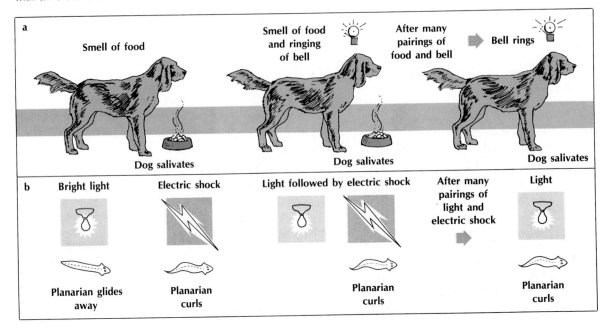

The other kind of conditioning involves trial and error learning. This type of conditioning is often used to teach animals to do tricks, such as teaching a dog to roll over. By rewarding the dog for doing the trick, the probability that the dog will repeat the trick is increased. Also, the dog learns by being punished for not doing the trick. (Punishment may be the absence of reward.) In this kind of conditioning, *the conditioned, or learned, response is the behavior that results in the desired effect* (in this case, the reward). Commands and gestures can be cues for the dog's behavior, but the dog has learned that the reward goes with certain behavior. Learning in this case is the result of the animal's own activities. Can you think of examples of both kinds of conditioning in humans?

FIGURE 31-8. Animals can be taught tricks by a kind of conditioning in which the animal voluntarily does certain behavior that results in the desired effect (reward).

In the second kind of conditioning, the animal learns a voluntary response that gets a desired effect.

31:9 Imprinting

Imprinting is a simple, very rapid, irreversible form of learning. Imprinting clearly shows the elements of innate and learned behavior. It often has been observed that many birds, such as ducks and geese, follow their mothers very soon after hatching. These observations led to the conclusion that birds have the innate behavior (instinct) to follow something they see within a critical time period after hatching. Normally, the object on which the young imprint is the mother. Young birds learn to recognize the mother rapidly. This behavior is clearly adaptive.

However, the young can imprint on another object or animal if that object or animal is substituted during the critical time. The young birds will then follow the substitute. The young birds have the innate behavior to follow but learn to recognize the substitute. Young geese

Imprinting is a form of learning related to instinctive behavior.

Young birds normally imprint on their mother.

FIGURE 31–9. These young Canada geese have imprinted on their mother. Imprinting is a following response and a form of learned behavior.

seeing a human within the critical time will imprint on that person. In one experiment, geese that had imprinted on a human would not later follow their natural mother when she was returned to them. The preference they formed for the human was permanent; as adults the geese still preferred the human over other geese.

Experiments with many birds have shown that birds will imprint on one of several types of objects. Exactly what certain birds will follow depends on the species. Many will follow a mechanical model shaped like a bird, or even a ball. The number of species that follow a mechanical model increases if a rhythmic sound is added. Some species will respond to the stimulus of sound alone. The period of time during which birds can be imprinted lasts for only a few days after they hatch. Once this critical time has passed, imprinting can no longer occur.

PEOPLE IN BIOLOGY

Konrad Zacharias Lorenz

(1903–   )

Konrad Lorenz's lifelong interest in animal behavior grew out of his boyhood pastime of collecting, observing, and caring for animals. The first person to study animals in their natural habitat instead of in the laboratory, Lorenz founded modern ethology, the study of animal behavior. Working mainly with geese, Lorenz developed the concept of imprinting, the process by which a young animal learns to identify its mother. He also discovered that the birds could be made to imprint on a human instead of their mother, and he became famous for his substitute "mother-goose" role as he was seen being followed by young geese. More recently, Lorenz has been studying territoriality and aggression in animal species, including humans.

Konrad Lorenz was one of the first researchers to state that many behaviors were genetically determined, not learned. For this work in animal behavior, Lorenz shared the 1973 Nobel Prize for physiology and medicine with two other ethologists, and he has earned many other degrees and honors. His research has been published in several popular books, including *On Aggression, King Solomon's Ring,* and *The Waning of Humaneness.*

FIGURE 31–10. (a) The chimpanzee first attempts to reach the bananas by jumping. (b) It then stacks boxes and (c) stands on them to reach the fruit. Insight is involved in the chimp's behavior.

31:10 Insight

Suppose a chimpanzee is placed in a large cage containing several boxes. Bananas have been hung from the top of the cage well out of the chimp's reach. At first, the chimp begins to jump for the bananas, but it cannot reach them. Later, instead of jumping, the chimp gathers the boxes, stacks them, and crawls up the stack of boxes reaching the bananas (Figure 31–10).

No past experience provided the chaimpanzee with a "plan of attack." Somehow it was able to "think" out the fact that stacking the boxes would allow it to reach the bananas. The chimp had insight to solve the problem. **Insight** is the ability to correctly plan a response to a new situation. This type of behavior reduces the amount of trial and error learning.

Insight eliminates some trial and error learning.

The chimpanzee formed a concept, or idea. The chimpanzee's concept was that added height would allow it to reach the bananas. This type of behavior is different from trial and error learning. Both forms involve experience, but insight or reasoning enables the animal to "plan."

Reasoning, insight, and concept forming are involved in much of behavior. Thus, learning in humans is very complex. Learning in early life is often by trial and error or imitation. However, as humans age, they rely more on reasoning, insight, and abstract idea formation. Humans draw on their own past experiences as well as those of others.

Most human learned behavior is complex, involving reasoning and insight.

COMMUNICATION

Communication is an adaptation in many animals.

Animals have a variety of ways of exchanging information. Communication is essential within species of animals. Communication among members of a species aids in activities such as locating food, reproduction, and defense against other animals. Courtship behavior (Section 31:5) involves communication such as visual displays and sounds. In the sections that follow, you will study some other examples of animal communication.

31:11 Honeybee "Talk"

Division of labor exists within a society.

Societies are common in some insects. The honeybee society has been studied carefully. One of the most important activities of bees is making honey from the nectar of flowers. As in all societies, a division of labor exists within an active beehive (Figure 31–11). Activity centers around a single queen bee who mates only once and produces thousands of offspring. Most numerous are the worker bees, bees

FIGURE 31–11. (a) A bee society is made up of (b) different kinds of bees with different roles.

a

b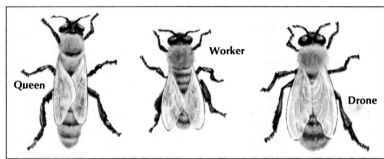

that do most of the work needed to maintain the hive. They feed larvae, produce honey, remove dead bees from the hive, and protect the hive from intruders. Worker bees are sterile females.

Drones are male bees that develop from unfertilized eggs. Several drones mate with a queen during her single mating flight. A drone mates with only one queen bee and then dies.

Karl von Frisch (1886–1982), a German scientist, experimented to learn if certain bees can communicate. He believed that a bee can "inform" other bees of the location of food. Von Frisch had observed that a bee, after finding a supply of food, returned to the hive; soon after, hundreds of bees swarmed to the food supply. Von Frisch assumed that the first bee had "informed" the other bees of the location of the food.

In order to study bees, von Frisch marked some bees and made a glass-walled hive. To attract the bees, he set out containers of a sugar-water solution.

How a bee informs other bees depends on distance. Suppose a bee discovers a food source at a distance of less than one hundred meters from the hive. The bee returns to the hive and performs what von Frisch called a round dance on the inside wall of the hive. The bee circles first in one direction and then in the other (Figure 31–12). Soon other bees leave the hive, return shortly, and perform the same dance. Once several bees have performed the dance, almost all the bees are aware of the distance to the food. The round dance must mean, "There is food somewhere within a radius of one hundred meters."

No directions are given during a round dance. Von Frisch determined this fact by placing the food source at various points around a one-hundred-meter circle. Regardless of the direction of the food, the dance was the same. How do the bees locate the food source? By using their antennae, they detect the scent of the food source on the dancer bee. The bees then seek the same scent when they leave the hive.

For distances greater than one hundred meters, a two-part dance called the waggle dance is performed by the bees. A bee returning from a food source begins the waggle dance on the wall of the hive. The bee does "figure eights" while wagging its abdomen. *Distance* from the hive is determined by the rate at which complete cycles of the dance are performed. The farther the food source from the hive, the fewer the number of figure eights. Sounds made by the dancing bee also may provide information about distance.

The waggle dance also gives the *direction* of the food. The angle (in comparison with an imaginary vertical line on the wall of the hive) at which each figure-eight dance is performed is the key. The vertical line represents the *present* direction of the sun. The angle from the vertical is the angle of the food source from the sun. If a bee begins its

ROUND DANCE

FIGURE 31–12. A round dance by one bee "tells" other bees that a food source is within 100 meters of the hive.

FIGURE 31–13. A waggle dance by one bee "tells" other bees both the direction of a food source and its distance from the hive. The angle of the dance with respect to a vertical corresponds to the angle of the food source from the sun. Frequency of the figure-eight movements indicates distance.

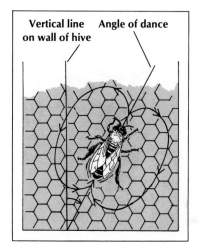

Vertical line Angle of dance
on wall of hive

dance pointing 30° right of vertical, it means that the food source is 30° to the right of the sun (Figure 31–13).

Round dances and waggle dances are complex instincts in certain bee species. These activities are governed by genes and are the products of natural selection. This type of innate behavior benefits the entire species. After returning to its hive, a bee does not dance for its own pleasure. Rather, the dance is a means of informing other bees of a food source.

31:12 Pheromones

Communication by "dancing" is only one way that information is passed along in animal societies. Another means of communication is by secretion of chemicals called **pheromones** (FER uh mohnz). Pheromones are chemicals that can be detected by their odor.

Pheromones are chemical signals.

One use of pheromones is for attraction of mates. In some species of insects, the female releases a pheromone so strong that it can be detected by a male more than a kilometer away. The male follows the scent and, as he gets closer, the scent becomes stronger. Thus, he can determine the precise location of the female and mate with her. Some female mammals secrete pheromones that inform the males that they are in estrus (Section 23:5) and are receptive to mating.

Some pheromones attract animals for mating.

In ant societies, some pheromones are used to inform others about the location of food. As an ant returns to its nest with food, it lays a trail of pheromone along the ground. Ants from the nest follow the scent to the food source and deposit their own trail as they return to the nest. Ants that do not find food do not release pheromones. In this way, there is no confusion about the location of food.

Other pheromones are produced by ants. One is a compound released by decomposing ants. The pheromone acts as a signal to other ants to remove the dead ant from the nest.

Some pheromones affect behavior over a long period of time. Male mice secrete a pheromone that is needed to initiate the female's estrous cycle. This cycle involves a series of changes in hormone levels and behavior. In some insect societies, pheromones eaten by the insects determine the insects' roles in the society, including their reproductive abilities. In honeybees, the queen bee orally passes a pheromone to worker bees. The pheromone inhibits ovarian development and the behavior needed to make a queen cell in the hive. Thus, there is only one queen per hive, a factor that contributes to the social organization of honey bees.

FIGURE 31–14. The queen honeybee passes out pheromone to worker bees. It inhibits their developing into queen bees.

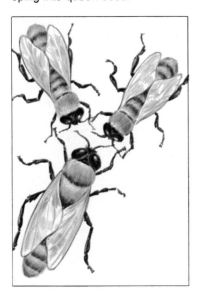

The search for a specific, single pheromone in humans is underway. Results of tests already carried out are unclear. However, human pheromones may be involved in human behavior as attractants between males and females and as attractants between mothers and their babies.

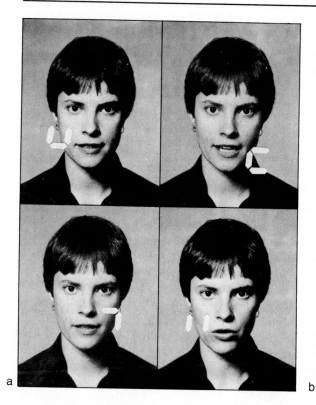

31:13 Language

Most scientists agree that language is a form of communication in which symbols are used to represent ideas. Human language involves both verbal and written symbols. Both spoken and written language provide young humans with the basics for survival and reproduction.

Humans have available to them a record of the experiences of all those who lived before them. Thus, humans do not have to experience everything themselves but can benefit from what others have learned. Humans can use that accumulated knowledge as a basis for their personal discoveries. In this way, knowledge keeps building upon itself and the species as a whole benefits.

Language can occur only in animals with complex nervous systems, memory, and insight. Chimps cannot speak words, but they can use symbols to communicate. Their language, though, is much simpler than that of humans. Chimp language does not seem to involve the many complex thought processes that are common to human language. It is difficult to tell whether all or part of a chimp's language is really just a form of learning in order to achieve rewards—a kind of trick. Whether chimps communicate in nature by a form of language also must be determined.

FIGURE 31–15. Special devices are being developed that help persons with hearing or sight problems communicate more easily. (a) An aid to persons with hearing problems, the computerized "autocuer," contained in special eyeglasses, "listens" to speech and displays cues that aid in lipreading. The photos show what the wearer would see as the girl pictured says, "He can go." (b) The Kurzweil reading machine reads typed text aloud, speaking in a clear voice that simulates human speech, and has been a great aid to blind persons.

Chimps can be taught to use symbols to communicate.

FIGURE 31–16. The gorilla, Koko, uses a keyboard to communicate. The scientist who taught Koko to use the keyboard is one of a group who argues that apes are capable of communication much like humans.

In experiments in which scientists study ape language, a computer and a keyboard with symbols are used. The apes recognize the symbols. By pushing different keys, they can put together words to make simple sentences. In some cases, they have put symbols together in new combinations to make meanings other than those they were taught. Some chimps and gorillas have been taught sign language used by people who are deaf. With sign language, apes can express many words and simple thoughts.

Marine mammals such as whales, dolphins, and porpoises are intelligent animals. Many studies show that they make a variety of sounds. Some of the sounds are used to aid the animals in navigation and location of objects by means of echoes. Other sounds, including moans, whistles, and barking noises, may be a form of communication. The sounds made by some whales can travel as far as 80 km. Most "talkative" of all these mammals is the humpback whale. It produces a great variety of sounds, and the sounds alternate between whales, suggesting a "conversation." Although there has been much work done, scientists do not know what these sounds represent.

REVIEWING YOUR IDEAS

12. How do some honeybees communicate?
13. What are pheromones?
14. What is language?

SOCIAL BEHAVIOR

Communication is one form of behavior that is important in animal societies. A **society** is a group of animals of one species living together in an organized way. Other behavior patterns are also found in many societies. Like communication, these other kinds of behavior are important to the organization of the society and the survival of its members.

31:14 Social Hierarchies

A social hierarchy is composed of levels of authority.

Another example of social behavior involves a pecking order, or social hierarchy (HI rar kee). A **social hierarchy** is a behavior pattern based on dominance relationships. Much human life involves this aspect of behavior. A social hierarchy is composed of levels of authority. Social hierarchies are present in most organizations, such as a school or a business. Such a chain of command provides order and organization within a group of people. A social hierarchy eliminates confusion and promotes efficiency of the organization.

In many respects, the same pattern exists in animal societies. The term "pecking order" was coined as a result of studies of dominance relationships in animals. A society of chickens forms a hierarchy by "pecking" at each other. Two chickens approach each other and may fight, or one may peacefully give way to the other (Figure 31–17). In either case, a dominance relationship results between these two chickens.

Dominance in animals is based on a system of authority relationships. Contacts similar to the one just described occur throughout a flock so that a total hierarchy emerges. One chicken becomes dominant to all the others. Another chicken may be dominant to all but the first one, and so on. Once the order is decided, it remains constant. Sometimes a chicken with lower status challenges a more dominant one, but these encounters rarely change the pecking order. This type of hierarchy occurs in many groups of animals such as birds, some reptiles, and nonhuman primates.

What are the factors involved in establishing a hierarchy? In general, males dominate females. Beyond this, the strong dominate the weak, and the old dominate the young. Size and other factors may be involved. Strangers usually lose to animals that are in familiar territory.

Why is a hierarchy adaptive? A hierarchy promotes survival in several ways. The overall result of dominance is increased order and reduced aggression throughout the society. After the first pecking is over, the less dominant learns to "accept its fate." There is less fighting over food and mates. Therefore, if conditions are such that not every individual can eat and mate, then at least the more dominant ones will be able to do so. These members are usually the better adapted. Thus, a social hierarchy is good insurance that the "best" genes are the ones passed on, and it prevents overbreeding. In any event, the pecking order promotes coordination and unity among the members of a society. Therefore, it is adaptive to the total group of animals.

FIGURE 31–17. Social hierarchy in chickens is established as a result of "conflicts" or pecks between members of the flock. A "pecking order" results.

Once a pecking order is established, order is increased within the society.

FIGURE 31–18. (a) Impalas and (b) walruses engage in combat to establish social order in their groups.

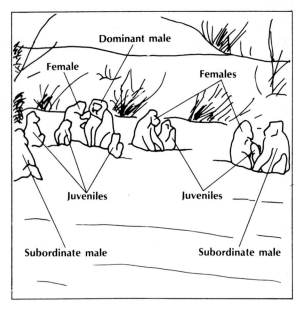

FIGURE 31–19. A troop of Anubis baboons shows a social hierarchy in which there is one dominant male, several females and their offspring, and subordinate males. Subordinate males work with the dominant male to provide defense for the troop.

Territoriality involves the occupation and defense of definite areas.

Territoriality often enables a male to attract a mate.

31:15 Territoriality

Many animals occupy and defend specific territories. This behavior, known as **territoriality** (ter uh tor ee AL ut ee), was first observed in birds. The songs of many male birds announce that their home has been established. The song tells other males to stay away, and it informs females of the species that a male has staked out his territory. Because having a territory "improves the status" of a male, territoriality is important to success in mating for many species of animals.

Studies of other kinds of animals in nature have shown many examples of territoriality. For example, howler monkeys travel in clans. They defend their territory against rival clans by howling and by bluffing attacks against the rival. Rarely is anything more serious needed. The intruding clan realizes that it is not on safe ground and usually retreats peacefully.

What is the advantage of having a territory? What does a territory do for an individual or, more importantly, for a society? Territoriality, like a social hierarchy, promotes survival. Possession of a territory helps attract a mate. Those that are most "fit" (best adapted) are those that win a territory and mate. Thus, their genes are most likely to be passed to the next generation. The weaker members of the group, those with no territory, rarely pass on their genes because they do not mate. In many primate societies, mateless males live on the edges of the territories of other males. These males enjoy none of the benefits of having a territory. Thus, territoriality is related to position in the hierarchy.

Territoriality spreads the members of a group over a large area. Wide spacing ensures a better food supply for all members. Overbreeding and overcrowding are dangerous to a species. With given areas in which to live (eat, breathe, mate), conflicts are reduced and survival is ensured. Thus, the society (and species) benefits.

Several biologists question whether territoriality exists in humans. However, because many animals exhibit territoriality, it seems reasonable to assume that humans might have this trait also. More research is needed to determine if humans are territorial. Perhaps a keener understanding of human biological nature will reveal some of the reasons for human behavior. How might such knowledge be useful?

FIGURE 31–20. The male wolf in the foreground defends a territory against another male wolf.

REVIEWING YOUR IDEAS

15. What is a society?
16. What is a social hierarchy?
17. What is territoriality? How is territoriality related to social hierarchy?

CHAPTER REVIEW

SUMMARY

1. Behavior, the pattern of responses made to environmental stimuli, has a biological basis. **p. 649**
2. Innate behavior is genetically controlled. It is predictable and adaptive. **p. 649**
3. Plant behavior includes tropisms and responses to touch. **31:1**
4. Reflexes are behavior patterns involving fixed pathways of the nervous system. **31:2**
5. Instinct is complex innate behavior involving a set of related responses. **31:3**
6. Migration of birds is mainly instinctive behavior, but it also may involve various degrees of learning. **31:4**
7. Courtship behavior is a complex instinct common in many animals. **31:5**
8. Behavior associated with time is thought to be controlled by biological clocks. **31:6**

9. Learned behavior is behavior that can be changed. **p. 654**
10. Trial and error is a simple form of learned behavior. Motivation, reward, and punishment are often necessary for learning. **31:7**
11. Conditioning involves learning by association. One form of conditioning involves involuntary behavior; the other results in voluntary behavior. **31:8**
12. Imprinting combines both innate and learned behavior. The animal follows an instinct to learn. **31:9**
13. Complex learning can involve insight and concept formation. **31:10**
14. Information about food sources is passed along by dancing in some honeybees. **31:11**
15. Pheromones are chemical signals important in many animal societies. **31:12**

16. Human language is written and spoken. It enables humans to learn from the experience of other humans. Other complex animals, such as chimps and whales, may have a form of language. **31:13**

17. A social hierarchy promotes survival in several ways, including reduced aggression and increased order. **31:14**

18. Territoriality also promotes survival. For example, it is important in mating, in even distribution of members of a society, and in reducing conflicts. **31:15**

LANGUAGE OF BIOLOGY

behavior	pheromone
biological clock	punishment
conditioning	releaser
imprinting	reward
innate behavior	social hierarchy
insight	society
instinct	territoriality
learned behavior	trial-and-error learning
motivation	

CHECKING YOUR IDEAS

On a separate paper, indicate whether each of the following statements is true or false. Do not write in this book.

1. Learning is behavior that can be changed.
2. A dog may learn a trick by conditioning.
3. A bee's waggle dance only gives information about the distance of a food source from the hive.
4. Behavior has a biological basis.
5. Migration of birds is an example of behavior that is mainly learned.
6. Insight enables an animal to plan a response to its environment.
7. Once "set," a biological clock cannot be "reset."

8. A reflex is a type of innate behavior.
9. Plant behavior usually occurs more slowly than animal behavior.
10. A newly hatched duckling will follow nothing but its mother.

EVALUATING YOUR IDEAS

1. What determines the complexity of an organism's behavior?
2. Why is innate behavior considered an adaptation? Why is innate behavior predictable?
3. Explain the reaction to touch of *Mimosa* plants. What causes the reaction?
4. How are reflexes important to humans? List some human reflexes.
5. What is the role of a releaser? Explain the behavior of a female goose after she sights an egg that has rolled from her nest. What does that behavior illustrate about instincts in general?
6. What are some of the factors involved in navigation by birds?
7. How are migration and courtship adaptive?
8. Describe several examples of activities probably controlled by biological clocks.
9. What feature of the nervous system may be important for learning?
10. How is the learning process of an earthworm in a T-maze trial-and-error learning?
11. How did Pavlov demonstrate conditioned behavior in dogs?
12. Distinguish between and give examples of the two types of conditioning.
13. Describe imprinting behavior in young ducks and geese. How is imprinting an adaptive behavior?
14. How does insight differ from learning by trial and error?
15. How does a bee inform other bees of a food source within 100 m of the hive?
16. Describe the waggle dance of bees. What information does it contain?

17. How are pheromones important in mating? In locating food?
18. How is language important to human learning?
19. Do animals such as chimps and whales have a language? Explain.
20. How is a social hierarchy formed in chickens? How is a hierarchy adaptive?
21. How is territoriality advantageous?

APPLYING YOUR IDEAS

1. List several examples of trial-and-error learning in humans.
2. A certain dog licks a person's face. The person in pushing the dog away rubs the dog's belly. The dog and the person then both repeat the behaviors many times. The dog gets the "belly rub" reward for face licking; the person gets the "dry face" reward for rubbing the dog's belly. Is the dog or the person conditioned in this example? Explain.
3. Might imprinting be a type of learning in humans? Explain.
4. Compile a list of some of the unique advantages of human learning.
5. How might pheromones be used to control insect pests?
6. Why is territory defense usually limited to gestures, bluffed attacks, or loud noises?
7. A queen bee leaves her old hive to start a new one. Why can a new queen then develop in the old hive?
8. How is reflex behavior related to a reflex arc?
9. Why might certain medicines be more effective at specific times of the day?
10. A *Paramecium* bumps into an object. What kind of behavior does the *Paramecium* exhibit in its response to the object?
11. Why do plants not have learned behavior?

EXTENDING YOUR IDEAS

1. Obtain a culture of the ciliated protozoan *Stentor*. Use a microscope to observe its behavior. Then, use a pipette to scatter India ink on the *Stentor*. Notice its response. Repeat several times. Does *Stentor* always respond in the same way? Explain.
2. Expose a pill bug to a light source from one side. Note the reaction of the bug. Is the bug exhibiting a tropism? Explain.
3. Construct a maze to test learning in white mice or white rats. Keep a record of the number of trails necessary for the animals to learn the maze.
4. How does a baboon society compare to an insect society? How does the baboon society compare to a human society?
5. Place two sets of white mice that have been reared apart together in a cage. Note any signs of a social hierarchy being formed.
6. Take your temperature at different times of the day. Is it always the same? Explain.

SUGGESTED READINGS

Crockett, Carolyn M., "Family Feuds." *Natural History,* Aug., 1984.

Ghiglieri, Michael P., "The Social Ecology of Chimpanzees." *Scientific American,* June, 1985.

Griffin, Donald R., *Animal Thinking.* Cambridge, MA, Harvard University Press, 1984.

Palmer, J. D., *Human Biological Rhythms.* Carolina Biology Reader. Burlington, NC, Carolina Biological Supply Co., 1983.

Rose, Kenneth Jon, "How Animals Think." *Science Digest,* Feb., 1984.

Topoff, Howard, "Invasion of the Body Snatchers." *Natural History,* Oct., 1984.

Wilson, E. O., "Altruism in Ants." *Discover,* Aug., 1985.

POPULATION BIOLOGY

S everal factors affect the sizes of populations of organisms. Some of these factors are the availability of food and space, and the presence of disease organisms. What are some of the factors that might limit the population of these fur seals? What factors might limit the population of other organisms? What factors limit the human population?

population is a group of organisms within a certain area that belong to the same species. As reproduction occurs over a period of time, population size increases. **Population size** is the number of organisms in the population. As populations live and reproduce, they interact with their environment. In this chapter, you will study ways in which the environment affects population size.

Objectives:
You will
- describe the growth of populations.
- identify and explain the effects of various limiting factors.
- discuss factors that have contributed to the growth of the human population.

GROWTH OF POPULATIONS

Charles Darwin (Section 10:8) recognized that organisms tend to produce large numbers of offspring. He also realized that not all offspring survive to reproduce, and that, as a result, population size tends to remain stable. A population's size, then, must be determined by two opposing "forces"—a tendency for parents to leave many offspring, balanced by factors of the environment that prevent many of those offspring from surviving.

32:1 Biotic Potential

Under ideal conditions—unlimited food, absence of disease, lack of predators, and so on—the size of a population would increase indefinitely. For example, under ideal conditions, certain species of bacteria reproduce and double their population size every twenty minutes.

FIGURE 32–1. Under ideal conditions, a population's size would continue to increase indefinitely. This increase can be shown graphically.

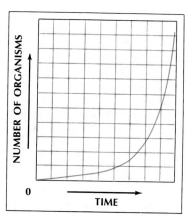

Populations tend to increase maximally under ideal conditions. Consider the bacteria that reproduce every twenty minutes. If no environmental factors limited their survival, in 36 hours one bacterium could give rise to enough descendants to cover Earth with a layer of bacteria 30 cm deep! Even organisms that reproduce more slowly could leave many offspring under ideal conditions. It has been estimated that, starting with one pair of elephants, there would be over 19 000 000 elephants in only 750 years. This highest rate of reproduction under ideal conditions is called a population's **biotic** (bi AHT ihk) **potential.**

Populations do not reproduce to their biotic potential. Earth is not overrun with bacteria, elephants, or any other organism. Factors exist that limit population size. Factors that hold the biotic potential in check are called **limiting factors.**

Biotic potential is balanced by limiting factors.

32:2 Population Growth Curves

Population growth is the increase in the size of a population with time. Growth of some populations can be studied in the laboratory. To study the population growth of yeast cells, a culture medium can be inoculated with a few yeast cells. Counting the yeast cells at regular time intervals would give data much like that shown in Table

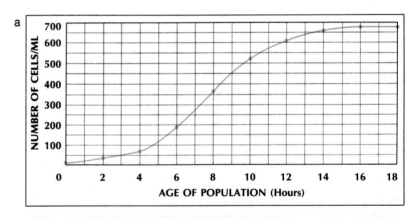

FIGURE 32–2. A yeast population can be grown under laboratory conditions. (a) When the growth is graphed, an S-shaped population growth curve results. (b) The liquid medium in which the yeast are grown becomes cloudy as the number of cells increases. The tube on the left corresponds to a population on the left side of the growth curve. The tube on the right corresponds to a population on the right side of the growth curve.

32–1. Note that the size of the population increases slowly at first. This period of slow growth is followed by a period of rapid growth, and then another period of slowed growth. Eventually, a point is reached when the population does not grow. At this point, the population size remains stable.

Growth of the yeast population can be graphed (Figure 32–2). The graph of such growth is called a population growth curve or **S-shaped curve.** Growth of the yeast population is similar to the growth of many other populations. Up to a point, the population size increases. The increase is slow at first, and then becomes faster. However, after a certain amount of time, the population growth levels off due to limiting factors. Can you think of what some of these factors might be?

How quickly a population grows can also be graphed (Figure 32–3). This graph, called a **population growth rate curve,** shows the *rate of increase* of population size per unit of time. The data for this graph come from the third column of Table 32–1. The graph shows that up to a point, the size of a population increases more and more rapidly. Then, the *rate* of increase slows down. The number of organisms still increases, but the rate of growth decreases. When the population growth levels off (at about hour 18), the population growth rate approaches zero.

When a population reaches the point at which its number is no longer increasing, it has reached the carrying capacity of the environment. The **carrying capacity** is the number of individuals that a given environment can support. At the carrying capacity, the number of organisms born or produced in a given period of time, the **birthrate,** balances the number of organisms that die during that time, the **death rate.** At this point, the size of the population remains fairly stable.

TABLE 32–1. GROWTH OF A YEAST POPULATION		
Population Age (h)	Number of Cells	Increase in Cell Number
0	10	
2	29	19
4	71	42
6	175	104
8	351	176
10	513	162
12	595	82
14	641	46
16	656	15
18	662	6

As a population grows, the rate of increase begins to decrease.

Carrying capacity is the number of organisms that can survive in a particular area.

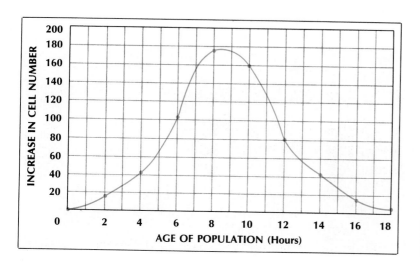

FIGURE 32–3. The growth rate of a population can be graphed. This growth rate curve is for the same yeast population whose growth is graphed in Figure 32–2(a).

Biotic potential is limited by environmental resistance.

The environment includes physical and biological limiting factors, both of which are subject to change. Thus, the carrying capacity is also subject to change. Different environmental factors produce different carrying capacities for a given population. The sum of the limiting factors of a population is called **environmental resistance.** The size of a population is determined by a balance between biotic potential and environmental resistance. Environmental resistance is the difference between the biotic potential and the actual size of a population.

REVIEWING YOUR IDEAS

1. What is biotic potential? What checks biotic potential?
2. Describe the growth of a yeast population. Distinguish a population growth curve from a population growth rate curve.
3. Distinguish between birthrate and death rate.
4. What is carrying capacity?
5. What is environmental resistance?

LIMITING FACTORS

FIGURE 32–4. Both (a) animal populations, such as gannets, and (b) plant populations, such as phlox, have limiting factors. Limiting factors also affect populations of organisms other than animals and plants.

Limiting factors, both physical and biological, control population size and growth. However, at any given time, only one factor limits a population. Limiting factors can change depending on environmental conditions. Factors affecting a population may be related to **population density,** the number of organisms per given area. Such factors are called density-dependent. Other factors not related to population density are called density-independent.

a

b

32:3 Predation and Food

Predation (prih DAY shun), the feeding of one organism on another, is a density-dependent factor that may limit the size of a population. Almost every organism is preyed on by other organisms. It is often hard to find out if predation is a limiting factor, but the size of most populations is at least partly checked by predators.

How may predation be density-dependent? Consider the prey population. The greater the number of prey organisms in a given area, the more likely it is that a predator will find and kill one. As the density of the prey population begins to fall as a result of predation, fewer prey are killed. Most often, a balance is reached in a predator-prey relationship. The size of one population affects the size of the other. An increase in size of the prey population leads to an increase in size of the predator population because more food is available for the predators. Later, the size of the prey population begins to fall. Soon, because less food is available, fewer predators survive, and the size of the predator population falls. This predator-prey interaction creates a cycle of population increase and decrease in both populations (Figure 32–5). Notice that changes in the size of the predator population follow those in the prey population. In a typical predator-prey relationship, predation is the limiting factor for the prey population and the availability of food is the limiting factor for the predator population.

Predation is "healthy" for a prey population. Very often, the prey trapped and killed are those that are very young, very old, or weak. Thus, the healthy, best-adapted individuals usually survive, reproduce, and pass their genes on to the next generation. In this sense, predation is a form of natural selection that benefits the population as a whole. Predation may also serve to maintain prey population size very near the carrying capacity while other factors would not. For example, if predators were removed from the environment, the prey population size would likely increase dramatically. However, food

FIGURE 32–5. (a) Predator-prey cycles can be represented graphically. After the number of prey increase, the number of predators increase, and after the number of prey decrease, the number of predators decrease. (b) A fox is a predator on rabbits (the prey). (c) Ants are predators on this grasshopper (the prey).

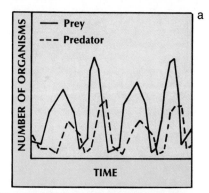

a

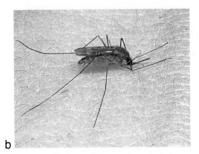
b

c

FIGURE 32-6. (a) Ticks are parasites on various animals. Here two are shown on a dog. Ticks live on blood sucked from the host. (b) A mosquito is a parasite living on blood from its host. (c) These parasitic mites have killed the host beetle on which they were living.

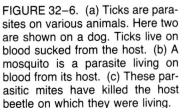

Predation is often more beneficial to a prey population than some other limiting factor.

Parasitism can be a limiting factor when the number of parasites carried by a host becomes too large.

The greater the population density, the more likely that parasitism will limit the population size.

might then become a limiting factor. The unusually large number of prey organisms might overbrowse, eating plants more quickly than the plants could reproduce. Eventually, almost all of the animals would starve to death and the population size would decline to a point well below the carrying capacity. Such a case happened in a moose population on an island in Lake Superior. Before wolves reached the island, many moose starved. In the absence of predators, the size of the moose population increased unchecked. Food then became a limiting factor, and the majority of the moose starved. Later, wolves reached the island and began to prey on the moose. Since that time, the size of the moose population has remained stable at a point closer to the carrying capacity.

32:4 Parasitism

Parasitism (PER uh suh tihz um) can be a limiting factor. In a parasitic relationship (Section 1:2), one organism (the parasite) lives in or on another organism (the host), which is usually harmed by the parasite. Almost all organisms have some parasites, but the parasites usually do not cause death because death of the host would result in death of the parasites. However, if a host has too many parasites, the host might die. Thus, the limitation is the *number* of parasites that a host is carrying.

This limitation is density-dependent. Parasitism usually causes death only when the parasites are in very dense host populations. The possibility that parasites will be passed from one organism to the next is greater in a dense host population. As the rate of transfer increases, the number of parasites in each host also increases. Too many parasites on one host may cause death by interfering with the host's nutrition or metabolism. The parasites may also reduce fertility of the host. As a result, the size and density of the host population decreases. Host organisms then have more space available to them, and transfer of parasites is less frequent. Since hosts have fewer parasites, the host population can survive and increase in number again. Thus, the cycle continues.

FIGURE 32–7. This dense population of trees is more subject to a severe attack by parasites than a scattered population would be.

32:5 Diseases and Populations

Population density is closely related to the spread of certain diseases. For example, malaria results from a parasitic relationship between *Plasmodium* and humans (Section 19:6). Malaria is density-dependent in that the denser the human population, the greater the chance of the disease being passed to others. *Plasmodium* is carried by female *Anopheles* mosquitoes and transferred to humans by the bites of these mosquitoes. Thus, the spread of this disease is also dependent on the density of the mosquito population.

Disease is a density-dependent limiting factor.

Malaria can be treated with certain drugs, but the best solution to malaria is prevention of the disease by population control of the mosquitoes. Population control has been carried out in many places by destroying the swampy breeding grounds of the *Anopheles* mosquito.

Spread of disease can be used to control the size of certain animal populations. The disease can spread rapidly in a dense population and kill many organisms in a matter of days. For example, rabbits are not native to Australia; they were brought there by humans. The result was nearly a disaster. Because there were no predators, the size of the rabbit population increased rapidly. Soon rabbits began competing with sheep for food. In order to reduce the rabbit population and preserve the sheep population, biologists injected some rabbits with a lethal virus that spread rapidly, killing large numbers of rabbits. The virus did not kill all the rabbits because disease is a density-dependent factor. The virus was transferred directly from rabbit to rabbit. As the number of rabbits decreased, the rate of transfer also decreased. As a result, the smaller population size became stable.

The size of pest populations may be decreased by the spread of disease.

Killing pests by using disease-causing organisms must be carefully controlled. Care must be taken that only the pest population is affected. If other organisms are killed, the natural balance could be upset, causing even more harm (Chapter 35).

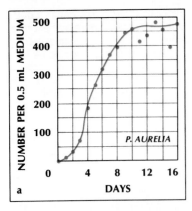

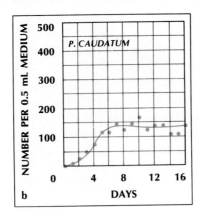

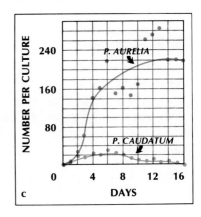

FIGURE 32–8. (a) A population growth curve shows *Paramecium aurelia* grows to a higher density than (b) *Paramecium caudatum* before leveling off. (c) In this particular experiment when the two are grown together, interspecific competition occurs and *P. aurelia* survives while *P. caudatum* dies.

Two populations cannot occupy the same ecological niche.

32:6 Interspecific Competition

Suppose a certain factor limits one population, and two populations of different organisms compete for it. The competition itself may then become a limiting factor. In the laboratory, two different populations can be made to compete for the same limiting factor. As a result, one population always dies. This fact led to the formulation of the **competitive exclusion principle:** *complete competitors cannot coexist.* This principle means that two populations cannot occupy the same ecological niche (Section 11:7), or place. Two different types of organisms cannot fulfill the exact role within a community.

In an experiment, two species of paramecia were introduced into separate culture media. These species were *Paramecium caudatum* (kaw DAYT um) and *Paramecium aurelia* (aw REEL yuh). The food source in both media was a species of bacterium. The growth rate of each population was recorded (Figure 32–8a,b). The graphs showed that the biotic potential of *P. aurelia* is greater than that of *P. caudatum.*

Next, the two species of paramecia were placed in the same culture medium. The same species of bacterium was used as the only food source. Thus, both species of paramecia competed for the same food source. As a result of the competition, *P. caudatum* died while *P. aurelia* survived (Figure 32–8c). Each species survived on the food source when in separate cultures. However, when the two species were together, the competition for food limited the population. Such competition between populations of different species is called **interspecific** (ihn tur spih SIHF ihk) **competition** and is density-dependent.

Interspecific competition is competition between populations of different species.

In this example, neither paramecium species harmed the other. *P. caudatum* died because it was more sensitive to wastes produced by the bacteria. *P. aurelia* was better adapted to living among the bacterial wastes, and it survived. In similar experiments using other food sources, *P. caudatum* was better adapted and became the surviving species.

32:7 Intraspecific Competition

Competition between members of the same species is called **intraspecific** (ihn truh spih SIHF ihk) **competition.** It is the most severe type of competition and is also density-dependent. The greater the population density, the greater the chance for competition because there are more contacts among organisms. Some degree of competition is good for a species because by natural selection the "more fit" survive. But too much competition is not good.

Excess intraspecific competition is avoided in many ways. Life cycles may be such that competition is avoided within a species. For example, adult frogs do not compete with tadpoles because the habitats and foods of frogs and tadpoles are different. The life span of an organism may also reduce competition. In many animal species, adults die shortly after the young are born. Thus, old and young do not compete for food or other factors. In animal species in which the parent does not die after the young are produced, parental care is common. This avoids competition because the offspring depend on the parents. In plants, dispersal of spores and seeds also prevents competition between members of the same species.

In social animals, social hierarchies (Section 31:14) and territoriality (Section 31:15) reduce conflict. Once a social hierarchy is set up, competition for food, mates, and space is greatly reduced. Also, nonproductive activity is reduced so animals have more time and energy to "make a living." After a male bird sets up a territory, he is able to attract a female. The male's territory also provides enough space and food for supporting the "entire family."

Societies in which members have definite roles also have reduced intraspecific competition. In insect societies such as bees, ants, and termites, roles are genetically determined. Usually there is one queen that produces all new members of the society. Other members

Intraspecific competition occurs among members of the same species. It is density-dependent.

Life cycles of plants and animals may prevent intraspecific competition.

Social hierarchies and territoriality reduce the chances for intraspecific competition.

Existence of different roles among members of a society reduces intraspecific competition.

FIGURE 32–9. The social hierarchy established by confrontations of elephant seals is a form of intraspecific competition. Once established, the hierarchy reduces conflict.

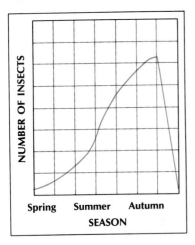

FIGURE 32–10. An insect population in the temperate zone increases each spring and summer and dies rapidly with the freezing weather of late fall. Temperature is therefore a density-independent limiting factor.

Temperature may be a density-independent limiting factor.

A decrease in oxygen supply can affect many species in an aquatic community.

of the society may have roles as workers, soldiers, or foragers. Spreading out the work and activities reduces conflict and competition among the members of the society.

Overcrowding can lead to behavioral and physiological changes among members of the same species. If rats are overcrowded and cannot escape, the amount of aggressive behavior they exhibit may increase, resulting in more deaths than usual. Also, nests may be poorly built or not built at all. Hormonal changes resulting from stress may cause a reduction in litter size. In some cases, no young are produced. In other cases, the young are abandoned or eaten by their parents.

As a result of these behavioral and physiological changes, birth-rate decreases and death rate increases. The result is a reduction in the size of the population to a number that the environment can support.

Another solution to overcrowding in some populations is **emigration,** the leaving of an area. For example, some kinds of locusts swarm to new areas when competition becomes intense as a result of increased population density. Other locusts do not emigrate. They are able to survive in their original environment under conditions of reduced population size and density.

32:8 Density-Independent Limiting Factors

Populations may be limited by a variety of factors. In fact, *almost anything affecting the lives of organisms could act to limit a population's size.* Often these factors are density-independent. Temperature and oxygen supply are two examples.

Temperature often is a factor that limits the size of insect populations. In temperate zones, the population size begins to grow slowly in late spring and then more rapidly through the summer. But if the growth of such a population were graphed, it would be different from an S-shaped graph (Section 32:2). Like the S-shaped graph, it would start slowly and then rise rapidly. But instead of leveling off, the graph would show a sharp decline in population size because the adults rapidly die in autumn as freezing weather begins (Figure 32–10). The population size declines before reaching the carrying capacity. Before they die, the insects deposit their fertilized eggs in some "safe" place. The following spring, the young adults emerge and begin another population "explosion."

Amount of oxygen can be a limiting factor in aquatic communities (Section 35:7). Oxygen is needed for respiration by almost all organisms. If the oxygen level drops too low, many animals cannot survive. Drop in oxygen content of the water can be caused by pollution. For example, improper sewage treatment can result in the dumping of much organic material into the water. Decomposition of this material by bacteria consumes huge amounts of oxygen. As a result, many animals die. When their dead bodies decay, the oxygen level is lowered even further. This lack of oxygen can affect an entire aquatic community.

Oxygen content of water is temperature related. More oxygen can dissolve in cold water than in warm water. Sometimes hot water is released into rivers, streams, or lakes from factories or power plants. The hot water may raise the temperature of the water to which it is added and cause the oxygen content to be lowered. Fish and other animals may not be able to survive.

REVIEWING YOUR IDEAS

6. What is population density?
7. How are limiting factors related to population density?
8. What is predation? Are humans predators?
9. How may parasitism be a limiting factor?
10. Distinguish between interspecific competition and intraspecific competition.
11. What is a density-independent factor?

THE HUMAN POPULATION

Complex intelligence and behavior of humans have had a major impact on all organisms, including humans themselves. Unlike other organisms, humans have learned to control the environment to some extent and can alter it according to their own needs. Advances in agriculture and technology have been especially important. These areas are related to the size and growth of the human population.

FIGURE 32–11. Advances in how land is farmed have influenced the size and structure of the human population. This aerial view of farmland shows the large scale on which farming takes place today.

32:9 Agriculture

Development of agriculture probably took place between 10 000 and 15 000 years ago. This development was one of the most profound steps in human progress. Crops were planted where they did not grow naturally; new strains of plants such as corn, wheat, and rye were developed. As a result, humans added variety to their diet and extended their range of habitats. Growing of crops reduced competition for food; thus, population size increased and permanent homes could be established.

Agriculture remained the central feature of human society for thousands of years, and all human societies still depend on agriculture. In the past, everyone lived directly "off the land," and farming is still the major way of life in many countries. But some people gradually began to live in towns and villages. Division of labor increased as people began to make their livings in other ways besides farming. From the smaller population centers emerged larger cities and a more complex society.

Agriculture added variety to the human diet, increased population size, and made a more permanent home possible.

32:10 Technology

As curious animals, humans investigate nature. The discovery of natural laws has enabled humans to become more productive, and they have learned to make many things with the help of tools and machines. Since the beginning of the Industrial Revolution in the early nineteenth century, society has become more technological than agricultural.

The products made by today's society have become important parts of our daily lives. Some advances add to our comfort. Others directly affect our health, and thus, our population. Technological

Human curiosity has led to the development of a society that relies on technology.

FIGURE 32–12. (a) Sewage treatment plants that clean area water supplies have greatly reduced water pollution. (b) A worker tests a water sample for chlorine, a chemical added to water to inhibit the growth of harmful bacteria.

a

b

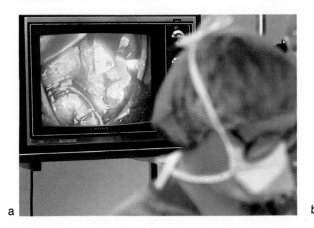

a b

FIGURE 32–13. Medical advances in patient care, professional training, and in getting aid to patients have greatly influenced the survival rate of humans. (a) A surgical operation is shown on closed circuit television as a method of training new doctors. (b) A mobile cancer therapy unit reaches areas of a New Mexico Indian reservation where patients are long distances from professional medical care.

advances have had a direct bearing on the size and growth of the human population.

Advances in sanitation have prevented many lethal diseases. For example, bubonic plague is a disease carried by fleas. In the 14th century, one fourth of the population of Europe died of this disease! Proper sanitation procedures have prevented recent outbreaks of diseases such as plague. Modern sewage treatment is an important technological advance in sanitation.

Technological advances have also improved agriculture (Chapter 35). Crop yields have been improved and more nutritious foods have been produced. These advances affect populations by reducing the death rate and lengthening the life span of population members.

Advances in medicine and surgery have also played a major role in decreasing the death rate and increasing life span. Many diseases that long ago might have resulted in an early death can now be treated or cured. Also, technological advances in disease detection have resulted in early treatment of some diseases and thus have increased the number of survivors. Medical advances have been especially important in reducing infant mortality. Survival of infants is a major way in which death rate has been reduced and by which the size of the human population has increased.

32:11 The Population Explosion

It is estimated that humans numbered about 0.25 billion in the year 1 A.D. By 1650, world population doubled. It took 200 more years to reach the billion mark. The population doubled again in the period between 1850 and 1930. In 1980, the world's population was over 4.0 billion. It is estimated that total population will reach 8.0 billion by 2022. Thus, the length of time it takes for the population to double is continuing to decrease (Figure 32–14).

The 1970s marked the first time in modern history that the world population growth rate decreased. During that time the growth rate

The time required to double the human population is decreasing.

FIGURE 32–14. The growth of the world's human population can be shown graphically. The dotted portion of the curve represents projected future growth.

declined from 2 percent to 1.7 percent. Even though the rate of 1.7 percent has held steady through most of the 1980s, the world population passed the 5.0 billion mark in 1987. Although growing more slowly, the world population is *still increasing.* Eighty to ninety million people are being added each year! It is estimated that world population will reach 6.1 billion by the year 2000.

An increase in birthrate has not caused the human population explosion. Instead, the death rate has decreased due primarily to improvements in sanitation, medicine, and food production. The most important consequence of the decline in death rate is that the population includes more and more people of reproductive age. Thus, a greater percentage of the population produce offspring, and total population size continues to increase.

The birthrate in the United States has dropped from a peak of nearly 3.8 births per woman of reproductive age in the late 1950s to 1.8 by the mid 1980s. However, the size of the population is still increasing. A rate of 2.1 births per woman for many generations is

The human population explosion has been caused by a decrease in the death rate.

FIGURE 32–15. Much of the world's human population is crowded together in cities. Crowding can have harmful effects.

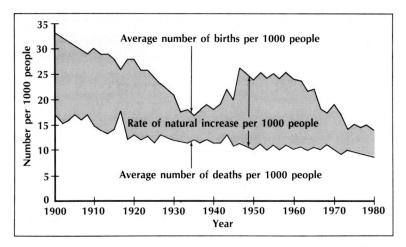

FIGURE 32–16. When the average number of births and deaths per 1000 people are compared, it becomes more apparent that because the number of births is greater than the number of deaths, the population is still growing.

necessary to attain zero population growth. **Zero population growth** is a condition in which the birthrate equals the death rate, so the rate of population growth equals zero. Because of the large proportion of people of reproductive age now in the United States, even a birthrate of 1.8 would cause the 1980 United States population of 222.5 million to double in 99 years.

The human population, like all other populations, interacts with its environment. It cannot continue to expand indefinitely because a point is reached in any population at which the environment cannot support additional growth. Most ecologists think that this point will be reached by humans. What do you think?

Most natural checks of population growth are increases in the death rate. An alternative for humans would be to decrease the birthrate. Decreasing the birthrate can be done as part of family planning methods. Biologists and physicians have developed several artificial ways of preventing pregnancy based on the biological aspects of reproduction. Natural methods to prevent pregnancy have also been learned.

Many countries around the world educate their citizens about family planning. Which, if any, family planning method is used depends on physical, emotional, religious, and moral factors. The results of this education may be a factor in the recent decline in growth rates of some human populations.

In zero population growth, the birthrate equals the death rate.

Family planning methods are a means of reducing birthrate.

Information on family planning is now available in many countries throughout the world.

REVIEWING YOUR IDEAS

12. How did agriculture affect early humans?
13. How do modern humans depend on technology?
14. How has the size of the human population changed during the last 2000 years? Is the birthrate or the death rate responsible for this change? Explain.
15. What is zero population growth?

INVESTIGATION

Problem: How do population changes alter population pyramids?

Materials:

2 population graphs

Procedure:

Part A. Stationary Population Graph Analysis

1. Make a copy of Table 32-4. Use this table to record your answers to questions 4a-g.
2. Obtain a population graph from your teacher marked "Stationary Population." This type of population shows little increase or decrease in size.
3. The age groups have been divided into three categories. Locate these three categories along the right side of the graph. They are defined as follows:
   (a) prereproductive—persons aged 0-14, portion of the population that will be reproducing in future years
   (b) reproductive—persons aged 15-44, portion of the population considered to be reproducing at the present time
   (c) postreproductive—persons aged 45+, portion of the population that generally does not reproduce and will not in the future
4. Part of the graph has been completed for you. Males are shaded in while females are left unshaded on the graph. The percentage of the total population for each age group by sex can also be read from the graph.
   (a) In age group 0-4, what percentage of the population is male?
   (b) In age group 0-4, what percentage of the population is female?
   (c) In age group 5-9, what percentage of the population is male?
   (d) In age group 5-9, what percentage of the population is female?
   (e) In age group 10-14, what percentage of the population is male?
   (f) In age group 10-14, what percentage of the population is female?
   (g) What percentage of the total population is of prereproductive age?
5. Complete the remainder of the pyramid using the data in Table 32-2. Remember to shade in males and leave females unshaded. It will be necessary to estimate the length of the line when given a decimal percent.

Part B. Growing Population Graph Analysis

1. Obtain a graph from your teacher marked "Growing Population." This type of population is increasing in size.
2. Complete the graph using the data in Table 32-3. Remember to shade in males and leave females unshaded.

TABLE 32-2. DATA FOR STATIONARY POPULATION GRAPH		
Age Group	% Male	% Female
15-19	3.4	4.0
20-24	3.4	3.9
25-29	3.2	3.9
30-34	3.0	3.8
35-39	2.8	3.6
40-44	2.6	3.0
45-49	2.6	3.0
50-54	2.4	2.8
55-59	2.1	2.0
60-64	1.6	1.8
65-69	1.2	1.3
70-74	1.0	1.0
75-79	0.6	0.8
80-84	0.3	0.5
85+	0.1	0.3

TABLE 32–3. DATA FOR GROWING POPULATION GRAPH		
Age Group	% Male	% Female
0–4	9.0	8.8
5–9	7.8	7.2
10–14	6.0	5.9
15–19	4.5	4.5
20–24	4.3	4.0
25–29	3.9	3.3
30–34	4.0	3.5
35–39	3.5	2.9
40–44	2.5	2.4
45–49	2.5	2.0
50–54	1.8	1.6
55–59	1.7	1.5
60–64	1.0	1.2
65–69	0.8	1.1
70–74	0.6	0.9
75–79	0.3	0.6
80–84	0.1	0.4
85+	0.1	0.3

Data and Observations:

TABLE 32–4. STATIONARY POPULATION	
Question	Answer
(a)	
(b)	
(c)	
(d)	
(e)	
(f)	
(g)	

Questions and Conclusion:

1. Define:
   (a) population
   (b) prereproductive group
   (c) reproductive group
   (d) postreproductive group

(e) stationary population
(f) growing population

2. The graphs that you have constructed are called population pyramid graphs. Describe and compare the general pyramid shapes for the two population graphs.

3. Compare the following categories for a stationary population to those of a growing population.
   (a) % of prereproductive groups
   (b) % of reproductive groups
   (c) % of postreproductive groups

4. Based on your percentages in question 3 and the general pyramid shapes of question 2, explain what problems or trends may be seen in the future in terms of food supply, housing, waste disposal, and impact on the environment:
   (a) for a stationary population
   (b) for a growing population
   (c) explain your answers to (a) and (b)

5. The following pyramid graph was prepared for a population.

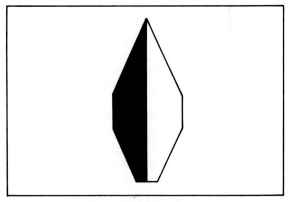

(a) Could it be considered a stationary, growing, or declining population?
(b) How does this population appear to differ from the others studied in regard to the % of prereproductive groups?

Conclusion: How do population changes alter population pyramids?

CHAPTER REVIEW

SUMMARY

1. The biotic potential of all populations is checked by limiting factors. **32:1**
2. The size of most populations usually increases rapidly and then levels off, producing an S-shaped population growth curve. **32:2**
3. In a predator-prey relationship, predation is usually a limiting factor on the prey population while food limits the predator population. **32:3**
4. Parasitism may be a density-dependent limiting factor, depending on the number of parasites a host is carrying. **32:4**
5. Disease may be a density-dependent limiting factor. **32:5**
6. Competition between two species for a requirement may itself become a limiting factor. **32:6**
7. Competition within a species may promote natural selection. Too much competition may be avoided by such factors as life cycles, territoriality, and emigration. **32:7**
8. Physical aspects of the environment, such as temperature and oxygen supply, can be density-independent factors that limit population size. **32:8**
9. Development of agriculture provided more food and contributed to an increase in population size. **32:9**
10. Technological advances have increased human life span and contributed to an increase in human population size. **32:10**
11. The human population is exploding as a result of a decrease in the death rate. One way of checking human population growth is by decreasing the birthrate. **32:11**

LANGUAGE OF BIOLOGY

biotic potential
birthrate
intraspecific competition
limiting factor
carrying capacity
competitive exclusion
    principle
death rate
emigration
environmental
    resistance
interspecific competition
parasitism
population density
population growth rate
    curve
population size
predation
S-shaped curve
zero population growth

CHECKING YOUR IDEAS

On a separate paper, complete each of the following statements with the missing term(s). Do not write in this book.

1. The _____ is the number of individuals which an environment can support.
2. An S-shaped curve is a graph that shows _____.
3. Experiments with *P. caudatum* and *P. aurelia* illustrate how _____ competition may be a limiting factor.
4. Parasitism is a density-_____ limiting factor.
5. _____ is the number of organisms in a population.
6. _____ is the number of organisms in a given area.
7. A social hierarchy is a way of reducing _____ competition.
8. _____ is the condition in which the birthrate equals the death rate.
9. _____ is the highest rate of reproduction of a population under ideal conditions.
10. The sum of limiting factors of a population is called _____.

EVALUATING YOUR IDEAS

1. Compare the growth of a population that has limits to its growth to a population with no limits.
2. Why does the population growth rate of an organism eventually decrease?

3. What is the rate of increase in population size when the carrying capacity has been reached?
4. Describe the interaction of a predator population and its prey population. What is the limiting factor for each population?
5. In what ways is predation beneficial to a prey population?
6. How are parasitism and disease related to population density?
7. Explain the competitive exclusion principle.
8. What happens when *Paramecium caudatum* and *Paramecium aurelia* are grown in the same culture medium? Why?
9. Why is some intraspecific competition good for a species? Why would too much be harmful?
10. Discuss several ways in which intraspecific competition is overcome.
11. Discuss two examples of density-independent limiting factors. Compare the effects of density-independent and density-dependent factors on a population.
12. How have agriculture and technology helped increase the human population size?
13. In what way has a decline in death rate contributed to the human population explosion?
14. Explain why, even if zero population growth were attained, it would still take a very long time for the human population size to stabilize. What role can family planning play in slowing the population explosion?

APPLYING YOUR IDEAS

1. How is the fact that all organisms have high biotic potentials adaptive?
2. Assuming that a human female has a "reproductive life span" of 30 years, how many offspring could a couple hypothetically produce during that time? If half of the offspring were females, how many children could they produce? What does this example indicate about the biotic potential of humans?

3. Explain how changes in overcrowded rat populations might be considered adaptive.
4. Discuss several ways in which technology has resulted in harmful effects on the environment.
5. The average height of Americans has increased several centimeters since the beginning of the twentieth century. How can this increase be explained? Is increased height an adaptation? Explain.

EXTENDING YOUR IDEAS

1. Place several male and female *Drosophila* into a vial containing a food source. Plug the vial with sponge rubber or cotton. Count new flies as they emerge each day. (Dispose of flies once they have been counted.) Prepare a population growth graph and a population growth rate graph of your data.
2. Conduct a class discussion on the limiting factors of the human population. Should the population be allowed to increase unchecked? How might it be controlled?
3. Use library resources to learn about the rate of population growth in other countries. Which countries seem to face the most severe problems? What is the standard of living in these countries?
4. Write a story about humans in the year 2050. Assume that the problems of pollution and population have not been solved.
5. How would you design an experiment to determine the population size of a given animal over a period of time?

SUGGESTED READINGS

Bernado, Stephanie, "Why Babies Die." *Science Digest*, Dec., 1985.
Strong, Donald R., "Banana's Best Friend." *Natural History*, Dec., 1984.

THE ECOSYSTEM

Organisms in an ecosystem interact with the physical environment and with one another. As it horns the mud prior to wallowing in the water, this water buffalo allows oxpeckers to settle on its back. What benefits might the oxpeckers provide the water buffalo? What benefits might the water buffalo provide the oxpeckers? How do organisms in other places interact with their physical environment and with other organisms?

opulations live together in communities (Section 1:1). A community is a group of populations living in the same area. The structure of a community is based on all the organisms living together. Within a community, organisms interact with members of their own species and with those of other species, and they interact with their physical environment as well. Interaction of a community with its environment is an ecological system, or **ecosystem** (EE koh sihs tum). Interactions between the living and nonliving parts of an environment are the functions of a community. Ecologists study a community in terms of its functions.

These functions are "give-and-take" relationships. Organisms "take" from the environment materials and energy needed for life. As they use these materials and energy, organisms "give," or transfer, materials and energy back to the environment. In a normal ecosystem, most materials are constantly recycled. But, energy cannot be recycled (Section 1:4). If an ecosystem is to survive, energy must be added constantly. The source of this energy is the sun. Energy is needed to maintain the organization of individuals and of entire ecosystems.

Objectives:
You will
- describe the transfer of energy in an ecosystem.
- discuss recycling of materials in an ecosystem.
- identify and explain the importance of various physical factors of the environment.

Organisms take energy and materials from their environment and release energy and materials back to the environment.

Materials can be recycled, but energy cannot.

BIOTIC FACTORS OF ENVIRONMENT

Many relationships exist among organisms. These relationships are called **biotic** (bi AHT ihk), or living, **factors** of the environment. In a normal ecosystem there is a delicate balance among the many forms of life.

Biotic factors are relationships among organisms.

693

FIGURE 33–1. (a) Autotrophs, such as these grassland plants, are producers and make up the first trophic level of an ecosystem. (b) Organisms such as these mountain goats that are herbivores, are called first-order consumers. (c) An otter eats sea urchins and thus is a second-order consumer. (d) Top level carnivores include animals such as lions.

In an ecosystem, materials and potential energy are transferred along a series of trophic levels.

First-order consumers are also called herbivores. Herbivores eat plants.

Carnivores eat meat.

Second-order and third-order consumers are carnivores.

33:1 Trophic Levels

Energy enters an ecosystem as sunlight. This solar energy can be directly used only by autotrophs such as grasses, trees, or algae (Section 1:2). In photosynthesis, autotrophs trap light energy and use it to make sugars and other organic compounds. Light energy is changed to chemical energy in these compounds.

Because only autotrophs can trap and change light energy to chemical energy, all other organisms depend on them. Materials and potential energy are transferred from one organism to another within an ecosystem. Each organism represents a step, or **trophic level,** in the passage of energy and materials. Autotrophs, the producers, make up the first trophic level of an ecosystem.

In order for an ecosystem to function, materials and energy must pass from producers to consumers. A consumer that feeds directly on a producer is called a **first-order consumer.** First-order consumers are also called **herbivores** (HUR buh vorz), or plant eaters. A cow grazing on grass, a deer browsing on foliage, and a tadpole eating algae are examples of herbivores. Herbivores make up the second trophic level in ecosystems.

Consumers that feed on other consumers are called **carnivores** (KAR nuh vorz), or meat eaters. A mouse is preyed on by snakes. Therefore, a snake is a carnivore and a **second-order consumer.** An owl that eats the snake is a **third-order consumer.** Second-order and third-order consumers make up higher trophic levels.

The transfer of materials and potential energy from organism to organism (trophic level to trophic level) forms a series called a **food chain** (Section 1:4). A food chain is represented using arrows.

Materials and potential energy are transferred along a food chain.

plant	$\rightarrow$	mouse	$\rightarrow$	snake	$\rightarrow$	owl
(producer)		(first-order consumer)		(second-order consumer)		(third-order consumer)

A food chain represents *one* possible route for the transfer of materials and energy through an ecosystem. Many other routes exist. In an ecosystem, many species are found at each trophic level, and one species does not always feed on the same food source. A snake may also eat a lizard or a toad; an owl may feed directly on a mouse. In addition, some animals are **omnivores** (AHM nih vorz), organisms that eat both plants and animals. Humans are omnivores. All the possible feeding relationships that exist in an ecosystem make up a **food web** (Figure 33–2).

Many interrelated food chains make up a food web.

FIGURE 33–2. A food web describes all the feeding relationships within an ecosystem. Some organisms can occupy various trophic levels.

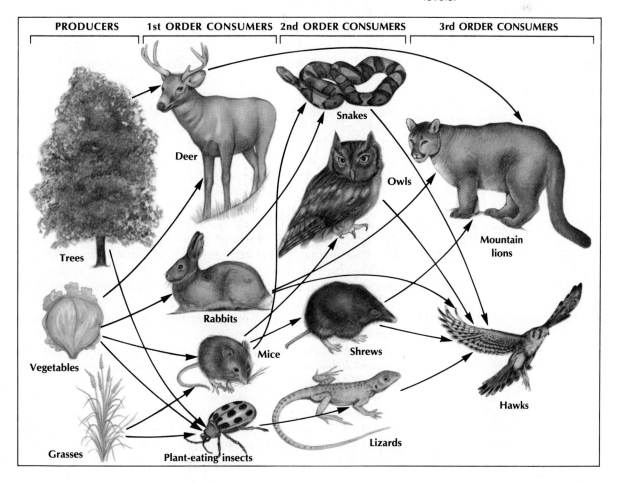

PRODUCERS	1st ORDER CONSUMERS	2nd ORDER CONSUMERS	3rd ORDER CONSUMERS

Trees

Vegetables

Grasses

Deer

Rabbits

Mice

Plant-eating insects

Snakes

Owls

Shrews

Lizards

Mountain lions

Hawks

FIGURE 33–3. A pyramid can be drawn to show the energy levels within an ecosystem. Little energy is available at the top of the pyramid because some is lost at each level. Graphs of the numbers and biomass of organisms at each trophic level also have pyramid shapes.

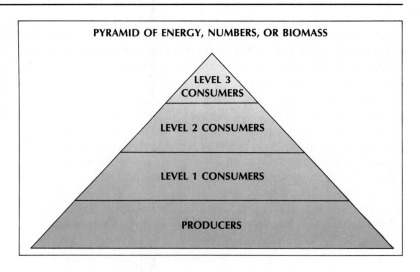

PYRAMID OF ENERGY, NUMBERS, OR BIOMASS

LEVEL 3 CONSUMERS

LEVEL 2 CONSUMERS

LEVEL 1 CONSUMERS

PRODUCERS

33:2 Pyramid of Energy

Food chains enable all organisms to obtain their share of the energy entering the ecosystem. However, energy is lost to organisms between each link in a food chain. The amount of energy actually transferred at each level depends on the ecosystem. Accurate data are difficult to obtain because of variables such as season, age and size of organisms, and amount of sunlight. A study of a river ecosystem in Silver Springs, Florida, was made by Howard Odum. It showed that only 17 percent of the potential energy in producers was transferred to herbivores. Second-order consumers obtained only 21 percent of the potential energy in herbivores. Only about 5 percent of the energy in second-order consumers was transferred to third-order consumers. The third-order consumers received only about 0.1 percent of the potential energy originally available in the producers. A graph of data such as these has a pyramid shape (Figure 33–3). Thus, ecologists refer to the transfer of energy as a **pyramid of energy.**

Loss of energy between each link in a food chain results in a pyramid of energy.

What happens to the energy in producers? Some of the potential energy that enters a food chain is used as each organism carries out its life functions. More than half of the potential energy of each food molecule is lost as heat during respiration. The usable energy from cellular respiration is used for life functions of growth and development, maintenance and repair, organization, reproduction, and obtaining food. During these functions, chemical energy is transformed to heat and is lost. This energy cannot be used again. Much of the potential energy at each trophic level never reaches the next level. Some organisms die and are acted on by decomposers. Also, members of a higher trophic level cannot consume every part of an organism. For example, only a small fraction of the organic material in a

Chemical energy from cellular respiration is used for life functions, but some energy is lost as heat.

tree is consumed by herbivores. Not all parts of the material eaten by an organism are digested. For all these reasons, each link in the food chain has less potential energy available to it than the previous link. How does this energy loss affect humans?

The pyramid of energy explains why an organism feeds on certain other organisms. A lion preys on large animals because it gets more energy from a large animal than from a small animal. The lion might use more energy catching a rodent than it would get by eating the rodent. Therefore, lions hunt and eat zebras, for example. One chase results in an energy "profit" for the lion.

33:3 Pyramids of Numbers and Biomass

Loss of energy in a food chain explains several ecological principles. A food chain rarely includes more than four links. So much energy is lost at each trophic level that usually not enough energy remains to support fourth- and fifth-order consumers. The loss of energy also explains why there are usually fewer organisms in each higher trophic level than in the previous level. For example, there are more mice than there are snakes that feed on those mice. This relationship is known as the **pyramid of numbers.** Each higher trophic level has fewer organisms than the previous trophic level. A pyramid of numbers could be drawn in the same way as a pyramid of energy (Figure 33:3).

The pyramid of numbers does not apply to all food chains, especially where a large organism is fed upon by smaller ones. For example, a single tree might be food for thousands of caterpillars, and a dog might be infested with many parasites.

Research has shown an expected relationship in the biomass in many ecosystems. **Biomass** (BI oh mas) is the total mass of dry organic matter per unit of area. Table 33–1 shows the amount of biomass in each trophic level in the Silver Springs, Florida, ecosystem. These data show a **pyramid of biomass** in the ecosystem. A pyramid of biomass can be drawn in the same way as the other pyramids. Each higher trophic level contains less biomass than the previous trophic level. The pyramid of biomass also results from the loss of energy along a food chain.

FIGURE 33–4. A timber wolf is at the "top of the pyramid." Top level consumers have less energy available to them than organisms at lower levels. Also, the number of wolves and their biomass is less than the number and biomass of animals on which they prey.

Because of energy loss along a food chain, each trophic level has fewer organisms than the preceding level.

Each level of a food chain has less biomass than the preceding level.

TABLE 33-1. BIOMASS IN AN ECOSYSTEM		
Trophic Level	Biomass (g/m^2)	Percent of Biomass Compared to Previous Level
Producers	809	—
Herbivores	37	4.6
Second-order consumers	11	29.7
Third-order consumers	1.5	7.3

FIGURE 33–5. (a) The caracara and (b) the hyena are scavengers. Scavengers aid in the recycling of materials in an ecosystem by feeding on dead animals.

a

b

33:4 Recycling of Materials

FIGURE 33–6. In an ecosystem, materials are constantly recycled. However, energy is lost and must be replaced.

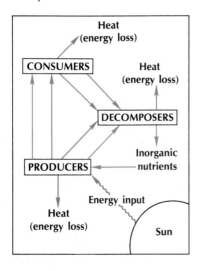

Although usable energy is lost as it is transferred through an ecosystem, it is replaced by energy of sunlight. But the amount of *matter is limited* and it must be recycled. Some carbon, hydrogen, and oxygen are recycled by the processes of photosynthesis and cellular respiration (Section 1:4). These materials are also released during the burning of fossil fuels such as coal and oil. But sooner or later, organisms die. If the matter in dead organisms were not recycled, all available matter would one day be used up. This problem does not occur because the matter in dead organisms and in waste materials is recycled in the ecosystem.

Decomposers (Section 1:4), mainly bacteria and fungi, release the materials in dead organisms and waste products, and these materials can be used again. Scavengers (SKAV un jurz) play a role in decomposition too. **Scavengers** are animals such as buzzards and jackals that feed on dead animals. Decomposers are consumers. As they feed on dead organisms and waste materials, decomposers produce carbon dioxide and water as by-products of respiration. These by-products are then recycled through the ecosystem. Decomposers not only recycle carbon dioxide and water, but also a variety of other materials. They convert organic matter to inorganic form. Inorganic materials are taken up and converted to organic molecules by autotrophs. Organic molecules are then passed along the food chain and so the cycle continues. Figure 33–6 summarizes the flow of matter and energy through an ecosystem.

33:5 Nitrogen Cycle

Nitrogen is an important element in all organisms. For example, it is an essential part of amino acids, proteins, and DNA. Nitrogen must be recycled if an ecosystem is to survive. Nitrogen moves through ecosystems in what is called the **nitrogen cycle** (Figure 33–7). The nitrogen cycle includes three major pathways, each of which involves microorganisms.

In one pathway, decomposer bacteria in soil or water act on proteins and other nitrogen-containing organic compounds in dead organisms or waste products. They convert these organic compounds into inorganic materials that can be absorbed by autotrophs. One group of bacteria converts the organic compounds to amino acids that are then metabolized, releasing ammonia (NH_3). Conversion of amino acids to ammonia is called **ammonification** (uh moh nuh fuh KAY shun). Ammonia reacts with water in the soil, forming ammonium ions (NH_4^+). These ions are then changed by other bacteria to nitrite ions (NO_2^-), which are then converted by other bacteria to nitrate ions (NO_3^-). These changes are called **nitrification** (ni truh fuh KAY shun). During this process, the bacteria obtain energy for their needs.

During the nitrogen cycle, nitrogen compounds of dead organisms and waste materials are converted to inorganic forms.

FIGURE 33–7. The nitrogen cycle is a complex pathway by which nitrogen moves through an ecosystem. It involves the interaction of organisms and nitrogen compounds.

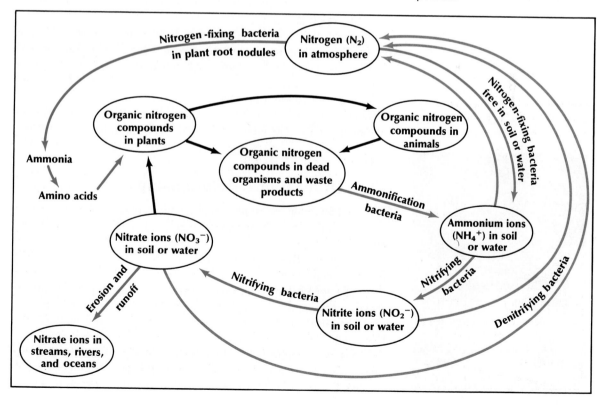

FIGURE 33–8. Nitrogen-fixing bacteria inhabit root nodules, shown here as white knobs on roots of a clover plant. These bacteria convert atmospheric nitrogen to nitrate ions and then to amino acids that the plant can absorb.

FIGURE 33–9. An orchid is an epiphyte that is a commensal with the trees in which it lives.

Most land plants absorb nitrogen from the soil in the form of nitrate ions. Plants use nitrate ions in making amino acids, proteins, and nucleotides (Section 9:2). These organic compounds become part of the food web. Thus, conversion of organic nitrogen compounds to nitrates is one way in which nitrogen is recycled.

Although the atmosphere contains about 78 percent nitrogen gas (by volume), most organisms cannot use nitrogen in that form (N_2). However, many forms of cyanobacteria (Section 13:4) and several kinds of other bacteria can use nitrogen gas. They change nitrogen gas to a usable form, a process called **nitrogen fixation.** Nitrogen fixation is another pathway in the nitrogen cycle.

All nitrogen-fixing organisms convert nitrogen gas to ammonia. The cyanobacteria and some of the other bacteria live free in soil or water, and nitrogen-fixation occurs there. The ammonia produced is then changed by nitrifying bacteria to nitrates. Still other nitrogen-fixing bacteria live in the roots of certain plants, especially legumes (peas, clover, and alfalfa). These bacteria enter the cortex cells (Section 22:1) of the root and cause them to become swollen areas called **nodules** (NAHJ ewlz). Some of the ammonia produced by these bacteria enters the soil and is converted to nitrates. Most of the ammonia is changed by the bacteria to amino acids, which then enter the plant cells. The amino acids are used by the plant to make proteins and other nitrogen compounds. Both the legumes and the bacteria benefit from this relationship.

A third pathway in the nitrogen cycle drains nitrogen from an ecosystem. By the process of **denitrification** (dee ni truh fuh KAY shun), some bacteria obtain energy by changing nitrates, nitrites, or ammonia to nitrogen gas. Thus, some nitrogen that could be used by plants escapes to the atmosphere. Also, erosion can wash away soil containing nitrates. These events affect the nitrogen cycle, but the overall amount of nitrogen in the soil usually remains fairly stable.

33:6 Other Biotic Relationships

The most obvious biotic relationships involve feeding, but organisms also interact in other ways. Some biotic interactions are highly specialized.

Commensalism (kuh MEN suh lihz um) is a relationship in which one organism benefits from a host, without aiding or harming the host. The benefitting organism is called a commensal. For example, in dense, tropical rain forests, the tree branches are so thick that the forest floor is dark. As a result, small plants cannot grow on the forest floor. Some small plants, such as orchids, live in the branches of trees and thus get enough light. Because of the high humidity in a rain forest, their roots can absorb water from the air. Plants like orchids that live in the tops of trees are called **epiphytes** (EP uh fites).

FIGURE 33–10. (a) A bromeliad is an epiphyte in rain forest trees. (b) Small tree frogs get moisture from plants, such as the bromeliad, that collect water.

The bromeliad (broh MEE lee ad), a member of the pineapple family, is another epiphyte of rain forests. Not only is a bromeliad a commensal, it is also a small ecosystem. The leaves of a bromeliad overlap to form a hollow in which water collects. Many arthropods and even tree frogs live in the water. Some of these frogs live entirely in the treetops.

Another commensal is "Spanish moss," again, a member of the pineapple family. It is an epiphyte of oak and other trees in the southern United States.

Animals exhibit commensalism also. A remora (REM uh ruh) is a small fish that attaches itself to the belly of a larger animal such as a shark. The shark provides a "free ride" for the remora, which feeds from the leftovers of the shark's meals.

An epiphyte living on a tree is an example of a commensal.

FIGURE 33–11. (a) These sea anemones exhibit commensalism with a hermit crab. Because the anemones are being transported on the crab's shell, they are better able to feed than if they were attached to a rock. (b) Barnacles may become attached to whales, which aids the sessile barnacles in obtaining food.

a

b

FIGURE 33–12. (a) Termites have protozoa within their intestines that enable them to digest wood. The relationship is mutualism. (b) Wasp larvae are parasites of the tomato horn worm.

Both organisms benefit in mutualism.

In parasitism, a parasite depends on the host, which it usually harms.

Mutualism (MYEW chuh lihz um) is an interaction in which two organisms depend on and benefit from each other. Termites are able to "eat" wood because their intestines contain certain protozoans that digest wood cellulose. In turn, termites supply the protozoans with materials for their metabolism.

Another example of mutualism is that of the yucca plant and yucca moth. Female moths transfer pollen from one yucca flower to another. These moths also deposit eggs in the developing seedpods of yucca plants. The eggs develop into larvae that feed on some of the seeds. The seeds not eaten can develop into yucca plants. Thus, the larvae have an adequate food supply, and pollination and reproduction of yucca plants are assured.

Parasitism is another relationship between organisms. In parasitism, one organism, the parasite, is completely dependent at some point in its life cycle on a host. That host is usually harmed. Parasites cause many plant and animal diseases. Parasites may also be important in limiting population size (Section 32:5).

Besides these specific relationships in ecosystems, many more general ones exist. Each ecological niche involves relationships among living and/or once living systems. For example, some birds build nests in trees, some insect larvae live in rotten logs, and some seeds are dispersed by animals. Humans use parts of plants and animals for clothing, chemicals, and tools.

REVIEWING YOUR IDEAS

1. Distinguish between a community and an ecosystem.
2. Define biotic factors of the environment.
3. Distinguish among herbivore, carnivore, and omnivore.
4. What is transferred in a food chain? Distinguish between food chain and food web.
5. What organisms are necessary for the recycling of materials through an ecosystem?
6. How are epiphytes examples of commensals?

ABIOTIC FACTORS OF ENVIRONMENT

In addition to the biotic factors of the environment, an ecosystem also is influenced by abiotic (ay bi AHT ihk) factors. **Abiotic factors** are the physical aspects of the surroundings. They can control the distribution and range of organisms and their reproduction, feeding, growth, and metabolism. An abiotic factor often can act as a limiting factor on a population (Chapter 32). Because populations live together in ecosystems, an entire ecosystem can be affected by abiotic factors. Study of abiotic factors and their importance is often difficult because they interact to affect the makeup and functions of the ecosystem. The following physical factors are presented separately, but keep in mind that they interact together.

Abiotic factors are the physical aspects of the environment. They interact to affect an ecosystem.

33:7 Water

All organisms need water. Water is important as a medium in which many organisms live. Water is also important to living systems in other ways. It is a major part of protoplasm, a solvent in which many metabolic reactions occur, and is a reactant in some reactions. How is water important to humans?

All living systems need water.

FIGURE 33–13. Water moves in a cycle. Water recycling depends on heat energy from the sun and metabolic activities of organisms.

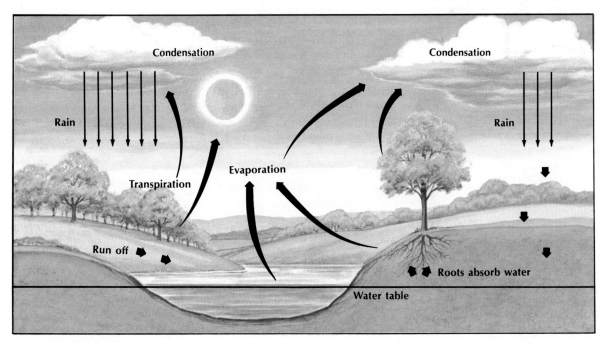

Condensation   Condensation

Rain   Rain

Transpiration   Evaporation

Run off   Roots absorb water

Water table

Water affects the distribution of land plants. All plants require water for photosynthesis, and plants such as mosses and ferns need water for reproduction. The amount of water on land is controlled by rainfall. The amount of rainfall, in combination with temperature, determines the type of plant that will be dominant in a given area. For example, forests are common in areas receiving 75 cm or more of rain per year. In regions where rainfall is from 25 to 75 cm per year, grasslands are common. Water is scarce in desert regions. Thus, desert plants and animals are adapted to obtain and conserve water (Section 34:10).

Amount of rainfall, in combination with temperature, determines the dominant plant life of an area.

33:8 Soil

Soil is very important in several ways to land organisms. Most plants are anchored in **soil.** Soil is the substrate upon which many animals move and in which many organisms live. Decomposers in soil return materials to the ecosystem.

Organisms aid the formation of soil. The beginning of soil formation is a mechanical process in which weathering (freezing, thawing, and erosion) breaks rocks into small particles that are the basis of soil. After small particles are formed, microorganisms and small plants begin to live there. The microorganisms' metabolism releases carbon dioxide that dissolves in water, forming a weak acid. The action of the acid and leaching continues the breakdown of the rock. Leaching is the dissolving of minerals out of rock or soil by water. Eventually, fine particles such as clay are formed.

Weathering and the action of simple organisms cause the breakdown of rock into smaller particles.

After the breakdown of rock into soil begins, organic material is added as the first organisms die. More plants begin to live in the soil, and then animals. As the plants and animals die, they build up on the surface of the soil and are decomposed. The decayed remains of organisms are called **humus** (HYOO muhs). Humus is an important part of soil because it contains organic material, and it enriches the soil. Thus, the remains of once-living systems provide the materials that other organisms will use later. Humus also increases the soil's capacity for holding the water and air necessary for plant growth. Humus is found in the upper, dark layers of soil called **topsoil.** Removal of topsoil by erosion (Section 35:10) can be a severe agricultural problem.

Soil fertility is determined by mineral content, organic content, and pH. Different plants are adapted for growth in different soils. Some plants can grow in acidic soils, while others grow best in neutral or basic soils. Lime, a basic material, often is added to acidic soils to make them neutral. Acid rain (Section 35:6) alters the pH of soil and may affect plant life. Acid rain is defined as rainwater with a pH of 5.6 or less.

FIGURE 33–14. Fertilizer is added to soil to restore minerals that have been lost. The three numbers indicated on fertilizer labels give the respective amounts of nitrogen, potassium, and phosphorus that the fertilizer contains.

FIGURE 33–15. Light is essential for (a) photosynthesis by green plants. (b) Vision also requires light. This biker could not be seen if light was not reflected.

Soil fertility is also determined by particle size. Few plants can grow well in loose, sandy soils. Most plants grow best in soil composed of small particles in which water and oxygen are plentiful.

33:9 Light

Light, or radiant energy, is the source of energy for all ecosystems. By photosynthesis, light energy is changed to the chemical energy needed by every living system.

Plant distribution is affected by the amount of light. Because light cannot penetrate deep into water, most algae live near the surface of aquatic communities. In a forest, trees are exposed to the greatest amount of light. Tree leaf arrangements are adaptations for maximum light exposure. Other forest plants are adapted to grow in the dim light of the forest floor. Some of these other plants begin to grow before leaves appear on the trees. Epiphytes receive enough light by growing in tree branches.

Only about two percent of the light striking Earth is used for photosynthesis. Most of the light is absorbed by the atmosphere, land, and seas and is converted to heat or thermal energy. Thus, light energy raises the temperature of land and water and in this way, maintains Earth's temperature. Also, heat has many specific roles. For example, heat is needed for the transpiration of water from plants. Heat also causes the evaporation of water from oceans and rivers. Without heat, water could not be recycled (Figure 33–13).

Light is also important to organisms in other ways. Light is required for vision. Some animals see well in dim light, but no animal can see in the complete absence of light. Light is also a factor in the migration of birds and the flowering of certain plants (Section 22:12). Light energy is needed to make vitamin D in humans (Table 25–2). Can you think of other examples of why light is important?

Light influences the distribution of plant life.

Light energy is transformed to heat that is necessary for many specific processes and also for life in general.

Light is necessary for vision and for triggering certain biological processes.

FIGURE 33–16. Life forms vary with altitude and latitude. Community types parallel one another in altitudinal and latitudinal succession.

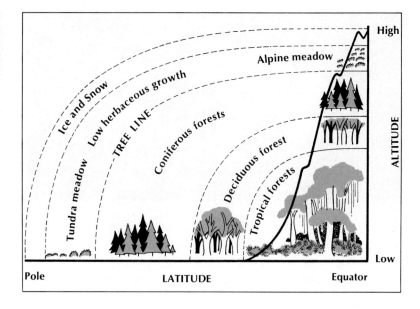

33:10 Temperature and Distribution of Organisms

Temperature of the environment has an important role in the distribution of organisms. In general terms, temperature changes with altitude and latitude. That is, temperature decreases from Earth's surface upward and from the equator north and south to the poles. The differences in temperatures, along with factors such as wind and moisture, affect the distribution of life forms. As a result, there is a continuous progression of community types from the equator to the poles. This is called **latitudinal** (lat uh TEWD nul) **succession.** The same pattern occurs with changes in altitude. This change is called **altitudinal** (al tuh TEWD nul) **succession.** The two types of successions parallel one another (Figure 33–16).

Latitudinal and altitudinal successions show similar changes in community types.

Temperature varies with time as well as with space. Temperatures differ from day to night and from season to season. Temperature differences also exist between times of sunshine and times of cloudiness. Such changes affect land organisms more than water organisms because air temperatures change more rapidly and to a greater degree than water temperatures. Air temperature can change as much as 40°C between day and night. In the same time period, the temperature of the surface (top ten meters) of the ocean changes no more than 4°C. Below ten meters, there is little change in ocean temperature. Thus, in temperature, water is a more stable environment than land.

Temperature changes affect life on land more than they affect life in the water.

Temperature, especially on land, is an important ecological factor. Bacteria can live within the greatest temperature range. Some bacteria form endospores and can withstand temperatures above the

FIGURE 33–17. Monarch butterflies are shown in flight above the trees. They migrate south in autumn to avoid the harsh winter and return north in spring.

boiling point and far below the freezing point of water (Section 13:2). For this reason, bacteria are found almost everywhere. Formation of spores or seeds is a feature of many plants. Some spores and seeds can survive temperatures that would kill a mature plant. In this way, the species survives extreme temperatures.

By forming endospores, some bacteria can withstand temperatures above the boiling point and below the freezing point of water.

Temperature often starts biological processes. For example, peach trees bloom when it is warm, but they must first be exposed to chilling during winter. The trees will not bloom without the chilling. Gypsy moth pupae will not develop into adults unless a freezing temperature has occurred. An adult, though, would die in freezing temperatures. These temperature requirements limit the distribution of the organisms. Peach trees and gypsy moths could not reproduce in the tropics.

Temperature often is responsible for the beginning of biological processes.

Temperature is among the factors that can affect the movement of organisms from place to place. Many birds and some mammals migrate in the winter. Often this migration of animals, such as bighorn sheep, caribou, and many birds, aids in finding food. This lack of food is brought on by the temperature (weather) changes.

Temperature changes can lead to a reduced food supply and the migration of animals.

33:11 Temperature and Metabolism

Metabolic rate in animals varies with temperature. Up to about 45° C, metabolic rate increases two or three times for every ten degrees the temperature is raised. Approaching 50° C, the rate of increase slows down. Then it falls off sharply to zero. At high temperatures, the enzymes needed for metabolism are destroyed. Therefore, most forms of life cannot survive when their body temperatures are above 50° C (Figure 33–18).

Ectotherms have body temperatures close to that of the environment. These animals have behavioral and physiological adaptations that prevent severe changes in body temperature and metabolism. A

Ectotherms have many behavioral and physiological adaptations that enable them to survive temperature extremes.

FIGURE 33–18. Metabolic rate is influenced by temperature. Above about 50°C, the rate decreases rapidly because enzymes are destroyed. Most organisms cannot live at temperatures above 50°C.

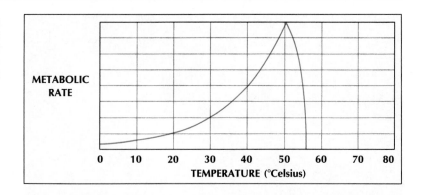

desert lizard is active in the early morning when it is cool. However, it will spend the hottest part of the day where it is cooler, such as behind or under a rock. A frog may spend the winter buried in the mud at the bottom of a pond. The frog's metabolism and energy use are greatly reduced. This condition is called hibernation. A frog may again go through a dormant period during the summer. This period of summer dormancy is called **estivation** (es tuh VAY shun). It allows an organism to "escape" from the heat or from drought.

Birds and mammals are endotherms. Their body temperatures are constant and are controlled internally.

A human has a body temperature of about 37° C. In a polar bear it is about 38° C, and in an elephant it is about 36.2° C. Why are these temperatures so similar? They are points at which enzymes operate efficiently and metabolism occurs rapidly. Thus, enough energy is made available for the animals' complex activities. Some endotherms have physiological adaptations to survive severely cold weather. For example, many small mammals, such as chipmunks and bats, hibernate during the winter. During hibernation, body temperature de-

Body temperature of endotherms is maintained at a constant level favorable for metabolism.

FIGURE 33–19. (a) Earthworms estivate and (b) a bat hibernates. During estivation and hibernation, body metabolism is slowed and less energy is expended.

a

b

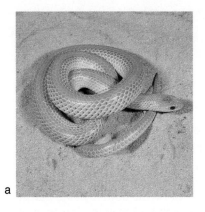

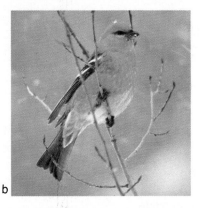

a   b

FIGURE 33–20. (a) The snake, *Sonora simiannulata,* is an ectotherm. Its internal temperature varies with its surroundings. (b) The pine grosbeak, *Pinicola enucleator,* an endotherm, is maintaining a fairly constant internal temperature.

creases as the metabolic rate decreases. Body temperature in vertebrates is monitored and regulated by the hypothalamus (Section 29:9). The hypothalamus is sensitive to its own temperature and to temperature information received from heat sensors located in other parts of the body.

The hypothalamus regulates body temperature in vertebrates.

Adjustments to external temperature involve both physiological and behavioral changes. For example, you may perspire when hot. As perspiration occurs, the skin is cooled. Also, some arteries in your skin expand and heat escapes. You also tend to be less active, so less heat is generated by the muscles. When you are cold, you may become more active and even shiver. Arteries in the skin contract. These changes conserve heat.

REVIEWING YOUR IDEAS

7. Define abiotic factors of the environment.
8. How is water necessary for life?
9. How is soil important to an ecosystem?
10. In general, how is light important to plants? How is it important to animals?
11. How do altitude and latitude affect temperature? What are several ways that temperature affects organisms?

FIGURE 33–21. As these runners perspire, their body temperatures are lowered. The hypothalamus is responsible for regulating body temperature.

Problem: How does one measure soil humus?

Materials:

graduated cylinder   large test tube
flask   ring stand
glass-marking pencil   water
rubber tube   hydrogen peroxide
clamp   5 soil samples
metric ruler   shallow pan
1-hole stopper with glass tube

Procedure:

1. Copy Table 33–2 onto a sheet of paper.
2. Complete the column marked "Type/location of soil sample." Your teacher will provide you with this information.
3. Use a glass-marking pencil to place marks 4 cm and 8 cm from the bottom of the large test tube.
4. Set up the apparatus as shown in Figure 33–22.
5. Fill a shallow pan with water to a height of about 3 cm.
6. Fill the graduated cylinder with water. Cover the cylinder with your hand and invert the cylinder into the pan of water. NOTE: The cylinder must be full of water after it is inverted. If it is not, repeat this step.
7. Slightly raise the cylinder so that its bottom is not resting directly on the bottom of the pan. It must, however, still be below the surface of the water in the pan. Secure the cylinder with the ring stand and clamp.
8. Feed the loose end of the rubber tube into the cylinder. Use Figure 33–23 as a guide. Be sure that the rubber tube is not pinched closed by the cylinder.

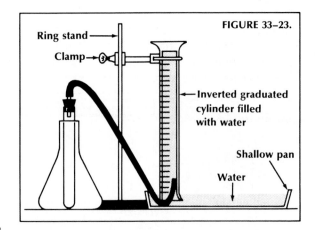

FIGURE 33–23.

Ring stand

Clamp

Inverted graduated cylinder filled with water

Shallow pan

Water

9. Remove the stopper from the test tube previously prepared in steps 1–4. Make sure that the rubber tube has not been pulled out from under the cylinder. Fill the tube to the 4 cm mark with soil sample A.
10. Fill the test tube to the 8 cm mark with hydrogen peroxide. Quickly stopper the tube.
11. After 10 minutes, record in Table 33–2 the volume of gas in milliliters that has collected in the cylinder. The gas collected is a measure of the amount of humus present in a

Glass tube
1-hole stopper
Large test tube
Rubber tube
8 cm mark
Flask used as test tube support
4 cm mark

FIGURE 33-22

sample of soil. Use Figure 33–24 as a guide.

12. Repeat steps 6–11 using soil sample B. Begin this test using a clean test tube and a cylinder totally filled with water.

13. Repeat steps 6–11 using soil sample C. Begin this test using a clean test tube and a cylinder totally filled with water.

14. Repeat steps 6–11 using soil samples D and E. Begin each test using a clean test tube and a cylinder totally filled with water.

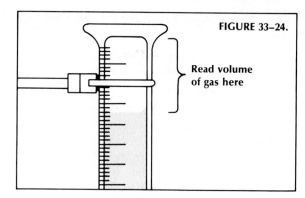

FIGURE 33–24.

Read volume of gas here

Data and Observations:

Soil sample	Type/location of soil sample	Volume of gas formed by humus
	TABLE 33-2. MEASUREMENT OF GAS PRODUCTION	
A		
B		
C		
D		
E		

Questions and Conclusion:

1. Define:
   (a) soil
   (b) topsoil
   (c) humus

2. Hydrogen peroxide (H_2O_2) reacts chemically with humus (mainly carbon) to form oxygen gas. In this investigation:
   (a) What gas was formed during the reaction?
   (b) What chemical was the source of this gas?
   (c) Where was the gas in (a) observed or collected?

3. The amount of gas produced by a soil sample is directly related to the amount of humus contained in it.
   (a) Which sample produced the most gas?
   (b) Which sample contained the most humus?
   (c) Which sample produced the least gas?
   (d) Which sample contained the least humus?

4. (a) Describe the source of humus in soil.
   (b) Considering your answer above, explain why samples A and B, even though they were both taken from the same place, had differing amounts of humus present.

5. Water retention and mineral availability to plants is improved by the amount of humus present in soil.
   (a) Which soil sample would possibly be best for growing plants?
   (b) Explain.
   (c) Which soil sample would possibly be worst for growing plants? Explain.

Conclusion: How does one measure soil humus?

CHAPTER REVIEW

SUMMARY

1. In an ecosystem, organisms interact with one another and with the physical factors of their environment. **p. 693**
2. Materials and potential energy are transferred through trophic levels within an ecosystem by food chains. **33:1**
3. Because energy is lost along a food chain, less energy is available at each trophic level. Loss of energy results in a pyramid of energy. **33:2**
4. Loss of energy also results in a pyramid of numbers and a pyramid of biomass. **33:3**
5. Because energy cannot be recycled, a constant energy source, the sun, is needed to maintain an ecosystem. Materials in an ecosystem are recycled. Carbon, hydrogen, and oxygen are recycled as a result of processes including respiration and photosynthesis, and through the activities of decomposers and scavengers. **33:4**
6. Decomposer bacteria and nitrogen-fixing bacteria are important organisms in the nitrogen cycle. **33:5**
7. Commensalism, mutualism, and parasitism are examples of biotic relationships that are specialized. **33:6**
8. Water is part of protoplasm, a solvent for cellular reactions, and a reactant in some cellular reactions. Amount of rainfall is important in the determination of dominant plant life in a given area. **33:7**
9. Type of soil influences the type of plant life and animal life in an ecosystem. **33:8**
10. Light influences photosynthesis. It also is changed to heat, which provides a suitable range of temperatures for life in air, on land, and in water. Heat also is necessary to "drive" the water cycle. **33:9**
11. Variation in temperature influences the distribution of organisms. Temperature also triggers many biological processes. **33:10**
12. Temperature affects metabolism. Animals have adaptations for survival in different temperature regions. **33:11**

LANGUAGE OF BIOLOGY

abiotic factor	mutualism
altitudinal succession	nitrogen cycle
biotic factor	nitrogen fixation
carnivore	nodule
commensalism	omnivore
ecosystem	pyramid of biomass
estivation	pyramid of energy
first-order consumer	pyramid of numbers
food chain	scavenger
food web	second-order consumer
herbivore	third-order consumer
humus	topsoil
latitudinal succession	trophic level

CHECKING YOUR IDEAS

On a separate paper, complete each of the following statements with the missing term(s). Do not write in this book.

1. A first-order consumer is also called a(n) _____.
2. _____ are animals that feed upon other animals.
3. _____ is a relationship in which one organism benefits and the other is neither aided nor harmed.
4. Because of its diet, a human is an example of a(n) _____.
5. All possible feeding relationships in an ecosystem are known as a(n) _____.
6. Changing nitrogen gas to a usable form is called _____.
7. Rate of metabolism is related to an organism's _____.
8. _____ is the progression of community types from the equator to the poles.

33

9. In _____, two organisms live in close association and both benefit.
10. _____ is the rich, organic portion of top-soil.

19. Explain some ways in which ectotherms respond to changes in temperature.
20. In what ways is constant body temperature adaptive?

EVALUATING YOUR IDEAS

1. Why is a constant supply of sunlight needed to maintain an ecosystem?
2. What are the general names for the links in a food chain?
3. Does each ecosystem have specific food chains? Explain.
4. What factors result in a pyramid of energy?
5. Why do large carnivores eat other large animals rather than small animals?
6. Why is there a pyramid of numbers?
7. Why do plants need nitrogen? Explain the pathways by which nitrogen becomes available to plants in the nitrogen cycle. Include the activities and importance of nitrogen-fixing bacteria.
8. Distinguish between commensalism and mutualism. Give examples.
9. How does parasitism differ from commensalism and mutualism?
10. How does availability of water determine the distribution of plants?
11. How is soil formed?
12. How does the nature of the soil help determine the structure of a community?
13. How does light affect the distribution of plants in a forest?
14. How is light energy converted to heat energy? How is heat energy important?
15. Distinguish between latitudinal succession and altitudinal succession.
16. In terms of temperature, which environment is more stable, water or land? Why?
17. How are spores and seeds an important adaptation to temperature?
18. How does temperature affect metabolism in animals?

APPLYING YOUR IDEAS

1. Explain several ways that humans fit into food chains. Are humans usually preyed on by carnivores? Explain.
2. Why is wood such a good source of energy?
3. In what ways does a dog stay cool in hot weather?
4. How may clearing land for agriculture affect a natural ecosystem?

EXTENDING YOUR IDEAS

1. Assume that a small group of humans is leaving Earth to colonize another planet. You are on a committee to develop plans for a well-organized ecosystem in the new environment. Report on your plans.
2. Prepare a report on the place of humans in the ecosystem. Discuss their needs and how these needs are met. Also, discuss how humans have affected the ecosystem.
3. Phosphorus is an important mineral needed for life. Find out how phosphorus is recycled in an ecosystem.

SUGGESTED READINGS

Mopper, Susan, and Whitham, Thomas, "Natural Bonsai of Sunset Crater." *Natural History,* Dec., 1986.

Morse, Douglass, H., "Milkweeds and their Visitors." *Scientific American,* July, 1985.

ORIGIN AND DISTRIBUTION OF COMMUNITIES

A community is a group of populations occupying the same area. The development of a community depends on a variety of physical factors that influence the kinds of producers that grow. The kinds of producers present, in turn, determine the types of consumers present. Because the physical factors in an area remain basically the same, characteristic types of organisms may be found. What factors affect the development of a community? How do communities differ? What changes may naturally occur in a community?

chapter 34

M any natural biological communities exist. They include lakes, forests, and open meadows. If you study any one of these communities, you can determine its common plants and animals. Also you can observe the effects of biotic and abiotic factors within it. In short, you can analyze the community as it exists today.

Have the same plants and animals always been in the community? How did the plants and animals get there? Were the same biotic and abiotic factors important today also important in the past? Can a community change with time? Are there communities in other parts of the country and the world similar to the ones where you live? What factors might cause similarities? Keep these questions in mind as you study this chapter.

Objectives:
You will
- describe the changes in community life that occur during ecological succession.
- compare and contrast primary succession and secondary succession.
- identify and describe various examples of biomes.

ECOLOGICAL SUCCESSION

Like any other biological level of organization, a community is a living system; it has a developmental history. The series of changes in a community during its development is called **ecological succession** (ee kuh LAHJ ih kul • suk SESH un). Succession may occur as a result of natural, orderly changes, or it may follow a disaster such as fire or disease. Each part of a community's history is well integrated and ordered. The "characters" and "stage" change with time, but the "plot" remains the same.

During its development, a community undergoes a series of changes called ecological succession.

34:1 Primary Succession

Succession that begins with bare rock is called **primary succession.** The first stage of primary succession involves "hardy" autotrophs that can grow under adverse conditions. The first stage of succession is called the **pioneering stage.**

Primary succession is ecological succession in an area that has not been previously occupied by a community.

FIGURE 34–1. The first stage in primary succession of a forest is lichens growing on bare rocks. As they die and form humus, larger plants such as mosses, grasses, ferns, and shrubs can grow.

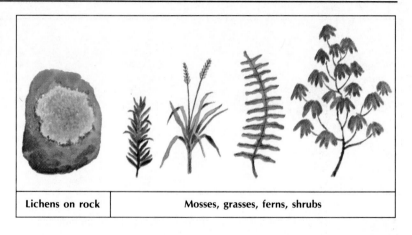

| Lichens on rock | Mosses, grasses, ferns, shrubs |

"Hardy" organisms are the first to live in a new area.

The types of plants and animals present in an area change during succession.

Each stage of an ecological succession has a more complex set of organisms. Each stage of succession "prepares" for more complex stages.

Consider a natural forest community in New England. This forest might have begun long ago with lichens growing on bare rock. Lichens produce acids that may cause rock to become corroded. Freezing and thawing of water in cracks in rock then cause it to break into smaller fragments. As lichens die, bacteria and other decomposers break them down. Small amounts of organic materials may collect in the cracks of the rock, thus beginning soil formation (Section 33:8). Mosses may then root in the "primitive" soil. As mosses die and decompose, they further enrich the soil. Later, more complex plants such as ferns, grasses, and shrubs appear. The decay of these plants further increases the humus content of the soil, making it thicker and richer.

While these changes in plant species are occurring, animal life is also changing. Animals such as nematodes and earthworms add to the richness of the soil, and insect life becomes abundant above the soil. The insects may help pollinate the plants and speed the population growth of the plants. Birds and small mammals may also enter the area.

The kinds of plants and animals present at each stage of succession not only change the soil, but also change other abiotic and biotic factors. For instance, larger plants cast shadows that cause subtle differences in temperature, moisture, and light in some small parts of the community. These slight differences create areas that are known as **microenvironments.** Differences in organisms also influence food chains and energy flow (Section 33:1). All these changes result in the eventual replacement of the organisms present at each stage. The organisms are replaced by different organisms in later stages.

In the New England community, pine trees gradually come to replace the grasses and shrubs. Seedlings of beech and maple trees grow in the shade of the pines. Gradually beeches and maples replace the pines by crowding them out.

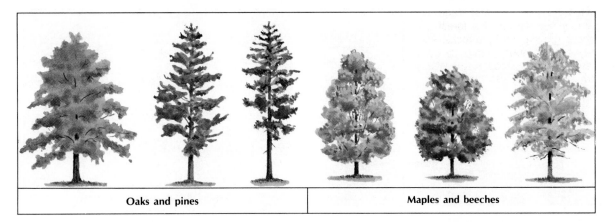

| Oaks and pines | Maples and beeches |

34:2 Climax Community

Beeches and maples are known as the **dominant species** in the New England forest. They are the final, stable stage in the development of the community called the **climax community.** The type of climax community that emerges in a given area depends on the interaction of all factors of the environment, both physical and biotic. The climax community in this example is a **beech-maple forest.** In the southeastern United States, the climax community might be an **oak-hickory forest.**

Because it is complex, a climax community tends to remain the same over a long period of time without further succession. But factors such as changes in climate, disease, a natural disaster, or human interference could be so severe that further succession might occur later.

34:3 Secondary Succession

New succession often begins as a result of natural destruction. It may also result from interference by humans. In either case, new succession begins when the dominant species of a plant community are removed. If the land is later left untended, **secondary succession** will occur.

During the early history of the United States, much of the forest land of New England was cleared and used for farming. As people moved westward, many of these farms were abandoned. Secondary succession then began in the abandoned fields. This type of succession is still occurring today in New England (and in other parts of the world). At first, hardy grasses and weeds appear in the fields. Later, shrubs grow and they are followed by junipers, poplars, and white pines. Eventually, beeches and maples begin to sprout in the shadow of the other trees so that what was once an open field becomes a beech-maple forest.

FIGURE 34–2. Succession leads to a climax community. Oaks and pines are replaced by beech and maple trees in a beech-maple climax community.

A climax community is stable and usually has a dominant plant species.

Certain factors may alter a climax community and lead to further succession.

Secondary succession may occur in an area in which there has already been a community.

FIGURE 34–3. Secondary succession occurs when natural growth is disrupted. (a) After a forest fire, secondary succession occurs eventually resulting in formation of the forest again. (b) This abandoned farm shows early signs of secondary succession.

In this example, the sequence of stages of community development is different from that beginning with bare rock, but the climax community is the same. However, most ecologists today believe that succession in a given area does not necessarily lead to the same climax community all the time. The characteristics of a climax community probably result from many chance occurrences including both biotic and abiotic environmental factors over long periods of time. In general, though, each climax community is more stable and productive than previous stages of the succession that produced it.

34:4 From Lake to Forest

A lake is a body of water surrounded by land. It is a type of ecosystem in which cyanobacteria and green algae are the main producers. Toward the shore, plants such as cattails also provide energy. Consumers consist of ciliates (SIHL ee ayts) and other protists, insect and other larvae, worms, crayfish, and fish. Study of a lake shows that it is a stable, well-organized ecosystem.

A lake is not a permanent ecosystem. It eventually undergoes succession.

A lake is organized, but it is not a climax community. A lake is not a permanent ecosystem; it undergoes succession. The life span of a lake depends on its size and the conditions of the environment.

Succession of a lake occurs from the edges toward the middle.

Consider a lake surrounded by a beech-maple forest. As the lake ages, sediments collect at its bottom. The shallow portions around the shoreline become filled first. In this marshy soil simple plants such as *Sphagnum* moss, insectivorous plants, and shrubs take root. Sediments continue to collect on the bottom of the lake. As land takes the place of water, simple plants continue to grow. Meanwhile, the areas around the old shore undergo further succession. Shrubs are replaced by trees, and finally, beeches and maples become dominant. This pattern continues until no more water remains (Figure 34–4). A

Eventually a lake is replaced by a climax community.

beech-maple forest, the climax community, gradually replaces the lake.

If you examine the area surrounding an old lake, you may see evidence of succession. Close to the lake are hardy, simple plants common to marshy areas. Areas farther from the lake support life of increasing complexity up to the climax community. Do human activities affect succession? Explain.

FIGURE 34–4. A lake is not a permanent ecosystem. As time passes, the lake becomes filled and the surrounding climax community replaces it.

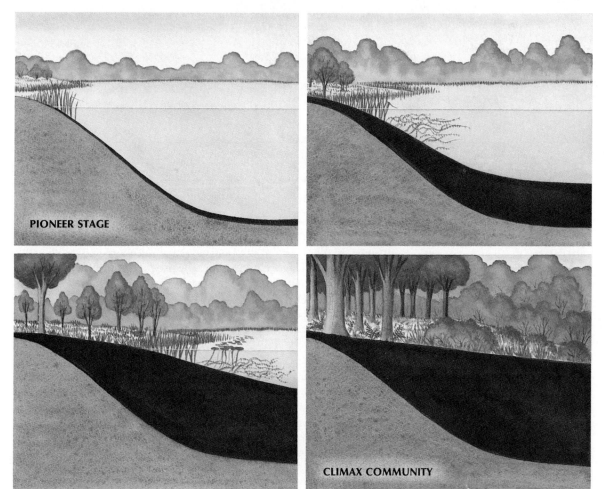

PIONEER STAGE

CLIMAX COMMUNITY

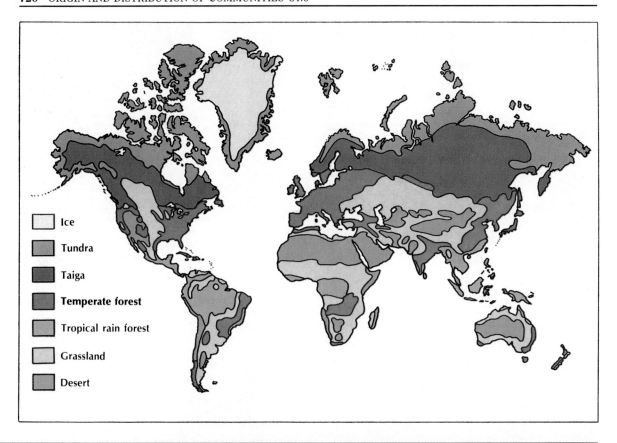

Ice

Tundra

Taiga

Temperate forest

Tropical rain forest

Grassland

Desert

FIGURE 34–5. The major world biomes can be shown on a map. Notice how areas with similar climate have the same biomes.

FIGURE 34–6. One major difference among the biomes is the average amount of annual rainfall.

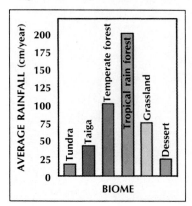

BIOMES

Several communities with the same major life forms make up a **biome** (BI ohm). Geography and climate determine the type of biome an area will be. These factors affect the producers in each community, which in turn affect the other life forms. Figure 34–5 shows the major world biomes. Distribution of these biomes shows a latitudinal succession. A similar altitudinal succession also occurs (Section 33:10).

34:5 Tundra

The **tundra** (TUN druh) biome is found at the northernmost latitudes. It has low average temperatures and low annual precipitation. The precipitation of the tundra is mostly in the form of snow. Converted to rain, the annual precipitation would be about 12 cm per year. Because the growing season is short and rainfall is low, forests cannot develop. In the winter, water in the soil is frozen. Although a slight amount of thawing occurs in summer, thawing extends down for

only a few meters. Thus, there is always a layer of frozen ground called **permafrost** (PUR muh frawst) that prevents the rooting of large plants.

Producers of a tundra are lichens, mosses, grasses, sedges, and some low, stunted shrubs and bushes. Rodents such as mice, moles, and lemmings are common in the tundra. Other animal life includes snowshoe hares, polar bears, reindeer (caribou), and foxes. Some animals live in the tundra only in warmer weather. Mosquitoes and flies are common in the summer. Many birds migrate to the tundra in summer, though few birds stay all year.

The tundra is a very fragile biome that is being affected by humans. Disturbance of the vegetation that insulates the permafrost can cause widespread erosion. Footpaths used for one season may be visible decades later. Vehicles have left tracks that have eroded into gullies three meters deep. The search for oil in the area is also posing grave threats to the tundra, reducing grazing ranges, and disturbing the migratory patterns of animals such as caribou.

FIGURE 34–7. (a) The tundra contains (b) low shrubs and bushes. Harsh winters with strong winds, permafrost, short summers, and lack of available water prevent growth of large plants.

Lichens, mosses, grasses, and some shrubs are producers that can exist in the permafrost of the tundra.

FIGURE 34–8. Animals of the tundra are adapted for life in cold zones. (a) Musk oxen have thick fur that protects them. (b) Ptarmigans have feathers that cover their feet and protect against cold.

FIGURE 34–9. The taiga, land characterized by coniferous forests, is the largest of the world's biomes.

34:6 Taiga

South of the tundra are the great coniferous (kuh NIHF rus) forests (Section 14:9) covering large portions of Canada, Alaska, and the Soviet Union. This biome, called **taiga** (TI guh), is probably the largest of the world's biomes. The more southern areas of the taiga have a rainfall of about 35 to 40 cm per year. The taiga also has a lot of fog and a low evaporation rate that results in a very wet area. The soil is full of water and is acidic. There is no permafrost because the temperature in the taiga is higher than that in the tundra.

Probably the largest of Earth's biomes is the taiga.

Trees can grow in taiga soil. Once spruce and firs begin to grow, they further change an area by creating microenvironments. Trees of the taiga provide shade, homes, and protection for many organisms. Trees also cause changes in the soil.

The taiga is slightly warmer than the tundra, has relatively little precipitation, and has no permafrost.

Animal life common in the taiga includes moose, caribou, weasels, mink, and ermine (UR mun). Crossbills are birds well adapted to the taiga. Because their upper and lower bills overlap like a pair of scissors, crossbills (Figure 34–10) are able to break apart cones and feed on the seeds.

FIGURE 34–10. Taiga animals include (a) the moose and (b) the crossbill, which are adapted to survive in coniferous forests.

a

b

34:7 Temperate Forest

Regions with a rainfall totaling about 100 cm per year make up the **temperate forest** biome. As the name suggests, this biome is found in temperate regions of North America, South America, Europe, and Asia. Temperate forests are usually characterized by definite seasons. The growing season, therefore, lasts at least six months.

Temperate (and other) forests have **vertical stratification** (strat uh fuh KAY shun), a series of **strata** (STRAYT uh), or layers, from top to bottom. The top layer is the **canopy** (branches) of the trees. The canopy shades the ground below and provides homes for animals such as insects, squirrels, and birds. Most of the food is produced in the canopy. Next is the **shrub** layer, under which is the **forest floor.** Light is scarce on the floor so only plants that can grow well in dim light are present. Insects, other arthropods, and snakes are common on the forest floor. A fallen, rotting log may contain a microcommunity of life. Beneath the ground, in the soil, is the lowest level. In this layer are many small arthropods, worms, and decomposers.

Each level of the forest has its own set of environmental conditions. Each is a separate zone. Although an organism usually lives in only one level, the organization of the community depends on the interaction of all organisms.

Most temperate forests in North America and Europe are composed of **deciduous** (dih SIHJ uh wus) trees. A temperate deciduous forest is one in which leaves regularly fall from the trees, and so at times, the trees are "bare." Most leaves fall in the autumn and new

FIGURE 34–11. Temperate deciduous forests are found in regions that get about 100 cm of rain per year.

Forests have vertical stratification.

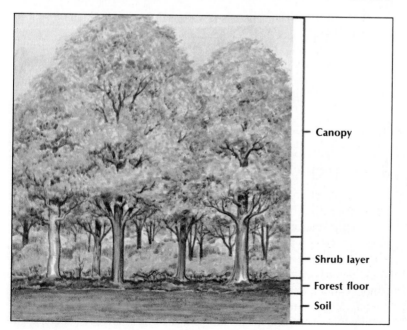

Canopy

Shrub layer

Forest floor

Soil

FIGURE 34–12. Forests show vertical stratification. Life forms in each layer vary.

a   b

FIGURE 34–13. Temperate deciduous forest areas appear much different in winter. (a) Spring brings the budding of trees and shrubs, and the area supports much animal life. (b) In winter, plant leaves are lost, and the area appears nearly lifeless.

During the winter, a temperate deciduous forest is dormant. The forest is full of life in the spring and summer months.

buds open in the spring. Most of the region east of the Mississippi River is the temperate deciduous forest biome of the United States.

A temperate deciduous forest changes with the seasons. The falling of leaves in the autumn affects the community in many ways. Most important is that the food supply is reduced because food sources such as leaves, seeds, and fruits are less abundant. Also, without leaves, the effects of cold temperatures and wind are more severe. The net effect of these changes is a dormant period during the winter months. The forest floor is covered with dead leaves. Also, animal life is scarce as hibernation and long winter sleeps are common among the animals. In the spring, there is a rapid burst of growth as plants on the forest floor begin to bloom and leaves form on the trees.

Some of the major animals of the temperate deciduous forest in the United States are Virginia deer, black bears, opossums, salamanders, and squirrels. Birds and insects are also abundant.

Fire can be a major problem for temperate forest communities. Fire has proved especially destructive in the western United States, destroying not only plant and animal life, but also human property.

34:8 Tropical Rain Forest

In the equatorial (ek wuh TOR ee ul) regions around the world is the **tropical rain forest** biome. These forests are found in Central and South America, central Asia, parts of Australia, and Africa. This biome is often called jungle, but a **jungle** is really a very dense growth resulting from increased sunlight along a riverbank or other area that has been cleared. In a tropical rain forest, the temperature is almost constant at 25°C and rainfall is heavy. A total of more than 200 cm of rain falls per year. Heavy rainfall and high temperature produce a humid environment much like a florist's greenhouse, and growth

FIGURE 34–14. Heavy rainfall and constantly warm temperatures contribute to the luxuriant growth of the tropical rain forest.

occurs constantly. The tropical rain forest has the most abundant growth of all the biomes.

Most prominent of the plants in the tropical rain forest are large trees reaching 30 to 40 m in height. Unlike coniferous and deciduous forests, a tropical rain forest has no dominant species of trees. In fact, there may be hundreds of species in the same forest. The branches of the trees form a large canopy, which is a major layer of the forest. The canopy is so thick that the forest floor may be dark, even at noon. Because the forest floor is dim, floor plant life is scarce. A few shrublike plants may survive, but often the soil is bare.

Common in this biome are vinelike plants called **lianas** (lee AHN uz). Lianas are rooted in the ground, but their leaves are in the canopy where they receive sunlight. Also in the canopy are epiphytes such as the bromeliad. These plants trap and store water in the hollow areas where their leaves are joined. This water is used by the bromeliad (Section 33:6) and by the animals that live in the canopy.

A tropical rain forest is a year-round "hothouse" of luxuriant growth.

A tropical rain forest has no dominant species of trees.

Lianas and epiphytes are commonly found in the canopy of a tropical rain forest.

a

b

FIGURE 34–15. (a) The slow loris is a tropical rain forest animal that is nocturnal and often sleeps hanging by its feet from a tree limb. (b) Lianas are vines that are rooted in the soil but have their leaves in the canopy, where they receive sunlight.

Although plants are more evident, animals abound. The forest is alive with many arthropods, such as beetles, and vertebrates, such as tree frogs. Most of the mammals in the rain forest live in the trees.

Many tropical rain forests exist in developing countries of the world. For economic reasons, huge tracts of these forests are being destroyed. Trees are sometimes slashed, dried, and then burned, and crops are planted right among the ashes. Some cleared areas have been planted with grass used to feed cattle. However, these cleared areas can be used in these ways for only a few years. Because high temperature and humidity cause rapid decay of organisms, little humus is formed. Minerals are leached from soil due to heavy rainfall. Thus, the soil is poor in nutrients to begin with and those that are present are quickly used up by the new plants. The land is abandoned as useless.

Tropical rain forests have also been cleared to provide a source of wood and to make space available for new roads and mines. In the process, a great many species have been destroyed. Biologists are concerned that potentially useful organisms (and chemicals produced by those organisms) are being eliminated before they can be discovered. It is also possible that removal of the forests may affect concentrations of both oxygen and carbon dioxide in the atmosphere and

The soil of tropical rain forests is not suitable for farming.

FIGURE 34–16. When tropical rain forests are cleared for farming and the land is abandoned after a few years, erosion results. In Colombia, erosion of this former forest has caused the large gorges.

contribute to an increase in temperature worldwide. It is also clear that succession will not lead to restoration of a climax community like the original. Once destroyed, a tropical rain forest cannot be replaced. Recent efforts by both scientists and governments offer hope that the rate of destruction will be greatly reduced in the years ahead.

34:9 Grassland

Some areas in both temperate and tropical regions receive between 25 and 75 cm of unevenly distributed rainfall per year. Grasses are the major species of the climax community in these areas. Grasses will also grow where there is more than 75 cm of rain per year, but they are often crowded out in the shade of trees. Biomes in which grass is predominant have different names in different regions. These names include steppe, prairie, plain, pampa, and veldt. A savannah (suh VAN uh) is a biome in which grasses predominate but in which there also are scattered trees. However, in general, these biomes are called **grasslands.**

In the United States, most grassland is found west of the Mississippi River to the foothills of the Rocky Mountains. Grasses are the natural climax species of this biome. The size of the grasses increases from west to east as average rainfall increases. At one time grasses close to two meters tall were common.

FIGURE 34–17. (a) An African savannah is a special type of grassland. (b) Although most of the U.S. grasslands are in the Great Plains, some are in other places, such as this area in California.

Once cleared, a tropical rain forest does not undergo secondary succession.

The grassland biome has a rainfall of between 25 and 75 cm per year.

FIGURE 34–18. (a) Giraffes and (b) zebras are among animals that live in the African savannah. Both live in herds of about a dozen.

FIGURE 34–19. This farm was abandoned because poor farming practices ruined the soil. Winds eroded the loose soil.

Animals of the grasslands in the United States once included grazing animals such as bison. Today, antelopes, jackrabbits, gophers, ground squirrels, and meadowlarks are common. Hunting of ground-nesting birds, such as the prairie chicken and ring-necked pheasant, is restricted to prevent their extinction.

Grasslands are very important to humans because of their use in agriculture. To take advantage of the high fertility of grassland soils, humans have changed this biome. They have replaced the natural grasses with cereal grains such as wheat and corn.

Natural grasses have been replaced with cereal grains, and bison with cattle.

Humans have been careless with this biome. Much of the rich topsoil has been removed as a result of wind and water erosion of plowed soil. Constant plowing and lack of rain change the rich soil into loose dust. In the 1930s, winds picked up the soil and carried it for great distances. The south central part of the United States became known as the "dust bowl," and many farms had to be abandoned. The tragedy of the "dust bowl" was a major factor in starting conservation practices.

34:10 Desert

Deserts occur in areas receiving less than 25 cm of rain per year.

Where rainfall is less than 25 cm per year and evaporation is rapid, there is little plant and animal life. Such areas are desert biomes. **Deserts** are found in the western United States, Africa, India, Asia, South America, and Australia. Depending on the amount of rainfall relative to the average temperature, deserts can be hot, cold, or temperate. The Mojave Desert of southern California is an example of a hot desert. The Northern Great Basin Desert in parts of Nevada, Utah, and California is a cold desert.

Desert plants are well spaced.

Deserts have little plant cover; bare ground is a common sight. However, endless stretches of barren sand dunes are not typical of most deserts. The plants that are present are widely spaced due to the small amount of water that is available.

Not only is the amount of water small, but also there is no regular pattern of rainfall. Long periods of drought are not uncommon and may last for many years. Also, when it does rain, much of the water quickly runs off and is not usable by the plants.

Organisms in desert biomes have many adaptations to dryness and temperature extremes. In plants these include reduction of the surface area of leaves, storage of water in fleshy parts, and large root systems for absorbing water. Cacti leaves have been modified into spines, and the stem of the plant carries out photosynthesis. Some plants mature quickly and produce seeds in a few days when water is available. During droughts some parts of plants may die, but the remaining parts can produce new growth when water becomes available. Seeds of some desert plants can remain dormant for fifty years or more so that a species can survive even under the most adverse conditions. Other seeds have a chemical that prevents germination until the chemical is dissolved by water. Because a large amount of water is needed to dissolve the chemical, the plant starts to grow only when water is plentiful.

Seeds of desert plants have adaptations for surviving during long dry spells.

Animals, too, are adapted for desert life. Conservation of water is important in their survival. Desert animals can obtain water from their metabolism. Water produced as a by-product of the breakdown of foods can be used for other purposes. Also, animals such as a kangaroo rat excrete highly concentrated urine (Section 27:12) and thus conserve water.

Temperature extremes are common in deserts. Many desert animals are small burrowing rodents. By staying in their burrows, they avoid the heat of the day and the cold of the night. Some burrowing animals undergo a period of estivation (Section 33:11). As in hibernation, metabolism is slowed down. In this way, an animal survives adverse conditions.

Estivation enables some desert animals to withstand temperature extremes.

Many desert animals, such as the jackrabbit and kangaroo rat, move by jumping. Jumping allows them to move quickly from place to place. Large animals do not usually live in a desert. They would be unable to find enough food, and they are not adapted to drought and temperature extremes.

FIGURE 34–20. (a) The horned toad is a desert animal that estivates if the temperature gets too high. (b) Cacti have modified leaves that enable the plant to survive in desert conditions.

a

b

a   b   c

FIGURE 34–21. (a) The moray eel, (b) sea urchin, and (c) squirrel fish are ocean animals that live close to the surface.

Also, some behavioral adaptations protect animals from temperature extremes. Many are active only at night. Others stay in shady places during the hottest part of the day (Section 33:11).

34:11 Ocean

You are familiar with life on land. However, the surface of Earth is more than two-thirds water, mostly in the form of **oceans.** The kinds and number of organisms living in the oceans far exceed those on land. How might this vast resource be best used?

Distribution of organisms in the oceans depends mostly on abiotic factors. The most important of these is light. Light intensity decreases with depth because light is both reflected and absorbed by the water. The penetration of light varies with the contents of the water and with latitude, but it generally does not reach a depth greater than 200 m. The oceans have an average depth of about 4000 m so the zone of water in which light is present is quite small. Producers, which affect the makeup of any ecosystem, only live where there is sufficient light for photosynthesis to occur. Thus, light is an important limiting factor of life in the ocean.

> Limited penetration of light restricts photosynthesis to about the top two hundred meters of the oceans.

Temperature is also important to life in water. However, temperature does not vary as much in the ocean as on land (Section 33:10). Thus, the sea is a more stable environment than the land. In the mid North Atlantic Ocean, for example, water temperature varies only about 10 C° from top to bottom.

> Plankton are organisms that float in water.

Ecologists classify marine organisms into three major groups. Those organisms that float in the water are called **plankton** (PLANG tun). These include many types of unicellular algae, heterotrophic protists, and small multicellular animals. Copepods (KOH puh pahdz), microscopic crustaceans, are a major form of animal plankton. Jellyfish and their relatives and the larval forms of other animals are also plankton.

> Nekton are animals that move freely through water.

Animals that can move freely through the water under their own power are called **nekton** (NEK tun). Nektonic animals include fish, whales, arthropods, and squid.

a
b

FIGURE 34–22. (a) The littoral zone is the ocean biome close to shore. A tide pool shows some of the life, such as sea anemones and starfish, found in the littoral zone. (b) A coral reef has organisms of the neritic zone.

Animals that live attached to or crawl on the ocean floor are called **benthos** (BEN thahs). Included in this group are organisms such as barnacles, starfish, and clams. Benthic organisms are mostly found close to shore or in shallow water.

34:12 Ocean: Littoral and Neritic Zones

Conditions vary greatly from one part of the ocean to another. For this reason, the ocean is made up of many biomes that have certain features and representative life forms. For example, the area close to the shore is the **littoral** (LIHT uh rul) **zone.** The littoral zone is subject to the action of tides and is a transitional area between land and sea. Organisms of this biome, such as many types of algae, clams, crabs, and barnacles, are adapted to changing conditions. These organisms can survive exposure to both air and water and changes in temperature.

Along the continental shelf, the land begins to slope down and the ocean becomes deeper. This area is known as the **neritic** (nuh RIHT ihk) **zone.** Life in the neritic zone is different from that in the littoral zone. The environment is more stable because there is no direct exposure to air. Organisms in the neritic zone are strictly aquatic. Plankton and nekton abound in the surface waters. Also, there are many bottom dwellers.

34:13 Ocean: Abyssal Zone

The deepest part of the ocean is the **abyssal** (uh BIHS ul) **zone.** Here light is absent so no photosynthesis occurs. Some nektonic animals, mostly fish, exist here. These fish feed on other fish and on dead organisms settling from above. Some organisms of this zone have organs that give off light. A certain angler fish has a "lure" that emits light and attracts prey. Giving off a light in the abyssal zone may also be needed for reproduction.

Benthos are bottom-dwelling animals.

Organisms inhabiting the littoral zone must be adapted to changing conditions.

Life in the neritic zone is more stable than life in the littoral zone.

In the abyssal zone, animals feed on other animals or dead organisms sinking from above.

Decomposition of ocean organisms occurs on the ocean bottom.

Some animals live at great depths in the abyssal zone. Experiments have been done in which bait (food), lights, and movie cameras were lowered to a depth of 7000 m. It was found that within a few hours small fish, crustaceans, and even sharks were attracted to the bait, probably by its odor. The animals quickly ate most of the food and mollusks and echinoderms fed on the leftovers.

Decomposition of organic materials on the ocean floor is done by bacteria. The molecules are broken down into simpler forms such as carbon dioxide, nitrates, and phosphates. After decomposition, materials must be made available to producers. If materials continued to build up on the ocean bottom, a point would be reached where life could no longer continue. An ecosystem can survive only if materials are recycled.

Materials produced by decomposition are constantly mixed and redistributed in the upper layers of the oceans as a result of temperature differences and wind.

Return of materials to the surface of the oceans is affected mostly by water temperature and wind. In cold climates, surface water becomes colder than water below the surface. Because cold water is more dense than warm water, cold surface water sinks and warm bottom water rises. Also, the wind causes currents, which move the surface water. Water below the surface rises to replace the water displaced by the wind. Although both of these processes return decomposed materials to the surface, water temperature differences account for almost all of the return.

REVIEWING YOUR IDEAS

6. What factors affect the type of biome an area will be?
7. Name the major world biomes. Where is each found? What is the dominant species of plant in each one?
8. What factor accounts for the different biomes east and west of the Mississippi River in the United States?
9. Distinguish between jungle and rain forest.
10. Why is there more life within a hundred meters of the ocean surface than on the ocean floor?

ADVANCES IN BIOLOGY

Unique Ocean Communities

Most of the ocean floor is a cold, barren, desertlike biome. Light does not reach the ocean floor; thus there can be no photosynthesis. Because of this, it had long been thought that few, if any, life forms could exist in this area. However, in recent years, scientists have discovered unique communities of animals living on the ocean floor at depths as great as 3000 meters.

Many of these unusual communities exist in the Pacific Ocean. The first was found in 1977 off the Galapagos Islands (Section 10:8), and others were later discovered off the coasts of Mexico and Peru. Although the water temperature on most of the ocean's floor is near freezing, each of the Pacific communities is in an area of warm water. The communities are located over areas of molten rock where new ocean floor is forming. Cold seawater seeps into cracks in the new crust of the ocean floor, comes into contact with the molten rock below, and becomes heated. The hot water later gushes up through hydrothermal vents, openings that look like small volcanoes. The water temperature may be as high as 350°C. These communities thrive in the warm water around the vents.

Animal life never before seen has been found to inhabit these communities. Brown mussels and white clams that average 30 cm long live among mounds of cooling lava. Clusters of orange and white crabs, starfish, octopi, and strangely-colored fish move around the vents. Strangest of all are colonies of giant tube worms, some of them as long as three meters, that live around the vents. Many other life forms collected from the sites differ so much from known animals that they have not yet been classified.

Every community must have an outside energy source and producers that trap it to make food. In most communities, light is the energy source, plants or algae are the producers, and glucose is made by photosynthesis. But, there is no light energy in these ocean floor communities. However, there is geothermal energy. **Geothermal energy** is heat energy due to the natural radioactivity of Earth. This energy, present in the molten rock, increases the temperature of the water. The very hot water is important because it reacts with minerals in the rocks to produce an energy-rich substance, hydrogen sulfide. Certain chemosynthetic bacteria (Section 13:4) are the producers of these unique communities. Like other producers, they make food. However, these bacteria make carbohydrates by using the energy in hydrogen sulfide. Thus, these bacteria are the basis of the community's food web! Animals such as mussels and fish feed upon the bacteria and are, in turn, preyed upon by other animals.

Tube worms and clams depend upon the bacteria as well, but in a unique way. Instead of feeding upon the bacteria, these animals live with them in a mutualistic relationship (Section 33:6). The bacteria live in different parts of the tube worm's body. In clams, the bacteria live in the gills. Both animals extract hydrogen sulfide from the water and transport it in their blood to the bacteria. In turn, the bacteria provide food for the animals.

In 1986, hydrothermal vents were discovered in the Atlantic Ocean. This discovery suggests that hot spring ocean communities may occur wherever the ocean floor is spreading. It may also mean that these vents have a marked effect on both the temperature and mineral content of the oceans.

FIGURE 34–23. Unusual forms of life live at the vents. Tube worms are shown here.

Ocean vent communities contain an abundance of newly discovered animal life.

FIGURE 34–24. Sulfide-blackened hot water comes from a "chimney" at the East Pacific Rise near the Gulf of California. Water from the vents is as hot as 350°C and contains minerals that nourish the biological community.

INVESTIGATION

Problem: How do biomes of North America differ?

Materials:

climatograms
map of North America
colored pencils: red, blue, orange, green,
yellow, black, brown

Procedure:

Part A. Preparing Climatograms for Biomes

1. Examine Figure 34–25. It is called a climatogram. It shows the monthly variation of temperature and rainfall during an entire year for a city in Kansas.

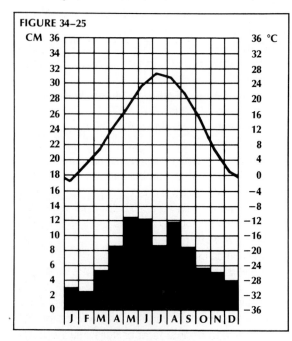

FIGURE 34–25

2. Note that the line for temperature corresponds to the degrees listed along the right side of the climatogram. Note that the vertical bar graph lines for rainfall correspond to the

centimeters listed along the left of the climatogram. The letters along the bottom correspond to the months of the year.

3. There are seven biomes located in North America. The city shown in Figure 34–25 (Lawrence, Kansas) is located in the grassland biome.

4. You are to prepare climatograms for the remaining six biomes of North America.

5. Obtain a series of blank climatograms from your teacher.

6. Use the climatogram in Figure 34–25 as a model as you complete the six charts using the data provided in Table 34–1.

7. Remember to identify each biome being graphed.

Part B. Locating Biomes in North America

1. Obtain a map of North America from your teacher. The outlines for the seven biomes have been marked for you.

2. Use colored pencils to color in the different biomes as follows:

Tundra	= black
Desert	= yellow
Grassland	= green
Taiga	= blue
Temperate forest	= red
Chaparral	= brown
Tropical rain forest	= orange

3. Complete the key at the bottom of the map by showing the colors that depict each biome.

4. Calculate an average annual temperature and precipitation for each biome and include this information on the map key.

5. Indicate in the space marked "Major plant forms" what main vegetation type one would expect to find in each biome. Omit this information for the chaparral. (Use your text for this information.)

Data and Observations:

TABLE 34-1. AVERAGE MONTHLY TEMPERATURE AND PRECIPITATION		J	F	M	A	M	J	J	A	S	O	N	D
Biome		J	F	M	A	M	J	J	A	S	O	N	D
Tundra	Temperature (in degrees C)	−24	−27	−25	−18	−6	2	4	3	−2	−6	−16	−24
Tundra	Precipitation (in centimeters)	0.2	0.2	0.1	0.1	0.1	0.2	2.1	2.0	0.7	0.6	0.3	0.2
Desert	Temperature	0	2	6	8	13	16	20	19	14	10	4	0
Desert	Precipitation	2.5	2.6	1.8	1	1	0.8	0.6	0.6	0.4	1.2	1.6	2.2
Taiga	Temperature	−10	−6	−2	2	8	13	14	12	8	2	−6	−10
Taiga	Precipitation	1.8	1.7	1.7	1	1.7	2.2	4	6.5	6.7	4.5	2.2	2
Temperate Forest	Temperature	4	4	10	14	20	26	28	27	24	18	12	4
Temperate Forest	Precipitation	12	10.2	13	9	9	8	10	8.2	6.8	6.2	8.5	10.2
Chaparral	Temperature	11.7	11.7	12.8	14.4	15.6	17.2	19	18.3	18.3	16.7	14.4	12.8
Chaparral	Precipitation	8.9	7.6	7.4	1.3	1.3	0	0	0	0.3	1.5	3.6	5.8
Tropical Rain Forest	Temperature	25	25.6	24.4	25	24.4	23.3	23.3	24.4	24.4	25	25.5	25.5
Tropical Rain Forest	Precipitation	26	25	31	16.5	25.4	19	16.8	11.7	22	18.3	21	29.2

Questions and Conclusion:

1. Define:
   (a) biome
   (b) climatogram
   (c) precipitation
2. Using your climatograms, describe how the pattern of precipitation differs during the seasons in a
   (a) chaparral versus a desert
   (b) grassland versus a chaparral
3. Make an estimate as to the length of the growing season for plants in a
   (a) tundra, (b) taiga, (c) chaparral,
   (d) tropical forest, and (e) grassland.
4. Which biome appears to have the:
   (a) highest annual average precipitation?
   (b) lowest annual average precipitation?
   (c) highest annual average temperature?
   (d) lowest annual average temperature?
   (e) largest area in North America?

(f) smallest area in North America?
5. The map of North America is marked off in degrees of latitude. In general,
   (a) how do the biomes change in regard to annual precipitation and temperature as one moves farther north in latitude?
   (b) how do these changes affect the type of plant life present?
6. (a) What biome do you live in?
   (b) Based on your climatograms and your completed map of North America, describe the average annual precipitation, high and low months of precipitation, average annual temperature, high and low months of temperature, and expected type of vegetation present in the biome where you live.

Conclusion: How do biomes of North America differ?

CHAPTER REVIEW

SUMMARY

1. The series of changes in the development of a community is ecological succession. **p. 715**
2. Each stage of primary succession from bare rock affects both biotic and abiotic factors and prepares the community for more complex stages that follow. **34:1**
3. Eventually, primary succession results in a stable climax community with a dominant species. **34:2**
4. Secondary succession occurs after a community has been affected in such a way that one of its dominant species is removed. **34:3**
5. A lake is a community in the process of succession. It will eventually be replaced by a terrestrial climax community. **34:4**
6. The type of biome in a given area is determined by geography and climate. **p. 720**
7. The tundra biome has permafrost and lacks large plants. **34:5**
8. The taiga biome is a wet region composed of evergreen forests. **34:6**
9. Temperate deciduous forests are found in areas of moderate rainfall. Most plant life is dormant during the winter. **34:7**
10. Tropical rain forests develop in regions with heavy rainfall and constant warm temperatures. **34:8**
11. Grassland biomes receive less rain than forests. Much of the grasslands has been cleared for agriculture. **34:9**
12. Deserts are biomes with sparse rainfall that support little plant or animal life. **34:10**
13. Oceans cover most of Earth. Many more kinds of organisms live in the oceans than on land. **34:11**
14. The littoral zone, an area close to shore, and the neritic zone, where the water is deeper, are two ocean biomes. **34:12**
15. The abyssal zone is the biome in the deepest part of the ocean. **34:13**
16. In ocean-floor communities powered by geothermal energy, producers use energy of hydrogen sulfide to make carbohydrates by chemosynthesis. **p. 732**

LANGUAGE OF BIOLOGY

abyssal zone	nekton
benthos	neritic zone
biome	ocean
climax community	plankton
deciduous	primary succession
desert	secondary succession
dominant species	taiga
ecological succession	temperate forest
geothermal energy	tropical rain forest
grassland	tundra
littoral zone	vertical stratification

CHECKING YOUR IDEAS

On a separate paper, indicate whether each of the following statements is true or false. Do not write in this book.

1. Conifers are dominant in the tundra.
2. A lake is a permanent climax community.
3. Abandonment of New England farms has led to secondary succession.
4. The first organisms to play a role in primary succession are small animals.
5. Each stage during an ecological succession modifies the environment and makes possible the next stage.
6. Plankton are free-swimming organisms.
7. All organisms in the unique ocean-floor communities obtain energy from hydrogen sulfide.
8. Wide spacing of plants is a characteristic of deserts.
9. No photosynthesis occurs in the abyssal zone.

34

10. Barnacles are examples of benthos.
11. Lianas are common in the taiga biome.

EVALUATING YOUR IDEAS

1. Compare the events of primary succession from bare rock in New England and secondary succession on an abandoned farm.
2. How does a lake become a forest?
3. What is permafrost? How does permafrost affect the type of plant life in the tundra?
4. Describe the animal life in the tundra.
5. What is the climate of the taiga? How does the climate affect plant life there?
6. How does the presence of microenvironments affect a community?
7. What kinds of animals live in the taiga?
8. Describe the climate of a temperate deciduous forest.
9. What is vertical stratification? Describe the vertical stratification of a temperate deciduous forest.
10. How does the climate of a tropical rain forest affect the plant life there?
11. Compare life in the canopy of a tropical rain forest to life on the forest floor.
12. Describe the climate of the grasslands.
13. How have the grasslands of the United States changed?
14. Why are the plants of a desert scattered rather than clumped together?
15. What kinds of animals live in a desert? How are they adapted for life there?
16. How does light affect the distribution of organisms in the oceans?
17. Compare the three kinds of ocean life.
18. Discuss the three major ocean zones.
19. How are materials of the ocean recycled?
20. Explain energy "flow" in most newly-discovered ocean-floor communities.

APPLYING YOUR IDEAS

1. What might happen to a climax community if a parasite severely damaged the dominant species of plant?
2. How might life in the canopy layer of a temperate deciduous forest affect the shrub, floor, and soil levels?
3. What kinds of animal life would be affected if a forest were cleared? How?
4. Which type of life is most abundant in an ocean?
5. Vertical stratification is common in forest biomes. Does vertical stratification also occur in oceans? Explain.
6. In what kind of biome do you live? How have human activities altered the characteristics of that biome?
7. Do you think that the unique ocean-floor communities are permanent? Explain.

EXTENDING YOUR IDEAS

1. Examine an area near your home for evidence of secondary succession. Are there different kinds of plants in different areas? Make a model or drawing of the area.
2. Examine a natural area for microenvironments. Make a list of microenvironments and the special conditions in them. How do they affect the life of the community?
3. Find out about biomes called estuaries. What unique conditions exist there? How are organisms adapted for life in estuaries?

SUGGESTED READINGS

Brownlee, Shannon, "Explorers of Dark Frontiers." *Discover*, Feb., 1986.

Bruemmer, Fred, "Life upon the Permafrost." *Natural History*, April, 1987.

Humans and the Environment

Humans depend on and interact with the environment. Here, a photographer "shoots" a Dall sheep in Alaska. How do you depend on the environment? How do you interact with the environment? What types of effects have humans had on the environment? What can you do to assure that a clean, healthy environment will be available to others in the future?

espite the fact that humans can avoid many of the forces of natural selection, they are part of the **biosphere,** the world of life. The biosphere is made up of ecosystems where organisms interact with their environment. The environment affects the lives of the organisms, and the organisms affect one another and the physical factors of their environment.

Humans depend on the environment, both directly and indirectly. They need food, fuel to run their machines, and other raw materials for modern-day life. Humans change their environment by clearing land, substituting one kind of life for another, and killing "pests." They also disrupt food chains and put waste materials into the air, water, and soil. The results of these actions are some of the most urgent problems of ecology. These changes that humans make in the environment are important to all organisms, including humans themselves.

Objectives:
You will
- discuss the pros and cons of using pesticides.
- describe various sources of pollution and their effects.
- explain means of conserving resources.

Humans have altered their natural surroundings in many ways.

PEST CONTROL AND POLLUTION

While agriculture and technological advances (Section 32:10) have been responsible for human successes and comforts, these advances also have produced unexpected and unwanted results. In growing crops and producing goods, many wastes have been released into the environment. Often these wastes cause **pollution,** or contamination, of air, water, and soil, and decrease the quality of the environment. A major source of pollution has resulted from the effort to control organisms that compete with or cause diseases in humans.

Introduction of wastes into the biosphere contributes to pollution.

a

b

FIGURE 35–1. Insect damage can ruin various crops. (a) Squash bugs harm squash plants and (b) Japanese beetles can ruin many kinds of fruit plants, such as this peach tree.

Pests destroy or damage man-made products, crops, forests, and livestock.

Some pests were accidentally brought to North America.

Changes in natural ecosystems may increase the population density of pest organisms.

35:1 Pests and Pesticides

Many organisms are pests to humans. Termites destroy wood, moths destroy woolen products, and Japanese beetles attack fruit trees and other plants, to name a few. In addition to pests that may directly affect humans in and around the home, there are pests that cause severe damage to agricultural crops, livestock, and forests. Weeds cause nearly ten percent of the crop losses each year by competing with and crowding out cultivated plants. Corn borers attack corn, and bollworms (larvae of a particular moth) invade cotton. Fungi such as rusts (Section 19:5) damage wheat plants, and other fungi cause Dutch elm disease (Section 13:15). Nematode worms attack and destroy a variety of plant parts.

Many pests, such as the Japanese beetle, gypsy moth, and Mediterranean fruit fly (medfly) were brought to North America from other parts of the world either on purpose or accidentally. The gypsy moth was brought from Europe in 1868 for use in scientific experiments. Japanese beetles first appeared in 1916, accidentally imported, probably in the form of larvae in soil around plants. The first invasion of medflies occurred in 1929 in fruit thought to be smuggled from Hawaii. With no natural predators, these insects flourished and became pests in the new environment.

Humans also have contributed to pest problems by altering natural ecosystems and planting large areas with single food crops. These activities remove natural predators and create a high population density of hosts suitable for transmission of parasites (Section 32:4).

In trying to control pests, humans often use chemicals called **pesticides,** poisons that kill the unwanted organisms. Insects are the most numerous pests. One of the most effective pesticides ever used to kill insects is a chemical known as DDT. DDT has been used to kill disease carriers such as *Anopheles* (uh NAHF uh leez) mosquitoes. These mosquitoes transmit the organism that causes malaria (Section 13:11).

FIGURE 35–2. These trees are being sprayed with pesticides to eliminate organisms that damage the crop.

DDT also has been used to kill insects around the home and those that destroy crops. Because DDT and other pesticides were effective and easily obtained, they became convenient weapons for farmers, orchard keepers, and homeowners.

DDT and other pesticides have been used to kill disease-causing and crop-destroying insects.

35:2 Pesticides: Some Problems

Use of pesticides to eliminate diseases and to protect crops is extensive, and large areas are sprayed with pesticides. But DDT and other pesticides kill many organisms besides pests. In applying chemicals to large areas, entire ecosystems are affected.

Suppose that an insect pest is the natural prey of a larger, carnivorous insect. When DDT is used to kill the pest, both insect populations are affected. The predators, fewer in number than the pests (Section 32:3), may be completely destroyed by the pesticide. But some of the pests may survive, even after the effects of the pesticide wear off. In the absence of natural predators, the pest population would increase rapidly. The pest problem may get worse than it was before (Figure 35–3). Thus, in this case, pesticide use would have made conditions worse.

Pesticides have caused another unforeseen problem. When DDT was widely used in the 1940s, houseflies were easily killed. However, some flies were not affected because they were resistant (genetically) to DDT. These flies survived and passed their genes to later generations. Thus, as a result of natural selection, a resistant strain of houseflies evolved. In the same way, mosquitoes that transmit malaria have become resistant to DDT. It is estimated that a total of 400 insect species and 50 species of fungi around the world are now resistant to currently used pesticides. Some organisms are resistant to only one type of poison, while others are resistant to several.

FIGURE 35–3. Dragonflies are natural predators of aphids. Spraying an area with DDT to kill aphids may kill all dragonflies. Some aphids may survive and, without the natural predator, the aphid population would grow even larger than it was originally.

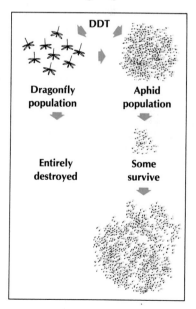

FIGURE 35–4. As a top level carnivore, the bald eagle can accumulate large concentrations of DDT. DDT and related pesticides are known to be involved in eggshell thinning in these and other birds.

35:3 Pesticides: Effects on Other Organisms

A biodegradable substance is one that can be broken down in organisms and thus be recycled through a community.

DDT is a synthetic product that cannot be metabolized, or broken down, by animals. Chemicals structurally similar to DDT, such as DDD, DDE, and dieldrin, also cannot be broken down. A compound that cannot be chemically broken down into harmless, inactive compounds is called **nonbiodegradable** (nahn bi oh dih GRAYD uh bul).

DDT is a nonbiodegradable substance. It accumulates in the fatty tissues of animals. As it is passed upward along food chains, it becomes concentrated and, in larger amounts, can kill animals.

DDT, insoluble in water, is soluble in fats. Thus, DDT builds up in the fatty tissues of animals and is transferred along food chains from animal to animal. Because it is not broken down, the concentration of DDT increases as it passes higher up the food chain. This increased concentration of chemicals along a food chain is called **biological magnification.**

In small doses, DDT is not harmful to animals other than insects. But, as it builds up in other animals, DDT can be harmful, even lethal, to animals such as fish, birds, and mammals. As these animals die, the ecosystems of which they were a part are disrupted.

Birds often are affected by pesticides such as DDT. Top carnivores, such as the bald eagle and golden eagle, build up the most DDT. Studies show that these birds are rapidly decreasing in number. In fact, they are in danger of extinction. DDT does not kill the adult birds, but instead affects the population's birthrate (Section 32:2). DDT causes eggshells to be very thin, so the eggs break easily. DDT also may interfere with breeding behavior. Ospreys, falcons, and pelicans are other birds affected by DDT. The sizes of their populations have been reduced by the use of pesticides.

Another group of pesticides, which includes parathion, malathion, and phosdrin, has different ecological effects. Unlike DDT and related substances, these pesticides tend to break down quickly. Thus, they do not spread through food chains and their effects are not magnified. However, these pesticides affect not only insects but also vertebrates, including humans, before they break down. When applied over a large area, they may kill many predators of pests, thus disturbing natural controls of pest populations. Malathion was widely used in California in 1981 to control a "foreign" pest, the medfly.

FIGURE 35–5. Because of damage done by medflies, samples of fruits shipped within, into, and out of the country are inspected for medfly larvae.

35:4 The Pesticide Dilemma

Use of pesticides such as DDT was widespread until their harmful ecological effects were fully realized. In 1972, use of DDT was banned in the United States. Since that time, the use of other dangerous pesticides also has been restricted.

But how does pesticide use affect the economy? In order to feed a growing world population, a high crop yield is needed. Without pesticides, pests would destroy a large portion of the crops and a lot of food would be lost.

Apart from the need for enough food for everyone, many shoppers in the United States expect a large selection of food in grocery stores. Consumers want to buy food that is pest free and that does not show signs of pest damage. For example, grocers must supply mold-free oranges and worm-free apples. Thus, pesticide use affects the economy in many ways.

Much research on pesticides is still necessary to find ways to use pesticides with less harmful consequences. For example, new ways of applying pesticides besides aerial spraying must be found. The chemicals should be applied more directly to reduce the risk of harming other organisms. It is estimated that if pesticides could be applied directly to pests, 99 percent of the pesticide use now employed could be eliminated. Scientists are studying the use of pesticides in plastic capsules to reduce the risks to other organisms. The capsules would be placed in the soil where microorganisms would degrade the plastic. Thus, the pesticide would be in the soil and would not directly enter the air. Also, development of pesticides that quickly break down to form harmless products is being pursued. Alternatives to control by pesticides are discussed on page 759.

Because pesticides can be very beneficial, what should be done about the problems they cause? Many people defend pesticide use because it helps to reduce crop loss and damage. Maximizing crop yield may be especially important in countries where people do not get enough food. What do you think?

Pesticide use can be beneficial at the same time it is harmful in other ways.

Pesticides should be applied to an ecosystem in such a way that organisms other than pests are not affected.

OTHER SOURCES OF POLLUTION

Pesticides and other chemicals are pollutants. A **pollutant** is any factor that damages or makes the environment unclean. Pollutants affect air, water, and soil and may endanger the health and lives of all forms of organisms.

35:5 Air Pollution

A major source of air pollution is the burning of fossil fuels.

Most air pollution is caused by the burning of fossil fuels. Burning of fuels is needed in industry, transportation, and homes. During this burning, substances that may damage organisms are released directly into the atmosphere.

One of the chief causes of air pollution is the burning of gasoline, mainly in automobiles. By-products of gasoline combustion include hydrocarbons (compounds of hydrogen and carbon), carbon monoxide, sulfur dioxide, nitrogen oxides, and lead compounds. Some hydrocarbons are known to cause cancer. Carbon monoxide deprives animals of oxygen because carbon monoxide binds readily to hemoglobin.

Nitrogen oxides are produced at high temperatures from the reaction between nitrogen and oxygen in the air. Nitrogen oxides, in

FIGURE 35–6. (a) The U.S. Environmental Protection Agency developed the Pollutant Standards Index (PSI) to convey air pollution information and associated health effects to the public in a simple, uniform way. (b) Often, people think of air pollution as something that is outside. Air pollution inside cars and houses from smoking and poor ventilation can pose major health hazards.

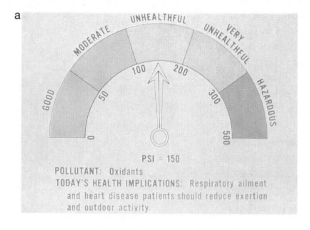

PSI = 150
POLLUTANT: Oxidants
TODAY'S HEALTH IMPLICATIONS: Respiratory ailment and heart disease patients should reduce exertion and outdoor activity.

turn, can combine in sunlight with hydrocarbons to form a compound called PAN, a chief ingredient in the kind of smog that often surrounds large cities. A major effect of this type of smog is damage to the lungs. Damage of this kind can be extremely dangerous to people who suffer from other respiratory illnesses.

Nitrogen dioxide, a brownish-colored gas, is a compound that can kill lung cells and, in high amounts, cause a buildup of fluids in the lungs. How would fluid buildup affect respiration?

Some automobile engines require a form of gasoline containing lead compounds. As this gasoline is burned, lead compounds are released into the air along with other pollutants. Lead is a poison that can interfere with cellular metabolism by combining with proteins and interfering with enzymes. Because of lead's harmful effects, gasoline-using cars made in the United States since 1975 are built to run on unleaded gasoline. U.S. cars manufactured before 1975 and some imported cars still use gasoline containing lead.

Another air pollutant is sulfur dioxide, a compound produced by burning sulfur-containing fuel oil and coal. Sulfur dioxide is a gas that can damage the lining of the respiratory tract and lungs. The ciliated cells of the air passages are weakened, and the respiratory tract becomes more easily damaged by pollutants such as dust and other chemicals.

Sometimes the air over a highly industrialized area becomes trapped because of atmospheric conditions. Pollutants then become even more concentrated. Such a situation is called an **inversion** (ihn VUR zhun). In 1952, an inversion occurred in London, England. A smog containing sulfur dioxide, smoke, and fog was trapped for four days and resulted in the deaths of more than 4000 people.

Burning of fossil fuels may have a long-range effect on average global temperatures. The amount of CO_2 in the atmosphere has increased greatly with the burning of fossil fuels in industry and transportation. Increased CO_2 in the atmosphere may cause some of the heat leaving Earth to be radiated back to the surface. Thus, some people feel that an increase in CO_2 in the environment may have an overall warming effect. One severe consequence of a prolonged warming effect would be melting of polar ice caps. The changing of ice to water would increase the depth of oceans, possibly submerging coastal regions. Other scientists feel that the increasing number of all particles in the air from pollution may have an opposite effect. They suggest that more sunlight will be reflected back into space before it reaches Earth's surface. This reflection could lead to a decline in temperatures on Earth. How might a decrease in temperature affect life on Earth?

Products of fuel combustion are not the only source of air pollution. Other dangerous chemicals include asbestos fibers, mercury, and a variety of organic compounds released during industrial pro-

FIGURE 35–7. Hydrocarbons are among the by-products of gasoline combustion, which contributes to the formation of the smog that surrounds many large cities.

In an inversion, a layer of air is trapped close to the ground. Because of atmospheric conditions, the concentration of pollutants continues to increase. Serious health problems may result.

Pollution leads to a greater number of particles in the air. An increase in these particles may affect temperatures world-wide.

cesses. One of the major difficulties in analyzing and dealing with air pollution is the fact that pollutants often interact with one another to form even more dangerous compounds. Thus, it is often difficult to know the effect of a pollutant. It can be concluded, though, that air pollution is a threat to health and requires a great deal of further study.

35:6 Acid Rain

Rain is normally slightly acidic, having a pH of 5.6 (Section 3:8). The acidity results from the dissolving of carbon dioxide in water forming weak carbonic acid.

Sulfur dioxide and nitrogen dioxide combine with water and/or oxygen in the air to form acids that lower the pH of rain.

Release of sulfur dioxide and nitrogen oxides into the atmosphere further lowers the pH of rain. These pollutants react with water and/or oxygen in the air to produce sulfuric and nitric acids. So much sulfur dioxide and nitrogen oxide are released into the atmosphere in urban areas that the pH of rain in some places may now be as low as 3! Rain with a pH below 5.6 is known as **acid rain.**

In the United States, acid rain is most damaging in the northeast where the average pH is between 4.0 and 4.5. Rain that is only slightly more acidic than normal may increase the yield of some crops. However, as acidity continues to increase (as pH decreases), plant tissues are damaged or destroyed and photosynthesis and nitrogen fixation in legumes are affected. Changing the pH of soil also affects the types of plants that may grow (Section 33:8).

Acid rain may damage plant tissues and interfere with photosynthesis and nitrogen fixation.

Acid rain severely affects lake ecosystems. A pH lower than normal may cause nutrients to be removed from the water. A low pH also may lead to increased solubility of dangerous metals, such as mercury. Greater acidity is lethal to a variety of organisms — plankton, eggs of fish and salamanders, frogs, and adult fish. Because many bacteria

FIGURE 35–8. (a) Salmon develop normally in neutral water and (b) abnormally in the presence of acid rain. (c) Radishes grow much larger at a neutral pH than (d) when grown at a low pH.

a

b

c

pH = 7.0

d

pH = 2.6

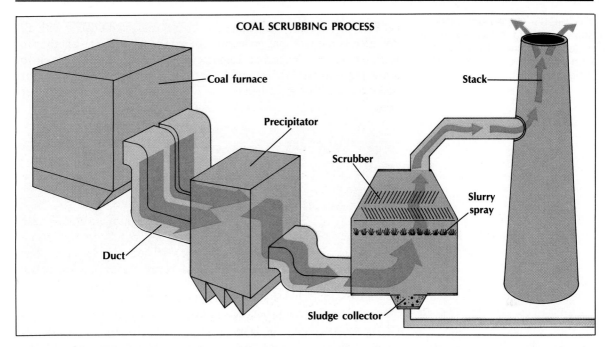

COAL SCRUBBING PROCESS

Coal furnace

Precipitator

Scrubber

Stack

Slurry spray

Duct

Sludge collector

FIGURE 35–9. When burned, high sulfur coal releases sulfur, which can be removed from the emitted gases by a coal scrubbing process.

also die, the rate of decomposition decreases and organic material accumulates at the bottom of the lake.

Acid rain is yet another problem that results from burning fossil fuels and is another reason why some people have suggested turning to other energy sources. Coal, also a fossil fuel, does not solve the problem because it contains sulfur, and when burned, releases sulfur dioxide. One answer may be the development of scrubbers and other devices that remove pollutants from gases escaping to the atmosphere. Alternate energy sources, such as nuclear, solar, tidal, and wind power, do not release substances that cause acid rain into the atmosphere.

Some sources of energy do not release substances that cause acid rain and other problems of air pollution.

35:7 Pollution of Surface Waters

Chemicals pollute water as well as air. Many industrial by-products are poured in large amounts into streams and rivers. Mercury is an example of an extremely dangerous by-product. The mercury settles to the bottom in a safe, insoluble form. But, bacteria in the bottom mud may change insoluble mercury to a soluble, poisonous form. This form of mercury enters the food chain where, like DDT, it becomes more concentrated along each link. People who eat mercury-poisoned fish may become ill and even die. Fishing has been prohibited in many areas because of high mercury levels. Swordfish, a large carnivore, was for a long time declared unsafe for human consumption in the United States because of its high mercury content.

Aquatic ecosystems are affected by poisons such as mercury produced by some industries.

FIGURE 35–10. An algal bloom is shown covering the surface of the water of this pond. Such an algal population explosion can lead to water stagnation.

PCBs are nonbiodegradable chemicals that harm both humans and animals.

PCBs (polychlorinated biphenyls) are chemicals once used in a variety of industrial products. Before their use was banned in 1979, they entered rivers and lakes from factories. Because they are non-biodegradable, PCBs remain in the environment and pose a health threat to humans. They may affect the nervous and muscular systems. Experiments suggest they also may cause cancer and birth defects in animals.

Water pollutants may play a role in the aging and death, or **eutrophication** (yoo troh fuh KAY shun), of a lake. In such a situation, the oxygen level of the lake becomes very low, and the ecological balance is destroyed. The following is an example of how eutrophication of a lake occurs.

Phosphates and nitrates can cause an algal bloom, which eventually may cause a body of water to become stagnant.

Phosphates from home detergents may enter a lake from sewage treatment plants. Phosphates and nitrates from fertilizers enter from soil runoff. These phosphates and nitrates are important nutrients for algae. Thus, with large amounts of these chemicals and warm temperatures, the algal populations thrive and grow rapidly in what is called an **algal bloom.** The algal population may become so large that the algae use up all available nutrients and begin to die. The dead algae settle to the bottom where they are decomposed. Huge amounts of oxygen are used up as decomposition occurs. As a result, less oxygen is available for other aquatic organisms and many die. Death of a large number of organisms causes a disrupted food web. Continued decomposition of dead organisms uses more oxygen. Consumers, such as insect larvae and fish, may die and other organisms, such as bacteria and mosquitoes, may thrive. The entire lake may become stagnant as the ecological balance is destroyed.

Pollution accelerates eutrophication and speeds the succession from a lake to dry land (Section 34:4). At one time Lake Erie was near destruction, but conservation efforts are saving it.

Water also can be polluted by heat. Addition of hot water to a natural body of water is called **thermal pollution.** Water from a nearby natural ecosystem is often used as a cooling agent in industries such as nuclear power plants. The temperature of the water increases and then the water is piped back to the lake, stream, or bay. One effect of thermal pollution is the lowering of the oxygen content of water (Section 32:8). Animal life in the water may not be able to survive if enough oxygen is not present. Thermal pollution may also interfere with the spawning (mating) habits of fish and may kill organisms that cannot adapt to the warmer waters. Organisms better adapted to the warm water may increase, thus changing the balance of the ecosystem.

Thermal pollution lowers the oxygen content of water, interferes with spawning of fish, and may kill some organisms.

PEOPLE IN BIOLOGY

When toxic wastes, pollutants that are harmful to humans and other life forms, began to destroy her beautiful Japanese fishing village, Michiko Ishimure became a determined ecologist.

Michiko Ishimure

(1927–   )

For over 25 years she had watched the "cat's dance disease," which made cats go mad and drown themselves, spread throughout the village. Humans then became affected, becoming crippled and disfigured, and finally dying. University studies showed that the people were suffering from a nervous disorder that resulted from eating fish contaminated by mercury wastes discharged into the bay by a large chemical plant.

Ishimure attacked the toxic waste problem by writing books about the people who suffered from the poisoning. Her first book, *Kukai Jodo (Sea of Suffering),* made her an outcast in her village because the people were so dependent on the chemical plant for jobs. After her second book, *Rumin no Miyako (City of Drifters),* was published, the Japanese government began to take an interest in the toxic waste problem. With her books, Michiko Ishimure has led the fight in Japan to control the types of wastes that can be dumped by industry. Because of her efforts, and the efforts of many like her, governments have passed guidelines to control toxic wastes.

INVESTIGATION

Problem: What evidence is there that the greenhouse effect is occurring?

Materials:

graph outline supplied by your teacher
ruler

Procedure:

Part A. Infrared Radiation and Temperature on Earth

1. Infrared radiation is responsible for heating the atmosphere. This form of energy is released by the sun and is not visible. When infrared radiation strikes the atmosphere, one of two things may happen. If much carbon dioxide is present, most of the infrared radiation will be absorbed. Absorption of infrared radiation by carbon dioxide causes atmospheric temperature to rise. If little carbon dioxide is present, little infrared radiation is absorbed. Atmospheric temperature does not rise. Radiation that is not absorbed bounces back into space.
2. Copy Figure 35–11.
3. The outlines in Figure 35–11 represent the following:
   (a) an atmosphere with no carbon dioxide (no shading of atmosphere)
   (b) an atmosphere with a small amount of carbon dioxide (light shading of atmosphere)
   (c) an atmosphere with much carbon dioxide (dark shading of atmosphere)
4. Draw six arrows on each outline to represent the pathway of incoming infrared radiation. Use a straight arrow (↓) to represent absorbed infrared radiation. Use a bent arrow (↘) to represent infrared radiation that bounces back into space. Reread statement 1 if you need help.
5. Included in each outline is a thermometer. Indicate the *relative* temperature expected in each atmosphere by shading the center of the thermometer. Reread statement 1 if you need help.

Part B. The Effect of Carbon Dioxide on Temperature

1. Table 35–1 shows the change in average temperature for the atmosphere on Earth since 1890.
2. Use the graph outline supplied by your teacher to construct a graph of the data in Table 35–1.
3. Using the edge of a ruler, draw a dashed line on your graph to show the change in average temperature from 1890 to 1980. Your teacher will help you with this step if you are not sure what to do.

TABLE 35-1. AVERAGE CHANGE IN TEMPERATURE			
Year	Increase °C	Year	Increase °C
1890	0.00°	1940	0.52°
1900	0.10°	1950	0.40°
1910	0.25°	1960	0.40°
1920	0.20°	1970	0.38°
1930	0.40°	1980	0.42°

TABLE 35-2. AVERAGE AMOUNT OF CO_2			
Year	Amount CO_2	Year	Amount CO_2
1958	315 ppm	1970	325 ppm
1960	317 ppm	1972	327 ppm
1962	318 ppm	1974	330 ppm
1964	319 ppm	1976	332 ppm
1966	321 ppm	1978	335 ppm
1968	323 ppm		

4. Extend the line drawn in Step 3 to the year 2000 so that you can predict the temperature in that year.
5. Table 35–2 shows the change in amount of carbon dioxide for the atmosphere on Earth since 1958. The amounts are stated in parts per million (ppm). This measurement indicates how many parts of carbon dioxide are in one million parts of air.
6. Use the graph outline supplied by your teach-er to construct a graph of the data in Table 35–2.
7. Using the edge of a ruler, draw a dashed line on your graph to show the average change in amount of CO_2 from 1958 to 1978. Your teacher will help you with this step if you are not sure what to do.
8. Extend the line drawn in Step 7 to the year 2000 so that you can predict the amount of carbon dioxide present in that year.

Data and Observations:

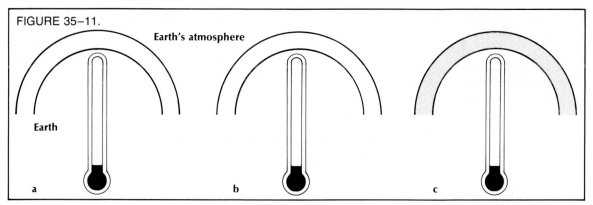

FIGURE 35–11.

Earth's atmosphere

Earth

a   b   c

Questions and Conclusion:

1. Explain how
   (a) Earth's temperature is affected by the amount of infrared radiation absorbed.
   (b) Earth's temperature is affected by the amount of carbon dioxide present in the atmosphere.
2. (a) Explain how your graphs support your answer to question 1(b).
   (b) Does the trend seen in years past seem to continue into the year 2000?
   (c) What is the expected change in average temperature by the year 2000?
   (d) What is the expected change of carbon dioxide in ppm by the year 2000?
3. There are several causes for increased carbon dioxide levels in our atmosphere. One is the burning of fossil fuels. A second cause is the continued destruction of forests for agricultural use.
   (a) What are fossil fuels?
   (b) List several ways in which the burning of fossil fuels may be lessened.
   (c) Explain how destruction of forests can increase the amount of carbon dioxide in the atmosphere.
   (d) List several ways to reduce the destruction of forests.
4. Predict several outcomes if Earth's average temperature were to rise even a few degrees because of the greenhouse effect.

Conclusion: What evidence is there that the greenhouse effect is occurring?

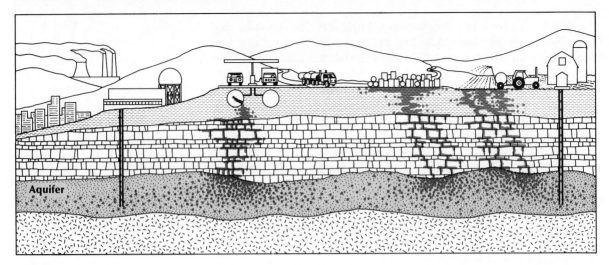

FIGURE 35–12. Large aquifers beneath the ground supply water to a large segment of the population. Aquifers become polluted when hazardous materials leach downward through the soil from the ground's surface. Once ground water has become contaminated, it is virtually impossible to clean it up.

35:8 Pollution of Ground Water

Not only surface waters, but also underground reservoirs of water, called aquifers, are becoming polluted. Pollution of underground water supplies poses a grave threat to segments of the population that obtain drinking water from wells. A great variety of potentially harmful chemicals have found their way into aquifers. Although many of the chemicals are present in small doses, repeated usage of contaminated well water over a long period of time may lead to severe health problems.

Hazardous wastes enter underground water supplies from disposal sites and from several other sources.

Aquifers become contaminated as pollutants move downward through layers of soil. Some pollutants get into the soil as they are leached from landfills where hazardous wastes have been deposited. Often the hazardous wastes have been stored in containers, only to begin leaking out later. Some of the pollutants enter from buried gasoline tanks, septic systems, and farms, and some enter as a result of illegal dumping of wastes.

Dioxins are known to be toxic to animals.

Radiation from radioactive isotopes can cause mutations and, in some cases, can be lethal.

Chemicals entering groundwater include PCBs and a variety of pesticides and other organic chemicals. Dioxins, a general name for a group of chemicals, also have tainted underground water supplies. A particular dioxin, known as TCDD, is a by-product of the manufacture of certain herbicides. TCDD is toxic to certain animals, but its long-term effects on humans are not yet known. Radioactive isotopes (Section 10:1), which result from manufacture of atomic devices and are produced industrially, also can enter groundwater (and air, too) as a result of improper storage. Radiation from some of these isotopes can be lethal, even in very small amounts. Other substances release radiation that increases the incidence of mutations; thus, the effects of contamination with radioactive compounds may extend to future generations.

Many scientists and government officials feel that it will be too expensive to cleanse large quantities of contaminated water. In many parts of the United States, wells have been shut down and people are using bottled water. Efforts are being made to clean up hazardous waste dump sites. This cleanup is also expensive. The most logical course for the future is to prevent further contamination by developing new and safer methods of storage and containment of hazardous wastes.

Because it is expensive to clean contaminated water, efforts must be made to improve methods of disposal of hazardous wastes.

REVIEWING YOUR IDEAS

5. What is a pollutant?
6. What is the major cause of air pollution?
7. How is acid rain formed?
8. What is eutrophication? Does this process occur under natural conditions?
9. How may streams and rivers become polluted?
10. How do underground water supplies become polluted?

CONSERVATION OF RESOURCES

Humans depend on the entire biosphere for their needs. All the things taken from the environment are called **resources.** In a broad sense, resources are either renewable or nonrenewable. **Renewable resources** can be replaced; **nonrenewable resources** cannot. Humans must realize how their use of these resources affects other organisms and their own future.

35:9 Food

One of the most important of all resources is food. Food provides the energy needed for life. Because crops can be replanted and livestock can be bred, food is a renewable resource.

However, food shortages exist. They are most critical in developing countries where the combination of dense populations, primitive farming methods, and poverty results in poor food supplies. Fuel for harvesting crops is difficult to get and is expensive, so food production is decreased even more. Prolonged drought in certain areas also has led to widespread famine.

By selective breeding, geneticists have produced new strains of plants such as rice and wheat that have large heads of grain. Also, these plants respond well to fertilizers. These new strains have produced enough food to feed many more people than could be fed with "wild" varieties of rice and wheat.

FIGURE 35–13. Millions of people around the world are malnourished. Food shortages in many areas cause survival problems, especially for children who often suffer from protein deficiency.

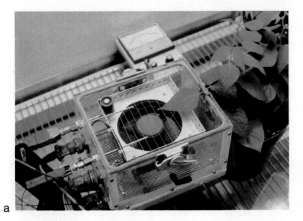

a

b

FIGURE 35–14. (a) The photosynthetic rate of a bean leaf is measured in this chamber. Photosynthetic rates are studied in an effort to breed and manage plants so that total seasonal photosynthesis increases and crop production improves. (b) Information obtained from satellites can be coupled with ground-based information to estimate crop yields. This statistician is mathematically combining Landsat satellite images with computer generated maps.

By means of genetic engineering and use of enzymes, synthetic food may be made in the future.

With food shortages, protein deficiency becomes a major problem. Protein is very important for growth and mental development. A strain of corn rich in protein has been developed. However, there are some poor traits, such as soft kernels and lack of resistance to pests, that must be bred out of this strain.

Using new sources of protein is another way of solving protein shortages. Many types of plant seeds are good sources of protein. Soybean plants, which resist drought and thrive in a variety of soils, are widely used. Soybean seeds can be used in fresh or dried form or can be ground to make meal. Oil can be extracted from soybeans. Impurities are removed from the oil, leaving a concentrated protein that can be used as a meat substitute or mixed with meat. Experiments are underway to develop crops of plants, the amaranths, as a food source. These plants, once a staple of the diet of the Aztecs, contain more protein than grain plants now used, and the protein contains amino acids needed by humans. The sea is also an excellent source of protein. Much protein is available in fish that are not usually eaten by humans because they are too small or too oily. However, a nutritious protein concentrate could be made from these fish and used as a meat substitute.

Experiments suggest that synthetic foods can be made using genetic and biochemical techniques. These foods would be the same as natural foods; they would just be made in a different way. For example, using recombinant DNA techniques, bacteria have been engineered that produce an enzyme normally produced in the stomachs of calves. The enzyme is used in producing puddings and cheeses. More of these foods could be produced with this source of the enzyme. Scientists predict that bacteria and yeasts can be engineered to produce proteins found in a variety of meats, fish, and other foods. The protein could be made into a powder and mixed with other foods. Another approach is using enzymes that carry out the same metabolic reactions that organisms do. For example, the enzymes used to make milk in a cow could be used to make milk in a lab.

a

b

FIGURE 35–15. (a) Both water and wind can erode soil. (b) Careful plowing, such as the contour plowing shown, can prevent or minimize erosion damage.

35:10 Soil

Food production on land depends on the soil. The growing of crops removes nutrients from the soil. Normally the nutrients of plants would be returned to the soil by decay of the plants or animals that eat the plants. However, when humans use crops as food, they remove organic material so some nutrients are not returned to the soil. As a result, soil fertility is reduced. This process is called **soil depletion** (dih PLEE shun). Fertilizers are used in an effort to replace the nutrients that are removed.

Often farmers alternate crops, a process called **crop rotation.** In crop rotation, a soil-enriching plant such as a legume is alternated with a soil-depleting crop such as corn or wheat. The legumes contain nitrogen-fixing bacteria that return nitrogen to the soil (Section 33:5). Crop rotation is also often effective in reducing the number of insect pests.

Soil resources can be lost as a result of **erosion** (ih ROH zhun). Erosion is a natural process in which topsoil is removed by the action of water and wind. Under the cover of plants, soil is held in place. However, careless farming methods increase the erosion rate (Section 34:8). Harvesting and plowing expose the topsoil by removing the plant cover. Under these conditions, water and wind carry topsoil away more easily (Figure 35–15).

Soil is a renewable resource because it is constantly being formed, but soil formation takes thousands of years. Thus, increased erosion due to human activity can cause soil to be a nonrenewable resource.

Erosion can be reduced in several ways. **Contour plowing,** plowing along the contour of the land, reduces water runoff. **Terracing,** the creation of level banks of land on a slope, also prevents runoff. Terraces slow the flow of water, which slows erosion. Wind erosion can be lessened by planting trees as windbreaks. Also, plowing at right angles to the wind reduces erosion.

Removal of materials from farmlands results in soil depletion.

Natural plant covers tend to minimize the effects of erosion. However, careless farming methods increase the rate of erosion.

Soil erosion can be prevented by contour plowing, terracing, proper planting of trees, and by plowing at right angles to the wind.

FIGURE 35–16. Humans depend on fossil fuels. (a) An offshore drill rig is used to get oil that is in deposits under the ocean waters. (b) The Alaskan pipeline moves oil from northern Alaska to ports in southern Alaska for transport to places where the oil is refined.

a

b

35:11 Fuels

Humans have used nearly all of the available fossil fuel resources.

Humans today rely on fossil fuels such as coal, oil, and natural gas, but supplies are rapidly dwindling. The supply of petroleum products required millions of years to form, but humans have almost totally used up the supply in a few hundred years. New sources of oil and natural gas are constantly being sought. Offshore drilling may provide new supplies of these energy sources. Shale is another good source of oil. However, even if new supplies are found, the demand cannot be met for long as these supplies will soon run out, too.

Nuclear energy is generated in the United States and other countries by nuclear reactors. In one type of nuclear reactor, radioactive isotopes of uranium are split apart in a process called **nuclear fission.** During nuclear fission, energy is released. In a nuclear power plant, the energy is used to generate electricity. Only a small part of the electricity generated in the United States comes from nuclear power plants. Use of nuclear reactors conserves fossil fuels, but the radioactive isotopes of uranium used to produce nuclear energy are nonrenewable resources. Radioactive wastes produced by nuclear reactors pose difficult disposal problems, and the water used for cooling the reactors may cause thermal pollution in rivers and lakes. In a properly functioning reactor, radiation does not escape as fission occurs. However, in the event of a large accident, some radiation could escape and affect the environment.

FIGURE 35–17. Fission reactions release energy in a nuclear fission reactor. Shown here is the reactor in which the reactions occur.

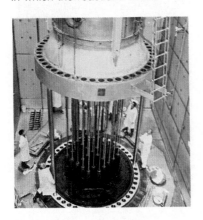

Use of solar energy is slowly increasing. Currently, solar energy is used mainly for heating and cooling buildings. Solar panels are installed on houses or other buildings to collect solar energy. Solar-powered cells are being used to operate calculators and watches, fuel a factory that makes fresh water out of salt water in Saudi Arabia, power satellites, and run field telephones for the military. As the cost of solar cells decreases, more uses for solar cells will be found.

What may the future hold? Hydrogen may be an energy source for the future. Hydrogen is readily available from water and, when burned, does not release pollutants. However, the process now being used to obtain hydrogen from water is very expensive. Therefore, hydrogen is not being used commonly as an energy source. However, it already has been used as a rocket fuel. Hydrogen also may be used as a nuclear fuel. Isotopes of heavy hydrogen (Section 3:2) may be used in a process called **nuclear fusion.** Nuclear fusion is a reaction in which small atoms collide to form larger atoms. As these collisions occur, a great deal of energy is released. Fusion is the reaction that fuels the sun. In order to "mimic" this reaction on Earth, large amounts of energy and extremely high temperatures are needed. Research is continuing to find an efficient way to control fusion and harness the energy it releases. Unless more energy is released from fusion than is used in the process, hydrogen fusion will not be an economical energy alternative.

Still other sources of energy are being investigated for use in the future. Possible sources of untapped energy include wind, geothermal energy (p. 732), and tidal power. These energy sources have two advantages — they are renewable, and they produce little or no pollution.

FIGURE 35–18. Solar collectors on this building help with heating by using energy from the sun. Water moving through the panels picks up heat and can be used as hot water or for heating the building.

35:12 Wildlife and Forests

Humans have often interfered with natural food chains and caused the extinction of many species of plants and animals. The dodo bird and passenger pigeon have become extinct because of human interference. Also, the hunting of animals, such as tigers, has nearly wiped out these animals. Today, the major cause of extinction is destruction of habitats.

People should understand the value of wildlife conservation. They should realize the long-range effects of extinction. It not only interferes with the natural balance of ecosystems but also removes forever a part of the world of life. Once an organism is gone, it is gone forever.

Many efforts to conserve wildlife have been undertaken. Setting up of wildlife preserves and imposing limits on hunting and fishing have proved successful. Population sizes of many nearly extinct animals have been increasing as a result of conservation efforts in the United States. Among these animals are whooping cranes, bison, sea otters, wild turkeys, and white-tailed deer. However, despite con-

FIGURE 35–19. The California condor is a bird that is nearly extinct. It is listed on the U.S. EPA endangered species list.

servation efforts, the California condor and the black-footed ferret remain highly endangered North American species.

Large areas of forests also have been wasted. At one time, forests covered most of the eastern and western United States. These forests were cleared for farmland and much of the timber was burned or used for other purposes. By the early 1900s, it was evident that too much of this natural resource had been removed.

Since that time, humans have begun to conserve forests and other plants and increase their value. Trees are planted every few years so that some will be ready for cutting at regular intervals. In this way, there is a continuous production of trees, and a sufficient supply of valuable, healthy trees is available each year. Forests are also conserved by not cutting an entire forest at once. Instead, blocks of trees are left to reseed the cut areas.

Other forest conservation practices include **selective cutting** and **improvement cutting.** In selective cutting, mature trees are cut while young trees are left to mature. In improvement cutting, old, crooked, or diseased trees are removed. Thus, healthy trees can grow into large, valuable trees.

Humans can conserve forests by planting trees every few years, by not cutting an entire forest at once, and by selective cutting and improvement cutting.

35:13 Cause for Optimism

Pollution of the environment, wasting of resources, and an expanding population are important problems facing humans. Humans must care about the environment now and in the future and do everything possible to solve these problems. Ignoring these problems will compound other problems related to human survival.

FIGURE 35–20. One cause for optimism is (a) formerly polluted areas that (b) are now being cleaned up. Lake Erie areas are shown.

a

b

Fortunately, many people realize the effects of pollution, poor conservation methods, and overpopulation. Scientists and nonscientists alike are aware of and concerned about these problems. Humans should continue to use their reason and intelligence to ensure their own survival and the survival of all that is living.

REVIEWING YOUR IDEAS

11. What are resources? How are resources classified?
12. How can food become a nonrenewable resource?
13. What is soil depletion? How does soil depletion occur?
14. What is erosion? Explain what humans can do to prevent erosion.
15. How are fuels used today? Are fuels a renewable resource? Explain.
16. Distinguish between nuclear fission and nuclear fusion.
17. What has caused the extinction or near extinction of many animals?
18. Why has much of the original forest land of the United States been cleared?

ADVANCES IN BIOLOGY

Alternative Solutions in Pest Control

Although the use of pesticides continues, it is not the only answer to pest control. Biological control methods also are being developed. **Biological control** is the use of organisms to check pest populations. Such methods retain a more natural, balanced ecosystem and avoid the disadvantages of chemicals.

One biological method used to control pests is to preserve their natural predators—insects, spiders, frogs, and birds. Predators of pests also may be introduced into a new area. This method was first used in 1888 when an Australian ladybird beetle was introduced in California to help reduce the numbers of a small insect pest that was destroying citrus crops. The beetle saved the citrus industry.

As predators or parasites, wasps (Figure 35–21) are especially useful in controlling other insects. Two species of parasitic wasps imported from India may prove useful in controlling gypsy moth caterpillars. Gypsy moth caterpillars destroy the leaves of millions of acres of U.S. trees each year. The wasps lay their eggs in the caterpillars. Within two to three weeks, the wasp larvae destroy the host caterpillars. Field tests are planned to determine how effective the

FIGURE 35–21. This wasp is laying eggs in a gypsy moth pupa. When the eggs hatch, the wasp larvae will feed on and kill the pupa. Introduction of the parasitic wasps into the area thus can control the moths.

FIGURE 35–22. Japanese beetles can be controlled by use of sex attractants. The attractants in the dark patch on the trap lure the beetles, which then fall into the trap.

Hormones are used to alter the developmental pattern of some insects.

Pheromones can be used as "bait" in traps or to interfere with the finding of mates.

wasps may be. Currently, several species of insects have been approved for pest control use in the United States. What precautions should be taken before introducing a "foreign" insect into an ecosystem?

Use of bacteria, viruses, fungi, and protozoa as agents of control is being studied and several products have been developed. Bacteria that control the larvae of Japanese beetles and a variety of other insects are already in use. Spores of bacteria that affect certain insects are applied over an area like an insecticide. The spores, once ingested by the insects, develop and become toxic, causing the insects to die. Spores of bacteria have been applied to a variety of crops, including lettuce and cabbage. Viruses are available for use against the cotton bollworm and gypsy moth. A problem with using microorganisms is that they often work slowly, so that insects may cause a great deal of damage before dying. However, microorganisms usually attack only one pest species, leaving other organisms intact.

Another form of pest control is the sterilizing of male insects by radiation. This method in which insects contribute to their own destruction is called **autocidal control.** Insects are raised in a laboratory and the males are sterilized at maturity. Then, these males are released into the ecosystem. Some of the sterile males will mate with females. When mating of this kind occurs, the eggs are not fertilized, and the size of the insect population is reduced. Scientists think that if there are ten times more sterile males than fertile ones in a population, many insect species could be eliminated within four generations. In California, use of sterile males was combined with malathion to control the spread of medflies. Other species of fruit flies have also been successfully controlled by sterilization procedures.

Hormonal control of insect pests is also in use. These hormones affect the insect's rate of metamorphosis (Section 28:9). Some hormones may reduce the time needed to reach maturity and are effective in cases where an insect causes damage as a larva. Other hormones retard development and are effective in cases where the insect causes damage as an adult or where reproduction must be controlled. Hormones such as these are being used to control the development of fleas and mosquitoes.

Pheromones (Section 31:12) also are being used to control some insects. Pheromones can be placed in traps to attract insects. Pheromones also may be spread over a large area, "confusing" insects in search of mates. Both uses of pheromones lead to a reduction in the size of the insect population. Pheromones are being used in an attempt to control the cotton bollweevil (a beetle) as well as the bark beetle that carries the fungus responsible for causing Dutch elm disease.

Cultural control is another means of dealing with pests. **Cultural control** involves the breeding, planting, and harvesting of crops by methods that reduce pest damage. Crop rotation is an example of a cultural control method. By alternating crops, pests are less likely to attack, since the same crop is not present every year. The time when crops are planted or harvested is also important. With proper timing of these events, crops may escape the effects of potential pests.

Genetics is an important aspect of cultural control. Breeding of pest-resistant crops is one way to combat damage and crop loss. Recombinant DNA techniques (Chapter 13) may be used in the future to produce pest-resistant plants. Such techniques, once developed, would change crops much more quickly than breeding, which often takes up to ten years.

Genetics can be used in the production of pest-resistant crops.

The problems of pest control are complex, and solutions must be thought out carefully. As more is learned about the relationships among organisms, it becomes clear that a pest problem rarely can be solved using only one approach. Many people feel that integrated pest management may be the most effective plan of attack. **Integrated pest management** is a system in which a combination of techniques is used to control pests. This system involves the use of chemical pesticides along with biological and cultural controls. In every case, the goal of a program designed to control pests should be to preserve the natural balance of an ecosystem.

Integrated pest management may be the most sensible approach to pest control.

CHAPTER REVIEW

SUMMARY

1. Modern technology has accelerated pollution of the environment. **p. 739**
2. Pesticides kill organisms that compete with or are harmful to humans. **35:1**
3. Pesticides may destroy natural predators of pests as well as the pests themselves. Some insect species have evolved that are resistant to pesticides. **35:2**
4. Many pesticides are nonbiodegradable and accumulate in the tissues of animals. Other pesticides are biodegradable but may be toxic to nonpest organisms, for which they are not intended. **35:3**
5. The use of pesticides is related to economy. New pesticides having less harmful consequences for the environment must be developed. **35:4**
6. Air pollution is caused mainly by the burning of fossil fuels. **35:5**
7. Release of nitrogen oxides and sulfur dioxide has contributed to formation of acid rain, which damages or kills a variety of organisms. **35:6**

CHAPTER REVIEW

8. Pollution of surface waters can lead to poisoning and death of organisms and to the destruction of entire ecosystems. **35:7**
9. Pollution of groundwater occurs as hazardous wastes work their way downward through layers of soil. **35:8**
10. In many areas of the world, food is a resource that cannot be produced quickly enough to meet the increase in population size. **35:9**
11. Soil quality can be improved by farming methods that decrease erosion. **35:10**
12. Fossil fuels are being used up rapidly as energy demands increase. New sources of energy must be developed. **35:11**
13. Wildlife can be conserved through human concern. Proper forestry techniques must also be continued. **35:12**
14. An awareness of ecological problems and dedication by some to solving these problems are reasons for optimism about the future. **35:13**
15. Pest control involves the use not only of pesticides but also of biological and cultural control methods. Integrated pest management may be the most effective means of controlling pests. **p. 759**

LANGUAGE OF BIOLOGY

acid rain
algal bloom
autocidal control
biological control
biological magnification
biosphere
crop rotation
cultural control
erosion
eutrophication
integrated pest
  management

inversion
nonbiodegradable
nonrenewable resource
nuclear fission
nuclear fusion
pesticide
pollutant
pollution
renewable resource
resources
soil depletion
thermal pollution

CHECKING YOUR IDEAS

On a separate paper, complete each of the following statements with the missing term(s). Do not write in this book.

1. A(n) _____ is any factor that makes the environment unclean.
2. _____ is the aging and death of a lake.
3. Substances such as DDT, which are not broken down, are said to be _____.
4. Fossil fuels are an example of a(n) _____ resource.
5. Large amounts of phosphates and nitrates may result in a population explosion of algae called a(n) _____.
6. _____ may be used to lure insect pests to traps or to prevent insects from finding mates.
7. _____ is the increased concentration of chemicals along a food chain.
8. _____ is a natural process that results in removal of topsoil.
9. _____ is a dangerous element present in some gasoline.
10. Alternation of crops to increase soil fertility is called _____.

EVALUATING YOUR IDEAS

1. Name some specific organisms that are pests to humans. What organisms do the pests affect?
2. How have humans contributed to pest problems?
3. What can happen if insects that prey on insect pests are also killed when exposed to a pesticide?
4. Why are some pesticides no longer effective against certain insects?
5. What are the effects of DDT on large birds?
6. How are pesticides such as malathion harmful to an ecosystem?

7. What are the advantages of using microorganisms to control pests? What are the disadvantages?
8. Explain how hormones and pheromones may be used to control pests.
9. How do the pollutants given off by burning fuels affect organisms?
10. Describe the sequence of events leading to an algal bloom and eutrophication.
11. Identify some hazardous wastes that reach aquifers. Why is pollution of aquifers a threat to humans?
12. How may radioactive isotopes act as pollutants?
13. Explain several approaches that have been or might be taken to increase the world's food supplies.
14. What is crop rotation? How does crop rotation prevent soil depletion?
15. How can erosion be prevented?
16. Why are fossil fuels considered to be nonrenewable? What are some possible new energy sources?
17. Discuss the pros and cons of generating electricity by nuclear power.
18. How is solar energy being used today?
19. How can wildlife be conserved?
20. How are forests conserved? How would recycling help conserve forests?

APPLYING YOUR IDEAS

1. Explain how insects have become resistant to DDT and other pesticides.
2. The amount of oxygen dissolved in water increases with lower temperatures. How might hot water produced by a factory affect life in a body of water?
3. Why must offshore oil-drilling procedures be well controlled?
4. In what ways would an energy shortage affect a technological society?

5. Which areas of the United States are most heavily polluted? Are other areas safe from pollution? Explain.

EXTENDING YOUR IDEAS

1. How can you as an individual help in the fight against pollution? Do you have an obligation to keep the environment free of pollution? How can you inform others about the need for a healthy environment?
2. Use mounds of soil to illustrate the principles of contour plowing and terracing. Use other mounds as controls. Add the same amount of water to each mound and note the differences in the water that runs off.
3. Increased use of electricity means more pollution by electric plants that operate on fossil fuels. Increased electrical usage could also lead to the use of more nuclear power plants that cause thermal pollution and pose the threat of nuclear contamination. Which would you prefer, less electricity or more electricity and these possible consequences? Explain.
4. Investigate the processes for recycling wastes such as glass, aluminum cans, and paper and uses for recycled materials.
5. Study your home and school for ways to conserve energy and start an energy conservation program based on your findings.

SUGGESTED READINGS

Grove, Noel, "Air—an Atmosphere of Uncertainty." *National Geographic,* April, 1987.

Schneiderman, H.A., "Altering the Harvest." *Science 85,* Nov., 1985.

Stranahan, Susan Q., "The Deadliest Garbage of All." *Science Digest,* April, 1986.

Yulsman, Tom, "The Threat from PCBs." *Science Digest,* Feb., 1985.

APPENDIX A

A Classification of Living Systems

Although biologists have devised several different classification schemes, the clues used to group organisms are universal ones. In general, taxonomists classify organisms on the basis of their evolutionary relationships. Evidence for determining such relationships comes from a variety of studies: the fossil record, homologous structures, comparative embryology and biochemistry, and modern genetics.

The following is a classification showing five kingdoms. Of the five kingdoms, only Kingdom Monera contains prokaryotic organisms. Major phyla of each kingdom are included. Several minor phyla have been omitted. Classification of the classes of several minor phyla and of important orders of insects and mammals is included. For more complete descriptions of most groups, refer to Unit 4 — Diversity.

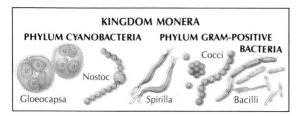

KINGDOM MONERA

Note: Most organisms in this kingdom consist of single cells, or of cells arranged in pairs, clusters, or chains. A few forms exist as multicellular filaments. Most forms are enclosed by a cell wall and reproduction is always asexual. Some forms can move.

EUBACTERIA (true bacteria)

Phylum Gram-Positive Bacteria: heterotrophic forms; chemicals in cell wall react with Gram's solution.

Phylum Gram-Negative Bacteria: heterotrophic forms; chemicals in cell wall do not react with Gram's solution.

Phylum Cyanobacteria: aerobic autotrophic forms; pigments located in flattened membranes within cytoplasm.

Phylum Green-Sulfur Bacteria and **Phylum Purple Bacteria:** anaerobic autotrophic forms having a unique form of chlorophyll.

Phylum Prochlorophyta: autotrophic forms having pigments much like those of eukaryotic autotrophs.

ARCHAEBACTERIA (ancient bacteria)

Phylum Methanogens: Methane-producing bacteria. Obtain energy by chemosynthesis, producing methane as a by-product.

Phylum Halophiles: Salt-loving bacteria. Carry out a unique form of photosynthesis without chlorophyll.

Phylum Thermoacidophiles: Heat- and acid-loving bacteria. Produce organic compounds by chemosynthesis.

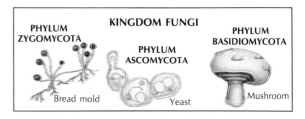

KINGDOM FUNGI

Phylum Zygomycota: Sporangium fungi. Multicellular; heterotrophic; spores produced in sporangia.

Phylum Basidiomycota: Club fungi. Multicellular; heterotrophic; spores produced on basidia.

Phylum Ascomycota: Sac fungi. Mostly multicellular; heterotrophic; spores produced in asci.

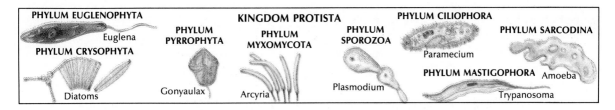

KINGDOM PROTISTA — PHYLUM EUGLENOPHYTA — Euglena; PHYLUM CRYSOPHYTA — Diatoms; PHYLUM PYRROPHYTA — Gonyaulax; PHYLUM MYXOMYCOTA — Arcyria; PHYLUM SPOROZOA — Plasmodium; PHYLUM CILIOPHORA — Paramecium; PHYLUM MASTIGOPHORA — Trypanosoma; PHYLUM SARCODINA — Amoeba

KINGDOM PROTISTA

Phylum Euglenophyta: Euglenoids. Unicellular algae; mostly autotrophic; usually one flagellum for locomotion; mainly freshwater forms; some animal parasites.

Phylum Chrysophyta: Golden algae. Yellow-brown color; mostly unicellular; marine.

Phylum Pyrrophyta: Dinoflagellates. Unicellular; two flagella for locomotion; marine and freshwater forms.

Phylum Sarcodina: Sarcodines. Unicellular; pseudopods (false feet) for locomotion and obtaining food; heterotrophic.

Phylum Ciliophora: Ciliates. Unicellular; many cilia for both locomotion and obtaining food; macronucleus controls basic cell activities; micronucleus is involved in reproduction; heterotrophic.

Phylum Mastigophora: Flagellates. Unicellular; have flagella; heterotrophic.

Phylum Sporozoa: Sporozoans. Unicellular; reproduce by spores; no adaptations for locomotion; parasitic.

Phylum Gymnomycota: Slime molds. Mostly colonial; some cells amoebalike; spores; multinucleate.

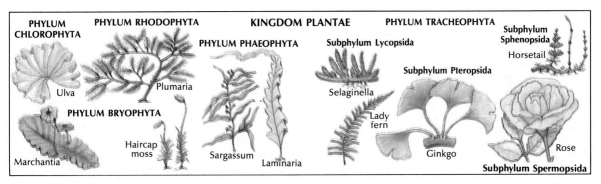

KINGDOM PLANTAE — PHYLUM CHLOROPHYTA — Ulva; PHYLUM RHODOPHYTA — Plumaria; PHYLUM BRYOPHYTA — Marchantia, Haircap moss; PHYLUM PHAEOPHYTA — Sargassum, Laminaria; PHYLUM TRACHEOPHYTA — Subphylum Lycopsida — Selaginella; Subphylum Pteropsida — Lady fern, Ginkgo; Subphylum Sphenopsida — Horsetail; Subphylum Spermopsida — Rose

KINGDOM PLANTAE

Phylum Chlorophyta: Green algae. Many unicellular; some filaments or colonies.

Phylum Phaeophyta: Brown algae. Multicellular; mostly sessile; mostly marine.

Phylum Rhodophyta: Red algae. Multicellular; filamentous and branchlike forms; some live deep in ocean.

Phylum Bryophyta: Mosses and liverworts. Very small; multicellular; no vascular system; most live in moist environments; gametophyte generation predominant.

Phylum Tracheophyta: Plants with vascular tissue.

Subphylum Psilopsida: Psilopsids. Very rare; vascular tissue in stem only.

Subphylum Lycopsida: Club mosses. Vascular tissue throughout plant; cones.

Subphylum Sphenopsida: Horsetails. Vascular tissue throughout; branches and leaves in a whorled pattern; cones.

Subphylum Pteropsida: Ferns. Sporophyte generation predominant; rhizomes; sori.

Subphylum Spermopsida: Seed plants. Most complex of the tracheophytes.

Class Gymnospermae: Conifers. Seeds in cones; needlelike leaves; sporophyte generation predominant.

Class Angiospermae: Flowering plants. Flowers are reproductive organs; seeds within fruits; well protected seeds; flat leaves; sporophyte generation predominant.

Subclass Monocotyledonae: Monocots. Seeds with one cotyledon.

Subclass Dicotyledonae: Dicots. Seeds with two cotyledons.

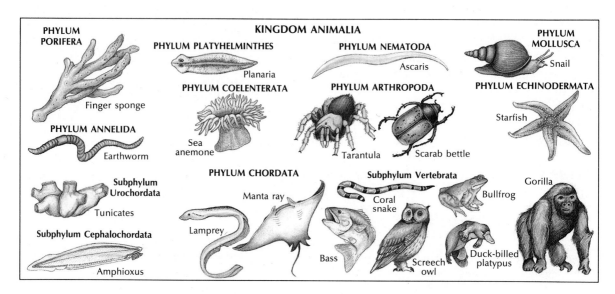

KINGDOM ANIMALIA

Phylum Porifera: Sponges. Specialized cells but no tissue layers; no symmetry; no locomotion; no nervous system; skeleton of mineral content; all aquatic.

Phylum Coelenterata (Cnidaria): Coelenterates. Two tissue layers; radial symmetry; nerve net; single body opening for mouth and anus; tentacles and stinging cells; all aquatic.

Phylum Ctenophora: Comb jellies. Two tissue layers; radial symmetry; rows of cilia along body for swimming; single body opening; marine.

Phylum Platyhelminthes: Flatworms. Three tissue layers; bilateral symmetry; flattened body; single body opening; many parasitic.

Phylum Nemertina: Proboscis worms. Three tissue layers; bilateral symmetry; flattened body; cilia; two body openings; circulatory system lacking a heart; proboscis for food getting and defense; mostly marine.

Phylum Nematoda: Roundworms. Three tissue layers; bilateral symmetry; round shape (cylindrical); two body openings; mostly parasitic.

Phylum Rotifera: Rotifers. Three tissue layers; bilateral symmetry; microscopic but multicellular; ring of cilia around mouth for drawing in food; freshwater and marine.

Phylum Acanthocephala: Spiny-headed worms. Three tissue layers; bilateral symmetry; no digestive system; young are parasites in arthropods; adults are parasites in vertebrates.

Phylum Bryozoa: Moss animals. Three tissue layers; bilateral symmetry; colonial; tentacles around mouth; secretion protects colony; primitive nervous system; freshwater and marine.

Phylum Brachiopoda: Lamp shells. Three tissue layers; bilateral symmetry; body in two shells like clam; two "arms" with tentacles; supported by stalk; many fossil forms; marine.

Phylum Annelida: Segmented worms. Three tissue layers; bilateral symmetry; segmented body; two body openings; ventral nervous system; primitive dorsal circulatory system; many are aquatic.

Phylum Mollusca: Mollusks. Three tissue layers; bilateral symmetry; often with shells; complex digestive, respiratory, circulatory, and excretory systems; mostly aquatic.

Phylum Arthropoda: Arthropods. Three tissue layers; bilateral symmetry; jointed appendages; tough exoskeleton; segmented; body in sections; all systems specialized.

Class Chilopoda: Centipedes. One pair of legs per segment.

Class Diplopoda: Millipedes. Two pairs of legs per segment.

Class Crustacea: Crustaceans. Five or more pairs of legs; two body sections; mandibles; mainly aquatic.

Class Insecta: Insects. Three pairs of legs; three body sections; mandibles; usually winged.

Order Coleoptera: Beetles. Membranous hind wings folded under hard front wings; chewing mouthparts.

Order Diptera: Flies and mosquitoes. One pair of wings; piercing and sucking mouthparts.

Order Hemiptera: True bugs. Wings thin out from base to tip; sucking mouthparts.

Order Homoptera: Cicadas, leafhoppers, and plant lice. Wingless or wings lifted above body; sucking mouthparts.

Order Hymenoptera: Ants, bees, and wasps. Front wings longer than hind wings (or wingless); chewing or sucking mouthparts; some with stingers.

Order Isoptera: Termites. Two pairs of wings of similar size; chewing mouthparts.

Order Lepidoptera: Butterflies and moths. Two pairs of scaly wings; sucking mouthparts.

Order Odonata: Damselflies and dragonflies. Two pairs of long wings.

Order Orthoptera: Grasshoppers. Membranous hind wings folded under long, tough front wings; chewing mouthparts.

Class Arachnida: Arachnids. Four pairs of legs; two body sections; chelicerae.

Phylum Echinodermata: Echinoderms. Three tissue layers; radial symmetry in adult; spiny body often with rays; all marine.

Phylum Hemichordata: Acorn worms. Burrow in mud; marine; feed on microorganisms.

Phylum Chordata: Chordates. Three tissue layers; bilateral symmetry; dorsal nervous system; ventral circulatory system; presence of gill slits and notochord during some stages of life; paired appendages.

Subphylum Cephalochordata: Lancelets. Fishlike; filter feeders; marine.

Subphylum Urochordata: Sea squirts. Sessile; chordate characteristics prominent only in larvae; marine; filter feeders.

Subphylum Vertebrata: Vertebrates. Backbone and general internal skeleton.

Superclass Pisces: Fish. Gills; skeleton of cartilage and/or bone.

Class Agnatha: Jawless fish. Lack paired fins; cartilaginous skeleton; notochord throughout life.

Class Chondrichthyes: Cartilaginous fish. Paired fins; cartilaginous skeleton; prominent gill openings; mostly marine.

Class Osteichthyes: Bony fish. Scaled; hard covering over gills; paired fins; marine and freshwater forms.

Superclass Tetrapoda: Vertebrates with two pairs of limbs.

Class Amphibia: Amphibians. Lungs in most adult forms; aquatic and terrestrial forms; external fertilization, development in water; moist, smooth skin.

Class Reptilia: Reptiles. Dry skin; lungs; internal fertilization; development within amniotic egg; most live on land.

Class Aves: Birds. Feathers; endotherms; usually winged; lungs; internal fertilization; development within amniotic egg.

Class Mammalia: Mammals. Endotherms; hair or fur; internal fertilization; usually internal development; mammary glands; lungs; mostly terrestrial.

Order Monotremata: Monotremes. Most primitive mammals; development in amniotic egg, primitive mammary glands.

Order Marsupialia: Marsupials such as kangaroos, koalas, opossums, wallabies. Young complete development in pouch of female; variety of body forms.

Order Insectivora: Moles and shrews. Small burrowing forms; eat insects and other small animals.

Order Chiroptera: Bats. Forelimbs adapted for flight; active at night.

Order Rodentia: Rodents such as beaver, gopher, mouse, squirrel. Chisel-like teeth for gnawing; little body specialization but forelimbs adapted for variety of niches.

Order Lagomorpha: Hares, rabbits, pikas. Tooth structure similar to that of rodents but more incisors; use side to side mouth movements to eat plants; long ears.

Order Edentata: Includes armadillo, sloth, great anteater. Most lack teeth but some have molars; eat small animals.

Order Cetacea: Dolphins, porpoises, whales. Streamlined body; hair vestigial or absent in adults; forelimbs modified as flippers; hind limbs absent; marine.

Order Sirenia: Manatees, dugongs, sea cows. Streamlined body; sparse amount of hair; forelimbs modified as flippers; hind limbs absent; flipperlike tail; freshwater.

Order Proboscidea: Elephants. Front teeth missing except for incisors (tusks); trunk; herbivorous.

Order Carnivora: Includes cat, dog, raccoon, ferret, seal. Powerful jaws; large canine teeth for eating flesh; sharp claws.

Order Ungulata: Includes cow, giraffe, horse, rhinoceros. Teeth adapted for grinding plants; feet modified as hoofs.

Order Primates: Primates such as lemurs, monkeys, orangutans, humans. Prehensile hands; large brain, erect or semi-erect posture, nails.

APPENDIX B

SI Measurement

SI is a convenient, widely used system of measurement that has the advantage of units based on ten and multiples and fractions of ten. The names of the units used in measuring length, volume, mass, and time differ. However, the prefixes used to indicate the fraction or multiple of ten are always the same. Table B–1 lists SI prefixes and their values. Information about converting a given measurement to a different form follows the table. The basic units for length, volume, and mass are the meter, liter, and kilogram respectively. The basic unit of time is the second.

TABLE B–1. SI PREFIXES

Prefix	Symbol	Meaning	Multiplier (Numerical)	Multiplier (Exponential)
Greater than 1				
tera	T	trillion	1 000 000 000 000	10^{12}
giga	G	billion	1 000 000 000	10^{9}
mega	M	million	1 000 000	10^{6}
kilo	k	thousand	1 000	10^{3}
hecto	h	hundred	100	10^{2}
deka	da	ten	10	10^{1}
Less than 1				
deci	d	tenth	0.1	10^{-1}
centi	c	hundredth	0.01	10^{-2}
milli	m	thousandth	0.001	10^{-3}
micro	μ	millionth	0.000 001	10^{-6}
nano	n	billionth	0.000 000 001	10^{-9}
pico	p	trillionth	0.000 000 000 001	10^{-12}
femto	f	quadrillionth	0.000 000 000 000 001	10^{-15}
atto	a	quintillionth	0.000 000 000 000 000 001	10^{-18}

APPENDIX C

Respiration and Photosynthesis
Cellular Respiration

During aerobic cellular respiration, the energy in the bonds of glucose (or other energy-rich compounds) is transferred to the bonds of ATP. How does this transfer of energy occur? The *net* equation for aerobic respiration is:

$$C_6H_{12}O_6 + 6O_2 + 36ADP + 36 \!-\!\!\textcircled{P} \rightarrow$$
$$6CO_2 + 6H_2O + 36ATP$$

Aerobic respiration occurs in four major stages (Figure C–1). As you study the stages, focus on the general point of energy transformation (bonds of glucose to bonds of ATP) and relate the details to the general equation above.

C:1 Glycolysis

The first stage, **glycolysis** (gli KAHL uh sus), occurs in the cytoplasm. As a result of many enzyme-controlled reactions, one molecule of glucose, a six-carbon compound (C_6), is changed to two molecules of pyruvic (pi REW vihk) acid (C_3). Two molecules of ATP are used during the process. As these reactions occur, the bonds of glucose are broken. In one important energy-releasing step, four hydrogen atoms are removed and join two each with a coenzyme, NAD^{+*}, forming 2 $NADH + 2H^+$. Part of the energy released is stored in the NADH molecules and H^+, whose importance will be discussed later. Most of the rest of the energy released is used to form 2 molecules of ATP. A later step releases more energy, used to form 2 more ATP. Thus, there is a net "profit" of 2 ATP. Note that glycolysis itself is an anaerobic process; it occurs in the absence of oxygen. In fact, oxygen is not used until the fourth stage of aerobic respiration.

*nicotinamide adenine dinucleotide

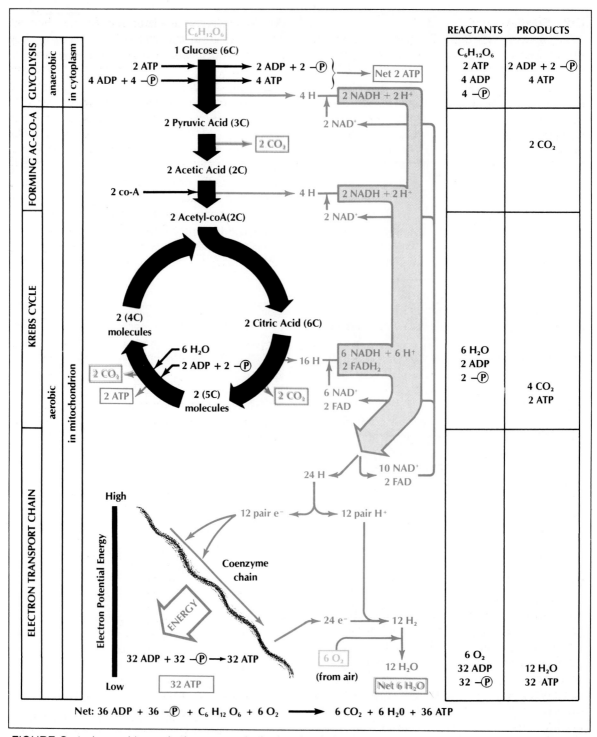

FIGURE C–1. In aerobic respiration, energy in the bonds of glucose is released and used to produce ATP. The process consists of four interrelated stages.

C:2 Formation of Acetyl-CoA

The pyruvic acids formed in glycolysis move into the mitochondria. During several steps, each pyruvic acid molecule is converted to acetic acid (C_2). The carbons removed form CO_2 (part of the CO_2 shown in the general equation above). Each acetic acid is joined to a molecule of coenzyme A (CoA), forming acetyl-CoA. As these steps occur, four more hydrogen atoms are removed and join with 2 NAD^+ to form 2 NADH + $2H^+$.

C:3 Krebs Cycle

Each acetyl-CoA formed combines with a C_4 molecule to form citric acid (C_6). In a series of steps known as the **Krebs cycle,** each citric acid is broken down to a C_5 molecule and then finally to the original C_4 molecule, giving off two CO_2. Since two citric acids enter the cycle for each original molecule of glucose, a total of four CO_2 molecules are given off. (These four plus those from the previous stage account for the total of six CO_2 in the general equation.) For each acetyl-CoA that enters the Krebs cycle, 3 H_2O are used and 8 H are removed. Because two acetyl-CoA enter for each glucose molecule being broken down, 6 H_2O are used and 16 H are removed in the Krebs cycle. Twelve H combine with 6 NAD^+ to form 6 NADH + 6 H^+, and 4 H combine with another coenzyme, FAD**, to form 2 $FADH_2$. Much energy is released during the Krebs cycle. Most is stored in the molecules of NADH and $FADH_2$. Some is used to produce 2 ATP molecules.

C:4 Electron Transport Chain

Note that at the end of the Krebs cycle, only 4 ATP molecules (net) have been produced. A great deal of the energy originally in glucose is now stored in 10 NADH + 10 H^+ and in 2 $FADH_2$. In the final stage, electrons from these energy-rich substances are passed along a series of coenzyme-like molecules located in the cristae and known as the **electron transport chain** or **respiratory chain.** A total of 12 pairs of electrons, 10 from NADH and 2 from $FADH_2$, travel along the chain. Some of the electrons do not enter the chain at the

first molecule, but at a later one. As the electrons are passed from one molecule to another, they release energy. The energy released is used to convert 32 ADP + 32 —(P) to 32 ATP. Each pair of electrons, at the end of the chain, combines with hydrogen ions and oxygen to form water. Six H_2O were used in the Krebs cycle; thus, a net of six H_2O are released during aerobic respiration. The oxygen organisms use during aerobic respiration comes from air (or that dissolved in water) and is necessary as the final electron acceptor in the electron transport chain.

In summary, aerobic respiration of one molecule of glucose produces 6 CO_2, 6 H_2O, and 36 ATP. Much of the potential energy of glucose is transferred to ATP; the rest is lost as heat.

C:5 Other Energy Sources

Fats and proteins can be used as energy sources. Fats are broken down into fatty acids and glycerol. Proteins are converted to amino acids. These simpler molecules are changed to acetyl-CoA or Krebs cycle compounds that enter the aerobic respiration pathway, giving off hydrogens that are sent, via coenzymes, to the electron transport chain. There, energy is released for ATP formation (Figure C–2).

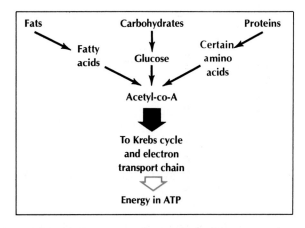

FIGURE C–2. Fats and proteins can be converted to simpler compounds that may enter the respiration pathway at several points and release energy for ATP formation.

**flavin adenine dinucleotide

C:6 Anaerobic Respiration

Some organisms, such as certain bacteria, cannot live in the presence of oxygen and always respire anaerobically. Most other cells, including yeast, some bacteria, and many plant and animal cells, respire anaerobically only when deprived of oxygen. The process of anaerobic respiration consists partly of glycolysis (Section C:1), resulting in the formation of 2 pyruvic acid molecules, 2 NADH

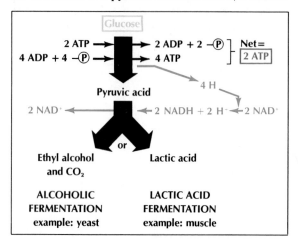

FIGURE C–3. Anaerobic respiration may result in the production of ethyl alcohol or lactic acid.

+ 2 H⁺, and a net of 2 ATP molecules. When no oxygen is available, the process of fermentation follows glycolysis.

In anaerobic respiration, NADH cannot donate electrons to the electron transport chain. There is no oxygen to accept them. Instead, the H^+ and the hydrogen carried by NAD are combined with pyruvic acid. In cells such as yeasts, addition of the four hydrogens to the two pyruvic acids results in the formation of two molecules of ethyl alcohol and two molecules of CO_2. Because alcohol is a product, this process is called **alcoholic fermentation.**

During heavy exercise, human skeletal muscle cells may not receive enough oxygen. These cells, under such conditions, undergo **lactic acid fermentation.** When the H^+ and hydrogens from NADH are combined with the two pyruvic acid molecules, two lactic acid molecules are formed.

Anaerobic respiration nets only 2 ATP for each glucose molecule. Thus, it releases much less of the potential energy of glucose than does aerobic respiration. Much of the original energy in glucose remains "locked" in molecules such as ethyl alcohol or lactic acid. Aerobic respiration is a far more efficient process.

Photosynthesis

A simple equation for photosynthesis shows the reactants and products. It also shows that enzymes, chlorophyll, and light energy are needed.

$$6CO_2 + 6H_2O \xrightarrow[\text{light energy}]{\text{enzymes, chlorophyll}} C_6H_{12}O_6 + 6O_2$$

It does not show the many, complicated reaction steps that were worked out by the efforts of many scientists over many years.

C:7 Source of Oxygen

It was known before 1900 that CO_2 and H_2O are reactants, that chlorophyll and light are needed, and that oxygen is given off when glucose, $[CH_2O]$, is made. These facts can be represented by a simple, unbalanced equation.

$$CO_2 + H_2O \rightarrow [CH_2O] + O_2$$

What is the source of the O_2 given off? Early researchers thought it came from CO_2 and that carbon joined water to form $[CH_2O]$. This idea was tested using an isotope (Section 3:2) of O_2 with ten neutrons. Because oxygen (^{16}O) and the heavy oxygen isotope (^{18}O) differ in mass, it is possible to distinguish between them. An experiment was done in which plants were given $C^{16}O_2$ and $H_2{}^{18}O$ and the products of photosynthesis were analyzed. The heavy oxygen appeared in the oxygen produced. Thus, the O_2 given off is not from CO_2.

$$C^{16}O_2 + H_2{}^{18}O \rightarrow [CH_2{}^{16}O] + {}^{18}O_2$$

Thus, ideas about photosynthesis changed.

C:8 The Light Reactions

How does light interact with chlorophyll, and how is radiant energy changed to chemical energy?

The reactions of photosynthesis that require light are called the light reactions. They occur in the

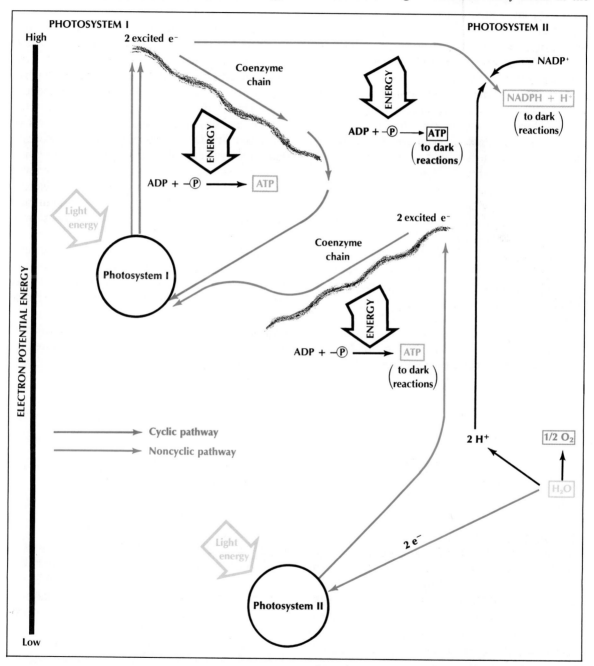

FIGURE C–4. In the light reactions of photosynthesis, light energy excites electrons. In the returning of these electrons to their ground state, energy is given off as ATP.

membranes of the grana of chloroplasts. The reactions in the grana involve two pathways that are referred to as **photosystem I** and **photosystem II.** Each photosystem consists of a group of chlorophyll and carotene molecules. The pigment molecules of photosystem I absorb light energy and transfer energy to one special chlorophyll molecule. Light moves certain electrons of the chlorophyll to levels farther from the atoms' nuclei — to a higher energy level. These energy-rich electrons are called **excited electrons.** Their energy originally came from light. In photosystem I, excited electrons are passed down a chain of coenzymes (like the electron transport chain in a mitochondrion). The electrons return to the chlorophyll molecule in the **ground state,** a low energy level (red path, Figure C–4). Energy is released as the excited electrons return to their ground state. That energy is used in forming ATP. Because the electrons return to the chlorophyll molecule of which they were originally a part, this passage of electrons is known as the **cyclic pathway.**

The cyclic pathway is the only form of photosynthesis that occurs in certain bacteria, and it does not involve water or release oxygen. In algae and plants, excited electrons follow a **noncyclic pathway** (blue path, Figure C-4), water is used, and oxygen is given off.

The noncyclic pathway involves photosystems I and II. Light hits photosystem I and electrons become excited. These electrons follow a different set of coenzymes and do not return to their original chlorophyll molecule. As they move along the coenzyme chain, they release energy used to make ATP. Since the electrons do not return, the chlorophyll molecule has electron vacancies. How are these vacancies filled? Light hitting photosystem II also results in release of excited electrons from a certain chlorophyll molecule. These electrons are passed along a transport chain leading to photosystem I. There they return to their ground state, replacing the "lost" electrons. Energy given off as the electrons move from molecule to molecule is used to make ATP. Now, the chlorophyll molecule from photosystem II has electron vacancies. How are they filled? Water molecules split into H^+ ions, O_2 molecules, and electrons. Each water molecule

provides two electrons, which are carried to the waiting chlorophyll molecule in photosystem II. Pairs of H^+ ions from water and pairs of electrons that came from photosystem I are picked up by the coenzyme NADP*, forming energy-rich NADPH + H^+. The NADPH + H^+ and the ATP formed during passage of electrons in the noncyclic pathway will be used to make glucose.

C:9   The Dark Reactions

Synthesis of glucose occurs in a second set of reactions, the dark reactions (so named because they do not require light). The dark reactions occur in the stroma of the chloroplast. Glucose is synthesized from CO_2. This synthesis requires energy and hydrogen. The energy (ATP) and hydrogen (NADPH + H^+) come from the noncyclic pathway of the light reactions.

In a series of reactions called the Calvin cycle, each CO_2 molecule combines with a C_5 molecule, RBP*, to form an unstable C_6 molecule that quickly breaks down to form 2 C_3 molecules. Using energy

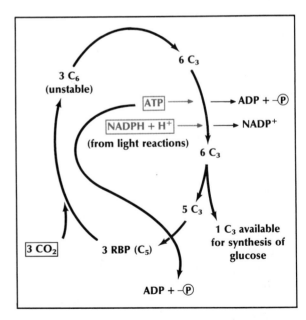

FIGURE C–5. In the dark reactions, CO_2 is converted to glucose. ATP and NADPH + H^+, both produced in the noncyclic pathway of the light reactions, are used. Two "spins" of the cycle are needed to produce one glucose molecule.

*nicotinamide adenine dinucleotide phosphate

*ribulose biphosphate

of ATP and hydrogen from NADPH + H$^+$, some of the C$_3$ molecules are changed and used to make glucose. The remaining C$_3$ molecules, using ATP energy, are converted back to RBP. Thus, the cycle can continue. Note that Figure C-5 shows 3 CO$_2$ molecules entering the cycle at a time. In all, 6 CO$_2$ molecules would enter for each molecule of glucose produced.

As you study photosynthesis, focus on how the noncyclic pathway of the light reactions prepares for the events of the dark reactions and relate the entire process to the general equation.

APPENDIX D

Biology-Related Careers

Careers in biology-related fields are many and varied. Some jobs in these fields may require only on-the-job training. Others may require up to eight years of formal college training plus on-the-job training. Below is a list of a few of the biology-related careers with brief descriptions and training requirements. The training requirements may vary from place to place. You will want to check with local companies, schools, and professional groups for details.

Training and education are indicated below using the following abbreviations: On-the-job training—JOB, High school diploma—HS, Vocational or technical school training—VT, Junior college (2 yr)—JC, Bachelor of Science degree—BS, Master of Science degree—MS, Doctor of Philosophy degree—PhD, Internship—IN.

LIFE SCIENCE. Life sciences often are divided into three broad categories—agriculture, biology, and medicine. Life scientists may do research to determine facts. They also may solve practical problems or teach. Improved plants, new drugs, and better food are some of the results of the work of life scientists.

Animal laboratory assistant (JOB, HS)—cares for lab animals

Biochemist (BS, MS, PhD)—studies substances such as foods and drugs and their changes in living systems

Biomedical engineer (BS, MS, PhD)—applies engineering technology to medical and health problems

Farmer (JOB, VT, BS)—cultivates land, raises crops and/or livestock

Horticulturist (BS)—breeds plants, raises flowers, fruits, vegetables, and decorative plants

Microbiologist (BS, MS, PhD)—studies microscopic organisms

Science teacher (BS, MS)—instructs students about general or specific areas of science

Soil scientist (BS)—studies biological, physical, and chemical properties of soil

ENVIRONMENTAL CONSERVATION. Persons in conservation and environmental occupations help us live within our physical environment. Some help protect, develop, and manage our forests, rangelands, wildlife, soil, and water. Others study our surroundings to improve the quality of life. All play an important role in solving environmental pollution problems.

Forester (BS)—manages, protects, and develops forests

Forestry technician (VT)—aids forester; prevents and controls fires; supervises woodcutting

Geologist (MS, PhD)—studies composition, structure, and history of Earth's crust

Geophysicist (BS, MS, PhD) — studies chemical and physical traits of Earth and other planets

Meteorologist (BS) — studies atmosphere and its effects; forecasts weather

HEALTH CARE. The availability of health service is important to people who are sick or injured. These services depend on people employed in health occupations. Certain careers in some professional health fields — medicine, dentistry, pharmacy — require not only several years of preprofessional college training but also professional school education and a passing score on a state board examination. However, other careers in these fields require less specialized training.

Dentistry

Dental assistant (JOB, VT) — prepares patients; helps dentist

Dental hygienist (VT or BS) — cleans teeth, gives oral hygiene instruction

Dental laboratory technician (JOB) — prepares dentures, inlays, and dental appliances

Dentist (Doctor of Dental Surgery degree-6 yr.) — examines and treats people with tooth-related problems

Medical secretary (VT, SC) — office work for doctors, insurance companies, and others

Medical Practitioners

Chiropractor (Doctor of Chiropractics degree-4 yr.) — treats human patients by manual manipulation of body parts; cannot prescribe medication.

Optometrist (Doctor of Optometry degree-6 yr.) — examines eyes for vision problems and disease

Osteopathic physician (Doctor of Osteopathy degree-7 yr., IN) — diagnoses and treats human diseases, especially by manipulation of skeletal and muscular systems; can use all accepted methods of medical care

Physician (Doctor of Medicine degree-7 yr., IN) — examines, diagnoses, and treats human disease and injury; often specialized

Podiatrist (Doctor of Podiatric Medicine degree-6 yr.) — treats foot injury and disease

Veterinarian (Doctor of Veterinary Medicine-6 yr.) — diagnoses and treats animal disease and injury

Nursing

Licensed practical nurse, LPN (HS, 1 yr. training) — provides nursing care to sick or injured patients

Nursing assistant (JOB) — serves meals; provides for patient comfort; frees registered nurse and licensed practical nurse for more critical work

Registered nurse, RN (Nursing school-2 to 5 yrs.) — gives medication ordered by physician; observes patient symptoms and progress; supervises nursing assistants; teaches

Dietician (BS, MS, PhD) — plans nutritious meals; supervises food service workers

Food technologist (BS) — investigates nature of food and applies this to processing, packaging, and storage of foods

Immunologist (MS, PhD) — studies how the body fights disease

Medical assistant (paramedic) (JOB, VT) — helps physician examine and treat patients; does clerical work

Medical lab worker: Assistant (JOB), Technician (College-2 yr.), Technologist (College-4 yr.) — works in laboratory performing various medical tests; responsibility and test complexity depend on training and experience

Pharmacist (Pharmacy degree-5 yr., IN) — dispenses drugs and medicines prescribed by medical practitioners

Physical therapist (Physical therapy degree-4 yr.) — provides training and helps rehabilitate persons with muscle, bone, and nerve disease or injury

Physical therapy aide (JOB, JC) — assists physical therapist

Further information about biology-related careers and a more complete listing of additional sources can be found in the *Occupational Outlook Handbook* and *Keys to Careers in Science and Technology*. Check also with your school guidance counselors for any information they may be able to provide.

GLOSSARY

A

abdomen: posterior body region of an arthropod; region in vertebrates housing many internal organs

abiotic (ay bi AHT ihk) **factors:** physical (nonliving) aspects that interact with the organisms of an ecosystem

abscisic acid: plant hormone that inhibits growth of root cells causing the root to grow downward

abscission (ab SIHZH un): the falling of leaves, flowers, or fruits from trees

abscission layer: group of cells that forms between the stem and the petiole of a leaf

absorption spectra (ub SORP shun • SPEK truh): spectra that have some wavelengths of light either reduced or absent

abyssal (uh BIHS ul) **zone:** deepest area of the ocean where light is absent

acceleration center: area in the medulla oblongata from which the accelerator nerves originate

accelerator nerves: nerves that increase the heartbeat by stimulating the S-A node

acetylcholine (uh seet ul KOH leen): transmitter that is produced by the vagus nerves and inhibits the rate of contraction of heart muscle in humans

acid: substance, which in solution, has a greater concentration of hydrogen ions than hydroxide ions

acid rain: rain with a low pH

acquired characteristic: change in an organism caused by use or disuse of certain body parts

acquired immune deficiency syndrome (AIDS): viral disease that is transmitted through body fluids and that attacks the immune system

actin (AK tun): protein that makes up part of the filaments in striated muscle

action potential: reversal of polarity and flow of ions in a neuron caused by a stimulus

activation energy: energy necessary to start chemical reactions

active immunity (ihm YEW nut ee): immunity or disease resistance resulting from production of antibodies by the host

active site: particular portion of an enzyme molecule that fits a substrate

active transport: energy-requiring process in which a membrane has an active role in the passage of materials across it

adaptation: inherited characteristic that promotes survival in a natural environment

adaptive radiation: the evolution of many new species from a common ancestor in a new environment

adenine (AD un een): one of the bases in nucleic acids

adenosine diphosphate (uh DEN uh seen • di FAHS fayt) **(ADP):** complex molecule containing adenine, ribose, and two phosphates

adenosine triphosphate (uh DEN uh seen • tri FAHS fayt) **(ATP):** complex molecule containing adenine, ribose, and three phosphates; used as an energy source in all organisms

adhesion (ad HEE zhun): attraction between unlike molecules

adrenal (uh DREEN ul) **cortex:** outer portion of the adrenal gland

adrenal gland: endocrine gland on top of each kidney

adrenal medulla (muh DUL uh): inner portion of the adrenal gland

adrenaline (uh DREN ul un): hormone secreted by the adrenal medulla; aids the body's response to emergencies; also called epinephrine

adrenocorticotrophic hormone (ACTH): hormone secreted by the anterior pituitary; stimulates the adrenal cortex to secrete its hormones

adventitious (ad ven TIHSH us) **roots:** roots that grow from the stems of plants allowing them to reproduce vegetatively

aerobe: organism that uses oxygen in respiration

aerobic (er ROH bihk) **respiration:** respiration that occurs in the presence of oxygen

African sleeping sickness: disease caused by a parasitic flagellate

afterbirth: part of the placenta that is expelled from the body after the birth of offspring

agar: substance made from red algae, often used to make culture media gel

agglutination: process in which cells, such as red blood cells or bacterial cells, clump together

agglutinin (uh GLEWT uh nun): antibody present in blood of people with certain blood types that causes clumping of red cells in blood of people having a different blood type

air sac: respiratory structure in birds that fills with air as the lungs fill

alcoholic fermentation: type of anaerobic respiration used by plant cells and microbes

aldosterone (al DAWS tuh rohn): hormone secreted by the adrenal cortex; controls salt level in the blood

algae: simple autotrophs; may be unicellular or multicellular

algal bloom: a very large and dense population of planktonic algae

alimentary (al uh MEN tree) **canal:** long, hollow tube in animals through which food passes during digestion

allantois (uh LANT uh wus): sac that collects metabolic waste products in the amniotic egg; also functions with the chorion in gas exchange

allele (uh LEEL): one of two or more alternate forms of a gene

all-or-none response: a muscle fiber contracts fully or does not contract at all; a neuron either carries an impulse or does not carry it at all

alternation of generations: life cycle in most plants in which multicellular diploid and haploid generations follow each other

altitudinal (al tuh TEWD nul) **succession:** continuous succession of community types from low to high altitudes

alveoli (al VEE uh li): small, moist sacs in the lungs; sites of gas exchange

amino (uh MEE noh) **acids:** compounds that are the building blocks of proteins

amino group: part of all amino acids that has the structural formula ($-NH_3$)

ammonia (NH_3): nitrogenous waste excreted by planaria and other organisms

ammonification (uh moh nuh fuh KAY shun): stage of the nitrogen cycle during which bacteria metabolize amino acids and produce ammonia

amniocentesis (am nee oh sen TEE sus): technique in which some of the amniotic fluid surrounding an embryo is removed for analysis

amnion (AM nee ahn): fluid-filled sac surrounding the embryos of reptiles, birds, and mammals; protects the embryo and keeps it moist

amniotic egg: embryo surrounded by a tough shell; protects embryo as it develops outside the mother; permits development away from water

amoebocyte (uh MEE buh site): type of cell in sponges that moves in amoeboid fashion

amoeboid (uh MEE boyd) **motion:** movement by pseudopodia; motion of amoebae and amoebalike cells

amphetamines (am FET uh meenz): stimulant drugs used for appetite control

amphibians: vertebrates that live both on land and in water; class Amphibia

amplexus: clasping of the female frog by the male that results in gametes being released at the same time

ampulla: bulb-shaped structure in the water-vascular system of starfish

anaerobe: organism not using oxygen in respiration

anaerobic (an uh ROH bihk) **respiration:** respiration in the absence of oxygen

anal pore: structure in ciliates that allows egestion of undigested materials

anaphase (AN uh fayz): phase of mitosis during which one strand of each chromosome is pulled to a pole of the cell; phase in meiosis I in which homologous chromosomes are separated and each is pulled to a pole of the cell

anatomy: study of the structures of organisms

androgens (AN druh junz): male sex hormones

anemia (uh NEE mee uh): disease in which the blood is low in red blood cells, hemoglobin, or total volume

angiosperms (AN jee uh spurmz): flowering plants

animal: multicellular, heterotrophic organism, usually mobile; ingests food

animal pole: top half of a frog egg composed of cytoplasm and a dark pigment

annelid: segmented worm; phylum Annelida

annual rings: circular lines in a woody stem that represent growth during a year

antennae: sensory organs of arthropods

anterior (an TIHR ee ur): front part of an animal

anterior pituitary: one of the lobes of the pituitary gland; many of its hormones control other endocrine glands

anther (AN thur): male sporangium at tip of stamens

antheridium (an thuh RIHD ee um): male sex organ in bryophytes and lower vascular plants such as ferns

anti-A agglutinin: antibody contained in the plasma of blood type B and type O

anti-B agglutinin: antibody contained in the plasma of blood type A and type O

antibiotic: drug that cures certain bacterial diseases

antibody (ANT ih bahd ee): protein formed in response to a chemical in the blood or tissues

anticodon: set of three bases at one end of tRNA; fits with specific codons of mRNA

antigen (ANT ih jun): specific chemical on the surfaces of red blood cells; any foreign chemical that causes antibody formation

antiserum: solution of antibodies and blood serum used to fight disease

anus (AY nus): opening through which undigested materials are expelled

aorta (ay ORT uh): large artery leading from the heart

aortic (ay ORT ihk) **arches:** five pairs of enlarged tubes, "hearts," in annelids

apical (AY pih kul) **dominance:** retardation of growth of lateral buds in the presence of the apical bud

apical meristem: plant growth tissue located at the tip of roots and stems

appendage (uh PEN dihj): limb or other projection attached to the main trunk of an animal

appendix: small organ attached to the large intestine; vestigial in humans

applied science: using scientific knowledge to solve real problems

aquatic (uh KWAHT ihk): having to do with water

arachnids (uh RAK nudz): class of arthropods; spiders, ticks, mites, and scorpions

archegonium (ar kih GOH nee um): female sex organ in bryophytes, gymnosperms, and lower vascular plants

archenteron (ar KENT uh rahn): internal cavity in the developing embryo; develops into the alimentary canal

artery: blood vessel that carries blood away from the heart to the body

arthropods (AR thruh pahdz): animals having an exoskeleton and jointed appendages; phylum Arthropoda

artificial selection: procedure in which humans choose organisms to breed that have desirable features

ascending colon: portion of the large intestine into which undigested materials from the small intestine pass

ascus: saclike structure in which spores are produced by meiosis in certain fungi

asexual (ay SEK shul) **reproduction:** reproduction in which a single parent produces one or more offspring by fragmentation, fission, or mitotic cell division

aster: structure that radiates outward from a centriole

asthma: respiratory condition in which the muscles of the bronchial system contract, reducing the diameter of the air passages

atherosclerosis (ath uh roh skluh ROH sus): condition in which the arteries become lined with fatty deposits

atom: smallest particle of an element; basic building block of all matter

ATP-ADP cycle: series of reactions by which ATP is converted to ADP and ADP is converted back to ATP

atrioventricular (ay tree oh ven TRIHK yuh lur) **node (A-V node):** small bundle of tissue between atria and ventricles that receives the current from the S-A node

atrium (AY tree uhm): heart chamber that receives blood in vertebrates; also called auricle

Australopithecus (aw stray loh PIHTH uh kus): genus of prehumans that lived more than three million years ago in southern Africa

autonomic nervous system: system that controls involuntary responses

autosome (AWT uh sohm): chromosomes not related to an organism's sex

autotroph (AWT uh trohf): organism that can produce its own food

auxin (AWK sun): one class of plant hormone

Aves (AY vayz): class of vertebrates characterized by feathers and lightweight bones for flight

axon (AK sahn): long part of a neuron leading away from the cell body; transmits impulses away from the cell body

B

bacillus (buh SIHL us): rod-shaped bacterium

bacteria: very small, prokaryotic, unicellular, heterotrophic organisms (lack nuclei and some organelles)

bacterial transformation: form of genetic recombination that occurs when one bacterium breaks open and part of its DNA enters another bacterium

bacteriophage (bak TIHR ee uh fayj): virus that reproduces in bacteria; also called phage

balanced equation: chemical equation in which the number of atoms of each element agree on both sides

ball and socket joint: type of joint that can rotate as well as move side to side and back and forth

barbiturates (bar BIHCH uh ruts): most powerful and most often abused depressants

bark: outermost layers of a woody stem

basal body: structure that attaches cilia and flagella to the inside of a cell

base: substance which, in solution, has a greater concentration of hydroxide ions than hydrogen ions; an alkaline solution; compound containing nitrogen that makes up part of a nucleic acid

basidium (buh SIHD ee uhm): club-shaped, spore-producing structure in certain fungi

B cell: lymphocyte that does not pass through the thymus gland; divides to form a clone and produces antibodies

beech-maple forest: climax community in northeastern United States

behavior: responses made by an organism to the stimuli of its environment

behavioral (bih HAY vyuh rul) **adaptation:** adaptation involving reactions to the environment

benthos (BEN thas): animals or plants that live attached to or crawl on the ocean floor

beta cell: type of islet of Langerhans cell that produces insulin

bicuspid (bi KUS pud) **valve:** valve that prevents blood from being forced back into the left atrium when the left ventricle contracts

bilateral (bi LAT uh rul) **symmetry:** body pattern in which only one specific longitudinal cut will produce two sides that are mirror images

bile: greenish liquid produced by the liver and stored in the gallbladder; emulsifies fat in the small intestine

bile salts: substances contained in the gall bladder that break up fat globules in the intestine

binomial nomenclature (bi NOH mee ul • NOH mun klay chur): two-term naming system used in classifying organisms

biogenesis (bi oh JEN uh sus): theory that states that at the present time and under present conditions on Earth, all life is produced from life

biological clock: mechanism that controls certain time-related activities of animals

biological control: using natural, biological methods of controlling pest populations

biological magnification: increased concentration of chemicals along a food chain

biologist: person who studies biology

biology: study of life

biomass (BI oh mas): total mass of a group of organisms per unit of area

biome (BI ohm): communities characterized by the same major life forms, such as tundra

biosphere (BY uh sfir): total world of life

biotic (bi AHT ihk) **factors:** relationships among living organisms in an ecosystem

biotic potential: tendency of all organisms to reproduce in large numbers; reproductive potential

bipedal (bi PED ul) **locomotion:** moving on two feet

birth: expulsion of fetus from the uterus

birthrate: number of organisms born in a given time

bivalves (BI valvz): mollusks that have two shells; includes clams, oysters, and mussels

blade: expanded part of a leaf

blastocoel (BLAS tuh seel): fluid-filled cavity inside the blastula

blastocyst: ball of cells formed as a zygote divides

blastopore (BLAS tuh por): opening formed by the dorsal lip cutting into the blastula

blastula (BLAS chuh luh): hollow mass of cells in the early development of an embryo

blood: liquid that serves as an exchange medium between the internal parts of an organism and its external environment

blood clot: network of fibers and other blood parts that covers a wound and prevents the loss of excess blood

blood pressure: pressure exerted by the blood on the walls of the blood vessels

body stalk: structure that connects the trophoblast and the embryo to the uterine wall

bone: living tissue that makes up the endoskeleton of most vertebrates

bone marrow: tissue in the hollow part of bones; site of production of blood cells or fat storage

bony layer: layer of bone that contains minerals

Bowman's capsule: cup-shaped part of a nephron

brain: major coordinating center in vertebrates; nerve center in invertebrates similar in position and function to the vertebrate brain

breathing: movement of air into and out of the lungs of complex animals

breathing center: area in the medulla oblongata that controls the breathing rate

bromeliad: epiphyte that lives in the canopy of the tropical rain forest

bronchi (BRAHN ki): two large tubes branching from the trachea in the lungs

bronchial tubes: small branches of the bronchi

bronchioles (BRAHN kee ohlz): tiny branches of the bronchial tubes

bryophytes (BRI uh fites): most primitive of true land plants; mosses and liverworts

bud: protected, dormant meristem that may develop into new stems, flowers, or leaves

bud scale: modified leaf that protects a bud

bud scale scar: where bud scales surrounded a bud

budding: type of asexual reproduction in which an outgrowth forms on the parent organism and later separates giving rise to a new organism

bulb: short underground stem surrounded by many scales (modified leaves)

C

Calvin cycle: cyclic pattern of reactions in the dark reaction of photosynthesis

cambium: special layer of cells that produces secondary growth as a plant grows in diameter

camouflage (KAM uh flahj): any means of blending with the environment

cancer: disease characterized by lack of control of cell division

canopy: top layer of a forest where most of the food is produced

cap: top, umbrellalike portion of a mushroom; guanine at the end of mRNA

capillaries: smallest branches of blood vessels where materials of the blood and cells are exchanged; connect arteries and veins

carbohydrates: organic compounds composed of carbon, hydrogen, and oxygen; used for energy sources in organisms

carbon-14: a radioactive isotope with a half-life of 5730 years

carbon compound: chemical compound containing carbon; organic compound

carboxyl (kar BAHK sul) **group:** part common to all fatty acids and amino acids; has a structural formula of (-COOH)

carboxyhemoglobin: molecule formed when carbon dioxide combines with hemoglobin in blood

carcinogen (kar SIHN uh jun): cancer-causing agent

cardiac (KARD ee ak) **muscle:** muscle of which the heart is composed

carnivore (KAR nuh vor): consumer that feeds on other consumers; meat eater

carotene (KER uh teen): yellow or orange pigment present in most autotrophs

carrying capacity: number of individuals of a population that a particular area can support

cartilage (KART ul ihj): tough, flexible, connective tissue that does not contain minerals

cell: smallest living unit of which all organisms are composed

cell body: portion of a neuron most resembling other cells; contains a nucleus and cytoplasm

cell cycle: events in the life of a cell, which include interphase and mitosis

cell membrane: structure that separates the components of a cell from its environment by regulating the passage of materials into and out of the cell

cell plate: structure completed in plants to separate the daughter cells formed in mitosis

cell theory: concept that applies to all organisms; the cell is the basic unit of structure and function and all cells are produced from other cells

cell wall: structure that surrounds the plasma membrane of bacterial, fungal, and plant cells

cellular respiration: process by which the energy of glucose is converted to the energy of ATP in a cell

cellulose: a polysaccharide of which the cell walls of plants and some algae are composed

centipede: member of class of arthropods characterized by one pair of legs per segment; class Chilopoda

central disc: center part of a starfish from which arms stick out

central nervous system: a brain and one or more longitudinal nerve cords

centrioles (SEN tree olz): pair of organelles that play an important role in division of animal cells

centromere (SEN truh mihr): special region of a chromosome at which chromatids are held together

cephalothorax (sef uh luh THOR aks): one of the two body sections of crustaceans; the result of fusion of the head and thorax

cerebellum (ser uh BEL um): convoluted region of the hindbrain at the back of the head below the cerebrum

cerebral cortex: convoluted surface of the cerebrum

cerebrospinal fluid: fluid in brain and spinal cord; protects brain from shock

cerebrum (suh REE brum): forebrain of humans

cervix (SUR vihks): hollow, narrow, neck area of the uterus leading to the vagina

chelicerae (kih LIHS uh ree): fanglike or pincerlike mouthparts in arachnids

cheliped (KEE luh ped): first pair of appendages, which are modified as claws in crustaceans

chemical bond: force that holds atoms together

chemical change: process by which atoms or molecules are rearranged; chemical reaction

chemical energy: energy within molecules

chemical equation: statement using formulas to describe a chemical reaction

chemical formula: combinations of symbols that represent the number and kind of each atom in a compound

chemical property: property of a substance that depends on how the substance reacts with other substances

chemical reaction: chemical change; rearranging of the atoms or molecules of substances to form completely new substances with new properties

chemical symbol: shorthand method of representing an element

chemosynthesis (kee moh SIHN thus sus): process in which energy released from inorganic materials such as iron, sulfur, and nitrogen compounds is used to fix carbon into organic compounds

chitin (KITE un): carbohydrate material forming the walls of most fungi; one of the main components of an arthropod's exoskeleton

chlorophyll: green pigment that is essential for the conversion of light energy to chemical energy in photosynthetic organisms

chloroplast (KLOR uh plast): cellular organelle in eukaryotic autotrophs that contains chlorophyll

cholinesterase (koh luh NES tuh rays): enzyme that breaks down acetylcholine into two smaller molecules

chordates (KOR dates): members of phylum Chordata, includes vertebrates

chorion (KOR ee ahn): thin membrane inside the shell of the egg of a reptile, bird, or monotreme that functions in gas exchange; a membrane in mammals contributing to the formation of the placenta

chromatid (KROH muh tid): individual strand in a double-stranded chromosome

chromatin (KROH mut un): mass of material inside the nuclear membrane that appears as chromosomes during cell division

chromatophore (kroh MAT uh for): cell containing one or more pigments

chromosome (KROH muh sohm): distinct body in the nucleus that appears during cell division; contains genes

chromosome theory of heredity: theory that genes are located on chromosomes

chunk feeder: animal that eats pieces of food material

chyme (KIME): acidic, liquid form of food after digestion in the stomach

cilia (SIHL ee uh): tiny, hairlike projections; used for locomotion in some one-celled organisms; also present on some cells of multicellular organisms

ciliary (SIHL ee er ee) **motion:** locomotion achieved by movement of cilia

ciliates (SIHL ee ayts): protozoa that have cilia

circulatory (SUR kyuh luh tor ee) **system:** system of blood and the structures through which it passes

cirri: large bundles of cilia found in some ciliates

class: division of classification system that represents a subdivision of a phylum

cleavage (KLEE vihj): series of many cell divisions in the developing embryo

climax community: final stage in the evolution of a community

cloaca (kloh AY kuh): a common chamber that receives the digestive and excretory wastes as well as products of the reproductive system

clone: strain of genetically identical cells or individuals

closed circulatory system: system in which blood is always in the blood vessels

coccus (KAHK us): round- or sphere-shaped bacterium

cochlea: fluid-filled canal in the inner ear

codominance: condition in which one allele is not dominant to the other; both alleles are expressed equally

codon (KOH dohn): sequence of three nucleic acid bases which represent a certain amino acid

coelenterates: simple animals characterized by tentacles with stinging cells, two tissue layers, and a single body opening; phylum Coelenterata

coelom (SEE lum): body cavity that houses the major organs in annelids and more complex animals

coenzyme: molecules with which enzymes temporarily join during a reaction

cohesion (koh HEE zhun): clinging together of the same kind of molecules

coleoptile (koh lee AHP tul): sheath surrounding apical meristem and young leaves in grass embryos

collar cells: flagellated endoderm cells that line the body cavity of sponges; function in trapping food

collecting duct: structure for passage of urine from tubules to kidney pelvis

colon (KOH lun): large intestine

commensalism (kuh MEN suh lihz um): relationship in which one organism (commensal) benefits from another organism (host) without affecting it

common bile duct: canal formed by the merging of the bile duct and the pancreatic duct; empties into the duodenum

communication: use of symbols to represent ideas

community: naturally occurring group of interacting organisms living in a certain area

companion cells: phloem cells with nuclei that lie next to sieve tubes in a stem

comparative anatomy: studies of similar structures among organisms

comparative biochemistry: studies of chemical similarities among organisms

comparative embryology: studies of similar development among different organisms

competitive exclusion principle: two different populations cannot occupy the same ecological niche

complement system: proteins in the bloodstream that attack foreign cells

complete metamorphosis: four-stage process of development in insects

compound: substance composed of different kinds of atoms joined by a chemical bond

compound leaf: leaf blade subdivided into leaflets

concussion: brain injury that may result from a severe blow

condensation: chemical reaction in which a large molecule is formed from smaller molecules by removing water

conditioning: involuntary response to a stimulus that does not normally cause the response; voluntary response to get a desired effect; depends on type of conditioning

conductor: structure that transmits nerve impulses

cone: seed-bearing or pollen-bearing structure of gymnosperms

conidium (kuh NIHD ee um): spore formed by mitosis at the tip of a conidiophore

conidiophore (kuh NIHD ee uh for): hypha found in some molds on which conidia are produced

conifers (KAHN uh furs): pines and their relatives; cone-bearing gymnosperms

conjugation: fusion of nuclear material of two cells

consumer: organism that obtains food from other organisms; heterotroph

continuous variation: existence of varying degrees of a characteristic; controlled by more than one pair of genes

contour plowing: conservation practice in which land is plowed along its natural contours, reducing erosion from water runoff

contractile (kun TRAK tul) **vacuole:** organelle that pumps excess water from a protozoan

control group: group that is identical to the experimental group except for the variable being tested

controlled experiment: experiment in which all factors are identical except the one being tested

convergence (kun VUR junts): evolutionary process in which distantly related species produce descendants that resemble each other

convolutions (kahn vuh LEW shunz): folds in the cerebral cortex

cork: cells that make up the outer bark of trees and woody shrubs

corpus luteum (KOR pus • LEWT ee um): yellow tissue that forms from the ruptured follicle

corpus luteum stage: stage in the menstrual cycle during which the corpus luteum produces progesterone to maintain the uterus for pregnancy

cortex: food storage area in roots and stems; outer part of kidney; outer part of adrenal gland; outer part of cerebrum

cortisol: hormone that causes fats and proteins to be converted to glucose; secreted by the adrenal cortex

cortisone: hormone secreted by the adrenal cortex, often administered in synthetic form to counteract pain and swelling

cotyledon (kaht ul EED un): seed leaf of plant embryo; often stores food

courtship behavior: type of complex behavior of animals

covalent (koh VAY lunt) **bonds:** force resulting from sharing of electrons between atoms

cranial nerves: nerves from the brain that carry impulses of the sensory-somatic system

creatine phosphate (CP): compound that stores a high energy phosphate and donates it to ADP to form ATP used as an energy source for muscle contraction

cretinism (KREET uhn ihz um): disease caused by a great deficiency of thyroxine in early childhood; victim is called a cretin

cristae (KRIS tee): folds on the inner layer of mitochondria

Cro-Magnon: type of modern human that lived 10 000 to 50 000 years ago

crop: food storage organ in the alimentary canal of an earthworm

crop rotation: conservation practice in which different crops are planted in a field from year to year

crossing-over: exchange of segments of chromosomal material between two strands of a tetrad; occurs during prophase of meiosis I

cross-pollination: pollination between flowers of separate plants

crustaceans (krus TAY shee uns): class of arthropods that includes lobsters, crayfish, shrimp, and crabs

cryptic coloration: type of camouflage in which the color of an organism blends with the environment

cultural control: the breeding, planting, and harvesting of crops by methods that reduce pest damage

cutin: waxy material covering the epidermis of a leaf that helps prevent water loss

cyclic AMP: "second messenger" formed as a result of certain hormones combining with receptors on target cells

cyclic pathway: passage of electrons from low to high to low energy levels in the light reactions of photosynthesis

cytoplasm (SITE uh plaz um): protoplasm outside the nucleus of the cell

cytoplasmic streaming: movement of cytoplasm within cell that aids in distribution of materials

cytosine (SITE uh seen): one of the bases in the nucleic acids

cytotoxic T cell: cells of the immune system that recognize antigens on certain host cells, such as cancer cells and transplanted cells

D

dark reaction: phase of photosynthesis in which the actual synthesis of glucose occurs

data: scientific facts

daughter cells: two cells formed as the result of mitotic cell division

day-neutral plant: plant whose flowering does not depend on the length of daylight or darkness

DDT: pesticide that kills insects by attacking the nervous system; nonbiodegradable, passed along food chains

death rate: percentage of a population that dies in a given period of time

deciduous (dih SIHJ uh wus) **forest:** forest composed of trees that periodically shed all their leaves

decomposer: organisms such as bacteria or fungi that cause decay

degenerative (dih JEN uh ruh tihv) **evolution:** process in which organisms lose some of their features as they evolve from a more complex form into a simpler one

dendrite (DEN drite): highly branched structure at one end of a neuron

denitrification (dee ni truh fuh KAY shun): process by which nitrogen is released from the soil by the action of denitrifying bacteria

deoxyribonucleic (dee AHK sih ri boh noo klay ihk) **acid (DNA):** nucleic acid with the "genetic message"

deoxyribose: a sugar that joins with a nitrogen base and a phosphate group to form a nucleotide

depressant: one of a major group of drugs that reduce nerve transmission in the brain

descending colon: section of the large intestine that leads to the rectum

desert: biome of arid regions where rainfall is less than 25 cm per year; sparse, widely-spaced vegetation

desmid: a form of green alga

development: series of changes an organism undergoes in attaining its mature form

diabetes insipidus (ihn SIHP ud us): disease caused by secretion of too little vasopressin; characterized by frequent urination and constant thirst

diabetes mellitus (MEL ut us): disease characterized by frequent urination and constant thirst as a result of a high blood sugar level; also called sugar diabetes

diaphragm: large, muscular sheet that separates the chest from the abdomen; important in breathing

diastole (di AS tuh lee): short period of "rest" after the heart contracts

diatomaceous earth: huge deposit of cell walls of dead diatoms

diatom (DI uh tahm): unicellular golden alga

dicot (DI kaht): flowering plants whose embryos bear two cotyledons

differentiation (dihf uh ren chee AY shun): series of changes that result in the formation of specialized cells and parts

diffusion (dihf YEW zhun): random movement of particles resulting in their movement from a region of greater concentration to one of lesser concentration

digestion: process of breaking large molecules into smaller ones by chemical and physical means

dinoflagellate: unicellular alga with two flagella for locomotion

dipeptide: two amino acids joined together by a peptide bond

diploid (DIHP loyd): having two of each chromosome

disaccharide: double sugar formed by the combination of two monosaccharides

dissociation: process in which the ions of ionic compounds separate in solution

divergence (di VUR junts): evolution of a species into two or more species with different characteristics

division of labor: organization of parts or organisms for specific jobs

dominant (DAHM un nunt) **species:** characteristic species of a community

dominant trait: genetic trait that dominates or prevents the expression of the recessive trait

dormancy (DOR mun see): period of inactivity of a seed or spore before germination; period of little or no activity in some types of animals

dorsal (DOR sul): top part of the body of an animal having bilateral symmetry

dorsal aorta: aorta that carries blood from the gills to all parts of a fish

dorsal blood vessel: major blood vessel on the top side of an annelid worm

dorsal lip: top part of the blastopore in some animal embryos

double fertilization: the joining of a sperm nucleus with the two polar nuclei and the joining of a sperm nucleus with an egg in the ovule of a flowering plant

Down syndrome: abnormality caused by nondisjunction, characterized by eye shape that looks like that of people of the Mongoloid race, mental retardation, short arms and legs, and internal malformations

drone: male bee with haploid chromosome number

duet rule: atoms having only one energy level must have a total of two electrons in that level in forming a chemical bond

dynamic equilibrium (di NAM ihk • ee kwul LIHB ree um): state in which the number of particles entering and leaving a cell is equal

E

ecdyson (EK duh sahn): hormone that causes molting

echinoderms: marine animals characterized by spiny skin and radial symmetry; phylum Echinodermata

ecological niche: the specific role an organism plays in a community

ecological succession (ee kuh LAHJ ih kul • suk SESH un): series of changes in a community during its evolution

ecology: study of the relationships between organisms and their environments

ecosystem (EE koh sihs tum): interaction of a community with its physical environment; an ecological system

ectoderm (EK tuh durm): outer tissue layer in an embryo or mature animal

ectotherm (EK tuh thurm): animal whose body temperature is not internally regulated but rather approximates that of the environment

effector (ih FEK tur): organ that carries out the appropriate response to a stimulus

egestion: the elimination of undigested food

egg: female reproductive cell; ovum

electron: negatively-charged particle of an atom

electron cloud: description of probable electron location around the nucleus of an atom

electron microscope: microscope that uses a beam of electrons instead of light

electron transport chain: series of special molecules that act as coenzymes; involved in cellular respiration

element: substance composed of only one type of atom

elongation region: cells of the meristem that grow only in length adding to the length of a young stem or root

embryo: a developing plant or animal after fertilization

embryology (em bree AHL uh jee): study of development of an embryo

embryonic induction (em bree AHN ihk • ihn DUK shun): hypothesis that a particular section of an embryo influences, or induces, the development of an adjacent region

emigration: movement of animals out of an area

emphysema: respiratory disease linked to smoking in which the bronchioles become inflamed

end brush: filaments forming the end of an axon

endergonic (en dur GAHN ihk) **reaction:** chemical reaction that requires energy

endocrine (EN duh krun) **gland:** ductless gland, secretes hormones directly into blood

endocrinology (en duh krih NAHL uh jee): study of endocrine glands and hormonal control

endocytosis (en duh si TOH sus): process in which a cell uses energy to surround and take in large particles

endoderm (EN duh durm): inner tissue layer in an embryo or mature animal

endodermis: a layer of cells surrounding the stele of a root

endoplasmic reticulum (en duh PLAZ mihk • rih TIHK yuh lum): network of tubelike structures in most cells; transports certain proteins within the cell

endorphin: neuropeptide that inhibits pain messages

endoscope: device that allows a fetus to be viewed while it is still in the uterus

endoskeleton (en doh SKEL ut un): internal skeleton

endosperm (EN duh spurm): mass of storage tissue found in many seeds, providing a food source for the germinating embryo plant

endosperm nucleus: $3n$ structure formed when the sperm nucleus joins with the polar nuclei in a flower; develops into endosperm

endospore: protective structure formed by certain bacteria during unfavorable conditions

endotherm (EN duh thurm): an animal that regulates and maintains a fairly constant body temperature

endotoxin: poisonous chemical released when certain bacteria die and break open

energy: ability to do work or cause motion

energy level: distance at which an electron travels around the nucleus of an atom

enkephalin: neuropeptide that reduces pain

environment: all factors that act upon an organism

environmental resistance: total of the environmental factors that check the biotic potential of a population

enzyme (EN zime): protein that activates a chemical reaction in a cell

enzyme induction: process in which a substance entering a cell causes the cell to start producing a certain enzyme

epicotyl (EP ih kaht ul): part of plant embryo above the cotyledon(s); composed of young apical meristem and young leaves

epidermis (ep uh DUR mus): outermost layer of cells of an organism

epiglottis (ep uh GLAHT us): structure that covers the opening to the trachea when a person swallows food

epiphyte (EP uh fite): plant that grows upon another plant

epitope (EP uh tohp): a specific antigen or peptide in a virus' coat that stimulates production of antibodies by a host

era: division of time used by geologists to divide Earth's history

erosion (ih ROH zhun): naturally-occurring process in which topsoil is removed by wind and water action

erythrocyte (ih RIHTH ruh site): red blood cell, produced in bone marrow; important in transport of oxygen and carbon dioxide

esophagus (ih SAHF uh gus): tube that passes food from the mouth or throat to other digestive organs

estivation (es tuh VAY shun): period of inactivity and low metabolism similar to hibernation but occurring in hot weather

estrogen (ES truh jun): female sex hormone

estrous (ES trus) **cycle:** entire series of chemical and physical changes leading up to the production of mature eggs

eukaryote (yew KER ee oht): organism with cells that have true nuclei

eustachian tube: a canal that connects the middle ear to the nasal passages and throat

eutrophication: natural aging and death of a lake; may be speeded up by upsetting the ecological balance

evolution: process of change with time during successive generations among organisms

excited electron: electron of chlorophyll that becomes energy-rich as it moves to higher energy levels

excretion: release and removal of harmful metabolic by-products

excretory canal: structure in a planarian that collects excess water from tissues

excretory duct: branch of excretory canal that opens as a pore on the surface of a planarian; excess water leaves through the pore

excretory system: an animal system specialized for removal of nitrogenous wastes and maintenance of osmotic balance

excurrent pore: opening in a sponge through which water leaves; osculum

exergonic (ek sur GAHN ihk) **reaction:** chemical reaction that releases energy

exocytosis (ek soh si TOH sus): process in which materials are expelled from a cell as a vesicle fuses with the cell membrane

exon: section of DNA or mRNA that carries a code for amino acids that will become part of a protein

exoskeleton (ek soh SKEL ut un): external skeleton

exotoxin: poisonous chemical released by certain living bacteria

experimental group: part of an experiment in which the variable factor is tested

experimentation: scientific testing of a hypothesis

expiration: process in which air is expelled from the lungs

extensor (ihk STEN sur): muscle that, when contracted, causes a joint to straighten

external fertilization: fertilization in animals in which the egg is fertilized outside the female's body

extinct: no longer existing form of life

extracellular (ek struh SEL yuh lur) **digestion:** digestion that occurs outside cells

extract: solution of a chemical obtained from cells

eyespots: two light-sensitive organs at the anterior end of planaria

F

facilitated diffusion: process in which permeases aid the movement of certain particles through cell membranes

fact: observed phenomenon agreed upon by a number of people; something about which there is no doubt

Fallopian (fuh LOH pee un) **tube:** oviduct

family: subdivision of an order in the complete classification of an organism

fatty acid: complex molecule containing a carboxyl group; part of a fat molecule

feces (FEE seez): solid waste material, egested through the anus

feedback mechanism: system in which control over the glands depends on information received from them

fermentation: anaerobic respiration

ferns: vascular plants that have roots, stems, leaves, and rhizomes and reproduce by spores

fertility factor: genetic information located on plasmid in certain bacteria that are designated F^+; bacteria without this factor are F^-

fertilization: fusion of sperm and egg

fetoscopy (fee TAHS kuh pee): medical test that allows direct observation of the fetus and surrounding tissues

fetus (FEET us): a human embryo that is at least eight weeks old; any vertebrate embryo in egg or uterus

fiber: a skeletal muscle cell

fibril: unit that makes up fibers of striated muscle

fibrin (FI brun): long strands of protein that form a blood clot

fibrous (FI brus) **root system:** system of many secondary roots and root hairs

filament: stalklike portion of a stamen; unit that makes up fibrils in striated muscle

filter feeder: animal that takes food from water as it flows through the animal

first filial (FIHL ee ul) **generation:** first generation of offspring produced from a parental cross

first-order consumer: consumer that feeds directly on a producer; herbivore

first polar body: one of the two cells produced from the primary oocyte during meiosis I

fission (FIHSH un): asexual reproduction in which one organism divides into two small organisms of equal size

fixed joint: joint at which bones are fused together and do not move

flagella (fluh JEL uh): long, whiplike projections of a cell used for locomotion

flagellar motion: locomotion achieved by movement of flagella

flagellate: protozoan that moves by means of flagella

flame cell: cell with flagella that moves water into and along the excretory canal in a planarian

flatworms: bilaterally symmetrical, aquatic, semiaquatic, or parasitic organisms; members of phylum Platyhelminthes

flexor (FLEK sor): muscle that causes the bending of a joint

florigen: hypothetical hormone that promotes flowering

flower: reproductive organ in angiosperms

flower scar: place on the stem where a flower was once attached

fluid mosaic model: model of membrane structure with a lipid bilayer that is a liquid with proteins scattered through it

follicle (FAHL ih kul): cell nest within an ovary; site of egg development

follicle stage: stage in the menstrual cycle during which an egg matures and the preparation of the uterus for a possible pregnancy begins

follicle-stimulating hormone (FSH): hormone that stimulates the ripening of eggs in follicle of ovary; hormone that stimulates sperm production in males

food: substance that provides energy and nutrients required by living systems

food chain: passage of energy and materials through a community

food vacuole: structure within certain cells that collects and digests food

food web: all the possible feeding relationships in an ecosystem

foot: muscular organ in mollusks used for locomotion

forebrain: one of the three general areas of the vertebrate brain

forest floor: bottommost layer of a forest

fossil: any part or trace of an organism that lived long ago

fossil fuels: fuels made from the remains of once-living organisms

fragmentation: form of asexual reproduction in which a part of an organism breaks off and grows into a new organism

fraternal (fruh TURN ul) **twins:** twins that develop when two different eggs are fertilized

frond (FRAHND): fern leaf, highly branched, often with a lacy appearance

fruit: enlarged ovary of a plant, which contains and protects the seeds

fruit scar: place on a stem where a fruit was once attached

fungi: plantlike, heterotrophic organisms; representative of kingdom Fungi; reproduce by spores

fusion: uniting of two sets of DNA; sexual reproduction

G

galactose: a simple sugar found in milk

galactosemia: a genetic disease in which galactose cannot be converted to glucose and instead builds up and causes nerve damage

gallbladder: organ beneath the liver in which bile is stored

gametes (GAM eets): sex cells; sperm and eggs

gametophyte (guh MEET uh fite) **generation:** the haploid part of the life cycle of a plant

ganglia (GANG glee uh): a mass of nerve cell bodies

gas exchange: movement of oxygen and carbon dioxide between organisms and their environment

gastric glands: glands in the stomach that secrete gastric juice

gastric juice: fluid secreted into the stomach by gastric glands that contains hydrochloric acid and pepsinogen

gastrovascular (gas troh VAS kyuh lur) **cavity:** internal area of coelenterates where digestion begins

gastrula: stage of the developing embryo following the blastula stage

gastrulation (gas truh LAY shun): movement of cells from the outside surface of the blastula through the blastopore into the interior of the blastula; process in which tissue layers are formed

gel: somewhat solidified state such as is characteristic of the cytoplasm at times

gemmule (JEM yewl): reproductive cell of a sponge enclosed by a tough, outer covering that is highly resistant to dryness and cold

gene: unit responsible for transmitting hereditary traits; segment of a DNA molecule that codes for synthesis of a particular protein

gene linkage: presence of genes for different traits on the same chromosome

gene pool: all the genes of a population; sum of genetic information that will be passed to each new generation

gene therapy: replacement of defective genes with normal ones to treat diseases

generative nucleus: nucleus of a pollen grain; produces two haploid sperm nuclei

genetic continuity: reproduction of offspring having the same set of features as the parent

genetic counseling: type of counseling in which parents can be advised of their chances of having offspring with a hereditary disease

genetic engineering: altering an organism's genetic makeup

genetic map: diagram indicating locations of genes on a chromosome

genetic recombination: rearranging of genetic instructions

genetics (juh NET ihks): science of heredity

genotype (JEE nuh tipe): particular combination of alleles of an organism

genus (JEE nus): classification division between family and species; the first name in the scientific name of an organism

geographic isolation: division or separation of populations by geographical features

geothermal energy: energy created by the natural heat of Earth

geotropism (jee oh TROH pihz um): plant growth response to gravity

germ theory of disease: theory proposed by Louis Pasteur; stated that bacteria can cause disease

germinate (JUR muh nayt): begin development

gestation (jeh STAY shun) **period:** period of internal development

gibberellin (jihb uh REL un): chemical regulator affecting plant growth

gill: respiratory organ in fish and other aquatic animals; spokelike structure in the cap of a mushroom where spores are produced

gill chamber: area in which gills are located in fish

gill filaments: double rows of thin-walled tissue in gills through which capillaries pass

gizzard: digestive organ in which food is ground up in earthworms and birds

gliding joint: joint in which bones move easily over one another in a back and forth motion

glomerulus (gluh MER uh lus): mass of capillaries in the center of each Bowman's capsule of a kidney

glucose: simple sugar produced by photosynthesis

glycerol: complex molecule that is part of fats

glycogen (GLI kuh jun): starchlike carbohydrate in animals

glycolysis: series of energy-releasing reactions in a cell in the absence of oxygen; stage one of aerobic respiration though it does not require oxygen; process of anaerobic respiration

goiter: disease characterized by enlargement of the thyroid gland; often caused by lack of iodine in the diet

Golgi (GAWL jee) **body:** organelle that functions in storage, modification, and packaging of certain proteins

gonads (GOH nadz): sex organs; ovaries and testes

gradualism: hypothesis that speciation occurs over very long periods of time

grana (GRAY nuh): membranous structures that contain chlorophyll in a chloroplast

grassland: biome of plains regions; characterized by grasses

gray matter: neuron cell bodies that make up the cerebral cortex and the interior of the spinal cord

ground state: low energy level for electrons in light reactions of photosynthesis

growth: increase in the amount of living material in an organism; increase in the number of individuals in a population

guanine (GWAHN een): one of the bases in nucleic acids

guard cells: cells that surround and control the size of the stomata in leaves of complex plants

gymnosperms (JIM nuh spurmz): class of seed plants; plants in which seeds develop unprotected on the scales of cones; phylum Gymnospermae

H

half-life: time necessary for one half of a radioactive material to decay

hand-eye coordination: ability to use sight and touch together for coordinated movements

haploid: having a set of chromosomes that consists of one chromosome from each of the homologous pairs in a diploid organism

Hardy-Weinberg Law: principle that under certain conditions, the genotypic ratios and frequency of appearance of alleles remain constant in a population generation after generation

Haversian (huh VUR zhun) **canals:** channels running throughout the bony layer that contain blood vessels and nerves that supply osteocytes

hazardous wastes: wastes dangerous to humans and other life forms

head-foot mollusk: mollusk that has tentacles originating in the head region such as squid and octopus

heart: muscular pump that circulates blood throughout an animal

helper T cell: T cells that stimulate B cells and macrophages by releasing certain chemicals

hemocoel (HEE muh seel): body cavity, spaces through which blood passes in an open circulatory system

hemoglobin (HEE muh gloh bun): complex protein in red blood cells containing iron that combines with oxygen; important in oxygen transport

hemophilia (hee muh FIHL ee uh): sex-linked hereditary disease in which the blood fails to clot properly

hepatic (hih PAT ihk) **portal vein:** vein that carries blood from the intestine to the liver

herbaceous (hur BAY shus) **stem:** soft, green stem of plants that live for only one growing season

herbivore (HUR buh vor): consumer that eats producers (plants); plant eater

heredity (huh RED ut ee): transmission of characteristics from one generation to the next

hermaphrodite (hur MAF ruh dite): organism that has both ovaries and testes

heterocyst (HET uh roh sihst): special cell in a cyanobacterium filament that has thick walls, no DNA, and appears empty; adapted for nitrogen fixation

heterotroph (HET uh ruh trohf): organism that cannot make its own food; depends on other living or dead organisms for food

heterotroph hypothesis: proposal by Alexander Oparin that life originated as prokaryotic heterotrophs that later gave rise to autotrophs

heterozygous (het uh roh ZI gus): having different alleles for a given character at the corresponding sites on homologous chromosomes

hibernation: state in which metabolism is reduced and energy expenditure is minimized; occurs in some animals during the winter months

high energy bond: bond that, when broken, provides a large amount of chemical energy

high energy phosphate: phosphate group containing high energy bonds

hindbrain: one of the three general areas of the vertebrate brain

hinge joint: joint that moves in one direction and does not rotate

histone: a chromosomal protein that forms spools around which DNA is wound

holdfast: special anchoring cell found in some algae

homeostasis (hoh mee oh STAY sus): the steady state of the internal operation of an organism regardless of external changes

hominid: early human

Homo erectus: prehuman that lived about 1.5 million years ago; may have led to the extinction of *Australopithecus*

Homo habilis: prehuman that lived 2 to 3 million years ago; thought to be a toolmaker

Homo sapiens (HOH moh • SAY pee unz): scientific name for modern humans

homogeneous (hoh muh JEE nee us): the same throughout

homologous (huh MAHL uh gus): parts of organisms that are inherited from a common ancestor

homologs (HOH muh lawgs): two chromosomes of a pair carrying genes that control the same trait; one received from each parent

homozygous (hoh muh ZI gus): having identical alleles for a given trait at the corresponding sites on homologous chromosomes

hormone: a chemical regulator that is produced in one part of the organism and affects other parts; complex molecules that direct chemical control within the body

host: organism from which another organism benefits

human growth hormone: anterior pituitary hormone that influences total growth of the body

human immunodeficiency virus (HIV): virus that causes AIDS

humus: rich layer of soil containing decayed remains of organisms

Huntington's disease: genetic disease that affects the nervous system and is fatal

hydrocarbons: compounds of hydrogen and carbon produced when fossil fuels are burned

hydrolysis (hi DRAHL uh sus): chemical reaction in which large molecules are broken into small molecules by the addition of water

hypertension: unusually high blood pressure

hyphae (HI fee): filamentous strands of cells belonging to fungi or certain funguslike protistans; the hyphae together comprise the mycelium

hypocotyl (HI puh kaht ul): embryonic stem of plant embryo; structure that occurs between the cotyledons and the radicle

hypothalamus (hi poh THAL uh mus): region of the brain that controls many of the body's internal activities; brain region involved in many feedback mechanisms of internal control

hypothesis: idea or statement that explains the relationship among observed facts

I

identical twins: twins that have the same genotypes, resulting from the splitting of a zygote into two separate parts

immune system: cells and tissues that identify and defend the body against foreign chemicals and organisms

imperfect flower: flower that contains only the male or female organs

implantation: attachment to and embedding of the embryo in the uterine wall

imprinting: simple, rapid, irreversible form of learning by which an animal forms a social attachment to an object or organism shortly after hatching or birth

improvement cutting: removing old, crooked, or diseased trees while leaving the healthy trees

impulse: series of action potentials along a neuron

incomplete dominance: one allele of a pair is not dominant to the other; three phenotypes result, the third being an intermediate form

incomplete metamorphosis: three-stage process of development in insects

incurrent pore: opening or perforation through which water enters a sponge

indoleacetic (IHN dohl uh seet ihk) **acid (IAA):** auxin necessary for the lengthening of plant cells

inducer: substance that causes enzyme induction; chemical that causes onset of embryonic development

industrial melanism (MEL uh nihz um): phenomenon in which a color change in a population of peppered moths in England evolved during a period of industrialization

infectious disease: disease caused by a pathogen

infusion: nutrient-rich solution in which microorganisms can live

ingestion: taking in of solid particles

inheritance of acquired characteristics: invalid hypothesis that states that characteristics that each individual acquires during its lifetime are passed on to the offspring of that individual

inhibition (ihn uh BIHSH un) **center:** area of the medulla oblongata from which the vagus nerves originate

innate (ihn AYT) **behavior:** behavior that is genetically passed from parent to offspring; behavior that does not change

inner bark: layer of a woody stem made up of cortex and phloem tissue

inner cell mass: one of the two regions of cells that form as the cells of a blastocyst divide

insectivorous plant: plant that feeds on insects

insects: largest class of arthropods with three pairs of legs, three body sections, and usually wings

insight: ability to plan a response to a new situation

inspiration: process in which air is taken into lungs

instinct: innate behavior that involves complicated responses to a stimulus

insulin: hormone that regulates the level of sugar in the blood; hormone secreted by beta cells of the islets of Langerhans

integrated pest management: system in which a combination of techniques is used to control pests

integument (ihn TEG yeh ment): outer covering of a seed; skin

interferon (ihnt ur FIHR ahn): protein produced by some cells after exposure to viruses that inhibits further virus growth

internal fertilization: process in animals in which sperm is deposited inside the female reproductive organs where fertilization occurs

International System of Measurement (SI): special language of measurements and their symbols used by scientists and by other people in most countries throughout the world

interneuron: neuron that connects sensory neurons and motor neurons in the brain and spinal cord

internode: space between nodes on a stem

interphase (IHNT ur fayz): part of cell cycle during which chromosomes are replicated

interspecific (ihn tur spih SIHF ihk) **competition:** competition between populations of different species

intestine: alimentary canal organ where some or most food is chemically digested and absorbed

intracellular (ihn truh SEL yuh lur) **digestion:** digestion that occurs within individual cells

intraspecific (ihn truh spih SIHF ihk) **competition:** competition between members of the same species

intron: segment of DNA or unprocessed mRNA that does not code for amino acids

invertebrates: animals without backbones

involuntary muscles: muscles that an organism cannot consciously control

ion: charged atom or group of atoms

ionic bond: attraction between ions of opposite charge

ionic compound: combination of different ions

islets (I lutz) **of Langerhans** (LAHNG ur hahnz): endocrine cells within the pancreas that secrete insulin and glucagon

isomers (I suh murz): compounds that have the same chemical formula but different structural formulas

isotopes (I suh tohps): two or more atoms of the same element differing only in the number of neutrons

J

Java man: variety of *Homo erectus* that lived between 300 000 and 1.6 million years ago

joint: where body segments of arthropods meet; where bones of vertebrates meet

jungle: dense tropical rain forest community along river banks or other cleared areas

juvenile hormone: hormone that plays a role in control of insect metamorphosis

K

K lymphocyte: cells of the immune system which can bind to bacteria, thus killing them

karyotype (KER ee uh tipe): characteristics of all chromosomes in a cell including shape and number

kelp: a medium to large benthic brown alga

kidneys: excretory organs in vertebrates

kinetic energy: energy of motion or energy at work

kingdom: broadest division in the classification of organisms; the five kingdoms are animals, plants, protists, fungi, and monerans

Klinefelter syndrome: set of abnormal symptoms resulting from nondisjunction that produces a male having an XXY chromosome pattern

Koch's postulates: a set of procedures used to determine whether a certain bacterium or virus causes a disease

Krebs cycle: a stage of aerobic respiration; stage in which 16 hydrogen atoms are released and 2 ATP molecules produced per glucose molecule entering glycolysis

L

labor: series of contractions of the uterine muscles prior to giving birth

lactase: enzyme that breaks down lactose

lacteal (LAK tee ul): lymph vessel in villi that absorbs fatty acids and glycerol

lactic acid fermentation: anaerobic respiration carried out by muscle cells

ladder-type nervous system: nervous system in planarians and other flatworms; system composed of two ganglia and two major nerve cords connected at regular intervals by nerves

language: form of communication in which symbols are used to represent ideas

large intestine: organ in humans that absorbs water from the undigestible materials; colon

larva: developing organism leading an independent existence

lateral aspects: side edges of a bilaterally symmetrical organism

lateral bud: bud along the sides of branches that gives rise to new branches

latitudinal (lat uh TEWD nul) **succession:** continuous succession of community types from the equator to the poles

law of conservation of energy: energy can neither be created nor destroyed, but can be transformed into other kinds of energy

law of conservation of mass: mass can neither be created nor destroyed

law of dominance (DAHM uh nuhns): the dominant form of a trait dominates or prevents the expression of the recessive form

law of independent assortment: genes for different traits located on different homologous pairs of chromosomes are sorted independently during gamete formation

law of segregation (seg rih GAY shuhn): during gamete formation the pair of genes responsible for each trait separate so each gamete receives only one gene for each trait

law of use and disuse: idea that an animal could strengthen part of its body by using it and that if a certain body part were not used, it would disappear; one of Lamarck's two laws used to explain evolution

leaching: dissolving of minerals out of rock and soil by water

leaf: major photosynthetic organ of complex plants

leaf scar: mark left on a stem after a leaf falls off

learned behavior: behavior that can be changed

lenticel (LENT uh sel): tiny pore on a stem; site of gas exchange between the stem and atmosphere

left ventricular assist device (LVAD): device that takes over for the heart until a donor heart becomes available for a transplant

lethal gene: an allele that causes death

leukemia: cancer of blood-forming tissue

leukocyte (LEW kuh site): general term for a variety of white blood cells produced in bone marrow, the spleen, and lymph nodes; protects the body against infection

levels of organization: different groups of biological parts (cells, tissues, and so on) organized to perform particular functions

liana (lee AHN uh): vinelike plant of tropical rain forests; rooted in the ground but with leaves in the canopy

lichen (LI kun): mutualistic combination of a fungus and an alga or cyanobacterium; often lives on rock or in other barren places

ligament (LIHG uh munt): connective tissue that connects most bones to one another

light: radiant energy

light reaction: series of photosynthetic reactions that depend on the presence of light; portion of photosynthesis in which light energy is converted to chemical energy

limiting factor: aspect of the environment that prevents an increase in population size at any given time

lipids: class of organic compounds that includes fats, waxes, and oils; composed of carbon, hydrogen, and oxygen

littoral (LIHT uh rul) **zone:** area of the ocean close to shore and subject to the action of tides

liver: abdominal organ; produces bile, destroys old red blood cells, stores glycogen; important in many processes of metabolism

lock and key hypothesis: each enzyme is specific for a given substrate because its shape matches that of the substrate

locomotion: movement of an organism from one place to another

long-day plant: plant whose flowering is dependent on long periods of light exposure

low energy phosphate: a phosphate group containing low energy bonds

luminescent: glows in the dark

lungs: moist internal organs for gas exchange in most adult amphibians and all reptiles, birds, and mammals

luteinizing (LEW teen i zing) **hormone (LH):** hormone that causes the follicle to rupture and the ruptured follicle to change to the corpus luteum; causes production of testosterone in males

lymph: excess tissue fluid that enters lymph vessels

lymph nodes: structures scattered throughout the body, remove harmful substances from lymph

lymph vessel: vessel that channels lymph throughout the body

lymphatic (lihm FAT ihk) **system:** network of lymph vessels and lymph nodes

lymphocyte (LIMH fuh site): type of white blood cell that forms clones of plasma cells and memory cells in response to a specific antigen

lysis (LI sus): destruction of bacteria or other cells by breaking of the membrane

lysogenic (li suh JEN ihk) **cycle:** viral life cycle in which the DNA of a host cell becomes part of the host's DNA and is replicated when the host cell reproduces

lysosome (LI suh sohm): organelle containing enzymes that hydrolyze large molecules brought into the cell

lytic cycle: viral life cycle in which viral DNA takes over a host cell, uses the host cell's materials to reproduce, and destroys the host cell

M

macronucleus: large structure in ciliates that controls the basic activities of a cell

macrophage (MAK ruh faj): kind of white blood cell that engulfs foreign material

malaria: disease caused by *Plasmodium* and transmitted by the female *Anopheles* mosquito; causes bursting of red blood cells

malnutrition: inadequate supply of one or more essential nutrients in a person's diet

Malpighian (mal PIHG ee un) **tubules:** cluster of string-like structures attached to the intestine of a grasshopper that function in excretion

maltase: enzyme that breaks down maltose

mammals: complex vertebrates that have hair or fur, mammary glands, and bear live young; animals in class Mammalia

mandible: mouthpart adapted for biting and chewing

mantle: folded tissue in mollusks covering the internal organs

marrow: tissue in the hollows of some bones; produces blood cells or stores excess fat

marsupials (mar SEW pee ulz): group of pouched mammals in which partial development of embryos occurs internally

matter: anything that has mass and occupies space

maturation region: area of a developing root or stem in which cells are large and develop into different types of specialized tissue

medulla (muh DUL uh): inside part of a kidney; inside section of an adrenal gland

medulla oblongata (muh DUL uh • ahb long GAHT uh): portion of the hindbrain; controls involuntary responses of internal organs

medusa: motile body form of coelenterate in which mouth and tentacles usually face downward

megaspore (MEG uh spor): in seed plants, the haploid cell that forms the female gametophyte generation within the developing ovule

meiosis (mi OH sus): process of nuclear division usually associated with the formation of haploid gametes or spores; reduction division of the nucleus

melanin (MEL uh nun): a dark pigment

membrane potential: condition in which the outside of the membrane is more positively charged than the inside; also called resting potential

memory cell: cells that remain in the body after an infection is over; memory cells confer active immunity

meninges (muh NIHN jeez): membranes that protect and nourish the spinal cord and brain

menopause: period when the menstrual cycle of human females ends, usually occurs in the 40s

menstrual (MEN strul) **cycle:** monthly series of hormonal changes leading to egg maturation and uterine preparation for a possible pregnancy

menstruation (men STRAY shun): stage of the menstrual cycle usually lasting from three to five days during which blood, some uterine tissue, and the unfertilized egg are expelled from the vagina

meristem (MER uh stem): special regions of plant tissue where cell division (growth) occurs

mesoderm (MEZ uh durm): tissue layer between the ectoderm and the endoderm

mesoglea (mez uh GLEE uh): jellylike layer between the two tissue layers of coelenterates

messenger RNA: RNA that carries the genetic code of DNA; functions in protein synthesis

metabolism (muh TAB uh lihz um): total of the chemical reactions that build up and tear down complex molecules within a cell

metamorphosis (met uh MOR fuh sus): series of changes in form during development of an immature form to an adult

metaphase (MET uh fayz): phase of mitosis or meiosis in which chromosomes move to the "equator" of the cell and become attached to spindle fibers at their centromeres; phase of meiosis in which tetrads, or pairs of chromosomes, attach to the spindle fibers

microbe: microscopic organism

microenvironment: small area in an ecosystem that differs from the rest of the area

microfilament: cellular structure made of protein that functions mainly in movement

micronucleus: small structure in ciliates involved in reproduction

microorganism: organism too small to be seen with the unaided eye; microbe

micropyle (MI kruh pile): opening in the ovule of a flowering plant through which sperm nuclei enter

microscope: a scientific instrument used to magnify small objects so that they can be easily seen

microspores: in seed plants, the haploid cells within the flower stamens or pollen cones that are transformed into pollen grains (the male gametophyte)

microtubule: tubular cellular structure composed of protein that functions in support and movement of, or within, a cell

midbrain: one of the three general areas of the vertebrate brain

migration: seasonal movement of animals

millipede: member of class of arthropods characterized by two pairs of legs per segment; class Diplopoda

mimicry (MIHM ih kree): protective adaptation in which one organism (the mimic) resembles another organism (the model)

mineral: essential nutrient needed for maintenance of an organism

mitochondrion (mite uh KAHN dree un): organelle in the cytoplasm; site of most of the stages of aerobic respiration

mitosis (mi TOH sus): process of nuclear division resulting in two genetically identical daughter nuclei

model: organism a mimic is similar to; image of an idea or object that helps simplify understanding the idea or object

modifier genes: genes that affect the expression of other alleles

molecule: a combination of two or more atoms joined by a covalent bond

mollusks (MAHL usks): soft-bodied, mostly marine animals, usually enclosed within a hard outer shell of calcium carbonate; phylum Mollusca

molting: process of shedding the exoskeleton in arthropods; process in some birds and reptiles in which the outer skin or feathers are shed and replaced by new growth

moneran: an organism in kingdom Monera; prokaryotic, usually unicellular

monoclonal antibody: antibody produced by a particular clone of hybridoma

monocot: flowering plant whose seeds contain only one cotyledon

monosaccharide (mahn uh SAK uh ride): simple sugar; basic building block for complex carbohydrates

monotreme: group of primitive egg-laying mammals; includes platypus and spiny anteater

morphogenesis (mor fuh JEN uh sus): series of embryonic changes in which migration of cells results in a change of form

morphological (mor fuh LAHJ ih kul) **adaptation:** adaptation that involves structures of organisms

mosaic evolution: hypothesis that basic internal functions may remain constant or change very slowly, while structures and processes related to environmental crises may evolve quickly

motivation (moht uh VAY shun): drive necessary for the learning process of most animals

motor neuron: neuron that transmits outgoing impulses from the brain or spinal cord to the effectors

mRNA processing: the capping, tailing, and removal of introns from mRNA that make mRNA functional for protein synthesis

multicellular: composed of many cells

multiple alleles: three or more alternate forms of a gene, such as A, B, and O genes for blood types

multiple genes: set of three or more different alleles controlling a trait

murein (MYOOR ee un): material that composes the cell wall of most prokaryotes

muscle: animal tissue specialized for contraction

muscle tone: condition in which muscles are always slightly contracted

mutagen (MYEWT uh jun): agent that causes mutations by altering the structure of DNA molecules

mutation (MYEWT ay shun): change in the genetic code of an organism

mutualism (MYEW chuh lihz um): relationship in which two organisms live in a mutually beneficial and usually necessary association

mycelium (mi SEE lee um): mass of hyphae

myelin (MI uh lun) **sheath:** fatty outer layer that encloses the axon of many neurons

myosin (MI uh sun): protein that makes up part of the filaments in striated muscle

N

NAD⁺: nicotinamide adenine dinucleotide; coenzyme that transports hydrogens in respiration

NADP⁺: nicotinamide adenine dinucleotide phosphate; coenzyme that transports hydrogens during light reactions of photosynthesis

natural selection: principle that explains evolution; idea that better adapted organisms tend to live longer and leave more offspring than organisms that are not well adapted

navel (NAY vul): point at which the umbilical cord was attached to the body of the fetus during its development in the uterus

Neanderthal (nee AN dur thawl): humans that lived between 35 000 to 125 000 years ago; classified as *Homo sapiens*

nekton (NEK tun): ocean animals that can move freely through the water under their own power

nematocyst (nih MAT uh sihst): stinging cell of coelenterates such as hydra; used in trapping food

nematodes: roundworms; phylum Nematoda

neoteny (nee AWT un ee): retaining immature traits in the adult form

nephridia (nih FRIHD ee uh): pairs of excretory units in almost every segment of an earthworm

nephridiopore (nih FRIHD ee up por): small opening of each bladder to the outside of an earthworm; releases wastes

nephron (NEF rahn): microscopic excretory unit in the human kidney

nephrostome (NEF ruh stohm): funnel-shaped part of nephridia

neritic (nuh RIHT ihk) **zone**: area of ocean along the continental shelf beyond the littoral zone

nerve: bundle of neuron fibers

nerve impulse: series of action potentials sweeping down an axon

nerve net: diffuse nerve system within the mesoglea of coelenterates

nervous system: system that controls rapid responses to stimuli

neural (NOOR ul) **folds**: raised cells of the neural plate in the developing embryo

neural plate: rapidly dividing ectoderm cells that form a flat section along the dorsal side of an embryo

neural tube: hollow groove formed by the fusion of the neural folds; becomes the brain and spinal cord of the organism

neurofibril (noor oh FIBE rul): structure that transmits impulses in some protists

neuron (NOO rahn): specialized cell of the nervous system that conducts impulses; nerve cell

neuropeptide: small protein that functions in the nervous system

neutral solution: solution neither acidic nor basic

neutron: neutral particle in the nucleus of an atom

nicotine: stimulant found in tobacco

nitrification (ni truh fuh KAY shun): process in which bacteria oxidize ammonium ions into nitrite ions and then to nitrate ions

nitrogen base: a compound that makes up part of a nucleic acid

nitrogen cycle: the movement of nitrogen through an ecosystem

nitrogen fixation: process in certain prokaryotes, either free-living or associated with vascular plants, in which nitrogen gas is fixed into organic compounds

nitrogen-fixing bacteria: bacteria that convert atmospheric nitrogen to ammonia, which is then 'nitrified' by other bacteria

nitrogenous (ni TRAHJ uh nus) **waste**: animal waste product, contains nitrogen

node: place on a stem where leaves develop; point where the myelin sheath does not cover the axon

nodule (NAHJ ewl): swollen area on roots of legumes; contains nitrogen-fixing bacteria

nonbiodegradable (nahn bi oh dih GRAYD uh bul) **substance**: material that cannot be converted by organisms to inactive compounds

noncyclic pathway: in light reactions of photosynthesis in algae and plants, the path of excited electrons in which water is used and oxygen given off

nondisjunction (nahn dihs JUNK shun): failure of homologs to segregate during meiosis

nonhistones: chromosomal proteins that may serve to activate genes in eukaryotes

nonrenewable resource: resource that, once used, cannot be replaced

nonvascular plants: plants that do not have vascular (conductive) tissues for transport of food, water, and minerals

noradrenaline (nor uh DREN ul un): nerve transmitter that excites heart muscle and inhibits alimentary canal muscles in humans

notochord (NOHT uh kord): stiff rod of cartilage along the dorsal side of chordates at some stage of their life cycle; replaced by the vertebral column in most adult chordates

nuclear fission: process in which radioactive isotopes of uranium are split

nuclear fusion: reaction in which small atoms collide to form larger atoms, releasing energy

nuclear membrane: membrane that surrounds the nucleus

nucleic (noo KLAY ihk) **acids:** large, complex molecules that control heredity (DNA and RNA)

nucleoid: region in which nuclear material of a prokaryote is located

nucleolus (new KLEE uh lus): a small body within the nucleus of most eukaryotic cells; site of synthesis of ribosomal RNA

nucleoplasm (NEW klee uh plaz um): protoplasm inside the nucleus

nucleoprotein (new klee oh PROH teen): combination of nucleic acid and protein; makes up chromosomes of eukaryotes

nucleotide (NEW klee uh tide): subunit of nucleic acids, composed of a sugar, a phosphate group, and a nitrogen base

nucleus: central, round body that is the control center of a cell; the central part of an atom

nutrient (NEW tree unt): in general, material required by cells as energy sources or for building protoplasm

nutrition: the study of foods and how they are used by the body

nymph (NIHMF): second stage in the process of incomplete metamorphosis

O

oak-hickory forest: climax community in the southeastern United States

observation: a noticed and recorded event

ocean: aquatic community covering over two thirds of the Earth's surface

octet rule: when atoms having more than one energy level combine their outer energy levels, they must have a total of eight electrons

omnivore (AHM nih vor): consumer that feeds on both producers and other consumers; consumer that eats both plants and animals

oncogene (AHN koh jeen): gene that has mutated into a cancer-causing gene

ootid: cell produced in meiosis II that develops into a mature ovum

open circulatory system: system in which blood is not always in the blood vessels

operator: region of DNA with which a repressor combines inhibiting enzyme formation

operculum (oh PUR kyuh lum): hard covering of the gill chamber of bony fish

operón: consists of the operator and three structural genes

oral groove: opening in *Paramecium* through which food is ingested

order: subdivision of a class in the complete classification of an organism

organ: group of specialized tissues performing a specific function

organelle (or guh NEL): specialized cell part

organelle DNA: DNA, found in the organelles, which synthesizes mRNA, tRNA, rRNA, and several proteins

organic chemistry: the study of chemistry of carbon compounds

organic compounds: molecules that contain the element carbon

organism: individual that is capable of carrying on life processes

organization: the orderly functioning of a living system; the physical structure of cells, organisms, or higher levels of life

organizer: tissue that induces other tissues of an embryo to differentiate

osculum (AHS kyuh lum): large opening in the body of sponges; excurrent pore

osmosis (ahs MOH sus): diffusion of water across a semipermeable membrane

osmotic (ahs MAHT ihk) **balance:** a kind of dynamic equilibrium; state when the number of water molecules entering a cell equals the number leaving

ossification (ahs uh fuh KAY shun): formation of bone; due to activity of osteocytes and the deposition of minerals

osteocyte (AHS tee uh site): living bone cell

ostia (AHS tee uh): tiny openings through which blood passes into the heart of a grasshopper

outer bark: outermost part of a woody stem made up of cork; contains old, nonfunctioning secondary phloem

oval window: membrane in the middle ear

ovary (OHV ree): swollen lower region of the pistil in flowering plants; female gonad in animals; where eggs are produced

oviduct (OH vuh dukt): tube close to each ovary in the mammalian female; conveys egg from ovary to uterus; Fallopian tube

ovipositor: posterior abdominal segments of a female grasshopper used to deposit fertilized eggs in the soil

ovisac: structure in female frogs that stores eggs

ovulation (ahv yuh LAY shun): short stage in the menstrual cycle in which the follicle bursts and the mature egg is released

ovule (OHV yewl): female sporangium within the ovary of a flowering plant

ovum: egg; female reproductive cell

oxygen debt: process of a cell obtaining ATP energy by anaerobic respiration; ATP is produced when glycogen stored in muscles is converted to lactic acid

oxyhemoglobin (ahk sih HEE muh gloh bun): compound containing oxygen and hemoglobin; form in which oxygen is transported by red blood cells

oxytocin (ahk sih TOHS un): hormone secreted by the posterior pituitary; stimulates contraction of uterine muscles during labor

ozone layer: layer in the Earth's atmosphere composed of O_3 molecules that blocks most incoming UV radiation

P

pacemaker: small bundle of tissue in the right atrium where the impulse for the heartbeat originates; sinoatrial node

palisade cells: long cells arranged vertically under the upper epidermis of a leaf; site of photosynthesis

palmate (PAL mayt): venation pattern in which large veins originate and spread out from the point where the petiole is attached to the blade

palmately compound: leaf pattern in which leaflets are all attached to the petiole at a central point

PAN: compound of nitrogen oxides and hydrocarbons that pollutes the air, damaging lungs and irritating eyes

pancreas (PAN kree us): organ that secretes enzymes for digestion in the small intestine and the hormones insulin and glucagon

pancreatic amylase: enzyme that changes starch to maltose

pancreatic duct: duct that transports digestive enzymes from the pancreas to the duodenum

parapodia (PER uh poh dee uh): paired, bristled, paddle-like projections that extend from the segments of some annelids and function in locomotion and respiration

parasite: organism that lives in or on a host and gets nourishment from the host

parasitism (PER uh suh tihz um): relationship in which one organism (parasite) is completely dependent on a host organism; the host is usually harmed

parasympathetic (per uh sihm puh THET ihk) **system:** division of the autonomic nervous system; counteracts the results of the sympathetic system's actions returning the body to normal after an emergency

parental cross: mating of two organisms to produce offspring; usually refers to first mating in a series

parthenogenesis: process in which male bees are hatched from unfertilized eggs

passive immunity: immunity to a disease brought about by the injection of an antiserum

passive transport: exchange of materials across the cell membrane in which the cell does not expend energy

pathogen (PATH uh jun): disease-causing organism

pathogenic: causing disease

pedipalp: a pair of appendages on spiders that hold and tear apart food

Peking man: variety of *Homo erectus* similar to Java man

pellicle (PEL ih kul): stiff covering on ciliates and euglenoids

pelvis: central, hollow cavity of the kidney; a set of bones enclosing a hollow space

penicillin: chemical produced by the mold *Penicillium;* kills certain disease-causing bacteria

penis: male reproductive organ in animals that have internal fertilization

pepsin: enzyme that hydrolyzes proteins into polypeptides

pepsinogen (pep SIHN uh jun): inactive form of the enzyme pepsin

peptidase: enzyme that converts small polypeptides to amino acids

peptide: name given to an amino acid when bonded to other amino acids in a chain

peptide bond: bond that holds amino acids together

perfect flower: flower that contains both male and female organs

pericycle: layer of root cells that gives rise to branch roots

period: subdivision of an era in the geologic timetable

periosteum (per ee AHS tee um): outer layer of bones, encloses long bones such as arm and leg bones

peripheral (puh RIHF rul) **nervous system:** system that sends information from the receptors to the brain and spinal cord and transmits impulses from them to the effectors

peristalsis (per uh STAHL sus): series of alternating muscular contractions and relaxations, moves food along the digestive tract

permafrost (PUR muh frawst): layer of frozen ground in the tundra that never thaws

pesticide: chemical such as DDT that kills pests

petiole (PET ee ohl): slender stalk of a dicot leaf, attaches leaf to the stem

pH scale: scale of numbers representing the concentration of hydrogen ions and hydroxide ions in a solution

phage: a virus that attacks a bacterium

phagocytosis (fag uh si TOH sus): engulfing of a solid particle by a cell

pharynx (FER ingks): tube through which food passes after it leaves the mouth in planarians and earthworms; in vertebrates, passageway for gases and food; throat

phenylketonuria (PKU): genetic disease in which a missing enzyme leads to severe mental retardation

phenotype (FEE nuh tipe): trait that a genotype determines

pheromone (FER uh mohn): substance secreted by an organism into the environment that affects the physiology or behavior of other members of the same species

phloem (FLOH em) **cells:** cells that function in the transfer of food and certain minerals within a vascular plant

photoperiodism: response of flowering plants to light and dark conditions

photosynthesis: process by which light energy is absorbed and then converted to the chemical energy of glucose

photosystem I: one of the pathways of the light reactions of photosynthesis

photosystem II: one of the pathways of the light reactions of photosynthesis

phototropism (foh toh TROH pihz um): plant growth response to light

phycocyanin: blue pigment found in cyanobacteria

phycoerythrin (fi koh ER ih thrun): red pigment in red algae

phylum: major classification division of a kingdom

physical change: any change in the physical properties of a substance

physical property: characteristic that describes a piece of matter

physiological adaptations: adaptations involved with the physical and chemical needs of organisms

physiology (fihz ee AHL uh jee): study of the functions of parts of organisms

phytochrome (FITE uh krohm): pigment in leaves thought to be involved in the flowering of plants

pinnate (PIHN ayt): venation pattern in which veins of a leaf branch from a central vein

pinnately compound: leaf pattern in which leaflets branch out along different parts of the petiole

pinocytosis (pi noh si TOH sis): a type of endocytosis in which liquids and small particles are taken into a cell

pioneering stage: first stage in ecological succession; usually involves hardy autotrophs

pistil (PIHS tul): the long, vase-shaped female organ of a flower

pith: central part of stems and monocot roots in angiosperms

pituitary (puh TEW uh ter ee) **gland:** endocrine gland at the base of the brain; secretes hormones, some of which control other endocrine glands

pivot joint: joint that moves in one direction as well as rotates

placenta (pluh SENT uh): mass of small blood vessels and associated tissues across which materials are exchanged between embryo and mother

placentals (pluh SENT ulz): group of mammals in which the entire development of the embryo occurs internally with material exchange occurring through the placenta

plankton (PLANG tun): marine organisms that float in the water

plant: autotrophic, complex organism with chlorophyll in chloroplasts; cell walls made of cellulose; mostly multicellular

plasma cells: cluster of cells that are genetically identical to one another; formed when activated B cell divides

plasmid: circular segment of DNA in bacteria that is independent of the main chromosome

plasmodium (plaz MOHD ee um): large mass of "naked" protoplasm composing the vegetative stage of slime molds; contains many nuclei but no cell walls

plasmolysis (plaz MAHL uh sus): separation in walled cells of the protoplast from the cell wall due to loss of water from the protoplast by osmosis

plastids (PLAS tudz): organelles in plant cells; chloroplasts are the most common kind

platelet (PLAYT lut): human blood cell fragments lacking nuclei; involved in blood clotting

point of insertion: site where a muscle attaches to a bone that moves during flexing

point of origin: site where a muscle attaches to a bone that does not move during flexing

polar nuclei: two haploid nuclei within the ovule of a flowering plant that fuse with one of the sperm nuclei to form the endosperm nucleus

pollen grain: structure that contains the male sex cells in seed plants; the male gametophyte of seed plants

pollen sac: male sporangium contained in flower anthers

pollen tube: structure formed by the tube nucleus that "digs" through the style to the ovary of a flower, delivering sperm nuclei to the egg in the ovule

pollination: process by which pollen is transferred to the ovule of gymnosperms or the stigma of flowering plants

pollutant: any substance that makes water, soil, or air unclean

pollution: introduction of materials into the environment that decreases the purity, or cleanliness, of the environment

polyp (PAHL up): one of the body forms of coelenterates; sessile, cylindrical body with mouth and tentacles facing upward

polypeptide: molecule formed by the bonding together of many amino acids

polysaccharide: compound composed of many monosaccharides bonded together

population: group of organisms that naturally interbreed

population density: number of individuals per given area

population genetics: study of gene pools and their evolution

population growth: change in size of population with time

population growth rate curve: a graph of the rate of increase in the size of a population with time

population size: number of individuals in a population

pore: a hollow permease in the cell membrane through which certain particles may pass; an opening in the nuclear membrane

posterior (pah STIHR ee ur): rear part of an organism

posterior pituitary: lobe of the pituitary gland; secretes oxytocin and vasopressin; does not directly influence other endocrine glands

postsynaptic (pohst suh NAP tihk) **cell:** cell that receives an impulse after it crosses the synapse

post-transcriptional level: point after transcription of DNA at which gene expression can be controlled

potential energy: energy due to position; stored energy

predation (prih DAY shun): feeding of one animal on another

predator (PRED ut ur): animal that preys on another animal

prehensile (pree HEN sul): able to grasp, like the human hand

pressure-flow hypothesis: explanation of food transport in plants in which pressure differences in sieve cells between leaves and roots account for translocation

presynaptic (pree suh NAP tihk) **cell:** cell that carries an impulse toward a synapse

primary oocyte: cell that divides in meiosis I to produce the secondary oocyte and first polar body

primary spermatocyte: cell that divides in meiosis I to produce two secondary spermatocytes

primary succession: ecological succession that may begin with bare rock; the first succession in an ecosystem

primary transcript: a molecule of mRNA from which introns have not been removed

primate: order of mammals that includes lemurs, monkeys, great apes, and humans

proboscis (pruh BAHS us): modified mouthparts of certain insects, used for obtaining food; long, flexible snout

producer: organism that can make its own food; autotroph

product: substance that results from a chemical reaction

product rule: rule that states that the probability of two or more events occurring together equals the product of the individual probabilities of each event occurring alone

progesterone (proh JES tuh rohn): hormone secreted by the corpus luteum; maintains the uterus in its prepared condition for pregnancy

prokaryote (proh KER ee oht): organism that does not have distinct nuclei; bacteria

promoter: a region of DNA to which the enzyme needed for mRNA synthesis binds

prophase (PROH fayz): first stage in mitosis and meiosis I during which the nucleolus and the nuclear membrane disappear and the chromosomes become clearly visible as separate bodies; in meiosis I, pairing of homologous chromosomes also occurs

prostaglandin: special group of hormones made from fatty acids

protease (PROHT ee ays): enzyme that breaks down proteins

protein: large, complex molecule; contains carbon, hydrogen, oxygen, and nitrogen; some have structural uses, while others play specific functional roles

prothallium (pro THAL ee um): small, heartshaped gametophyte of the fern

prothrombin: chemical needed for the clotting of blood

protist: simple, eukaryotic organism; includes simple algae, protozoa, and slime molds; kingdom Protista

proton: positively charged particle in the nucleus of an atom

protonema (proht uh NEE muh): filamentous, algal-like stage of the young moss gametophyte

protoplasm (PROHT uh plaz um): the jellylike cellular substance in which much metabolism occurs

protoplasmic streaming: flow of protoplasm in a cell that distributes cellular material and, in some cases, aids locomotion

protozoan: unicellular, animallike protist

provirus: a virus that has become part of its host's DNA

pseudopodium (sewd uh POHD ee um): structure of locomotion and food getting; false foot in amoeba

puberty (PYEW burt ee): onset of the development of secondary sexual characteristics

pulmonary (PUL muh ner ee) **artery:** artery in birds and mammals that takes deoxygenated blood from the right atrium to the lungs

pulmonary veins: veins in amphibians, reptiles, birds, and mammals that take oxygenated blood from the lungs to the left atrium

punctuated equilibrium: idea that those organisms that survive a crisis rapidly evolve into a new species

Punnett (PUN ut) **square:** chart used to determine the expected genotypes of the offspring of a cross

punishment: negative reinforcement

pupa (PYEW puh): stage of an insect life cycle in which the tissues of the organism are completely reorganized during complete metamorphosis

pyramid of biomass: relationship showing the decrease of per unit area biomass with each higher trophic level

pyramid of energy: relationship showing the loss of energy as it is transferred along a food chain

pyramid of numbers: relationship showing the decrease in number of organisms with each higher trophic level

pyrenoid (pi REE noyd): area of the chloroplast where starch is formed and stored

Q

queen bee: central figure in a beehive; mates only once and produces thousands of offspring

R

radial symmetry (SIH muh tree): body plan in which a cut lengthwise through the middle in any direction produces two identical halves

radicle (RAD ih kul): embryonic seed structure, becomes the primary root

radiant energy: heat and light energy

radioactive elements: isotopes with unstable nuclei that disintegrate, releasing nuclear energy

radioactive isotope: an isotope with unstable nuclei

radiocarbon dating: method of dating fossils that uses the half-life of radioactive materials

radula (RAJ uh luh): structure with toothlike projections used by univalves and certain other mollusks to obtain food

ray: cells that transport materials horizontally between pith, wood, and bark

reactant: substance that enters into a chemical reaction

receptor (rih SEP tur): nerve structure that detects stimuli; cell membrane protein that "recognizes" and combines with certain other molecules (antigens, hormones)

recessive (rih SES ihv) **trait:** form of a trait that is dominated by another form

recombinant DNA: DNA that results from the combination of DNA from two organisms; recombined DNA

recombination gametes: gametes that contain a set of linked genes different than that found in the parent

rectum (REK tum): end of large intestine; storage area for solid wastes in mammals

red blood cell: blood cell that transports oxygen and some carbon dioxide; erythrocyte

red-green colorblindness: sex-linked genetic disorder in which a person has trouble identifying red or green

red tide: large population of dinoflagellates that present a danger to seafood and fish

reflex: innate behavior pattern resulting from the fixed pathways of the nervous system; simple response that involves no conscious control

reflex arc: pathway of nerves followed by an impulse in a reflex

refractory (rih FRAK tree) **period:** interval of time in which the membrane potential of a neuron is being restored

regeneration (rih jen uh RAY shun): regrowing of missing parts

regulator gene: gene that controls the production of repressor proteins

releaser: stimulus that causes the onset of an instinctive behavior

releasing factor: hormone that travels from the hypothalamus to the anterior pituitary where it causes release of another hormone

renal (REEN ul) **arteries:** blood vessels that bring blood containing wastes to the kidneys

renal veins: blood vessels that return purified blood from the kidneys to the general circulation

renewable resource: resource that can be replaced

replication: duplication, as in DNA

repressor: special protein that binds to the operator region and prevents transcription of related structural genes

reproduction: process in which new organisms are produced

reproductive isolation: the prevention of interbreeding

reptiles: vertebrates characterized by dry, scaly skin, internal fertilization and development within amniotic shelled eggs; class Reptilia

resource: anything humans take from the environment

resolving power: ability to distinguish two objects as being separate

respiration: exchange of oxygen and carbon dioxide between an organism and the environment

respiratory system: system made up of special regions for gas exchange and methods of transporting gases

resting potential: membrane potential

retrovirus: kind of virus in which RNA serves as a pattern for producing DNA; some known to cause cancer in animals; one form causes AIDS

reverse transcription: process in which DNA is made from RNA

reward: positive reinforcement for a behavior

Rh factor: antigens on red cells; first discovered in Rhesus monkeys

Rh negative blood: blood in which red cells do not have the Rh factor

Rh positive blood: blood in which red cells have the Rh factor

rhizoid (RI zoyd): rootlike structure that anchors certain fungi to the food source and secretes enzymes into it; rootlike structure in bryophytes

rhizome (RI zohm): underground fern stem from which roots and leaves develop

ribonucleic (ri boh noo KLAY ihk) **acid (RNA):** large, complex molecule that works with DNA in carrying out the instructions of the genetic code

ribose: sugar contained in RNA

ribosomal RNA: type of RNA that makes up ribosomes

ribosome (RI buh sohm): organelle in cytoplasm or on surface of endoplasmic reticulum that is the site of protein synthesis

rickets: disease caused by lack of vitamin D that results in the improper development of bone

root cap: protective cells covering the tip of a growing root

root hair: outgrowth of a root epidermis cell

root pressure: pushing force caused by water in the xylem

round dance: bee behavior that means there is food within 100 m

rust: type of club fungus that is a serious pathogen on many economically important crops; has a complex life cycle, often involving alternate hosts

S

saliva (suh LI vuh): liquid secretion of the salivary glands; moistens food and contains an enzyme that breaks starch into maltose

salivary (SAL uh ver ee) **glands:** three pairs of glands that secrete saliva

saprophyte (SAP ruh fite): organism that obtains its food from dead organisms or from waste products of living ones

sarcomere: a contractile unit of a skeletal muscle fiber composed of two Z-lines and the region between

scanning EM: electron microscope that passes a beam of electrons through a live sample, forming an image that can be viewed on a screen

scavenger (SKAV un jur): animal that feeds on dead animals

Schwann cells: cells that wrap around an axon and aid in the nutrition and regeneration of axons

science: a body of organized knowledge about nature; method of solving problems

scrotum (SKROHT um): external sac that encloses the testes of a mammal

second filial generation: generation of offspring produced from interbreeding offspring of the first filial generation

second messenger hypothesis: idea that some peptide hormones cause target cells to produce a second compound that acts as a messenger and causes chemical changes in the target cells

second-order consumer: consumer that eats a first-order consumer

second polar body: cell produced from the secondary oocyte during meiosis II that dies

secondary oocyte: cell that undergoes meiosis II to produce the ootid and a second polar body

secondary spermatocytes: two cells produced by meiosis I that undergo meiosis II to produce spermatids

secondary succession: series of ecological changes that occur when species of a climax community are removed

secretion: process by which an organism or some part of an organism releases a material; the material released

seed: specialized structure containing a young plant embryo, usually with a food storage source, and surrounded by integument

seed coat: hardened outer covering of a seed

seed dispersal: the scattering of seeds

segmentation: division of a body plan into more or less similar, repeating parts

segmented: divided into units

selective breeding: mating of animals or plants to produce offspring with desired features

selective cutting: cutting of mature trees while leaving the young trees to mature

self-pollination: pollination that occurs in a single flower or between flowers of the same plant

semen (SEE mun): combination of sperm and fluid

semilunar (sem ih LEW nur) **valve:** valve at the base of the pulmonary artery or aorta; prevents blood from flowing back into the ventricles after they contract

semipermeable (sem ih PER mee uh bul) **membrane:** membrane that allows only certain materials to pass through it

sensory neuron: cell that transmits incoming impulses from receptors to a coordinating center such as the brain or spinal cord

sensory-somatic system: division of the peripheral nervous system; controls the exchange of information between receptors, the central nervous system, and the skeletal muscles

septum (SEP tum): in some vertebrates, wall dividing ventricle that keeps oxygenated and deoxygenated blood separated

sessile (SES ul): nonmotile

setae (SEE tee): bristles on each segment of some segmented worms; aid in locomotion

sex chromosomes: chromosomes (X and Y) that determine sex

sex-linked characteristic: trait whose genes are carried on the X chromosome

sexual reproduction: union of two sets of DNA; fusion process

sexually-transmitted disease: disease transmitted during sexual contact

short-day plant: plant whose flowering depends on short periods of light exposure

shrub: low, woody plant

sickle-cell anemia: hereditary disease in which hemoglobin is abnormal and red blood cells are shaped like sickles

sieve tube: phloem cell in which the end walls are perforated, resembling a sieve

simple formula: chemical formula showing the kinds and numbers of atoms per molecule

simple leaf: leaf composed of a single blade

sinoatrial (si noh AY tree ul) **node (S-A node):** small bundle of tissue in the right atrium in which the impulse for the heartbeat originates; pacemaker

sinus (SI nus): space in the body cavity through which blood passes in animals with an open circulatory system; a cavity or depression

sinus venosus (SI nus • vih NOH sus): saclike structure in fishes and frogs that collects blood as it returns to the heart

skeletal muscles: striated muscles

skeletal system: specialized support system in most complex animals

sliding filament hypothesis: explanation of muscle contraction

slime capsule: outer protective covering that encloses some bacterial cells

slime mold: funguslike protist; phylum Gymnomycota

small intestine: digestive organ in humans in which most chemical digestion occurs

smooth ER: endoplasmic reticulum without ribosomes

smooth muscle: muscle that moves many of the internal parts of the body; nonstriated or involuntary muscle

smut: type of club fungus that is a serious plant pathogen; has a relatively simple life cycle

social hierarchy (HI rar kee): levels of authority in animals; pecking order

society: group of animals of the same species living together in an organized way

sodium-potassium pump: process occurring in a neuron where sodium ions are actively transported out of and potassium ions into a cell

soil: the upper layer of earth in which plants are anchored

soil depletion: reduction of soil fertility as a result of nutrients being removed by crops

sol (SAHL): liquified state that the cytoplasm may take

solute (SAHL yewt): portion of a solution in lesser quantity; that which is dissolved in a solvent

solution: homogeneous material with variable composition

solvent (SAHL vunt): portion of a solution in greater quantity; that which dissolves a solute

sorus (SOR us): in ferns, small cluster of sporangia, usually on lower surface of frond

speciation: evolution of a new species

species: classification division after genus; group of organisms that normally interbreed in nature to produce fertile offspring; second name in scientific name

sperm: male reproductive cell

sperm duct: area in the reproductive system of a male grasshopper where fluids are added to the sperm

sperm nuclei: two haploid nuclei produced by mitosis from the generative nucleus in the pollen tube

sperm receptacle: special sac attached to the vagina of a grasshopper used for sperm storage

spermatids: four haploid cells produced during meiosis II that mature into sperm cells

spicule (SPIHK yewl): minerallike structure that supports a sponge

spinal cord: major nerve cord running along the dorsal side of vertebrates

spinal nerves: thirty-one pairs of nerves that carry impulses between body parts and the spinal cord in humans

spindle: oval-shaped structure composed of microtubules, to which chromosomes become attached during mitosis and meiosis

spinneret: structures in a spider's body that release threads of silk for making webs

spiracle (SPIHR ih kul): small opening through which air enters a terrestrial arthropod; external opening of a trachea

spirillum (spi RIHL um): spiral-shaped bacterium

spleen: saclike abdominal organ that mainly stores blood, destroys dead blood cells, and produces lymphocytes

sponges: simple, sessile, mostly marine animals; phylum Porifera

spongy layer: layer of photosynthetic cells beneath the palisade cells in leaves

spontaneous (spahn TAY nee us) **generation:** mistaken belief that organisms can be produced from nonliving sources

sporangiophore (spuh RAN jee uh for) **hypha:** upright structure upon which sporangia are located; present in common bread mold

sporangium: case in which spores are produced and stored

spore: specialized reproductive cell that gives rise to a new organism; may be asexual or sexual

sporophyte (SPOR uh fite) **generation:** generation of a plant in which the cells are diploid and spores are produced

sporozoans (spor uh ZOH unz): phylum of protozoa that have no way of moving and reproduce by means of spores

spring wood: wood produced in the spring; consists mainly of large xylem vessels

S-shaped curve: population growth curve

stamen: male reproductive organ of a flower

starch: large, complex molecule made up of hundreds of monosaccharides bonded together

stele (STEEL): the vascular cylinder, inside the cortex, of roots and stems

stem: main stalk of vascular plants; supports the plant and transports materials

sterile: free of life

steroids (STIHR oydz): group of organic compounds that includes cholesterol, sex hormones, hormones from the adrenal cortex, and vitamin D

stigma: structure on *Euglena* sensitive to light; sticky part of a female flower; sticky area of pistil, receives pollen

stimulant: one of a major group of drugs that works by increasing the reactions of the nervous system

stimulus: anything that causes activity or change in an organism

stipe: stalklike section of a mushroom

stolon (STOH lun): hypha that grows along the surface of the food supply

stoma (STOH muh): tiny pore in a leaf

stomach: large, hollow organ where protein digestion begins in humans

strata: layers

striated (STRI ayt ud) **muscle:** muscle attached to bones; skeletal muscle or voluntary muscle

stroma: the ground substance of a chloroplast, where actual glucose synthesis occurs

structural formula: formula that shows the arrangement of the atoms of a molecule

structural gene: section of DNA that codes for a polypeptide

style: stalklike portion of a pistil

suberin (SEW buh run): oily material that coats cork cells and protects against water loss

subspecies: division of species; capable of interbreeding with other subspecies of the same species; race

substrate (SUB strayt): a reactant in a reaction controlled by an enzyme

sucrase: enzyme that breaks down sucrose

summer wood: wood produced in the summer; made up of vessels that are smaller in diameter than those in spring wood

suppressor T cell: cells that counteract the effects of helper T cells by releasing chemicals that inhibit B cell and macrophage activity

survival of the fittest: natural selection

swarm cells: flagellated cells of slime molds that can fuse to form a zygote

swimmeret: abdominal appendage on some crustaceans that functions in both locomotion and reproduction

symbiosis: a relationship in which two organisms live in close association; may be mutualistic, parasitic, or commensalistic

symbiotic theory: theory that explains the origin of eukaryotes through internal symbiosis of prokaryotes

sympathetic (sihm puh THET ihk) **system:** division of the autonomic nervous system; initiates responses that prepare the body for emergencies

synapse (SIHN aps): gap between two neurons or between a neuron and effector

synapsis (suh NAP sus): process in which homologous chromosomes pair during prophase of meiosis I

synfuels (SIHN fyewlz): fuels made using artificial processes

synthesize (SIHN thuh size): build up

system: group of organs working together to perform a specific function

systole (SIHS tuh lee): contraction of the heart chambers

T

taiga (TI guh): biome characterized by coniferous forests

taproot system: root system in which plants have one large, primary root

target organ: organ affected by a particular hormone

taxa: categories used to classify organisms

taxis: movement of an organism toward or away from a stimulus

taxonomy: science of classification

Tay-Sachs: genetic disease in which the nervous system fails to develop properly

T cell: lymphocyte that passes through the thymus gland; involved in immune response to foreign or infected cells

technology: applying scientific knowledge to real problems

telophase (TEL uh fayz): last phase of mitosis in which the events are opposite those of prophase; stage in meiosis I and meiosis II

temperate deciduous forest: temperate forest in which leaves are periodically all shed from the trees

temperate forest: biome characterized by an even distribution of rain totaling about one hundred centimeters per year

tendon (TEN dun): tough, elastic connective tissue that attaches muscle to bone

tentacles (TENT ih kulz): structures that surround the mouth of some animals, used in food getting

terminal bud: bud at the tip of a stem or one of its branches

terracing: conservation practice in which banks of land are created on slopes to prevent erosion from water runoff

territoriality (ter uh tor ee AL ut ee): adaptation in many animals that leads to occupation and defense of a specific territory

testis (TES tus): male gonad

testosterone (teh STAHS tuh rohn): androgen that stimulates the formation of secondary sexual characteristics in males

tetanus: bacterial disease resulting in continuous muscle contraction

tetrad (TEH trad): a pair of the homologous, double-stranded chromosomes present during later prophase I of meiosis

thallus: the entire body of an alga that lacks roots, stems, and leaves

theory: fundamental hypothesis that is supported by extensive experimentation

third-order consumer: consumer that eats a second-order consumer

thoracic duct: major tube in the chest region for collecting lymph

thorax: chest region

threshold: degree of intensity of a stimulus necessary to cause a nerve impulse

thromboplastin: chemical needed for the clotting of blood

thymine (THI meen): one of the bases in DNA

thymus gland: gland beneath the breastbone; thought to control the production of certain antibodies in juveniles

thyroid (THI royd) **gland:** small endocrine gland on the trachea; secretes thyroxine

thyroid-stimulating hormone (TSH): hormone secreted by the anterior pituitary that stimulates thyroxine secretion

thyroxine (thi RAHK sun): hormone that controls the metabolic rate of the body cells; produced by the thyroid gland

tissue: a group of cells similar in structure that are organized to perform a certain function

tissue fluid: fluid that passes out of the blood vessels and exchanges materials with tissue cells

T-maze: maze, shaped like a T, in which an animal has a choice of one of two turns

topsoil: dark, upper layer of soil; most fertile layer

toxin (TAHK sun): poisonous chemical

trachea (TRAY kee uh): windpipe; tube leading from mouth to bronchi; tube in insects and spiders that opens to the outside, functions in gas exchange

tracheal (TRAY kee ul) **system:** system made up of spiracles, tracheae, and tracheoles; respiratory system in terrestrial arthropods

tracheid (TRAY kee ud): type of xylem cell

tracheole (TRAY kee ohl): tiny air sac that contains water; part of the tracheal system in terrestrial arthropods; site of gas exchange with cells

tracheophytes: plants characterized by vascular tissue; phylum Tracheophyta

tranquilizer: drug that depresses the central nervous system

transcription (trans KRIHP shun): transferring the genetic code from DNA to RNA in protein synthesis

transduction (trans DUK shun): transfer of DNA from one organism to another by a virus

transfer RNA: RNA that brings amino acids to messenger RNA in protein synthesis

transformation: accidental means of genetic recombination in bacteria that results as one cell breaks open and its DNA enters another cell

transforming principle: chemical that causes transformation; DNA

translation: synthesis of a protein from the code carried by a mRNA molecule

translocation (trans loh KAY shun): transport of food and certain ions in a vascular plant; occurs within phloem cells

transmission EM: electron microscope that passes a beam of electrons through a very thinly sliced sample

transmitter: hormonelike chemical stored and released from vesicles at the ends of presynaptic cells; secreted across synapses

transpiration (trans puh RAY shun): evaporation of water from a plant

transpiration-cohesion theory: theory that states that water is pulled up a stem because water loss through the leaf pulls on the water column in the xylem

transpiration pull: tension created by cohesion of water molecules

transport system: system that carries needed materials to and wastes away from body cells; links cells to external environment

transverse colon: section of the large intestine

trial and error learning: simple learning in which an organism repeatedly tries a task and learns by its mistakes

trichocyst: structure in *Paramecium* used for protection

tricuspid (tri KUS pud) **valve:** valve between the right atrium and right ventricle; prevents blood from returning to the right atrium when the right ventricle contracts

triploid (TRIHP loyd): 3*n* structure

trophic level: level of energy transfer in an ecosystem

trophoblast: outer layer of cells that develops as the cells of the blastocyst divide

tropical rain forest: biome of equatorial regions, characterized by heavy rainfall, constant warm temperature, dense growth of many different species of trees

tropism (TROH pihz um): plant growth response caused by unequal stimulation on opposite sides of the plant

trypsins: protein-digesting enzymes secreted by the pancreas

tube feet: bulblike structures on the underside of a starfish; used in locomotion and food getting

tube nucleus: one of the two haploid nuclei formed in the anther of a flower

tuber: modified, swollen underground stem

tubule: tube that leads to the bladder in each nephridium of an earthworm; long, coiled tube leading from each Bowman's capsule in a kidney

tundra (TUN druh): biome characterized by low average temperature, permafrost, low average rainfall, lack of large plants

turgidity (tur JIHD ut ee): stiffness caused by the pressure of water in cells

Turner syndrome: set of abnormal characteristics in a female who has one X and no other chromosomes; results from nondisjunction

tympanic membrane: eardrum

U

ultrasonography (ul truh suh NAHG ruh fee): technique used to determine the position and anatomy of a fetus

umbilical (um BIHL ih kul) **cord:** structure that contains blood vessels; transports blood between the embryo and the placenta

unicellular: made up of only one cell

univalve (YEW nih valv): one-shelled mollusk

universal donor: person having type O blood

universal recipient: person having type AB blood

upper epidermis: upper layer of a leaf

uracil (YOOR uh sihl): a base in RNA but not in DNA

urea (yoo REE uh): nitrogenous waste in amphibians and mammals; component of urine

ureter (YOOR ut ur): tube that transports urine from the kidney to the urinary bladder

urethra: canal from which urine is expelled from the body and in males also transports sperm

uric acid: nitrogenous waste produced by insects, snails, reptiles, and birds

urinary bladder: hollow organ that stores urine

urine: combination of urea, excess salts, and water

uterus (YEWT uh rus): thick-walled, muscular organ in female mammals; organ in which the embryo develops; womb

V

vaccine: solution of weakened or killed microorganisms administered to prevent disease by producing active immunity

vacuole (VAK yuh wohl): a cellular organelle used for storage

vagina (vuh JI nuh): the organ in the female that receives sperm from the male

vagus (VAY gus) **nerves:** pair of nerves that originate in the medulla oblongata that decrease the heartbeat rate

valve: mollusk shell

variable factor: factor being tested in an experiment

vascular bundle: group of xylem and phloem tissue

vascular cambium: meristem that produces cells that later develop into xylem and phloem tissue

vascular plants: plants that have specialized tissue for the transport of food, water, and minerals

vascular tissue: tissue that transports food, water, and minerals throughout vascular plants

vasopressin (vay zoh PRES un): hormone secreted by the posterior pituitary gland that controls water balance in the body

vegetal (VEJ ut ul) **pole:** ventral part of a frog egg; contains the yolk

vegetative reproduction: reproduction from a nonsexual, or vegetative, part of a plant

vein: conducting vessel of leaves (xylem and phloem); blood vessel that carries blood toward the heart from the body

vena cava: the main vein that carries blood to the right atrium

venation (ve NAY shun): vein pattern in a leaf

ventral (VEN trul): underside of a bilaterally symmetrical animal

ventral aorta: aorta that carries blood from the heart to the gills of a fish

ventral blood vessel: major blood vessel on the underside of an earthworm

ventricle (VEN trih kul): chamber that pumps blood away from the heart; cavity or space within the brain

vertebrae (VUR tuh bray): bony structures that connect to form the spinal column in vertebrates

vertebrates (VURT uh brayts): animals with backbones

vertical stratification: layers of a forest from top to bottom

vesicle: small, saclike organelle

vessel: type of xylem cell

vestigial organ: body structure having no known function

villi (VIHL i): structures on the intestinal lining that extend into the hollow of the intestine; increase the surface area for absorption of digested food

viroid: small segment of RNA known to cause disease in a variety of plants

virus: microscopic particle of nucleic acid and protein that depends on a specific host cell for its reproduction

visible spectrum: band of colors created when white light is separated into different wavelengths

vitamin: an organic substance necessary in small amounts for the proper metabolism

voluntary muscles: muscles that can be controlled; skeletal muscles

W

waggle dance: bee behavior that indicates distance (over 100 m) and direction to food source

warning coloration: the display of various bright colors that announce rather than conceal the presence of an animal

water-vascular system: system that controls movement and food getting in a starfish

wavelength: distance between consecutive crests of two waves

weathering: mechanical process of freezing, thawing, and erosion

white light: radiant energy with wavelengths of about 400 to 700 nm

white matter: axons with myelin sheaths in the interior of the human brain and exterior of the spinal cord

womb: uterus

woody stem: stem that contains woody tissue derived from xylem

worker bee: sterile female bee that performs most of the tasks necessary for maintaining a beehive

X

X chromosome: one of the chromosomes that determines sex; a sex chromosome

xylem (ZI lum): vascular tissue in plants; carries water and minerals from roots to the leaves

Y

Y chromosome: one of the chromosomes that determines sex; a sex chromosome

yolk sac: structure in the shelled amniotic egg that contains yolk, the food source for the developing embryo

Z

zero population growth: condition in which the birthrate equals the death rate

Z line: crossband to which thin (actin) microfilaments are attached

zoospore: a motile spore produced in a sporangium

zygospore: diploid zygote with tough outer wall

zygote (ZI goht): fertilized egg resulting from the union of a sperm and an egg

INDEX

A

Abdomen, 330, 334, 335
Abiotic factor, 703–708, 718
Abscisic acid, 462
Abscission, 463
Absorption of food, 530
Absorption spectrum, 436
Abyssal zone, 731
Acceleration center, 543, 617
Acetylcholine, 609, 616, 621, 639
Acid, 48–50; *table,* 50
Acid rain, 704, 746–747; *illus.,* 746
Acorn worm, 337; *table,* 337
Acquired characteristic, 206–207
Acquired immune deficiency syndrome, *See* AIDS
ACTH, 620; *table,* 595
Actin, 639; *illus.,* 639
Action potential, 607, 609; *illus.,* 607
Activation energy, 91, 93, 95
Active site, 93, 94
Active transport, 71–72; *illus.,* 684
Adaptation, 18, 181, 221–227; animal, 490, 491, 495; behavioral, 224; bird, 347; ciliate, 383; eel, 342; evolution of, 222; euglenoid, 267; fungal, 273; innate, 650; integumentary, 339; morphological, 224, 225; origin of, 221–222; physiological, 224, 225; plant, 290, 291, 292, 293, 295, 296, 297, 298, 456, 650; types of, 224–225; *illus.,* 18; 224; 229; camouflage, 225; melanism, 223; wasp-flower, 298
Adaptive radiation, 231–232; *illus.,* 231
Adenine, 173, 174, 175, 179; *illus.,* 173; 174
Adenosine diphosphate (ADP), 96, 641
Adenosine triphosphate (ATP), 95–96, 98, 439, 440, 606, 640, 641; *illus.,* 98
Adhesion, 454
ADP, *See* Adenosine diphosphate
Adrenal cortex, 589
Adrenal gland, 589; *illus.,* 589
Adrenal medulla, 589, 617
Adrenaline, 589, 596, 609; *table,* 595
Adventitious root, 424
Aerobic prokaryote, 215
Aerobic respiration, 76, 96–99, 262
African sleeping sickness, 271, 395
Agar, 290
Agglutination, 398–399
Agglutinin, 399
Agnatha, 340; *table,* 379

AIDS (acquired immune deficiency syndrome), 392; transmission of, 401
AIDS virus, 401–403; *illus.,* 401
Air pollution, 744–746
Air sac, 347; *illus.,* 347
Alcohol, 99, 266, 443, 610
Aldosterone, 589; *table,* 595
Algae, 7, 376, 379, 453, 702, 748; brown, 289–290; digestion in, 376; golden, 268; green, 287–289, 413–414; life cycle, 413–415; red, 290; *illus.,* green, 287; 413–415; kelp, 289; red, 290; *Spirogyra,* 413; *Ulothrix,* 413; *Ulva,* 413; *Volvox,* 289
Algal bloom, 748; *illus.,* 748
Algal plant, 287–290
Alimentary canal, 518–519
Allele, 133, 135, 137, 138, 141, 150, 155, 164, 188, 222, 551
Allantois, 493, 496
Allergy, 570
All-or-none response, 608, 640
Alternation of generations, 413, 414
Altitudinal succession, 706
Aluminum, 40
Alveoli, 567
Amanita, 275; *illus.,* 275
Amino acid, 55, 92, 176, 213, 397, 585, 700; *table,* 179
Amino group, 55
Ammonia, 212, 380, 573, 574, 576, 699
Ammonification, 699
Amniocentesis, 166; *illus.,* 166
Amnion, 493
Amniotic egg, 493–494
Amniotic fluid, 493; *illus.,* 500
Amniotic sac, 499
Amoeba, 7, 269, 361; digestion, 377; disease caused by, 269, 395; locomotion of, 381; phagocytosis in, 73; reproduction of, 361; response of, 383; structure of, 270; *illus.,* 270; 381; digestion, 377; movement, 382
Amoebic dysentery, 269, 395
Amoebocyte, 516; *illus.,* 516
Amoeboid motion, 269, 381; *illus.,* 381
Amphetamine, 610, 611
Amphibia, 343; *table,* 348
Amphibian, 343–344; circulatory system, 538–539; development of, 491, 502, 505–506, 508; excretory system of, 579; respiratory system of, 566, 567; *illus.,* 343; circulatory system of, 538; frog, 489; 491; 502; 503; 505; respiratory system, 566. *See also* Frog

Amphioxus, 338
Amplexus, 477; *illus.,* 477
Ampulla, 635
Anabaena, 362
Anaerobic respiration, 264, 641
Anal pore, 377; *illus.,* 270; 377
Anaphase, 113; *illus.,* 113
Anatomy, 39, 201
Androgen, 589; *table,* 595
Anemia, 523, 549
Angiosperm, 296, 297–298, 300–301; structure of, 449–453; *illus.,* 297; *table,* 303
Angler fish, 224, 342; *illus.,* 224
Animal, 54, 55; asexual reproduction in, 474–475; dominance, 667; ectotherms, 340, 345; endotherms, 346, 349; food gathering, 515–520; invertebrate, 309–332, 329–337; life cycle of, 476–478; sexual reproduction in, 476–482; vertebrate, 338–350; *illus.,* dominant, 667; *tables,* arthropod characteristics, 336; classifications of, 251; invertebrate characteristics, 323; major characteristics, 350; vertebrate characteristics, 348
Animal behavior, 348, 650–651
Animal communication, 662–666; *illus.,* 662; 663; 664; 665; 666
Animal development, 488–509
Animal pole, 490, 502
Animalia, 33, 252; *table,* 323
Annelid, 319, 330; circulatory system of, 535–536; respiration of, 564; *illus.,* 564; respiratory system, 564
Annelida, 319; *table,* 323
Anopheles mosquito, 395–396, 679, 740
Antennae, 331, 332, 334; *illus.,* 331
Anterior, 314; *illus.,* 314
Anterior lobe, 591
Anther, 419, 420, 422; *illus.,* 419; 420
Antheridium, 416, 417; *illus.,* 417
Anthrax, 389
Anti-A agglutinin, 551
Anti-B agglutinin, 551
Antibiotics, 400
Antibody, 397–399, 405, 551, 556, 557; *illus.,* 397
Anticodon, 182
Antidote, 346
Antigen, 397–399, 403, 551, 552, 556; 552
Antiserum, 400
Anus, 318, 505

Anvil, 619
Aorta, 536, 540; *illus.,* 575
Aortic arch, 536
Apical bud, 462–463
Apical dominance, 462
Appendage, 330
Aquifer, 752
Arachnid, 335; *illus.,* 335; *table,* 335
Archaebacteria, 261, 265–266
Archaeopteryx, 346; *illus.,* 346
Archegonium, 416, 417; *illus.,* 416
Archenteron, 505
Aristotle, 107, 247
Artery, 537, 544, 575, 563; aorta, 540;
  disease of, 543; pulmonary, 539; renal,
  575, 576; *illus.,* 575
Arthropod, 329, 330–331; circulatory
  system of, 536–537; digestive system
  of, 519–520; growth of, 628;
  locomotion of, 636–637; skeletal system
  of, 628; *illus.,* 520; circulatory system,
  536; muscle, 637; skeletal system, 628;
  tables, characteristics, 336; 350
Artificial heart, 557–559
Artificial pacemaker, 542, *illus.,* 542
Artificial selection, 209
Artificial vegetative propagation, 424
Ascaris, 318
Ascending colon, 530
Ascomycota, 275; *table,* 278
Asexual reproduction, 359–363, 474
Asthma, 570
Atherosclerosis, 554, 571
Athlete's foot, 393
Atmosphere, early, 212–213
Atom, 39–40; *illus.,* 40
ATP, *See* Adenosine triphosphate
Atrioventricular node, 542
Auditory canal, 619
Auditory nerve, 620
Aurelia, 313, 474; *illus.,* 313
Auricle, 537
Australopithecine, 235–236, 237
Australopithecus afarensis, 236, 239;
  illus., 236
Autocidal control, 760
Autoclave, 266
Autonomic nervous system, 617, 638
Autosome, 149, 151
Autotrophic eubacteria, 264–265
Autotroph, 214, 433, 436, 693;
  digestion, 376; gas exchange, 379;
  nutrient needs of, 376; transport, 379
Auxin, 459–463
A–V node, 542
Aves, 346; *table,* 348
Axon, 605; *illus.,* 605
AZT, 402

B

Bacillus, 262; *illus.,* 262

Bacteria, adaptation of, 222; asexual
  reproduction in, 360; beneficial, 266;
  chemosynthesis in, 266; defense
  against, 396–400; digestion in, 376;
  environment of, 706; enzyme induction
  in, 189; gas transport in, 379; genetic
  recombination in, 364; in intestine, 530;
  macrophage attacking, 398; and
  nitrogen cycle, 699–700; pathogenic,
  390–391, 570; penicillin and, 25, 26,
  204; photosynthetic, 266; structure of,
  262–263; virus attacking, 367; *illus.,*
  25; 280; 364; *table,* 278
Bacterial pneumonia, 570
Bacterial transformation, 171–173
Bacteriophage, 366–368; *illus.,* 279;
  366
Ball and socket joint, 633
Barbiturate, 609–610
Barnacle, 333; *illus.,* 333
Base, 48–50, 178; *table,* 50
Basidia, 274
Basidiomycota, 274; *table,* 499
Basidiospores, *illus.,* 275
B-cell, 399; *illus.,* 399
Beak adaptations, *illus.,* 229
Behavior, 648–669; animal, 650–651;
  complex, 334, 349; courtship, 653;
  innate, 649–654; learned, 654–669;
  plant, 650; social, 666–669; *illus.,* 650;
  651; 653; 658; 659; 660; 661; 667; 668;
  669. *See also* Adaptation; Response
Benthos, 731
Biceps, 636; *illus.,* 636
Bicuspid valve, 540
Bilateral symmetry, 314; *illus.,* 314
Bile, 526
Biloba, 297
Binomial nomenclature, 247–248
Biological clock, 654
Biology, 5
Biomass, 697; *table,* 697
Biome, 720–732; desert, 728; grassland,
  727; ocean, 730; temperate forest, 723;
  taiga, 722; tropical rain forest, 724–
  726; tundra, 720; *illus.,* 720; 721; 722;
  723; 724; 725; 726; 727; 731
Biosphere, 739
Biotic factor, 693, 703–709, 718;
  environmental, 693–736
Biotic potential, 673–674; *illus.,* 673
Bipedal locomotion, 629
Bipedalism, 234; evolution of, 234
Bird, 346–347, 578, 742; circulatory
  system of, 539–540; development of,
  493–494; migration of, 651; *illus.,* 346;
  347; 502; migration, 651
Bird behavior, 651–652; *illus.,* 650;
  651; 653
Birth, 499–501
Birthrate, 686, 742; decrease in, 682
Bivalve, 321; *illus.,* 321

Blastocoel, 502; *illus.,* 505
Blastocyst, 496–497
Blastula, 502, 505
Blood, 548–557; cholesterol in, 555;
  earthworm, 320, 535; fish, 537; frog,
  538; gas transport by, 548, 564, 567;
  glucose in, 163, 586, 587; grasshopper,
  536, 537; human, 539–553; Rh factor
  in, 552–553; sugar level of, 587;
  tranfusion of, 552; *illus.,* 548; 549;
  tables, 548; 576
Blood clot, 159, 550, 554; *illus.,* 550
Blood contents, *table,* 548
Blood disease, 159–160, 164; *illus.,*
  159; 160; 164
Blood plasma, 551
Blood pressure, 543–544
Blood sugar, 163, 586, 587, 589; *illus.,*
  163
Blood type, 551–552; *tables,* 141; 551
Boiling point, 707
Bone, 629, 631–633; structure of, 632;
  illus., 629; human, 630; 631; 632; 633
Bowman's capsule, 576; *illus.,* 576
Brain, 612–613; human, 613–615; *illus.,*
  591; 612; 613; 614; *table,* 614
Bread mold, 273, 274, 363, 378; *illus.,*
  273; 365; 366
Breathing, 568–569; rapid, 641
Breathing rate, 569
Breeding, of plants, 130–141; *See also*
  Reproduction; Trait
Bridges, Calvin, 151–152
Bromeliad, 701, 725
Bronchi, 567
Bronchiole, 567
Bronchitis, 570, 571
Brown, Robert, 64
Bryophyta, 291–293; *table,* 303
Bubonic plague, 390, 392; *illus.,* 390
Bud, 301, 425, 463, 464; apical, 462;
  illus., 517
Bud formation, 424
Bud scale, 301; *illus.,* 301
Budding, 361–362, 475; *illus.,* 362; 475
Butterfly, 492

C

Calcium, 608, 640
Calcium compound, 631
Calvin, Dr. Melvin, 443
Camouflage, 225–227; by color, 225–
  226; *illus.,* 225
Cancer, 402, 403–404; *illus.,* 404
Capillary, 530, 536, 537, 544–545
Carbohydrate, 54–55, 443; *illus.,* 54;
  table, 56
Carbon, 45
Carbon compound, 52

Carbon dioxide, 456, 457, 458, 704, 726, 745; in the blood, 571; while breathing, 569
Carbon monoxide, 744
Carbonic acid, 746
Carboniferous Period, 293
Carboxyhemoglobin, 549
Carboxyl group, 55, 57
Carcinogen, 403–404
Cardiac muscle, 638; *illus.*, 638
Cardiac sphincter, 525
Carnivore, 694; *illus.*, 694; 697
Carotene, 264, 436; yellow, 287
Carrier molecule, 70
Cartilage, 631, 632
Casting, 319
Caterpillar, 492, 759
Cave artwork, *illus.*, 238
Cell, 63–83; animal, 113, 114, 117; B-, 397, 398; basic unit of living organisms, 111–120; beta, 489; characteristics of, 64–65; clone, 398; collar, 469; cork, 451; daughter, 113, 115, 117; discovery of, 63; disease, 391; division of, 106–123; enzyme induction, 189; epithelial, 64; eukaryotic, 76, 113, 188, 190–191; flame, 573; follicle, 590; gland, 519; guard, 434, 450; and heredity, 111–121; heterocyst, 362; holdfast, 288, 289; interaction in, 82; islet, 587; memory, 399; haploid, 413; nerve, 392; nucleus of, 63, 81, 111; nutrition, 375–378; organization of, 75; osteocyte, 631; palisade, 433; plant, 113, 433; plasma, 397, 398, 399; prokaryotic, 111, 188–190, 261–267, 360; protein in, 176–186; red blood, 69, 396, 548–549, 552, 553; reproduction of, 111–116; Schwann, 605; sex, 117–121; stinging, 312; swarm, 270, 271; T-, 397, 398; white blood, 74, 397, 550; *illus.*, 17; 111; blood, 548
Cell cycle, 111; *illus.*, 111
Cell division, 111–121; meiosis, 117–121; mitosis, 112–116; *table*, 119
Cell membrane, 65–73, 113, 114, 118, 269, 376, 379; composition of, 66–68; and diffusion, 68–69; and facilitated diffusion, 70; interaction of, 82–83; and osmosis, 69; permeability of, 607; and plasmosis, 69; respiration in, 261; *illus.*, 68; 70; 73
Cell plate, 113, 114
Cell theory, 63–64
Cell wall, 79, 113, 114; murein in, 261, 273; *illus.*, 413
Cellular energy, 91–99
Cellular respiration, 10–11, 74, 94, 112, 379, 433, 451, 696; with oxygen, 97–98; without oxygen, 98–99; *illus.*, 98; 99
Cellulose, 54, 79

Centipede, 331–332; *illus.*, 331; *table*, characteristics, 336
Central disc, 337
Central nervous system, 612
Centrifuge, 81–82
Centriole, 80, 114; *illus.*, 80
Centromere, 113
Cephalothorax, 333, 335
Cerebellum, 615; *illus.*, 591; 612; 614; *table*, 614
Cerebral cortex, 614
Cerebrospinal fluid, 613
Cerebrum, 614
Cervix, 480, 500; *illus.*, 480
Chambered nautilus, 322; *illus.*, 322
Chelicerae, 335, 336
Cheliped, 333
Chameleon, 225
Chemical bond, 42; energy in, 90
Chemical change, 51
Chemical energy, 95
Chemical equation, 51–52
Chemical formula, 46, 53
Chemical reaction, 51; effects on nervous system, 609–611; endergonic, 95, 96; exergonic, 95, 97; regulation of, 68
Chemosynthesis, 265
Chemosynthetic bacteria, 733
Childbirth, 594
Chilopoda, 331; *table*, characteristics, 336
Chinese liver fluke, 316
Chitin, 273, 628
Chlamydomonas, 288; *illus.*, 64
Chlorine, 44
Chlorophyll, 10, 287, 436, 438; outside of chloroplast, 261
Chlorophyta, 287; *table*, 303
Chloroplast, 79, 188, 215, 262, 439; *illus.*, 79; 215; 216; 252; 413
Cholera, 392
Cholesterol, 55, 67, 555
Cholinesterase, 609
Chondrichthyes, 341; *table*, 348
Chordata, 252, 329, 337, 338–343; characteristics of, 338–339; *illus.*, 351; *table*, 350
Chorion, 493
Chromatid, 113, 118, 154
Chromatin, 81, 112, 114
Chromatophore, 225, 436
Chromosome, 81, 111, 114, 175, 359; autosome, 149, 151; and disease, 165–167; and movement of genes, 186; mutation of, 186–190; and nondisjunction, 151–153; sex, 147–167; *illus.*, 116
Chromosome mutation, 186–188
Chromosome number, 115–116
Chromosome problem, 165
Chromosome theory, 147–167

Chrysophyta, 268; *illus.*, 278
Chunk feeder, 517
Chyme, 525, 526
Cigarette, and heart disease, 555
Cilia, 80, 377, 382, 396, 571; *illus.*, 80; 382
Ciliary motion, 382
Ciliate, 270, 718; *illus.*, 270; *table*, 278
Ciliophora, 270; *table*, 278
Circulation, placental, 498; portal, 530
Circulatory system, 535–545; adaptive, 538; amphibian, 538–539; annelid, 535–536; arthropod, 536–537; bird, 539–540; closed, 536; disease of, 554, 555; earthworm, 535; fish, 537–538; frog, 538; grasshopper, 536–537; human, 540–558; insect, 536; mammal, 539–540; open, 536; reptile, 538–539; *illus.*, blood, 548; 549; 550; 551; 552; 553; earthworm, 536; fish, 537; grasshopper, 536; human, 540; 541; 542; 543; 554; lymphatic, 545; salamander, 538; turtle, 539
Cirri, 383
Class, 251
Classification, 244–253; bases for, 248–250; system of, 250–253; the need for, 246; theory of, 245–246; *illus.*, 247; 253; *table*, 251
Classification groups, 251
Cleavage, 501–505; *illus.*, 502; 503
Climax community, 717
Cloaca, 476, 579
Clone, 507–508
Clostridium, 392; *illus.*, 266
Clotting, 549, 554
Club fungi, 274–275, 365, 393; *illus.*, 274; 275; *table*, 278
Cobb, William Montague, 253; *illus.*, 253
Coccus, 262; *illus.*, 262
Cochlea, 619, 620
Codeine, 610
Codon, 178–179, 187; *table*, 178
Coelenterata, 312; *table*, 322
Coelenterate, 312–313; nervous system of, 611; *illus.*, 312; 313; *table*, 323
Coelom, 320, 476; *illus.*, 320
Coenzyme, 94, 376, 522
Cohesion, 454–455
Coleoptile, 459
Collar cell, 515; *illus.*, 516
Collecting duct, 575
Colon, 530
Colony, 289, 310
Color blindness, 164; *illus.*, 164
Combination, 136
Commensalism, 700; *illus.*, 700
Common bile duct, 526; *illus.*, 525
Communication, 662–669; language, 665–666; pheromone, 664; *illus.*, 662; 663; 664; 665; 666

Community, 5–11, 693; climax, 717; forest, 716; lake, 718, 748; *illus.,* 717
Companion cell, 457; *illus.,* 452
Comparative anatomy, 201–202
Comparative biochemistry, 203, 249
Comparative embryology, 202, 249, 321; *illus.,* 202
Competitive exclusion principle, 680
Competitor, complete, 680
Complement, 398
Compound, 42–43, 45, 46, 52–57
Concept, 661
Concussion, 613
Condensation, 57
Condensation reaction, 93
Conditioning, 658–659
Conduction, 603–611
Cone, 296, 418, 722
Conidia, 363
Conidiophore, 363
Conifer, 296; life cycle of, 418; *illus.,* 419; *table,* 303
Coniferous forest, 722, 725
Conjugation, 363, 364, 414; *illus.,* 364; 365
Conjugation tube, *illus.,* 413
Conservation, 753, food, 753–754; forest, 757; fuel, 756; soil, 755; wildlife, 757
Conservation of energy, law of, 90
Conservation of mass, law of, 52
Consumer, 694
Continuous variation, 155; *illus.,* 155
Contour plowing, 755; *illus.,* 691
Contractile vacuole, 381
Control group, 27
Controlled experiment, 27
Convergence, 232–233, 322
Convolutions, 614
Copa-iba tree, 443
Copepod, 730
Coral, 312
Coronary bypass surgery, 554
Corpora allata, 597
Corpus luteum, 482, 591
Corpus luteum stage, 482
Cortex, 449, 451, 457, 575; *illus.,* 450; 451; 452; 574
Cortex cell, 700
Cortisol, 589; *table,* 595
Cortisone, 589; *table,* 595
Cosmarium, 288
Cotid, 119
Cotyledon, 298, 426
Coughing, 571
Courtship behavior, 477, 653; *illus.,* 477; 653
Covalent bond, 42, 53; *illus.,* 47
CP, *See* Creatine phosphate
Cranial nerve, 612, 617
Cranium, 633; *illus.,* 613
Crayfish, *illus.,* 332

Creatine phosphate (CP), 641
Cretinism, 588
Crick, Francis, 174
Cristae, 76
Cro-Magnon, 238, 239; *illus.,* 238
Crop, 755
Crop rotation, 755, 761
Cross-pollination, 423; *illus.,* 130
Crossbill, 720; *illus.,* 720
Crossing-over, 154–155; *illus.,* 154
Crustacea, 332; *table,* characteristics, 336
Crustacean, 332–333; *illus.,* 332
Cryptic coloration, 225
Cultural control, 761
Cup fungi, *illus.,* 275
Cutin, 292; *illus.,* 292
Cutting (of plant stem), 424
Cuttlefish, 322, *illus.,* 322
Cyanobacterium, 287
Cyclic AMP, 596
Cytoplasm, 74, 83, 111, 179; *illus.,* 413
Cytoplasmic streaming, 379
Cytosine, 173, 179; *illus.,* 173
Cytoskeleton, 79

D

Dark reaction, 439
Dart, Raymond, 236
Darwin, Charles, 207–211, 221; and speciation, 231
Data, 25
DDT, 741, 743, 747
Death rate, 682
Deciduous forest, 723–724, 725; *illus.,* 723; 724
Decomposer, 11, 704
Decomposition, 197, 732
Degenerative evolution, 279
Dehydration synthesis reaction, 57, 93, 74; *illus.,* 57
Dehydrogenase, 92
Delbrück, Max, 366
Dendrite, 605
Denitrification, 700
Density-dependent limiting factor, 676, 678
Density-independent limiting factor, 676, 682–683
Deoxygenated blood, 538, 539, 540
Deoxyribonucleic acid (DNA), 56, 81, 111, 173–176, 249; and classification, 248–249; code, 178–179; exon, 191; and genetics, 189; intron, 191; in macrophage, 367, 369; model of, 173–179; organelle, 188; outside the nucleus, 188; and phenotype, 188; replication of, 176; the role of, 176; and sexual reproduction, 363, 364; structure of, 171; transcription of, 179; transduction of, 368; translation of,

181–186; in virus, 368–370; *illus.,* 56; 173; 174; replication, 175; transcription, 179; translation, 181, 182
Deoxyribose, 179; *illus.,* 179
Depressant, 609
Descending colon, 530
Desert, 728–730; *illus.,* 720; 729
Desmid, 288
Development, 14–15; bird, 493–495; external, 491; fish, 490; frog, 491, 502, 505, 506, 508; human, 496–509; insect, 492–493; marsupial, 495–496; mechanism of, 501–509; placental, 496–499; *illus.,* bird, 494, 502; fish, 490; frog, 489; 491; 505; human, 497; 498; 499; insect, 492; 493; marsupial, 495
Diabetes insipidus, 586–587, 594
Diabetes mellitus, 281, 586, 594
Diaphragm, 568–569
Diastole, 543
Diatom, 268, 381, *illus.,* 268
Diatomaceous earth, 268
Dicot, 298–299, 450; *illus.,* 299; 426; *table,* 303
Diet, 522–524
Differentiation, 506; control of, 507–508
Diffusion, 68–69, 70, 453; facilitated, 70; *illus.,* 69
Digestion, 376; in autotroph, 376; extracellular, 378, 517; fungi, 378; intracellular, 376, 377, 517; patterns of, 515; in plant, 441–443; protozoan, 377; *illus.,* fungal, 378; protozoan, 377; *table,* 376
Digestive system, 516–527; arthropod, 519–520; earthworm, 518; human, 521–527; hydra, 516–517; insect, 519–520; planaria, 518; sponge, 516; starfish, 520; *illus.,* earthworm, 519; human, 524; 525; 526; 530; sponge, 516; starfish, 520; *table,* 527
Dinoflagellate, 268, 381; *illus.,* 268; *table,* 278
Dioxin, 752
Dipeptide, 56, 57
Diphtheria, 392
Diploid, 116
Diploid cell, 418, 420, 473
Diplopoda, 331; *table,* characteristics, 336
Disaccharide, 54, 93
Disease, 739; African sleeping sickness, 270, 379, 395; amoebic dysentery, 270, 395; anemia, 523, 549; asthma, 570; atherosclerosis, 544, 554, 571; athlete's foot, 393; bacterial pneumonia, 570; bacteriological, 389–393; bronchitis, 570, 571; cancer, 402, 403–404; cretinism, 588; defenses against, 396–404; diabetes, 163, 586, 587, 594; Down syndrome, 165; Dutch elm, 276,

740; elephantiasis, 318; emphysema, 571–572; fungal, 393–394; germ theory of, 389–390; goiter, 588; heart, 524, 554–557; human genetic, 159–167; infectious, 389; influenza, 569–570; as a limiting factor, 679; lung cancer, 572; malaria, 679, 740; malnutrition, 523; plague, 685; plant, 391; pneumonia, 171–173, 570, 571; and population, 679–680; protozoan, 395–396; provirus, 367; respiratory, 569–570; rickets, 522, 631; sex-linked, 164; spread of, 390–391; strep throat, 570; tetanus, 637; tobacco mosaic, 279; tuberculosis, 570; venereal, 390; viral, 389–393; vitamin deficiency, 94, 522; *illus.*, blood, 159; 160; 164; pathogens, 390; 395

Dissociation, 47
Distemper, 392
Distribution of organism, 706–707
Divergence, 230
DNA, *See* Deoxyribonucleic acid
Dominance, 156, 667
Dominant trait, 131
Dormancy, 425, 708
Dorsal aorta, 537
Dorsal blood vessel, 536
Dorsal lip, 505; *illus.*, 505
Dorsal side, 314; *illus.*, 314
Double bond, 55
Double helix, 175
Down syndrome, 165; *illus.*, 165
Drone, 663; *illus.*, 662
Drosophilia melanogaster, 155, 187; genetic study of, 147–150; *illus.*, 148, 155
Drug, 609–611
Duckbill platypus, 494
Duet rule, 42
Duodenum, 526
"Dust bowl," 728
Dutch elm disease, 276, 740, 760
Dwarfism, 592
Dynamic equilibrium, 69

E

E. coli, 189, 280; *illus.*, 155
Ear, 619
Eardrum, 619
Earthworm, 319, 474; circulatory system of, 535; digestive system of, 518; excretory system of, 574; locomotion of, 634; respiratory system of, 564; *illus.*, 319
Ecdyson, 597
Echinoderm, 329, 337; locomotion of, 635
Echinodermata, 337
Ecological niche, 232
Ecology, 18–19

Ecosystem, 692–709
Ectoderm, 312, 315, 502, 508
Ectotherm, 340, 345; and temperature, 707–708
Eel, 342
Effector, 603
Egestion, 318
Egg, 16, 117, 119, 120, 311, 415, 416, 418, 419, 420, 421, 422, 423, 478, 490; fertilized, 117, 131; shelled, 493–494; 742; *illus.*, 117
Eggshell, 344, 493–494, 742
Eldridge, Niles, 231
Electron, 40; sharing, 42; transfer, 44
Element, 41
Elephantiasis, 318
Elodea, 70
Elongation region, 426
Embryo, 202, 317, 396, 418, 422, 482; development of, 489–509; human, 553; protection of, 489–490
Embryology, 489
Embryonic induction, 508–509; *illus.*, 509
Emigration, 677
Emphysema, 571
Emulsification, 526
Encephalitis, 392
End brush, 605
Endergonic reaction, 95–96
Endocrine control, 591–597
Endocrine gland, 586, 589
Endocrinology, 586
Endocytosis, 72–73, 83
Endoderm, 312, 315, 505
Endodermis, 450, 454
Endoplasmic reticulum (ER), 77, 113, 114; *illus.*, 77
Endorphin, 621
Endoskeleton, 337, 629–630
Endosperm, 421
Endosperm nucleus, 421
Endospore, 266
Endotherm, 346, 349, 494, 708
Endotoxin, 392
Energy, 8–9, 89–99; of activation, 91, 92; cellular, 91–99; chemical, 90, 95; conservation of, 90; def. of, 89; kinetic, 90; and mitochondria, 76; potential, 90; pyramid of, 696–697; radiant, 435; relationships, 439–440; solar, 756–757; transfer of, 9–11; transformation of, 89–90; trapping of, 433–436; uses of, 14–17
Energy level, 40, 42
Enkephalin, 621
Environment, 18–19; abiotic factors of, 703–709; biotic factors of, 693–736; cell in, 65–67; and evolution, 222–223; and pesticides, 740–743; plant response to, 383, 459–465; and pollution, 739, 744–750; reproduction, 478; response

to, 383
Environmental resistance, 675
Enzyme, 68, 77, 91–94, 97, 160, 176, 189, 369, 376, 378, 396, 441, 524, 708; and activation energy, 91, 92, 95; active site on, 93, 94; properties of, 92
Enzyme induction, 190
Epicotyl, 425
Epidermis, 433, 449
Epiglottis, 567
Epinephrine, 589
Epiphyte, 700, 705, 724
Epithelial cell, 64
Epitope, 404–405
Equation, 51–52
ER, *See* Endoplasmic reticulum
Erosion, 704, 755
Erosion rate, 755
Error catastrophe theory, 755
Erythrocyte, 548
Esophagus, 519, 524, 525; and digestion, 524–525
Estivation, 708
Estrogen, 482, 590, 591
Estrus cycle, 478, 664
Eubacteria, 261
Euglena, 252, 288, 361, 383
Euglenoid, 267
Euglenophyta, 267
Euglenoid, 266; digestion in, 377
Eukaryote, 76, 111, 175, 265, 438, 439; asexual reproduction of, 360–361; cell parts of, 76–80; origin of, 215; RNA synthesis in, 180–181; *illus.*, 76
Euphorbia, 443
Euplotes, 383
Eustachian tube, 619
Eutrophication, 748, 749
Evolution, 197; anatomical evidence, 201–211; Darwin's concept of, 210; degenerative, 280; embryonic evidence, 202; and environment, 222–223; fossil evidence, 199–200; genetic evidence, 203; human, 234–239; Lamarck's work on, 206, 210–211; of life, 212–216; natural selection, 208–209; observation of, 204–205; of species, 228–230
Excretion, 380–381, 562–579, 573–579
Excretory canal, 573
Excretory duct, *illus.*, 320
Excretory system, amphibian, 579; earthworm, 574; fish, 578; grasshopper, 574; human, 575–577; planaria, 573; vertebrate, 578–579
Excurrent pore, 310
Exercise, 555
Exergonic reaction, 95–96
Exocytosis, 72–73, 83, 377, 608; *illus.*, 73
Exon, 191
Exoskeleton, 330, 597, 628
Exotoxin, 391–392

Experimental group, 26
Experimentation, 26–28
Expiration, 568–569
Extension, 636
Extensor, 636
Extinct, 200

F

F₁ generation, 130
F₂ generation, 130
Facilitated diffusion, 70
Fact, 24–25, 26
Fallopian tube, 479; *illus.,* 479
Family, 251
Family planning, 686
Fat, 55, 71, 545
Fat emulsification, 526
Fatty acid, 596
Feces, 530
Feedback mechanism, 592–593; *illus.,* 593
Femur, 633
Fermentation, 98–99; *illus.,* 98
Fern, 294–295, 417; life cycle of, 417; *illus.,* 294; 295; 417; *table,* 303
Fernandez-Moran, Humberto, 605; *illus.,* 605
Fertility factor, 364
Fertilization, 423, 476–478; external, 476–477, 489; internal, 476, 478, 489, 493–494, 495; *illus.,* 421
Fertilizer, 755; *illus.,* 704
Fetoscopy, 167, *illus.,* 167
Fetus, 119, 166, 499, 632; *illus.,* 499
Fiber, 638, 609
Fibril, 639; *illus.,* 639
Fibrin, 550
Fibrous root system, 300
Filament, 419, 639, 640; *illus.,* 639
Filaria worm, 318
Filial generation, 130
Filter feeding, 515–516
First polar body, 119
Fish, angler, 342; bony, 342–343; cartilage, 341; circulatory system of, 537–538; excretory system of, 578–579; development of, 490; jawless, 340; respiratory system of, 564–565; *illus.,* 340; 341; 342; 490; circulatory system, 537; respiratory system, 565
Fission, 360; *illus.,* 360
Flagella, 80, 119, 573
Flagellar motion, 382
Flagellate, 270–271; *illus.,* 270; *table,* 278
Flatworm, 315–316; digestive system of, 518; *illus.,* 315; 518; 573; *table,* 323
Fleming, Alexander, 25, 27, 28
Flexor, 636
Florigen, 465

Flower, 298, 299, 419–422; imperfect, 419; perfect, 419; *illus.,* 419; 420; 421; 423; 465
Flower scar, 301
Flowering, control of, 464–465
Fluid mosaic model, 68
Fly, 115, 116, 740, 741, 743, 760; genetic study of, 147–152; life cycle of, 108; *illus.,* 107; 148; 151; 155
Follicle, 479
Follicle stage, 482
Follicle-stimulating hormone (FSH), 482, 591; *table,* 595
Food, 7–8; absorption of, 530; as a limiting factor, 677–678; conservation of, 753–755; production, 9–11
Food absorption, 530
Food chain, 11, 199, 695, 716, 742, *illus.,* 11
Food getting, 514–527
Food group, 522–523
Food shortage, 753
Food transport, 453
Food vacuole, 83
Food web, 700; *illus.,* 700
Foot-and-mouth disease, 392
Forebrain, 612, 613–614; *illus.,* 614; *table,* 614
Forest, conservation of, 757–758; deciduous, 723–724; temperate, 723–724; tropical rain, 724–726
Forest floor, 723; *illus.,* 723
Formaldehyde, 187
Formula, 45, 53
Fossil, 197–200; formation and dating, 197–199; human, 235; interpretation of, 199–200; *illus.,* 199; 200; carbon dating, 200; *table,* geologic time, 200
Fossil fuel, 756
Fragmentation, 361–362, 413, 475
Fraternal twins, 504; *illus.,* 504
Freezing point, 707
French, Charles, 438, *illus.,* 438
Frisch, Karl von, 663
Frog, 344, 477; circulatory system, 538; development of, 491, 502, 505–506; reproductive system of, 566, 567, *illus.,* 343; 489; 491; 502; 505; 566; *table,* 348
Frond, 295, 417
Frontal lobe, *illus.,* 613; *table,* 614
Fructose, 93
Fruit fly, 116; genetic study of, 146–152; *illus.,* 148; 151; 155
Fruit scar, 301
FSH, *See* Follicle-stimulating hormone
Fuel, 744, 747; conservation of, 756–757; from plants, 442–443
Fungi, 33, 273–276, 363, 378, 379, 702; club, 274–275; cup, 274–275; in lichen, 277; sac, 275–276; sexual reproduction of, 364–365; sporangium, 273–274;

illus., 273; 274; 275; 278; 378; *table,* 278
Fusion, 363

G

Galactose, 161
Galactosemia, 161
Galapagos finch, 228–230, 231–232
Galapagos Islands, 208, 732–733
Gamete, 117, 130, 132, 133, 135, 138, 149, 155, 186, 206, 365, 413, 414, 415, 478; *illus.,* 139
Gametophyte, 415, 417, 418, 420, 421, 422; *illus.,* 415; 417
Gametophyte generation, 422
Ganglia, 611
Gas exchange, 379–380, 453, 457–458, 563
Gas transport, 548–549
Gasohol, 443
Gasoline, 443, 744
Gastric gland, 525
Gastric juice, 525
Gastrin, 525
Gastrovascular cavity, 517
Gastrula, 505, 506
Gastrulation, 505
Geckos, 345
Geiger counter, 198
Gel, 382
Gemmule, 311
Gene, 131, 133, 151, 154–155, 155–156, 159, 280, 364; in asexual reproduction, 359; and continuous variation, 155–156; cross-over, 154–155; dominant vs. recessive, 131–133; expression of, 157–158, 186–191; incomplete dominance of, 137–140; jumping, 186, 191; lethal, 159, 188; location of, 155; movement of, 186; mutation of, 186–190; nondisjunction, 151–152, 165; operator, 190; problem-causing, 159–161; and protein synthesis, 176–186; in provirus, 367; recessive vs. dominant, 131–133; regulator, 188–189; structural, 188, 189; switched on, 508; theory of, 149–150; and transduction, 368; *table,* codons, 178
Gene linkage, 153–154; *illus.,* 153
Gene pool, 229
Gene segregation, 132
Gene therapy, 281
Generative nucleus, 420
"Genetic code," 111, 170–195; *table,* codons, 178
Genetic continuity, 115
Genetic control, in bacteria, 189–190; in eukaryote, 190–191; in prokaryote, 188–190
Genetic counseling, 159

Genetic engineering, 186, 280–281
Genetic map, 155; *illus.*, 155
Genetic recombination, 154, 203, 211, 368; *illus.*, 123
Genetics, 129–142, 203, 266, 761; evolution, 203–204; origin of, 129; probability of, 134–135; *illus.*, genetic ratio, 136; 138; 140; homologous chromosome, 133; Punnett square, 132; traits, 137; 139; *table*, blood types, 141
Genotype, 133, 135, 136, 149; *table*, 141
Genotypic ratio, 136; *illus.*, 136; 138; 139
Genus, 247–248, 251
Geographic isolation, 229
Geologic timetable, *table*, 200
Geotropism, 459, 462; *illus.*, 462
Germ theory of disease, 389–390
Germination, 422, 425–427; *illus.*, 425
Gestation period, 499–500
Giantism, 592; *illus.*, 592
Gibberellins, 463–464
Gila monster, 345
Gill, 275, 344, 537, 564–565; *illus.*, 344; 564; 565
Gill arch, 565
Gill chamber, 565
Gill filament, 565
Ginkgo biloba, 297; *illus.*, 296
Ginkgo tree, 296–297
Gizzard, 519
Gland cell, 519
Gliding joint, 633
Globular protein, 92
Glomerulus, 575, 576; *illus.*, 575
Glucagon, *table*, 595
Glucose, 54, 93, 97, 98, 70, 80, 190, 434, 440, 587, 596; formation of, 439; in the blood, 530; *illus.*, 54; *table*, 548
Glycerol, 55; *illus.*, 55
Glycogen, 54, 587, 596, 641
Goiter, 588
Golden algae, 268
Golgi body, 77, 82; *illus.*, 77
Gonad, 474, 590
Gould, Stephen Jay, 231
Gradualism, 231
Gram's solution, 264
Grana, 79, 438
Grasshopper, circulatory system of, 536–537; digestive system of, 519–520; excretory system of, 574; reproductive system of, 478; respiratory system of, 566; skeletal system of, 628; *illus.*, 478; 493; 520; excretory system, 574; respiratory system, 566; skeletal system, 628
Grassland, 727–728; *illus.*, 720; 727
Gray matter, 614, 615
Griffith, Fred, 172
Ground pollution, 752–753
Growth, 14–15; *illus.*, 14

Guanine, 173, 179; *illus.*, 173
Guttation, 454
Gymnomycota, 271
Gymnosperm, 296–297; life cycle of, 418–419; *illus.*, 296; 419
Gymnospermae, *table*, 303
Gypsy moth, 759

H

Hagfish, 340; *table*, 348
Half-life, of an isotope, 198
Hammer, 619
Hand-eye coordination, 235
Haploid cell, 117, 473
Haploid spore, 418
Hardy, Godfrey, 211
Hardy-Weinberg Law, 211–212
Haversian canal, 632; *illus.*, 632
Hay fever, 570
Head, 334
Head-foot mollusk, 322
Health, 522–524
Hearing, 619–620
Heart, 320; artificial, 557–559; disease of, 554–555; earthworm, 535, 536; fish, 537, 538; frog, 538; grasshopper, 536–537; human, 539–558; *illus.*, artificial, 558; fish, 537; human, 540; 541; 554; rabbit, 539; salamander, 538; turtle, 539
Heart attack, 571
Heart disease, 524, 554–557
Heart transplant, 556–557
Heartbeat, 542–543
Heart lung machine, 556
Heat energy, 91
Heavy hydrogen, 757
Helium, 40
Hemichordata, 337
Hemichordate, 338
Hemocoel, 574
Hemoglobin, 159, 176, 548, 568; and pollution, 744; *illus.*, 160
Hemophilia, 164, 187; *illus.*, 164
Hepatic vein, 530
Herbaceous, 300
Herbivore, 694, 696
Heredity, cellular basis of, 106–127; chemical change of, 172–173; chromosome theory of, 147–152; genetics of, 129–141; principles, 128–141; probability of, 134–135, 139; *table*, blood types, 141; *See also* Trait
Hermaphrodite, 311, 316, 474
Heroin, 610
Herpes, 392
Heterocyst, 362; *illus.*, 362
Heterotroph, 214, 252, 264, 515; nutrient needs of, 376
Heterotroph hypothesis, 214

Heterotrophic eubacteria, 264; of anaerobic respiration, 264
Heterozygote, 160
Heterozygous genotype, 137
HGH, *See* Human growth hormone
Hibernation, 708–709; *illus.*, 708
High blood pressure, 523–524
Hierarchy, 666, 667, 668, 682
Hindbrain, 612, 613, 615; *illus.*, 614; *table*, 614
Hinge joint, 633
Histone protein, 261
Histoplasmosis, 393; *illus.*, 393
Hodgkin, Dorothy Crowfoot, 29; *illus.*, 29
Homeostasis, 15, 585, 618; *illus.*, 15
Hominidae, 235, 350
Hominids, 235
Homo erectus, 237, 239; *illus.*, 237
Homo habilis, 237, 239
Homo sapiens, 238, 239, 350
Homogeneous substance, 46–47
Homolog, 116
Homologous chromosome, 117, 118, 148, 154; *illus.*, 133
Homologous pair, 117
Homologous structure, 249
Homozygous, 133
Honeybee, 654, 662–664; *illus.*, 662; 663; 664
Hooke, Robert, 63–64
Hookworm, 318; *illus.*, 318
Hormone, 225, 603, 621, 760; human growth, 585–596; and metamorphosis, 597; as a pesticide, 760; *tables*, 548; 595
Hormone action, 595–596
Horsetail, 294; *illus.*, 294; *table*, 303
Housefly, *illus.*, 519
Human blood, 539–553
Human blood type, 141, 551–553
Human brain, 613–614; *illus.*, 591; 612; 614; *table*, 614
Human chromosome number, 116
Human circulatory system, 539–560; *illus.*, 540; 541; 542; 543; 554
Human digestive system, 521–530; *table*, 506
Human embryo, fertilization, 496
Human environment, 739; and pesticides, 740–743; pollution of, 739, 744–750
Human evolution, 234–240; *illus.*, 235; 236; 237; 239; *table*, human tree, 240
Human excretory system, 575–577; *illus.*, 575; 577
Human genetic disease, 159
Human growth hormone, 592; *table*, 595
Human hormones, 585–597; *table*, 595
Human immunodeficiency virus, 401–403

Human insulin, 281
Human locomotion, 636–642
Human nerve, 603–611; *illus.,* 604; 605; 606; 607; 608; 609
Human nervous system, 613–620; *illus.,* 613; 614; 615; 616; 617; 618; *table,* 614
Human origin, 233–239
Human population, 683–687
Human reproduction, 479–482, 496–509; and Rh factor, 553; *illus.,* 497; 498; 499; 504; female, 479; male, 480; menstrual, 483
Human respiratory system, 567–570; diseases of, 569–570; smoking effect on, 571–572; *illus.,* 567
Humerus, 636
Humus, 704, 726
Huntington's disease, 161
Hybridoma, 405
Hydra, 312; digestive system of, 516–517; locomotion of, 634; nervous system of, 611; reproduction of, 475; *illus.,* 320; 473; 517; 611; 634; nervous system, 611
Hydrocarbon, 443, 744
Hydrochloric acid, 525
Hydrogen, 40, 45, 212, 757
Hydrogen chloride, 48
Hydrogen sulfide, 733
Hydrolysis, 57, 521
Hydrolysis reaction, 93, 376, 515; *illus.,* 57
Hydrothermal vent, 733
Hydroxide ion, 48, 49; *table,* 50
Hyman, Libbie Henrietta, 316; *illus.,* 316
Hypertension, 543–544, 554
Hypha, 273; minus, 364; plus, 364
Hypocotyl, 425
Hypothalamus, 482, 577, 592, 594, 614, 620, 709; *illus.,* 591
Hypothesis, formation of, 26, 92–94

I

Identical twins, 504; *illus.,* 504
Ileum, 526
Immigration, 682
Immune system, 550–551, 556; function, 397–399; structures of, 397
Immunity, active, 399; to disease, 399; passive, 400
Immunology, techniques in, 404–405
Implantation, 497
Imprinting, 659–660; *illus.,* 660
Improvement cutting, 758
Incomplete dominance, 156
Incurrent pore, 310
Indian pipe, 298; *illus.,* 298

Indoleacetic acid, 461
Inducer, 190
Induction, *illus.,* 189
Industrial melanism, 223; *illus.,* 222
Infection, 549–550
Inferior vena cava, *illus.,* 575
Influenza, 392, 569–570
Infusion, 109
Ingestion, 252
Inhibition center, 543
Inheritance of acquired characteristics, 205–206
Inhibition center, 543
Innate behavior, 649–650
Inner bark, 451; *illus.,* 452
Inner cell mass, 497
Inner ear, 619
Insect, 334; circulatory system of, 536–537; development of, 492–493; digestive system of, 519, 520; *illus.,* 334; 335; 492; 493; 520; circulatory system, 536
Insecta, 334; *table,* 336
Insectivorous plant, 441–442; *illus.,* 441
Insight learning, 661; *illus.,* 661
Inspiration, 568; *illus.,* 568
Instinct, 651–652, 659
Insulin, 161, 587, 627; *illus.,* 281; *table,* 595
Integrated pest management, 761
Integumentary system, 339
Interbreeding, 229, 230
Interferon, 400–401
Interneuron, 616
Internode, 301; *illus.,* 301
Interphase, 112; *illus.,* 112
Interpretation, 25
Interspecific competition, 680
Intraspecific competition, 681–682
Intron, 191
Inversion, 745
Invertebrate, 309; nervous system of, 611–612
Iodine, 588
Ion, 44–45, 47, 68, 442; mineral, 454
Ionic bond, 45
Ionic compound, 45, 47; *illus.,* 47
Ishimure, Michiko, 749; *illus.,* 749
Islets of Langerhans, 586; *illus.,* 586
Isomer, 53; *illus.,* 53
Isotope, 41, 198

J

Japanese beetle, 740, 760
Java, *illus.,* 237
Jawless fish, 340
Jejunum, 526
Jellyfish, 312; *illus.,* 314
Jet lag, 654

Joint, 633; ball and socket, 633; gliding, 633; hinge, 633; pivot, 633; *illus.,* 633
Jungle, 724
Junk food, 523
Just, Earnest Everett, 75; *illus.,* 75
Juvenile hormone, 597

K

Kangaroo, 495
Kaposi's sarcoma, 402
Karyotype, 116, 166; *illus.,* 116; 165
Katydid, 226
Kelp, 289
Khorana, Har Gobind, 152; *illus.,* 152
Kidney, 575, 578, 589, 594, 596; *illus.,* 575
Kinetic energy, 90; *illus.,* 90
Kingdom, 251, 252–253; *table,* 323
Klinefelter syndrome, 165
Koch, Robert, 389–390
Koch's postulates, 389–390

L

Labor, 500
Lactase, 190, 527
Lacteal, 530
Lactic acid, 641
Lactose, 54, 189
Ladder-type nervous system, 611
Lamarck, Jean Baptiste de, 206–207, 210–211, 222, 226
Lamarckism, 206
Lamprey, 340; *illus.,* 340; *table,* 348
Lancelet, 338; *illus.,* 338
Language, 665–666; *illus.,* 665; 666
Large intestine, and digestion, 530–531; *table,* 506
Larva, 491, 492, 493
Lateral area, 314
Lateral bud, 301; *illus.,* 301
Latitudinal succession, 706
Law of Conservation of Energy, 91
Law of Conservation of Mass, 52
Law of Dominance, 131
Law of Independent Assortment, 138
Law of Segregation, 135
Law of Use and Disuse, 206
Leaching, 704
Lead, 744, 745
Leaf, 301–302, 433–434; compound, 302; palmately compound, 302; pinnate, 302; simple, 302; *illus.,* 301; 302; 434
Leaf scar, 301; *illus.,* 301
Leaflet, 302
Learned behavior, 650, 654–661
Learning, trial and error, 659
Leech, 319–320; *illus.,* 319
Leeuwenhoek, Anton van, 109

Leeuwenhoek's microscope, *illus.,* 109

Left ventricular assist device (LVAD), 558

Legume, 700, 755

Lenticel, 301, 458; *illus.,* 301; 458

Leopard frog, 490

Lethal gene, 159, 188

Leukocyte, 550; *table,* 548

Li, Choh Hao, 593; *illus.,* 593

Liana, 725; *illus.,* 725

Lichen, 276–277, 702, 716, 721

Life, kingdoms of, 32–33; origin of, 212–214

Life cycle, animal, 473–474; bacteriophage, 366; diploid phase of, 291, 295; haploid phase of, 291

Ligament, 633

Light, 435; as a biotic factor, 705; reactions to, 437–438

Light reaction, 438

Limiting factor, 676–683; *illus.,* 677; 680

Linnaeus, Carolus, 247, 249

Lipase, 526

Lipid, 55, 69, 72, 585

Lithium, 40

Littoral zone, 731

Liver, 526, 530, 549, 577, 641; *illus.,* 525

Liver damage, 523

Liverwort, 291, 453; *illus.,* 291; 292; *table,* 303

Lizard, 345; *illus.,* 345; *table,* 348

Lobe, 614

Lobster, 333; *illus.,* 333

Lock and key hypothesis, 92–94

Locomotion, 626–641; amoeboid, 381–382; arthropod, 636–637; bipedal, 629; ciliary, 382; coelenterate, 634; earthworm, 634; echinoderm, 635; flagellar, 382; hydra, 634; mollusk, 635; starfish, 635; vertebrate, 636–637; *illus.,* 269; 634; 635; 636

Lorenz, Konrad, 660, *illus.,* 660

LSD, 611

Lumbricus, 319; *See also* Earthworm

Luminence, 342

Lundegardh, Henrik, 461; *illus.,* 461

Lung, 344, 345, 347, 566–568; book, 336; *illus.,* 567; 568

Lung cancer, 572

Luteinizing hormone (LH), 482, 591; *illus.,* 595

Lyell, Charles, 207

Lympanic membrane, 619

Lymph, 397, 545, 574

Lymph node, 397, 545

Lymph vessel, 545

Lymphatic system, 530, 544–545

Lymphocyte, 397, 405

Lysis, 367, 399

Lysogenic cycle, 367; *illus.,* 367

Lysosome, 78, 377; functions of, 82–83; *illus.,* 78; 82

Lytic cycle, 367; *illus.,* 367

M

Macronucleus, 270, 361

Macrophage, 397; *illus.,* 397

Maggot, 492

Maggot development, 108

Magnesium, 442

Magnification, 30

Maintenance, 15

Malaria, 271, 362, 395–396, 679; *illus.,* 395

Malathion, 743, 760

Male gamete, 130

Malnutrition, 523

Malpighian tubule, 574

Maltase, 527

Maltose, 54

Malthus, Thomas, 208

Mammal, 348–350; circulatory system of, 539–540; reproductive system of, 479–482, 495–510; *illus.,* 348; 349; human, 497; 498; 499; *table,* 348; main tissues, 506

Mammalia, 348; *table,* 348

Mammary gland, 349, 496

Mandible, 332, 334, 335

Mangold, Hilde, 508

Mantle, 321

Marijuana, 611

Marrow, 633

Marsupial, 348; development of, 495–496; *illus.,* 348; 495

Mastigophora, 270; *table,* 278

Matching, 552

Material, recycling of, 698

Mating, 478, 668

Matter, classification of, 39–41; properties of, 50–51

Maturation region, 426

McClintock, Barbara, 186; *illus.,* 186

Measles, 392

Medfly, 740, 743

Medium, 703

Medulla, 575; *illus.,* 575

Medulla oblongata, 543, 569, 615; *illus.,* 591; 612; 614; *table,* 614

Medusa, 312, 633; *illus.,* 313

Medusa stage, 474

Megaspore, 420; *illus.,* 420

Meiosis, 117–123, 148, 203, 365, 413, 418, 420, 473; in females, 119–121; importance of, 121; in males, 118; *illus.,* 117; 118; 119; *table,* 119

Meiosis I, 117, 118, 119; *illus.,* 117; 119

Meiosis II, 119; *illus.,* 117

Meiosis-mitosis cycle, *illus.,* 121

Melanin, 188; *illus.,* 223

Melanism, 223

Membrane, 112–113

Membrane potential, 606–607

Mendel, Gregor, 129–133, 135, 138, 147, 156, 171

Meninges, 613, 615; *illus.,* 614

Meningitis, 390

Menopause, 590

Menstrual cycle, 481–483, 590, 593; *illus.,* 483

Menstruation, 482, 590

Mercury, 747

Meristem tissue, 426, 450, 451, 462; apical, 426

Mesoderm, 315, 508; *illus.,* 320; 505

Mesoglea, 312; *illus.,* 517

Messenger RNA, 179, 188, 190, 191

Metabolic rate, 588

Metabolism, 74, 588, 707–709; *illus.,* 707

Metamorphosis, 490, 597, 760; complete, 492; incomplete, 493; in insects, 492–493; stages of, 492; *illus.,* 491; 492; 493

Metaphase, 113; *illus.,* 113

Metaphase chromosomes, *illus.,* 122

Methane, 53, 212, 265

Mica, 460

Microenvironment, 716

Microfilament, 78, 114, 382, 638; *illus.,* 78

Micronucleus, 270, 361

Micropyle, 421

Microscope, 29–31

Microspore, 420, 422; *illus.,* 420

Microtubule, 78, 80, 112, 113, 382; *illus.,* 78

Midbrain, 612, 613, 615; *illus.,* 614; *table,* 614

Middle ear, 619

Migration, 652–653, 707; *illus.,* 651

Mildew, *illus.,* 274

Miller, Stanley, 213

Millipede, 331–332; *illus.,* 331; *table,* characteristics, 336

Mimicry, 227; *illus.,* 227

Mimosa, 650

Mineral, 521, 726; *table,* 521

Minoka-Hill, Lillie Rosa, 523; *illus.,* 523

Mitochondrial DNA, 249–250

Mitochondrion, 76, 97, 188, 215; *illus.,* 76

Mitosis, 111, 112–115, 360, 362, 413, 420, 424, 473, 508; analysis of, 115; *illus.,* 114; *table,* 119

Model, 227

Modifier gene, 157

Mold, 25; bread, 273, 274, 363, 378; slime, 270, 271; *illus.,* 273; 365; 378; slime, 270; 271

Molecule, 42

Mollusk, 321–322; locomotion of, 635; respiratory system of, 564; *illus.*, respiratory system, 564; *table*, 323

Molting, 597, 628; *illus.*, 346

Monera, 32, 261; asexual reproduction in, 360; illus., 278; 360; *table*, 278

Monoclonal antibody, 557

Monocot, 298–299; herbaceous, 451; *illus.*, 299; 426

Monosaccharide, 54, 93

Monotreme, 348; *illus.*, 348

Morgan, Thomas Hunt, 148–150

Morphine, 610

Morphogenesis, 505–506; *illus.*, 505

Morphological adaptation, 224

Mosaic evolution, 231

Mosquito, 395, 678–679; *table*, characteristics, 336

Moss, 291, 416; club, 294; life cycle of, 416; Spanish, 701; *illus.*, 291; 292; 416; *table*, 303

Moth, 223, 225, 492, 760; *illus.*, 222, 226; 331

Motivation, 655

Motor neuron, 604, 612, 639

Mouth, and digestion, 524–525; *illus.*, 517; 518; 524

"Mouthbreeders," 490; *illus.*, 490

Mouthparts, 330, 331, 334

Mucin, 524

Mucus, 524

Multicellular organism, 65

Multiple allele, 141

Mumps, 39

Murein, 262

Muscle, 636–637; biceps, 636; cardiac, 638; contraction of, 641; involuntary, 638; nonstriated, 638; pectoralis, 347; skeletal, 638, 639–641; smooth, 638; striated, 638–639; triceps, 636; vertebrate, 638–641; *illus.*, 636; 637; 638; 639; 640

Muscle tone, 637, 641, 615

Mushroom, 274, 365, 378; *illus.*, 274

Mutagen, 187

Mutation, 186–188, 203, 211, 223, 229; and vaccine development, 392; *illus.*, 186; 188

Mutualism, 271, 277, 531, 702

Mutualistic relationship, 733

Mycelia, 273, 365, 378

Myelin sheath, 605, 614

Myosin, 639, 640; *illus.*, 639

Myxomycota, 270–271; *table*, 278

N

Natural selection, 208, 221–222, 231, 232

Nautilus, *illus.*, 322

Navel, 500

Neanderthal, 238; *illus.*, 238

Nekton, 730

Nematocyst, 312, 516–517; *illus.*, 516

Nematode, 318, 320, 378; *illus.*, 320

Neoteny, 344

Nephridia, 574

Nephridiopore, 574

Nephron, 575

Nephrostome, 574

Nereis, 319, 634; respiratory system of, 564; *illus.*, 319; 564

Neritic zone, 731

Nerve, 603–604; accelerator, 543; action potential of, 607; auditory, 620; cranial, 612; drugs affecting, 609–610; response of, 608

Nerve cord, 612; *illus.*, 320

Nerve impulse, 607, 609

Nerve net, 611

Nervous control, 602–621

Nervous system, 589, 611–613, 650; autonomic, 617; central, 612; human, 613–620; invertebrate, 611–612; ladder-type, 611; peripheral, 612, 616–618; vertebrate, 612–613; *illus.*, 604; 605; 606; 607; 608; 609; 612; 613; 614; 615; 616; 617; 618; *table*, 614

Nervous tissue, 315

Neural fold, 506, *illus.*, 507

Neural plate, 506

Neural tube, 506, 508; *illus.*, 507

Neurofibril, 383

Neuron, 604–606; interneuron, 605, 606, 616; motor, 605, 616; sensory, 603; *illus.*, interneuron, 604; 616; motor, 604; 616; sensory, 604; 616

Neuropeptide, 620–621

Neutral solution, 48; *table*, 50

Neutron, 40, 41

New groups, origin of, 228–233

Nicotine, 555

Night blindness, 522

Nitrate, 748

Nitrate ion, 699

Nitrification, 699

Nitrite ion, 699

Nitrogen, 45, 55, 442, 573, 699–700, 755

Nitrogen cycle, 699–700; *illus.*, 699

Nitrogen dioxide, 745

Nitrogen fixation, 700, 746

Nitrogen gas, 212, 700

Nitrogen oxide, 744, 746

Nitrogen-fixing bacteria, 700, 702, 756

Nitrogenous waste, 380, 573, 575, 576, 578

Node, 301, 605; *illus.*, 301

Nodule, 700; *illus.*, 700

Nonbiodegradable, 742

Nondisjunction, 151–152, 165, 187; *illus.*, 151

Nonhistone, 191

Nonstriated muscle, 638

Noradrenaline, 609, 621

Normal red blood cells, *illus.*, 160

Nostoc, 362

Notochord, 338, 340

Nuclear energy, 756

Nuclear fission, 756

Nuclear fusion, 757

Nucleic acid, 56, 179

Nucleoli, 81

Nucleoplasm, 74, 81

Nucleotide, 173–175, 187; *illus.*, 174

Nucleus, 40, 63, 81, 116; composition of, 112; importance of, 111; *illus.*, 81; 111; 413

Nutrient, 375, 521, 522, 755; plant, 442; requirements, 375–376

Nutrition, 380–381, 521–522; *tables*, 376; 521; 522

Nymph, 493

O

Oak-hickory forest, 717

Obesity, 523, 555

Observation, 24–25, 29; *illus.*, 24

Occipital lobe, *illus.*, 612; *table*, 614

Ocean, 730–733; and light, 730; and temperature, 730; *illus.*, 720; 730; 731

Octet rule, 42

Octopus, 322; *illus.*, 322

Odum, Howard, 696

Offshore drilling, 756

Olduvai Gorge, 236

Omnivore, 695

Oncogene, 404

Oocyte, 481, 482; primary, 119, 120; secondary, 119

Oparin, Alexander, 212

Operator, 190

Operculum, 343, 565

Opossum, 495; *illus.*, 495

Optic lobe, 612; *illus.*, 612

Oral groove, 361, 377

Order, 251

Organ, 17; *illus.*, 17

Organ of Conti, 619

Organelle, 75, 82

Organelle DNA, 188

Organic chemistry, 52

Organic compound, reactions of, 56–57

Organic reaction, 56–57; *illus.*, 57

Organism, 5–7; adaptation of, 18; change in, 197–215; classifying, 247–248; development of, 15; disease-causing, 679; distribution of, 706–707; environment, 18–19; grouping, 32; growth of, 14; homeostasis in, 15; organization of, 16–17; photosynthetic, 436; reproduction of, 16

Organization, 16–17; *illus., 17*
Osculum, 310, 516
Osmosis, 69–70, 292, 454, 457, 544
Osmotic balance, 69, 573; control of, 577; *illus.,* 69; 573
Ossification, of bone, 631–633
Osteichthyes, 342; *table,* 348
Osteocyte, 631; *illus.,* 632
Ostia, 536
Outer bark, 451; *illus.,* 452
Outer ear, 619
Outer flap, 619
Oval window, 619
Ovary, 418, 422, 474, 482, 590; *illus.,* 419; 482
Overcrowding, 682
Oviduct, 476, 479
Ovipositor, 478
Ovisac, 476
Ovulation, 482
Ovule, 419, 421; *illus.,* 420
Ovum, 117
Oxygen, 45, 96–98, 379, 434, 564, 565, 726, 748; as a limiting factor, 682
Oxygen debt, 641; *illus.,* 641
Oxygenated blood, 538, 540
Oxyhemoglobin, 548
Oxytocin, 594, 620; *table,* 595
Ozone layer, 215

P

Pacemaker, 542; *illus.,* 542
Pain killer, 621
Palisade cell, 433
Palmate venation, 302
PAN, 744
Pancreas, 526, 586; *illus.,* 525; 587
Pancreatic amylase, 526
Pancreatic duct, 526; *illus.,* 525
Paramecium, 7, 270, 361, 680; digestion in, 377; growth rates of, 681; *illus.,* 64; 270; 361; digestion, 377; 383
Parapodia, 319
Parasite, 8, 316, 318, 319, 395, 678, 759; *illus.,* 317
Parasitism, 678, 702; *illus.,* 678
Parasympathetic system, 617; *illus.,* 617
Parathion, 743
Parathormone, *table,* 595
Parental cross, 130
Parietal lobe, *illus.,* 613; *table,* 614
Parthenogenesis, 475
Passive immunity, 501
Passive transport, 70–71; *illus.,* 70
Pasteur, Louis, 109–110
Pasteur's experiment, 110; *illus.,* 110
Pasteurization, 392
Pathogenic bacteria, 531

Pathogen, 389, 399–401; defenses against, 396–397; *illus.,* 390
Pavlov, Ivan, 658
PCB, 748
Pea plant traits, *illus.,* 130
Peat moss, 291
Pecking, 666; *illus.,* 666
Pedipalp, 335
Peking people, *illus.,* 237
Pellicle, 270
Pelvis, 575, 633
Penicillin, 26, 27, 28, 29, 205, 276; *illus.,* 26
Penicillium mold, 25, 276, 363; *illus.,* 276; 363
Penis, 478; *illus.,* 480
Pepsin, 92, 525
Pepsinogen, 525
Peptidase, 526
Peptide bond, 56, 404–405
Pericycle, 450; *illus.,* 450
Periosteum, 632; *illus.,* 632
Peripatus, 330; *illus.,* 330
Peripheral nervous system, 612, 616, 617–618
Peristalsis, 525, 638
Permafrost, 721
Permeability, 68, 71
Permeases, 70, 71, 72
Perspiration, 577
Pertussis, 390
Pest, 740–741; control of, 739, 740–743, 759–761; alternative solutions to, 759–761
Pesticide, 741–743; and their effects on organisms, 742
Petiole, 301, 463; *illus.,* 301
Petite mutation, 188
Petrochemical, 442
pH scale, 49–50; *table,* 50
Phaeophyta, 289; *illus.,* 303
Phagocytosis, 72, 83, 271, 377; *illus.,* 83
Pharynx, 518; and digestion, 524–525; *illus.,* 518
Phenotype, 133, 135, 137, 188; *illus.,* 141
Phenotypic ratio, 136; *illus.,* 136; 138
Pheromone, 664, 760
Phloem, 434, 450, 456–457; *illus.,* 434; 450; 451; 452; 457
Phosdrin, 743
Phosphate group, 67, 173, 174, 748
Phospholipid, 67
Photoperiodism, 465
Photosynthesis, 7, 10–11, 74, 79, 199, 265, 433, 437–440, 449, 456, 457, 458, 694, 704, 746; dark reactions of, 439, 440; of lichen, 277; light reactions of, 437–438; of salt-loving bacteria, 266; *illus.,* 439; *table,* 440
Photosynthetic organism, 214

Phototropism, 459, 460–461; *illus.,* 459; 460
Phycocyanin, 264
Phycoerythrin, 290
Phylum, 251, 252, 291
Phylum Cyanobacteria, 264–265
Physical change, 50
Physical property, 50; *illus.,* 51
Physiological adaptation, 224
Physiology, 39
Phytochrome, 465
Pigment, 290, 436; carotene, 268, 287, 289, 436; chlorophyll, 80, 287, 289, 376, 436; *illus.,* 261; 437
Pine tree, life cycle of, 296, 418–419; *illus.,* 418
Pinnate, 301
Pinocytosis, 73; *illus.,* 73
Pioneering stage, 715
Pipefish, 226; *illus.,* 226
Pistil, 419; *illus.,* 419
Pitcher plant, 441; *illus.,* 441
Pith, 450, 453; *illus.,* 453
Pituitary gland, 482, 577, 591–592; anterior lobe of, 591–592; posterior lobe of, 594–595; *illus.,* 591
Pivot joint, 633
PKU, 161; *illus.,* 161
Placenta, 496–499, 500; *illus.,* 499
Placental, 348; *illus.,* 348
Plague, 685
Planarian, 315–316, 475; digestive system of, 518; excretory system of, 573; nervous system of, 611–612; respiratory system of, 563, 564; *illus.,* 315; 320; 518; excretory system, 573; nervous system, 611; 658; respiratory system, 563
Plankton, 730
Plant, 54, 55, 63; adaptation, 423, 434; aquatic, 287–289; cycles of, 413–427; day-neutral, 464; digestion in, 441; diploid, 413; disease of, 390, 393, 394; evolving from water to land, 291; flowering, 419–427; gas exchange in, 453, 457, 458; haploid, 413; land, 290–299; life cycle of, 413–427; long-day, 464; 414; nonvascular, 291–292; response to environment, 459–465; seed, 296–303; short-day, 464; structure of, 433–434, 449–453; transport in, 453–458; used for fuel, 442–443; *illus.,* adaptation, 18; flower parts, 419; 420; 421; 423; germination, 425; insectivorous, 441; life cycle, 413; 414; 415; 417; 418; 420; seed, 426; seed dispersal, 422; seed types, 423; *table,* characteristics, 303
Plant and animal cell comparison, *illus.,* 79
Plant behavior, 650; *illus.,* 650
Plant distribution, affected by light, 705

Plant genetics, 129–141; *illus.,* genetic ratio, 136; 138; 140; homologous chromosome, 133; Punnett square, 132; traits, 130, 137, 139
Plantae, 33, 252; *illus.,* 302
Plasma cell, 398
Plasmid, 263, 281
Plasmodium, 362, 395, 679; *illus.,* 395
Plasmodium, 271–272, 395, 679; *illus.,* 271; 272
Plasmolysis, 70
Plastid, 79
Plastron, 629
Platelet, 549; *illus.,* 550; *table,* 548
Platyhelminthes, 315; *illus.,* 323
Pneumococcus, 171, 364
Pneumocystis carinii pneumonia, 402
Pneumonia, 390, 392, 402
Point of insertion, 636
Polar, nuclei, 420, 421; *illus.,* 420
Polar body, 119
Polarization, 606; *illus.,* 606
Pole, 112, 113, 118
Polio, 392
Pollen grain, 298, 418, 420, 421, 423; *illus.,* 419; 420
Pollen sac, 419, 420; *illus.,* 420
Pollen tube, 421
Pollination, 421–422, 423–424
Pollution, 739, 744–753; air, 744–745; ground, 752–753; thermal, 749; water, 747–748; *illus.,* 744; 745; 746; 747
Polychlorinated biphenyl, 748
Polyp, 312, 633; *illus.,* 313
Polyp stage, 474
Polypeptide, 56, 176, 585; *illus.,* 56
Polysaccharide, 54
Pongidae, 350
Population, 120, 200, 221, 229, 673–683; and biotic potential, 673; competition in, 681–682; growth of, 673–683; human, 683–687; limiting factors of, 676–683; pest, 741; *table,* 674
Population biology, 672–687
Population density, 677, 679
Population explosion, 685–687; *illus.,* 685; 686
Population genetics, 211
Population growth curve, 674–676; *illus.,* 674; 675; 681; 682; 685
Pore, 70; *table,* 323
Porifera, 310
Portuguese man-of-war, 312–313; *illus.,* 312
Posterior, 314
Posterior lobe, 594–595
Posterior pituitary, 621
Postsynaptic fiber, 609
Potassium ion, 606
Potential energy, 90, 696, 697; *illus.,* 90

Predation, 677–678; *illus.,* 677
Predator, 677, 741, 759; def. of, 8; *illus.,* 677
Prehensile hand, 234; *illus.,* 234
Pressure-flow hypothesis, 451
Presynaptic fiber, 608, 609
Prey, 677
Primate, 350
Probability, 134–136; *illus.,* 122
Proboscis, 519
Producer, def. of, 7, 268, 694, 696, 721, 730, 731
Product rule, 134
Progesterone, 482, 590; *table,* 595
Prokaryote, 76, 213, 215, 360; aerobic respiration in, 262; characteristics of, 261–262; RNA synthesis in, 180–181; *illus.,* 76
Prolactin, *table,* 595
Prophase, 112; *illus.,* 112
Prophase I, 118, 154
Prostaglandin, 596
Prostate gland, *illus.,* 480
Protease, 92, 526
Protein, 55–56, 65, 67, 68, 111, 189, 585; in cell, 176–186; and genes, 176; *illus.,* 55; 56; hemoglobin, 176; *tables,* 56; amino acid, 178; 548
Protein digestion, 525
Protein synthesis, 178, 215; transcription, 179; translation, 181; *illus.,* 183; transfer, 181
Prothallium, 417; *illus.,* 417
Protista, 33, 267–272; *table,* 278
Protococcus, 288
Proton, 40
Protonema, 416
Protoplasm, 74
Protozoa, 267, 362, 379, 380; digestion in, 377; excretion, 380; pathogenic, 395; *illus.,* digestion, 377; 395
Provirus, 367
Pseudopodia, 269, 381, 382; *illus.,* 269; 382
Puberty, 590
Puffballs, *illus.,* 274
Pulmonary artery, 539
Pulmonary vein, 538
Pump, 71
Punctuated equilibrium, 231
Punishment, 655
Punnett square, 132, 136; *illus.,* 132; 135; 149; 153
Pupa, 493
Pupal stage, 492
Pus, 551
Pyloric sphincter, 526
Pyramid of biomass, 697; *table,* 697
Pyramid of energy, 696–697; *illus.,* 696
Pyramid of numbers, 697
Pyrenoid, 267

Pyrrophyta, 268; *illus.,* 278

Q

Queen bee, 662; *illus.,* 662; 664

R

Rabies, 392
Radial symmetry, 314; *illus.,* 314
Radiant energy, 705
Radiation, 188
Radicle, 425
Radioactive isotope, 198–199, 752, 756
Radiocarbon dating, 199
Radiolaria, 269
Radius, 636
Radula, 321
Rain forest, 701, 724–726; *illus.,* 724; 725; 726
Random mating, 211
Ratio, 136
Ray, 311; *illus.,* 341; *table,* 348
Reactant, 51
Receptor, 603, 616, 650; *illus.,* 604; 616
Recessive allele, 131
Recombinant DNA, 280, 761; *illus.,* 280; 281
Recombination gamete, 155
Rectum, 530
Recycling, 698
Red blood cell, 548–549, 633
Red-green color blindness, 164
"Red tide," 268, 332; *illus.,* 268
Redi, Francesco, 108
Redi's experiment, *illus.,* 108
Reflex, 616, 650–651
Reflex arc, 616; *illus.,* 616
Reflex response, 617
Refractory period, 607
Regeneration, 475, 508
Regulator gene, 189, 190
"Reindeer moss," 277, 721
Remora, 701
Renal artery, 575
Repair, 15
Replication, 111, 112, 114, 115, 118; of DNA, 175–176; *illus.,* 175
Repressor, 189, 190
Reproduction, 16, 473–476; asexual, 359–361, 413, 414, 474–475; cellular basis of, 111–121; human, 479–483; nonsexual, 424; sexual, 117–121, 292, 311, 317, 363–365, 413–423, 476–483; vegetative, 424; in virus, 366–370; *illus.,* human, 497; 498; 499; 504; menstrual cycle, 483
Reproductive isolation, 230
Reptile, 344–346; circulatory system, 538–539; *illus.,* 345; 346; circulatory system, 539; *table,* 348
Resolving power, 30
Resources, conservation of, 753

Respiration, 96–98, 457–458, 562–579; aerobic, 76, 96–98; anaerobic, 98–99; on cell membrane, 261; external, 564; internal, 564; muscle, 641–642; in water, 563–572; *table,* 440

Respiratory disease, 569–570

Respiratory system, 345; annelid, 564; aquatic, 564; disease of, 569–570; land animal, 566–569; smoking effects on, 571, 572; *illus.,* gills, 564–565; human, 567–568; 570; lungs, 566–567

Response, all-or-none, 608, 640; animal, 383, 650–669; conditioned, 658–659; immune, 397; involuntary, 658; nervous, 603–611; reflex, 616, 617; by target organ, 596; *illus.,* 604; 650; 651; 653; 658; 659; 660; 661; 663; 667; 668; 669

Resting potential, 607

Retrovirus, 369, 404

Reverse transcription, 369

Reward, 655

Rh factor, 552–553

Rhesus, 552

Rhizoid, 274, 292; *illus.,* 274

Rhizome, 295

Rhizopus, 273, 363, 378; *illus.,* 273

Rhodophyta, 290; *table,* 303

Ribonucleic acid (RNA), 56, 179, 280, 368, 369; messenger, 179, 188, 190, 191; processing, 191; transcription of, 179, 190–192; transfer, 181; translation of, 181; *illus.,* transcription, 179; transfer, 181

Ribose, 179; *illus.,* 179

Ribosome, 77, 179; *illus.,* 77

Rickets, 522, 631

Ringworm, 393

RNA, *See* Ribonucleic acid

RNA codon, 181

RNA processing, *illus.,* 181

RNA synthesis, *illus.,* 180

RNA transcriptase, 368–369

Root, 71, 426, 427, 449–450, 455, 457, 462; dicot, 450; fibrous, 300; monocot, 450; *illus.,* 71; 300; 426; 450

Root cap, 449, 454, 458; *illus.,* 450; 454

Root hair, 449, 454, 458; *illus.,* 450; 454

Root pressure, 454; *illus.,* 454

Round dance, 663, 664; *illus.,* 663

Roundworm, 318; *illus.,* 318; *table,* 323

Rust, 394, 740; *illus.,* 395

S

S-A node, 542, 543, 607

Sabin, Florence Rena, 541; *illus.,* 541

Sac fungi, 275

Salamander, 344, 508; *illus.,* 343; 344; circulatory system, 538; *table,* 348

Saliva, 524

Salivary amylase, 524

Salivary gland, 524

Salt, 44, 46, 47, 524, 575, 576, 578; *illus.,* 47

Sand dollar, 337

Sanitization, 685

Saprophyte, 264, 274, 298

Sarcodine, 269; *illus.,* 269; *table,* 278

Sargasso Sea, 289

Sargassum, 289

Savannah, 727

Scanning electron microscope, 30–31

Scapula, 636

Scavenger, 698; *illus.,* 698

Schleiden, Matthias, 64

Schwann, Theodor, 64

Schwann cell, 605

Science, 24

Scorpion, 336; *illus.,* 336; *table,* 336

Scrotum, 480; *illus.,* 480

Sea anemone, 312

Sea horse, 490; *illus.,* 490

Sea urchin, 337

Seaweed, 289

Second filial generation, 130

Second messenger hypothesis, 596

Second polar body, 119

Secretion, 77

Sedimentary rock, 197; *illus.,* 198

Seed, 130, 298, 418, 419, 421, 422, 425, 426

Seed coat, 419, 421, 422

Seed dispersal, 422; *illus.,* 422

Segmented worm, 319–320; *illus.,* 323

Segregation, 136; the test of, 132–133

Selective breeding, 204, 209, 753; *illus.,* 204

Selective cutting, 758

Self-pollination, 423

Semen, 480

Semilunar valve, 540

Semipermeable membrane, 68, 69, 71

Senses, 618–620

Sensory neuron, 604, 612, 616

Sensory organ, 315, 330

Septum, 539

Sequoia, 296

Serum, 405

Sessile, 310

Setae, 634; *illus.,* 634

Sewage treatment plant, 748

Sex determination, 148–149; *illus.,* 148; 149; 150

Sex hormone, 589, 590

Sex-linked characteristic, 590

Sex-linked disease, 164

Sex organ, 474; *See also* Gonads

Sexual characteristic, 150

Sexual reproduction, 117–122, 292, 311, 317, 363–365, 413–423, 476–483; *illus.,* 120

Sexually-transmitted disease, 390

Shark, 341; *illus.,* 341; *table,* 348

Shelf fungi, *illus.,* 274

Shrimp, *illus.,* 332; *table,* characteristics, 336

Shrub layer, 723; *illus.,* 723

Sickle-cell anemia, 159–160, 177; *illus.,* 160

Sickled red blood cells, *illus.,* 160

Sieve tube, 456

Sinoatrial node, 542

Sinus, 536

Sinus venosus, 537

Skate, 341; *illus.,* 348

Skeletal system, 627–633; external, 628; internal, 629–632; *illus.,* endoskeleton, 629; exoskeleton, 628; human, 630; 631; 632; 633

Skeleton, 269, 343, 347

Skin, 320, 396, 566

Slime mold, 271–272; *illus.,* 272

Small intestine, and digestion, 526–527; *table,* 506

Smallpox, 392

Smell, 619

Smog, 744–745

Smoking, 555, 571–572

Smut, 394; *illus.,* 394

Snail, 321; *illus.,* 321

Snake, 346; *illus.,* 345; 346; *table,* 353

Social behavior, 666–669

Social hierarchy, 666–667, 681

Society, 662, 664, 666, 681, 684

Sodium, 44

Sodium chloride, 44–45, 46

Sodium ion, 606

Sodium-potassium pump, 606

Soil, 704, 716, 723; *illus.,* 723

Soil depletion, 755

Soil fertility, 704, 755

Sol, 382

Solar energy, 756–757

Solute, 47

Solution, 46–47; *illus.,* 46; 47

Solvent, 47

Sori, 417; *illus.,* 417

Sowbug, 333; *illus.,* 332

Space, as a limiting factor, 677

Spanish moss, 701

Specialization, 423–424

Speciation, 229; convergence, 232; divergence, 230, 233, 234; evolution of, 228–230; and geographic isolation, 229; human, 234–239; the tempo of, 231; *illus.,* 228; 229; 230; 236; 237; 238; 239; adaptive radiation, 231; convergence, 232; prehensile hand, 234; *table,* human tree, 240

Species, 247, 251

Spectrum, absorption, 436

Spemann, Hans, 508

Sperm, 16, 65, 117, 120, 131, 311, 415, 417, 418, 419, 421, 478, 480–481; *illus.,* 117; 416; 417
Sperm duct, 478
Sperm formation, *illus.,* 118
Sperm receptacle, 478
Spermatid, 119
Spermatocyte, 118; primary, 118; secondary, 114, 115
Spermatophore, 343
Sphagnum moss, 291
Spicule, 310; *illus.,* 516
Spinal cord, 612, 615–617
Spinal nerve, 612, 616, 617
Spindle fiber, 113, 115, 118, 186, 361; *illus.,* 113
Spinneret, 336
Spiny anteater, 494; *illus.,* 494
Spiracle, 566
Spirillum, 262; *illus.,* 262
Spirochete, 264
Spirogyra, 288, 413–414; *illus.,* 414
Spleen, 397, 549
Split gene, 191
Sponge, 310–311, 515; digestive system of, 515; glass, 310–311; sexual reproduction of, 311; *illus.,* 311; 516; *table,* 323
Spongy layer, 433; *illus.,* 434
Spontaneous generation, 107–110
Sporangium, 273, 363, 419; *illus.,* 270
Sporangiophore, 273
Spore, 109, 110, 117, 266, 270, 362–363, 413, 414, 415; diploid, 413; *illus.,* 109; 266; 274; 293; 362; 415
Sporophyte, 415, 417, 418, 421, 422; *illus.,* 416; 417
Sporophyte generation, 414
Sporozoa, 270–271, 362; *illus.,* 278
Spring wood, 452
Squid, *illus.,* 322
Stamen, 419; *illus.,* 419
Staphylococcus, 25
Starch, 54
Starfish, 337; digestive system of, 520; locomotion of, 635; *illus.,* 337; 520; 635
Stele, 449
Stem, 450–453; herbaceous, 300, 450; woody, 301, 450, 453, 458; *illus.,* dicot, 451; monocot, 451; woody, 301
Sterile, 110
Sterilization, 760
Sternum, 633
Steroid, 585, 596
Stickleback fish, 489
Stigma, 419
Stimulant, 608
Stimulus, 585, 618; internal, 649
Stipe, 275
Stirrup, 619
Stolon, 273

Stomach, 520; and digestion, 525; *illus.,* 525; *table,* 506
Stomata, 434, 455, 45, 458; *illus.,* 434; 456
Strata, 723
Strep throat, 390, 570
Stress, 682
Striated muscle, 638
Stroma, 439
Structural gene, 189
Style, 419, 421
Suberin, 451
Subspecies, 251
Substance P, 621
Substrate, 92, 273
Succession, 749; altitudinal, 706; ecological, 715–719; latitudinal, 706; primary, 715–716; secondary, 717–718; *illus.,* 715; 716
Sucrase, 527
Sugar, 9, 47, 54, 93, 173, 174, 376, 524, 587; *illus.,* 47
Sulfur, 55
Sulfur dioxide, 744, 745, 746
Survival of the fittest, 209
Swimmeret, 333
Swine influenza, 392
Symbiosis, 215; *illus.,* 215
Symbiotic theory, 215; *illus.,* 215
Symbol, 45
Symmetry, 314; *illus.,* 314
Sympathetic system, 617
Synapse, 608–609, 616; *illus.,* 608; 616
Synapsis, 118
Synfuel, 443
Systole, 543

T

Tadpole, 490; *illus.,* 491
Taiga, 722; *illus.,* 720; 722
Tapeworm, 317; *illus.,* 317
Taproot, 300; *illus.,* 300
Tarantula, *illus.,* 336
"Target organ," 591, 595, 596
Taste, 619
Taxonomy, 245, 253
Tay-Sachs disease, 161; *illus.,* 161
TCDD, 752
T cell, 399
Technology, 684–685
Teeth, and digestion, 524
Telophase, 114; *illus.,* 114
Telophase I, 114
Temperate forest, 723–724
Temperature, 188, 480, 682, 705–707; as a biotic factor, 706–707; as a limiting factor, 682; for development, 494; and metabolism, 707–709; *illus.,* 707
Temporal lobe, *illus.,* 613; *table,* 614
Tentacle, 312; *illus.,* 517
Terminal bud, 301; *illus.,* 301

Termite, 702
Terracing, 756
Territoriality, 668–669, 681
Testes, 474, 480; *illus.,* 480
Testosterone, 590
Tetanus, 392, 637; *illus.,* 390
Tetrad, 118
Thalamus, 614
Thallus, 287, 289
Theory, 29
Theory of biogenesis, 110
Thermal pollution, 749
Thoracic duct, 545
Thorax, 334
Threshold, 608
Thymine, 173, 179; *illus.,* 173; 181
Thymus gland, 397
Thyroid, 588, 591, 592, 593
Thyroid-stimulating hormone (TSH), 591–592; *table,* 595
Thyroxine, 588, 593, 597; *table,* 595
Tick, *illus.,* 336; *table,* characteristics, 336
Tidal pool, *illus.,* 731
Tissue fluid, 545
Tissue layer, *table,* 506
Toad, 344; *illus.,* 344; *table,* 348
Toadstool, 275
Tobacco mosaic disease, 279
Topsoil, 704, 755
Touch, 619
Touch receptor, 619
Toxin, 391
Tracheae, 332, 566, 567, 588
Tracheal system, 566
Tracheid, 452
Tracheole, 566
Tracheophyta, 291; characteristics of, 293; *illus.,* 303
Trait, 129, 507, 754; adaptation of, 221–227; and bacterial transformation, 171–172; codominance, 137; convergence of, 233, 234; divergence of, 233; dominant, 131, 155; expression of, 158; important human, 234; multiple, 138–140; mutation of, 186–189; recessive, 131, 159, 160, 161; sex-linked, 150–152; *illus.,* 130; 137; 139; adaptive radiation, 231; convergence, 232; divergence, 230; genotypic ratio, 136; 138; 140; geographic isolation, 229; homologous chromosomes, 123; mutation, 234; prehensile hand, 234; Punnett square, 132
Tranquilizer, 609
Transaminase, 92
Transcriptase, reverse, 369
Transcription, 180, 191; reverse, 369; *illus.,* 180
Transduction, 368, 404; *illus.,* 368
Transfer RNA, 181; *illus.,* 181
Transformation, 364

Transforming principle, 173
Transfusion, 552
Translation, 191; *illus.*, 191
Translocation, 456–457
Transmission electron microscope, 30
Transmitter, 608–609, 621; *illus.*, 609
Transpiration, 454, 455–456; *illus.*, 703
Transpiration-cohesion theory, 454; *illus.*, 454
Transpiration pull, 455
Transport, 379, 453–457, 534–545
Transport system, animal, 535–559
Transverse colon, 530
Trial and error learning, 655, 659
Triceps, 636; *illus.*, 636
Trichinella worm, 318
Trichinosis, 318
Trichocyst, 383
Tricuspid valve, 540
Triploid, 421
Trophic level, 694–695
Trophoblast, 496–497
Tropical rain forest, 724–727
Tropism, 459, 650; *illus.*, 462
Trypanosoma, 271; *illus.*, 271
Trypsin, 526
Tsetse fly, 271
Tube feet, 635; *illus.*, 635
Tube nucleus, 420
Tuber, 424
Tuberculosis, 392, 570
Tubule, 575
Tumor, 403
Tundra, 720–721; *illus.*, 720; 721
Tunicate, 338
Turgidity, 301
Turner syndrome, 165
Turtle, 346
Twins, 159, 404; *illus.*, 404
Tympanic membrane, 619

U

Ulna, 636
Ulothrix, 288, 413–414; *illus.*, 413
Ultrasonography, 167
Ultraviolet radiation, 213
Ulva, 414–415; *illus.*, 414; 416
Umbilical cord, 497; *illus.*, 497; 499
Unicellular organism, vs. multicellular, 65; *See also* Monera, Protista
Univalve, 321
Universal donor, 552
Universal recipient, 552
Uracil, 179
Uranium, 756
Urea, 576, 579; *table*, 548
Ureter, 577; *tables*, 480; 575
Urethra, 480; *illus.*, 479; 480
Uric acid, 574, 579

Urinary bladder, 577; *illus.*, 479; 480; 499; 575
Urine, 576; *table*, 576
Uterus, 480; *illus.*, 479; 499

V

Vaccination, 399–400
Vaccine, 399–400; synthetic, 404
Vacuole, 78; contractile, 361, 381; food, 269, 377; *illus.*, 78; 267; 270; 381
Vagina, 478, 480, 500; *illus.*, 480; 500
Vagus nerve, 543
Variable factor, 27
Variation, 210, 211
Vascular bundle, 298, 451, 453, 456
Vascular cambium, 427, 451; *illus.*, 451; 452
Vascular plant, 454
Vascular tissue, 291
Vasopressin, 577, 594; *illus.*, 595
Vegetal pole, 490, 502, 505
Vein, 434, 537, 538, 540; *illus.*, 301; 575
Vena cava, 539
Venation, 301, 302; *illus.*, 302
Venom, 346
Ventral aorta, 537
Ventral blood vessel, 535, 536
Ventral nerve cord, 331
Ventral side, 314; *illus.*, 314
Ventricle, 537; of the brain, 613
Venus's-flytrap, 441; *illus.*, 441
Vertebrae, 339, 629, 633
Vertebrate, 338–350; excretory system of, 578–579; locomotion of, 636–637; nervous system of, 612–613; *illus.*, 339; locomotion, 636; 637; 638; 639; 641; 642; *table*, 348
Vertical stratification, 723; *illus.*, 723
Vesicle, 82
Vessel, 451, 535, 536
Vestigial organ, 202; *illus.*, 202
Villi, 526–527; *illus.*, 526
Viral pneumonia, 392
Viroid, 393
Virus, 279–280; and cancer, 368–369; characteristics of, 279–280; controlling population, 680; defenses against, 397–401; and disease, 392; pathogenic, 392; reproduction of, 366–368; *illus.*, 279; 280; tobacco mosaic, 279; 367
Visible spectrum, 435
Vision, 619; three-dimensional, 234
Vitamin, 94, 376; *table*, 522
Vitamin D, 631
Vitamin K, 550
Vocal sac, 344
Voluntary response, 617
Volvox, 289

W

Waggle dance, 663; *illus.*, 663
Warning coloration, 226; *illus.*, 226
Waste, 71, 573
Waste product, 11
Water, 14, 42, 43, 46, 47, 57, 69, 93, 376, 381, 438, 439, 440, 442, 577; as a biotic factor, 703–704; and diffusion, 68; as a nutrient, 521; and osmosis, 69, 454; and plants, 454, 456–458; and reproduction, 478
Water balance, 573, 577; *illus.*, 577
Water cycle, *illus.*, 703
Water pollution, 747–749
Water vapor, 212
Water-vascular system, 635
Watson, James, 174
Watson and Crick model of DNA, 174, 175
Wavelength, 435, 436
Weinberg, Wilhelm, 211
White blood cell, 550–551, 633
White light, 435
White matter, 615
Wildlife, conservation of, 757–758
"Woodpecker finch," 232
Worker bee, 662; *illus.*, 662; 664
Worm, segmented, 319–320
Wright, Jane, 403; *illus.*, 403

X

X chromosome, 148
X-ray crystallography, 29
Xylem, 296, 434, 450, 456–457; *illus.*, 434; 450; 451

Y

Y chromosome, 148
Yeast, 276, 362, 365; *illus.*, 362; 365
Yolk plug, *illus.*, 505
Yolk sac, 493
Yucca moth, 702
Yucca plant, 702

Z

Zero population growth, 687
Z-line, 639
Zoospore, 413
Zygomycota, 273; *table*, 278
Zygospore, 414
Zygote, 117, 132, 271, 311, 317, 418, 421, 422, 482; development of, 489–510; diploid, 413, 415; human, 481, 482; *illus.*, 416; 417; chicken, 494; 502; 503; frog, 491; 502; 505; 508; human, 497; 498; 499; 504; insect, 492; 493; marsupial, 495

COVER: Frank Cocco

2–3, Dr. Betty Linthicum; **4,** William Weber; **6(l)** Gene Frazier, **(r)** Richard Brommer; **7(l)** William Ferguson, **(r)** Rick McIntyre/Tom Stack & Assoc.; **8(l)** Helen Rhode, **(c)** Richard Brommer, **(r)** Alvin Staffan; **14,** Grant Heilman; **15,** Greg Sailor; **16,** Scripps Institute of Oceanography; **18(l)** Richard Brommer, **(r)** Sharon Kurgis; **19(l)** R. B. Satterthwaite, **(r)** Robert Mayer; **22,** Chip Clark © 1984; **24,** James Westwater; **25(t)** David M. Dennis, **(b)** C. Findel, J. Gnau, and A. Ottolenghi at OSU; **26,** Doug Martin; **27,** Tom Stack/Tom Stack & Assoc.; **28,** Historical Pictures Service; **30,** © Carroll H. Weiss, 1979/Camera M.D. Studios, Inc.; **31(l)** Hans Pfletschinger/Peter Arnold, Inc., **(c)** David M. Dennis, **(r)** David Scharf/Peter Arnold, Inc.; **38,** Steve Lissau; **40,** Johns Hopkins University for Photo Researchers; **45,** Gerard Photography; **46,** Hickson-Bender Photography; **49(l)** Lightforce, **(r)** Herm Beck; **50, 51, 54,** Gerard Photography; **56,** J. Somers/Science Photo Library International/Taurus; **62,** Lennart Nilsson, © Boehringer Ingelheim Itl, GmbH, THE INCREDIBLE MACHINE, National Geographic Society © 1986; **64(l)** Elaine Shay, **(r)** Tom Stack & Assoc.; **66(l)** Don Fawcett, M.D. for Photo Researchers, **(r)** Dr. Judie Walton, Lawrence Livermore National Laboratory; **68,** Gerard Photography; **70(t)** Earth Scenes/Breck P. Kent, **(b)** Grant Heilman; **74,** M. Abbey/Photo Researchers; **76(tl)** Don Fawcett, M.D. for Photo Researchers, **(tr)** Carolina Biological Supply Company; **76(b)** file photo; **77(t)** Barbara Stevens, **(b)** Dr. Judie Walton, Lawrence Livermore Laboratory; **78(t)** file photo, **(b)** Barbara Stevens; **80(t)** Carey Calloway/Centers for Disease Control, **(b)** Omikron/Photo Researchers; **81,** Science Photo Library/Photo Researchers; **88,** E. R. Degginger; **90,** Thomas Zimmerman/FPG; **94,** Tim Dietz; **95(l)** Roger K. Burnard, **(r)** Commercial Image; **97,** Thomas Zimmerman/FPG; **98(b)** Bruce Charlton, **(r)** Ted Rice; **104–105,** Townsend P. Dickinson/Photo Researchers; **106,** © Lennart Nilsson, THE INCREDIBLE MACHINE, National Geographic Society © 1986; **109,** Dr. Judie Walton, Lawrence Livermore Laboratory; **112(t)** Runk/Schoenberger from Grant Heilman, **(b)** Alfred Owczarzak/Taurus Photos; **113(t)** Ward's Natural Science Establishment, Inc., **(b)** Alfred Owczarzak/Taurus Photos; **114,** Alfred Owczarzak/Taurus Photos; **115,** Runk/Schoenberger from Grant Heilman; **116(l)** Alfred Hochrein/Berg and Associates, **(r)** Russ Lappa; **117(t)** Manfred Kage/Peter Arnold, Inc., **(b)** Photo Researchers; **118,** Dr. Judie Walton, Lawrence Livermore Laboratory; **128,** Jan & Des Bartlett/Bruce Coleman, Inc.; **137(l)** W. Atlee Burpee Company, **(r)** George W. Parke Seed Company; **146,** Pictures Unlimited; **157,** Allan Roberts; **158(t)** Carolina Biological Supply Company, **(bl)** USDA, **(br)** Steve Lissau; **160(tl)** Kessel/Shih/Springer-Verlag © 1976, **(tr)** SEM Laboratory, Morris Brown College, **(br)** Doug Martin; **161,** Image Workshop; **164,** Russ Kinne for Photo Researchers; **165(l)** Medical Genetics Laboratory, Children's Hospital, Columbus, OH, **(r)** Image Workshop; **166,** Kenneth Garrett; **167(t)** Gerard Photography, **(bl)** Yale University, School of Medicine, **(br)** Alexander Tsiaras/Science Source/Photo Researchers; **170,** Breck P. Kent; **174(l)** Ira Wyman/Sygma; **(r)** McCoy/Langridge/UCSF/Rainbow; **177,** Science Source/Photo Researchers; **187(l)** Richard P. Smith/Tom Stack & Assoc., **(c)** Grant Heilman, **(rt)** Larry Hamill, **(rb)** Lightforce; **188,** Richard Brommer; **194–195,** Dr. R. B. Rickards; **196,** Larry Agenbroad/Hot Springs Mammoth Site, SD.; **197,** University of Houston; **198(tl)** Frank Balthis, **(tr)** TASS from Sovfoto, **(bl)** Michael Collier, **(br)** Larry Roberts; **204(tl)** Hickson-Bender Photography, **(bl)** L & M Photo, **(r)** Robert Mischka; **208(tl)** Animals Animals/Bertram G. Murray, Jr., **(tr)** Virginia Crowl, **(bl)** Roger K. Burnard, **(br)** William Ferguson; **209(l)** Earth Scenes/Raymond A. Mendez, **(c)** Animals Animals/Z. Leszcyznski, **(r)** Lightforce; **212,** Stephen J. Krasemann/DRK Photo; **215(t)** Biophoto Associates/Science Source/Photo Researchers, **(b)** Dr. Jeremy Burgess/Science Photo Library for Photo Researchers; **220,** Johnny Johnson; **222,** Stan Wayman for Photo Researchers; **223, 224(l)** Animals Animals/Breck P. Kent, **(r)** Animals Animals/Z. Leszczynski; **225(t)** Roger K. Burnard, **(b)** Latent Image; **226(tl)** Animals Animals/Oxford Scientific Film, **(tr)** Dave Woodward/Tom Stack & Assoc., **(bl)** Robert Fridenstine, **(br)** J. R. Schnelzer; **228(l)** Jean Wentworth, **(c)** Animals Animals/Mark Chappell, **(r)** Grant Heilman; **230,** Tom & Pat Leeson; **231,** William Ferguson; **232,** Miguel Castro for Photo Researchers; **233(tl)** Animals Animals/J. C. Stevenson, **(tr)** Alvin Staffan, **(bl)** Animals Animals/Stouffer Enterprises, **(br)** Joey Jacques; **234(l)** Ruth Dixon, **(r)** Gary Milburn/Tom Stack & Assoc.; **235,** American Museum of Natural History; **236(l)** The Cleveland Museum of Natural History, **(r)** American Museum of Natural History; **237(l)** American Museum of Natural History, **(r)** Reconstruction by Harry Shapiro, American Museum of Natural History, **(b)** David L. Brill, Courtesy National Geographic Society; **238(l)(r)** American Museum of Natural History, **(b)** French Government Tourist Office; **244, 247,** Latent Image; **249,** Animals Animals/Lynn M. Stone; **250,** W. H. Hodge/Peter Arnold, Inc.; **251,** Richard Brommer; **258–259,** Breck P. Kent; **260,** Carolina Biological Supply Company; **262,** Ward's Natural Science Establishment; **263(t)** American Society for Microbiology, **(b)** Biophoto Associates; **265,** Courtesy M. D. Moore/R. S. Wolfe, Department of Microbiology, University of Illinois at Urbana-Champaign; **266,** Dannon Yogurt, **(inset)** Lightforce; **267,** Lester V. Bergman; **268(t)** Manfred Kage/Peter Arnold, Inc., **(b)** Dr. Carleton Ray/Photo Researchers; **269(tr)** Michael Abbey/Photo Researchers, **(bl)** Manfred Kage/Peter Arnold, Inc., **(bc)** Alfred Pasieka/Taurus Photos, **(br)** Paola Koch/Photo Researchers; **270(tl)** SEM Laboratory/Morris Brown College, **(b)** Roger K. Burnard; **271(l)** E. R. Degginger/Bruce Coleman, Inc., **(r)** Runk/Schoenberger for Grant Heilman; **274(t)** Dan Gurovich/Photo Researchers, **(bl)** Ward's Natural Science Establishment, **(bc)** James Westwater, **(br)** Alvin Staffan; **275(tl)** Veryl Scheibner/Photo Researchers, **(bl)** Ken Hayes, **(br)** William Ferguson; **276(tl)** Courtesy of Charles Pfizer and Co., Inc., **(tr)** Stephen J. Krasemann/DRK Photo, **(ml)** Larry Roberts, **(bl)** Dwight R. Kuhn; **277,** James Westwater; **279(t)** Dr. C. Wayne Ellet, Professor Emeritus/OSU, **(bl)** J. Somers/Science Photo Library International/Taurus Photos, **(br)** Centers for Disease Control; **281,** Steven D. Northrup/TIME Magazine; **286,** Stephen J. Krasemann/DRK Photo; **289(tr)** Manfred Kage/Peter Arnold, Inc., **(bl)** Eileen Tanson for Photo Researchers, **(br)** Tom Stack/Tom Stack & Assoc.; **290,** file photo; **291(t)** Alvin Staffan, **(b)** Ruth Dixon; **293(l)** Roger K. Burnard, **(r)** SEM Laboratory, Morris Brown College; **294(tl)** Tom Stack/Tom Stack & Assoc., **(tr)** Ward's Natural Science Establishment, **(b)** Robert Neulieb; **295(l)** Gary Retherford, **(r)** Roger K. Burnard; **296(t)** Doug Martin, **(c)** William J. Jahoda/Photo Researchers, **(r)** Runk/Schoenberger from Grant Heilman; **297(l)** Alvin Staffan, **(c)** John Shaw/Tom Stack & Assoc., **(r)** Richard Brommer; **298(t)** Earth Scenes/Oxford Scientific Film, **(ml)** Runk/Schoenberger from Grant Heilman, **(bl)** Breck P. Kent; **300,** Image Workshop; **302(l)** Kodansha, **(lc)** Grant Heilman, **(rc)(r)** Robert Ashworth/Photo Researchers; **310,** Douglas Faulkner; **312(l)** Terry Ashley/Tom Stack & Assoc., **(c)(r)** Joey Jacques; **313(l)** Florida Department of Commerce, **(r)** Lightforce; **314(l)** Courtesy CCM: General Biological, Inc., **(r)** Sharon Kurgis; **315(tr-A.)** Runk/Schoenberger from Grant Heilman, **(tr-B.)** Bill W. Curtsinger/Photo Researchers, **(bl)** Dennis L. Crow/FPG, **(br)** Animals Animals/Oxford Scientific Film; **316,** Runk/Schoenberger from Grant Heilman; **317,** Grant Heilman; **319,** Harris Biological Supplies, Ltd.; **320(l)** Edward J. Webster, **(c)** Lester V. Bergman & Associates, **(r)** Carolina Biological Supply Company; **321(tl)** Allan Roberts, **(tr)** Jeff Foott, **(br)** Breck P. Kent; **323(t)** Gene Frazier, **(b)** Ward's Natural Scientific Establishment, Inc.; **324(tl)** Geri Murphy, **(tr)** Peter David for Photo Researchers, **(bl)(br)** Tom McHugh for Photo Researchers; **330,** W. Perry Conway/Tom Stack & Assoc.; **332,** Lester V. Bergman & Assoc.; **333(tl)** David Scharf/Peter Arnold, Inc., **(tr)** Mrs. George T. Butts, **(bl)** Alvin Staffan, **(br)** H. Armstrong Roberts; **334(l)** William Ferguson, **(c)** Alvin Staffan, **(r)** Michael DiSpezio; **335,** Gwen Fidler; **337 (A.)** Courtesy CCM: General Biological, Inc., **(B.)** Wallace Kirkland for Tom Stack & Assoc., **(C.)** J. W. Thompson, **(D.)** Gene Frazier, **(E.)** Stephen Dalton for Photo Researchers, **(F.)** Roger K. Burnard, **(G.)** Alvin Staffan, **(H.)** J. W. Thompson; **338(l)** Breck P. Kent, **(c)** Courtesy CCM: General Biological, Inc., **(r)** Michael DiSpezio; **339(l)**